# Preface

When this book was originally conceived, its primary objective was to present a comprehensive body of chemical principles in a manner that would be understandable without being oversimplified. Since that time, there have been significant advances in chemistry, the career goals of students enrolled in introductory chemistry have become more diverse, and the scientific and mathematical backgrounds of freshman students have became more varied than ever before.

This new edition has been prepared with these changes in mind, although the overall intent remains the same — to provide the student with a comprehensive text for the introductory chemistry course. To that end, the organization and scope of this revised edition are essentially the same as those of the earlier editions.

The purpose of this book is to present a balanced treatment of modern theoretical and descriptive chemistry while avoiding the use of calculus. New sections have been added and many topics have been rewritten to bring the material up to date or to improve the clarity of the presentation.

The chapter on thermodynamics has been rewritten and placed after the chapter on kinetics and equilibrium. The order of these two chapters has been reversed so that the concept of chemical equilibrium can be used in the discussion of Gibbs free energy. From a chemist's viewpoint, no thermodynamic function is more important than the standard free energy change of a chemical reaction. Interpretation of $\Delta G°$ values in terms of "reaction spontaneity" alone, without reference to the equilibrium state, can lead to serious misunderstanding.

The growing acceptance of the International System of Units (SI) makes it important that students be introduced to this modern metric system. Immediate and complete transition to SI units, however, may not be possible. Much of the data found in the chemical literature has been recorded in units that are not SI units, for example, thermodynamic values in kilocalories (not kilojoules), crystal distances in ångstrom units (not nanometers), vapor pressures in torr (not pascals). Consequently, students must be familiar with both the old and the new units, and both are presented in this book. Marginal notes are used to show the relation between SI units and the older but still commonly accepted units. In many chapters the problems are presented in such a manner that they can be solved in joules or calories.

Many of the changes in this edition were made in response to comments received from instructors and students who have used earlier editions. For example, a mathematical review has been included in the Appendix, a section on osmosis has been introduced, and the section on molecular orbitals and the discussion of intermolecular attractive forces have been rewritten. The discussion of oxidation numbers has been moved forward to serve as the basis for an expanded section on chemical nomenclature. Most of the chapter-end problems are new; over 850 are included.

The scope and format of the book are intended to permit the instructor relatively wide latitude in course content and organization. A number of instructors prefer to begin their courses with a consideration of stoichiometry (Chapter 5), and several incorporate topics from Chapter 21 (Nuclear Chemistry) into the discussion of Chapter 2 (Atomic Structure). It is gratifying to know that the text has successfully served different approaches.

I sincerely thank the following for their comments: Professors Myron G. Berry, James J. Bohning, Joseph F. Chiang, Alan C. Dickinson, Bruce K. Dietrich, Thomas B. Eames, Lowell P. Eddy, Michael F. Farona, Russell Ham, John Kaczynski, Gerard L'Heureux, Sister Maria Maurice Harrington, Robert W. Miller, John W. Moore, Lawrence C. Nathan, Charles R. Rhyner, Nancy J. Sell, and William J. Wasserman.

Suggestions for the improvement of this edition will be welcomed.

Charles E. Mortimer

# CHEMISTRY
## A Conceptual Approach
### THIRD EDITION

## Charles E. Mortimer
MUHLENBERG COLLEGE

**D. VAN NOSTRAND COMPANY**
*New York   Cincinnati   Toronto   London   Melbourne*

*To*
J. S. M. *and* C. E. M.[II]

D. Van Nostrand Company Regional Offices:
New York   Cincinnati   Millbrae

D. Van Nostrand Company International Offices:
London   Toronto   Melbourne

Published by D. Van Nostrand Company
450 West 33rd Street, New York, N.Y. 10001

Published simultaneously in Canada by
Van Nostrand Reinhold Ltd.

10  9  8  7  6  5  4

The illustration on the cover shows the result of laser impact upon an old bronze art object. The laser removes a sample of bronze slightly larger than a human hair, $50 \times 10^{-6}$ m to $80 \times 10^{-6}$ m in diameter and to a depth of $80 \times 10^{-6}$ m to $100 \times 10^{-6}$ m. This sample is then analyzed to identify the elements present in the bronze. Scientific techniques such as this can be used to examine and authenticate works of art. The illustration is reproduced courtesy of the Research Laboratory of the Museum of Fine Arts, Boston, Massachusetts.

# OF THE ELEMENTS

| | | | III A | IV A | V A | VI A | VII A | 0 |
|---|---|---|---|---|---|---|---|---|
| | | | | | | | | 2<br>**He**<br>Helium<br>4.00260 |
| | | | 5<br>**B**<br>Boron<br>10.81 | 6<br>**C**<br>Carbon<br>12.011 | 7<br>**N**<br>Nitrogen<br>14.0067 | 8<br>**O**<br>Oxygen<br>15.9994 | 9<br>**F**<br>Fluorine<br>18.99840 | 10<br>**Ne**<br>Neon<br>20.179 |
| | I B | II B | 13<br>**Al**<br>Aluminum<br>26.98154 | 14<br>**Si**<br>Silicon<br>28.086 | 15<br>**P**<br>Phosphorus<br>30.97376 | 16<br>**S**<br>Sulfur<br>32.06 | 17<br>**Cl**<br>Chlorine<br>35.453 | 18<br>**Ar**<br>Argon<br>39.948 |
| 28<br>**Ni**<br>Nickel<br>58.71 | 29<br>**Cu**<br>Copper<br>63.546 | 30<br>**Zn**<br>Zinc<br>65.38 | 31<br>**Ga**<br>Gallium<br>69.72 | 32<br>**Ge**<br>Germanium<br>72.59 | 33<br>**As**<br>Arsenic<br>74.9216 | 34<br>**Se**<br>Selenium<br>78.96 | 35<br>**Br**<br>Bromine<br>79.904 | 36<br>**Kr**<br>Krypton<br>83.80 |
| 46<br>**Pd**<br>Palladium<br>106.4 | 47<br>**Ag**<br>Silver<br>107.868 | 48<br>**Cd**<br>Cadmium<br>112.40 | 49<br>**In**<br>Indium<br>114.82 | 50<br>**Sn**<br>Tin<br>118.69 | 51<br>**Sb**<br>Antimony<br>121.75 | 52<br>**Te**<br>Tellurium<br>127.60 | 53<br>**I**<br>Iodine<br>126.9045 | 54<br>**Xe**<br>Xenon<br>131.30 |
| 78<br>**Pt**<br>Platinum<br>195.09 | 79<br>**Au**<br>Gold<br>196.9665 | 80<br>**Hg**<br>Mercury<br>200.59 | 81<br>**Tl**<br>Thallium<br>204.37 | 82<br>**Pb**<br>Lead<br>207.2 | 83<br>**Bi**<br>Bismuth<br>208.9804 | 84<br>**Po**<br>Polonium<br>(210)a | 85<br>**At**<br>Astatine<br>(210)a | 86<br>**Rn**<br>Radon<br>(222)a |

metals ← → nonmetals

| 63<br>**Eu**<br>Europium<br>151.96 | 64<br>**Gd**<br>Gadolinium<br>157.25 | 65<br>**Tb**<br>Terbium<br>158.9254 | 66<br>**Dy**<br>Dysprosium<br>162.50 | 67<br>**Ho**<br>Holmium<br>164.9304 | 68<br>**Er**<br>Erbium<br>167.26 | 69<br>**Tm**<br>Thulium<br>168.9342 | 70<br>**Yb**<br>Ytterbium<br>173.04 | 71<br>**Lu**<br>Lutetium<br>174.97 |
|---|---|---|---|---|---|---|---|---|
| 95<br>**Am**<br>Americium<br>(243)a | 96<br>**Cm**<br>Curium<br>(247)a | 97<br>**Bk**<br>Berkelium<br>(249)a | 98<br>**Cf**<br>Californium<br>(251)a | 99<br>**Es**<br>Einsteinium<br>(254)a | 100<br>**Fm**<br>Fermium<br>(253)a | 101<br>**Md**<br>Mendelevium<br>(256)a | 102<br>**No**<br>Nobelium<br>(254)a | 103<br>**Lr**<br>Lawrencium<br>(257)a |

# THE ELEMENTS

| | Symbol | Atomic Number | Atomic Weight | | Symbol | Atomic Number | Atomic Weight |
|---|---|---|---|---|---|---|---|
| Actinium | Ac | 89 | (227)[a] | Molybdenum | Mo | 42 | 95.94 |
| Aluminum | Al | 13 | 26.98154 | Neodymium | Nd | 60 | 144.24 |
| Americium | Am | 95 | (243)[a] | Neon | Ne | 10 | 20.179 |
| Antimony | Sb | 51 | 121.75 | Neptunium | Np | 93 | 237.0482[b] |
| Argon | Ar | 18 | 39.948 | Nickel | Ni | 28 | 58.71 |
| Arsenic | As | 33 | 74.9216 | Niobium | Nb | 41 | 92.9064 |
| Astatine | At | 85 | (210)[a] | Nitrogen | N | 7 | 14.0067 |
| Barium | Ba | 56 | 137.34 | Nobelium | No | 102 | (254)[a] |
| Berkelium | Bk | 97 | (249)[a] | Osmium | Os | 76 | 190.2 |
| Beryllium | Be | 4 | 9.01218 | Oxygen | O | 8 | 15.9994 |
| Bismuth | Bi | 83 | 208.9804 | Palladium | Pd | 46 | 106.4 |
| Boron | B | 5 | 10.81 | Phosphorus | P | 15 | 30.97376 |
| Bromine | Br | 35 | 79.904 | Platinum | Pt | 78 | 195.09 |
| Cadmium | Cd | 48 | 112.40 | Plutonium | Pu | 94 | (242)[a] |
| Calcium | Ca | 20 | 40.08 | Polonium | Po | 84 | (210)[a] |
| Californium | Cf | 98 | (251)[a] | Potassium | K | 19 | 39.098 |
| Carbon | C | 6 | 12.011 | Praseodymium | Pr | 59 | 140.9077 |
| Cerium | Ce | 58 | 140.12 | Promethium | Pm | 61 | (145)[a] |
| Cesium | Cs | 55 | 132.9054 | Protactinium | Pa | 91 | 231.0359[b] |
| Chlorine | Cl | 17 | 35.453 | Radium | Ra | 88 | 226.0254[b] |
| Chromium | Cr | 24 | 51.996 | Radon | Rn | 86 | (222)[a] |
| Cobalt | Co | 27 | 58.9332 | Rhenium | Re | 75 | 186.2 |
| Copper | Cu | 29 | 63.546 | Rhodium | Rh | 45 | 102.9055 |
| Curium | Cm | 96 | (247)[a] | Rubidium | Rb | 37 | 85.4678 |
| Dysprosium | Dy | 66 | 162.50 | Ruthenium | Ru | 44 | 101.07 |
| Einsteinium | Es | 99 | (254)[a] | Samarium | Sm | 62 | 150.4 |
| Erbium | Er | 68 | 167.26 | Scandium | Sc | 21 | 44.9559 |
| Europium | Eu | 63 | 151.96 | Selenium | Se | 34 | 78.96 |
| Fermium | Fm | 100 | (253)[a] | Silicon | Si | 14 | 28.086 |
| Fluorine | F | 9 | 18.99840 | Silver | Ag | 47 | 107.868 |
| Francium | Fr | 87 | (223)[a] | Sodium | Na | 11 | 22.98977 |
| Gadolinium | Gd | 64 | 157.25 | Strontium | Sr | 38 | 87.62 |
| Gallium | Ga | 31 | 69.72 | Sulfur | S | 16 | 32.06 |
| Germanium | Ge | 32 | 72.59 | Tantalum | Ta | 73 | 180.9479 |
| Gold | Au | 79 | 196.9665 | Technetium | Tc | 43 | 98.9062[b] |
| Hafnium | Hf | 72 | 178.49 | Tellurium | Te | 52 | 127.60 |
| Helium | He | 2 | 4.00260 | Terbium | Tb | 65 | 158.9254 |
| Holmium | Ho | 67 | 164.9304 | Thallium | Tl | 81 | 204.37 |
| Hydrogen | H | 1 | 1.0079 | Thorium | Th | 90 | 232.0381[b] |
| Indium | In | 49 | 114.82 | Thulium | Tm | 69 | 168.9342 |
| Iodine | I | 53 | 126.9045 | Tin | Sn | 50 | 118.69 |
| Iridium | Ir | 77 | 192.22 | Titanium | Ti | 22 | 47.90 |
| Iron | Fe | 26 | 55.847 | Tungsten | W | 74 | 183.85 |
| Krypton | Kr | 36 | 83.80 | Uranium | U | 92 | 238.029 |
| Lanthanum | La | 57 | 138.9055 | Vanadium | V | 23 | 50.9414 |
| Lawrencium | Lr | 103 | (257)[a] | Xenon | Xe | 54 | 131.30 |
| Lead | Pb | 82 | 207.2 | Ytterbium | Yb | 70 | 173.04 |
| Lithium | Li | 3 | 6.941 | Yttrium | Y | 39 | 88.9059 |
| Lutetium | Lu | 71 | 174.97 | Zinc | Zn | 30 | 65.38 |
| Magnesium | Mg | 12 | 24.305 | Zirconium | Zr | 40 | 91.22 |
| Manganese | Mn | 25 | 54.9380 | | | | |
| Mendelevium | Md | 101 | (256)[a] | | | | |
| Mercury | Hg | 80 | 200.59 | | | | |

[a] Mass number of most stable or best known isotope.

[b] Mass of most commonly available, long-lived isotope.

# Contents

# 1

# Introduction

**Chemistry** may be defined as the science that is concerned with the characterization, composition, and transformations of matter. This definition, however, is far from adequate. The interplay between the branches of modern science has caused the boundaries between them to become so vague that it is almost impossible to stake out a field and say "this is chemistry." Not only do the interests of scientific fields overlap, but concepts and methods find universal application. Moreover, this definition fails to convey the spirit of chemistry, for it, like all science, is a vital, growing enterprise, not an accumulation of knowledge. It is self-generating; the very nature of each new chemical concept stimulates fresh observation and experimentation that lead to progressive refinement as well as to the development of other concepts. In the light of scientific growth, it is not surprising that a given scientific pursuit frequently crosses artificial, human-imposed boundaries.

Nevertheless, there is a common, if somewhat vague, understanding of the province of chemistry, and we must return to our preliminary definition; a fuller understanding should emerge as this book unfolds. Chemistry is concerned with the composition and the structure of substances and with the forces that hold the structures together. The physical properties of substances are studied since they provide clues for structural determinations, serve as a basis for identification and classification, and indicate possible uses for specific materials. The focus of chemistry, however, is probably the chemical reaction. The interest of chemistry extends to every conceivable aspect of these transformations and includes such considerations as: detailed descriptions of how and at what rates reactions proceed, the conditions required to bring about desired changes and to prevent undesired changes, the energy effects that accompany chemical changes, the syntheses of substances that occur in nature and those that have no natural counterparts, and the quantitative mass relations between the materials involved in chemical changes.

**1.1 MATTER AND ENERGY**

Science interprets nature in terms of energy and matter. **Matter,** the material of which the universe is composed, may be defined as anything that occupies space and has mass. Mass is a measure of quantity of matter. A body that is not being acted upon by some external force has a tendency to remain at rest or, when it is in motion, to continue in uniform motion in the same direction. This property is known as **inertia;** the mass of a body is proportional to the inertia of the body.

The mass of a body is invariable; the weight of a body is not. **Weight** is the gravitational force of attraction exerted by the earth on a body; the weight of a given body varies with the distance of that body from the center of the earth. The weight of a body is directly proportional to its mass as well as to the earth's gravitational attraction. Therefore, at any given place, two objects of equal mass have equal weights.

The mass of an object may be determined by means of an analytical balance; standard masses are placed on one pan of the instrument to balance the object of unknown mass that is placed on the other pan. When the instrument is "in balance," the gravitational force on one pan equals that on the other and the weights on the two pans are equal; therefore, the masses on the two pans are equal.

The terms mass and weight have long been used interchangeably in chemistry. Such usage is strictly incorrect but is firmly established through such terms as atomic weight, molecular weight, and weight-percent. Mass determinations are made through weight comparisons and, in writing and speaking, the verb "weigh" is usually less awkward than the phrase "determine the mass." In general, no significant errors arise through this loose usage; the precise meaning is almost invariably evident from the context in which the terms are used.

Matter exists in three physical states: gaseous, liquid, and solid. A **gas** maintains neither volume nor shape; it completely fills any container into which it is introduced and may be expanded or compressed within relatively wide limits. A **liquid** also has no fixed shape; however, a liquid assumes the shape of the container only within the limit of the volume that the sample occupies. Volumes of liquids do not vary greatly with changes in temperature and pressure. A **solid** under ordinary conditions maintains both volume and shape. Observed variations in the volumes of solids with changes in temperature and pressure are very small.

Changes in state (such as the melting of a solid and the vaporization of a liquid), as well as changes in shape or state of subdivision, are examples of **physical changes**—changes that do not involve the production of new chemical species. Physical means (such as filtration and distillation) may be used to separate the components of a mixture, but a substance that was not present in the original mixture is never produced by these means. **Chemical changes,** on the other hand, are transformations in which substances are converted into other substances.

A material may be characterized by its **intrinsic properties**—attributes that distinguish it from all other types of matter. Properties such as density, color, physical state, melting point, and electrical conductivity are known as physical properties because they can be observed without causing any change in the chemical composition of the specimen. The **chemical properties** of a substance are those that describe the chemical changes (chemical reactions) that the substance undergoes. Properties that are not characteristic of any particular type of matter, such as mass, length, and temperature, are known as **extrinsic properties.**

A distinct portion of matter that is uniform throughout in composition and in intrinsic properties is called a **phase.** If a material consists of only one phase, it is said to be **homogeneous;** if it consists of more than one phase, it is classified as **heterogeneous.** Iron, salt, and a solution of salt in water are each homogeneous; wood, granite, and a mixture of ice and water are each heterogeneous.

The phases of heterogeneous matter have distinct boundaries and are usually easily discernible. In granite, for example, it is possible to identify pink feldspar crystals, colorless quartz crystals, and shiny black mica crystals. When the number of phases in a sample is being determined, all portions of the same kind are counted as a single phase. Thus, granite is said to consist of three phases.

Heterogeneous materials do not have fixed compositions and are, therefore, **mixtures.** The relative proportions of the three phases of granite vary from specimen to specimen. An infinite number of heterogeneous mixtures is possible, since there are no restrictions on the proportions in which the various components may be present.

Homogeneous mixtures are called **solutions;** the single phase in which a solution occurs may be gaseous, liquid, or solid. Air, salt dissolved in water, and a silver–gold alloy are examples of gaseous, liquid, and solid solutions. All gases mix with one another in all proportions, but in solid and liquid solutions there is usually a limit to the solubility of one substance in another. Nevertheless, the compositions of solutions are variable; more than one solution is possible for each group of components from which a solution may be prepared; solutions are, therefore, mixtures.

Each phase of a heterogeneous mixture has its own properties. A homogeneous mixture has a single set of intrinsic properties, but each of these properties depends upon the composition of the mixture. The density, taste, boiling point, and electrical conductivity, for example, of aqueous salt solutions vary with the proportion of salt and water present.

**Pure substances** are homogeneous materials that have fixed compositions and invariable intrinsic properties; two types exist. An **element** is a pure substance that cannot be decomposed into any simpler pure substance. A **compound** is a pure substance that is composed of two or more elements in fixed and definite proportion. Thus, hydrogen and oxygen are elements; water is a compound composed of 11.19% hydrogen and 88.81% oxygen. Preparation of a compound from the constituent elements is known as **synthesis,** and decomposition of a compound into the component elements is called **analysis.** The classification of matter is summarized in Figure 1.1.

**Figure 1.1** Classification of matter.

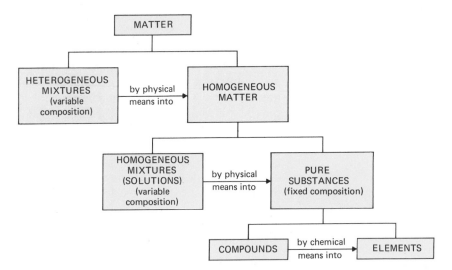

**Energy** is usually defined as the capacity to do work. Work is done when a body is displaced by a force, and the amount of work is equal to the product of the displacement, $d$, and the magnitude of the force in the direction of the displacement, $f$.

$$w = fd$$

A given force is directly proportional to the acceleration, $a$, it causes in a body (increase in velocity per unit time); the mass of the body, $m$, is the proportionality constant.

$$f = ma$$

There are many forms of energy (Section 14.1)—heat energy, electrical energy, kinetic energy (energy of motion), and potential energy (intrinsic energy of an object due to its position) are examples. Under appropriate conditions, one form of energy can be converted into another form; in these transformations energy is conserved (Section 14.1).

**1.2  MEASUREMENT**     The metric system of measurement is used in all scientific studies. As a result of a treaty signed in 1875, metric conventions are established and modified when necessary by international agreement. From time to time, an international group, the General Conference of Weights and Measures, meets to ratify improvements in the metric system. Some aspects of the currently approved **International System of Units** (*Le Système International d'Unites,* officially abbreviated **SI**) are summarized inside the back cover of this book. This system is a modernization and simplification (fewer units are used than previously) of the old metric system, which developed from one proposed by the French Academy of Science in 1790.

The International System is founded on seven **base units** and two **supplementary units** (see the table inside the back cover). The selection of primary standards for the base units is arbitrary. For example, the primary standard of mass, the kilogram, is defined as the mass of a cylinder of a platinum–iridium alloy that is kept at the International Bureau of Weights and Measures at Sèvres, France. The criteria for the selection of a primary standard are that it be reproducible, unchanging, and capable of being used for precise measurement. Throughout the years, the primary standards for some base units have been changed (when a new standard appeared to be superior to an old one).

Fractions or multiples of base units are indicated by the use of prefixes (see Table 1.1). The base unit of length is the meter. Since the prefix *centi-* means 0.01 ×, 1 centimeter is 0.01 meter.

$$1 \text{ cm} = 0.01 \text{ m}$$

Table 1.1 Prefixes used to modify unit terms in the metric system

| PREFIX | ABBREVIATION | FACTOR | |
|---|---|---|---|
| tera- | T- | 1 000 000 000 000 × | $(10^{12})$ |
| giga- | G- | 1 000 000 000 × | $(10^{9})$ |
| mega- | M- | 1 000 000 × | $(10^{6})$ |
| kilo- | k- | 1 000 × | $(10^{3})$ |
| hecto- | h- | 100 × | $(10^{2})$ |
| deka- | da- | 10 × | $(10^{1})$ |
| deci- | d- | 0.1 × | $(10^{-1})$ |
| centi- | c- | 0.01 × | $(10^{-2})$ |
| milli- | m- | 0.001 × | $(10^{-3})$ |
| micro- | $\mu$- | 0.000 001 × | $(10^{-6})$ |
| nano- | n- | 0.000 000 001 × | $(10^{-9})$ |
| pico- | p- | 0.000 000 000 001 × | $(10^{-12})$ |
| femto- | f- | 0.000 000 000 000 001 × | $(10^{-15})$ |
| atto- | a- | 0.000 000 000 000 000 001 × | $(10^{-18})$ |

In like manner, the prefix *kilo-* means 1000 ×, and 1 kilometer is 1000 meters.

$$1 \text{ km} = 1000 \text{ m}$$

All other SI units, called **derived units,** are derived from the base units by algebraic combination. For example, the base units for mass (kilogram, kg), length (meter, m) and time (second, sec) are combined to give the derived unit of force, the newton, N. The newton is defined as the force that gives a mass of 1 kg an acceleration of 1 m/sec². 

$$f = ma$$

$$f = (1 \text{ kg})(1 \text{ m/sec}^2) = 1 \text{ kg m/sec}^2 = 1 \text{ N}$$

The derived unit of work, or energy, is the joule, J (which is a newton–meter).

$$w = fd$$

$$w = (1 \text{ N})(1 \text{ m}) = 1 \text{ kg m}^2/\text{sec}^2 = 1 \text{ J}$$

The newton and the joule are examples of SI units that are given special names. Some derived units are named in terms of base units; for example, the SI unit of volume is the cubic meter, m³.

The use of certain units that are not a part of SI but are widely employed is sanctioned. The liter, for example, which is defined as the volume of 1 cubic decimeter (and hence is 1000 cm³), may be used in addition to the official SI unit of volume, the cubic meter. Certain other units that are not a part of SI are to be retained for a limited time; the standard atmosphere (atm, a unit of pressure) and the ångström unit (1 Å = $10^{-8}$ cm = $10^{-10}$ m) fall into this category. The International

Committee of Weights and Measures considers it preferable, in general, to avoid the use of certain other units that are outside the International System; the use of the calorie (cal, an energy unit) and the torr (760 torr = 1 atm) is therefore discouraged.

Not all scientists have adopted SI units, but their use appears to be growing. Many scientific groups and agencies (for example, the National Bureau of Standards) have recommended their use and have adopted them. The advantages of uniformity in usage among scientists of all nations are obvious. Strict adherence to the International System, however, poses a problem since the system eliminates some units that have previously been widely used. For example, chemists have commonly used the calorie (which is the amount of heat required to raise the temperature of 1 g of water from 14.5° to 15.5°C) as a unit of energy. The International System incorporates the joule as the unit for all energy measurements, and the International Committee of Weights and Measures recommends that the use of the calorie be discontinued. Thermodynamic values, however, have been recorded almost exclusively in terms of calories and kilocalories in the past; therefore, one must be familiar with both units (Section 14.1).

Marginal notes appear throughout this book to show the relations between SI units and the older but still commonly accepted units. The SI abbreviations for units, as well as the units themselves, are specified. The text of this book does not strictly follow SI conventions. For example, in the interest of clarity, the abbreviation sec is used for second — the official SI abbreviation is s. The marginal notes, however, employ the approved abbreviations.

● The Fahrenheit temperature scale is commonly used. When we speak of 98.6° as normal body temperature or 72° as a comfortable room temperature, we mean 98.6°F and 72°F. This scale, however, is not a part of the International System. The official SI temperature scale is the Kelvin scale; however, temperatures may also be expressed in degrees Celsius.

The **Celsius** (also called centigrade) temperature scale, named for Anders Celsius, is employed in scientific studies and is a part of the International System. This scale is based on the assignment of 0° to the normal freezing point of water and 100° to the normal boiling point of water. On the **Fahrenheit** temperature scale (named for G. Daniel Fahrenheit, a German instrument maker), the normal freezing point of water is 32° and the normal boiling point of water is 212°. Since there are 100 Celsius degrees and 180 Fahrenheit degrees between the freezing and boiling points of water, 5 Celsius degrees equal 9 Fahrenheit degrees. In Figure 1.2 these two temperature scales are compared and the formula for converting a reading on one scale to the other is given. The thermodynamic temperature scale, called the Kelvin scale, is described in Section 6.4.

**1.3 SIGNIFICANT FIGURES**

In scientific studies the number of figures recorded for a measured quantity must reflect the precision with which the measurement has been made and the degree of certainty that can be attributed to the reported

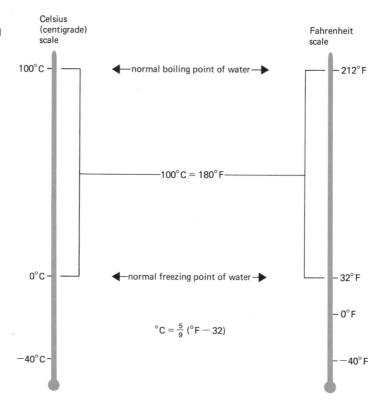

**Figure 1.2** Comparison of the Celsius (centigrade) and Fahrenheit temperature scales.

value. In short, all the figures must be **significant;** the value must provide as much usable information as is possible but must also avoid giving misinformation.

For example, let us suppose that the mass of an object is determined to be 12.3456 g. The analytical balance commonly employed in chemistry is capable of measuring the mass of an object to a sensitivity of 0.0001 g; therefore, all the reported figures are significant. There is some uncertainty in the last figure (6), but it has value and is significant because it tells us that the true mass lies closer to 12.3456 g than to 12.3455 g or 12.3457 g. This is the most precise value we can secure; it is impossible to derive a value that contains seven significant figures for the mass of this particular object by using a balance with a sensitivity of 0.0001 g.

If we add a zero to the measurement, we indicate a value of *seven* significant figures (12.34560), which is incorrect and misleading. The conclusion drawn from the value 12.34560 g is that the true mass lies between 12.34559 g and 12.34561 g, a much closer tolerance than that indicated by the properly reported value. In fact, we have no idea of the magnitude of the integer of the fifth decimal place. A zero does not indicate that the fifth decimal place is unknown or undetermined; rather, the zero is interpreted to be as significant as any other figure in the number. A properly determined and recorded mass, such as 42.5630 g, may legitimately include a zero as a significant figure.

Zeros used to locate the decimal place, however, are not significant. The number 0.0005030 has four significant figures; two of the zeros are significant, whereas those preceding the numeral 5 are not. Expressing the diameter of a particle as 0.00000001 cm rather than 1 Å (an ångström, Å, is $10^{-8}$ cm) does not render the measurement any more precise.

Occasionally, difficulty arises in interpreting the number of significant figures in a term such as 6000. Are the zeros significant, or do they merely locate the decimal point? One solution to the problem is to indicate the position of the decimal point as a power of 10 and include only those zeros that are significant in the other portion of the number. Thus, if the value 6000 has been determined to four significant figures, it is indicated as $6.000 \times 10^3$; otherwise, fewer zeros are placed after the decimal point.

Certain values, such as those that arise from the definition of terms, are exact. Thus, there are *exactly* 1000 ml in 1 liter, and the number 1000 may be considered to possess an infinite number of significant zeros following the decimal point. Other values are not so precise as might normally be expected. For example, if a balance with a sensitivity of 0.0002 g is used to determine 12.3456 g as the mass of an object, the measurement should be recorded as $12.3456 \pm 0.0002$ g. The uncertainty range ($\pm 0.0002$ g) gives the reliability of the determination. The last figure of a value recorded without an uncertainty range is understood to be precise plus or minus one digit.

Frequently the values used in a calculation have different numbers of significant figures. The method used to determine the proper number of significant figures in the answer depends upon the mathematical operation performed. The result of an **addition** or **subtraction** should be reported to the same number of decimal places as that of the term with the least number of decimal places. The answer for the addition

$$\begin{array}{r} 161.032 \\ 5.6 \\ \underline{32.4524} \\ 199.0844 \end{array}$$

should be recorded as 199.1 since the number 5.6 has only one digit following the decimal point.

The answer to a calculation involving **multiplication** or **division** is rounded off to the same number of significant figures as is possessed by the least precise term used in the calculation. Thus, the result of the multiplication

$$152.06 \times 0.24 = 36.4944$$

should be reported as 36, a value which conforms to the limitations imposed by the term 0.24. In reality, this answer is slightly more precise than the least precise term. The value 0.24 indicates a precision of 1 part in 24, whereas the answer indicates the slightly greater precision of 1 part in 36.

The values used in a calculation involving several steps should be rounded off before the mathematical operations are performed. Each value that has an excess number of significant figures is rounded off to *one more* significant figure than the number of significant figures required to express the answer. For example, the answer to the computation

$$1.267 \left( \frac{4.353178}{56} \right)$$

should contain only two significant figures since this is the level of precision of the value 56. Therefore, the values to be used in the calculation should be rounded off to *three* significant figures, and the answer should be rounded off to *two* significant figures. Thus,

$$1.27 \left( \frac{4.35}{56} \right) = 0.099$$

# 2

# Atomic Structure

One of the theories that has, in large measure, provided the impetus for the rapid advance of chemistry is the modern atomic theory. An understanding of atomic structure, the ways in which atoms interact, and the nature of the products of these interactions is central to an understanding of chemistry. Furthermore, atomic theory makes the quantitative aspects of chemistry intelligible.

Individual atoms cannot be weighed, measured, or examined directly. Indirect evidence has been used, and of necessity must be used, to develop the atomic theory to its present complex (but still incomplete) state. Many nineteenth-century scientists (Ludwig Boltzmann, for example) believed in the physical existence of atoms; many others (Michael Faraday, for example) thought the atomic theory to be a mental construct that may or may not represent reality; still others (Wilhelm Ostwald, for example) rejected atomism completely. Whether atoms are real or imaginary was an open question as late as 1904. In that year a debate, attended by noted international scientists, was held at the St. Louis World Fair; Ostwald argued against the existence of atoms and Jacobus van't Hoff and Boltzmann defended the other side.

**2.1 THE ATOM**

Through the ages men have been intrigued with the problem of the ultimate constitution of matter. Credit for the first "atomic" theory is usually given to the ancient Greeks; however, this concept may have had its origins in even earlier civilizations. Aristotle (fourth century B.C.) held that matter was continuous and hence, hypothetically, could be divided and subdivided endlessly into smaller and smaller particles. The atomistic school, principally of Leucippus, Democritus, and Epicurus (sixth to fourth centuries B.C.), believed that the subdivision of matter ultimately would yield atoms (from the Greek word *atomos,* meaning "uncut" or "indivisible"), which could not be further reduced. In the first century B.C. the atomic concept was eloquently presented by the Roman poet–philosopher Lucretius in his *De Rerum Natura* (*The Nature of Things*).

The ancients' theories were based on abstract thinking and not on planned experimentation. For nearly 2000 years the atomic theory was mere speculation. It formed a part of the thinking of Robert Boyle in his book *The Sceptical Chymist* (1661) and of Isaac Newton in his works *Principia* (1687) and *Opticks* (1704). It remained, however, for John Dalton to propose an atomic theory (1803–1807) that became a landmark in the progress of chemistry. Dalton's theory, based on experimentation and chemical laws known at that time, assigned weights and combining capacities to the postulated atoms. Because of its quantitative nature, his theory convincingly interpreted many observed facts and stimulated new work and thought. Dalton's theory in its broad outline is still valid; however, some of the particulars have been modified in the light of modern discovery.

Today we know that the division of matter (any type from any source) into the smallest particles that are capable of independent, prolonged existence would result in the separation of about 300 different kinds of atoms (naturally occurring **nuclides**). Some of these nuclides are very similar chemically; if the 300 nuclides are grouped according to chemical characteristics, about 100 sets result. The atoms of a given set are isotopes of the same element (Section 2.8).

**2.2 SUBATOMIC PARTICLES**

Although Dalton, like the Greeks, regarded the atom as indivisible, it is now known that the atom consists of even smaller particles into which it may be divided. In fact, certain radioactive atoms spontaneously disintegrate into smaller fragments.

At the present time approximately 35 different **subatomic particles** have been identified from the subdivision of atoms. Many of these particles are unstable and exist for only a split second; their status relative to the atom is not completely understood.

Three types of particles—the **electron, proton,** and **neutron** (see Table 2.1)—are regarded as "fundamental." Whether this term is wisely chosen is questionable since these particles may arise from lesser species. The neutron breaks down spontaneously when isolated. For purposes of the study of chemistry, however, atomic structure is adequately explained on the basis of these three fundamental particles.

The neutron is not electrically charged. The electron and proton have electrical charges of equal magnitude (the unit charge) but of opposite sign. The electron is negative; the proton, positive. The unit electrical charge is $1.6022 \times 10^{-19}$ coulomb or $4.8033 \times 10^{-10}$ esu. A **coulomb** is the quantity of charge that passes a given point in an electrical circuit in 1 second when the current is 1 ampere. One coulomb is $2.998 \times 10^9$ esu.

● The coulomb (C) is the SI derived unit of electric charge. It is equal to an ampere second.

$$C = A \cdot s$$

The ampere (A) is the SI base unit of electric current and the second (s) is the SI base unit of time.

The definition of the **electrostatic unit** (esu) is derived from **Coulomb's law.** The force of attraction or repulsion, $F$, between two charges $q_1$ and $q_2$ separated by a distance $d$ is

$$F = \frac{q_1 q_2}{D d^2}$$

where $D$ is the dielectric constant. The value of $D$ depends upon the medium in which the charges are dispersed; for a vacuum, $D = 1$. The electrostatic unit is defined as the charge that will repel another identical charge by a force of exactly 1 dyne when the charges are separated by exactly 1 cm in a vacuum.

● A dyne is the force that gives a mass of one gram an acceleration of 1 cm/s$^2$. The dyne is related to the newton (the SI derived unit of force).

$$1 \text{ dyne} = 10^{-5} \text{ N}$$

The dyne, however, is not a unit of the International System. Use of this unit with SI units is discouraged. The esu is also not a part of the International System.

1 esu = $3.33564 \times 10^{-10}$ C

Table 2.1 Fundamental subatomic particles

| | MASS | | |
|---|---|---|---|
| PARTICLE | GRAMS | UNIFIED ATOMIC MASS UNITS[a] | CHARGE[b] |
| electron | $9.1096 \times 10^{-28}$ | 0.00054859 | 1 − |
| proton | $1.6726 \times 10^{-24}$ | 1.007277 | 1 + |
| neutron | $1.6749 \times 10^{-24}$ | 1.008665 | 0 |

[a] The unified atomic mass unit (u) is 1/12 the mass of a $^{12}$C atom (Section 2.9).
[b] The unit charge is $1.6022 \times 10^{-19}$ coulomb or $4.8033 \times 10^{-10}$ esu.

Figure 2.1 **Cathode-ray tube.**

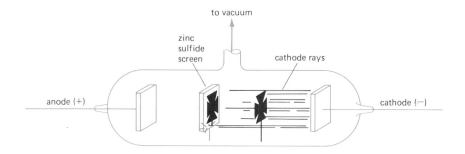

### 2.3 THE ELECTRON

In the early 1800's the English chemist Humphry Davy studied chemical electrolysis—the decomposition of substances brought about by electric currents. These studies led Davy to the theory that the elements of a chemical compound are held together by bonds that are electrical in nature. The quantitative relations between the amount of electricity used in an electrolysis and the chemical effect of the electrolysis were determined by Michael Faraday, Davy's protégé and successor, in 1832 and 1833 (Chapter 10). In interpreting Faraday's work, George Johnstone Stoney proposed in 1874 that units of electrical charge are associated with atoms, and in 1891 he suggested that these units be called electrons (from the Greek, meaning "amber"—so named because of the effect of friction on this substance).

Much of the information about electrons comes from the study of **cathode rays.** When a high voltage is impressed across two electrodes sealed in a glass tube from which the air is almost completely withdrawn, rays emanate from the negative electrode (Figure 2.1). These rays are negatively charged, travel in straight lines, and cause the walls opposite the cathode to glow.

In the latter part of the nineteenth century, the nature of these cathode rays was extensively investigated. Experiments showed the rays to be streams of fast-moving electrons. These electrons, which originate from the metal that composes the negative pole, are the same no matter what material is employed for the electrode.

In 1897 Joseph J. Thomson determined the ratio of the charge of the electron to its mass, $e/m$, by studying the deflection of cathode rays in magnetic and electric fields. The methods of Thomson, however, are incapable of giving either the charge, $e$, or the mass, $m$, alone.

Since they are electrically charged, electrons in motion constitute a current of electricity; magnetic fields interact with electric currents. (The electric motor is based on this principle.) When an electron of charge $e$ and mass $m$, moving with a velocity $v$, enters a magnetic field, it is deflected from its usual straight-line path into a circular path of radius $r$ (Figure 2.2). The deflection occurs at a right angle to the field, and the electron continues in a straight line when it leaves the field.

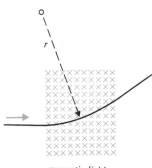

magnetic field (perpendicular to paper)

Figure 2.2 Path of an electron in a magnetic field.

The value of $r$ is given by

$$r = \frac{mv}{He} \qquad (1)$$

where $H$ is the magnetic field strength. The larger the field strength or the larger the charge on the electron, the greater the deflection from a straight-line path. A large deflection, however, means that the value of $r$ is small. Therefore, $H$ and $e$ are *inversely* proportional to $r$. On the other hand, a comparatively large value of $r$, indicating little deviation from a straight line, is characteristic of a particle with a large mass or one that is moving rapidly. Hence, $m$ and $v$ are *directly* proportional to $r$.

Thomson measured the values of $r$ for the deflections of cathode rays in magnetic fields of known strengths and was able to obtain a value for $e/mv$. From (1),

$$\frac{e}{mv} = \frac{1}{Hr} \qquad (2)$$

In order to get $e/m$, however, the velocity of the electron, $v$, had to be determined. One of the methods Thomson used to find $v$ involved the study of cathode rays in a tube around which magnetic and electric fields, perpendicular to each other, could be simultaneously applied (Figure 2.3).

Electrons describe a parabolic path in an electric field; deflection occurs toward the positive plate. Under the influence of the electric field alone, the electron would strike the tube at point (c). The force causing the deflection of a charged particle in an electric field depends only upon the strength of the electric field, $E$, and the magnitude of the charge ($e$, for the electron). The force is the same whether the particle is at rest or in motion. Hence,

$$F = Ee \qquad (3)$$

Under the influence of the magnetic field alone, the cathode ray in Figure 2.3 would strike the tube at point (a). A magnetic field has no

Figure 2.3 Thomson's determination of $e/m$ for the electron.

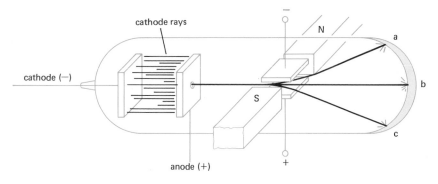

effect on a charged particle at rest. Magnetic fields interact with electric currents — charges in motion. The force causing the magnetic deflection depends, therefore, not only upon the magnetic field strength and the charge of the electron but also upon the velocity of the electron.

$$F = Hev \qquad (4)$$

Equations (3) and (4) refer only to the *forces* that cause deflections. The magnitude of a deflection in either a magnetic or an electric field also depends upon the mass of the electron and the length of time that the electron spends in the field. Thomson balanced the intensities of the electric and magnetic fields, when they were applied *simultaneously,* so that no net deflection occurred and the impact of the ray in Figure 2.3 was at point (b). Under these conditions, the electron, which has a constant mass, is subjected to the effect of the magnetic field for the same length of time that it is subjected to the effect of the opposing electric field. Since there is no net deflection, the two opposing forces must be equal. Therefore, from equations (3) and (4),

$$Hev = Ee \qquad (5)$$

$$v = \frac{E}{H} \qquad (6)$$

Multiplication of the value of $e/mv$ (determined by means of equation 2) by $v$ (determined by means of equation 6) gives the value of $e/m$.

$$e/m = -5.2728 \times 10^{17} \text{ esu/g}$$

The first precise measurement of the charge of the electron, which is the unit of electrical charge, was made by Robert A. Millikan in 1909.

**Figure 2.4** Millikan's determination of the charge on the electron.

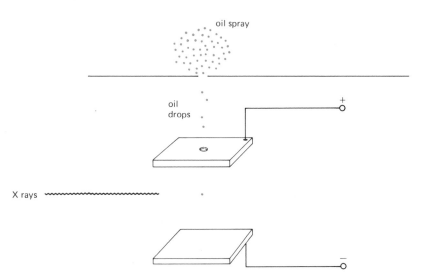

By the use of X rays, very small oil drops are made to acquire electric charges. These drops are allowed to settle between two horizontal plates, and the mass of a single drop is determined by measuring its rate of fall (Figure 2.4).

When the plates are charged, the rate of fall of the charged drop is altered. Measurement of the velocity of the drop in these circumstances enables calculation of the charge on the drop. The charges thus calculated are not identical; they are, however, all simple multiples of the same value, $4.8033 \times 10^{-10}$ esu, the unit electrical charge. Therefore,

● In SI units the charge of the electron is $-1.6022 \times 10^{-19}$ C, and the mass of the electron is $9.1096 \times 10^{-31}$ kg.

$$e = -4.8033 \times 10^{-10} \text{ esu}$$

Division of this value for $e$ by the value for $e/m$ gives $m$, the mass of the electron.

$$m = 9.1096 \times 10^{-28} \text{ g}$$

**2.4 THE PROTON**   If one or more electrons are removed from a neutral atom, the residue, which is called an **ion** (from the Greek, meaning "to go"), has a charge equal to the sum of the charges of the electrons removed but opposite in sign. When the lightest of all atoms, hydrogen, loses its lone electron, the ion produced is a fundamental particle called the **proton** (from the Greek, meaning "first"). The proton has a mass 1836 times that of the electron, and it has a positive charge of exactly the same magnitude as the negative charge on the electron (1+ and 1− in terms of the unit electrical charge).

In an electric discharge tube the cathode rays form ions by ripping electrons away from atoms of the gas present in the tube. These ions are positively charged and move in a direction contrary to that of the cathode rays (which are negatively charged); they move away from the positive pole toward the cathode. Many of these ions pick up electrons and again become neutral atoms. Some of them, however, reach the cathode; if holes have been bored in this electrode, the ions pass through them and beyond. These streams of positive ions, called **positive rays** or **canal rays,** were first observed by Eugen Goldstein in 1886 (Figure 2.5).

The deflections of positive rays in electric and magnetic fields were

Figure 2.5 Positive rays.

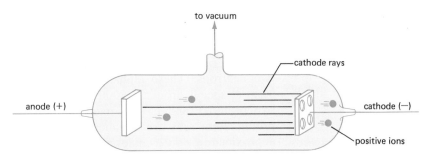

studied by Wilhelm Wien (1898) and J. J. Thomson (1906), and values of $e/m$ for the positive ions that constitute these rays were determined (Section 2.8). The ions differ depending upon the gas used in the discharge tube.

The value of $e/m$ for a positive ion depends upon the charge of the ion (which is proportional to the number of electrons lost by the atom in forming the ion) and upon the mass of the particular ion under study. For ions with the same charge, the value of $e/m$ is comparatively large when the mass of the ion is comparatively small. When hydrogen gas is used in the discharge tube, positive ions are formed that have the largest $e/m$ value observed for positive ions, $+2.872 \times 10^{14}$ esu/g. These ions, which are protons, have the smallest mass observed for any positive ion. Since each proton has a unit positive charge, $+4.8033 \times 10^{-10}$ esu, the mass of the proton can be calculated from the $e/m$ value; it is $1.6726 \times 10^{-24}$ g.

● In SI units the charge of the proton is $+1.6022 \times 10^{-19}$ C, and the mass of the proton is $1.6726 \times 10^{-27}$ kg.

**2.5 THE NEUTRON**  Since atoms are electrically neutral, a given atom must contain as many electrons as protons. To account for the total masses of atoms, Ernest Rutherford in 1920 postulated the existence of an uncharged particle. Since this particle is uncharged, it was difficult to detect and characterize. In 1932, however, James Chadwick published the results of his work which established the existence of the **neutron** (from the Latin, meaning "neutral"). He was able to calculate the mass of the neutron from data on certain nuclear reactions (see Chapter 21) in which neutrons are produced. By taking into account the masses and energies of all particles used and produced in these reactions, Chadwick determined the mass of the neutron. It has a mass very slightly larger than that of the proton; the neutron has a mass of $1.6749 \times 10^{-24}$ g and the proton, $1.6726 \times 10^{-24}$ g.

**2.6 THE NUCLEAR ATOM**  Certain atoms are unstable combinations of the fundamental particles. These atoms spontaneously emit rays and are thereby **transmuted** into atoms of different chemical identity. This process, **radioactivity,** was discovered by Henri Becquerel in 1896, and the nature of the rays produced was subsequently elucidated by Rutherford. The three types of rays emitted by naturally occurring radioactive materials (others have now been identified from man-made atoms) were named alpha ($\alpha$), beta ($\beta$), and gamma ($\gamma$).

Alpha rays consist of particles that are composed of two protons and two neutrons. These alpha particles are ejected from the radioactive atom at speeds around 10,000 miles/sec, carry a 2+ charge (from their two protons), and have a mass approximately four times that of the proton. Beta rays are streams of electrons (1− charge) that travel at approximately 80,000 miles/sec. Gamma radiation is essentially a highly energetic form of light; these rays are uncharged and are similar to X rays.

● The mile is not a unit of the International System. The SI base unit of length is the meter (m). Since 1 mile equals approximately 1.6 kilometers, $\alpha$ rays are ejected at speeds around $1.6 \times 10^4$ km/s, and β rays travel at approximately $12.8 \times 10^4$ km/s. Note that the unit of length may be spelled either "meter" or "metre."

By the use of α particles, Rutherford in 1911 presented convincing evidence for a nuclear construction of the atom. A radioactive source of α rays was placed behind a lead shield. A thin beam of α rays emerged through a hole bored in the thick lead shield and was directed against very thin (ca. 0.0004 cm) foils of gold, platinum, silver, or copper. About 99.9% of the α particles went directly through the foil in use; however, some were deflected from their straight-line path, and a few were sent back toward their source.

Rutherford explained the results of these scattering experiments by postulating the existence of a nucleus in the center of the atom. Most of the mass and all of the positive charge of the atom (the protons and neutrons) are concentrated in this nucleus. The electrons, which occupy most of the total volume of the atom, are outside this nucleus (extra-nuclear) and are in rapid motion around it.

It is important to grasp the scale of this model. If an atom had a diameter of 1 mile, its nucleus would have a diameter on the order of $\frac{2}{3}$ inch. Since most of the volume of an atom is empty space, most α particles pass undeflected through the target foils, even though the foils are approximately 1000 atoms thick.

If an α particle scores a direct hit on a nucleus, it recoils in the direction of its source; such recoils, though few in number, were observed. A near miss—a very close approach of a positively charged α particle to a positively charged nucleus—results in repulsion of the α particle and deflection from its straight-line path (Figure 2.6). The comparatively

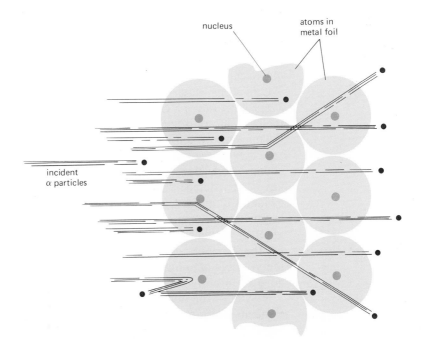

**Figure 2.6** Deflections and recoil of α particles by nuclei of metal foil in Rutherford's experiment (not to scale).

nucleus

atoms in metal foil

incident α particles

light electrons do not cause the deflection of the heavier, fast-moving $\alpha$ particles.

The composition and stability of the nucleus is discussed in Chapter 21.

**2.7** ATOMIC SYMBOLS    An atom may be identified by two numbers: the atomic number, $Z$, and the mass number, $A$. The atomic number (Section 2.12) represents the number of unit positive charges on the nucleus and is equal to the number of protons in the nucleus since each proton carries a $1+$ charge. An atom, since it is electrically neutral, must have as many electrons (at $1-$ each) as it has protons (at $1+$ each). Therefore, the atomic number also indicates the number of electrons in the neutral atom under consideration.

The mass number of an atom is the total number of neutrons and protons (**nucleons**) in the nucleus of the atom. The number of neutrons in a given nucleus can be calculated by subtracting the atomic number (the number of protons) from the mass number (the number of neutrons and protons).

number of neutrons $= A - Z$

The mass *number* represents the total *number* of nucleons in a nucleus and not the mass of the nucleus. However, since the mass of both the proton and the neutron are approximately equal to 1 u (Section 2.9), the mass number is generally a whole number approximation of the atomic masses expressed in atomic mass units.

An atom of sodium with an atomic number of 11 and a mass number of 23˙ would have 11 protons and 12 neutrons in the nucleus and 11 extranuclear electrons. Since chemical reactions involve only extranuclear electrons, all neutral atoms with 11 electrons behave the same chemically; all such atoms are given the same chemical name—sodium. We shall see later that it is possible to have, within limits, atoms with 11 protons and 11 electrons but a varying number of neutrons. Nevertheless, all atoms with 11 protons in the nucleus and 11 electrons around the nucleus are sodium atoms; they constitute the *element* sodium.

Each element is given a symbol, a practice introduced by the Swedish chemist Jöns Berzelius in the early 1800's. The names of the elements have been decided upon by international agreement; the symbols are usually the initial letters of these names. Where more than one element begins with the same letter, a second letter (not necessarily the second letter of the name) is added to distinguish them. Most of the symbols for the elements correspond closely to their English names. However, the symbols of some of the elements, known and characterized long ago, have been assigned on the basis of their Latin names (see Table 2.2). The symbol for tungsten, W, is derived from the German name for the element, wolfram.

Table 2.2 Symbols of elements derived from Latin names

| ENGLISH NAME | LATIN NAME | SYMBOL |
|---|---|---|
| antimony | stibium | Sb |
| copper | cuprum | Cu |
| gold | aurum | Au |
| iron | ferrum | Fe |
| lead | plumbum | Pb |
| mercury | hydrargyrum | Hg |
| potassium | kalium | K |
| silver | argentum | Ag |
| sodium | natrium | Na |
| tin | stannum | Sn |

A specific type of atom (nuclide) is designated by using the chemical symbol for the element with the atomic number placed at the lower left corner of the symbol and the mass number placed at the upper left corner. (The other two corners are reserved for other designations: the upper right for the charge if the atom has lost or gained one or more electrons and become an ion; the lower right corner for the number of atoms present in a formula unit of a compound.) Thus, the sodium atom previously described would be designated $_{11}^{23}\text{Na}$. An atom with the symbol $_{17}^{35}\text{Cl}$ has 17 protons and 18 neutrons in the nucleus and 17 extranuclear electrons. An atom of uranium with the designation $_{92}^{238}\text{U}$ has 92 protons and 146 neutrons in its nucleus and 92 electrons in motion about this nucleus.

**2.8 ISOTOPES AND ISOBARS**

One of the postulates of Dalton's atomic theory is that every atom of a given element is identical in all respects (including mass) to every other atom of that element. By the early twentieth century, however, it was clear that an element may consist of several varieties of atoms that differ from one another in atomic mass. Frederick Soddy proposed the name **isotopes** (from the Greek, meaning "same place") for atoms of the same element that differ in mass.

The **mass spectrograph** is used to study isotopes. Instruments of this type were first constructed by Francis W. Aston (1919) and Arthur J. Dempster (1918), who followed principles developed by J. J. Thomson in 1912. If an element consists of several types of atoms that differ in mass (isotopes), this difference is reflected in the $e/m$ values of the several types of positive ions derived from this element (Section 2.4). The mass spectrograph separates ions according to their charge-to-mass ratios; different positive ions are caused to impinge at different positions on a photographic plate.

When the instrument is operated, atoms of the vaporized material to be studied are converted into positive ions by bombardment with streams of electrons. These ions are accelerated by passage through an electric field of several thousand volts. Provided the voltage is held constant, all ions that have the same $e/m$ value enter a magnetic field with about the same velocity. This velocity, the $e/m$ value, and the strength of the magnetic field determine the radius of the path of the ion in the magnetic field (equation 1 in Section 2.3).

If the strength of the magnetic field and the accelerating voltage are held constant, ions that have the same value of $e/m$ will focus at the same position on the photographic plate. This position may be altered by changing the potential used to accelerate the ions. Ions that have different $e/m$ values, however, will focus at different positions on the plate. When an electrical device that measures the intensities of the ion beams is substituted for the photographic plate, the instrument is known as a **mass spectrometer** (Figure 2.7). Use of the mass spectrometer enables determination of both the exact atomic masses of the isotopes and the isotopic composition of the elements (the types of isotopes present and the relative amount of each).

Isotopes are atoms with the same atomic number but different mass numbers. They have very similar (in most cases, indistinguishable) chemical properties. In nature there are two types of atoms of chlorine: $^{35}_{17}Cl$ and $^{37}_{17}Cl$. Both these atoms have 17 protons and 17 electrons; however, $^{35}_{17}Cl$ has 18 neutrons and $^{37}_{17}Cl$ has 20 neutrons. Isotopes, therefore, differ in the number of neutrons in the nucleus. Some elements exist in nature in only one isotopic form (e.g., sodium, beryllium, and fluorine); however, most elements have more than one isotope—tin has ten. The term **nuclide** is used for atomic species in general.

**Isobars** are nuclides that have the same mass number but different

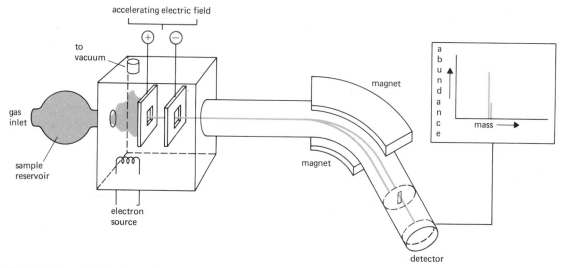

**Figure 2.7** Essential features of a mass spectrometer.

atomic numbers. Thus, $^{36}_{16}S$ and $^{36}_{18}Ar$ are isobars. The total numbers of nucleons in each of these isobars is the same, but the numbers of protons and neutrons are different. The sulfur isotope has 16 protons and 20 neutrons (a total of 36 nucleons), whereas the argon isotope has 18 protons and 18 neutrons (also a total of 36 nucleons). Isobars are not alike chemically since chemical characteristics depend principally upon the number of electrons, which is determined by the atomic number.

**2.9 ATOMIC WEIGHTS**

● The symbol u is the SI symbol for the atomic mass unit; amu, another symbol for the atomic mass unit, is also widely used.

● Prior to 1961 oxygen was used as the standard. Chemists based their scale on the assignment of exactly 16 to the *average* mass of oxygen atoms as they occur in nature. Physicists, however, based their scale on the *single* isotope $^{16}_8O$. Since naturally occurring oxygen consists of three isotopes (mass numbers 16, 17, 18), the average value for oxygen was 16.0044 on the physical scale. To eliminate this duality, the unified scale based on $^{12}_6C$ was developed. The SI symbol u was selected so that it would indicate measurements made on the *unified* scale.

The usual standards of mass are too large to be convenient for the measurement of the very small nucleons and nuclides. Therefore, a standard of mass has been adopted for the measurement of these particles that is on a scale commensurate with them.

The **atomic mass unit** (u) is defined as one-twelfth the mass of the nuclide $^{12}_6C$. Note that this definition gives an approximate value of unity to the mass of the proton and to the mass of the neutron. The assignment of the value of 12 u to this isotope of carbon is quite arbitrary; other nuclides could have been used, and, indeed, others have been used in the past (Dalton used hydrogen as his standard). At the present time, the masses of all nuclides are expressed as a ratio of their masses to the mass of $^{12}_6C$ taken as exactly 12 u.

On this scale the mass of the proton is 1.007277 u, the mass of the neutron is 1.008665 u, and the mass of the electron is 0.0005486 u. Given these values, we might expect that the mass of any nuclide could be calculated from its atomic number and mass number; such, however, is not the case. The mass of $^{35}_{17}Cl$, for example, might be expected to equal the sum of the masses of 17 protons, 18 neutrons, and 17 electrons.

$$17(1.007277) \text{ u} + 18(1.008665) \text{ u} + 17(0.0005486) \text{ u} = 35.289005 \text{ u}$$

The mass of $^{35}_{17}Cl$ has been accurately determined, however, to be 34.96885 u. The difference between the two values is

$$35.28901 \text{ u} - 34.96885 \text{ u} = 0.32016 \text{ u}$$

This difference expressed in its energy equivalent is called the **binding energy** of the nuclide in question.

Einstein has shown that energy and mass are equivalent:

$$E = mc^2$$

where $E$ is energy (in ergs), $m$ is mass (in grams), and $c$ is the speed of light ($3.00 \times 10^{10}$ cm/sec). Thus, the energy equivalent of 1 g of matter is

$$E = (1.00 \text{ g})(3.00 \times 10^{10} \text{ cm/sec})^2$$
$$= 9.00 \times 10^{20} \text{ g cm}^2/\text{sec}^2 = 9.00 \times 10^{20} \text{ ergs}$$

Since 1 cal $= 4.184 \times 10^7$ ergs,

$$E = 2.15 \times 10^{13} \text{ cal} = 2.15 \times 10^{10} \text{ kcal}$$

● Since the SI derived unit of energy, the joule (J), is equal to $10^7$ ergs, one gram of matter is equal to $9.00 \times 10^{13}$ J of energy.

Therefore, 1 g of matter is equal to over 20 billion kcal of energy.

Binding energy is generally interpreted in terms of the particles of the nucleus. If it were possible to pull the nucleus apart, the binding energy would be the energy required to do the job. The reverse process, the condensation of nucleons into a nucleus, would release the binding energy with an attendant decrease in mass. The removal or addition of electrons to an atom also involves energy changes which, of course, have mass equivalents. These mass equivalents, however, are extremely small and may be neglected.

For every atom except $^1_1$H (the nucleus of which consists of a single proton), the sum of the masses of the constituent particles exceeds the actual mass of the nuclide. Thus, every nucleus that consists of two or more nucleons has a binding energy. Binding energy is discussed more fully in Section 21.6.

Most naturally occurring elements consist of a mixture of isotopes. Chemical calculations would be hopelessly complicated except for the fact that, with very few exceptions, these mixtures have constant compositions. The element chlorine consists of a mixture of 75.53% $^{35}_{17}$Cl and 24.47% $^{37}_{17}$Cl. Any sample of chlorine from a natural source consists of these two isotopes in this proportion.

The weighted average of the atomic masses of the natural isotopes of chlorine is called the **atomic weight** of chlorine. This average is calculated by multiplying the mass of each isotope by its fractional abundance and by adding the values obtained.

$$0.7553(34.97 \text{ u}) + 0.2447(36.95 \text{ u}) = 35.45 \text{ u}$$

The accepted value for chlorine is $35.453 \pm 0.001$.

There is no atom of chlorine with a mass of 35.453 u, but it is convenient to think in terms of one. In even a small sample of matter there is a huge number of atoms—there are more atoms in a drop of water than there are people on the earth. In most calculations no mistake is made by assuming that a sample of an element consists of only one type of atom with the average mass, the atomic weight.

**2.10** ELECTRO-MAGNETIC RADIATION

Much has been learned about atomic structure through studies of the absorption and emission of radiant energy by atoms. An appreciation of these investigations requires that the nature of radiant energy be understood.

Radiant energy may be considered to consist of fluctuating electric and magnetic fields that are propagated through space in a wave motion

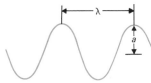

Figure 2.8 Wavelength, λ, and amplitude, *a*, of a wave.

● The International System uses the name hertz (Hz) for a reciprocal second.

$$1 \text{ Hz} = 1 \text{ s}^{-1}$$

This unit of frequency is the same as the cycle/sec used in the text (cycle need not be specified).

● The ångström (Å) is equal to $10^{-10}$ m. It is not an official SI unit, but it

(Figure 2.8). The distance from crest to crest (or trough to trough) is called a **wavelength,** λ (lambda). The height of the wave, called its **amplitude,** *a*, is related to the **intensity** of the radiation.

Matter is not needed for the propagation of these waves. In a vacuum all waves, regardless of wavelength, move with a velocity of $2.9979 \times 10^{10}$ cm/sec. This velocity, called the speed of light, is usually given the symbol *c*.

If the speed of light (in cm/sec) is divided by the wavelength of a given wave (in cm/cycle), the result gives the number of waves of that type that pass a given spot in a second (in cycles/sec). This value is the **frequency,** ν (*nu*), of the radiation. Thus,

$$\nu = \frac{c}{\lambda}$$

or

$$c = \lambda \nu$$

The electromagnetic spectrum is shown in Figure 2.9. Seemingly diverse types of radiation are in reality electromagnetic radiations of different wavelength. Radio waves have very long wavelengths, infrared rays (which are radiant heat) have moderate wavelengths, and γ rays (from radioactive decay) have extremely short wavelengths. White light consists of radiations with wavelengths in the approximate range of $4 \times 10^{-5}$ cm to $7.5 \times 10^{-5}$ cm (4000 Å to 7500 Å).

The wave theory successfully interprets many properties of electromagnetic radiation. Other properties, however, require that such radiation

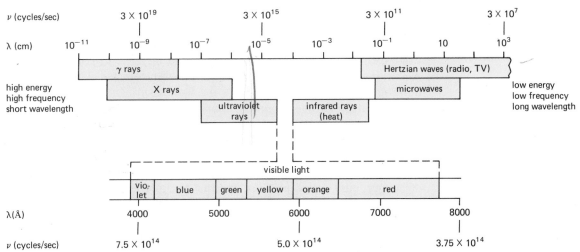

Figure 2.9 Electromagnetic radiation. Note that the approximate ranges of electromagnetic radiations are plotted on a logarithmic scale in the upper part of the diagram; the spectrum of visible light is not plotted in this way.

may be used with SI units for a limited period of time. The preferred unit is the nanometer (nm), which is $10^{-9}$ m. White light, therefore, consists of radiation with wavelengths in the approximate range of 400 nm to 750 nm.

be considered as consisting of particles, for example, the radiation emitted by hot bodies. The temperature of a hot body determines the range of wavelengths of the light emitted; at each temperature a maximum intensity is observed at a characteristic wavelength. As the temperature of the incandescent body increases, the maximum shifts to shorter wavelengths and the range of the light emitted includes shorter wavelengths.

The body glows red (the longest wavelengths in the visible region) at the first appearance of visible light. As the temperature increases, the emitted light appears orange since the shorter wavelengths characteristic of orange and yellow are included. Finally at "white heat," wavelengths of the entire visible band are included in the emitted radiation, which appears white. According to the wave theory, however, the energy of electromagnetic radiation is proportional to the square of the amplitude of the wave and is not dependent upon the wavelength.

In 1900 Max Planck proposed the **quantum theory** of radiant energy to explain these observations. Planck suggested that radiant energy could be absorbed or emitted only in discrete quantities (called **quanta**). The energy of each quantum is directly proportional to the frequency of the radiation, $\nu$. Since the frequency is equal to the speed of light, $c$, divided by the wavelength, $\lambda$,

$$E = h\nu = h \frac{c}{\lambda}$$

● In SI units Planck's constant, h, is $6.6262 \times 10^{-34}$ J · s. The use of a dot is used to indicate the product of two or more units. The dot may be dispensed with when there is no risk of confusion with another unit symbol.

where the proportionality constant, $h$, is Planck's constant, $6.6262 \times 10^{-27}$ erg sec. The higher the energy of an emitted quantum, the higher the frequency. and the shorter the wavelength. Therefore, as the temperature of a glowing body increases, the emitted radiation increasingly includes shorter and shorter wavelengths.

In 1905 Einstein used and extended Planck's quantum theory to explain the photoelectric effect, which consists of the emission of electrons from the surface of a metal when the metal is irradiated by light. According to the wave theory, the energies of the ejected electrons should be proportional to the intensity of the light beam. Instead, the energies of the photoelectrons are proportional to the frequency of the radiation, and each metal requires a characteristic frequency before any electrons are ejected. Increasing the intensity of the light dislodges more electrons but does not affect their energies.

Einstein proposed that Planck's quanta represented discontinuous bits of energy, later called **photons,** and that the interaction of a photon with an electron of the metal caused the photoelectric effect. In a collision the photon vanishes and all its energy is given to the electron. In order to dislodge the electron, the photon must have a certain minimum energy, $E = h\nu_0$, where $\nu_0$ is called the threshold frequency. If the energy of the incident photon ($h\nu$) is greater than that needed to disengage the electron ($h\nu_0$), the difference is imparted to the electron as kinetic energy ($mv^2/2$, where $m$ is the mass of the electron and $v$ its velocity).

$$h\nu - h\nu_0 = \frac{mv^2}{2}$$

Increasing the intensity of light of a given frequency increases the number of photons but does not increase the energies of the photons. Hence, when the intensity of the light is increased, more electrons are ejected, but the energies of these electrons are not altered.

Radiant energy, therefore, may be described in two ways: as waves of energy or as streams of photons. Each concept has the support of experimental observation. Which model is used for a given purpose depends upon which is more convenient. In Section 2.14 the dual character of matter is discussed.

**2.11  ATOMIC SPECTRA**

Since the extranuclear electrons of an atom determine its chemical characteristics, the arrangement of these electrons is of primary importance. Rutherford, as well as others, postulated that electrons are in rapid motion in spherical orbits around the nucleus (similar to the way the planets travel around the sun but with the important distinction that the units of the solar system are not charged). In 1913 Niels Bohr proposed a theory for the electronic structure of atoms that is based upon the nuclear model of Rutherford, the quantum theory of Planck, and data derived from the study of atomic spectra.

When a ray of electromagnetic radiation is passed through a prism, the ray is bent. The degree of refraction depends upon the wavelength of the ray; short wavelengths are refracted more than long wavelengths. Thus, a prism can be used to resolve radiation consisting of different wavelengths into its component wavelengths. Since ordinary white light consists of all wavelengths in the visible range, it is spread out into a wide band of colors called a **continuous spectrum.** This spectrum is a rainbow of colors with no discontinuities—violet merges into blue, blue merges into green, and so on.

When gases or vapors of chemical substances are heated by an electric arc, a spark, or a Bunsen flame, light is emitted. The resolution of a ray of this light by a prism produces a **line spectrum** (Figure 2.10). This spectrum is not continuous; it consists of a number of lines of various colors, each line representing a definite wavelength or energy. Each element has its own characteristic line spectrum; there are similarities, but each is unique. The frequencies of the lines emitted by hydrogen in the visible region of the spectrum (the Balmer series) can be expressed by the formula

**Figure 2.10 The spectro-
scope.**

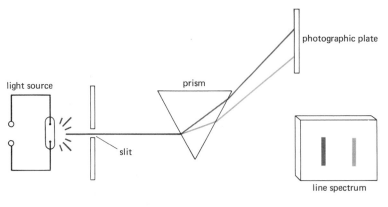

$$\nu = 3.29 \times 10^{15} \text{ cycles/sec} \left( \frac{1}{2^2} - \frac{1}{n^2} \right)$$

where $n$ is an integer equal to, or greater than, 3.

Bohr based his theory on the spectrum of hydrogen, which he interpreted in terms of the quantum concept of Planck. The radiation emitted in a spectral study is thought to arise from electronic transitions within atoms. Since the line spectrum of hydrogen shows that radiation is emitted only in discrete quanta, it is logical to assume that the energy of the electron of hydrogen is quantized and that a transition from a high energy state to a lower energy state results in the emission of the energy difference in the form of a quantum of light.

According to Bohr, the electron of hydrogen can exist only in certain spherical orbits (called energy levels or shells), which are arranged concentrically around the nucleus. These orbits are subject to a quantum restriction: the angular momentum of an electron in an orbit ($mvr$, where $m$ is the mass of the electron, $v$ is its velocity, and $r$ is the radius of the orbit) must be an integral multiple of the quantity $h/2\pi$.

$$mvr = n\frac{h}{2\pi} \qquad n = 1, 2, 3, \ldots$$

Thus, an electron moving in any orbit has a definite energy that is characteristic of that orbit. As long as an electron remains in a given orbit, it neither absorbs nor radiates energy.

Thus, the $K$ level ($n = 1$), the shell closest to the nucleus, has the smallest radius and lowest energy. The next shell ($L$; $n = 2$) has a larger radius, and electrons moving in it have higher energies. With increasing distance from the nucleus ($K, L, M, N, O$; $n = 1, 2, 3, 4, 5$), the radius of the shell increases, and the energy of electrons in the shell increases. No electron in an atom can have an energy that would place it between the permissible orbits.

If an electron is in an orbit of radius $r$, the force of attraction between the nucleus (with a charge of $+Ze$) and the electron (with a charge of $-e$) is given by Coulomb's law as $-Ze^2/r^2$. This force of attraction causes the electron to deflect from a straight-line path to a circular path around the nucleus. The coulomb force, therefore, is a centripetal force. The formula for centripetal force is $-mv^2/r$, where the minus sign indicates attraction inward. Therefore,

$$\frac{mv^2}{r} = \frac{Ze^2}{r^2} \tag{7}$$

Thus,

$$v^2 = \frac{Ze^2}{mr} \tag{8}$$

If we solve Bohr's quantum condition for $v$, we get

$$mvr = \frac{nh}{2\pi} \qquad n = 1, 2, 3, \ldots \tag{9}$$

$$v = \frac{nh}{2\pi mr} \tag{10}$$

Squaring equation (10) and combining the result with equation (8) gives

$$\frac{Ze^2}{mr} = \frac{n^2 h^2}{4\pi^2 m^2 r^2} \tag{11}$$

$$r = \frac{n^2 h^2}{4\pi^2 m Z e^2} \qquad n = 1, 2, 3, \ldots \tag{12}$$

If we solve this expression for the smallest orbit ($n = 1$) of hydrogen ($Z = 1$), we get

$$r = 0.529 \times 10^{-8} \text{ cm} = 0.529 \text{ Å}$$

● The Bohr radius is 0.0529 nm.

This quantity is called the **Bohr radius** and is sometimes given the symbol $a_0$.

The total energy of an electron is equal to the sum of its kinetic energy ($mv^2/2$), which is attributable to its motion, and its potential energy ($-Ze^2/r$), which results from its position in the field of the nucleus.

$$E = \frac{mv^2}{2} - \frac{Ze^2}{r} \tag{13}$$

From equation (8),

$$mv^2 = \frac{Ze^2}{r} \tag{14}$$

Therefore,

$$E = \frac{Ze^2}{2r} - \frac{Ze^2}{r} = -\frac{Ze^2}{2r} \tag{15}$$

Substitution of the value of $r$ from equation (12) into equation (15) gives

$$E = -\frac{2\pi^2 m Z^2 e^4}{n^2 h^2} \qquad n = 1, 2, 3, \ldots \tag{16}$$

By means of equation (16) it is possible to calculate the energy of an electron in each of the Bohr orbits.

When the electrons of an atom are arranged as close as possible to the nucleus (in the case of hydrogen, one electron in the $K$ level), they are in the condition of lowest possible energy; this condition is called the **ground state** or **ground level.** In a spectroscope one or more electrons of an atom absorb energy from the electric arc or spark and jump to outer, more energetic levels; this condition is called an **excited state.**

When an electron falls back to a lower orbit, it emits a definite amount of energy—the difference between the energy required by the outer level, $E_o$, and the energy required by the inner level, $E_i$. This energy is emitted as radiation; since it is a definite amount, it has a characteristic frequency (and wavelength) and produces a characteristic spectral line.

The energy of the photon radiated by the atom is

$$h\nu = E_o - E_i \qquad E_o \rightarrow E_i$$

Substitution of the values of $E$ from equation (16) into this expression gives

$$h\nu = \left(-\frac{2\pi^2 m Z^2 e^4}{h^2}\right)\left(\frac{1}{n_o^2}\right) - \left(-\frac{2\pi^2 m Z^2 e^4}{h^2}\right)\left(\frac{1}{n_i^2}\right) \tag{17}$$

$$= \frac{2\pi^2 m Z^2 e^4}{h^2}\left(\frac{1}{n_i^2} - \frac{1}{n_o^2}\right) \tag{18}$$

and

$$\nu = \frac{2\pi^2 m Z^2 e^4}{h^3}\left(\frac{1}{n_i^2} - \frac{1}{n_o^2}\right) \tag{19}$$

If the constants of this equation are evaluated for hydrogen ($Z = 1$), the frequencies of photons emitted by transitions from higher levels to the $n = 2$ level is given by

$$\nu = 3.29 \times 10^{15} \text{ cycles/sec} \left(\frac{1}{2^2} - \frac{1}{n^2}\right) \tag{20}$$

which agrees with the empirically derived equation describing the spectral lines of the Balmer series.

The relation between some of the electronic transitions of the hydrogen atom and the spectral lines is illustrated in Figure 2.11. Since electron transitions to the $n = 1$ level (Lyman series) release more energy than those to the $n = 2$ level, the wavelengths of the lines of this series

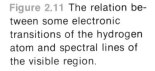
Figure 2.11 The relation between some electronic transitions of the hydrogen atom and spectral lines of the visible region.

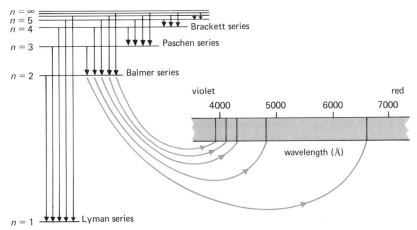

are shorter than those of the Balmer series; the lines of the Lyman series occur in the ultraviolet region. On the other hand, the lines of the Paschen series, which represent transitions to the $n = 3$ level, occur at wavelengths longer than the Balmer series; the Paschen lines appear in the infrared region. Certain wavelengths of light are not emitted because there are no corresponding energy drops for an electron in the atom. The removal of an electron to an infinite shell results in **ionization;** the **ionization energy** is the minimum amount of energy required for this process (Section 3.2).

Since atoms differ in size, charge on the nucleus, number of electrons, and so on, the spectrum of each element is different. The Bohr theory successfully interprets the spectra of atoms that contain only one electron, such as H, $He^+$, and $Li^{2+}$.

**2.12 ATOMIC NUMBERS AND THE PERIODIC LAW**

Early in the nineteenth century, chemists became interested in the relationships among the elements. In 1817 and 1829 Johann W. Döbereiner published articles in which he examined certain sets of elements that he called triads (Ca, Sr, Ba; Li, Na, K; Cl, Br, I; and S, Se, Te). The elements of each set have similar properties, and the atomic weight of the middle element is approximately the arithmetic mean of the other two.

In the years following, many chemists made attempts to discern regularities in the properties of elements, and several charts and arrangements of the elements were proposed. The work of John A. R. Newlands is significant. In the years 1863–1866 Newlands proposed and developed his law of octaves, in which he compared the elements to the notes of octaves in music. He stated that if the elements are arranged in order of increasing atomic weight, any given element is similar to the eighth element that follows it. Newlands assigned a serial number, based on

| | Group | | | | | | | | | | | | | | | |
|---|---|---|---|---|---|---|---|---|---|---|---|---|---|---|---|---|
| Period | I | | II | | III | | IV | | V | | VI | | VII | | VIII | |
| | a | b | a | b | a | b | a | b | a | b | a | b | a | b | a | b(0) |
| 1 | H 1.0 | | | | | | | | | | | | | | | He 4.0 |
| 2 | Li 6.9 | | Be 9.0 | | B 10.8 | | C 12.0 | | N 14.0 | | O 16.0 | | F 19.0 | | | Ne 20.2 |
| 3 | Na 23.0 | | Mg 24.3 | | Al 27.0 | | Si 28.1 | | P 31.0 | | S 32.1 | | Cl 35.5 | | | Ar 39.9 |
| 4 | K 39.1 | | Ca 40.1 | | Sc 45.0 | | Ti 47.9 | | V 50.9 | | Cr 52.0 | | Mn 54.9 | | Fe 55.8  Co 58.9  Ni 58.7 | |
| | Cu 63.5 | | Zn 65.4 | | Ga 69.7 | | Ge 72.6 | | As 74.9 | | Se 79.0 | | Br 79.9 | | | Kr 83.8 |
| 5 | Rb 85.5 | | Sr 87.6 | | Y 88.9 | | Zr 91.2 | | Nb 92.9 | | Mo 95.9 | | Tc | | Ru 101.1  Rh 102.9  Pd 106.4 | |
| | Ag 107.9 | | Cd 112.4 | | In 114.8 | | Sn 118.7 | | Sb 121.8 | | Te 127.6 | | I 126.9 | | | Xe 131.3 |
| 6 | Cs 132.9 | | Ba 137.3 | | La * 138.9 | | Hf 178.5 | | Ta 180.9 | | W 183.9 | | Re 186.2 | | Os 190.2  Ir 192.2  Pt 195.1 | |
| | Au 197.0 | | Hg 200.6 | | Tl 204.4 | | Pb 207.2 | | Bi 209.0 | | Po | | At | | | Rn |
| 7 | Fr | | Ra | | Ac ** | | | | | | | | | | | |

| * | Ce 140.1 | Pr 140.9 | Nd 144.2 | Pm | Sm 150.4 | Eu 152.0 | Gd 157.3 | Tb 158.9 | Dy 162.5 | Ho 164.9 | Er 167.3 | Tm 168.9 | Yb 173.0 | Lu 175.0 |
|---|---|---|---|---|---|---|---|---|---|---|---|---|---|---|
| ** | Th 232.0 | Pa | U 238.0 | Np | Pu | Am | Cm | Bk | Cf | Es | Fm | Md | No | Lr |

Figure 2.12 Periodic table based on the Mendeleev chart of 1871. Elements that were not known in 1871 appear in the colored squares.

increasing atomic weight, to each element with hydrogen assigned the number 1. These numbers may be regarded as early atomic numbers, although no physical significance could be attached to them at that time. Newlands' values do not correspond to modern values since some of the elements had not yet been discovered and since some of the atomic weights he used were incorrect.

The periodic classification of the elements was first proposed in polished form in 1869 by both Julius Lothar Meyer and Dmitri Mendeleev working independently. The modern **periodic law** states that the chemical and physical properties of the elements are periodic functions of atomic number; when the elements are arranged in order of increasing atomic number and the properties of each element are studied in turn, similarities recur periodically (Section 2.13). The original statement of the periodic law was based on atomic weights, not atomic numbers. Even in this form, however, it was a promising and useful generalization, since atomic numbers and atomic weights increase in a parallel manner with only three exceptions (K, Ni, and I). Mendeleev prepared a periodic table in which similar elements were listed under one another and in which blanks were left for missing elements (see Figure 2.12). On the basis of his system, Mendeleev was able to predict the existence, as well as the properties, of some of the elements that had not been discovered at that time (e.g., Sc, Ga, and Ge).

Subsequent study of Mendeleev's periodic classification convinced many that some fundamental atomic property other than atomic weight is the cause of the observed periodicity. It was proposed that this fundamental property is in some way related to atomic number, which at that time was only a serial number derived from the periodic classification. The work of Henry G. J. Moseley in the years 1913 and 1914 solved this problem. To explain his results, Moseley drew upon Bohr's theory, which had just been published.

When high-energy cathode rays are focused on a target, X rays are produced (Figure 2.13). This X radiation can be resolved into its com-

**Figure 2.13** X-ray tube.

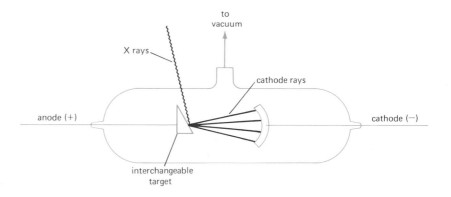

to
vacuum

X rays

cathode rays

anode (+)

cathode (−)

interchangeable
target

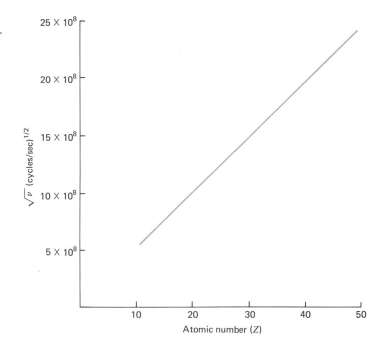

**Figure 2.14** Relationship between frequency of characteristic X-ray lines and atomic number.

ponent wavelengths, and the line spectra thus obtained can be recorded photographically. Different X-ray spectra result when different elements are used as targets; each spectrum consists of only a few lines.

Moseley studied the X-ray spectra of 38 elements with atomic numbers between 13 (aluminum) and 79 (gold). Using a corresponding spectral line for each element, he found that there is a linear relationship between the square root of the frequency of the line and the atomic number of the element (Figure 2.14). In other words, the square root of the frequency of the spectral line increases by a constant amount from element to element when the elements are arranged by increasing atomic number.

Moseley was able, therefore, to assign the correct atomic number to any element on the basis of its X-ray spectrum. In this way he settled the problem involving the classification of elements that have atomic weights out of line with those of their neighbors (K, Ni, and I). He also stated that there should be 14 elements in the series from $_{58}$Ce to $_{71}$Lu (found at the bottom of the chart in Figure 2.12), and he established that these elements should follow lanthanum in the periodic table. Moseley's diagrams indicated that 4 elements before $_{79}$Au remained to be discovered (numbers 43, 61, 72, and 75).

In the production of X rays, the cathode rays rip electrons from the inner shells of the target atoms. When outer electrons fall into the vacancies thus produced, photons of high energy are emitted, which have the high frequencies and short wavelengths characteristic of X rays. Since the Bohr equation (number 19 of Section 2.11) shows that the frequency of a spectral line is directly proportional to the square of the number of unit charges on the nucleus, $Z^2$, Moseley proposed that the atomic number of an element is the number of units of positive charge on the atomic nucleus.

## 2.13 THE PERIODIC LAW AND ELECTRONIC STRUCTURES

The most popular form of the periodic table is the long form that is found inside the front cover of this book. The **periods** consist of those elements that are arranged in horizontal rows in the table. The organization is such that elements of similar chemical and physical properties (called **groups** or **families**) appear in vertical columns. In this table the set of elements (the lanthanides) that appears at the bottom should actually appear in the body of the chart in proper order by atomic number so that the sixth period has its requisite 32 elements. For convenience in reproduction this is not done; but, properly, the chart should be vertically cut, the sections separated, and the lanthanides inserted in their proper positions. For the actinides, which appear below the lanthanides at the bottom of the periodic table, the same considerations pertain; they should be inserted in the seventh period (as yet incomplete).

The first period consists of only two elements, hydrogen and helium. Subsequent periods have 8, 8, 18, 18, and 32 elements. With the exception of the first period, each complete period begins with an **alkali metal** (group I A)–a highly reactive, light, silvery metal–and ends with a **noble gas** (group 0)–an unreactive, colorless gas. The element before the noble gas of each complete period (except for the first period) is a **halogen** (group VII A)–a very reactive nonmetal.

A general pattern exists for each period after the first. Starting with an alkali metal, the properties change from element to element–the metallic properties fade and are gradually replaced by nonmetallic characteristics. After a highly reactive nonmetal (a halogen) is reached, each complete period ends with a noble gas.

Since the periodic table classifies atoms by chemical behavior, and since chemical behavior is determined by electronic configuration, the periodic table may also be said to classify atoms by electronic configuration. The maximum number of electrons that each energy level may hold is given by $2n^2$, where $n$ is the principal quantum number of the shell in question. For the $n = 1$ shell, the maximum number is $2(1)^2$, or 2; for the $n = 2$ level, this number is $2(2)^2$, or 8; the maximum electron populations for the first five shells are given in Table 2.3.

The ground-state electronic configurations of the first 18 elements (the first three periods) are shown in Table 2.4. The first element, hydrogen, has one electron in the $K$ level; the configurations of the following

Table 2.3 Maximum number of electrons for each shell

| SHELL | $n$ | $2n^2$ |
|-------|-----|--------|
| $K$ | 1 | 2 |
| $L$ | 2 | 8 |
| $M$ | 3 | 18 |
| $N$ | 4 | 32 |
| $O$ | 5 | 50 |

Table 2.4 Electronic configurations by shells for the first three periods

| ELECTRONIC LEVELS | GROUP NUMBER | | | | | | | |
|---|---|---|---|---|---|---|---|---|
| | I A | II A | III A | IV A | V A | VI A | VII A | 0 |
| *K* | H 1 | | | | | | | He 2 |
| *K* | Li 2 | Be 2 | B 2 | C 2 | N 2 | O 2 | F 2 | Ne 2 |
| *L* | 1 | 2 | 3 | 4 | 5 | 6 | 7 | 8 |
| *K* | Na 2 | Mg 2 | Al 2 | Si 2 | P 2 | S 2 | Cl 2 | Ar 2 |
| *L* | 8 | 8 | 8 | 8 | 8 | 8 | 8 | 8 |
| *M* | 1 | 2 | 3 | 4 | 5 | 6 | 7 | 8 |

elements can be derived by successively adding electrons to the lowest available unfilled level. Comparison of the configurations of the third-period elements with those of the second-period elements yields an important fact. The number of electrons in the outermost shells (called the **valence shells**) is the same for elements of the same group, and this number is the same as the group number (with the exception of group 0, the noble gases). This similarity in number of valence electrons accounts for the similarities in properties.

The electronic configurations of the elements in the fourth and subsequent periods are not easily understood in terms of the electronic shells alone; a number of questions arise that may be answered only in terms of concepts that will be discussed later. The fourth period begins with potassium, number 19, which is a member of group I and has one valence electron in the $N$ ($n = 4$) level; potassium has a configuration of 2, 8, 8, 1. This configuration poses a question. Since the $M$ level can hold a maximum of 18 electrons, why should the final electron of potassium be placed in the $N$ level before the $M$ level has been filled?

Calcium, number 20, has the configuration 2, 8, 8, 2. The next element, scandium ($Z = 21$), has the electronic configuration 2, 8, 9, 2; it differs from calcium by the addition of an electron to the level next to the outer level—the $M$ level—rather than to the outer level—the $N$ level. Scandium is not a member of group III A because it does not have three valence electrons; therefore, it is not placed below aluminum in group III A. Scandium is called a **transition element;** since the "last" electron was added to the shell adjacent to the outermost shell, it is said to undergo **inner building.** The transition elements (numbers 21 through 30 in the fourth period) are placed in the center of the periodic table, and their groups are given B designations. Inner building cannot be explained on the basis of electronic shells alone.

The electronic configurations of the noble gases also pose a question. The first noble gas, helium, has a completed first shell; the second noble gas, neon ($Z = 10$), has a configuration of 2, 8 (two completed shells, $K$ and $L$); the third noble gas, argon ($Z = 18$), has the configuration 2, 8, 8. We might suppose that these elements of extremely low chemical reactivity would have all their electronic shells complete; this is not a requisite for electronic stability, however, since the third level in argon is not complete but has only an octet of electrons. We shall see later that all the noble gases have eight electrons in their outer shells with the exception of helium which has the maximum number of electrons that the $K$ level can hold—two.

**2.14 DUAL NATURE OF THE ELECTRON**

In the same way that electromagnetic radiation has both a wave and a corpuscular (particle-like) character, matter has a dual nature. We have seen that the energy of a photon, $E$, is related to its frequency, $\nu$, and to its wavelength, $\lambda$.

$$E = h\nu = h\frac{c}{\lambda} \tag{21}$$

Using Einstein's equation, $E = mc^2$, we derive

$$mc^2 = \frac{hc}{\lambda} \tag{22}$$

or

$$\lambda = \frac{h}{mc} \tag{23}$$

The term $mc$ is equal to the momentum of the photon (mass times velocity).

In 1925 Louis de Broglie postulated that a wavelength can be assigned to an electron:

$$\lambda = \frac{h}{m\upsilon} \tag{24}$$

where $m$ is the mass of an electron and $\upsilon$ is its velocity. This postulate has been confirmed by a variety of experimental data.

Consideration of the electron as a wave is important because the Heisenberg **uncertainty principle** (1927) demonstrates the ultimate futility of any attempt to make the Bohr model more exact and comprehensive. The precise prediction of the path of a moving body requires that both the position and the velocity (the latter denoting direction as well as speed) of the body be known at a given time. Werner Heisenberg showed that it is impossible to measure simultaneously the exact position and exact velocity of a body as small as an electron.

The determination of the position of an object depends upon our ability to observe it, either by noting the interference in the light rays that illuminate the object or, more indirectly, by noting the variation in some other test signal. According to the principles of optics, there is an uncertainty in the position of an object, $\Delta x$, approximately equal to the wavelength, $\lambda$, of the light used to illuminate it.

$$\Delta x \approx \lambda$$

Radiation of extremely short wavelength would be needed to detect the position of an object as small as an electron with any degree of precision. Such radiation is very energetic (the shorter the wavelength, the more energetic it is); therefore, when it impinges upon an electron, the impact causes the direction and speed of the tiny electron to change.

Solving equation (23) for the momentum of a photon, we obtain

$$mc = \frac{h}{\lambda}$$

In the collision of a photon with an electron, a part, or all, of the momentum of the photon is imparted to the electron. Therefore, the uncertainty in the momentum of an electron, $\Delta mv$, after collision is

$$\Delta mv \approx \frac{h}{\lambda}$$

Photons of longer wavelength (less energetic) have less effect on the momentum of the electron but, because of their longer wavelength, are less precise in indicating the position of the electron. The two uncertainties, therefore, are related. An approximate mathematical statement of the Heisenberg uncertainty principle is

$$\Delta x \, \Delta mv \approx \lambda \frac{h}{\lambda} = h$$

The mass of the electron is approximately $9 \times 10^{-28}$ g; according to the Bohr model, the velocity of the electron in the $n = 1$ shell of hydrogen is approximately $2 \times 10^8$ cm/sec. The momentum of the electron obtained by multiplying these values is

$$(9 \times 10^{-28} \text{ g})(2 \times 10^8 \text{ cm/sec}) = 2 \times 10^{-19} \text{ g cm/sec}$$

If the measured momentum is permitted to be in error by 100%,

$$\Delta x \, \Delta mv \approx h$$

$$\Delta x (2 \times 10^{-19} \text{ g cm/sec}) \approx 6.6 \times 10^{-27} \text{ g cm}^2/\text{sec}$$

$$\Delta x \approx \frac{6.6 \times 10^{-27} \text{ g cm}^2/\text{sec}}{2 \times 10^{-19} \text{ g cm/sec}} = 3 \times 10^{-8} \text{ cm} = 3 \text{ Å}$$

This value is approximately three times the Bohr diameter of the $n = 1$ shell of the hydrogen atom.

It appears, then, that a precise description of the path of an electron in a Bohr orbit is impossible. The wave postulate of de Broglie was used by Erwin Schrödinger to develop a differential equation that describes the electron in terms of its wave character.

The **Schrödinger equation** (1926) is the keystone of **wave mechanics.** It reconciles the idea of the quantum restriction on the energy of the electron (originally imposed by Bohr) with the idea of the electron as a standing wave (which may be described mathematically in much the same way as the waves of a vibrating guitar string). The Schrödinger equation mathematically relates a wave function for an electron, $\psi$, to the coordinates in space where it may be found as well as to its energy. Many different wave functions are obtained as solutions of this equation, and each $\psi$ corresponds to an allowed energy state for the system.

The intensity of a wave is proportional to the square of its amplitude (Section 2.10). The wave function, $\psi$, is an amplitude function, and the value of $\psi^2$ for a small segment of volume in space is proportional to the electron charge density of that volume segment. If one imagines a corpuscular electron in motion around the nucleus, $\psi^2$ is proportional to the probability of finding the electron in a given element of volume in space. The chance of finding an electron is high in a region where the cloud is dense. This interpretation does not fix electron position nor does it describe electronic paths; it merely predicts where an electron is most likely to be found. Insofar as a corpuscular interpretation of the electron is concerned, this is the best description that wave mechanics has to offer.

For an electron in the $n = 1$ state of the hydrogen atom, the charge cloud has its greatest density near the nucleus and becomes thinner as the distance from the nucleus increases (Figure 2.15). More information about this electronic distribution can be obtained from an examination of the probability curves of Figure 2.16.

In curve (a) the probability of finding an electron in a small segment of volume in space is plotted against distance from the nucleus. This probability is greatest near the nucleus, but there is a finite probability at all points in space that are finite distances from the nucleus.

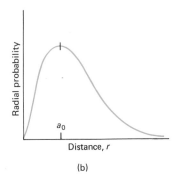

Figure 2.15 Cross section of the charge cloud of an electron in the $n = 1$ state of the hydrogen atom.

Figure 2.16 Probability curves for an electron in the $n = 1$ state of the hydrogen atom. (a) Probability of finding the electron per unit volume versus distance from the nucleus. (b) Probability of finding the electron at a given distance versus distance from the nucleus.

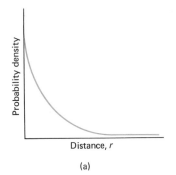

(a)

(b)

Curve (b) is a radial probability distribution. Imagine a group of spherical shells of very small thickness arranged concentrically around the nucleus; the total probability of finding the electron in such a shell is plotted against the distance of the shell from the nucleus in curve (b) of Figure 2.16. The probability per unit volume is greatest close to the nucleus, but the volume of a shell that is close to the nucleus is less than the volume of a shell that is farther out. Hence, the plot shows a maximum at 0.529 Å (the Bohr radius, $a_0$). The total probability of finding an electron at all points of distance $r$ from the nucleus is greatest when $r$ is equal to $a_0$. In terms of the charge cloud concept, a greater amount of the total charge of the electron is found in the spherical shell radius $a_0$ than is found in any other shell.

Since there is a probability of finding an electron at all finite distances from the nucleus, the three-dimensional representation of an $n = 1$ electron of the hydrogen atom poses a problem. Such representations, however, are important aids for interpreting chemical phenomena. It is not possible to depict a geometric shape that will encompass a region of 100% probability. However, a surface can be drawn that connects points of equal probability and encloses a volume in which there is a high probability (e.g., 90%) of finding the electron (Figure 2.17). Alternatively, the figure can be interpreted as representing a contour that encloses a high percentage (90%) of the electronic charge.

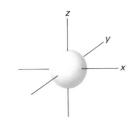

**Figure 2.17** Boundary surface representation of an electron in the $n = 1$ state of the hydrogen atom. Volume encloses 90% of the electron density. Nucleus is at the origin.

**2.15 QUANTUM NUMBERS**

Bohr explained spectral lines on the basis of electronic energy shells and assigned each shell a **principal quantum number,** $n$. Wave mechanics uses this quantum number as well as three additional numbers to describe electrons. We can consider that Bohr's shell represents a region where the probability of finding an electron is high. The value of $n$ gives an indication of the position of the shell relative to the nucleus; the larger the value of $n$, the farther the shell is from the nucleus. These shells must be regarded as discrete in terms of energy, and the only permissible electron transitions are between these shells. According to Bohr, these transitions account for the spectral lines.

The use of better spectroscopes, however, showed that the spectral lines first observed actually consist of groups of fine lines (fine structure). To account for all these lines, more permissible electron states are necessary; Arnold Sommerfeld in 1916 proposed that the Bohr shells with $n$ greater than 1 consist of subshells of slightly different energies — that these levels of Bohr are actually groups of sublevels. Sommerfeld proposed this in terms of the Bohr quantum theory; however, his idea is incorporated into the modern wave mechanical treatment.

Each principal shell has as many subshells as its value of $n$; there is only one sublevel for the $n = 1$ level, but there are two sublevels for the $n = 2$ level, three for the $n = 3$ level, and so on. A **subsidiary quantum**

**number,** *l,* is introduced, and each subshell is assigned an *l* value. For each quantum level, *n,* there are values of *l* corresponding to every term in the series

$$l = 0, 1, 2, 3, \ldots (n - 1)$$

When $n = 1$, the only value of *l* is 0, and there is only one subshell; when $n = 2$, there are two sublevels that have *l* values of 0 and 1, respectively; and when $n = 3$, the three sublevels have *l* values of 0, 1, and 2.

By convention, other letters are used at times to denote sublevels. Thus, the $l = 0$ sublevel is designated as an *s* sublevel; $l = 1$, *p;* $l = 2$, *d;* and $l = 3$, *f.* These spectroscopic notations are the initial letters of adjectives formerly used to describe spectral lines: sharp, principal, diffuse, and fundamental. For *l* values larger than 3, the letters used proceed alphabetically — *g, h, i,* and so on; however, for the ground states of the known elements, no value of *l* higher than 3 (*f*) is necessary. Combining the principal quantum number with these spectroscopic symbols gives a convenient way to denote the subshells (see Table 2.5 for the spectroscopic notation for the first four shells).

The maximum number of electrons in each level is given by the formula $2n^2$. How these electrons are distributed among the level's sublevels is seen by examining the remaining two quantum numbers.

The **spin quantum number,** *s,* characterizes the spin of an electron on its own axis and may have a value of either $+\frac{1}{2}$ or $-\frac{1}{2}$ (only two directions of spin are possible). Since a spinning charge generates a magnetic field, an electron has a magnetic moment associated with it. The magnetic moments of two electrons with different spins ($+\frac{1}{2}$ and $-\frac{1}{2}$) oppose and cancel each other.

The wave function, $\psi$, associated with an electron is termed an **orbital** to distinguish it from the path or orbit of the Bohr theory. Each orbital may hold two electrons; electrons paired in a single orbital have opposed spins. The $n = 1$ level has a maximum electron population of two but only one sublevel (1*s*) to hold them. The 1*s* sublevel, therefore, must consist of a single orbital.

The radial probability distribution for the 2*s* orbital is shown in Figure 2.18a. Notice that at the point where *r* is equal to $2a_0$, the radial

**Table 2.5** Subshell notations

| *n* | *l* | SPECTROSCOPIC NOTATIONS |
|---|---|---|
| 1 | 0 | 1*s* |
| 2 | 0 | 2*s* |
| 2 | 1 | 2*p* |
| 3 | 0 | 3*s* |
| 3 | 1 | 3*p* |
| 3 | 2 | 3*d* |
| 4 | 0 | 4*s* |
| 4 | 1 | 4*p* |
| 4 | 2 | 4*d* |
| 4 | 3 | 4*f* |

**Figure 2.18** Diagrams for the 2*s* orbital. (a) Radial probability distribution. (b) Cross section of electron charge cloud. (c) Boundary surface diagram.

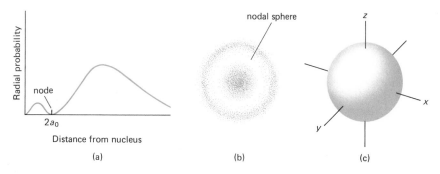

(a)  (b)  (c)

Figure 2.19 Boundary
surface diagrams for the 2*p*
orbitals.

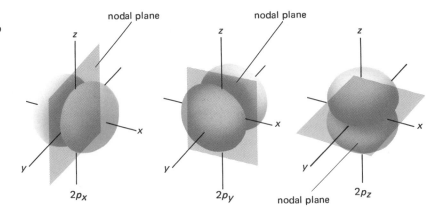

Figure 2.19 Boundary surface diagrams for the 2*p* orbitals.

probability drops to zero (a node). In the cross section of the electron density of this orbital (Figure 2.18b), this point shows up as a nodal sphere where electron density is zero. However, the conventional boundary surface of the 2*s* orbital, shown in Figure 2.18c, appears the same as that of the 1*s* orbital (Figure 2.17) except for size.

There are three 2*p* orbitals (Figure 2.19), and each has a nodal plane passing through the nucleus at the origin. Each *p* orbital consists of two lobes that may be considered to be oriented along the *x, y,* or *z* axis; hence, they may be given the designations $2p_x$, $2p_y$, and $2p_z$.

The number of orbitals in a sublevel may be inferred from the fourth quantum number—the **magnetic quantum number,** *m*. A magnetic field has no effect on an *s* electron since an *s* orbital is spherically symmetrical. If such an orbital were turned in a magnetic field, no effect would be observed because a sphere presents the same aspect toward the lines of force no matter what its orientation. This observation is true for all of the *s* orbitals—1*s,* 2*s,* 3*s,* and so on.

However, a *p* orbital is not spherically symmetrical. Each *p* sublevel consists of three differently oriented *p* orbitals. These *p* orbitals are identical in terms of energy, and no distinction may be noted between electrons that occupy different *p* orbitals in the absence of a magnetic field. However, if spectral studies are run with the source of emission in a magnetic field, certain spectral lines are split into several lines (Zeeman effect)—an effect that disappears when the magnetic field is removed.

For any sublevel, values of the magnetic quantum number, *m*, are given by the terms in the series

$$m = +l, +(l-1), \ldots, 0 \ldots, -(l-1), -l$$

Since *m* values indicate possible orientations in space for nonspherically symmetrical orbitals, the number of these values for a given sublevel reflects the number of orbitals in the sublevel. Each orbital is capable of

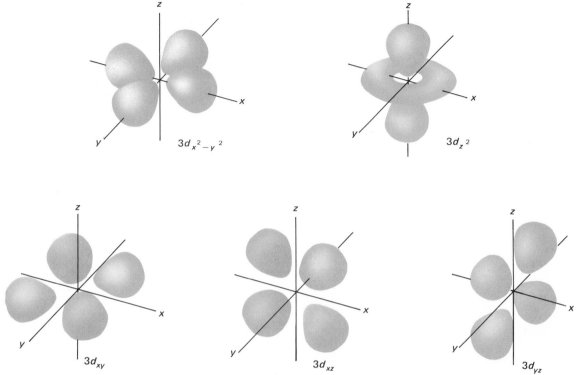

**Figure 2.20** Boundary surface diagrams for the 3d orbitals.

holding two electrons of opposed spin. Thus, for $l = 0$, $m = 0$. For $l = 1$, the $m$ values are $+1$, $0$, and $-1$, and three orbitals are indicated. When $l = 2$ ($d$ subshell), the $m$ values ($+2$, $+1$, $0$, $-1$, and $-2$) describe five orbitals that are oriented differently in space. Notice that the values of $m$ depend upon the value of $l$ for the subshell under consideration.

The conventional boundary surfaces of the five $3d$ orbitals are shown in Figure 2.20. The shape of the $d_{z^2}$ orbital is different from the others, but all these orbitals are energetically equivalent.

Each electron, therefore, may be described by a set of four quantum numbers — $n$ gives the relative distance of the electron from the nucleus, $l$ gives the subshell and the shape of the orbital for the electron (each orbital of a given subshell is equivalent in energy), $m$ designates the orientation of the orbital in space, and $s$ states the spin of the electron. The **exclusion principle** of Wolfgang Pauli states that no two electrons in the same atom may have identical sets of all four quantum numbers; electrons with the same set of quantum numbers would be indistinguishable. Even if two electrons have the same values for $n$, $l$, and $m$, they will differ in their $s$ values. The quantum numbers for the 32 electrons of the $n = 4$ level are given in Table 2.6.

Table 2.6 Quantum numbers for the electrons of the $n = 4$ level

| $n$ | $l$ | $m$ | $s$ | NUMBER OF ELECTRONS BY SUBSHELLS |
|-----|-----|-----|-----|-----|
| 4 | 0 ($s$) | 0 | $\pm\frac{1}{2}$ | 2 |
| | 1 ($p$) | +1 | $\pm\frac{1}{2}$ | |
| | | 0 | $\pm\frac{1}{2}$ | 6 |
| | | −1 | $\pm\frac{1}{2}$ | |
| | 2 ($d$) | +2 | $\pm\frac{1}{2}$ | |
| | | +1 | $\pm\frac{1}{2}$ | |
| | | 0 | $\pm\frac{1}{2}$ | 10 |
| | | −1 | $\pm\frac{1}{2}$ | |
| | | −2 | $\pm\frac{1}{2}$ | |
| | 3 ($f$) | +3 | $\pm\frac{1}{2}$ | |
| | | +2 | $\pm\frac{1}{2}$ | |
| | | +1 | $\pm\frac{1}{2}$ | |
| | | 0 | $\pm\frac{1}{2}$ | 14 |
| | | −1 | $\pm\frac{1}{2}$ | |
| | | −2 | $\pm\frac{1}{2}$ | |
| | | −3 | $\pm\frac{1}{2}$ | |

**2.16 MAGNETIC MOMENT**

There are a number of types of magnetic behavior; **diamagnetism, paramagnetism,** and **ferromagnetism** are the most important. Diamagnetic substances are weakly repelled by a magnetic field, while paramagnetic materials are drawn into a magnetic field. Ferromagnetism, exhibited by iron, is a comparatively rare phenomenon and is an extreme form of paramagnetism.

A single spinning electron behaves like a small magnet (Section 2.15). Two electrons that are paired in an orbital have opposed spins, and their magnetic moments oppose each other and cancel. The magnetic properties of *unpaired electrons* (electrons which occupy orbitals singly) cause atoms that contain these electrons to be paramagnetic. The larger the number of unpaired electrons, the greater the magnetic moment.

Two effects contribute to the paramagnetism of an atom: the spin of the unpaired electrons and the orbital motion of the electrons. The magnetic moment associated with the orbital motion of electrons is related to the orientation of the orbitals with respect to the magnetic field; therefore, it is related to the magnetic quantum number, $m$. In the same way that a pair of electrons has no net spin magnetic moment (the sum of the $s$ values equals zero), a filled or half-filled subshell has no resultant *orbital* magnetic moment (the sum of the values of $m$ for all the electrons of the subshell is zero). A half-filled subshell, however, has a magnetic moment attributable to the spins of its unpaired electrons.

The magnetic properties of some atoms are the result of both effects;

this is true for the lanthanides, in which the unpaired $f$ electrons lie deep within the atom and are well shielded from the surface. For many atoms, however, the effect of the orbital motion is quenched by the environment of the atom, and this contribution is negligible. In such cases the total magnetic moment is equal to the spin magnetic moment. The number of unpaired electrons ($n$ in the following equation) and the spin magnetic moment (in Bohr magnetons) are approximately related by the expression:

$$\text{spin magnetic moment} = \sqrt{n(n + 2)}$$

A material is diamagnetic if all its electrons are paired. Diamagnetism is a universal property of matter, but it is obscured by the stronger paramagnetic effect when unpaired electrons are present. Diamagnetism results from the interaction of the magnetic field with filled orbitals and is a function of the electronic charge density.

Magnetic measurements have helped to determine the electronic configurations of many elements and thereby to establish the hypothetical order in which the orbitals may be assumed to fill. The electronic configurations (by orbitals) of the first five elements are given in Table 2.7. A question arises concerning the electronic distribution of the sixth element, carbon. Since there are three $2p$ orbitals, does the next electron belong in the $2p$ orbital that already holds one electron, or does it belong in another $2p$ orbital?

**Hund's rule of maximum multiplicity,** which has been confirmed by magnetic measurements, provides an answer for this and similar questions. Hund's rule states that electrons are distributed among the orbitals of a subshell in a way that gives the maximum number of unpaired electrons with parallel spins (all $s$ values have the same sign). In carbon, therefore, each of the two $2p$ electrons occupies an orbital separately rather than pairing with the other in a single orbital. The six electrons of carbon may be described by the sets of quantum numbers indicated in Table 2.8. The next electron added, for nitrogen, has the quantum numbers $2, 1, -1, +\frac{1}{2}$.

In Figure 2.21 orbitals are represented by circles, and the order of filling for the first ten electrons is shown by the numbers entered in these

Table 2.7 Electronic configurations of the first five elements

|   | $1s$ | $2s$ | $2p$ |
|---|------|------|------|
| $_1$H | 1 |   |   |
| $_2$He | 2 |   |   |
| $_3$Li | 2 | 1 |   |
| $_4$Be | 2 | 2 |   |
| $_5$B | 2 | 2 | 1 |

Table 2.8 Quantum numbers for the electrons of carbon

| $n$ | $l$ | $m$ | $s$ |
|-----|-----|-----|-----|
| 1 | 0 | 0 | $+\frac{1}{2}$ |
| 1 | 0 | 0 | $-\frac{1}{2}$ |
| 2 | 0 | 0 | $+\frac{1}{2}$ |
| 2 | 0 | 0 | $-\frac{1}{2}$ |
| 2 | 1 | $+1$ | $+\frac{1}{2}$ |
| 2 | 1 | 0 | $+\frac{1}{2}$ |

Figure 2.21 Order in which the orbitals of the $n = 1$ and $n = 2$ shells are filled.

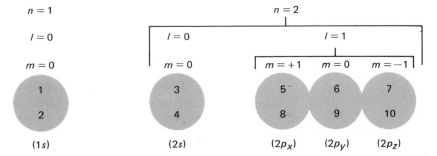

circles. Since electrons are negatively charged and repel each other, they spread out and occupy the $2p$ orbitals singly before they begin to pair. After each of the $2p$ orbitals holds one electron, pairing occurs because less energy is required to overcome interelectronic repulsion and add an electron to an orbital that already holds an electron than is required to place an electron in the next-higher empty orbital—a $3s$ orbital. This general order of orbital filling is observed for all subshells.

The spectroscopic notations for electronic configurations convey this information in a different form. The spectroscopic terms for the subshells are used with superscripts to indicate the number of electrons in each subshell. Thus, the electronic configuration of carbon may be indicated as $1s^2\ 2s^2\ 2p^2$. This notation clearly presents the first two quantum numbers—the number stands for $n$ and the letter relates to $l$— but the second two quantum numbers must be inferred.

Someone unfamiliar with this system must guard against interpreting all even-numbered superscripts as indicating a situation in which all electrons are paired. This is not the case. In carbon, for example, we know that there are two unpaired electrons despite the fact that all the superscripts are even numbers.

A way of circumventing this difficulty is to repeat the designation for each orbital of an incompete subshell. Thus, the electronic configuration of carbon would be indicated as $1s^2\ 2s^2\ 2p_x^1\ 2p_y^1$; that of the following element, nitrogen, would be $1s^2\ 2s^2\ 2p_x^1\ 2p_y^1\ 2p_z^1$; and that of the next element, oxygen, would be $1s^2\ 2s^2\ 2p_x^2\ 2p_y^1\ 2p_z^1$. After some experience with this spectroscopic system of notation, such a device is unnecessary.

**2.17 ELECTRONIC STRUCTURES OF THE ELEMENTS**

The quantum number $n$ for an orbital indicates the relative distance of the maximum electron density of the orbital from the nucleus. As $n$ increases, the electron spends more time at distances farther from the nucleus; therefore, the value of $n$ is important in determining the energy of an electron in an orbital.

In multielectron atoms the value of $l$ is also important. In such an atom, the positive charge of the nucleus is partially screened from an outer electron by the negative cloud of the intervening electrons; the electron does not "feel" the full nuclear charge. The screening, however, is not uniform for all the orbitals of a given level. The orbitals of an atom overlap in the region surrounding the nucleus, and there is at least a slight probability that an electron of any type orbital will be found close to the nucleus where the electron is more fully subject to the nuclear charge (see Figure 2.18). The more an orbital penetrates the negative cloud of the screening electrons, the more strongly the electron is attracted and the lower is its energy. For orbitals of the same level (same value of $n$), the smaller the value of $l$, the higher the probability of finding the electron near the nucleus. Thus, in a given atom a $2s$ electron ($l = 0$) is more penetrating and has a lower energy than a $2p$ electron ($l = 1$). For a given level the order by energy is: $s < p < d < f$.

Therefore, the energy of an electron in an orbital depends upon: the number of protons in the nucleus ($Z$), the average distance of the orbital from the nucleus (given by $n$), the screening of inner electrons, the degree to which the orbital under consideration penetrates the charge cloud of the other electrons and approaches the nucleus (given by $l$), as well as repulsions and interactions with the other electrons of the atom. There is no standard order of sublevels, based on energy, that pertains to all atoms. In the potassium atom ($Z = 19$), the $4s$ orbital has a lower energy than a $3d$ because the $4s$ penetrates the screening electrons more than the $3d$. In atoms of higher $Z$ (at approximately $Z = 29$, Cu), this situation reverses and a $3d$ orbital has a lower energy than the $4s$ presumably because the higher nuclear charge makes the penetration effect (related to $l$) less important than the distance of maximum electron probability (given by $n$).

The deduction of the electronic configurations of the elements by the successive addition of electrons to the Bohr energy levels was attempted in Section 2.13. A similar procedure, called the **aufbau** (from the German, meaning "building-up") **method**, takes into account the existence of sublevels and is therefore more successful in deriving electronic con-

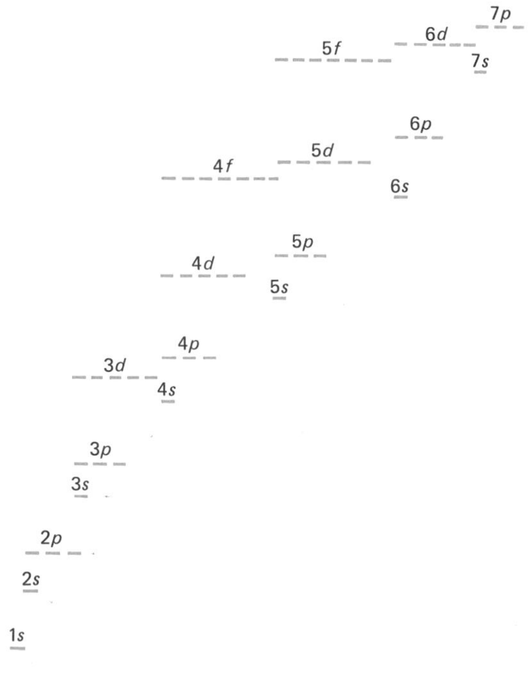

**Figure 2.22** Aufbau order of atomic orbitals.

figurations. In this *hypothetical process*, the electron that is added in going from element to element (the so-called differentiating electron that makes the electronic configuration of an element different from that of the preceding element) is entered into the orbital of lowest energy available to it in conformance with the Pauli exclusion principle and Hund's rule.

Since the order of orbitals, based on energy, changes as one proceeds through the elements, the *aufbau* order reflects *only* the orbital position that the differentiating electron assumes in this procedure. The aufbau order is shown in Figure 2.22; electronic configurations can be derived from this diagram by successively filling the orbitals starting at the bottom of the chart and proceeding upward. Remember that there are three orbitals in a *p* sublevel, five in a *d,* and seven in an *f.*

The aufbau order may be remembered by using the device illustrated in Figure 2.23. The notations are written in indented fashion; when these symbols are joined by downward vertical lines, the proper sequence emerges. However, the periodic table provides the best means of deriving electronic configurations, as well as correlating them.

In Figure 2.24 the type of differentiating electron is related to the position of the element in the periodic chart; this figure can be used to illustrate how the aufbau method works. The first period consists of hydrogen ($1s^1$) and helium ($1s^2$). The second period begins with lithium ($1s^2\ 2s^1$) and beryllium ($1s^2\ 2s^2$), which add electrons to the $2s$ orbital; the six elements that complete the period—boron ($1s^2\ 2s^2\ 2p^1$) to the noble gas neon ($1s^2\ 2s^2\ 2p^6$)—gradually add electrons to the three $2p$ orbitals.

The pattern of the second period is repeated in the third period. The two "*s* block" elements are sodium ($1s^2\ 2s^2\ 2p^6\ 3s^1$) and magnesium

**Figure 2.23** Aufbau order for filling sublevels.

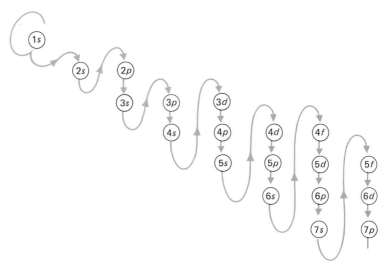

**Figure 2.24** Type of differentiating electron related to position of the element in the periodic chart.

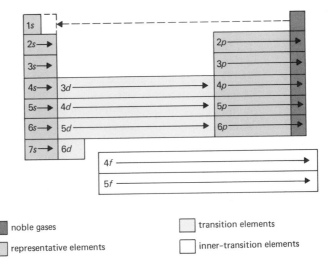

☐ noble gases      ☐ transition elements

☐ representative elements      ☐ inner-transition elements

$(1s^2 \ 2s^2 \ 2p^6 \ 3s^2)$; the six "$p$ block" elements go from aluminum $(1s^2 \ 2s^2 \ 2p^6 \ 3s^2 \ 3p^1)$ to the noble gas argon $(1s^2 \ 2s^2 \ 2p^6 \ 3s^2 \ 3p^6)$.

In the discussion of the configurations of the remaining elements, only outer orbitals will be indicated. The first overlap of orbital energies is observed with potassium $(Z = 19)$, the first element of the fourth period; its configuration is . . . $3s^2 \ 3p^6 \ 4s^1$ despite the fact that $3d$ orbitals are vacant. In like manner, calcium $(Z = 20)$ has the configuration . . . $3s^2 \ 3p^6 \ 4s^2$. With scandium the $3d$ sublevel comes into use (. . . $3s^2 \ 3p^6 \ 3d^1 \ 4s^2$), and in the series from scandium to zinc, this sublevel is gradually filled. The configuration of zinc $(Z = 30)$ is . . . $3s^2 \ 3p^6 \ 3d^{10} \ 4s^2$. The elements 21 to 30 are the first transition series (B families); they are said to exhibit inner building, since in each case the last electron is added to the shell $(3d)$ adjacent to the outermost shell $(4s)$. With element 31, gallium (. . . $3s^2 \ 3p^6 \ 3d^{10} \ 4s^2 \ 4p^1$), the $4p$ sublevel begins to be filled, and the fourth period ends with krypton (. . . $3s^2 \ 3p^6 \ 3d^{10} \ 4s^2 \ 4p^6$).

The fifth period starts with rubidium $(Z = 37)$ and the additional electron beyond the krypton core is added to the $5s$ subshell. A second transition series, which follows Sr $(Z = 38)$, starts with yttrium $(Z = 39,$ . . . $4s^2 \ 4p^6 \ 4d^1 \ 5s^2)$ and electrons are added to the $4d$ sublevel. The fifth period ends with the series from indium to xenon with electrons being added to the $5p$ sublevel. Xenon has the configuration . . . $4s^2 \ 4p^6 \ 4d^{10} \ 5s^2 \ 5p^6$.

The sixth period is much more complicated as far as orbital overlap is concerned. The first element, cesium $(Z = 55)$, has one electron in the $6s$ sublevel: . . . $4d^{10} \ 5s^2 \ 5p^6 \ 6s^1$; the second element, barium $(Z = 56)$, has an additional electron that completes the $6s$ sublevel. We here come upon one complication of the aufbau procedure. The $4f$ and $5d$ sublevels are so close in energy that the next electron (for lanthanum, $Z = 57$) is added to the $5d$ sublevel (thus lanthanum is a transition element), but the next succeeding electron (for cerium, $Z = 58$) is added to the $4f$ sublevel and the electron added for lanthanum falls back into the $4f$ sublevel. Thus, for La the configuration is . . . $4d^{10}$ $5s^2$ $5p^6$ $5d^1$ $6s^2$; and for Ce, . . . $4d^{10}$ $4f^2$ $5s^2$ $5p^6$ $6s^2$. For the elements 58 to 70 (cerium to ytterbium), the electrons are added to the $4f$ sublevel.

These elements are called inner-transition elements; for these elements, electron addition occurs in the third subshell ($4f$) from the outermost subshell ($6s$). After the $4f$ subshell has been filled, the next electron is added to the $5d$ subshell. Thus, for lutetium ($Z = 71$) the configuration is . . . $4d^{10}$ $4f^{14}$ $5s^2$ $5p^6$ $5d^1$ $6s^2$. This third transition series is completed and the $5d$ subshell filled with element 80, mercury (. . . $4d^{10}$ $4f^{14}$ $5s^2$ $5p^6$ $5d^{10}$ $6s^2$). The period ends with the filling of the $6p$ sublevel in elements 81 to 86

The seventh period is incomplete and includes many artificial, manmade elements. This period follows the pattern established by the sixth period: elements 87 and 88 have electrons added to the $7s$ sublevel, for element 89 an electron is added to the $6d$ sublevel, elements 90 to 103 constitute a second inner-transition series and exhibit an electron buildup of the $5f$ sublevel, and the transition elements 104 and 105 (as yet unnamed) have electrons added to the $6d$ sublevel.

To determine the electronic configuration of any element, we start with hydrogen and on the basis of the periodic table account for every electron added until the desired element is reached. Thus, for tungsten ($Z = 74$), the first period gives $1s^2$; the second period, $2s^2$ $2p^6$; the third, $3s^2$ $3p^6$; the fourth, $4s^2$ $3d^{10}$ $4p^6$; and the fifth, $5s^2$ $4d^{10}$ $5p^6$. The sixth period—in which tungsten is found—starts with $6s^2$ for cesium and barium, adds $4f^{14}$ for the inner-transition series, and concludes with $5d^4$, since tungsten is the fourth element in the $5d$ transition series. Rearranging these terms in sequence gives: $1s^2$ $2s^2$ $2p^6$ $3s^2$ $3p^6$ $3d^{10}$ $4s^2$ $4p^6$ $4d^{10}$ $4f^{14}$ $5s^2$ $5p^6$ $5d^4$ $6s^2$.

The aufbau order cannot be used to interpret processes that involve the loss of electrons (ionizations). The configuration of iron is $1s^2$ $2s^2$ $2p^6$ $3s^2$ $3p^6$ $3d^6$ $4s^2$, and that of the $Fe^{2+}$ ion is $1s^2$ $2s^2$ $2p^6$ $3s^2$ $3p^6$ $3d^6$; thus, ionization results in the loss of the $4s$ electrons even though $3d$ electrons are the last added by the aufbau procedure. The loss of two electrons in the formation of the $Fe^2$ ion reduces the screening effect, and hence the outer electrons experience the nuclear charge more fully. Therefore, for the ion the fact that electrons in $3d$ orbitals are on the average nearer to the nucleus than the $4s$ electrons is more important than the superior ability of the $4s$ electrons to penetrate the screening electrons. In general, the first electrons lost in an ionization are those with the highest value of $n$; therefore, spectroscopic notations should be written by increasing value of $n$ and not by any hypothetical order of filling.

**2.18 HALF-FILLED AND FILLED SUBSHELLS**

In Table 2.9 the correct electronic configurations of the elements are listed. The configurations predicted by the aufbau procedure are confirmed by spectral and magnetic studies for most elements. There are a few, however, that exhibit slight variations from the standard pattern. In certain instances, it is possible to explain these variations on the basis of the enhanced stability of a filled or half-filled subshell.

The predicted configuration for the $3d$ and $4s$ subshells in the chromium atom is $3d^4\,4s^2$, whereas the experimentally derived configuration is $3d^5\,4s^1$. Presumably the stability gained by having one unpaired electron in each of the five $3d$ orbitals (a half-filled subshell) accounts for the fact that the $3d^5\,4s^1$ configuration is the one observed. The existence of a half-filled subshell also accounts for the fact that the configuration for the $4d$ and $5s$ subshells of molybdenum is $4d^5\,5s^1$ rather than the predicted $4d^4\,5s^2$.

For copper the predicted configuration for the last two subshells is $3d^9\,4s^2$, whereas the accepted structure is $3d^{10}\,4s^1$. The explanation for this deviation lies in the superior stability of the $3d^{10}\,4s^1$ configuration that results from the completed $3d$ subshell. Silver and gold also have configurations with completely filled $d$ subshells instead of the $(n-1)d^9$ $ns^2$ configurations predicted. In the case of palladium, two electrons are involved—the only case with a difference of more than one electron. The predicted configuration for the last two subshells of palladium is $4d^8$ $5s^2$; the observed configuration is $4d^{10}\,5s^0$.

That half-filled and filled subshells contribute to the stability of atoms is also borne out in cases where the aufbau order is followed. Thus nitrogen, phosphorus, and arsenic show unusual properties (e.g., unexpectedly high ionization potentials) because they all have half-filled $p$ subshells. Zinc, cadmium, and mercury have properties that can be traced to the fact that all their subshells are filled. The noble gases have unusually stable configurations with all their subshells complete.

Some deviations from predicted configurations are observed other than those that can be accounted for on the basis of filled or half-filled subshells, particularly among elements of higher $Z$. For our purposes such exceptions are not important. In general, the chemistry of the elements is satisfactorily explained on the basis of the predicted configurations.

**2.19 TYPES OF ELEMENTS**

We may classify the elements according to their electronic configurations into four types (see Figure 2.23).

1. *The noble gases.* In the periodic table the noble gases are found at the end of each period in group 0. They are colorless gases, chemically unreactive, and diamagnetic. With the exception of helium (which has the configuration $1s^2$), all the noble gases have outer configurations of $ns^2\,np^6$, a very stable arrangement.

**Table 2.9.** Electronic configurations of the elements

| ELEMENT | Z | 1s | 2s | 2p | 3s | 3p | 3d | 4s | 4p | 4d | 4f | 5s | 5p | 5d | 5f | 6s | 6p | 6d | 7s |
|---|---|---|---|---|---|---|---|---|---|---|---|---|---|---|---|---|---|---|---|
| H  | 1  | 1 | | | | | | | | | | | | | | | | | |
| He | 2  | 2 | | | | | | | | | | | | | | | | | |
| Li | 3  | 2 | 1 | | | | | | | | | | | | | | | | |
| Be | 4  | 2 | 2 | | | | | | | | | | | | | | | | |
| B  | 5  | 2 | 2 | 1 | | | | | | | | | | | | | | | |
| C  | 6  | 2 | 2 | 2 | | | | | | | | | | | | | | | |
| N  | 7  | 2 | 2 | 3 | | | | | | | | | | | | | | | |
| O  | 8  | 2 | 2 | 4 | | | | | | | | | | | | | | | |
| F  | 9  | 2 | 2 | 5 | | | | | | | | | | | | | | | |
| Ne | 10 | 2 | 2 | 6 | | | | | | | | | | | | | | | |
| Na | 11 | 2 | 2 | 6 | 1 | | | | | | | | | | | | | | |
| Mg | 12 | 2 | 2 | 6 | 2 | | | | | | | | | | | | | | |
| Al | 13 | 2 | 2 | 6 | 2 | 1 | | | | | | | | | | | | | |
| Si | 14 | 2 | 2 | 6 | 2 | 2 | | | | | | | | | | | | | |
| P  | 15 | 2 | 2 | 6 | 2 | 3 | | | | | | | | | | | | | |
| S  | 16 | 2 | 2 | 6 | 2 | 4 | | | | | | | | | | | | | |
| Cl | 17 | 2 | 2 | 6 | 2 | 5 | | | | | | | | | | | | | |
| Ar | 18 | 2 | 2 | 6 | 2 | 6 | | | | | | | | | | | | | |
| K  | 19 | 2 | 2 | 6 | 2 | 6 | | 1 | | | | | | | | | | | |
| Ca | 20 | 2 | 2 | 6 | 2 | 6 | | 2 | | | | | | | | | | | |
| Sc | 21 | 2 | 2 | 6 | 2 | 6 | 1 | 2 | | | | | | | | | | | |
| Ti | 22 | 2 | 2 | 6 | 2 | 6 | 2 | 2 | | | | | | | | | | | |
| V  | 23 | 2 | 2 | 6 | 2 | 6 | 3 | 2 | | | | | | | | | | | |
| Cr | 24 | 2 | 2 | 6 | 2 | 6 | 5 | 1 | | | | | | | | | | | |
| Mn | 25 | 2 | 2 | 6 | 2 | 6 | 5 | 2 | | | | | | | | | | | |
| Fe | 26 | 2 | 2 | 6 | 2 | 6 | 6 | 2 | | | | | | | | | | | |
| Co | 27 | 2 | 2 | 6 | 2 | 6 | 7 | 2 | | | | | | | | | | | |
| Ni | 28 | 2 | 2 | 6 | 2 | 6 | 8 | 2 | | | | | | | | | | | |
| Cu | 29 | 2 | 2 | 6 | 2 | 6 | 10 | 1 | | | | | | | | | | | |
| Zn | 30 | 2 | 2 | 6 | 2 | 6 | 10 | 2 | | | | | | | | | | | |
| Ga | 31 | 2 | 2 | 6 | 2 | 6 | 10 | 2 | 1 | | | | | | | | | | |
| Ge | 32 | 2 | 2 | 6 | 2 | 6 | 10 | 2 | 2 | | | | | | | | | | |
| As | 33 | 2 | 2 | 6 | 2 | 6 | 10 | 2 | 3 | | | | | | | | | | |
| Se | 34 | 2 | 2 | 6 | 2 | 6 | 10 | 2 | 4 | | | | | | | | | | |
| Br | 35 | 2 | 2 | 6 | 2 | 6 | 10 | 2 | 5 | | | | | | | | | | |
| Kr | 36 | 2 | 2 | 6 | 2 | 6 | 10 | 2 | 6 | | | | | | | | | | |
| Rb | 37 | 2 | 2 | 6 | 2 | 6 | 10 | 2 | 6 | | | 1 | | | | | | | |
| Sr | 38 | 2 | 2 | 6 | 2 | 6 | 10 | 2 | 6 | | | 2 | | | | | | | |
| Y  | 39 | 2 | 2 | 6 | 2 | 6 | 10 | 2 | 6 | 1 | | 2 | | | | | | | |
| Zr | 40 | 2 | 2 | 6 | 2 | 6 | 10 | 2 | 6 | 2 | | 2 | | | | | | | |
| Nb | 41 | 2 | 2 | 6 | 2 | 6 | 10 | 2 | 6 | 4 | | 1 | | | | | | | |
| Mo | 42 | 2 | 2 | 6 | 2 | 6 | 10 | 2 | 6 | 5 | | 1 | | | | | | | |
| Tc | 43 | 2 | 2 | 6 | 2 | 6 | 10 | 2 | 6 | 6 | | 1 | | | | | | | |
| Ru | 44 | 2 | 2 | 6 | 2 | 6 | 10 | 2 | 6 | 7 | | 1 | | | | | | | |
| Rh | 45 | 2 | 2 | 6 | 2 | 6 | 10 | 2 | 6 | 8 | | 1 | | | | | | | |
| Pd | 46 | 2 | 2 | 6 | 2 | 6 | 10 | 2 | 6 | 10 | | | | | | | | | |
| Ag | 47 | 2 | 2 | 6 | 2 | 6 | 10 | 2 | 6 | 10 | | 1 | | | | | | | |
| Cd | 48 | 2 | 2 | 6 | 2 | 6 | 10 | 2 | 6 | 10 | | 2 | | | | | | | |
| In | 49 | 2 | 2 | 6 | 2 | 6 | 10 | 2 | 6 | 10 | | 2 | 1 | | | | | | |
| Sn | 50 | 2 | 2 | 6 | 2 | 6 | 10 | 2 | 6 | 10 | | 2 | 2 | | | | | | |
| Sb | 51 | 2 | 2 | 6 | 2 | 6 | 10 | 2 | 6 | 10 | | 2 | 3 | | | | | | |

Table 2.9 Continued Electronic configurations of the elements

| ELEMENT | Z | 1s | 2s | 2p | 3s | 3p | 3d | 4s | 4p | 4d | 4f | 5s | 5p | 5d | 5f | 6s | 6p | 6d | 7s |
|---|---|---|---|---|---|---|---|---|---|---|---|---|---|---|---|---|---|---|---|
| Te | 52 | 2 | 2 | 6 | 2 | 6 | 10 | 2 | 6 | 10 |   | 2 | 4 |   |   |   |   |   |   |
| I  | 53 | 2 | 2 | 6 | 2 | 6 | 10 | 2 | 6 | 10 |   | 2 | 5 |   |   |   |   |   |   |
| Xe | 54 | 2 | 2 | 6 | 2 | 6 | 10 | 2 | 6 | 10 |   | 2 | 6 |   |   |   |   |   |   |
| Cs | 55 | 2 | 2 | 6 | 2 | 6 | 10 | 2 | 6 | 10 |   | 2 | 6 |   |   | 1 |   |   |   |
| Ba | 56 | 2 | 2 | 6 | 2 | 6 | 10 | 2 | 6 | 10 |   | 2 | 6 |   |   | 2 |   |   |   |
| La | 57 | 2 | 2 | 6 | 2 | 6 | 10 | 2 | 6 | 10 |   | 2 | 6 | 1 |   | 2 |   |   |   |
| Ce | 58 | 2 | 2 | 6 | 2 | 6 | 10 | 2 | 6 | 10 | 2 | 2 | 6 |   |   | 2 |   |   |   |
| Pr | 59 | 2 | 2 | 6 | 2 | 6 | 10 | 2 | 6 | 10 | 3 | 2 | 6 |   |   | 2 |   |   |   |
| Nd | 60 | 2 | 2 | 6 | 2 | 6 | 10 | 2 | 6 | 10 | 4 | 2 | 6 |   |   | 2 |   |   |   |
| Pm | 61 | 2 | 2 | 6 | 2 | 6 | 10 | 2 | 6 | 10 | 5 | 2 | 6 |   |   | 2 |   |   |   |
| Sm | 62 | 2 | 2 | 6 | 2 | 6 | 10 | 2 | 6 | 10 | 6 | 2 | 6 |   |   | 2 |   |   |   |
| Eu | 63 | 2 | 2 | 6 | 2 | 6 | 10 | 2 | 6 | 10 | 7 | 2 | 6 |   |   | 2 |   |   |   |
| Gd | 64 | 2 | 2 | 6 | 2 | 6 | 10 | 2 | 6 | 10 | 7 | 2 | 6 | 1 |   | 2 |   |   |   |
| Tb | 65 | 2 | 2 | 6 | 2 | 6 | 10 | 2 | 6 | 10 | 9 | 2 | 6 |   |   | 2 |   |   |   |
| Dy | 66 | 2 | 2 | 6 | 2 | 6 | 10 | 2 | 6 | 10 | 10 | 2 | 6 |   |   | 2 |   |   |   |
| Ho | 67 | 2 | 2 | 6 | 2 | 6 | 10 | 2 | 6 | 10 | 11 | 2 | 6 |   |   | 2 |   |   |   |
| Er | 68 | 2 | 2 | 6 | 2 | 6 | 10 | 2 | 6 | 10 | 12 | 2 | 6 |   |   | 2 |   |   |   |
| Tm | 69 | 2 | 2 | 6 | 2 | 6 | 10 | 2 | 6 | 10 | 13 | 2 | 6 |   |   | 2 |   |   |   |
| Yb | 70 | 2 | 2 | 6 | 2 | 6 | 10 | 2 | 6 | 10 | 14 | 2 | 6 |   |   | 2 |   |   |   |
| Lu | 71 | 2 | 2 | 6 | 2 | 6 | 10 | 2 | 6 | 10 | 14 | 2 | 6 | 1 |   | 2 |   |   |   |
| Hf | 72 | 2 | 2 | 6 | 2 | 6 | 10 | 2 | 6 | 10 | 14 | 2 | 6 | 2 |   | 2 |   |   |   |
| Ta | 73 | 2 | 2 | 6 | 2 | 6 | 10 | 2 | 6 | 10 | 14 | 2 | 6 | 3 |   | 2 |   |   |   |
| W  | 74 | 2 | 2 | 6 | 2 | 6 | 10 | 2 | 6 | 10 | 14 | 2 | 6 | 4 |   | 2 |   |   |   |
| Re | 75 | 2 | 2 | 6 | 2 | 6 | 10 | 2 | 6 | 10 | 14 | 2 | 6 | 5 |   | 2 |   |   |   |
| Os | 76 | 2 | 2 | 6 | 2 | 6 | 10 | 2 | 6 | 10 | 14 | 2 | 6 | 6 |   | 2 |   |   |   |
| Ir | 77 | 2 | 2 | 6 | 2 | 6 | 10 | 2 | 6 | 10 | 14 | 2 | 6 | 7 |   | 2 |   |   |   |
| Pt | 78 | 2 | 2 | 6 | 2 | 6 | 10 | 2 | 6 | 10 | 14 | 2 | 6 | 9 |   | 1 |   |   |   |
| Au | 79 | 2 | 2 | 6 | 2 | 6 | 10 | 2 | 6 | 10 | 14 | 2 | 6 | 10 |   | 1 |   |   |   |
| Hg | 80 | 2 | 2 | 6 | 2 | 6 | 10 | 2 | 6 | 10 | 14 | 2 | 6 | 10 |   | 2 |   |   |   |
| Tl | 81 | 2 | 2 | 6 | 2 | 6 | 10 | 2 | 6 | 10 | 14 | 2 | 6 | 10 |   | 2 | 1 |   |   |
| Pb | 82 | 2 | 2 | 6 | 2 | 6 | 10 | 2 | 6 | 10 | 14 | 2 | 6 | 10 |   | 2 | 2 |   |   |
| Bi | 83 | 2 | 2 | 6 | 2 | 6 | 10 | 2 | 6 | 10 | 14 | 2 | 6 | 10 |   | 2 | 3 |   |   |
| Po | 84 | 2 | 2 | 6 | 2 | 6 | 10 | 2 | 6 | 10 | 14 | 2 | 6 | 10 |   | 2 | 4 |   |   |
| At | 85 | 2 | 2 | 6 | 2 | 6 | 10 | 2 | 6 | 10 | 14 | 2 | 6 | 10 |   | 2 | 5 |   |   |
| Rn | 86 | 2 | 2 | 6 | 2 | 6 | 10 | 2 | 6 | 10 | 14 | 2 | 6 | 10 |   | 2 | 6 |   |   |
| Fr | 87 | 2 | 2 | 6 | 2 | 6 | 10 | 2 | 6 | 10 | 14 | 2 | 6 | 10 |   | 2 | 6 |   | 1 |
| Ra | 88 | 2 | 2 | 6 | 2 | 6 | 10 | 2 | 6 | 10 | 14 | 2 | 6 | 10 |   | 2 | 6 |   | 2 |
| Ac | 89 | 2 | 2 | 6 | 2 | 6 | 10 | 2 | 6 | 10 | 14 | 2 | 6 | 10 |   | 2 | 6 | 1 | 2 |
| Th | 90 | 2 | 2 | 6 | 2 | 6 | 10 | 2 | 6 | 10 | 14 | 2 | 6 | 10 |   | 2 | 6 | 2 | 2 |
| Pa | 91 | 2 | 2 | 6 | 2 | 6 | 10 | 2 | 6 | 10 | 14 | 2 | 6 | 10 | 2 | 2 | 6 | 1 | 2 |
| U  | 92 | 2 | 2 | 6 | 2 | 6 | 10 | 2 | 6 | 10 | 14 | 2 | 6 | 10 | 3 | 2 | 6 | 1 | 2 |
| Np | 93 | 2 | 2 | 6 | 2 | 6 | 10 | 2 | 6 | 10 | 14 | 2 | 6 | 10 | 4 | 2 | 6 | 1 | 2 |
| Pu | 94 | 2 | 2 | 6 | 2 | 6 | 10 | 2 | 6 | 10 | 14 | 2 | 6 | 10 | 6 | 2 | 6 |   | 2 |
| Am | 95 | 2 | 2 | 6 | 2 | 6 | 10 | 2 | 6 | 10 | 14 | 2 | 6 | 10 | 7 | 2 | 6 |   | 2 |
| Cm | 96 | 2 | 2 | 6 | 2 | 6 | 10 | 2 | 6 | 10 | 14 | 2 | 6 | 10 | 7 | 2 | 6 | 1 | 2 |
| Bk | 97 | 2 | 2 | 6 | 2 | 6 | 10 | 2 | 6 | 10 | 14 | 2 | 6 | 10 | 8 | 2 | 6 | 1 | 2 |
| Cf | 98 | 2 | 2 | 6 | 2 | 6 | 10 | 2 | 6 | 10 | 14 | 2 | 6 | 10 | 10 | 2 | 6 |   | 2 |
| Es | 99 | 2 | 2 | 6 | 2 | 6 | 10 | 2 | 6 | 10 | 14 | 2 | 6 | 10 | 11 | 2 | 6 |   | 2 |
| Fm | 100 | 2 | 2 | 6 | 2 | 6 | 10 | 2 | 6 | 10 | 14 | 2 | 6 | 10 | 12 | 2 | 6 |   | 2 |
| Md | 101 | 2 | 2 | 6 | 2 | 6 | 10 | 2 | 6 | 10 | 14 | 2 | 6 | 10 | 13 | 2 | 6 |   | 2 |
| No | 102 | 2 | 2 | 6 | 2 | 6 | 10 | 2 | 6 | 10 | 14 | 2 | 6 | 10 | 14 | 2 | 6 |   | 2 |
| Lr | 103 | 2 | 2 | 6 | 2 | 6 | 10 | 2 | 6 | 10 | 14 | 2 | 6 | 10 | 14 | 2 | 6 | 1 | 2 |

2. *The representative elements.* These elements comprise the A families of the periodic table used in this book and include metals and nonmetals. They exhibit a wide range of chemical behavior and physical characteristics. Some of the elements are diamagnetic and some are paramagnetic; the compounds of these elements, however, are generally diamagnetic and colorless. These elements all have their electronic shells either complete or stable (e.g., $ns^2 np^6$) except the outer shells to which the last electron may be considered as having been added. This outer shell is termed the valence shell; electrons in it are valence electrons. The number of valence electrons for each atom is the same as the group number. The chemistry of these elements depends upon these valence electrons.

3. *The transition elements.* These elements are found in the B families of the periodic table. They are characterized by inner building — the "last" electron added by the aufbau procedure is an inner $d$ electron. Electrons from the two outermost shells are used in chemical reactions. All these elements are metals; most of them are paramagnetic and form highly colored, paramagnetic compounds.

4. *The inner-transition elements.* These elements are found at the bottom of the periodic table but should properly follow group III B. The sixth-period series that follows lanthanum (14 elements) is called the lanthanide series; the seventh-period series is known as the actinide series. The last electron added to each element is an $f$ electron. Since this electron is added to the third shell from the outermost shell, the outer three shells may be involved in the chemistry of these elements. All inner-transition elements are metals. The elements, in general, are paramagnetic; their compounds are paramagnetic and highly colored.

## Problems

**2.1** Complete the following table.

| Symbol | Z | A | Protons | Neutrons | Electrons |
|--------|---|---|---------|----------|-----------|
| Se | 34 | 80 | | | |
| Hg | | 202 | | | |
| | 48 | 114 | | | |
| Ce | | | | 82 | |
| Bi | | 209 | | 126 | |
| $Cr^{3+}$ | | | | 29 | |
| | 7 | | | 7 | 10 |

**2.2** Which unit would be larger: an atomic mass unit based on the mass of a $^{19}F$ atom set at exactly 19 u or an atomic mass unit based on the current standard? Only one isotope of fluorine occurs in nature, $^{19}F$.

**2.3** There are two naturally occurring isotopes of rubidium: $^{85}Rb$, which has a mass of 84.91 u, and $^{87}Rb$, which has a mass of 86.92 u. The atomic weight of rubidium is 85.47. What is the percent abundance of each of the isotopes?

**2.4** Naturally occurring bromine consists of two isotopes: $^{79}Br$, which has a mass of 78.918 u, and $^{81}Br$, which has a mass of 80.916 u. The atomic weight of bromine is 79.904. What is the percent abundance of each of the isotopes?

**2.5** If element X consists of 78.7% of atoms with a mass of 24.0 u, 10.1% of atoms with a mass of 25.0 u, and 11.2% of atoms with a mass of 26.0 u, what is the atomic weight of X?

**2.6** What is the atomic weight of element Y if Y consists of 57.25% of atoms with a mass of 120.90 u and 42.75% of atoms with a mass of 122.90 u?

**2.7** The $^{12}_{6}C$ atom is assigned a mass of exactly 12 u. What is the binding energy (in calories) of a $^{12}_{6}C$ atom? The masses of the fundamental particles are: electron, 0.0005 u; proton, 1.0073 u; and neutron, 1.0086 u. One u equals $1.49 \times 10^{-3}$ erg and 1 cal equals $4.18 \times 10^{7}$ erg.

**2.8** The atomic weight of beryllium

is 9.0122 and only one isotope of beryllium occurs in nature, $^{9}_{4}Be$. What is the binding energy (in ergs) of an atom of $^{9}_{4}Be$? Use the data given in Problem 2.7.

**2.9** The binding energy of a $^{29}_{14}Si$ atom is $3.90 \times 10^{-4}$ erg. What is the actual mass of this atom in u? Use the data given in Problem 2.7.

**2.10** The binding energy of a $^{23}_{11}Na$ atom is $2.97 \times 10^{-4}$ erg. What is the actual mass of this atom in u? Use the data given in Problem 2.7.

**2.11** What is the frequency and energy per quantum of (a) yellow light with a wavelength of 5500 Å and (b) blue light with a wavelength of 4500 Å?

**2.12** What is the frequency and energy per quantum of (a) an X ray with a wavelength of 1.50 Å and (b) a radio wave with a wavelength of $5.00 \times 10^{2}$ m ($5.00 \times 10^{12}$ Å)?

**2.13** What are the mass equivalents of the quanta described in Problem 2.12?

**2.14** If light with a wavelength of 4000 Å falls on the surface of sodium metal, electrons with a velocity of $5.37 \times 10^{7}$ cm/sec are ejected. (a) What is the energy (in ergs) of a 4000 Å wavelength photon? (b) What is the kinetic energy (in ergs) of the ejected electron? (c) What is the minimum frequency and the corresponding wavelength of light required to release an electron from sodium? Planck's constant, $h$, is $6.63 \times 10^{-27}$ erg sec; the speed of light,

● Since 1 u $= 1.49 \times 10^{-3}$ erg and 1 erg $= 10^{-7}$ joules (J), 1 u $= 1.49 \times 10^{-10}$ J. The erg is not an SI unit; its use is to be discouraged.

● In the International System a solidus (oblique stroke, /), a horizontal line, or negative powers may be used to express a derived unit formed from two or more other units by division. For example, 1 g cm²/sec in problem 2.14 is written

$$1 \text{ g} \cdot \text{cm}^2/\text{s}^2, \ 1 \ \frac{\text{g} \cdot \text{cm}^2}{\text{s}^2},$$

$$\text{or g} \cdot \text{cm}^2 \cdot \text{s}^{-2}$$

The solidus must not be repeated on the same line unless ambiguity is avoided by parentheses. Thus,

$$\text{g} \cdot \text{cm}^2/\text{s}/\text{s}$$

is not acceptable.

$c$, is $3.00 \times 10^{10}$ cm/sec; the mass of an electron is $9.11 \times 10^{-28}$ g; and 1 erg is 1 g cm²/sec².

**2.15** If ultraviolet light with a wavelength of 2480 Å falls on the surface of gold metal, electrons with a velocity of $2.50 \times 10^7$ cm/sec are ejected. Use the data given in Problem 2.14 to find the minimum frequency and the corresponding wavelength of light needed to release electrons from gold.

**2.16** The photoelectric ejection of electrons from calcium requires that light with a minimum frequency of $6.54 \times 10^{14}$/sec be used. If light with a wavelength of 2500 Å is used to irradiate calcium, what is the velocity of the emitted photoelectrons? Use the data given in Problem 2.14.

**2.17** The spectral lines of hydrogen in the visible region (the Balmer series) represent electron transitions to the $n = 2$ level from higher levels. What is the electron transition that corresponds to the 6563 Å spectral line?

**2.18** What is the electron transition that corresponds to the 4340 Å line in the visible region of the hydrogen spectrum?

**2.19** The Lyman series of lines in the hydrogen spectrum results from electron transitions from higher levels to the $n = 1$ level. (a) Compare equation (20) for the Balmer series with equation (19) for the general case (Section 2.11), and derive an equation for the Lyman series. (b) What wavelengths correspond to transitions from $n = 2$ to $n = 1$ and from $n =$

$\infty$ to $n = 1$? (c) In what region of the electromagnetic spectrum do the Lyman lines occur?

**2.20** The Paschen series of lines in the hydrogen spectrum results from electron transitions from higher levels to the $n = 3$ level. (a) Use equations (19) and (20) to derive an equation for the Paschen series. (b) What wavelengths correspond to transitions from $n = 4$ to $n = 3$ and from $n = \infty$ to $n = 3$? (c) In what region of the electromagnetic spectrum do the Paschen lines occur?

**2.21** (a) According to the Bohr theory, what is the radius of the H atom in the ground state? (b) The radii of many nuclei can be calculated by means of the formula $r = (1.3 \times 10^{-13} \text{ cm})A^{1/3}$, where $A$ is the mass number. What is the radius of the H nucleus? (c) Use your answers from parts (a) and (b) to calculate the percentage of the total volume of the Bohr H atom that is occupied by the nucleus.

**2.22** Moseley found that the frequency, $\nu$, of a characteristic line of the X-ray spectrum of an element is related to the atomic number, $Z$, of the element by the formula $\sqrt{\nu} = a(Z - b)$, where $a$ is approximately $5.0 \times 10^7/\sqrt{\text{sec}}$ and $b$ is approximately 1.0. What is the atomic number of an element for which the corresponding line in the X-ray spectrum occurs at a wavelength of 0.675 Å? What is the element?

**2.23** What is the atomic number of an element for which the line in the X-ray spectrum conforming to the formula given in Problem 2.22 occurs at a wavelength of 0.564 Å? What is the element?

**2.24** What is the wavelength of the line in the X-ray spectrum of Zr that conforms to the formula given in Problem 2.22?

**2.25** What is the wavelength of the line in the X-ray spectrum of Fe that conforms to the formula given in Problem 2.22?

**2.26** (a) What is the de Broglie wavelength of an electron moving at one-tenth the speed of light? (b) What is the de Broglie wavelength of a proton moving at the same velocity? The mass of an electron is $9.11 \times 10^{-28}$ g and that of a proton is $1.67 \times 10^{-24}$ g.

**2.27** If the de Broglie wavelength of an electron is 1.00 Å, at what velocity is the electron moving? The mass of an electron is $9.11 \times 10^{-28}$ g.

**2.28** (a) Use the approximate expression for Heisenberg's uncertainty principle, $\Delta x \, \Delta mv \approx h$, to determine the uncertainty in the velocity of a particle with a mass of 1.0 g when the uncertainty in the particle's position is 1.0 Å. Planck's constant is $6.6 \times 10^{-27}$ g cm²/sec. (b) Perform the same calculation for the electron (mass $= 9.1 \times 10^{-28}$ g).

**2.29** Give the values for all four quantum numbers for each electron in oxygen ($Z = 8$).

**2.30** Each electron in an atom may be characterized by a set of four quantum numbers. For each of the following parts, tell how many sets of quantum numbers are possible such that each set contains all the values listed. (a) $n = 5$, $l = 4$, (b) $n = 4$, $l = 1$, (c) $n = 4$, $l = 0$, $m = 0$, (d) $n = 2$, $l = 0$, $m = +2$, (e) $n = 3$, $l = 2$, $m = +1$, (f) $n = 2$, $l = 2$, (g) $n = 5$.

**2.31** In the ground state of Xe: (a) how many electrons have $m = -1$ as one of their quantum numbers? (b) how many electrons have $l = 2$ as one of their quantum numbers?

**2.32** In the ground state of Rb: (a) how many electrons have $m = 0$ as one of their quantum numbers? (b) how many electrons have $m = +2$ as one of their quantum numbers?

**2.33** The magnetic moments (in Bohr magnetons) of the ions of the fourth period transition elements conform fairly well to the formula $\sqrt{n(n + 2)}$, where $n$ is the number of unpaired electrons. Determine the magnetic moment of: (a) $Co^{2+}$, (b) $Cu^{2+}$, (c) $Fe^{2+}$, (d) $Sc^{3+}$, (e) $Mn^{2+}$, (f) $V^{3+}$.

**2.34** The magnetic moments of the ions derived from the fourth period transition metals (in Bohr magnetons) conform fairly well to the formula $\sqrt{n(n + 2)}$, where $n$ is the number of unpaired electrons. What is the charge on an ion of Cr that has a magnetic moment of 4.90 Bohr magnetons?

**2.35** Identify the atoms that have the following ground-state electronic configurations in their outer shell or shells: (a) $3s^2 \, 3p^6 \, 3d^8 \, 4s^2$, (b) $4s^2 \, 4p^4$, (c) $4s^2 \, 4p^6 \, 4d^2 \, 5s^2$, (d) $5s^2 \, 5p^3$, (e) $5s^2 \, 5p^6 \, 6s^1$, (f) $6s^2 \, 6p^6$.

**2.36** Identify the atoms that have the following ground-state electronic configurations in their outer shell or shells: (a) $4s^2 \, 4p^5$, (b) $4s^2 \, 4p^6 \, 4d^1 \, 5s^2$, (c) $4s^2 \, 4p^6 \, 4d^{10} \, 4f^3 \, 5s^2 \, 5p^6 \, 6s^2$, (d) $5s^2 \, 5p^6 \, 6s^2$, (e) $5s^2 \, 5p^6 \, 5d^4 \, 6s^2$, (f) $6s^2 \, 6p^5$.

**2.37** Write the notations for the ground-state electronic configurations of the following atoms: (a) $_{30}$Zn, (b) $_{52}$Te, (c) $_{61}$Pm, (d) $_{72}$Hf, (e) $_{82}$Pb, (f) $_{88}$Ra.

**2.38** State the number of unpaired electrons in each of the atoms listed in Problem 2.37.

**2.39** Write the notations for the ground-state electronic configurations of the following atoms: (a) $_{37}$Rb, (b) $_{50}$Sn, (c) $_{54}$Xe, (d) $_{67}$Ho, (e) $_{76}$Os, (f) $_{80}$Hg.

**2.40** State the number of unpaired electrons in each of the atoms listed in Problem 2.39.

**2.41** Write the notations for the ground-state electronic configurations of the following ions: (a) $_{20}$Ca$^{2+}$, (b) $_{24}$Cr$^{2+}$, (c) $_{27}$Co$^{3+}$, (d) $_{46}$Pd$^{2+}$, (e) $_{47}$Ag$^+$, (f) $_{53}$I$^-$.

**2.42** (a) State the number of unpaired electrons in each of the ions listed in Problem 2.41. (b) Which of these ions would you predict to be diamagnetic and which paramagnetic?

# 3
# Chemical Bonding

The electrons of atoms are responsible for chemical bonding. **Ionic** (also called **electrovalent) bonding** is characterized by electron transfer; one of the reacting atoms loses one or more electrons, and the other atom gains one or more electrons. In **covalent bonding,** electrons are not transferred but are shared; a covalent bond consists of a pair of electrons shared by two atoms. Pure ionic bonding and pure covalent bonding are seldom encountered. Most bonds have intermediate character, although many bonds are predominantly either ionic or covalent.

An understanding of electronic configuration, a principal topic of the previous chapter, is fundamental to an understanding of chemical bonding. Other properties of the elements, such as atomic size, are also important to a discussion of chemical bonding and will be considered in the opening sections of this chapter.

## 3.1 ATOMIC SIZES

Determination of atomic sizes poses a problem. If the atom is viewed as a sphere, the radius of the atom should be the distance from the center of the nucleus to the outer reaches of the last electron. But the electron cloud of an atom has a varying intensity, and the probability of finding an electron extends to infinity. It is impossible to isolate and measure a single atom.

It is possible, however, to measure the distance between the nuclei of two covalently bonded atoms (the so-called bond distance). **Atomic** (also called **covalent) radii** are determined from these bond distances by apportioning them between the bonded atoms. For example, one-half the Cl—Cl bond distance, 1.98 Å, gives a value of 0.99 Å for the atomic radius of chlorine. In turn, the atomic radius of chlorine, 0.99 Å, can be subtracted from the C—Cl bond distance, 1.76 Å, to derive the atomic radius of carbon, 0.77 Å.

Since changing the environment of an atom may cause its effective size to vary, discrepancies sometimes arise when this method of assigning atomic radii is used. Nevertheless, data of this type are useful in establishing trends and making generalizations.

Atomic radius is plotted against atomic number in Figure 3.1. Within a group of the periodic table, an increase in atomic radius is generally observed from top to bottom. The alkali metals (group I A) and the halogens (group VII A) are labeled in Figure 3.1, and the increase in size within each group is clearly evident. This trend is expected since the larger atoms of a group employ more electron levels than the smaller atoms.

**Figure 3.1** Atomic radii of the elements.

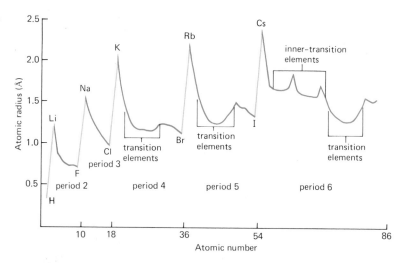

Figure 3.1 also shows a general decrease in atomic radius from left to right across any period. The outermost electrons of the elements of a given period all have the same principal quantum number. As the atomic number increases from element to element across a period, the additional electron is added to the same level or to an inner level. Since the positive charge on the nucleus also increases (the number of protons increases), the nucleus draws in the electron shells and the atoms decrease in size as a result.

The transition elements and inner-transition elements show some minor variations. For the transition elements the differentiating electrons fill inner $d$ orbitals. The effect of nuclear charge on outer, size-determining electrons is reduced by the screening effect of inner electrons. For a transition series, therefore, the gradual buildup of electrons in inner $d$ orbitals at first retards the rate of decrease in atomic radius and then, toward the end of the series when the inner $d$ subshell nears completion, causes the radius to increase.

In general, the lanthanides (elements 58 to 71) exhibit a slow but significant decrease in atomic radius called the **lanthanide contraction.** The differentiating electrons of these elements are added to the third level from the outside (the $4f$ sublevel). These electrons screen the increasing nuclear charge from the outer $6s$ electrons but are not completely efficient, and therefore the steady reduction in size is observed. The transition elements following the lanthanides exhibit a typical transition-element pattern; however, the effect of the lanthanide contraction causes these elements to have approximately the same size as the corresponding elements of the preceding period. Thus, hafnium ($Z = 72$) has an atomic radius of 1.44 Å; and zirconium ($Z = 40$), 1.45 Å.

The general trends in atomic radius are summarized in Figure 3.2.

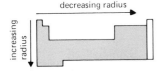

decreasing radius

increasing radius

**Figure 3.2** General trends in atomic radius in relation to the periodic classification.

**3.2 IONIZATION POTENTIALS**

The amount of energy required to remove the most loosely held electron from an isolated gaseous atom in its ground state is the **ionization energy** of the element.

$$A(g) \rightarrow A^+(g) + e^-$$

The symbol (g) indicates a gaseous species.

Conventions regarding the assignment of signs to energy terms must be noted. When a system absorbs energy, the term is given a positive sign. If the system evolves energy, the value is given a negative sign. Since energy is absorbed in the ionization process described, ionization energies have positive signs.

Ionization energies are usually determined by spectroscopic methods (Section 2.11). Figure 2.11 shows that the electronic energy levels converge to a limit. The ionization energy is the energy required to raise an electron to this limit, or it is the energy emitted when an electron falls from this limit back to the ground state. The limit is readily identified from a spectrum since beyond this point, in the direction of shorter wavelengths, the spectrum no longer consists of lines but is continuous. An electron is not restricted to quantized energy states after it has been removed from the atom and hence produces a continuous spectrum.

The **ionization potential** of an element is the minimum potential (measured in volts) required to effect electron removal. The ionization potential is numerically equal to the ionization energy expressed in electron volts (eV), which are energy units. One electron volt is the energy acquired by an electron when it falls through a potential difference of 1 V (Section 10.1); 1 eV equals $3.829 \times 10^{-20}$ cal. For one mole $(6.022 \times 10^{23})$ of electrons, 1 eV is 23.06 kcal/mol. Ionization energies in electron volts are customarily called ionization potentials, even though this terminology is not strictly correct.

Certain atoms produce negative ions by the addition of one or more electrons. The terms ionization energy and ionization potential pertain *only* to the production of positive ions by the *removal* of an electron. Some of the positive ions thus described (such as ions derived from the noble gases or from elements that commonly form only negative ions) are never produced in ordinary chemical reactions.

The factors that influence the magnitude of an ionization potential are those that control the orbital energies (Section 2.17). In Figure 3.3 the ionization potentials of the elements are plotted against atomic number. The ionization potential of helium is larger than that of hydrogen since the nuclear charge of He $(2+)$ is twice that of H $(1+)$ and the electron removed in each case is a $1s$ electron. The ionization potential of lithium, however, is much lower than that of He despite the fact that the

● Use of the calorie (cal) is discouraged; however, the calorie is defined in terms of the joule (J), which is the derived unit of heat, energy, and work.

$$1 \text{ cal} = 4.184 \text{ J}$$

One electron volt, therefore, is equal to $1.602 \times 10^{-19}$ J. For one mole $(6.022 \times 10^{23})$ of electrons, 1 eV is 96.49 kJ/mol.

**Figure 3.3** First ionization potentials of the elements versus atomic number.

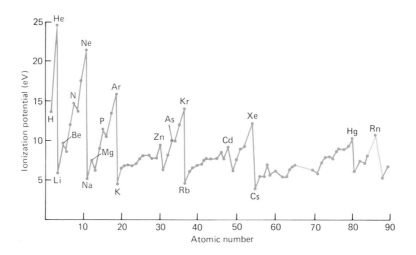

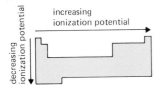

Figure 3.4 General trends in ionization potential in relation to the periodic classification.

nuclear charge of Li (3+) is higher than that of He (2+). In He, one electron is not effectively screened from the nuclear charge by the other since they both occupy the same subshell; however, the Li atom is larger than the He atom and has two $1s$ electrons that strongly shield the nuclear charge from the electron that is removed (a $2s$ electron).

In the second period (Li to Ne), the nuclear charge increases from 3+ to 10+ and the atomic radius decreases. Since electrons are being added to the same shell (the outer shell), the extent to which they augment the screening effect of the $1s$ electrons is slight, and the increasing nuclear charge causes the ionization potential to increase. In general, ionization potential increases across a period paralleling an increase in nuclear charge and a decrease in atomic radius (Figure 3.4).

The ionization potentials of the transition elements do not increase across a period as rapidly as those of the representative elements; the ionization potentials of the inner-transition elements remain almost constant. Therefore, these elements have low enough ionization potentials to react as metals—through electron loss. In these series, electrons are being added to inner shells, and the resulting increase in screening accounts for the effects noted.

For the representative elements, the ionization potential decreases as we go from the lighter to the heavier members of a group—notice the values for the noble gases (group 0) or the alkali metals (group I A) in Figure 3.3. The effect of increasing nuclear charge is greatly reduced by the screening effect of the increasing number of inner electrons, and ionization potential decreases as atomic radius increases (Figure 3.4).

Several features of the graph of Figure 3.3 are significant because they give insights into how certain electronic configurations affect the reactivity of atoms (Section 2.18). The high ionization potentials of the noble gases are an indication that the electronic configurations of the noble gases are very stable; it is difficult to remove an electron from these "closed" configurations.

The elements Be, Mg, Zn, Cd, and Hg (which have filled subshells) have ionization potentials that are high in comparison to the ionization potentials of the elements that follow them in the same period. In the ionization of each of the elements listed, an electron is removed from a filled $s$ subshell, and in each case the following element loses a $p$ electron. Since a $p$ electron is less able to penetrate screening electrons than an $s$ electron of the same shell, the $p$ electron is held less tightly, and less energy is required to remove it (Section 2.17).

In the curve of Figure 3.3 there are slight drops in ionization potential after N, P, and As (each of which has a half-filled $p$ subshell). In N, P, and As atoms, the three electrons of the half-filled $p$ subshell occupy orbitals singly, whereas in atoms of each of the following elements (O, S, and Se) one of the $p$ orbitals holds a pair of electrons. Since electrons repel one another, the removal of one of the $p$ electrons from the pair (in O, S, or Se) is facilitated.

The discussion so far has been concerned only with *first* ionization potentials. The *second* ionization potential of an element relates to the energy required to remove *one* electron from a 1+ ion of that element.

$$A^+(g) \rightarrow A^{2+}(g) + e^-$$

The *third* ionization potential pertains to a process in which *one* electron is removed from a 2+ ion. Even higher ionization potentials may be determined. When the term ionization potential is not further qualified, it is understood that the first ionization potential is meant.

Predictably, for any given element, the third ionization potential is higher than the second, and the second is higher than the first. It is more difficult to remove an electron (1− charge) from a 2+ ion than from a 1+ ion; likewise, it is more difficult to remove an electron from a 1+ ion than from a neutral atom. The high values observed for third and fourth ionization potentials indicate that few highly positive ions exist under ordinary conditions.

In terms of chemical reactivity, remember that ionization potential relates to the energy *required* to remove an electron. Thus, the elements most active in losing electrons (metallic behavior) are those with the *lowest* ionization potentials and are found in the lower left of the periodic table. This reactivity in terms of electron loss decreases as one moves upward or to the right from this corner of the chart.

**3.3 ELECTRON AFFINITIES**

The energy effect accompanying the process in which an electron is added to an isolated atom in its ground state is called an **electron affinity.**

$$e^- + A(g) \rightarrow A^-(g)$$

In this process, energy is usually evolved, and therefore most electron affinities have negative signs. Electron affinity, however, is frequently defined as energy released; in sources that employ such a definition, electron affinity values that relate to the liberation of energy are given positive signs. This sign convention is the reverse of the one that we employ; we shall consistently use a negative sign to indicate that the total energy of the system decreases and a positive sign to indicate that the system absorbs energy.

The direct determination of electron affinity is difficult and has been accomplished for only a few elements. Other values have been determined indirectly by calculation from thermodynamic data (Section 5.9). Electron affinity values, therefore, are available for only a limited number of elements, and many of these values are not highly accurate. Some electron affinities are listed in Table 3.1. Notice that some of these values

Table 3.1 Electron affinities (eV)

| H<br>−0.77 | | | | | | | He<br>+0.6$^a$ |
|---|---|---|---|---|---|---|---|
| Li<br>−0.6$^a$ | Be<br>+0.6$^a$ | B<br>−0.2$^a$ | C<br>−1.25 | N<br>+0.3$^a$ | O<br>−1.47<br>(+7.28)$^b$ | F<br>−3.45 | Ne<br>+1.0$^a$ |
| Na<br>−0.2$^a$ | Mg<br>+0.3$^a$ | Al<br>−0.3$^a$ | Si<br>−1.4$^a$ | P<br>−0.6$^a$ | S<br>−2.07<br>(+3.44)$^b$ | Cl<br>−3.61 | |
| | | | | | | Br<br>−3.36 | |
| | | | | | | I<br>−3.06 | |

$^a$ Calculated values.
$^b$ Values in parentheses are calculated and pertain to the total energy effects for the addition of two electrons.

have positive signs indicating that work must be done to force the atom under consideration to accept an additional electron.

A small atom should have a greater tendency to add an electron than a large atom since the added electron is, on the average, closer to the positively charged nucleus in a small atom. This trend is roughly followed from left to right in any period. In the second period, however, exceptions may be noted for beryllium (filled $2s$ subshell), nitrogen (half-filled $2p$ subshell), and neon (all subshells filled). Similar exceptions may be noted for analogous elements of the third period. Elements with relatively stable electronic configurations do not accept additional electrons readily. Since the electronic configuration of each of the group VII A elements is one electron short of a noble gas configuration, each element of group VII A has the greatest tendency toward electron gain of any element in its period.

The values for the halogens illustrate the typical group trend. Electron attracting ability, except for that of fluorine, tends to increase from bottom to top within the group in conformity with decreasing atomic size. The effect of small size, however, may be partially negated by the repulsion of electrons already present in the atom. The exception noted for fluorine, as well as for other second period elements, may be, in part, explained by these repulsions. The concentration of negative charge in a very small shell is greater than the concentration obtained when the same number of electrons is placed in a larger shell.

Some second electron affinities have been determined. These values refer to processes in which an electron is added to a negative ion; for example,

$$e^- + O^-(g) \rightarrow O^{2-}(g)$$

Since there is an electrostatic repulsion between a negative ion and an electron, energy is required, not released. All second electron affinities have positive signs.

**3.4 THE IONIC BOND**    When a metal reacts with a nonmetal, electrons are transferred from the atoms of the metal to the atoms of the nonmetal, and an **ionic (or electrovalent) compound** is produced. For example, a sodium atom loses an electron to a chlorine atom in the reaction of these two elements, and charged particles (called ions) result. Sodium has a comparatively low ionization potential; relatively little energy is required to remove an electron from a sodium atom. The sodium ion that forms has a 1+ charge since the sodium nucleus contains 11 protons (11+ charge) and the ion has only 10 electrons (one having been lost). On the other hand, the value for the electron affinity of chlorine indicates that the chlorine atom has a strong tendency toward electron gain. The chloride ion that forms has a 1− charge since the chlorine nucleus contains 17 protons (17+ charge) and the ion has 18 electrons (one having been gained). Positive ions are called **cations,** and negative ions are called **anions.** These names are derived from the terminology of electrochemistry (Chapter 10).

The reactions and compounds of the A family elements are frequently depicted by using the symbols of the elements under consideration together with dots to indicate valence electrons—the only electrons involved in the chemical reactions of these elements.

$$\text{Na} \cdot + \cdot \overset{..}{\underset{..}{\text{Cl}}} \colon \rightarrow \text{Na}^+ + \colon \overset{..}{\underset{..}{\text{Cl}}} \colon^-$$

The complete electronic configurations of the atoms and ions of this reaction are

$$\text{Na } (1s^2\, 2s^2\, 2p^6\, 3s^1) \rightarrow \text{Na}^+ (1s^2\, 2s^2\, 2p^6) + e^-$$

$$e^- + \text{Cl } (1s^2\, 2s^2\, 2p^6\, 3s^2\, 3p^5) \rightarrow \text{Cl}^- (1s^2\, 2s^2\, 2p^6\, 3s^2\, 3p^6)$$

The sodium ion has an electronic configuration identical to that of neon, and the chloride ion has the same configuration as argon. The ions may be said to be **isoelectronic** (the same in electronic configuration) with neon and argon, respectively.

In ionic reactions most A family elements lose or gain electrons in such a way as to produce ions that are isoelectronic with a noble gas (of the A group elements, only the post-transition metals, such as Ga and Sn, do not). Most of these noble-gas ions have an octet of electrons in the outer shell (an $s^2p^6$ **configuration**); a few (e.g., Li$^+$ and H$^-$) have the $1s^2$ configuration of helium.

In the reaction of sodium and chlorine, the total number of electrons

*Na⁺ Cl⁻* *(handwritten)*

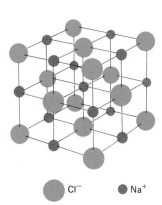

○ Cl⁻    ● Na⁺

**Figure 3.5** Sodium chloride crystal lattice.

lost by sodium must equal the total number of electrons gained by chlorine. Thus, the number of sodium ions produced is the same as the number of chloride ions produced, and the formula, NaCl, gives the simplest ratio of ions present in the compound (1 to 1). These ions attract one another to form a **crystal** (Figure 3.5).

In a sodium chloride crystal, no ion may be considered as belonging exclusively to another. Rather, each sodium ion is surrounded by six chloride ions and each chloride ion is surrounded by six sodium ions. The arrangement of ions in the crystal is such that the repulsion of like-charged ions is more than compensated for by the attraction of oppositely charged ions. This mutual attraction of positive and negative ions holds the crystal together; indeed, the ionic bond may be considered to be just this.

When oxygen undergoes an ionic reaction, each oxygen atom (electron configuration, $1s^2\, 2s^2\, 2p^4$) accepts two electrons and attains the neon configuration ($1s^2\, 2s^2\, 2p^6$); the oxide ion has a charge of 2−. In the reaction of sodium with oxygen, two atoms of sodium are required for every atom of oxygen since the number of electrons lost must equal the number of electrons gained.

$$2\text{Na}\cdot + \ddot{\underset{\cdot\cdot}{\text{O}}}\text{:} \;\rightarrow\; 2\text{Na}^+ + \text{:}\ddot{\underset{\cdot\cdot}{\text{O}}}\text{:}^{2-}$$

The simplest ratio of ions present in the product, sodium oxide, is indicated by the formula of the compound, $Na_2O$. The arrangement of ions in a crystal of sodium oxide, which is different from that of the sodium chloride crystal, is such that the ions are accommodated in the ratio of two sodium ions to one oxide ion.

The **electrovalence number** of an element is the charge (including sign) of the ion derived from the element after an electrovalent reaction. If we follow the octet principle, the electrovalence number of a group I A element should be 1+; group VII A, 1−; group II A, 2+; group VI A, 2−. These charges are derived from the number of valence electrons that is characteristic of any element of a given group and the number of electrons that must be lost (positive electrovalence) or gained (negative electrovalence) in order for the element to attain a noble-gas configuration.

From these ionic charges, formulas may be derived on the basis of the fact that the *total* positive ionic charge of any compound must equal the *total* negative ionic charge of that compound. Calcium is a group II A metal, has two valence electrons, and has an electrovalence number of 2+. We have already determined that the electrovalence number of chlorine is 1− and that the electrovalence of oxygen is 2−. Thus, the formula of calcium chloride is $CaCl_2$ and that of calcium oxide is CaO. In a similar manner one can derive the formulas of aluminum oxide ($Al_2O_3$), sodium sulfide ($Na_2S$), and potassium nitride ($K_3N$) from the predicted electrovalence numbers of aluminum (3+), oxygen (2−), sodium (1+), sulfur (2−), potassium (1+), and nitrogen (3−). We shall see later that not every compound predicted by this method exists and that not every one which does exist is truly ionic in character.

The energy effect associated with the condensation of positive and negative ions into a crystal is called the **crystal energy** or **lattice energy** (Section 7.13). Since energy is *evolved* in these processes, lattice energies have negative signs. The lattice energy (with a positive sign) of a given crystal may also be viewed as the energy required to separate the ions of the crystal.

The driving force of the ionic reaction is the electrostatic attraction of the ions for one another (this attraction results in the liberation of the lattice energy). The ionization of one mole of sodium ($6.02 \times 10^{23}$ atoms) requires +118 kcal (the ionization energy).

$$Na(g) \rightarrow Na^+(g) + e^-$$

Energy is released when gaseous chlorine atoms gain electrons (−83 kcal for one mole of chlorine atoms),

$$e^- + Cl(g) \rightarrow Cl^-(g)$$

but the amount released is insufficient to supply the energy required for the ionization of sodium. The deficit is more than compensated for by the lattice energy (−188 kcal/mol),

$$Na^+(g) + Cl^-(g) \rightarrow NaCl(crystal)$$

so that the process as a whole is energetically favorable.

This analysis is not complete since it pertains to the reaction between gaseous atoms of sodium and chlorine, and under ordinary conditions, these elements do not occur in this form (see Section 5.9 for a more complete analysis). Nevertheless, we can reach some valid conclusions by this approach.

Since electrostatic attraction is greater between multicharged ions than between singly charged ions, the lattice energies of compounds that contain multicharged ions are larger than those that contain only 1+ and 1− ions. The lattice energy of MgO (−948 kcal/mol), for example, is large enough to supply the energy required for the production of the $Mg^{2+}$ ion (+523 kcal/mol, the sum of the first *two* ionization energies of Mg) as well as the energy required for the formation of the $O^{2-}$ ion (+168 kcal/mol, the energy *absorbed* when *two* electrons are added to each O atom).

If each Mg atom lost only one electron, less energy would be required to produce *two* moles of $Mg^+$ ions for $Mg_2O$ (+352 kcal) than is required to produce *one* mole of $Mg^{2+}$ ions for MgO (+523 kcal). However, the energy released by the formation of the hypothetical $Mg_2O$ crystal (probably about −670 kcal/mol) would be so much lower than the energy

● Since 1 cal = 4.184 J, 1 kcal = 4.184 kJ. Therefore, the lattice energy of NaCl is −787 kJ/mol.

released by the formation of the MgO crystal (−948 kcal/mol) that the formation of MgO is favored.

Why, then, does sodium not form a $Na^{2+}$ ion in the oxide of this element? Table 3.2 lists the first to fourth ionization energies of some A family metals. For each element, the ionization energies increase from the first to the fourth, as expected (Section 3.2). However, in each case, a jump in the required energy occurs after all the valence electrons have been removed and it is necessary to break into the $s^2p^6$ noble-gas arrangement of the shell underlying the valence shell to remove the next electron (this place is marked with a vertical line in the table). The removal of two electrons from each sodium atom requires +1210 kcal/mol. Since the formation of the $O^{2-}$ ion requires +168 kcal/mol, over 1378 kcal would have to be supplied by the lattice energy in order to form the crystal. The lattice energy cannot supply this amount of energy. The lattice energy of the hypothetical $Na^{2+}O^{2-}$ would probably be of the same order of magnitude as that of MgO (−948 kcal/mol).

Table 3.2 Ionization energies of the third period metals

| METAL | GROUP | IONIZATION ENERGIES (kcal/mol) | | | |
|---|---|---|---|---|---|
| | | FIRST | SECOND | THIRD | FOURTH |
| Na | I A | +118 | +1091 | +1652 | +2280 |
| Mg | II A | +176 | +347 | +1848 | +2521 |
| Al | III A | +138 | +434 | +656 | +2767 |

The foregoing analysis shows why so many metals form cations with $s^2p^6$ (noble-gas) electronic configurations (see Table 3.3). All anions derived from atoms of the nonmetals also have $s^2p^6$ configurations; the addition of electrons is possible until the $s^2p^6$ limit is reached, and energy considerations favor the attainment of this limit. Energy is evolved by the formation of the anions of the VII A elements, and under appropriate conditions the lattice energy can supply the energy required for the formation of $O^{2-}$, $S^{2-}$, $Se^{2-}$, and $Te^{2-}$ of group VI A as well as $N^{3-}$ and $P^{3-}$ of group V A.

However, if an electron were to be added beyond the $s^2p^6$ configuration, the additional electron would be very loosely bound since it would be added to a higher quantum level, screened from the nuclear charge by a closed $s^2p^6$ shell, and repelled by the electrons of the negatively charged ion to which it was added. Hence, the addition would require a large amount of energy, more than any lattice energy could supply, and the formation of such anions is not observed.

Certain metals engage in ionic reactions but cannot possibly produce cations that are isoelectronic with noble gases. The ions $Sc^{3+}$, $Y^{3+}$, and $La^{3+}$ have $s^2p^6$ configurations, but the ions of the other transition metals, as well as the post-transition metals, do not. For these latter metals, a comparatively large number of electrons would have to be lost to achieve a noble-gas structure. Zinc $(\ldots 3s^2\,3p^6\,3d^{10}\,4s^2)$, for example, would have to lose 12 electrons. Electron losses of more than three are never observed; the energy required to remove four or more electrons is not obtainable in any ordinary ionic reaction. In its reactions zinc loses two electrons to form the $Zn^{2+}$ ion $(\ldots 3s^2\,3p^6\,3d^{10})$.

In the reactions of the transition elements, inner $d$ electrons can be used as well as outer $s$ electrons. Examples of various types of ions are given in Table 3.3; the configurations listed are those of the outer subshell(s). Notice that a number of ions (including $Zn^{2+}$) have the configuration $ns^2\,np^6\,nd^{10}$ in their outer shell (listed as $d^{10}$), and a number of others have a $(n-1)s^2\,(n-1)p^6\,(n-1)d^{10}\,ns^2$ configuration (listed as $d^{10}s^2$). Some metals have a tendency to achieve these two closed configurations in which all sublevels are filled (although it is not so pronounced as the tendency to attain a $s^2p^6$ configuration), but it is impossible for most transition elements to produce ions with any of the "regular" configurations $(s^2p^6, d^{10}, \text{ or } d^{10}s^2)$.

Table 3.3 Types of monatomic ions

| CONFIGURATION | EXAMPLES |
|---|---|
| $s^2$ | $Li^+$, $Be^{2+}$, $H^-$ |
| $s^2p^6$ | $Na^+$, $K^+$, $Rb^+$, $Cs^+$, $Mg^{2+}$, $Ca^{2+}$, $Sr^{2+}$, $Ba^{2+}$, $Al^{3+}$, $Sc^{3+}$, $Y^{3+}$, $La^{3+}$, $F^-$, $Cl^-$, $Br^-$, $I^-$, $O^{2-}$, $S^{2-}$, $Se^{2-}$, $N^{3-}$, $P^{3-}$ |
| $d^1$ | $Ti^{3+}$ |
| $d^2$ | $V^{3+}$ |
| $d^3$ | $Cr^{3+}$, $Mo^{3+}$ |
| $d^4$ | $Cr^{2+}$ |
| $d^5$ | $Mn^{2+}$, $Fe^{3+}$, $Ru^{3+}$ |
| $d^6$ | $Fe^{2+}$, $Co^{3+}$, $Rh^{3+}$ |
| $d^7$ | $Co^{2+}$ |
| $d^8$ | $Ni^{2+}$, $Pd^{2+}$, $Pt^{2+}$, $Au^{3+}$ |
| $d^9$ | $Cu^{2+}$ |
| $d^{10}$ | $Cu^+$, $Ag^+$, $Au^+$, $Zn^{2+}$, $Cd^{2+}$, $Hg^{2+}$, $Ga^{3+}$, $In^{3+}$, $Tl^{3+}$ |
| $d^{10}s^2$ | $Ga^+$, $In^+$, $Tl^+$, $Ge^{2+}$, $Sn^{2+}$, $Pb^{2+}$, $As^{3+}$, $Sb^{3+}$, $Bi^{3+}$ |

Many transition elements form more than one type of cation; iron, for example, forms either $Fe^{2+}$ or $Fe^{3+}$ depending upon reaction conditions. More energy is required to produce the $Fe^{3+}$ ion than the $Fe^{2+}$ ion, but the lattice energies of $Fe^{3+}$ compounds are larger than those of $Fe^{2+}$ compounds. These factors balance to the point that the preparation of compounds of both ions is feasible.

**3.5 IONIC RADIUS** The radii of ions have been determined by X-ray diffraction of ionic crystals. Such studies give only the distances between the centers of adjacent ions; apportioning these distances is a problem. To apportion

Figure 3.6 Determination
of ionic radii. (See text for
further discussion.)

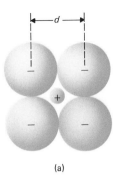

(a)

(b)

bond distances of covalent molecules to secure covalent radii (Section
3.1), one starts with a bond between two like atoms, which can be
divided equally. In ionic crystals, however, the sizes of the positive and
negative ions are of necessity different.

One solution to this problem is to study a crystal of a compound with
a very small cation and a very large anion, such as lithium iodide (Figure
3.6a). The assumption is then made that the iodide ions touch each other,
and the iodide–iodide distance ($d$ in the illustration) is divided in half to
obtain the radius of the iodide ion. This method of apportionment of
anion–anion distance ($d$), however, is impossible for a crystal of the more
common type illustrated in Figure 3.6b since there is no anion–anion
contact. Once the radius of the iodide ion has been fixed as a standard,
other ionic radii can be calculated by subtracting the radius of the standard ion from the cation–anion distance ($d'$ in Figure 3.6b) of another
crystal. For example, the radius of the potassium ion could be calculated
by subtracting the radius of the iodide ion from the $K^+$ to $I^-$ distance in
the potassium iodide crystal. Table 3.4 lists some ionic radii.

A positive ion is always smaller than the neutral atom from which it is

Table 3.4 Ionic radii (in ångström units, Å)

| I A | II A | III A | VI A | VII A |
|-----|------|-------|------|-------|
| $Li^+$ 0.60 | $Be^{2+}$ 0.31 | | $O^{2-}$ 1.40 | $F^-$ 1.36 |
| $Na^+$ 0.95 | $Mg^{2+}$ 0.65 | $Al^{3+}$ 0.50 | $S^{2-}$ 1.84 | $Cl^-$ 1.81 |
| $K^+$ 1.33 | $Ca^{2+}$ 0.99 | $Ga^{3+}$ 0.62 | $Se^{2-}$ 1.98 | $Br^-$ 1.95 |
| $Rb^+$ 1.48 | $Sr^{2+}$ 1.13 | $In^{3+}$ 0.81 | $Te^{2-}$ 2.21 | $I^-$ 2.16 |
| $Cs^+$ 1.69 | $Ba^{2+}$ 1.35 | $Tl^{3+}$ 0.95 | | |

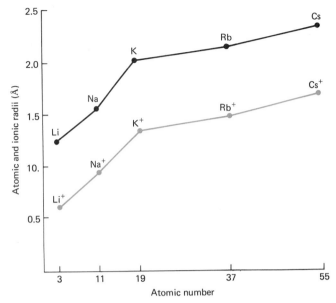

**Figure 3.7** Atomic and ionic radii of the group I A elements.

derived (Figure 3.7). Thus, the radius of Na is 1.57 Å, and the radius of Na⁺ is 0.95 Å. In the formation of the sodium ion, the loss of an electron represents the loss of the entire $n = 3$ shell of the sodium atom. Furthermore, the loss of an electron creates an imbalance in the proton–electron ratio; since the protons outnumber the electrons in the positive ion, the electrons of the ion are drawn in closer to the nucleus.

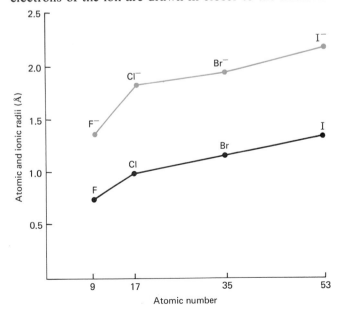

**Figure 3.8** Atomic and ionic radii of the group VII A elements.

For similar reasons a 2+ ion is larger than a 3+ ion. Thus,

$$\text{Fe} = 1.17 \text{ Å} \qquad \text{Fe}^{2+} = 0.75 \text{ Å} \qquad \text{Fe}^{3+} = 0.60 \text{ Å}$$

A negative ion is always larger than its parent atom. Thus, the radius of the fluorine atom is 0.72 Å, and that of the fluoride ion is 1.36 Å (Figure 3.8). In the formation of the fluoride ion, the addition of an electron to the $2p$ subshell causes an enhanced repulsion between electrons, and the $n = 2$ shell expands. The fluoride ion has nine protons and ten electrons.

Within a period, isoelectronic positive ions show a decrease in ionic radius from left to right because of the increasing nuclear charge. The same trend is observed for the isoelectronic negative ions of a period; ionic radius decreases from left to right.

Within a group of the periodic table, similarly charged ions increase in size from top to bottom. The ions of the heavier members of the group employ more electron shells than those of the lighter members.

**3.6 THE COVALENT BOND**

When atoms of nonmetals interact, molecules are formed which are held together by covalent bonds. Since these atoms are similar in their attraction for electrons (identical when two atoms of the same element are considered), electron transfer does not occur; instead, electrons are shared. A **covalent bond** consists of a pair of electrons (with opposite spins) that is shared by two atoms.

As an example, consider the bond formed by two hydrogen atoms. An individual hydrogen atom has a single electron that is symmetrically distributed around the nucleus in a $1s$ orbital. When two hydrogen atoms form a covalent bond, the atomic orbitals overlap in such a way that the electron clouds reinforce each other in the region between the nuclei, and there is an increased probability of finding an electron in this region. According to the Pauli exclusion principle, the two electrons of the bond must have opposite spins. The strength of the covalent bond comes from the attraction of the positively charged nuclei for the negative cloud of the bond (Figure 3.9).

Figure 3.9 Representation of the electron distribution in a hydrogen molecule.

The hydrogen molecule can be represented by the symbol H:H or H—H. Although the electrons belong to the molecule as a whole, each hydrogen atom can be considered to have the noble-gas configuration of helium (two electrons in the $n = 1$ level). This consideration is based on the premise that both shared electrons contribute to the stable configuration of each hydrogen atom.

The formula, $H_2$, describes a discrete unit – a molecule – and hydrogen gas consists of a collection of such molecules. There are no molecules in strictly ionic materials. The formula $Na_2Cl_2$ is incorrect because sodium chloride is an ionic compound and the simplest ratio of ions in a crystal of sodium chloride is 1 to 1; a molecule of formula $Na_2Cl_2$ does not exist.

For covalent materials, however, a formula such as $H_2O_2$ can be correct; this formula describes a molecule containing two hydrogen atoms and two oxygen atoms.

The hydrogen molecule can be described as being **diatomic** (containing two atoms). Certain other elements also exist as diatomic molecules. An atom of any group VII A element, for example, has seven valence electrons. By the formation of a covalent bond between two of these atoms, each atom attains an octet configuration characteristic of the noble gases. Thus, fluorine gas consists of $F_2$ molecules.

$$:\overset{..}{\underset{..}{F}}\cdot \; + \; \cdot\overset{..}{\underset{..}{F}}: \; \rightarrow \; :\overset{..}{\underset{..}{F}}:\overset{..}{\underset{..}{F}}:$$

Only the electrons between the two atoms are shared and form a part of the covalent bond (although the molecular orbital theory considers that all of the electrons affect the bonding — Section 4.4).

More than one covalent bond may form between two atoms. A nitrogen atom (group V A) has five valence electrons.

$$:\overset{.}{\underset{.}{N}}\cdot \; + \; \cdot\overset{.}{N}: \; \rightarrow \; :N:::N:$$

In the molecule, $N_2$, six electrons are shared in three covalent bonds (usually called a triple bond). Notice that, as a result of this formulation, each of the nitrogen atoms can be considered to have an octet of electrons.

Nonmetallic elements that exist as diatomic molecules are $H_2$, $F_2$, $Cl_2$, $Br_2$, $I_2$, $N_2$, and $O_2$. (Oxygen is a special case and will be discussed in Section 4.4.) These elements are always indicated in this way in chemical equations.

The electron-dot formulas we have been using are called **valence-bond structures** or **Lewis structures,** named after Gilbert N. Lewis who proposed this theory of covalent bonding in 1916 (more recent theories of covalent bonding are discussed in Chapter 4). The Lewis theory emphasizes the attainment of noble-gas configurations on the part of atoms in covalent molecules. Since the number of valence electrons is the same as the group number for the nonmetals, one might predict that VII A elements, such as Cl, would form one covalent bond to attain a stable octet; VI A elements, such as O and S, two covalent bonds; V A elements, such as N and P, three covalent bonds; and IV A elements, such as C, four covalent bonds. These predictions are borne out in many compounds containing only simple covalent bonds.

$$H\cdot \; + \; \cdot\overset{..}{\underset{..}{Cl}}: \; \rightarrow \; H:\overset{..}{\underset{..}{Cl}}:$$
hydrogen<br>chloride

$$2H\cdot \; + \; \cdot\overset{.}{\underset{..}{O}}: \; \rightarrow \; H:\overset{H}{\overset{..}{\underset{..}{O}}}:$$
water

$$3H\cdot + \cdot\overset{\cdot\cdot}{\underset{\cdot\cdot}{N}}\cdot \rightarrow H:\overset{\overset{H}{\cdot\cdot}}{\underset{\cdot\cdot}{N}}:H$$

ammonia

$$4H\cdot + \cdot\overset{\cdot}{\underset{\cdot}{C}}\cdot \rightarrow H:\overset{\overset{H}{\cdot\cdot}}{\underset{\cdot\cdot}{C}}:H$$

methane

Notice that in these molecules, each hydrogen atom can be considered to have a complete $n = 1$ shell; the other atoms have characteristic noble-gas octets.

The covalent bonding of compounds can also be indicated by dashes; each dash represents one bond, a pair of electrons.

$$:\overset{\cdot\cdot}{\underset{\cdot\cdot}{Cl}}\text{—}\overset{\cdot}{\underset{\overset{|}{:\overset{\cdot\cdot}{Cl}:}}{P}}\text{—}\overset{\cdot\cdot}{\underset{\cdot\cdot}{Cl}}: \qquad :\overset{\cdot\cdot}{\underset{\cdot\cdot}{Cl}}\text{—}\overset{\cdot\cdot}{\underset{\cdot\cdot}{O}}\text{—}\overset{\cdot\cdot}{\underset{\cdot\cdot}{Cl}}: \qquad \overset{\overset{H}{|}}{\underset{\underset{H}{|}}{H\text{—}C}}\text{—}\overset{\overset{H}{|}}{\underset{\underset{H}{|}}{C\text{—}H}}$$

phosphorous trichloride     dichlorine oxide         ethane

The following are examples of molecules that contain double and triple bonds.

$$:\overset{\cdot\cdot}{\underset{\cdot\cdot}{O}}: + :\overset{\cdot}{\underset{\cdot}{C}}: + :\overset{\cdot\cdot}{\underset{\cdot\cdot}{O}}: \rightarrow :\overset{\cdot\cdot}{\underset{\cdot\cdot}{O}}::C::\overset{\cdot\cdot}{\underset{\cdot\cdot}{O}}: \quad (\text{or} \quad :\overset{\cdot\cdot}{\underset{\cdot\cdot}{O}}{=}C{=}\overset{\cdot\cdot}{\underset{\cdot\cdot}{O}}:)$$

carbon dioxide

$$2H\cdot + \cdot\overset{\cdot}{\underset{\cdot}{C}}: + :\overset{\cdot}{\underset{\cdot}{C}}\cdot + 2H\cdot \rightarrow H:\overset{\overset{H}{\cdot\cdot}}{C}::\overset{\overset{H}{\cdot\cdot}}{C}:H \quad (\text{or} \quad \overset{\overset{H}{|}}{H\text{—}C}{=}\overset{\overset{H}{|}}{C\text{—}H})$$

ethylene

$$H\cdot + \cdot\overset{\cdot}{\underset{\cdot}{C}}: + :\overset{\cdot}{\underset{\cdot}{C}}\cdot + \cdot H \rightarrow H:C:::C:H \quad (\text{or} \quad H\text{—}C{\equiv}C\text{—}H)$$

acetylene

Notice that in each compound the number of covalent bonds on each atom agrees with the number predicted.

**3.7 FORMAL CHARGE** In the formation of certain covalent bonds, *both* of the shared electrons are furnished by *one* of the bonded atoms. For example, in the reaction of ammonia with a proton (a hydrogen atom stripped of its electron), the unshared electron pair of the nitrogen atom of $NH_3$ is used to form a new covalent bond.

$$H:\overset{\cdot\cdot}{\underset{\underset{H}{}}{N}}:H + H^+ \rightarrow \left[H:\overset{\overset{H}{\cdot\cdot}}{\underset{\underset{H}{}}{N}}:H\right]^+$$

A bond formed in this way is frequently called a "coordinate covalent" bond, but it is probably unwise to do so. The coordination *process* of covalent-bond formation is a useful distinction — particularly in the study of coordination complexes (Chapter 19) and in the interpretation of certain acid–base reactions (Section 15.7). However, labeling a specific *bond* as a "coordinate covalent" bond implies that it is different from other covalent bonds and has little justification.

All electrons are alike no matter what their source. All the bonds in $NH_4^+$ are identical; it is impossible to distinguish between them. Furthermore, the mode of formation of a covalent bond does not affect its nature. We can *imagine* any covalent bond as having been formed by electron pairing

$$2H\cdot + \cdot \overset{\cdot}{\underset{\cdot\cdot}{O}}: \rightarrow H_2O$$

or electron-pair donation

$$2H^+ + \left[:\overset{\cdot\cdot}{\underset{\cdot\cdot}{O}}:\right]^{2-} \rightarrow H_2O$$

Notice, however, that the number of covalent bonds on the N atom of $NH_4^+$ does not agree with the number predicted in the preceding section. Since a nitrogen atom has five valence electrons (group V A), it would be expected to satisfy the octet principle through the formation of three covalent bonds. This prediction is correct for $NH_3$; it is not correct for $NH_4^+$.

An answer to this question may be obtained by calculating the formal charges of the atoms in $NH_4^+$. A **formal charge** is calculated by apportioning the bonding electrons equally between the bonded atoms, one electron to each atom for each covalent bond, and then comparing the number of electrons that a given atom has in the structure with the number of valence electrons that this atom would have when electrically neutral.

In $NH_4^+$ the N atom may be considered to have four electrons — one from the division of each of the covalent bonds. Since a neutral N atom has five valence electrons, the N atom in $NH_4^+$ is assigned a formal charge of 1+. Each H atom of the $NH_4^+$ ion has the same number of electrons in the structure that it has as a neutral atom and hence carries no formal charge. We can indicate the formal charge on the N atom of $NH_4^+$ as

$$
\begin{array}{c}
\text{H} \\
| \\
\text{H}-\overset{\oplus}{\text{N}}-\text{H} \\
| \\
\text{H}
\end{array}
$$

Thus, the discrepancy between the predicted and actual number of bonds on the N atom in $NH_4^+$ may be explained on the basis of formal charge since the hypothetical $N^+$ would have four unpaired electrons and be able to form four covalent bonds.

In the assignment of formal charges, the assumption is made that the electron pair of any covalent bond is shared equally by the bonded atoms. Such an assumption is usually not true (Section 3.8), and formal charges must be interpreted carefully. The electron density on the N atom in $NH_4^+$ is less than that on the N atom in $NH_3$, but the actual charge is not a full positive charge because the bonding electrons are not equally shared. In addition, molecular polarity occurs in some molecules in which each atom has a formal charge of zero (Section 4.6).

Lewis structures may be assigned without reference to the source of the bonding electrons. For example, we could proceed in the following way to diagram the structure of the sulfur dioxide molecule, $SO_2$. This theory of covalent bonding offers no way to predict the arrangement of atoms in a molecule; the arrangement must be derived from experimental evidence. In the $SO_2$ molecule the O atoms are joined to a central S atom, and the molecule is angular.

$$\begin{array}{ccc} & S & \\ O & & O \end{array}$$

We next determine the number of valence electrons to be used in writing the structure. For a molecule this number is simply the sum of the numbers of the valence electrons of the atoms present. If the structure of an ion is being considered, this sum is increased (for an anion) or decreased (for a cation) to take into account the charge of the ion. For $SO_2$ there is a total of eighteen valence electrons to be distributed (twelve from the two O atoms plus six from the S atom). If we place single bonds between the atoms, four of the eighteen electrons are used. The fourteen electrons that are left must be distributed so that they give an octet to each of the atoms.

In an attempt to accomplish this division, we derive the structure

$$:\!\overset{..}{\underset{..}{O}}\!\!\overset{\overset{..}{S}}{\diagup}\!\!\diagdown\overset{..}{\underset{..}{O}}\!:$$

Both the O atoms in this structure have the proper number of electrons associated with them, but the S atom has only six. We can remedy this situation by switching an electron pair from one of the O atoms in such a way as to make a double bond between it and the S atom.

$$:\!\overset{..}{\underset{..}{O}}\!\!\overset{\overset{..}{S}}{\diagup}\!\!\diagdown\!\!=\!\overset{..}{\underset{..}{O}}\!:$$

For the calculation of formal charge, the S atom may be considered to have five electrons—two that are not engaged in bonding and three from the division of the three covalent bonds on that atom. Since a neutral S atom has six valence electrons (group VI A), the S atom is assigned a formal charge of 1+. The right-hand O atom has the same number of electrons in the structure that it has as a neutral atom, and hence it carries no formal charge. The O atom on the left has seven valence electrons in the structure (six unshared electrons plus one from the covalent bond). Since the neutral O atom has six valence electrons, this O atom is assigned a formal charge of 1−. Notice that the sum of the formal charges of all the atoms in the $SO_2$ molecule equals zero. The sum of the formal charges of the atoms of a molecule always equals zero; the sum of the formal charges of the atoms of an ion, however, equals the charge on the ion.

A structure for $SO_2$ may be diagramed as

$$\ominus \ddot{\underset{\cdot\cdot}{O}} \text{---} \overset{\cdot\cdot}{\underset{}{S}}{}^{\oplus} = \ddot{\underset{\cdot\cdot}{O}} \text{:}$$

The bonding in $SO_2$, however, is not adequately represented by a single structure of this type, and, as we shall see in Section 4.1, a further refinement is necessary.

**3.8** TRANSITION BETWEEN IONIC AND COVALENT BONDING

The bonding in most compounds is intermediate in character between purely ionic and purely covalent. The best examples of ionic bonding are found in compounds formed by a metal with a low ionization potential (for example, Cs) and a nonmetal with a strong tendency toward electron gain (for example, F). In a compound such as CsF, electron transfer is definite and complete; the bonding of the $Cs^+$ and $F^-$ ions in a crystal is solely the result of electrostatic attraction between the ions.

A purely covalent bond is found only in molecules formed from two *identical* atoms, such as $Cl_2$. The electron-attracting ability of one chlorine atom is exactly the same as the electron-attracting ability of another; the bonding results from an electron cloud that is distributed symmetrically with respect to the two atoms. In other words, the bonding electrons are shared equally by the two identical atoms.

The bonding in most compounds lies somewhere between these two extremes. The bonding of intermediate character that occurs in compounds that contain a metal and a nonmetal can be interpreted in terms of the interactions between the ions. The positively-charged ion of such a compound is believed to attract and deform the electron cloud of the anion; the electron cloud of the negative ion is drawn toward the cation. In extreme cases, such ion deformation leads to the formation of molecules and predominantly covalent compounds (Figure 3.10). The degree

ionic bond

distorted ions

polarized
covalent bond

covalent bond

**Figure 3.10** Transition between ionic and covalent bonding.

of covalent character that a compound exhibits can be considered to correspond to the extent of ion distortion in the compound.

The ability of a cation to distort the electron cloud of a neighboring anion depends upon the charge concentration of the cation; a small cation with a high positive charge is most effective. On the other hand, large multicharged anions are the most readily distorted since the outer electrons are relatively far from the nucleus in ions of this type.

Thus, in any group of metals in the periodic table, the member that forms the smallest cation (for example, Li in group I A) has the greatest tendency toward the formation of compounds with a high degree of covalent character. All compounds of beryllium ($Be^{2+}$ is the smallest cation of any group II A element) are significantly covalent.

Within a group of nonmetals, the tendency toward increasing covalence parallels increasing atomic number (or increasing anion size). Thus, in the case of the halides of aluminum, $AlF_3$ is completely ionic; $AlCl_3$ is intermediate in character; and both $AlBr_3$ and $AlI_3$ are essentially covalent.

In the series formed by the fourth period metals, KCl, $CaCl_2$, $ScCl_3$, and $TiCl_4$, the degree of covalent character increases with increasing charge and decreasing size of the "cation," each of which is isoelectronic with Ar; KCl is strongly ionic and $TiCl_4$ is definitely covalent. Truly ionic compounds that contain cations with a charge of 3+ or higher are rare and exist only when the cation is a large one. Thus, such compounds as $SnCl_4$, $PbCl_4$, $SbCl_5$, and $BiF_5$ are covalent. Boron forms only covalent compounds; the hypothetical $B^{3+}$ ion would have a comparatively high charge combined with a very small size and would cause extensive anion distortion leading to covalent bonding. Certain highly-charged cations exist, however, as hydrated species in water solution (Sections 9.2 and 9.3).

Another approach to bonds of intermediate character considers the polarization of covalent bonds. A purely covalent bond results only when two *identical* atoms are bonded. Whenever two *different* atoms are joined by a covalent bond, the electron density of the bond is not symmetrically distributed with respect to the two bonded atoms—the electrons of the bond are not shared equally. No matter how similar the atoms may be, there will be some difference in electron-attracting ability.

For example, chlorine has a greater attraction for electrons than bromine has. In the BrCl molecule the electrons of the covalent bond are more strongly attracted by the chlorine atom than by the bromine atom; the electron cloud of the bond is denser in the vicinity of the chlorine atom. Therefore, the chorine end of the bond has a partial negative charge, and since the molecule as a whole is electrically neutral, the bromine end is left with a partial positive charge of equal magnitude. Such a bond, with positive and negative poles, is called a **polar covalent**

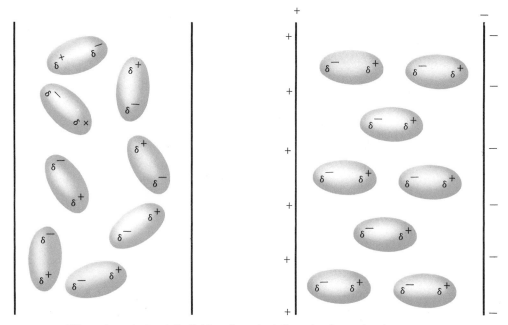

**Figure 3.11** Effect of an electrostatic field on the orientation of polar molecules.

**bond.** The partial charges of the bond are indicated by the symbols $\delta^+$ and $\delta^-$ to distinguish them from full ionic charges.

The greater the difference between the electron-attracting ability of two atoms joined by a covalent bond, the more polar the bond, and the larger the magnitude of the partial charges. If the unequal sharing of electrons were carried to an extreme, one of the bonded atoms would have all of the bonding electrons, and separate ions would result.

A polar covalent molecule orients itself in the electric field between the plates of a condenser in such a way that the negative end is toward the positive plate and the positive end is toward the negative plate (Figure 3.11). Polar molecules arranged in this way affect the amount of charge that a pair of electrically-charged plates can hold; as a result, measurements can be made that allow the calculation of a value called the dipole moment.

**Dipole moment,** $\mu$, is defined as the product of the distance separating equal charges of opposite sign and the magnitude of the charge. The dipole moments of nonpolar molecules, such as $H_2$, $Cl_2$, and $Br_2$, are zero. The more polar the bond of a diatomic molecule, the larger the dipole moment of the molecule. Linus Pauling has used the dipole moment of a compound to calculate the **partial ionic character** of its covalent bond. If hydrogen chloride were completely ionic, the $H^+$ and $Cl^-$ ions would each bear a unit charge ($4.80 \times 10^{-10}$ esu). Since the bond distance between the H and Cl atoms of hydrogen chloride is

1.27 Å, the dipole moment of the hypothetical, completely ionic HCl would be

$$\mu = (4.80 \times 10^{-10} \text{ esu})(1.27 \times 10^{-8} \text{ cm})$$
$$= 6.10 \times 10^{-18} \text{ esu cm} = 6.10 \text{ D}$$

● The debye unit is not an SI unit.

The debye unit, D, is $10^{-18}$ esu cm. The experimentally determined dipole moment of hydrogen chloride is 1.03 D. Thus, the observed dipole moment is

$$1.03 \text{ D}/6.10 \text{ D} = 0.17$$

times the value calculated for the hypothetical, ionic compound. Based on dipole moment measurements, the HCl bond appears to be 17% ionic.

There is a question as to whether the relationship between dipole moment and bond polarity is this simple. It has been demonstrated that the dipole moment of some molecules depends not only upon the bonding in the molecule but also upon the arrangement of the electrons not involved in the bonding (Section 4.6).

**3.9 ELECTRONEGA-TIVITY**

**Electronegativity** is a measure of the ability of an atom in a molecule to attract electrons to itself. Thus, the polarity of the HCl molecule may be said to arise from the difference between the electronegativity of chlorine and that of hydrogen. Since chlorine is more electronegative than hydrogen, the chlorine end of the molecule is the negative end of the dipole.

The concept of electronegativity is useful but inexact. There is no simple, direct way to measure electronegativities, and various methods of determining them have been proposed. Indeed, since this property depends not only upon the structure of the atom under consideration, but also upon the number and nature of the atoms to which it is bonded, the electronegativity of an atom is not invariable. The electronegativity of phosphorus would be expected to be different in $PCl_3$ than in $PF_5$. This concept, therefore, should be regarded as only semiquantitative.

The electronegativity scale of Pauling is the one most frequently used (Table 3.5) and is based upon experimentally derived values of **bond energies.** The bond energy of the Br—Br bond, for example, is the energy required to separate the $Br_2$ molecule into Br atoms; for one mole of $Br_2$ molecules ($6.022 \times 10^{23}$ molecules), +46 kcal is required. The bond energy of the H—H bond is +104 kcal/mol. Both the Br—Br and H—H bonds are nonpolar, and the arithmetic mean of the bond energies

$$\frac{46 \text{ kcal} + 104 \text{ kcal}}{2} = 75 \text{ kcal}$$

may be considered to describe a "nonpolar H—Br bond" in which the electrons are shared equally.

Table 3.5 Electronegativities of the elements[a]

| 1 | 2 | 3 | 4 | 5 | 6 | 7 | 8 | 9 | 10 | 11 | 12 | 13 | 14 | 15 | 16 | 17 | 18 |
|---|---|---|---|---|---|---|---|---|---|---|---|---|---|---|---|---|---|
| 1<br>H<br>2.1 | | | | | | | | | | | | | | | | | 2<br>He<br>— |
| 3<br>Li<br>1.0 | 4<br>Be<br>1.5 | | | | | | | | | | | 5<br>B<br>2.0 | 6<br>C<br>2.5 | 7<br>N<br>3.0 | 8<br>O<br>3.5 | 9<br>F<br>4.0 | 10<br>Ne<br>— |
| 11<br>Na<br>0.9 | 12<br>Mg<br>1.2 | | | | | | | | | | | 13<br>Al<br>1.5 | 14<br>Si<br>1.8 | 15<br>P<br>2.1 | 16<br>S<br>2.5 | 17<br>Cl<br>3.0 | 18<br>Ar<br>— |
| 19<br>K<br>0.8 | 20<br>Ca<br>1.0 | 21<br>Sc<br>1.3 | 22<br>Ti<br>1.5 | 23<br>V<br>1.6 | 24<br>Cr<br>1.6 | 25<br>Mn<br>1.5 | 26<br>Fe<br>1.8 | 27<br>Co<br>1.8 | 28<br>Ni<br>1.8 | 29<br>Cu<br>1.9 | 30<br>Zn<br>1.6 | 31<br>Ga<br>1.6 | 32<br>Ge<br>1.8 | 33<br>As<br>2.0 | 34<br>Se<br>2.4 | 35<br>Br<br>2.8 | 36<br>Kr<br>— |
| 37<br>Rb<br>0.8 | 38<br>Sr<br>1.0 | 39<br>Y<br>1.2 | 40<br>Zr<br>1.4 | 41<br>Nb<br>1.6 | 42<br>Mo<br>1.8 | 43<br>Tc<br>1.9 | 44<br>Ru<br>2.2 | 45<br>Rh<br>2.2 | 46<br>Pd<br>2.2 | 47<br>Ag<br>1.9 | 48<br>Cd<br>1.7 | 49<br>In<br>1.7 | 50<br>Sn<br>1.8 | 51<br>Sb<br>1.9 | 52<br>Te<br>2.1 | 53<br>I<br>2.5 | 54<br>Xe<br>— |
| 55<br>Cs<br>0.7 | 56<br>Ba<br>0.9 | 57–71<br>La–Lu<br>1.1–1.2 | 72<br>Hf<br>1.3 | 73<br>Ta<br>1.5 | 74<br>W<br>1.7 | 75<br>Re<br>1.9 | 76<br>Os<br>2.2 | 77<br>Ir<br>2.2 | 78<br>Pt<br>2.2 | 79<br>Au<br>2.4 | 80<br>Hg<br>1.9 | 81<br>Tl<br>1.8 | 82<br>Pb<br>1.8 | 83<br>Bi<br>1.9 | 84<br>Po<br>2.0 | 85<br>At<br>2.2 | 86<br>Rn<br>— |
| 87<br>Fr<br>0.7 | 88<br>Ra<br>0.9 | 89–<br>Ac–<br>1.1–1.7 | | | | | | | | | | | | | | | |

[a] Based on Linus Pauling, *The Nature of the Chemical Bond, Third Edition,* © 1960 by Cornell University Press. Used with the permission of Cornell University Press.

The actual HBr bond is polar, and the measured bond energy is +88 kcal/mol. The difference between the measured and calculated values, $\Delta$,

$$\Delta = 88 \text{ kcal} - 75 \text{ kcal} = 13 \text{ kcal}$$

may be considered to represent the portion of the measured bond energy that is required to separate the partial charges produced by the unequal sharing of the bonding electrons. Hence, a bond for which the value of $\Delta$ is large would be expected to be highly polar and to be formed from atoms that have a large difference in electronegativity. Consequently, the $\Delta$ values for a number of bonds can be used to derive relative electronegativity values for the elements.[1]

[1] The square root of a $\Delta$ value expressed in electron volts (1 eV = 23 kcal/mol) is assumed to equal the difference in electronegativity between the two bonded atoms. The electronegativity assigned to each atom is the value that leads to the best agreement among all the data pertaining to that atom; the assignment is made on a scale with the electronegativity of fluorine set equal to 4.0. Thus, for HBr

**Figure 3.12** The relation between position in the periodic classification and metallic or nonmetallic reactivity.

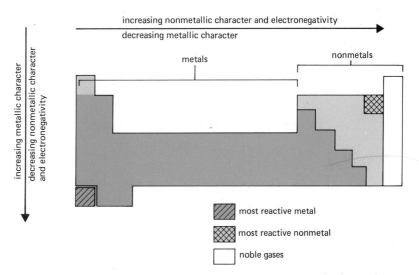

In general, electronegativity increases from left to right across any period (with increasing number of valence electrons) and from bottom to top in any group (with decreasing size). Thus, the most highly electronegative elements are found in the upper right corner of the periodic table (ignoring the noble gases) and the least electronegative elements are found in the lower left corner of the chart. These trends parallel those noted for ionization potentials and electron affinities.

Metals are elements that have small attractions for valence electrons (low electronegativities); nonmetals, except for the noble gases, have large attractions (high electronegativities). Thus, electronegativities (as well as ionization potentials and electron affinities) can be used to rate metallic reactivities and nonmetallic reactivities. The positions of the elements in the periodic table are helpful in making predictions concerning chemical reactivity (Figure 3.12).

Electronegativities can be used to predict the type of bonding that a compound will have. When two elements of widely different electronegativity combine, an ionic compound results; thus, the electronegativity difference between sodium and chlorine is 2.1, and NaCl is an ionic compound.

Covalent bonding occurs between nonmetals when the electronegativity differences are not very large. In such cases the electronegativity differences give an indication of the degree of polarity of the covalent bonds. If the electronegativity difference is zero or very small, an essentially nonpolar bond with equal or almost equal sharing of electrons can

$$\sqrt{\Delta} = \sqrt{(13/23)} \text{ eV} = 0.7$$

which is equal to the difference in the assigned electronegativities of Br and H (2.8 − 2.1 = 0.7, see Table 3.5).

Table 3.6 Some properties of the hydrogen halides

| HYDROGEN HALIDE | DIPOLE MOMENT (D) | BOND ENERGY (kcal/mol) | ELECTRONEGATIVITY OF HALOGEN | ELECTRONEGATIVITY DIFFERENCE BETWEEN HYDROGEN AND HALOGEN |
|---|---|---|---|---|
| HF | 1.91 | 135 | F = 4.0 | 1.9 |
| HCl | 1.03 | 103 | Cl = 3.0 | 0.9 |
| HBr | 0.78 | 87 | Br = 2.8 | 0.7 |
| HI | 0.38 | 71 | I = 2.5 | 0.4 |

be assumed. The larger the electronegativity difference, the more polar is the covalent bond (the bond being polarized in the direction of the atom with the larger electronegativity). Thus, from electronegativities we can predict that HF is the most polar, and has the largest bond energy, of any hydrogen halide (see Table 3.6).

The type of bonding that occurs between two metals (the metallic bond), in which electronegativity differences are also relatively small, will be considered in Section 4.7.

**3.10 OXIDATION NUMBERS**

In the assignment of formal charges to the atoms of a covalent molecule (Section 3.7), the bonding electrons are apportioned *equally* between the bonded atoms, and any bond polarity caused by unequal sharing of electrons is ignored. The charges thus derived are useful in interpreting the structure and some of the properties of covalent molecules (as we shall see in later sections), but the concept itself is merely a convention.

Another arbitrary but useful convention is the concept of oxidation numbers. **Oxidation numbers** are charges (fictitious charges in the case of covalent species) assigned to the atoms of a compound according to arbitrary rules that take into account bond polarity. The oxidation number of an atom in a binary, electrovalent compound is the same as the charge on the ion derived from that atom (the electrovalence number); thus, in NaCl the oxidation number of sodium is 1+ and that of chlorine is 1−. The oxidation numbers of the atoms in a covalent compound can be derived by assigning the electrons of each bond in the molecule to the more electronegative of the bonded atoms (rather than by dividing them equally as is consistently done in deriving formal charges). Thus, for the molecule

$$H \colon \overset{\cdot \cdot}{\underset{\cdot \cdot}{Cl}} \colon$$

both electrons of the covalent bond are assigned to the chlorine atom (chlorine is more electronegative than hydrogen), and the chlorine atom is said to have an oxidation number of 1− and the hydrogen atom 1+. In the case of a nonpolar bond between identical atoms, in which there is no electronegativity difference, the bonding electrons are divided equally between the bonded atoms in deriving oxidation numbers. Thus, the oxidation numbers of both chlorine atoms are zero in the molecule

$$:\overset{\cdot\cdot}{\underset{\cdot\cdot}{Cl}}:\overset{\cdot\cdot}{\underset{\cdot\cdot}{Cl}}:$$

The following rules, based on these ideas, can be used to assign oxidation numbers.

1. Any uncombined atom or any atom in a molecule of an element is assigned an oxidation number of zero.

2. The sum of the oxidation numbers of the atoms in a compound is zero since compounds are electrically neutral.

3. The oxidation number of a monatomic ion is the same as the charge on the ion (electrovalence number). The sum of the oxidation numbers of the atoms that constitute a polyatomic ion equals the charge on the ion.

4. The oxidation number of fluorine, the most electronegative element, is $1-$ in all fluorine-containing compounds.

5. In most oxygen-containing compounds, the oxidation number of oxygen is $2-$. There are, however, a few exceptions. In peroxides each oxygen has an oxidation number of $1-$. For example, the two oxygens of the peroxide ion, $O_2^{2-}$, are equivalent; each must be assigned an oxidation number of $1-$ so that the sum equals the charge on the ion. In the superoxide ion, $O_2^-$, each oxygen has an oxidation number of $\frac{1}{2}-$. In $OF_2$ the oxygen has an oxidation number of $2+$ (see rule 4).

6. The oxidation number of hydrogen is $1+$ in all its compounds except the metallic hydrides (e.g., $CaH_2$ and $NaH$) in which hydrogen is in the $1-$ oxidation state.

The oxidation states of the constituent elements of most compounds can be determined through the use of these rules. The oxidation numbers of the atoms of $H_3PO_4$, for example, must add up to zero. If each hydrogen is assigned $1+$ (total, $3+$) and each oxygen is assigned $2-$ (total, $8-$), the phosphorus must have an oxidation number of $5+$. The same conclusion can be reached by examining the ion derived from phosphoric acid, the phosphate ion, $PO_4^{3-}$. The atoms of a polyatomic ion, such as the phosphate ion, are covalently bonded; the ion exists as an entity and the sum of the oxidation numbers of the atoms that constitute the ion equals the charge of the ion. Since the charge on the $PO_4^{3-}$ ion is $3-$, and since each oxygen atom has an oxidation number of $2-$ (a total of $8-$ for the four oxygen atoms), the phosphorous atom must have an oxidation number of $5+$.

In the dichromate ion, $Cr_2O_7^{2-}$, the seven oxygen atoms have a combined oxidation number of $14-$. Since the charge on the ion is $2-$, the oxidation numbers of the two chromium atoms must add up to $12+$. The oxidation number of chromium in this ion, therefore, is 6+. Usual practice is to report the oxidation state of an element in a compound on the basis of the oxidation number of a single atom. It would be misleading to say that oxygen is in an oxidation state of 2− in $H_2O$ and 4− in $SO_2$; in both compounds, oxygen has an oxidation number of 2−.

What are the oxidation states of the elements in calcium perchlorate, $Ca(ClO_4)_2$? Since calcium is a member of group II A, its oxidation number (which is the same as its electrovalence number) is 2+. Since there are two perchlorate ions for every $Ca^{2+}$ ion in the compound, the charge on the perchlorate ion must be 1−. Each oxygen atom of a $ClO_4^-$ ion has an oxidation number of 2− (a total of 8− for the four oxygen atoms), and therefore the oxidation number of the chlorine atom is 7+. These values must be interpreted with care. The ionic charges have physical significance; the oxidation numbers of oxygen and chlorine are merely conventions.

Fractional oxidation numbers are possible. Calculation of the oxidation number of iron in $Fe_3O_4$ gives a value of $2\frac{2}{3}+$. (Since the oxygens add up to 8−, the irons must add up to 8+; 8+ divided by 3 is $2\frac{2}{3}+$.) This compound can be considered to be formed from $FeO$ and $Fe_2O_3$. Hence two iron atoms at 3+ each and one iron atom at 2+ average out to $2\frac{2}{3}+$. When an element has a fractional oxidation number in a particular compound, this often means that the compound contains two (or more) atoms of the element that are not perfectly equivalent.

Frequently an element displays a range of oxidation states in its compounds. For example, in the compounds of nitrogen, this element exhibits oxidation numbers from 3− (e.g., in $NH_3$) to 5+ (e.g., in $HNO_3$). Since the number of valence electrons of an A family element is the same as its group number, the highest positive charge (even a hypothetical one) that can logically be assigned to an A family element is the same as its group number. The highest oxidation number of an A family element is, therefore, its group number.

The lowest oxidation number of an A family element is its electrovalence number. Thus, the highest oxidation number of sulfur (group VI A) is 6+ (e.g., in $H_2SO_4$) and the lowest oxidation number of sulfur is 2− (e.g., in $H_2S$). The highest oxidation number of sodium (group I A) is the same as the lowest oxidation number, 1+. There are, however, exceptions to these generalizations (e.g., fluorine and oxygen).

## 3.11 NOMENCLATURE OF INORGANIC COMPOUNDS

The name of a **binary compound,** a compound formed from only two elements, is usually derived from the names of the elements of which it is composed. Ionic binary compounds are named with the name of the metal first and the nonmetal second; covalent binaries are named with the name of the less electronegative element first. For all binary compounds, the ending -*ide* is substituted for the usual ending of the name of the element that appears last in the name. All compounds ending in -*ide*, however, are not binaries; there are a few exceptions, such as the cyanides (e.g., NaCN) and the hydroxides (e.g., NaOH).

If two elements form only one compound with each other, the preceding guidelines are sufficient to name that compound. Thus, the name of $Na_2O$ is sodium oxide; $AlCl_3$, aluminum chloride; AgBr, silver bromide; ZnS, zinc sulfide; $CaH_2$, calcium hydride; and HF, hydrogen fluoride.

Some pairs of elements, however, form several compounds with each other (e.g., $FeCl_2$ and $FeCl_3$). Each compound must have a name that identifies it. In a system of nomenclature proposed by Albert Stock, the distinction between such compounds is made by indicating the oxidation state of the first element in Roman numerals immediately after the *English* name of that element. Thus, $FeCl_2$ is iron(II) chloride [$Fe^{2+}$ is called the iron(II) ion]. Notice that the Roman numeral applies to the oxidation state of the first element of the name (the one in a positive oxidation state) and not to numbers that appear as subscripts in the formula.

Another method to distinguish between the compounds formed by an element that exhibits more than one positive oxidation state is to change the suffix of the name of the element (*Latin* names are employed when the symbol derives from the Latin). The ending -*ous* is used to indicate the lower oxidation state of a pair, and -*ic* is used for the higher oxidation state. Thus, $FeCl_2$ is ferrous chloride (and $Fe^{2+}$ is the ferrous ion); $FeCl_3$ is ferric chloride (and $Fe^{3+}$ is the ferric ion). Difficulties arise in the use of this method since some pairs of elements form more than two compounds. The two systems of nomenclature that have been presented are compared in Table 3.6.

A system of nomenclature that is particularly useful for series of compounds formed from two nonmetals employs Greek prefixes: *mono-* (1), *di-* (2), *tri-* (3), *tetra-* (4), *penta-* (5), *hexa-* (6), *hepta-* (7), *octa-* (8),

Table 3.6 Names of some binary compounds

| FORMULA | NAME | |
|---|---|---|
| FeO | iron(II) oxide | ferrous oxide |
| $Fe_2O_3$ | iron(III) oxide | ferric oxide |
| $Cu_2S$ | copper(I) sulfide | cuprous sulfide |
| CuS | copper(II) sulfide | cupric sulfide |
| $SnF_2$ | tin(II) fluoride | stannous fluoride |
| $SnF_4$ | tin(IV) fluoride | stannic fluoride |
| $Hg_2Br_2{}^a$ | mercury(I) bromide | mercurous bromide |
| $HgBr_2$ | mercury(II) bromide | mercuric bromide |

$^a$ The mercury(I) ion (or the mercurous ion) is unusual since it is diatomic ($Hg_2^{2+}$). The two atoms of mercury are covalently bonded. Since the charge on the ion is 2+, each atom has an oxidation number of 1+.

Table 3.7 Names of the nitrogen oxides

| FORMULA | | NAME |
|---------|---|------|
| $N_2O$ | dinitrogen oxide | nitrogen(I) oxide |
| NO | nitrogen oxide | nitrogen(II) oxide |
| $N_2O_3$ | dinitrogen trioxide | nitrogen(III) oxide |
| $NO_2$ | nitrogen dioxide | nitrogen(IV) oxide |
| $N_2O_4$ | dinitrogen tetroxide | dimer of nitrogen(IV) oxide[a] |
| $N_2O_5$ | dinitrogen pentoxide | nitrogen(V) oxide |

[a]A dimer is a molecule formed from two identical, simpler molecules (monomers). In this case an $N_2O_4$ molecule is formed from two $NO_2$ molecules.

and so forth. One of the prefixes is added to the name of each element to indicate the number of atoms of that element in the molecule that is being named. The prefix *mono-* is frequently omitted. The names of the oxides of nitrogen, according to this system as well as to the Stock system, appear in Table 3.7.

Certain binary compounds have acquired non-systematic names by which they are known exclusively. The list of such substances includes water ($H_2O$), ammonia ($NH_3$), hydrazine ($N_2H_4$), and phosphine ($PH_3$). Notice that the formulas of the last three compounds are customarily inverted; the symbol H should appear first in the formula since hydrogen is the element that occurs in a positive oxidation state.

**Acids** are covalent compounds of hydrogen that dissociate in water to produce $H^+$(aq) ions, in which each $H^+$ is bonded to at least one water molecule. This type of ion is frequently indicated $H_3O^+$ (which is $H^+ \cdot H_2O$) and is called the *hydronium ion*. When hydrogen chloride dissolves in water, an acidic solution is produced that is called hydrochloric acid.

$$HCl(g) + H_2O \rightarrow H_3O^+(aq) + Cl^-(aq)$$

Aqueous solutions of *binary compounds* that function as acids are named by adding the prefix *hydro-* and the suffix *-ic* to the root of the name of the element that is combined with hydrogen and then adding the word *acid*. Hence, hydrogen fluoride (HF) forms hydrofluoric acid, and hydrogen sulfide ($H_2S$) forms hydrosulfuric acid. Although hydrogen cyanide (HCN) is not a binary compound, it is named as though it were; an aqueous solution of HCN is called hydrocyanic acid.

**Alkalies** are compounds that contain the hydroxide ion ($OH^-$) and are named accordingly [e.g., NaOH is sodium hydroxide and $Ca(OH)_2$ is calcium hydroxide]. Since they contain three elements, the hydroxides are classed as **ternary compounds.**

When an acid and an alkali are mixed, a neutralization reaction occurs between the $H_3O^+$ ion from the acid and the $OH^-$ ion from the alkali.

$$H_3O^+(aq) + OH^-(aq) \rightarrow 2H_2O$$

A **salt** is a compound derived from the combination of an alkali and an acid; it consists of the cation of the alkali and the anion of the acid. The salts of binary acids are themselves binary compounds, and the names of these salts have the customary *-ide* ending. Thus, sodium chloride (NaCl) is the salt derived from sodium hydroxide and hydrochloric acid.

Oxygen is one of the three constituent elements of most ternary acids. If a given element (sometimes called the central element) forms only one oxyacid, the acid is named by changing the ending of the name of the element to *-ic* and adding the word *acid*. Thus, $H_3BO_3$ is boric acid. If there are two common acids in which the same central element occurs in different oxidation states, the ending *-ous* is used in naming the acid of the element in its lower oxidation state; the ending *-ic* is used to denote the higher oxidation state (see Table 3.8).

There are a few series of oxyacids for which two names are not enough. The prefix *hypo-* may be added to the name of an *-ous* acid to indicate an oxidation state lower than that of the *-ous* acid. The prefix *per-* may be added to the name of an *-ic* acid to indicate an oxidation state higher than that of the *-ic* acid (see the names of the oxyacids of chlorine in Table 3.8).

The oxyacids are covalent molecules that produce covalently bonded polyatomic anions through the loss of protons ($H^+$ ions) in salt formation. The names of the anions produced by the loss of *all* possible protons are derived from the name of the parent acid by retaining the prefix (if any) and changing the suffix *-ic* to *-ate* or the suffix *-ous* to *-ite*. The salts of these anions are named by combining the name of the cation present (using Stock nomenclature or the *-ous/-ic* convention if necessary) with the name of the anion (see Table 3.8). Thus, $Fe(ClO_4)_3$ is iron(III)

Table 3.8 Names of some ternary acids and salts

| OXIDATION STATE OF CENTRAL ELEMENT | ACID | | SODIUM SALT | |
| --- | --- | --- | --- | --- |
| | FORMULA | NAME | FORMULA | NAME |
| 5+ | $HNO_3$ | nitric acid | $NaNO_3$ | sodium nitrate |
| 3+ | $HNO_2$ | nitrous acid | $NaNO_2$ | sodium nitrite |
| 6+ | $H_2SO_4$ | sulfuric acid | $Na_2SO_4$ | sodium sulfate |
| 4+ | $H_2SO_3$ | sulfurous acid | $Na_2SO_3$ | sodium sulfite |
| 7+ | $HClO_4$ | perchloric acid | $NaClO_4$ | sodium perchlorate |
| 5+ | $HClO_3$ | chloric acid | $NaClO_3$ | sodium chlorate |
| 3+ | $HClO_2$ | chlorous acid | $NaClO_2$ | sodium chlorite |
| 1+ | $HOCl$ | hypochlorous acid | $NaOCl$ | sodium hypochlorite |

Table 3.9 Common polyatomic anions

| FORMULA | NAME | FORMULA | NAME |
|---------|------|---------|------|
| $C_2H_3O_2^-$ | acetate | $ClO^-$ | hypochlorite |
| $AsO_4^{3-}$ | arsenate | $OH^-$ | hydroxide |
| $AsO_3^{3-}$ | arsenite | $NO_3^-$ | nitrate |
| $CO_3^{2-}$ | carbonate | $NO_2^-$ | nitrite |
| $ClO_3^-$ | chlorate | $ClO_4^-$ | perchlorate |
| $ClO_2^-$ | chlorite | $MnO_4^-$ | permanganate |
| $CrO_4^{2-}$ | chromate | $PO_4^{3-}$ | phosphate |
| $CN^-$ | cyanide | $SO_4^{2-}$ | sulfate |
| $Cr_2O_7^{2-}$ | dichromate | $SO_3^{2-}$ | sulfite |

perchlorate or ferric perchlorate. The names and formulas of some common polyatomic anions are given in Table 3.9.

Acids that can lose more than one proton per molecule are called **polyprotic acids.** Sulfuric acid ($H_2SO_4$, specifically a diprotic acid) can lose one or two protons. Phosphoric acid ($H_3PO_4$, specifically a triprotic acid) can lose one, two, or three protons. The salts formed by the loss of all possible protons are called **normal salts.** Salts formed by the incomplete neutralization of polyprotic acids are called **acid salts;** the anions of these salts retain one or more hydrogen atoms of the parent acid. Acid salts are named so as to indicate the number of hydrogen atoms present (see Table 3.10). In practice, the prefix *mono-* is usually omitted. In addition, the monohydrogen salt of a diprotic acid, such as $NaHSO_3$, may also be named by use of the prefix *bi-*. Thus, $NaHSO_3$ is sodium monohydrogen sulfite, sodium hydrogen sulfite, or sodium bisulfite.

Table 3.10 Names of the sodium salts of some polyprotic acids

| FORMULA | NAME |
|---------|------|
| $Na_3PO_4$ | sodium phosphate |
| $Na_2HPO_4$ | sodium monohydrogen phosphate (sodium hydrogen phosphate) |
| $NaH_2PO_4$ | sodium dihydrogen phosphate |
| $Na_2CO_3$ | sodium carbonate |
| $NaHCO_3$ | sodium monohydrogen carbonate (sodium hydrogen carbonate) (sodium bicarbonate) |

A large number of complex polatomic cations, anions, and neutral molecules exists. The chemistry of these substances is the topic of Chapter 19, and their nomenclature is discussed in Section 19.3. There are not many simple polyatomic cations of importance, but the ammonium ion ($NH_4^+$) and the mercury(I) ion ($Hg_2^{2+}$) deserve mention.

**Problems**

**3.1** Which member of each of the following pairs would you predict to be larger? (a) P or Cl, (b) P or Ge, (c) Mg or Ca, (d) Na or Mg, (e) K or Mg, (f) Co or Ga

**3.2** Which member of each of the following pairs would you predict to be larger? (a) Sn or Pb, (b) Sn or As, (c) As or Se, (d) Fe or Zn, (e) Al or Cl, (f) La or Hf

**3.3** Explain why a minimum is observed in each curve obtained by plotting atomic radius against atomic number for the members of each transition series.

**3.4** Of the elements of the third period (Na through Ar): (a) which has the largest atomic radius, (b) which has the highest first ionization potential, (c) which is the most electronegative, (d) which is the most reactive metal, (e) which is the most reactive nonmetal, (f) which is chemically the least reactive?

**3.5** Which of the Group IIA elements, Be, Mg, Ca, Sr, and Ba, (a) has the largest atomic radius, (b) has the highest first ionization potential, (c) is the most electronegative, (d) is the most reactive element?

**3.6** Which member of each of the following pairs would you predict to have the higher first ionization potential? (a) Cl or Ar, (b) Se or Cl, (c) S or Cl, (d) Cl or Br, (e) K or Ca

**3.7** Which member of each of the following pairs would you predict to have the higher first ionization potential? (a) Sr or Ba, (b) Cs or Ba, (c) In or Sn, (d) F or Ne, (e) Xe or Kr

**3.8** The first ionization potentials of the B family elements, Cu (7.7 eV), Ag (7.6 eV), and Au (9.2 eV), do not decrease with increasing atomic size in the way typical of the A family elements. Explain why this is true.

**3.9** Using arguments based on the energy effects of ion formation, explain why: (a) Na forms $Na^+$ and not $Na^{2+}$, (b) Cl forms $Cl^-$ and not $Cl^{2-}$, (c) it is possible for Fe to form both $Fe^{2+}$ and $Fe^{3+}$.

**3.10** Explain why the first ionization potential of O is lower than that of N.

**3.11** Give reasons why some of the electron affinity values appearing in Table 3.1 are positive.

**3.12** For each of the following, give the formulas of three ions that are isoelectronic with the atom or ion given. (a) Ne, (b) $Cd^{2+}$, (c) $Br^-$, (d) $Sc^{3+}$, (e) Hg.

**3.13** For each of the following, give the formulas of three ions that are isoelectronic with the atom or ion given. (a) Ni, (b) $Au^+$, (c) $Cs^+$, (d) Cd, (e) Kr

**3.14** Which member of each of the following pairs would you predict to be larger? (a) Br or $Br^-$, (b) Cs or $Cs^+$, (c) $O^{2-}$ or $F^-$, (d) $Au^+$ or $Au^{3+}$, (e) $In^{3+}$ or $Tl^+$, (f) $In^+$ or $Sn^{2+}$

**3.15** Which member of each of the following pairs would you predict to be larger? (a) $Cs^+$ or $Ba^{2+}$, (b) S or $S^{2-}$, (c) $S^{2-}$ or $Cl^-$, (d) $Cr^{2+}$ or $Cr^{3+}$, (e) Ag or $Ag^+$, (f) $Cu^{2+}$ or $Ag^+$

**3.16** Draw Lewis structures for the

following molecules (include formal charges): $OCF_2$, $OSF_2$, and $OPF_3$. In each case, each O and F atom is directly bonded to the central atom (C, S, and P respectively).

**3.17** Draw Lewis structures for the following molecules (include formal charges): $H_2NOH$, FNNF, $NF_3$, HCCH, $H_2CCH_2$, $H_2NNH_2$, HONO.

**3.18** Draw Lewis structures for the following ions (include formal charges): $SO_4^{2-}$, $SO_3^{2-}$, $ClO_2^-$, $CN^-$, $PO_4^{3-}$, and $O_2^{2-}$.

**3.19** The following species are isoelectronic: CO, $NO^+$, $CN^-$, and $N_2$. (a) Draw Lewis structures for each and include formal charges. (b) Each of the four reacts to form complexes with metals or metal cations. In the formation of a complex, we can assume that an electron pair from one of the four species is used to form a covalent bond by occupying an empty orbital of the metal. For each heteronuclear species, tell which atom is bonded to the metal atom.

**3.20** On the basis of anion deformation, predict which compound of each of the following pairs is the more covalent. (a) $CuI$ or $CuI_2$, (b) $Al_2O_3$ or $Al_2S_3$, (c) $SnBr_2$ or $PbBr_2$, (d) $CdBr_2$ or $CdI_2$, (e) CrS or $Cr_2S_3$, (f) $BeI_2$ or $MgI_2$.

**3.21** The bond distance in HF is 0.917 Å, and the dipole moment of HF is 1.91 D. Calculate the partial ionic character of HF.

**3.22** The bond distance in BrCl is 2.13 Å, and the dipole moment of BrCl is 0.57 D. Calculate the partial ionic character of BrCl.

**3.23** The bond in BrF is estimated to be 15% ionic. If the Br—F bond distance is 1.76 Å, what should the dipole moment of BrF be?

**3.24** The bond in HI is estimated to be 4.9% ionic. If the dipole moment of HI is 0.38 D, what is the H—I bond distance?

**3.25** (a) Use electronegativities to estimate the degree of polarity of the bond in each of the molecules: ICl, BrCl, BrF, and ClF. (b) List the molecules in decreasing order according to the magnitude of their dipole moments.

**3.26** For each of the following pairs, use electronegativities to determine which bond is the more polar; in each case tell the direction in which the bond is polarized. (a) O—H bond, O—F bond, (b) S—H bond, S—F bond, (c) O—H bond, S—H bond, (d) O—F bond, S—F bond.

**3.27** On the basis of electronegativity, state whether you think the bonds formed between the following pairs of elements would be ionic or covalent; if covalent, estimate the degree of polarity of the bond. (a) Mg—O, (b) P—Cl, (c) P—H, (d) Hg—Cl, (e) C—Br, (f) Cs—Br, (g) N—Cl, (h) B—H, (i) Al—I, (j) N—H.

**3.28** Arrange the following bonds in order of increasing ionic character. (a) C—H, (b) Si—H, (c) C—Cl, (d) Si—Cl, (e) C—O, (f) Si—O, (g) C—S, (h) Al—C, (i) Si—C.

**3.29** The bond energy of the C—C bond is 83 kcal/mol, and the bond energy of the O—O bond is 33 kcal/mol. The electronegativity of C is 2.5 and that of O is 3.5. Given that the difference in electronegativity between two elements approximately equals $\sqrt{\Delta/23}$, calculate a bond energy for the C—O bond.

**3.30** State the oxidation number of the atom other than O in each of the following ions: (a) $CrO_4^{2-}$, (b) $Cr_2O_7^{2-}$, (c) $AsO_4^{3-}$, (d) $MnO_4^-$, (e) $SO_3^{2-}$, (f) $ClO_4^-$, (g) $NO_3^-$.

**3.31** State the oxidation number of (a) Cl in $Cl_2O_4$, (b) Sn in $K_2SnO_3$, (c) N in $H_2N_2O_2$, (d) B in $CaB_2O_4$, (e) V in $VO^{2+}$, (f) Br in $BrF_6^-$, (g) U in $Li_2U_2O_7$, (h) S in $SO_3F_2$.

**3.32** State the oxidation number of (a) P in $OPF_3$, (b) Xe in $CsXeF_7$, (c) B in $LiBF_4$, (d) Te in $H_6TeO_6$, (e) P in $P_4O_8$, (f) Sb in $Sb(OH)_2^+$, (g) Bi in $(Bi_6O_6)^{6+}$, (h) Ge in $(Ge_3O_9)^{6-}$.

**3.33** Give the formulas for the sulfides, fluorides, arsenates, permanganates, and sulfates of potassium, calcium, and iron(III).

**3.34** (a) Give names for the oxy-acids of bromine: HOBr, $HBrO_2$, $HBrO_3$, and $HBrO_4$. (b) What is the name of the potassium salt of each of these acids?

**3.35** Give formulas for (a) iron(III) sulfate, (b) chromium(II) sulfite, (c) cupric sulfide, (d) silver(I) dihydrogen phosphate, (e) tellurium hexafluoride, (f) mercury(I) acetate, (g) disulfur dichloride, (h) ammonium nitrate.

**3.36** Give formulas for (a) cobalt(III) sulfide, (b) bromine pentafluoride, (c) ferrous carbonate, (d) gold(I) sulfate, (e) calcium bicarbonate, (f) tetraphosphorus trisulfide, (g) cuprous cyanide, (h) lanthanum(III) phosphate.

**3.37** Name the following: (a) $N_2F_4$, (b) $(NH_4)_2Cr_2O_7$, (c) $Co(CN)_2$, (d) $Cl_2O_6$, (e) $KMnO_4$, (f) $FeSO_3$, (g) $Sr(HSO_4)_2$, (h) TlOH.

**3.38** Name the following: (a) InN, (b) $PbCrO_4$, (c) $SF_4$, (d) $Cr(NO_3)_3$, (e) $NaNO_2$, (f) $Ni_3(AsO_3)_2$, (g) $Ba(HPO_4)$, (h) $HNO_2$.

# 4

# Molecular Geometry and the Covalent Bond

In 1853 Friedrich Wöhler wrote to Jöns Berzelius: "Organic chemistry just now is enough to drive one mad. It gives me the impression of a primeaval tropical forest . . . into which one may well dread to enter." Wöhler, who had at that time made significant contributions to organic chemistry (the chemistry of the compounds of carbon), subsequently turned his attention exclusively to the chemistry of minerals.

Organic chemistry is almost completely concerned with covalent compounds. Wöhler's frustration stemmed from the inadequate bonding theories of his time. Without a knowledge of the manner in which atoms are bonded into covalent structures, a satisfying interpretation of the complex natural compounds with which Wöhler worked is almost impossible.

The Lewis theory was a tremendous advance toward the solution of this problem. However, Lewis structures cannot be derived for some molecules. Furthermore, in many instances the observed properties of a molecule (such as bond distance and molecular polarity) are not satisfactorily represented by the Lewis structure assigned to the molecule. Since a characteristic of the covalent bond is that it is directed in space, the lack of correlation between a Lewis structure and the geometry of the molecule that it represents is a serious shortcoming. More recent theories supply answers to these problems and are more informative; they represent, therefore, better descriptions of covalent bonding.

**4.1 RESONANCE**    In some cases the properties of a molecule or ion are not adequately represented by a single Lewis structure. For example, the structure for $SO_2$ that was derived in Section 3.7,

is unsatisfactory in two respects. The structure depicts the S atom to be bonded to one O atom by a double bond and to the other O atom by a single bond. Double bonds are characteristically shorter than single bonds, yet experimental evidence shows that both S to O linkages have the same length. The structure also shows one of the O atoms to be more negative than the other, but it is known that both of the O atoms are equivalent in this respect.

In a case of this type, two or more valence-bond structures can be used in combination to depict the molecule. The molecule is said to be a **resonance hybrid** of the electronic structures, which are called **resonance forms.** For $SO_2$,

There is only one type of $SO_2$ molecule; the above notation does not mean that this substance is a mixture of two types of molecules. Nor does the symbolism denote a situation in which electrons oscillate so that the molecule appears in one form at one moment and in the other at the next. The symbol (↔) does not denote motion. Neither resonance form alone adequately represents the molecule, but the *single* structure of the $SO_2$ molecule can be imagined by superimposing both resonance forms.

Thus, resonance shows the O atoms of the molecule to be equally negative and the S to O linkages of $SO_2$ to be identical — intermediate between a single and a double bond. The problem of representation within the framework of the classical valence-bond approach arises because the approach is limited — not because the $SO_2$ molecule is unusual.

All resonance forms of a given species must have the *same configuration of nuclei* and be *energetically similar*. The resonance forms of the linear molecule dinitrogen oxide, $N_2O$, are

$$\overset{\ominus}{:}\!\ddot{N}\!=\!\overset{\oplus}{N}\!=\!\ddot{O}: \;\leftrightarrow\; :N\!\equiv\!\overset{\oplus}{N}\!-\!\ddot{\underset{..}{O}}:^{\ominus}$$

From measurements of the structures of other molecules, a N to N bond distance of 1.20 Å would be expected for a double bond and 1.09 Å for a triple bond. The N to N bond distance in $N_2O$ is 1.13 Å, a value that is intermediate between the double and triple bond values and in agreement with the bonding described by a hybrid of the two resonance forms shown. Similarly, the N to O bond distance in the $N_2O$ molecule (1.19 Å) falls between that expected for a N to O double bond (1.15 Å) and a N to O single bond (1.36 Å).

The dipole moment of $N_2O$ is close to zero. The formal charges appearing in the two resonance forms indicate that in the resonance hybrid both the N to N bond and the N to O bond are polar but that the polarities oppose each other. Since the molecule is linear, the negative charge center for the *molecule* falls in the middle of the molecule and approximately coincides with the positive charge center. Hence, the molecule shows no strong tendency to orient itself in a specific way in an electric field (Section 3.8).

Another possible resonance form of $N_2O$ is rejected.

$$\overset{2\ominus}{:}\!\ddot{N}\!-\!\overset{\oplus}{N}\!\equiv\!O:^{\oplus}$$

A form in which a comparatively high negative charge is placed on the end N atom and a positive charge is simultaneously forced on the more highly electronegative O atom has comparatively high energy. In addition, the O atom and the adjacent N atom are shown with formal charges of the same sign — a condition that would tend to disrupt the bond between these two atoms. This form, therefore, does not contribute appreciably to the resonance hybrid. Resonance tends to delocalize charge, not build up centers of high charge.

The delocalization of charge in an ion is illustrated by the resonance of the carbonate ion, $CO_3^{2-}$. This ion is planar, all bonds are equivalent (between single and double bonds), and all oxygen atoms are equally negative. The formal charges add up to the charge on the ion.

As depicted by resonance, the charge is delocalized; it is impossible to locate the exact position of the "extra" two electrons that give the ion its negative charge.

It is also possible, although usually not necessary, to use the concept of resonance to indicate the charge distribution of a polar molecule. The structure of hydrogen chloride can be illustrated as

$$H : \overset{\cdot\cdot}{\underset{\cdot\cdot}{Cl}} : \;\leftrightarrow\; H^+ : \overset{\cdot\cdot}{\underset{\cdot\cdot}{Cl}} :^-$$

structure I    structure II

Calculations from dipole moment measurements show that HCl has a 17% partial ionic character (Section 3.8). We conclude that structures I and II contribute to the resonance hybrid in a ratio of 83 to 17.

**4.2 HYBRID ORBITALS**    If properly applied, the concept of resonance makes the valence-bond theory a more realistic approach; however, additional extensions must be made to the valence bond view. We have seen that certain ions exist that do not have noble-gas configurations and that some of these ions are relatively stable. Some molecules also exist in which atoms have configurations other than those that the octet principle would lead us to expect.

There are a few molecules (such as NO and $NO_2$) that have an odd number of valence electrons to be distributed among the constituent atoms (Section 4.4). It is impossible to divide an odd number of electrons so that each atom of the molecule has a configuration of eight electrons (an even number). There are not many stable odd electron molecules; usually odd electron species are very reactive and consequently short-lived.

More common are the molecules that have an even number of valence electrons but contain atoms with valence shells of less than, or more than, eight electrons. Examples of this are $BF_3$, in which the central boron atom has six valence electrons

$$
\begin{array}{c}
: \overset{\cdot\cdot}{F} : \\
| \\
B \\
\diagup \quad \diagdown \\
: \overset{\cdot\cdot}{\underset{\cdot\cdot}{F}} \qquad \overset{\cdot\cdot}{\underset{\cdot\cdot}{F}} :
\end{array}
$$

and $PCl_5$ and $SF_6$, in which the phosphorus and sulfur atoms have ten and twelve valence electrons respectively.

For the elements of the second period, only four bonding orbitals are available ($2s$ and $2p$). The maximum covalence of these elements is therefore limited to four. The elements of the third and subsequent periods, however, have more orbitals available in their outer electron shells:

**Figure 4.1** Overlap of the $1s$ orbitals of two hydrogen atoms.

compounds are known in which these elements form four, five, six, or (infrequently) an even higher number of covalent bonds. In valence-bond structures representing compounds containing elements of the third and subsequent periods, therefore, the octet principle is frequently violated. Apparently the criteria for covalent bond formation should be centered on the electron pair rather than on the attainment of an octet.

In the formation of a covalent bond, orbitals from two atoms inter-penetrate, or overlap (Figure 4.1). The electron clouds reinforce each other in the bonding region between the two nuclei, and electronic charge is concentrated in this region. The strength of the bond comes from the attraction of the positively charged nuclei for the negative cloud of the bond; the strongest bonding occurs when the orbital overlap is at a maximum.

In many cases, therefore, the orbitals of the two atoms must be aligned in a specific way in order for a strong bond to form. The question of orbital alignment does not arise in the formation of $H_2$ since each H atom uses a $1s$ orbital in bonding and all $s$ orbitals are spherically sym-metrical. The $p$ orbitals of an atom, however, can be considered to be directed in space at right angles to one another along $x$, $y$, and $z$ axes (Figure 2.19). The electronic configuration of fluorine, for example, is $1s^2\ 2s^2\ 2p^2\ 2p^2\ 2p^1$; in the formation of a covalent bond between two fluorine atoms, the axes of the $p$ orbitals used in the bonding must be aligned. The atoms are directed along this axis in the resulting $F_2$ mole-cule.

In the vapor state beryllium chloride consists of linear Cl—Be—Cl molecules. Both covalent bonds of the molecule are equivalent in strength and bond length. A $p$ orbital of each Cl atom is used to form a bond with the central Be atom, but what type of Be orbitals are employed? The electronic configuration of Be is $1s^2 2s^2$, and it appears that in the ground state beryllium should not form covalent bonds at all. The excited state $1s^2\ 2s^1\ 2p^1$ provides two unpaired electrons for bonding, but it too is inadequate in explaining the bonding of the molecule.

If the $2p$ orbital of the excited state of Be is used to form a covalent bond with a $3p$ orbital of a Cl atom, these two atoms would be aligned along the bonding axis of the $p$ orbitals. For the $2s$ orbital of the Be atom, all directions are equivalent. The Be—Cl bond formed by utilizing this orbital, therefore, would be essentially undirected. Even if this second bond were assumed to take a position 180° away from the other bond, thus producing a linear arrangement, the two bonds would not be similar. The bond lengths would be different since the bonds would be produced by a $p$ orbital of a Cl atom overlapping a $2s$ Be orbital in one instance and a $2p$ Be orbital in the other. In addition, the two bonds would not be energetically equivalent, since the $2s$ and $2p$ orbitals of Be are not energetically equivalent.

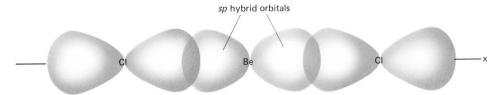

**Figure 4.2** Formation of $BeCl_2$. Shapes of orbitals have been simplified for clarity.

The structure of the molecule can be explained by assuming that the $2s$ orbital and a $2p$ orbital of Be are blended, or hybridized, into two **hybrid orbitals** that are used in the bonding (Figure 4.2). These hybrid orbitals (called *sp* **orbitals**) have one-half *s* character and one-half *p* character, are equivalent, and are directed 180° from one another (Figure 4.3). The *sp* orbitals extend farther along the *x* axis than either the *s* orbital or the *p* orbital. Greater orbital overlap results when an *sp* hybrid orbital is used for bonding than when a pure *s* or a pure *p* orbital is used. Calculations show that stronger bonds result from the use of hybrid orbitals.

The hypothetical formation of $BeCl_2$ may be summarized as

$$2p \ \underline{\ \ } \ \underline{\ \ } \ \underline{\ \ } \qquad 2p \ \underline{\uparrow} \ \underline{\ \ } \ \underline{\ \ } \qquad 2p \ \underline{\ \ } \ \underline{\ \ }$$
$$sp \ \underline{\uparrow} \ \underline{\uparrow}$$

$$2s \ \underline{\uparrow\downarrow} \qquad\qquad 2s \ \underline{\uparrow}$$

$$1s \ \underline{\uparrow\downarrow} \qquad\qquad 1s \ \underline{\uparrow\downarrow} \qquad\qquad 1s \ \underline{\uparrow\downarrow}$$
Be (ground state)     Be (excited state)     Be (hybrid state)

The energy required to produce the excited state as well as that required for the hybridization is more than supplied by the energy released when two comparatively strong covalent bonds are formed. The concept of hybridization is a theoretical device that permits correlation of the ground-state electronic configuration of an atom with the observed structural properties of a molecule that it forms. The preceding analysis

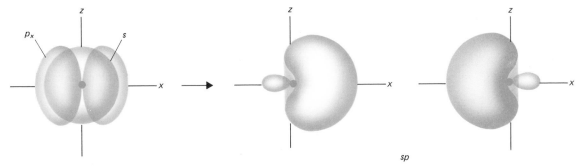

**Figure 4.3** Hybridization of *s* and *p* orbitals to produce *sp* hybrid orbitals. Nuclei at origins; *sp* orbitals should be superimposed.

**Figure 4.4** Directional characteristics of hybrid orbitals.

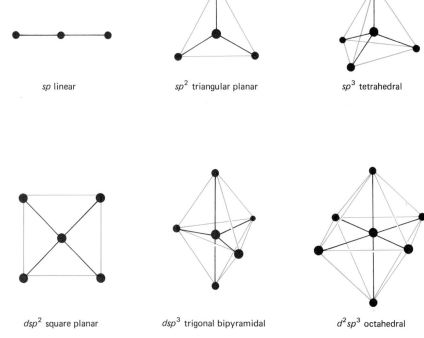

| | | |
|---|---|---|
| *sp* linear | *sp*$^2$ triangular planar | *sp*$^3$ tetrahedral |
| *dsp*$^2$ square planar | *dsp*$^3$ trigonal bipyramidal | *d*$^2$*sp*$^3$ octahedral |

(and others that follow) should not be interpreted as a description of an actual physical process by which a molecule is formed.

Other types of hybrid orbitals are known (Figure 4.4). In the description of molecules of the type $AB_n$, where $n$ is two or more, attention is directed to the bonding orbitals of the central atom, A, since these orbitals determine the overall geometry of the molecule.

The atoms of boron trifluoride lie in the same plane, the angle formed by any two bonds is 120°, and the three bonds are equivalent.

$$\begin{array}{c} F \\ | \\ F \diagdown B \diagdown F \end{array}$$

It is assumed that $sp^2$ **hybrid orbitals** of B are used in the formation of the molecule and that each orbital has $\frac{1}{3}s$ character and $\frac{2}{3}p$ character (Figure 4.5).

| $2p$ $\uparrow$ __ __ | $2p$ $\uparrow$ $\uparrow$ __ | $2p$ __ |
|---|---|---|
| | | $sp^2$ $\uparrow$ $\uparrow$ $\uparrow$ |
| $2s$ $\uparrow\downarrow$ | $2s$ $\uparrow$ | |
| $1s$ $\uparrow\downarrow$ | $1s$ $\uparrow\downarrow$ | $1s$ $\uparrow\downarrow$ |
| B (ground state) | B (excited state) | B (hybrid state) |

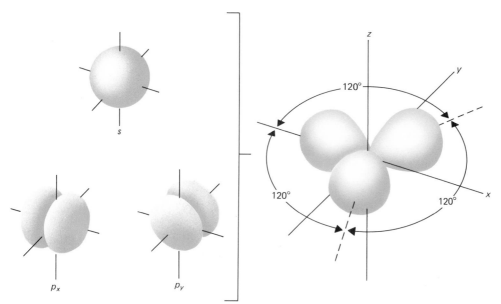

**Figure 4.5** Formation of hybrid $sp^2$ orbitals. The hybrid orbitals are coplanar. Shapes of the hybrid orbitals have been simplified for clarity.

All molecules in which the central atom uses $sp^2$ hybrid orbitals for bonding are triangular (trigonal) planar, and all $sp^2$ hybrid orbitals have the orientation shown in Figure 4.5.

An important and relatively common type of hybridization is found in methane, $CH_4$. In this molecule an $s$ and three $p$ orbitals are mixed to produce four equivalent $sp^3$ **hybrid orbitals,** each one with $\frac{1}{4}s$ character and $\frac{3}{4}p$ character.

$$2p \;\underline{\uparrow}\;\underline{\uparrow}\;\underline{\;\;}\qquad 2p\;\underline{\uparrow}\;\underline{\uparrow}\;\underline{\uparrow}$$

$$\qquad\qquad\qquad\qquad\qquad\qquad sp^3\;\underline{\uparrow}\;\underline{\uparrow}\;\underline{\uparrow}\;\underline{\uparrow}$$

$$2s\;\underline{\uparrow\downarrow}\qquad\qquad 2s\;\underline{\uparrow}$$

$$1s\;\underline{\uparrow\downarrow}\qquad\qquad 1s\;\underline{\uparrow\downarrow}\qquad\qquad 1s\;\underline{\uparrow\downarrow}$$

C (ground state)   C (excited state)   C (hybrid state)

The $sp^3$ hybrid orbitals are directed to the corners of a regular tetrahedron ·(Figure 4.6), and the $CH_4$ molecule has this geometry. In $CH_4$, all H—C—H bond angles are 109° 28′, the so-called tetrahedral angle. Chemical and physical evidence show that all the C—H bonds are equivalent in methane.

It might seem logical that carbon would form a $CH_2$ molecule in which two pure $p$ orbitals of the ground state of C would be used to bond two H atoms. The expected H—C—H bond angle for this molecule would be 90°, in conformity with the orientation of the $p$ orbitals of C

Figure 4.6 Orientation of the sp³ hybrid orbitals.

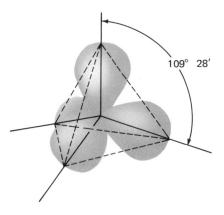

109° 28'

used in the bonding. The formation of such a molecule would not require energy to promote electrons and hybridize orbitals; $CH_2$, however, does not exist.

The bonds formed by using the hybrid orbitals of C are stronger than any formed by using simple atomic orbitals – a greater overlap of bonding orbitals is possible. Furthermore, *four* of the comparatively stronger bonds are formed in the preparation of $CH_4$ and only *two* of the weaker bonds are formed when $CH_2$ is prepared. Therefore, on balance, it is energetically more favorable to supply the energy required for the promotion and hybridization in order to get the larger energy return when $CH_4$ forms.

The nitrogen atom of $NH_3$ may be assumed to use $sp^3$ hybrid orbitals for bonding and to have one of these orbitals occupied by an unshared pair of electrons. The atoms of $NH_3$ form a trigonal pyramid, but the four N orbitals are arranged approximately in the form of a tetrahedron (see Figure 4.8 given subsequently).

$2p$ ↑ ↑ ↑

$sp^3$ ↑↓ ↑ ↑ ↑

$2s$ ↑↓

$1s$ ↑↓     $1s$ ↑↓

N (ground state)    N (hybrid state)

The H—N—H bond angle of $NH_3$ (107.3°) is close to the tetrahedral angle (109.5°). The N atom could conceivably use its three $2p$ orbitals for bonding. In this instance, however, the bond angle would be 90°, since $p$ orbitals are directed in space at right angles to one another along *x, y* and *z* axes. The observed bond angle is closer to the tetrahedral angle than it is to 90°.

Similarly, the oxygen atom of the $H_2O$ molecule may be assumed to use $sp^3$ hybrid orbitals in the bonding of the molecule and to have two of these hybrid orbitals occupied by unshared electron pairs.

$2p$ ⥮ ↑ ↑

$sp^3$ ⥮ ⥮ ↑ ↑

$2s$ ⥮

$1s$ ⥮          $1s$ ⥮

O (ground state)     O (hybrid state)

The atoms of the $H_2O$ molecule are arranged in angular form with a bond angle of $104.5°$ — an angle closer to the tetrahedral angle of $109.5°$ than to the $90°$ angle expected if pure $p$ orbitals were used.

The orbitals employed in a given type of hybridization must be comparatively close in energy. The ones most commonly used in the bonding of molecules are the $s$, $p$, and $d$ orbitals of the outer shell and, in some cases, the $d$ orbitals of the shell next to the outer shell. With the exception of these inner $d$ orbitals, electrons of inner orbitals are usually bound too tightly to be involved in covalent bond formation.

The sulfur atom of sulfur hexafluoride, $SF_6$, employs $d^2sp^3$ **hybrid orbitals** in the bonding of the molecule (Figure 4.7).

$3d$ _ _ _ _ _     $3d$ ↑ ↑ _ _ _     $3d$

$d^2sp^3$ ↑ ↑ ↑ ↑ ↑ ↑

$3p$ ⥮ ↑ ↑     $3p$ ↑ ↑ ↑

$3s$ ⥮     $3s$ ↑

S (ground state)     S (excited state)     S (hybrid state)

The $d^2sp^3$ hybrid orbitals are equivalent and are directed to the corners of a regular octahedron (Figure 4.4).

Phosphorus pentafluoride, $PF_5$, is a trigonal bipyramidal molecule (Figure 4.4), and $dsp^3$ **hybrid orbitals** are used in its formation.

$3d$ _ _ _ _ _     $3d$ ↑ _ _ _ _ _     $3d$ _ _ _ _

$3p$ ↑ ↑ ↑     $3p$ ↑ ↑ ↑     $dsp^3$ ↑ ↑ ↑ ↑ ↑

$3s$ ⥮     $3s$ ↑

P (ground state)     P (excited state)     P (hybrid state)

The $d_{x^2-y^2}$ atomic orbital is not used in the formation of $dsp^3$ hybrid orbitals, but the other atomic orbitals shown in Figure 4.7 are blended in the formation of this type of hybrid. The five orbitals of a $dsp^3$ set, however, are not equivalent (see Section 4.3).

The $Ni(CN)_4^{2-}$ ion is square planar (Figure 4.4) and its bonding is best described by means of a slightly different approach than the one we

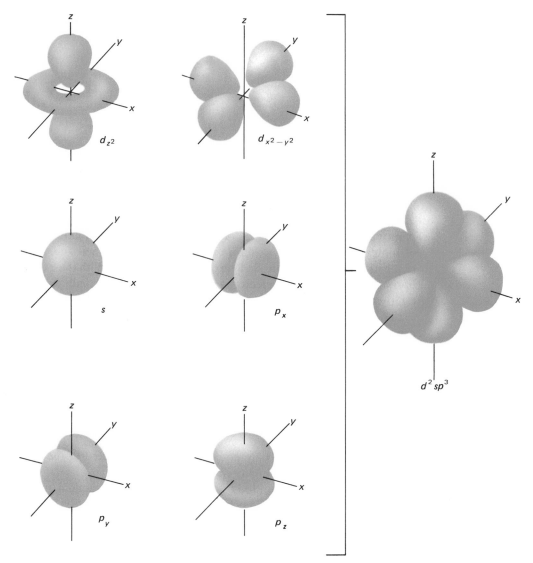

**Figure 4.7** Formation of $d^2sp^3$ hybrid orbitals. Shapes of the hybrid orbitals have been simplified for clarity.

have been using. The complex ion is prepared by the reaction of the $Ni^{2+}$ ion with excess cyanide ion.

$$Ni^{2+} + 4CN^- \rightarrow Ni(CN)_4^{2-}$$

The bonding in this complex may be considered to arise from the occupation of *empty $dsp^2$* **hybrid orbitals** of $Ni^{2+}$ by unshared electron pairs of the C atoms of the four cyanide ions, $:C\equiv N:^-$. The energies of the $3d$, the

$4s$, and the $4p$ orbitals are similar, and a $3d$ orbital is used for the formation of the hybrid.

$4p$ __ __ __                   $4p$ __
                              $dsp^2$ __ __ __ __   (used in bonding)
$4s$ __

$3d$ ⇅ ⇅ ⇅ ↑ ↑          $3d$ ⇅ ⇅ ⇅ ⇅
$Ni^{2+}$ (ground state)      $Ni^{2+}$ (hybrid state)

In the $Ni(CN)_4^{2-}$ ion, each of the $dsp^2$ orbitals is occupied by a pair of electrons (from a $CN^-$ ion), and Hund's rule is not violated. The actual steps in the physical process by which the electrons form the bonding of the complex ion are not known; this discussion is meant only to show how the atomic orbitals of nickel could be imagined to produce the square planar bonding of the $Ni(CN)_4^{2-}$ ion. It is assumed that $d_{x^2-y^2}$, $s$, $p_x$, and $p_y$ orbitals are used in the formation of four equivalent orbitals (a $dsp^2$ set) that are coplanar and are directed at 90° angles from one another toward the corners of a square (compare Figure 4.7).

## 4.3 ELECTRON PAIR REPULSIONS AND MOLECULAR GEOMETRY

The orientation that a set of atoms assumes in the formation of a given molecule may be explained qualitatively on the basis of electrostatic repulsions between electron pairs. In the application of this concept, attention is directed to the valence level of the central atom of the molecule. Diagrams for the examples that follow show only the electrons of this level of the central atom.

Mercury has two electrons in its valence level ($6s^2$); these electrons are used to form two covalent bonds with two chlorine atoms in the mercury (II) chloride molecule, $HgCl_2$. The molecule is *linear:*

Cl:Hg:Cl

The $HgCl_2$ molecule assumes a configuration such that the two electron-pair bonds are as widely separated as possible, thus minimizing the electrostatic repulsion between them. Molecules of the type B:A:B are invariably linear. Berylium, zinc, cadmium, and mercury form molecules of this type.

The boron trifluoride molecule is *triangular* and *planar.*

F
··
·B·
F· ·F

Each of the bond angles of the molecule is 120°. This arrangement provides the greatest possible separation between the three electron pairs.

Tin(II) chloride vapor, $SnCl_2$, consists of angular molecules.

$$\overset{\displaystyle ..}{\underset{Cl \quad\quad Cl}{.\,Sn\,.}}$$

The tin atom has four electrons in its valence shell (Sn is a member of group IV A), and each chlorine atom supplies one electron to form a bond. These six electrons constitute three electron pairs (two bonding and one nonbonding); these pairs assume a triangular planar configuration because of electrostatic repulsion. The shape of a molecule is described in terms of the positions of its constituent atoms, not its electrons; $SnCl_2$ is therefore described as *angular*.

In the methane molecule, $CH_4$, the carbon atom has four bonding pairs of electrons. The Lewis structure, which does not depict the molecular geometry, is

$$\begin{array}{c} H \\ .. \\ H:C:H \\ .. \\ H \end{array}$$

The electrostatic repulsions between the bond pairs are at a minimum when the bonds are directed toward the corners of a regular *tetrahedron* (Figure 4.8). In this configuration, all of the bonds are equidistant from one another, and all H—C—H bond angles are 109° 28′. The tetrahedral configuration is a common and important one. Many molecules and ions (e.g., $ClO_4^-$, $SO_4^{2-}$, and $PO_4^{3-}$) are tetrahedral.

The structure of ammonia

$$\begin{array}{c} .. \\ H:N:H \\ .. \\ H \end{array}$$

can also be related to the tetrahedron (Figure 4.8). The nitrogen has three bonding electron pairs and one nonbonding electron pair. The four electron pairs assume a slightly distorted tetrahedral configuration that causes the atoms of the molecule to have a *trigonal pyramidal* arrangement. The unshared electron pair, which is under the influence of only one positive charge center, spreads out over a larger volume than a bonding pair, which is under the influence of two nuclei. As a result, the

**Figure 4.8** Geometries of methane ($CH_4$), ammonia ($NH_3$), and water ($H_2O$) molecules.

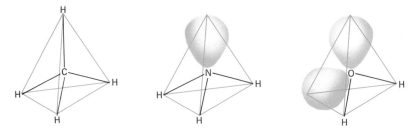

three bonds of the $NH_3$ molecule are forced slightly closer together than is normal for the tetrahedral arrangement. Thus, each of the H—N—H bond angles is 107° rather than the tetrahedral angle of 109°.

The water molecule has two bond pairs and two unshared pairs.

$$:\overset{..}{\underset{..}{O}}:H$$
$$\overset{..}{H}$$

The four electron pairs are arranged in an approximately tetrahedral manner (Figure 4.8) so that the atoms of the molecule have a *V-shaped* (*angular*) configuration. Since there are two nonbonding electron pairs in the molecule, the bonds are forced together even more than those of the $NH_3$ molecule. Hence, the H—O—H bond angle of $H_2O$ (105°) is less than the H—N—H bond angle of $NH_3$.

The central atoms of some ions and molecules have six electron pairs (bonding and nonbonding), and the geometries of these species are related to the octahedron. An example is the $SF_6$ molecule in which the central S atom has six bond pairs.

A S atom has six valence electrons; in $SF_6$ each of these electrons forms a covalent bond with an electron from a F atom. The molecule adopts the geometry of a regular *octahedron* (Figure 4.9a) in which the bonds are equidistant and all positions are equivalent. In this configuration, the electrostatic repulsions are minimized, and all bond angles are 90°.

The I atom of $IF_5$ has five bonding pairs and one nonbonding pair.

**Figure 4.9** Geometries of molecules and ions in which the central atom has six electron pairs.

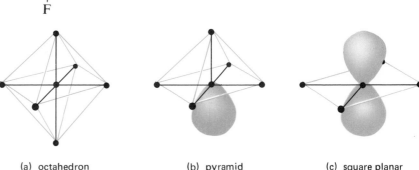

(a) octahedron     (b) pyramid     (c) square planar

An I atom has seven valence electrons; in $IF_5$, five of these electrons are engaged in bonding five F atoms, and the other two constitute a nonbonding pair. The electron pairs are directed to the corners of an octahedron. Since all the positions of an octahedron are equivalent, the atoms of the molecule form a *square pyramid* (Figure 4.9b), which may be slightly distorted.

The Br atom of the $BrF_4^-$ ion has four bonding pairs and two nonbonding pairs.

To account for the charge on the ion, we can consider that the central Br atom gains an electron to become a $Br^-$ ion with eight valence electrons. Four of these electrons are used to form four covalent bonds with F atoms (four bond pairs), and the other four electrons constitute two nonbonding pairs. Since the total number of electron pairs is six, the electron pairs of this ion also assume octahedral positions. The nonbonding pairs could conceivably assume a position with their axes at an angle of 90° or one with their axes at an angle of 180°. Since a nonbonding pair occupies a larger volume than a bonding pair, the 180° configuration minimizes electrostatic repulsion and is the one adopted (Figure 4.9c). The ion, therefore, has a *square planar* geometry.

The central atoms of some molecules and ions have five electron pairs (bonding and nonbonding). The configuration that minimizes electron-pair repulsions in cases of this type is the *trigonal bipyramid* (Figure 4.10a). This structure is assumed by the $PCl_5$ molecule in which the five valence electrons of a P atom form five bonding pairs with electrons from five Cl atoms. These five bonds, however, are not equivalent. The axial positions (numbers 1 and 3 in the diagram) are different from the equatorial positions (numbers 2, 4, and 5). Equatorial atoms are

**Figure 4.10** Geometries of molecules in which the central atom has five electron pairs.

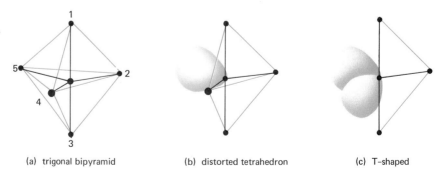

(a) trigonal bipyramid    (b) distorted tetrahedron    (c) T-shaped

coplanar; any bond angle formed by two equatorial atoms and the central atom is 120°. The axial atoms are on an axis that is at right angles to the equatorial plane; any bond angle defined by an axial atom, the central atom, and an equatorial atom is 90°. In addition, the P—Cl axial bond length (2.19 Å) is slightly longer than the P—Cl equatorial bond length (2.04 Å).

When one or more nonbonding electron pairs are substituted for bonding pairs in molecules with a trigonal bipyramidal geometry, the nonbonding pairs occupy equatorial positions. The charge concentration of an unshared pair is closer to the nucleus of the central atom than that of a shared pair, which is influenced by two nuclei. In a molecule containing five bond pairs (such as $PCl_5$ or $PF_5$), equatorial bonds are shorter than axial bonds; therefore, in molecules containing unshared pairs, the unshared pairs assume equatorial positions.

In $TeCl_4$, four of the six valence electrons of Te (group VI A) are used to form four bonding pairs; the other two electrons constitute a nonbonding pair. This nonbonding pair occupies an equatorial position, and a molecular configuration described as *distorted tetrahedral* (Figure 4.10b) results.

Chlorine has seven valence electrons (group VII A). In $ClF_3$, three of the seven electrons form bonding pairs with F atoms, and the remaining four make up two nonbonding pairs. Placement of these unshared pairs in equatorial positions produces a *T-shaped* molecule (Figure 4.10c).

The Xe atom of $XeF_2$ has two bonding pairs and three nonbonding pairs since six of the eight valence electrons of Xe remain unshared after two bonds have formed. These three nonbonding pairs occupy equatorial positions, and $XeF_2$ is therefore a *linear* molecule (compare Figures 4.10 and 11.2). The relationships between molecular shape and valence-shell electron pairs are summarized in Table 4.1.

The concept of electron pair repulsions can be extended to molecules and ions that contain double bonds. The double bond is counted as a unit; however, a double bond takes up more room than a single bond. For example, the molecule

has a *triangular planar* structure (similar to $BF_3$), but the presence of the double bond forces the Cl—C—Cl bond angle to contract from 120° (the angle typical of the triangular planar arrangement) to 111°.

Table 4.1 Number of electron pairs in the valence shell of the central atom and molecular shape

| TOTAL | BONDING | NONBONDING | SHAPE OF MOLECULE OR ION | EXAMPLES |
|-------|---------|------------|--------------------------|----------|
| 2 | 2 | 0 | linear | $HgCl_2$, $CuCl_2^-$ |
| 3 | 3 | 0 | triangular planar | $BF_3$, $HgCl_3^-$ |
| 3 | 2 | 1 | angular | $SnCl_2$, $NO_2^-$ |
| 4 | 4 | 0 | tetrahedral | $CH_4$, $BF_4^-$ |
| 4 | 3 | 1 | trigonal pyramidal | $NH_3$, $PF_3$ |
| 4 | 2 | 2 | angular | $H_2O$, $ICl_2^+$ |
| 5 | 5 | 0 | trigonal bipyramidal | $PCl_5$, $SnCl_5^-$ |
| 5 | 4 | 1 | distorted tetrahedral | $TeCl_4$, $IF_4^+$ |
| 5 | 3 | 2 | T shaped | $ClF_3$ |
| 5 | 2 | 3 | linear | $XeF_2$, $ICl_2^-$ |
| 6 | 6 | 0 | octahedral | $SF_6$, $PF_6^-$ |
| 6 | 5 | 1 | square pyramidal | $IF_5$, $SbF_5^{2-}$ |
| 6 | 4 | 2 | square planar | $BrF_4^-$, $XeF_4$ |

NUMBER OF ELECTRON PAIRS

Predictions may also be made for resonance hybrids. The nitrite ion, $NO_2^-$,

has an *angular* structure similar to $SnCl_2$.

**4.4 MOLECULAR ORBITALS**

Molecular structure has so far been depicted by forms that relate back to atomic orbitals that are centered on the constituent atoms. The **method of molecular orbitals** is an alternative approach in which orbitals are associated with the molecule as a whole and then electrons are entered into these molecular orbitals in an appropriate aufbau order. Corresponding to the practice of indicating atomic orbitals by the letters *s*, *p*, and so on, molecular orbitals are assigned the Greek letter designations σ (sigma), π (pi), and so on.

If two waves of the same wavelength ($\lambda$) and amplitude ($a$) are combined and are in phase, they reinforce each other (Figure 4.11a). The wavelength of the resultant wave remains the same, but the amplitude of the resultant wave is $a + a = 2a$. Two waves that are completely out of phase, however, cancel each other (Figure 4.11b); the "amplitude" of the resultant is $a + (-a) = 0$. Other combinations are, of course, possible; the waves do not have to be completely in phase or completely

**Figure 4.11** (a) Reinforcement of in-phase waves. (b) Cancellation of out-of-phase waves.

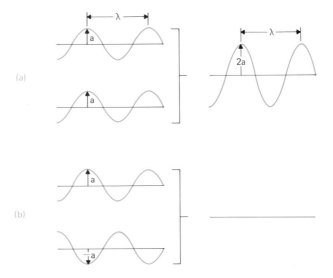

out of phase. The point is, however, that the combination of waves can be either *additive* or *subtractive*.

Molecular orbitals can be developed by assuming that they are **linear combinations of atomic orbitals.** The simplest example of molecular bonding is found in the hydrogen molecule ion, $H_2^+$, which consists of two protons (H nuclei) and one electron. Its role in the development of molecular orbital theory is analogous to the role of the hydrogen atom (the simplest atom) in the derivation of the theory of atomic orbitals. If the two hydrogen nuclei, A and B, are located at the proper internuclear distance for $H_2^+$, wave functions that pertain to the molecular orbitals of $H_2^+$, $\Psi$, can be derived by combining the wave function for an electron in a $1s$ orbital centered around nucleus A, $\psi_A$, and a similar wave function for an electron in a $1s$ orbital centered around nucleus B, $\psi_B$. Thus,

$$\Psi = c\psi_A + d\psi_B$$

where $c$ and $d$ are weighting factors that indicate the proportion of each atomic orbital in a given combination.

Since two atomic orbitals are used, equations for two molecular orbitals must be obtained. The two hydrogen nuclei are identical; in $H_2^+$, therefore, the electron density around nucleus A must be the same as that around nucleus B. As a result, $c$ must have the same *absolute value* as $d$ (the value without regard to sign, which we shall assume equals 1 in the discussion that follows), but in one equation $c = d$ and in the other $c = -d$. The first (additive) molecular orbital wave function may be imagined to result from the in-phase combination of waves

$$\Psi_\sigma = \psi_A + \psi_B$$

and the second (subtractive) from the out-of-phase combination of waves[1]

$$\Psi_{\sigma^*} = \psi_A - \psi_B$$

Just as in the case of an atomic orbital, $\Psi$ has no physical meaning, but the value of $\Psi^2$ at a given point in space is equal to the probability of finding the electron at that point. Hence, $\Psi^2$ may be considered to describe the electron density of the molecular orbital. For the molecular orbitals of $H_2^+$,

$$\Psi_\sigma^2 = (\psi_A + \psi_B)^2 = \psi_A^2 + \psi_B^2 + 2\psi_A\psi_B$$

$$\Psi_{\sigma^*}^2 = (\psi_A - \psi_B)^2 = \psi_A^2 + \psi_B^2 - 2\psi_A\psi_B$$

In each of these equations, the sum $\psi_A^2 + \psi_B^2$ can be thought of as representing the distribution of one electron around two nuclei that are not bonded together (i.e., infinitely separated). Since the term $2\psi_A\psi_B$ includes both $\psi_A$ and $\psi_B$, its value is appreciable only when the values of both $\psi_A$ and $\psi_B$ are appreciable, and such is the case only in the region between the nuclei. At points beyond this region, either $\psi_A$ or $\psi_B$ is negligible, and consequently the product $2\psi_A\psi_B$ is negligible.

In the equation pertaining to the first molecular orbital, the term $2\psi_A\psi_B$ is *added* to $\psi_A^2 + \psi_B^2$, which means that electron density is enhanced in the region between the nuclei. The attraction of the nuclei for this enhanced internuclear electronic charge holds $H_2^+$ together. This orbital is called a **sigma bonding orbital ($\sigma$)**.

In the equation for the second molecular orbital, the term $2\psi_A\psi_B$ is *subtracted* from $\psi_A^2 + \psi_B^2$, which indicates that internuclear electron density is depleted. In this case the nuclei repel each other since the low charge density in the region between the nuclei does little to counteract this repulsion. The net effect is disruptive. This orbital is called a **sigma antibonding orbital ($\sigma^*$)**.

These points are illustrated in Figure 4.12, in which electron density is plotted against distance along a line joining the nuclei of $H_2^+$. In (a), the electron density of the $\sigma$ orbital is compared with the hypothetical density

---

[1] In a rigorous derivation each of the molecular orbitals wave functions is normalized — each is derived so that its square ($\Psi^2$) when summed over all space adds up to a total electron probability density equal to one electron, no more nor less. If normalization is taken into account, the two sets of weighting factors are $c = d = N_b$ and $c = -d = N_a$. The equations then become

$$\Psi_\sigma = N_b(\psi_A + \psi_B) \qquad \text{and} \qquad \Psi_{\sigma^*} = N_a(\psi_A - \psi_B)$$

The factors $N_b$ (for the bonding molecular orbital) and $N_a$ (for the antibonding molecular orbital) are called normalizing factors. For this discussion the use of normalized functions is not necessary.

**Figure 4.12** Electron density against distance along an axis joining the two nuclei of $H_2^+$. Positions of the two nuclei, A and B, indicated. (a) A sigma bonding molecular orbital and one electron distributed between two nonbonded 1s atomic orbitals. (b) A sigma antibonding molecular orbital and one electron distributed between two nonbonded 1s atomic orbitals.

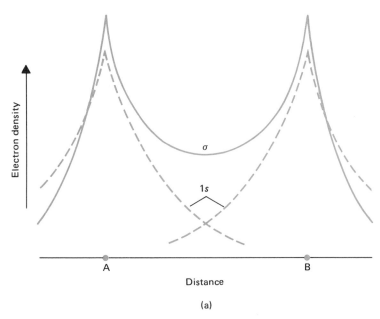

(a)

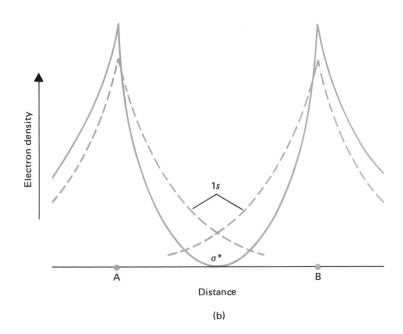

(b)

Figure 4.13 Formation of
σ and σ* molecular orbi-
tals from 1s atomic orbitals.

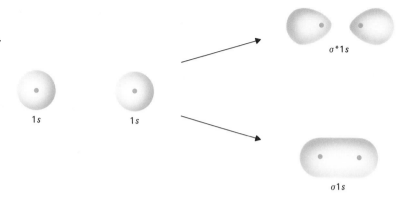

atomic        molecular      atomic
orbital       orbitals       orbitals

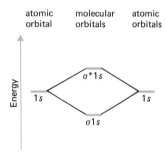

Figure 4.14 Energy-level
diagram for the formation
of σ and σ* molecular orbitals from the 1s orbitals
of two atoms.

of one electron distributed between two nonbonded atomic orbitals. In (b), the same comparison is made for the $\sigma^*$ orbital. The enhancement ($\sigma$) and depletion ($\sigma^*$) of internuclear electronic charge density is evident. Boundary surface diagrams for these molecular orbitals are shown in Figure 4.13; their formation can be imagined to be the result of in-phase ($\sigma$) and out-of-phase ($\sigma^*$) overlap of atomic orbitals. Sigma orbitals (both $\sigma$ and $\sigma^*$) are symmetrical about an axis joining the two nuclei; rotation of the molecule about this axis causes no observable change in orbital shape.

A qualitative energy-level diagram for the formation of $\sigma 1s$ and $\sigma^* 1s$ molecular orbitals from the 1s atomic orbitals of two atoms is shown in Figure 4.14. The energy of the $\sigma$ bonding orbital is lower than that of either atomic orbital from which it is derived, whereas the energy of the $\sigma^*$ antibonding orbital is higher. When two atomic orbitals are combined, the resulting bonding molecular orbital represents a decrease in energy, and the antibonding molecular orbital represents an increase in energy.

Any orbital can hold two electrons of opposed spin. The single electron of $H_2^+$ is placed in the $\sigma 1s$ orbital. In the hydrogen molecule, $H_2$, two electrons occupy the $\sigma 1s$ orbital. In both, the $\sigma^* 1s$ orbital is unoccupied. One-half the difference between the number of bonding electrons and the number of antibonding electrons gives the number of bonds in the molecule (the **bond order**). For $H_2^+$, $\frac{1}{2}(1-0) = \frac{1}{2}$ bond, whereas for $H_2$, $\frac{1}{2}(2-0) = 1$ bond.

If an attempt is made to combine two helium atoms, a total of four electrons must be placed in the two molecular orbitals that result from the combination of the two 1s orbitals of the He atoms. Since the $\sigma 1s$ orbital is filled with two electrons, the other two must be placed in the higher $\sigma^* 1s$ orbital. The number of bonds in $He_2$, then, would be $\frac{1}{2}(2-2) = 0$, and $He_2$ does not exist. However, the helium molecule ion, $He_2^+$ has been experimentally detected. Two of the three electrons of $He_2^+$ enter the $\sigma 1s$ orbital, and the other electron enters the $\sigma^* 1s$ orbital. Thus, the bond order of $He_2^+$ is $\frac{1}{2}(2-1) = \frac{1}{2}$.

The combination of two $2s$ orbitals produces $\sigma$ and $\sigma^*$ molecular orbitals similar to those formed from $1s$ orbitals, but the molecular orbitals derived from $2p$ atomic orbitals are slightly more complicated. The three $2p$ orbitals of an atom are directed along the Cartesian coordinates $x$, $y$, and $z$. If we consider that a diatomic molecule is formed by the atoms approaching each other along the $x$ axis, the $p_x$ atomic orbitals approach each other head on and overlap to produce $\sigma 2p$ bonding and $\sigma^* 2p$ antibonding molecular orbitals (Figure 4.15). All sigma orbitals are symmetrical about the internuclear axis; the $\sigma 2p$ and $\sigma^* 2p$ orbitals have shapes that are similar to those of the $\sigma 1s$ and $\sigma^* 1s$ orbitals.

In the formation of a diatomic molecule (Figure 4.15), the $p_z$ atomic

**Figure 4.15** Formation of molecular orbitals from $p$ atomic orbitals.

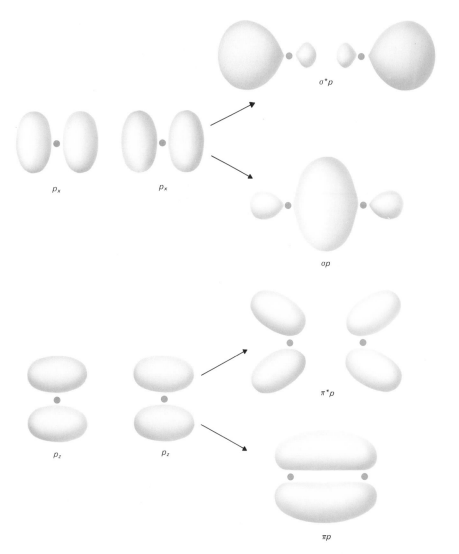

orbitals approach each other side to side and produce **pi ($\pi$) bonding and antibonding molecular orbitals.** Pi orbitals are not symmetrical about the internuclear axis. The electron density of a $\pi$ bonding orbital is zero in a plane that includes both nuclei of the molecule, but charge is concentrated in two regions that lie above and below this plane and between the two nuclei. The $\pi^*$ orbital reduces electronic density in the internuclear region.

The $p_y$ orbitals, which are not shown in Figure 4.15, also approach each other sideways. Consequently another set of $\pi$ and $\pi^*$ molecular orbitals, which lie at right angles to the set resulting from the overlap of $p_z$ orbitals, can be produced. The two $\pi 2p$ orbitals are **degenerate** (have equal energy), and the two $\pi^*2p$ orbitals are degenerate. Six molecular orbitals, therefore, arise from the two sets of $p$ orbitals — one $\sigma 2p$, one $\sigma^*2p$, two $\pi 2p$, and two $\pi^*2p$.

In what order are these molecular orbitals filled in the formation of a diatomic molecule? An aufbau order is based on orbital energies (Figure 4.16); the energy of a molecular orbital depends upon the energies of the atomic orbitals that compose it and upon the degree and the type of atomic-orbital overlap that occurs in its formation. Thus, both the $\sigma 2s$ and $\sigma^*2s$ molecular orbitals are lower in energy than any molecular orbital derived from $2p$ atomic orbitals because $2s$ atomic orbitals are lower in energy than $2p$ orbitals. The energy of the $\sigma 2p$ orbital should also be lower than the energy of either of the $\pi 2p$ orbitals (even though

**Figure 4.16** Energy-level diagram for the formation of molecular orbitals between identical atoms in which there is a large difference in energy between the $2s$ and $2p$ orbitals.

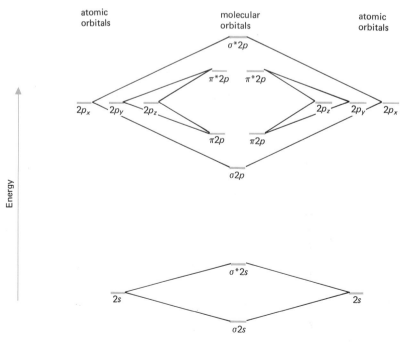

**Figure 4.17** Aufbau orders for homonuclear diatomic molecules of elements of the second period. (a) $O_2$ and $F_2$ (b) $Li_2$ to $N_2$.

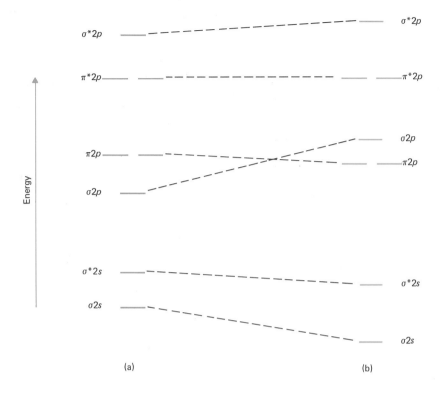

all the constituent $p$ orbitals have the same energy) because the $2p_x$ orbitals of the two atoms (the ones that form the $\sigma 2p$ orbital) overlap to a higher degree than the two $2p$ orbitals that form a $\pi 2p$ orbital ($p_y$ and $p_z$). In addition, the energy difference between a bonding orbital and the corresponding antibonding orbital is expected because of the difference in the type of atomic-orbital overlap. By use of reasoning such as this, the order shown in Figure 4.16 and Figure 4.17 (a) can be derived.

Of the homonuclear, diatomic molecules formed by elements of the second period, however, it is believed that this order is followed only by $O_2$ and $F_2$. In the derivation of this order, it was assumed that a $2s$ orbital of one atom would overlap *only* a $2s$ orbital of the other atom (in the formation of a $\sigma 2s$ molecular orbital). However, a $2s$ orbital of the first atom also overlaps a $2p_x$ orbital of the second (the type that forms $\sigma 2p$ molecular orbitals). This additional overlap, called $s$–$p$ interaction, strengthens the $\sigma 2s$ bond (lowers its energy) at the expense of the $\sigma 2p$ bond (the energy of which increases); see Figure 4.17 (b). Similarly, $s$–$p$ interaction causes the energy of the $\sigma *2s$ orbital to decrease and the energy of the $\sigma *2p$ to increase.

The importance of this $s$–$p$ interaction, which is always present, depends upon how close the energies of the $2s$ and $2p$ atomic orbitals are.

**Figure 4.18** Molecular orbital energy-level diagrams for $O_2$ and $F_2$.

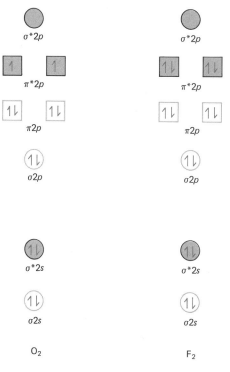

If they are far apart (as is the case for $O_2$ and $F_2$), this additional interaction causes no problem; the order of Figure 4.17 (a) is followed. On the other hand, if the energies of the $2s$ and $2p$ orbitals are close (as is the case for $Li_2$ to $N_2$), the added interaction will be large enough to cause the order of energies to shift to that shown in Figure 4.17 (b), in which the $\sigma 2p$ orbital is above the two $\pi 2p$ orbitals.

In Figure 4.18 molecular orbital energy-level diagrams for $O_2$ and $F_2$ are shown. The order shown is that of Figure 4.17 (a). Each diagram is obtained by successively placing the valence electrons of each molecule (twelve for $O_2$ and fourteen for $F_2$) into the lowest molecular orbitals available. The electrons of the inner shell ($1s$ electrons) do not appreciably affect the bonding.

For $F_2$ the number of bonds is $\frac{1}{2}(8-6)=1$; for $O_2$ the bond order is $\frac{1}{2}(8-4)=2$. The Lewis structure for $O_2$, $:\ddot{O}::\ddot{O}:$, depicts the double bond but does not show that oxygen is paramagnetic (has two unpaired electrons). The energy level diagram, however, shows two unpaired electrons for $O_2$ in the $\pi^*2p$ orbitals. Since the two $\pi^*2p$ orbitals have equal energies, the last two electrons to be added do not pair but enter the $\pi^*2p$ orbitals singly according to Hund's rule.

In Figure 4.19 energy level diagrams are given for $B_2$, $C_2$, and $N_2$, which utilize the aufbau order of Figure 4.17 (b). One confirmation of

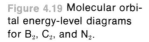

Figure 4.19 Molecular orbital energy-level diagrams for $B_2$, $C_2$, and $N_2$.

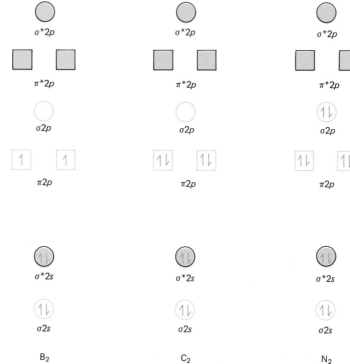

this order is afforded by measurement of the magnetic susceptibility of $B_2$. This molecule has two unpaired electrons (one in each of the two $\pi 2p$ orbitals) and is paramagnetic. Utilization of the aufbau order employed for $F_2$ would have given a diamagnetic $B_2$ molecule with the last two electrons paired in a $\sigma 2p$ orbital.

Table 4.2 Diatomic molecules of the elements of the second period

| MOLECULE | NUMBER OF ELECTRONS IN MOLECULAR ORBITALS | | | | | | BOND ORDER | BOND LENGTH (Å) | DISSOCI- ATION ENERGY (eV) | UNPAIRED ELECTRONS |
|---|---|---|---|---|---|---|---|---|---|---|
|  | $\sigma 2s$ | $\sigma^* 2s$ | $\pi 2p$ | $\sigma 2p$ | $\pi^* 2p$ | $\sigma^* 2p$ |  |  |  |  |
| [a]$Li_2$ | 2 |  |  |  |  |  | 1 | 2.67 | 1.1 | 0 |
| [b]$Be_2$ | 2 | 2 |  |  |  |  | 0 | — | — | 0 |
| [a]$B_2$ | 2 | 2 | 2 |  |  |  | 1 | 1.59 | 3.0 | 2 |
| [a]$C_2$ | 2 | 2 | 4 |  |  |  | 2 | 1.31 | 6.5 | 0 |
| $N_2$ | 2 | 2 | 4 | 2 |  |  | 3 | 1.09 | 9.8 | 0 |
|  | $\sigma 2s$ | $\sigma^* 2s$ | $\sigma 2p$ | $\pi 2p$ | $\pi^* 2p$ | $\sigma^* 2p$ |  |  |  |  |
| $O_2$ | 2 | 2 | 2 | 4 | 2 |  | 2 | 1.21 | 5.1 | 2 |
| $F_2$ | 2 | 2 | 2 | 4 | 4 |  | 1 | 1.43 | 1.6 | 0 |
| [b]$Ne_2$ | 2 | 2 | 2 | 4 | 4 | 2 | 0 | — | — | 0 |

[a] Exists only in the vapor state at elevated temperatures.
[b] Does not exist.

A summary of the diatomic molecules of the second period elements is given in Table 4.2. As the number of bonds increases, the bond distance shortens and the bonding becomes stronger. The molecule bonded most strongly is $N_2$, which is held together by a triple bond (Figure 4.19). Molecules for which this method of molecular orbitals assigns a bond order of zero ($Be_2$ and $Ne_2$) do not exist.

In heteronuclear, diatomic molecules the clouds of the bonding orbitals are distorted so that more electronic charge is located around the more electronegative atom than around the other atom; antibonding orbitals have their largest electron density in the regions close to the less electronegative atom. In the construction of the wave functions for molecular orbitals of molecules of this type,

$$\Psi_{MO} = c\psi_A + d\psi_B$$

the absolute values of the weighting factors, $c$ and $d$, are not equal.

For molecules such as CO and NO, the same types of molecular orbitals, although slightly distorted, are formed as those that occur in homonuclear diatomic molecules. Either aufbau order may be used in most cases with the same qualitative results; the actual order, however, is uncertain.

Since CO is isoelectronic with $N_2$ (each molecule has ten valence electrons), the molecular orbital energy-level diagram for CO is similar to that for $N_2$ (Figure 4.19). Carbon monoxide, therefore, has a bond order of 3. The dissociation energy of the CO molecule is about the same as that of the $N_2$ molecule.

We have said that it is impossible to diagram a valence-bond structure utilizing the octet principle for a molecule with an odd number of valence electrons. Nitrogen oxide, NO, is such a molecule. Since five valence electrons are contributed by the N atom and six valence electrons by the O atom, the total number of valence electrons is eleven. An energy-level diagram for the molecular orbitals of NO is given in Figure 4.20. Since there are eight bonding electrons and three antibonding electrons shown in the diagram, a bond order of $\frac{1}{2}(8 - 3)$, or $2\frac{1}{2}$, is indicated. Nitrogen oxide is paramagnetic; the NO molecule has one unpaired electron in a $\pi*2p$ orbital.

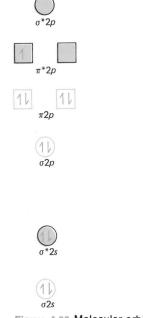

$\sigma*2p$

$\pi*2p$

$\pi2p$

$\sigma2p$

$\sigma*2s$

$\sigma2s$

**Figure 4.20** Molecular orbital energy-level diagram for NO.

**Figure 4.21** Rotation about the C—C bond in ethane.

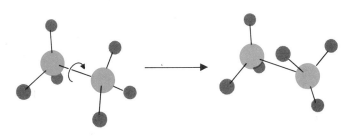

Molecular orbitals can be derived for molecules that contain more than two atoms, such as $H_2O$ and $NH_3$. In each case the number of molecular orbitals derived equals the number of atomic orbitals used, and the molecular orbitals encompass the whole molecule. For many purposes, however, it is convenient to think in terms of molecular orbitals that are localized between adjacent atoms. Consider the series

$$\begin{array}{ccc} \underset{\underset{\displaystyle H\ \ \ H}{\textstyle |\ \ \ \ |}}{\overset{\overset{\displaystyle H\ \ \ H}{\textstyle |\ \ \ \ |}}{H-C-C-H}} & \overset{H}{\underset{H}{}}C=C\overset{H}{\underset{H}{}} & H-C\equiv C-H \\ \text{ethane} & \text{ethylene} & \text{acetylene} \end{array}$$

Each C atom in ethane utilizes $sp^3$ hybrid orbitals in the formation of $\sigma$ bonds with the other C atom and the three H atoms (Figure 4.21). Thus, all bond angles are 109.5°, the tetrahedral angle. Since $\sigma$ bonding orbitals are symmetrical about the internuclear axis, free rotation about each bond is possible. Rotation about the C—C bond causes a changing atomic configuration (Figure 4.21).

There can only be one $\sigma$ bond between any two bonded atoms; any additional bonds must be $\pi$ bonds. The $\sigma$ bonding skeleton of ethylene, therefore, is

$$\overset{H}{\underset{H}{}}C-C\overset{H}{\underset{H}{}}$$

The C atoms utilize $sp^2$ hybrid orbitals to form this skeleton. This type of hybridization leads to a planar configuration of hybrid orbitals in which the orbitals form angles of 120° with each other. The ethylene molecule is planar with H—C—H bond angles of 118° and H—C—C bond angles of 121° (Figure 4.22).

**Figure 4.22** Geometric configuration of ethylene. Shapes of the $p$ orbitals that overlap to form a $\pi$ bond are simplified.

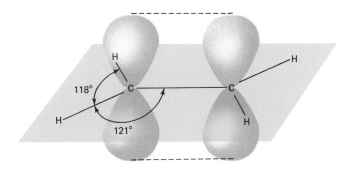

**Figure 4.23** Formation of the π bonds of acetylene. Orbital shapes simplified.

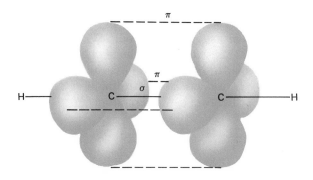

Each C atom of ethylene has one electron in a $p_z$ orbital that is not involved in $\sigma$ bond formation. The $p_z$ orbitals overlap to produce a $\pi$ bonding orbital. The electron density of the $\pi$ bond is above and below the plane of the molecule; free rotation about the C=C linkage is impossible without breaking this $\pi$ bond.

The bonding skeleton of acetylene is

H—C—C—H

Each C atom forms two $\sigma$ bonds, and *sp,* or linear, hybrid orbitals of C are employed in the formation of these $\sigma$ bonds. The $2p$ carbon orbitals that are not involved in the hybridization and $\sigma$ bond formation overlap to form two $\pi$ bonding molecular orbitals (Figure 4.23). The structure is much like that of $N_2$.

The multiple bonds of ethylene and acetylene are localized between two nuclei. In some molecules and ions **multicenter** (or **delocalized**) **bonding** exists in which some bonding electrons bond more than two atoms. The description of these species by the valence bond approach requires the use of resonance structures.

An example of delocalized bonding is found in the bonding in the carbonate ion (Figure 4.24). The ion is triangular planar, and each O—C—O bond angle is 120°. The C atom may be assumed to use $sp^2$ hybrid orbitals to form the $\sigma$ bonding skeleton; one $2p$ orbital remains unhybridized. This unhybridized $p$ orbital of C overlaps similar $p$ orbitals of the O atoms to form a $\pi$ bonding *system* that encompasses all the atoms of the ion.

The $CO_3^{2-}$ ion has a total of twenty-four valence electrons. Six are used in the formation of the three $\sigma$ bonds of the ion; placement of two pairs of unshared electrons on each O atom accounts for twelve more. The remaining six electrons are placed in the molecular orbitals of the $\pi$ bonding system. It must be understood that a molecular orbital, like an atomic orbital, can hold only two electrons. A $\pi$ bonding system, therefore, involves two or more contiguous molecular orbitals. The structures of sulfur trioxide, $SO_3$, and the nitrate ion, $NO_3^-$, are similar to the struc-

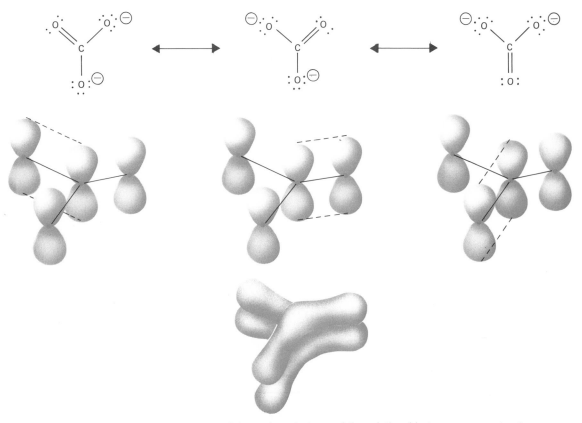

**Figure 4.24** Multicenter $\pi$ bonding system of the carbonate ion and the relationship to resonance structures.

ture of the carbonate ion, and the resonance forms of these species are also similar.

The S atom of the sulfur dioxide molecule, $SO_2$, may be assumed to employ $sp^2$ (triangular planar) hybrid orbitals in the formation of its $\sigma$ bonding framework; an unshared electron pair occupys one of these hybrid orbitals. The $\pi$ bonding system of this molecule is shown in Figure 4.25; ozone, $O_3$, and the nitrite ion, $NO_2^-$, have similar structures.

**4.6 INTERMOLECULAR FORCES OF ATTRACTION**

Atoms are held together in molecules by covalent bonds, but what forces attract molecules to each other in the liquid and solid states? Two types of intermolecular attractive forces are described in this section.

**Dipole–dipole** forces occur between polar molecules; such molecules have permanent dipoles and tend to line up in an electric field (Section 3.8). The attraction of the negative pole of one molecule for the positive pole of another molecule constitutes this type of intermolecular inter-

**Figure 4.25** Relation of the
π bonding system of sulfur
dioxide to the resonance
forms of the molecule.
Orbital shapes simplified.

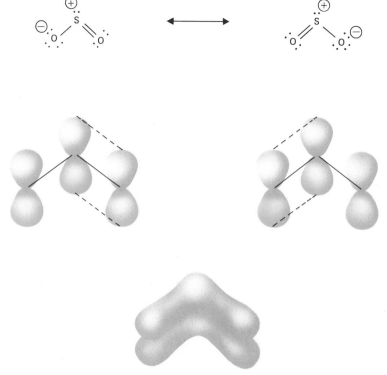

**Figure 4.26** Orientation of
polar molecules in a crystal.

action. In a crystal of a polar molecular substance, the molecules are
oriented in a way that reflects the dipole–dipole forces (Figure 4.26).

Electronegativities can be used to predict the degree of polarity of
a diatomic molecule as well as the orientation of the positive and negative
poles. Predictions concerning the polarity of a molecule that contains
more than two atoms, however, must be based upon a knowledge of the
geometry of the molecule *and the arrangement of unshared electron
pairs.*

Consider the three molecules ($CH_4$, $NH_3$, and $H_2O$) represented in
Figure 4.27. The dipole moment of a molecule is the result of the individual dipoles ascribed to the bonds and to the unshared electron
pairs of the molecule. In each of the molecules under consideration, the
central atom is more electronegative than the H atoms bonded to it;
the negative end of each *bond* dipole, therefore, points toward the
central atom. In the case of $CH_4$ the tetrahedral arrangement of the
four polar C—H bonds produces a molecule that is not polar; $CH_4$ has
no dipole moment. The center of positive charge of the *molecule* (derived
by considering all four bonds) falls in the center of the C atom and
coincides with the center of negative charge for the molecule.

On the other hand, the trigonal pyramidal $NH_3$ molecule is polar

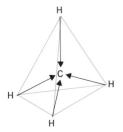

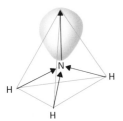

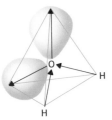

Figure 4.27 Analysis of the polarities of the methane ($CH_4$), ammonia ($NH_3$), and water ($H_2O$) molecules. Arrows point toward the negative end of the individual dipoles that compose the dipole moment of the molecule.

($\mu = 1.49$ D). The three polar bonds and the unshared electron pair are arranged so that the molecule has a dipole with the negative end directed toward the apex of the trigonal pyramid and the positive end toward the base. Similarly, the angular $H_2O$ molecule is polar ($\mu = 1.85$ D); the polar bonds and the unshared electron pairs contribute to a dipole with the negative end directed toward the O atom and the positive end directed toward a point halfway between the two H atoms.

The influence that an unshared pair of electrons has on the dipole moment of a molecule is seen in the case of $NF_3$. The $NF_3$ molecule has a structure similar to that of $NH_3$ (Figure 4.27), but the direction of the polarity of the bonds is the reverse of that in $NH_3$ (since F is more electronegative than N). Nitrogen trifluoride has a dipole moment of 0.24 D, a surprisingly low value in view of the highly polar nature of the N—F bonds. The N—F bond dipoles combine to give the molecule a dipole with the negative end in the direction of the base of the pyramid, but the contribution of the unshared electron pair works in the opposite direction and reduces the total polarity of the molecule.

What intermolecular forces attract *nonpolar* molecules to each other in the liquid and solid state? Such molecules do not have permanent dipoles, but nevertheless, all nonpolar substances can be liquefied. Some type of intermolecular force, therefore, must exist in addition to the dipole–dipole force.

The existence of **London forces (dispersion forces)** is postulated.[2] These forces are thought to arise from the motion of electrons. At one instant of time, the electron cloud of a molecule may be distorted so that an instantaneous dipole is produced in which one part of the molecule is slightly more negative than the rest. At the next instant, the orientation of the instantaneous dipole will be different because the electrons have moved. Over a period of time (a very short period of time—electrons move rapidly), the effects of these instantaneous dipoles cancel so that a nonpolar molecule has no permanent dipole moment.

The instantaneous, fluctuating dipoles of a molecule, however, induce matching dipoles in neighboring molecules (aligned in the same way that permanent dipoles are aligned), and the motion of the electrons of neighboring molecules is synchronized. The force of attraction between these mutually-induced, instantaneous dipoles constitutes the London force. The strongest London forces occur between large, complex molecules, which have large electron clouds that are easily distorted, or polarized;

[2] Johannes van der Waals postulated the existence of intermolecular attractive forces between gas molecules in 1873 (section 6.13). The explanation of the origin of this type of intermolecular force was proposed by Fritz London in 1930. Although there is a lack of uniformity in the use of terms, current usage appears to favor calling these specific forces *London forces* and intermolecular forces in general *van der Waals forces*.

these substances have comparatively high melting points and boiling points.

Since all molecules contain electrons, London forces also exist between polar molecules; in the case of nonpolar molecular substances, London forces are the *only* intermolecular forces that exist. The values listed in Table 4.3 show that London forces are the principal intermolecular forces for most molecular substances. The hydrogen bond, a special type of dipole–dipole interaction that is discussed in Section 8.9, is responsible for the magnitude of the dipole–dipole energy listed for $H_2O$, $NH_3$, and (to a lesser extent) HCl.

**Table 4.3** Intermolecular attractive energies in some simple molecular crystals

| | | ATTRACTIVE ENERGIES (kcal/mol) | |
| MOLECULE | DIPOLE MOMENT, D | DIPOLE–DIPOLE | LONDON |
| --- | --- | --- | --- |
| Ar | 0 | 0 | 2.03 |
| CO | 0.12 | 0.0001 | 2.09 |
| HI | 0.38 | 0.006 | 6.18 |
| HBr | 0.78 | 0.164 | 5.24 |
| HCl | 1.03 | 0.79 | 4.02 |
| $NH_3$ | 1.49 | 3.18 | 3.52 |
| $H_2O$ | 1.85 | 8.69 | 2.15 |

**4.7 THE METALLIC BOND**   The most characteristic chemical and physical properties of metals result from their low ionization potentials and low electronegativities. The outer electrons of metal atoms are relatively loosely held (see Table 3.2), and move freely throughout a metallic crystal. The remainder of the metal atoms, positive metal ions, occupy fixed lattice positions in the crystal. The negative cloud of the freely moving electrons binds the crystal together.

The wave functions of the atomic orbitals of the electrons used in the metallic bonding can be combined into wave functions for molecular orbitals that extend over the entire crystal. The mathematical process is complicated because a large number of atomic orbitals must be combined (from each of the atoms of the crystal), and an equal number of molecular orbitals must be obtained.

If only two Li atoms are brought together (in the formation of $Li_2$— Section 4.4), the $2s$ orbitals of the two atoms form $\sigma 2s$ and $\sigma^* 2s$ molecular orbitals (only the $\sigma 2s$ orbital is occupied in the ground state of $Li_2$). The $2s$ orbitals of two widely separated Li atoms are degenerate (have equal energy); as the atoms are brought together, however, this degeneracy is split in the formation of the $\sigma 2s$ molecular orbital (lower

Figure 4.28 Energy of the *2s* orbitals of two lithium atoms as a function of interatomic distance.

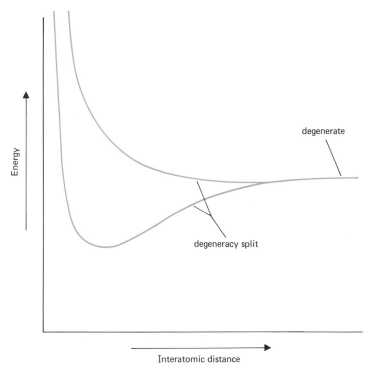

curve of Figure 4.28) and the $\sigma*2s$ molecular orbital (upper curve of Figure 4.28).

If seven lithium atoms are brought together, the degeneracy of the orbitals of the isolated atoms is split into seven levels (Figure 4.29). Each atom contributes one level to the aggregate, which is called a **band** (Figure 4.29b). In a large number ($N$) of isolated lithium atoms, the $2s$ orbitals have equal energy; this degeneracy is removed, however, as the atoms are brought together, giving rise to $N$ levels of slightly different energy. Each level of this band may accommodate two electrons of opposed spin.

If a fairly large number of atoms is considered, the spread between the levels is not significantly widened by an increased number of atoms. When interatomic spacing is uniform, orbital overlap is not significantly altered when the number of atoms is increased; this orbital interaction produces the effect on the energy levels. The gradual addition of atoms causes the increasing number of levels of the band to be more and more crowded together but does not affect the width of the band. As a result, the separation in energy between the levels of a band is very small, even for samples of extremely small size. Even at very low temperatures the electrons have enough energy to move from level to level within a

Figure 4.29 (a) Energy of $2s$ orbitals versus interatomic distance for a collection of seven atoms. (b) Energy-level diagram corresponding to the interatomic distance $R_0$.

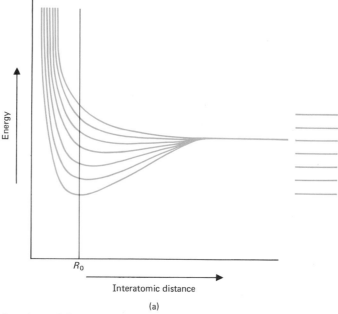

Energy

$R_0$

Interatomic distance

(a)

band, and for all practical purposes the band represents a continuum of energy.

Each level of a band may accommodate two electrons of opposed spin. If $N$ atoms contribute $N$ levels to a band, the band is capable of holding $2N$ electrons. Since $N$ lithium atoms have a total of $N$ valence electrons, it would appear that the band should be one-half occupied; however, the band derived from the $2p$ orbitals of lithium must also be considered.

In an isolated lithium atom the three $2p$ orbitals are only slightly higher in energy than the $2s$ orbital. The band derived from the $2p$ orbitals fans out and overlaps the band from the $2s$ orbitals (Figure 4.28). At the normal equilibrium interatomic distance, $R_0$, the two bands may be considered to form one. Since each atom contributes three $2p$ orbitals and one $2s$ orbital, the band resulting from the combination of $N$ atoms would consist of $4N$ levels that are capable of holding $8N$ electrons. The lithium band, therefore, is only one-eighth filled.

If this band overlap did not occur, the valence band of beryllium (electronic configuration, $1s^2 2s^2$) would be full, and beryllium would not conduct electricity. Beryllium is a metallic conductor because the band resulting from the overlap of the $2s$ and $2p$ orbitals is only one-quarter filled.

Metallic bonding generally arises from the interaction of the orbitals of the valence shell only. The interatomic spacing of a metallic crystal is determined by the overlap of the valence orbitals; the orbitals of the inner shells do not interact appreciably at this distance (Figure 4.30).

Figure 4.30 Overlap of 2s and 2p bands.

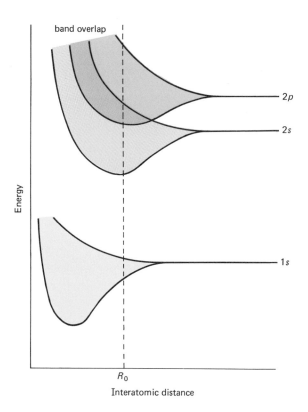

In a metallic crystal, therefore, the valence electrons are held by all the atoms of the crystal and are highly mobile. The valence band, which may be only partially filled, is overlapped by an unfilled conduction band. Transference of an electron to a higher level within a band requires the addition of very little energy since the levels are close together. Thus the valence electrons of a metal can move to higher levels by absorbing light of a wide range of wavelengths. When these electrons fall to lower energy levels, light is radiated. The lustrous appearance of metals is caused by these electron transitions.

The high degree of electron mobility of metallic crystals accounts for the high thermal and electrical conductivities of metals. Valence electrons of a metal absorb heat as kinetic energy and transfer it rapidly to all parts of the metal since their motion is relatively unrestricted. Solids in which electrons are localized, however, exhibit low thermal conductivities since conduction of heat can take place in these materials only through the motion of ions or molecules — a much slower process.

Energy-level diagrams for **conductors, insulators,** and **intrinsic semiconductors** are shown in Figure 4.31. In the diagram for a conductor (such as Li), the valence band (from the 2s orbitals) is overlapped by

Figure 4.31 Energy-level band diagrams for three types of solids.

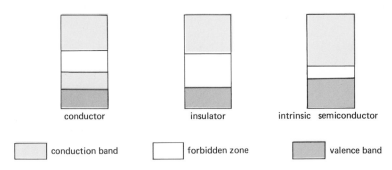

an empty conduction band (from the $2p$ orbitals); these two bands are separated from an empty, upper conduction band (derived from orbitals of the third shell) by a **forbidden energy zone.** Electrical conduction takes place by means of the motion of electrons within the lower conduction band; there is no need to supply the energy required to bridge the forbidden zone and use the upper conduction band.

The diagram for the insulator, however, shows a completely filled valence band that is widely separated from a conduction band by a forbidden energy zone. Electron motion, and hence electrical conductivity, is only possible if energy is provided to promote electrons across the comparatively large forbidden zone to the conduction band. Normally such promotion does not occur, and hence the conductivities of insulators are extremely low.

An intrinsic semiconductor is a material that has a low electrical conductivity that is intermediate between that of a conductor and an insulator; this conductivity increases markedly with increasing tempera-

Figure 4.32 Effect of deformation on (a) a metallic crystal and (b) an ionic crystal.

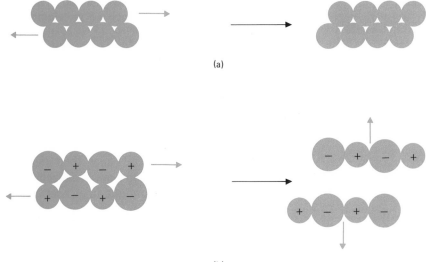

ture. For a semiconductor, the forbidden zone is sufficiently narrow that electrons can be promoted from the valence band to the conduction band by thermal excitation (see Figure 4.31). The vacancies left by the removal of electrons from the valence band permit the electrons remaining in the valence band to move under the influence of an electric field; conduction takes place through the motion of electrons in the valence, as well as in the conduction, band.

The electrical conductivity of a metal is not dependent upon thermal excitation of electrons. Whereas the conductivity of an intrinsic semiconductor increases with increasing temperature, the electrical conductivity of a metal decreases with increasing temperature. Presumably the increased temperature causes increased vibration of the metal ions of the crystal lattice which, in turn, impedes the flow of conduction electrons. Other properties of semiconductors are discussed in Section 7.14.

## 4.8 PROPERTIES AND STRUCTURE

The properties of ionic compounds are best understood on the basis of the structure of the ionic crystal (Section 3.4, Figure 3.5). The electrostatic forces between the ions in such a crystal are moderately strong forces. Thus ionic crystals are hard. In the process of melting, thermal energy must be supplied to overcome these forces sufficiently for the crystal to liquefy, and therefore ionic substances have relatively high melting points. The boiling points of ionic materials are high since the vaporization of ionic liquids is opposed by the strong electrostatic attractions between ions.

The ions in a crystal are held in a definite geometric pattern. The structure is such that the electrostatic attractions between positive and negative ions more than compensate for the repulsions that exist between ions of like charge. Thus, **ionic crystals** are brittle because the stability of the crystals depends upon the preservation of their geometric patterns (Figure 4.32). Ionic compounds are good conductors of electricity when molten or in solution (but not in the crystalline state where the ions are not free to move); also, ionic materials are very soluble in polar solvents (Chapter 9).

Molecules occupy lattice points in crystals of covalent compounds. The intermolecular forces that hold the molecules in the crystal lattice are not nearly so strong as the electrostatic forces encountered in ionic crystals. Hence, **molecular crystals** are soft and have low melting points; the liquids derived from them have low boiling points.

London forces hold nonpolar molecules in the lattice; in crystals of polar molecules, London forces are augmented by dipole-dipole forces. Therefore, polar compounds generally melt and boil at slightly higher temperatures than nonpolar compounds *of comparable molecular size and shape*. We must interpret this generalization carefully. For most compounds the London forces are stronger than the dipole-dipole forces. Even though HCl is a much more polar molecule than HI (and the dipole–dipole forces of HCl are stronger than those of HI), the boiling point of HCl ($-85\,°C$) is lower than that of HI ($-35\,°C$) because the London forces between the HCl molecules are weaker than those between the comparatively large HI molecules (see Table 4.3).

In general, liquids of covalent substances do not conduct electricity. Nonpolar molecules lack charges to respond to an electric field, and in polar molecules both the positive and negative charges of the dipole are contained in the same particle (causing it to be equally attracted toward the two poles of the electric field). However, some polar molecules in the liquid state dissociate to a slight extent, producing low concentrations of ions (Chapter 16); these particular substances are very poor electrical conductors. Liquids derived from polar molecular substances are the best solvents for ionic compounds.

**Network** (also called **atomic**) **crystals** are crystals in which the lattice

Table 4.4 Types of crystalline solids

| CRYSTAL | PARTICLES | ATTRACTIVE FORCES | PROPERTIES | EXAMPLES |
|---|---|---|---|---|
| ionic | positive and negative ions | electrostatic attractions | high m.p. hard, brittle good electrical conductor in fused state | $NaCl$, $BaO$, $KNO_3$ |
| molecular | polar molecules | London and dipole–dipole | low m.p. soft nonconductor or extremely poor conductor of electricity in liquid state | $H_2O$, $NH_3$, $SO_2$ |
| | nonpolar molecules | London | | $H_2$, $Cl_2$, $CH_4$ |
| network | atoms | covalent bonds | very high m.p. very hard nonconductor of electricity | C (diamond) SiC, AlN, $SiO_2$ |
| metallic | positive ions and mobile electrons | metallic bonds | fairly high m.p. hard or soft malleable and ductile good electrical conductor | Ag, Cu, Na, Fe, K |

points are occupied by atoms that are joined by a network of covalent bonds. It is impossible to distinguish molecules or ions; in fact, the entire crystal may be visualized as one giant molecule. An example of such a crystal is the diamond in which carbon atoms are joined by covalent bonds in a three-dimensional structure (Figure 4.33). Materials of this type have high melting points, high boiling points, low volatilities, and are extremely hard because of the large number of covalent bonds that would have to be ruptured to destroy the crystal structure. They are also nonconductors of electricity.

The **metallic crystal** has been described in the preceding section. The metallic bond is strong as evidenced by the close-packed arrangement of the positive ions in the crystal lattice and the high densities and high melting points of most metals (but not all metals—there are other factors to be considered, such as atomic radius and number of vacant orbitals). Unlike ionic crystals, the positions of the positive ions can be altered without destroying the crystal because of the uniform charge distribution provided by the freely moving electrons (Figure 4.32). Thus, metallic crystals are usually easily deformed, and most metals are malleable (capable of being pounded into shapes) and ductile (capable of being drawn into wire). The freely moving electrons are responsible for the high thermal and electrical conductivities of metals. The properties of the various types of crystals are summarized in Table 4.4.

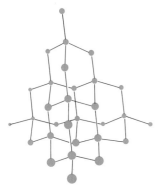

**Figure 4.33** Arrangement of atoms in a diamond crystal.

## Problems

**4.1** Diagram the resonance forms for the linear cyanate ion, $OCN^-$.

**4.2** In the $O_2NF$ molecule, the N atom is the central atom to which the others are bonded. Diagram the resonance forms of $O_2NF$. What is the shape of the molecule?

**4.3** Diagram the resonance forms of the bicarbonate ion, $HCO_3^-$. The C atom is the central atom to which two O atoms and one —O—H group are bonded. What is the shape of the ion?

**4.4** Diagram the resonance forms of nitramide, $H_2NNO_2$ (include

formal charges). Is a resonance form with a double bond between the two N atoms a valid contributor to the resonance hybrid?

**4.5** In dinitrogen tetroxide, $N_2O_4$, the two N atoms are bonded together and each N atom has two O atoms bonded to it. The structure is symmetrical and all N—O bond distances are equal. Diagram the resonance forms of $N_2O_4$.

**4.6** In the hyponitrite ion, $N_2O_2^{2-}$, the two N atoms are bonded to each other by a double bond and each N atom is bonded to an O atom by a single bond. Draw the

Lewis structure (complete with formal charges) for $N_2O_2^{2-}$ and predict the shape of the ion by using electron-pair repulsion theory. A structure in which one of the O—N linkages is a double bond and the other bonds of the molecule (the N—N and other N—O) are single does *not* contribute to a resonance hybrid even though it conforms to the octet principle. Why not?

**4.7** The structure of borazine, $B_3N_3H_6$, consists of a six-membered ring, composed of alternating B and N atoms, with a H atom bonded to each atom of the ring. Draw the resonance forms of borazine.

**4.8** For each of the following molecules, indicate the type of hybrid orbitals employed by the central atom and state the geometric configuration of the molecule. (a) $CdBr_2$, (b) $GeF_2$, (c) $SCl_2$, (d) $PCl_3$, (e) $GaI_3$, (f) $SnH_4$, (g) $SbCl_5$, (h) $SeF_6$.

**4.9** Interpret the configurations of the molecules given in Problem 4.8 on the basis of electron-pair repulsions. Which of the molecules would have dipole moments?

**4.10** The following species have been identified in aqueous solutions of mercury(II) chloride: $Hg^{2+}$, $HgCl^+$, $HgCl_2$, $HgCl_3^-$, and $HgCl_4^{2-}$. What type of hybrid orbitals are employed and what is the geometric configuration of $HgCl^+$, $HgCl_2$, $HgCl_3^-$, and $HgCl_4^{2-}$?

**4.11** The following molecules and ions of antimony are known: $SbCl_3$, $SbCl_4^-$, $SbCl_5^{2-}$, $SbCl_5$, and $SbCl_6^-$.

(a) On the basis of electron-pair repulsions, predict the shape of each of these species. (b) What type of hybrid orbitals does Sb employ in the bonding of each of these?

**4.12** Draw boundary surface diagrams to show how the component atomic orbitals are hybridized into a $dsp^2$ set.

**4.13** Use the concept of electron-pair repulsions to predict the geometric configuration of (a) $TlCl_2^+$, (b) $AsF_2^+$, (c) $IBr_2^-$, (d) $SnCl_3^-$, (e) $ClF_2^+$, (f) $ICl_3$, (g) $ClF_4^-$.

**4.14** Use the concept of electron-pair repulsions to predict the geometric configuration of (a) $SiF_5^-$, (b) $SbF_5^{2-}$, (c) $SeF_4$, (d) $NH_4^+$, (e) $NH_2^-$, (f) $SnCl_6^{2-}$, (g) $TlBr_4^-$.

**4.15** Use the concept of electron-pair repulsions to predict the geometric configuration of (a) $BeCl_2$, (b) $BeCl_4^{2-}$, (c) $BeF_3^-$, (d) $SF_3^+$, (e) $TeF_5^-$, (f) $AsF_4^-$, (g) $SiF_6^{2-}$.

**4.16** Use the concept of electron-pair repulsions to predict the geometric configuration of (a) $HgI_4^{2-}$, (b) $SeBr_2$, (c) $AsF_5$, (d) $BrF_3$, (e) $ClF_2^-$, (f) $BrF_5$, (g) $ICl_4^-$.

**4.17** Draw Lewis structures for the following and predict the geometric shape of each on the basis of electron-pair repulsions. (a) $SO_4^{2-}$, (b) $SO_3^{2-}$, (c) $NO_3^-$, (d) $ClO_2^-$, (e) $CO_2$, (f) $H_2CO$ (C is the central atom).

**4.18** In electron-pair repulsion theory, the effective volume of a bond pair is assumed to decrease with increasing electronegativity of the atom bonded to the central atom. In the light of this generalization, would you expect the Cl atoms of the trigonal bipyramidal $PCl_2F_3$ to occupy axial or equatorial positions?

**4.19** In view of the opening statement of Problem 4.18, predict which of the $PX_3$ molecules (where $X = F$, Br, Cl, or I) would have the smallest X—P—X bond angles.

**4.20** By means of molecular orbital energy-level diagrams, describe the bonding of the homonuclear diatomic molecules of the first ten elements.

**4.21** Oxygen forms compounds containing the dioxygenyl ion, $O_2^+$ (e.g., $O_2PtF_6$); the superoxide ion, $O_2^-$ (e.g., $KO_2$); and the peroxide ion, $O_2^{2-}$ (e.g., $Na_2O_2$). (a) Draw molecular orbital energy-level diagrams for $O_2$, $O_2^+$, $O_2^-$, and $O_2^{2-}$. (b) State the bond order for each species. (c) Which of the four are paramagnetic?

**4.22** The bond distances for $N_2$, $N_2^+$, $O_2$, and $O_2^+$ are: $N_2$, 1.09 Å; $N_2^+$, 1.12 Å; $O_2$, 1.21 Å; and $O_2^+$, 1.12 Å. Draw molecular orbital energy-level diagrams for these four species and explain why the bond distances vary in the way described.

**4.23** The anion of calcium carbide, $CaC_2$, should properly be called the acetylide ion, $C_2^{2-}$. Draw molecular orbital energy-level diagrams for $C_2$ and $C_2^{2-}$ and state the bond order of each.

**4.24** Numerous compounds are known in which the nitrosonium ion, $NO^+$, occurs (e.g., $NO^+HSO_4^-$ and $NO^+ClO_4^-$). By means of molecular orbital energy-level dia-

grams, compare the bonding in NO and $NO^+$. Which species has the stronger bonding? Why?

**4.25** In the derivation of equations describing electronic density by the molecular orbital theory, $2\psi_A\psi_B$ is sometimes called the overlap term. Explain why.

**4.26** Discuss the structure of the sulfur trioxide molecule, $SO_3$, in terms of resonance and the concept of delocalized $\pi$ bonding.

**4.27** Discuss the structure of the ozone molecule, $O_3$, in terms of resonance and the concept of delocalized $\pi$ bonding.

**4.28** How could measurement of the dipole moment of the trigonal bipyramidal molecule $PCl_2F_3$ help to determine whether the Cl atoms occupy axial or equatorial positions?

**4.29** The dipole moment of $PF_3$ is 1.03 D, whereas $PF_5$ has no dipole moment. Explain.

**4.30** The dipole moment of $NH_3$ (1.49 D) is greater than that of $NF_3$ (0.24 D). On the other hand, the dipole moment of $PH_3$ (0.55 D) is less than that of $PF_3$ (1.03 D). Explain these results.

**4.31** Explain why the dipole moment of SCO is 0.72 D, whereas the dipole moment of $CO_2$ is zero. Would $CS_2$ have a dipole moment?

**4.32** In Table 3.5 the electronegativity of C is given as 2.5 and that of O is given as 3.5. On the other hand, in Table 4.3 the dipole moment of CO is listed as only 0.12 D (the dipole–dipole forces of CO are negligible). Draw the Lewis structure of CO and offer an explanation for the low dipole moment of CO.

**4.33** Explain why the following melting points fall in the order given: $F_2$ ($-233°C$), $Cl_2$ ($-103°C$), $Br_2$ ($-7°C$), and $I_2$ ($113.5°C$).

**4.34** Consider the following molecules, each of which is tetrahedral with the C atom as the central atom: $CH_4$, $CH_3Cl$, $CH_2Cl_2$, $CHCl_3$, $CCl_4$. In which of these compounds would dipole–dipole forces exist in the liquid state? In what order would you expect the boiling points of the compounds to fall?

**4.35** Which molecules (not ions) given as examples in Table 4.1 would you expect to have a dipole moment of zero?

**4.36** What type of forces must be overcome in order to melt crystals of the following? (a) Si, (b) Ba, (c) $F_2$, (d) $BaF_2$, (e) $BF_3$, (f) $PF_3$.

**4.37** What type of forces must be overcome in order to melt crystals of the following? (a) $O_2$, (b) $Br_2$, (c) $Br_2O$, (d) Ba, (e) $BaBr_2$, (f) BaO

**4.38** Which substance of each of the following pairs would you expect to have the higher melting point? Why? (a) ClF or BrF, (b) BrCl or $Cl_2$, (c) CsBr or BrCl, (d) Cs or $Br_2$, (e) C (diamond) or $Cl_2$

**4.39** Which substance of each of the following pairs would you expect to have the higher melting point? Why? (a) Sr or $Cl_2$, (b) $SrCl_2$ or $SiCl_4$, (c) $SiCl_4$ or $SiBr_4$, (d) $SiCl_4$ or $SCl_4$, (e) SiC (carborundum) or $SiCl_4$.

**4.40** Which substance of each of the following pairs would you expect to have the higher melting point? Why? (a) Li or $H_2$, (b) LiH or $H_2$, (c) Li or LiH, (d) $H_2$ or $Cl_2$, (e) $H_2$ or HCl

**4.41** Use energy-level diagrams to explain the difference between conductors, insulators, and semiconductors.

# 5
# Chemical Equations and Quantitative Relations

As a result of the scientific revolution of the seventeenth century (and notably the work of Newton), science became increasingly quantitative and mathematical. Some scientists bemoaned the change. "The object of chemistry is the general, and particular understanding, not of the *masses* of bodies, nor of facts connected with those masses, but rather of the very nature of these bodies, the properties connected with them, and also their interrelations," wrote Jean-Baptiste Lamarck, the noted French naturalist, in the late eighteenth century. Nevertheless, chemists began increasingly to ask questions that could be answered quantitatively, and as a result, chemistry developed rapidly. **Stoichiometry** is the branch of chemistry that deals with the weight relations between elements and compounds in chemical reactions. The energy effects that accompany chemical changes are also quantitatively related to the substances involved in these changes.

**5.1 THE MOLE**

● The mole is the SI base unit of amount of substance. Its SI symbol is mol.

A **mole** is defined as the amount of a substance that contains the same number of units as the number of atoms in exactly 12 g of $^{12}_{6}C$. The number of atoms in 12 g of $^{12}_{6}C$, and hence the number of units in a mole, is $6.022 \times 10^{23}$ (Avogadro's number). Therefore, a mole of Be atoms contains $6.022 \times 10^{23}$ Be atoms, and a mole of He atoms contains $6.022 \times 10^{23}$ He atoms.

The standard for the atomic mass unit, u, is the $^{12}_{6}C$ atom; one atom of $^{12}_{6}C$ weighs 12 u. According to this standard, one atom of Be weighs 9.0 u ($\frac{3}{4}$ as much as the $^{12}_{6}C$ atom) and one atom of He weighs 4.0 u ($\frac{1}{3}$ as much as the $^{12}_{6}C$ atom).[1] If we take $6.022 \times 10^{23}$ atoms of $^{12}_{6}C$, the sample weighs 12 g (by definition). The same number of Be atoms (a mole) weighs $\frac{3}{4}$ of 12 g (9.0 g, the atomic weight of Be in grams), and a mole of He atoms weighs $\frac{1}{3}$ of 12 g (4.0 g, the atomic weight of He in grams). A mole of atoms, therefore, consists of Avogadro's number of atoms and has a mass in grams numerically equal to the atomic weight of the element.

A mole of *molecules* consists of Avogadro's number of *molecules* and has a mass in grams numerically equal to the molecular weight of the compound. The molecular weight of $H_2O$, for example, is 18.0 (twice the atomic weight of H plus the atomic weight of O), and 18.0 g of $H_2O$ (a mole of water) contains $6.022 \times 10^{23}$ molecules. Since there are two atoms of H and one atom of O in a molecule of water, a mole of water (18.0 g) contains two moles of H atoms (2.0 g) and one mole of O atoms (16.0 g).

The term *mole* can be applied to a wide variety of items—for example, atoms, molecules, ions, collections of ions, electrons, or chemical bonds. Hence one must be careful to specify what is being measured when the term mole is employed. A mole of O *atoms,* for example, contains $6.022 \times 10^{23}$ *atoms* and weighs 16.0 g, but a mole of $O_2$ *molecules* contains $6.022 \times 10^{23}$ *molecules* and weighs 32.0 g. A mole of an ionic substance contains $6.022 \times 10^{23}$ formula units. A mole of $CaCl_2$, for example, contains $6.022 \times 10^{23}$ $CaCl_2$ units; in reality, a mole of $CaCl_2$ consists of $6.022 \times 10^{23}$ $Ca^{2+}$ ions (1 mol of $Ca^{2+}$ ions) and $2(6.022 \times 10^{23})$ $Cl^-$ ions (2 mol of $Cl^-$ ions).

The number of units in a mole, **Avogadro's number,** was named for Amadeo Avogadro, who first interpreted the behavior of gases in terms of the numbers of reacting molecules (Section 6.7). The value of Avo-

---

[1] For most problem work atomic weights will be rounded off to the first figure following the decimal point.

gadro's number has been determined by electrochemical (Section 10.4, Example 10.1) and crystallographic (Section 7.9, Example 7.3) techniques.

The magnitude of Avogadro's number is difficult to imagine. A mole of people would be approximately $1.7 \times 10^{14}$ times the population of the entire world. The atomic weight of an element usually is not the mass of any individual atom; rather, the atomic weight of most elements is an average that takes into account the isotopic composition of the element (Section 2.9). Since there is an extremely large number of atoms in even a very small sample of an element, the chance of getting a collection of atoms with an average mass equal to the atomic weight is virtually assured. Therefore, we may think in terms of atoms with masses equal to the atomic weight of the element; the relations thus derived are perfectly valid for samples of a realistic size.

**5.2 CHEMICAL CALCULATIONS**

Units of measure must be included as an integral part of any measurement. It makes little sense to say that an object weighs 9.0, since this could mean 9.0 kg, 9.0 g, or 9.0 mg, and so on. Any terms employed in a calculation should also include proper unit labels; these labels should undergo the same mathematical operations as the numbers. In any calculation, if the units that appear in both the numerator and the denominator are canceled, those remaining will appear as a part of the answer and will provide the true dimension. If the answer does not have the units sought, then a mistake has been made in the calculation setup.

The use of dimensions in solving problems is illustrated in the examples that follow.

---

*Example 5.1*

Given the relation: 4.184 J = 1.000 cal, how many calories equal 1.000 kJ (which is 1000 joules)?

*Solution*

We might reason as follows:

$$4.184 \text{ J} = 1.000 \text{ cal}$$

Therefore,

$$1.000 \text{ J} = \left(\frac{1}{4.184}\right) \text{ cal}$$

and

$$1000 \text{ J} = 1000 \left(\frac{1}{4.184}\right) \text{ cal} = 239.0 \text{ cal}$$

The same line of reasoning can be followed by the use of a properly labeled conversion factor. If both sides of the equality

$$1.000 \text{ cal} = 4.184 \text{ J}$$

are divided by 4.184 J, we derive the equation

$$\left(\frac{1.000 \text{ cal}}{4.184 \text{ J}}\right) = 1$$

The factor (1.000 cal/4.184 J) is equal to unity since the numerator and denominator are equivalent. Our problem can be stated:

$$? \text{ cal} = 1000 \text{ J}$$

By use of the factor, we can solve the problem.

$$? \text{ cal} = 1000 \text{ J} \left( \frac{1.000 \text{ cal}}{4.184 \text{ J}} \right) = 239.0 \text{ cal}$$

Notice that the joule labels cancel and leave the answer in the desired units, calories.[2]

A second conversion factor can be derived from the equality

$$1.000 \text{ cal} = 4.184 \text{ J}$$

by dividing both sides of the equation by 1.000 cal

$$1 = \left( \frac{4.184 \text{ J}}{1.000 \text{ cal}} \right)$$

This factor, which is also equal to unity, can be used to convert calories to joules. For example, the number of joules that equals 50.00 cal can be found in the following way:

$$? \text{ J} = 50.00 \text{ cal} \left( \frac{4.184 \text{ J}}{1.000 \text{ cal}} \right) = 209.2 \text{ J}$$

A single equality, therefore, can be used to derive two factors. In the solution to a problem, the appropriate factor is the one that will lead to cancellation of the unit that must be eliminated.

---

*Example 5.2*

How many atoms are present in a 1.00 g sample of Na?

*Solution*

We can state the problem in the following way:

$$? \text{ Na atoms} = 1.00 \text{ g Na}$$

The atomic weight of Na is 23.0; therefore, 1.00 mol Na = 23.0 g Na. By use of the factor (1 mol Na/23.0 g Na), we can convert grams of Na into moles of Na.

$$? \text{ Na atoms} = 1.00 \text{ g Na} \left( \frac{1.00 \text{ mol Na}}{23.0 \text{ g Na}} \right)$$

This step, however, does not complete the solution.

Since 1.00 mol contains $6.02 \times 10^{23}$ atoms, use of the factor $(6.02 \times 10^{23} \text{ Na atoms}/1 \text{ mol Na})$ completes the solution.

$$? \text{ Na atoms} = 1.00 \text{ g Na} \left( \frac{1.00 \text{ mol Na}}{23.0 \text{ g Na}} \right) \left( \frac{6.02 \times 10^{23} \text{ Na atoms}}{1.00 \text{ mol Na}} \right)$$

$$= 2.62 \times 10^{22} \text{ Na atoms}$$

---

**5.3** THE LAW OF
CONSTANT
COMPOSITION

The law of constant composition (the law of definite proportions) was first proposed by Joseph Proust in 1799; it states that a pure compound always consists of the same elements combined in the same proportions by weight. In the preceding sections the assumption is tacitly made that compounds have definite compositions; indeed, this is a logical consequence of atomic and molecular theory (for exceptions, see Section 7.14).

The composition of a given compound can be determined by chemical analysis; the formula of the compound is usually derived from these analytical data (Section 5.4). Calculation of the percent composition

---

[2] Cancellation will not be indicated in future examples.

from a formula is sometimes necessary and is readily accomplished. In addition, other simple stoichiometric problems can be solved by use of the proportions represented by formulas.

---

*Example 5.3*

What is the percent composition by weight of $Fe_2O_3$?

*Solution*

One mole of $Fe_2O_3$ contains

$$2 \text{ mol Fe} = (2 \times 55.8) \text{ g} = 111.6 \text{ g}$$

$$3 \text{ mol O} = (3 \times 16.0) \text{ g} = \underline{\ \ 48.0 \text{ g}}$$
$$159.6 \text{ g}$$

The percent Fe is

$$\frac{111.6 \text{ g Fe}}{159.6 \text{ g Fe}_2\text{O}_3} \times 100\% = 69.9\% \text{ Fe in Fe}_2\text{O}_3$$

The percent O is

$$\frac{48.0 \text{ g O}}{159.6 \text{ g Fe}_2\text{O}_3} \times 100\% = 30.1\% \text{ O in Fe}_2\text{O}_3$$

(The same result, of course, could have been obtained by subtracting 69.9% from 100.0%.)

---

*Example 5.4*

How many moles of Au and grams of Au are present in 500 g of $Au_2O_3$?

*Solution*

The molecular weight of $Au_2O_3$ is 442.0, and 1 mol of $Au_2O_3$ (442 g of $Au_2O_3$) contains 2 mol of Au.

$$2 \text{ mol Au} = 1 \text{ mol Au}_2\text{O}_3$$

$$2 \text{ mol Au} = 442 \text{ g Au}_2\text{O}_3$$

The factor we need is (2 mol Au/442 g $Au_2O_3$).

$$? \text{ mol Au} = 500 \text{ g Au}_2\text{O}_3 \left( \frac{2 \text{ mol Au}}{442 \text{ g Au}_2\text{O}_3} \right)$$
$$= 2.26 \text{ mol Au}$$

Two moles of Au weigh 394 g (which is 2 times the atomic weight of Au, 197.0). Therefore

$$2 \text{ mol Au} = 1 \text{ mol Au}_2\text{O}_3$$

$$394 \text{ g Au} = 442 \text{ g Au}_2\text{O}_3$$

The factor needed to solve the second part of the problem is (394 g Au/442 g $Au_2O_3$)

$$? \text{ g Au} = 500 \text{ g Au}_2\text{O}_3 \left( \frac{394 \text{ g Au}}{442 \text{ g Au}_2\text{O}_3} \right)$$
$$= 446 \text{ g Au}$$

---

*Example 5.5*

A 1.215 g sample of nicotine (a compound that contains only C, H, and N) is burned in oxygen, and 3.300 g of $CO_2$, 0.945 g of $H_2O$, and 0.210 g of $N_2$ are the products of the combustion. What is the percent composition of nicotine?

*Solution*

We first calculate the quantity of each element in the sample. Since 1 mol of $CO_2$ (44.0 g) contains 1 mol of C (12.0 g),

$$12.0 \text{ g C} = 44.0 \text{ g CO}_2$$

and we derive the factor (12.0 g C/44.0 g $CO_2$). Therefore,

$$? \text{ g C} = 3.300 \text{ g CO}_2 \left( \frac{12.0 \text{ g C}}{44.0 \text{ g CO}_2} \right)$$
$$= 0.900 \text{ g C}$$

In 1 mol of $H_2O$ (18.0 g) there are 2 mol of H atoms (2.0 g). Hence,

2.0 g H = 18.0 g $H_2O$

Therefore,

$$? \text{ g H} = 0.945 \text{ g } H_2O \left( \frac{2.0 \text{ g H}}{18.0 \text{ g } H_2O} \right)$$

$$= 0.105 \text{ g H}$$

In a combustion such as that described, the nitrogen does not combine with oxygen but evolves as $N_2$. Hence the sample contains

0.210 g N. Since the sample weighed 1.215 g, the percents are

$$\frac{0.900 \text{ g C}}{1.215 \text{ g nicotine}} \times 100\% = 74.1\% \text{ C in nicotine}$$

$$\frac{0.105 \text{ g H}}{1.215 \text{ g nicotine}} \times 100\% = 8.6\% \text{ H in nicotine}$$

$$\frac{0.210 \text{ g N}}{1.215 \text{ g nicotine}} \times 100\% = 17.3\% \text{ N in nicotine}$$

---

**5.4 DETERMINATION OF FORMULAS**

The **empirical formula** (also called the simplest formula) of a compound reflects the simplest ratio of atoms in the compound. The **molecular formula** of a compound is based on the actual numbers of atoms that comprise a molecule of that compound. For some compounds the molecular and the empirical formulas are identical, for example, $H_2O$, $CO_2$, $NH_3$, and $SO_2$. For many compounds, however, the molecular and empirical formulas are different. The molecular formulas $N_2H_4$, $B_3N_3H_6$, $C_6H_6$, and $C_6H_{12}O_6$ correspond to the empirical formulas $NH_2$, $BNH_2$, CH, and $CH_2O$. Notice that the atomic ratio for an empirical formula may be obtained by reducing the atomic ratio of the molecular formula to the lowest possible set of whole numbers.

Since an ionic compound has no molecules, there is no molecular formula that describes it. Formulas of ionic compounds are empirical formulas that state the simplest ratio of ions in the compound.

An experimentally derived analysis of a compound is needed for the determination of the empirical formula of the compound.

---

*Example 5.6*

What is the empirical formula of an oxide of phosphorus that contains 43.6% P and 56.4% O?

*Solution*

We must convert the *weight* ratio of P to O (given by the percentages) to the *atomic* ratio of P to O (needed to write the formula). Since 1 mol of P (31.0 g) contains the same number of atoms as 1 mol of O (16.0 g), the ratio by moles is the same as the ratio by atoms.

On the basis of the percents given, a 100 g sample of the compound contains 43.6 g P and 56.4 g O.

$$? \text{ mol P} = 43.6 \text{ g P} \left( \frac{1 \text{ mol P}}{31.0 \text{ g P}} \right) = 1.41 \text{ mol P}$$

$$? \text{ mol O} = 56.4 \text{ g O} \left( \frac{1 \text{ mol O}}{16.0 \text{ g O}} \right) = 3.53 \text{ mol O}$$

Hence, there are 1.41 atoms of P for every 3.53 atoms of O in the compound. We need, however, the *simplest, whole-number* ratio in order to write the formula. By dividing the two values by the smaller value, we get

for P, $\dfrac{1.41}{1.41} = 1.00$     for O, $\dfrac{3.53}{1.41} = 2.50$

We still do not have a whole-number ratio, but we can get one by multiplying these two values by 2. Hence, the simplest whole-number ratio is 2 to 5 and the empirical formula is $P_2O_5$.

---

*Example 5.7*
A sample of caffeine contains 0.624 g C, 0.065 g H, 0.364 g N, and 0.208 g O. What is the empirical formula of caffeine?

*Solution*
The results of an analysis, such as the one described in Example 5.5, are usually reported in terms of percent composition. However, any weight ratio can be converted into a mole ratio and, in this form, used to derive an empirical formula. There is no need to convert the data given in this example into percentages.

$$? \text{ mol C} = 0.624 \text{ g C} \left( \frac{1 \text{ mol C}}{12.0 \text{ g C}} \right)$$
$$= 0.0520 \text{ mol C}$$

$$? \text{ mol H} = 0.065 \text{ g H} \left( \frac{1 \text{ mol H}}{1.0 \text{ g H}} \right)$$
$$= 0.065 \text{ mol H}$$

$$? \text{ mol N} = 0.364 \text{ g N} \left( \frac{1 \text{ mol N}}{14.0 \text{ g N}} \right)$$
$$= 0.0260 \text{ mol N}$$

$$? \text{ mol O} = 0.208 \text{ g O} \left( \frac{1 \text{ mol O}}{16.0 \text{ g O}} \right)$$
$$= 0.0130 \text{ mol O}$$

Division of these values by the smallest value (0.013) gives the ratio

4 (for C): 5 (for H): 2 (for N): 1 (for O)

and the empirical formula of caffeine, therefore, is $C_4H_5N_2O$.

---

Additional information is necessary to assign a molecular formula to a compound. The molecular weight of the compound can be experimentally determined or the number of atoms that constitute a molecule can be ascertained by diffraction techniques.

---

*Example 5.8*
What is the molecular formula of the compound described in Example 5.6 if the compound has a molecular weight of 284?

*Solution*
The weight obtained by adding the atomic weights indicated by the empirical formula,

$P_2O_5$, is 142. Since the actual molecular weight is 284, there must be twice as many atoms present in a molecule as is indicated by the empirical formula. Therefore the molecular formula is $P_4O_{10}$.

---

*Example 5.9*
The molecular weight of caffeine is 194.0, and the empirical formula is $C_4H_5N_2O$. What is the molecular formula of caffeine?

*Solution*
The formula weight indicated by $C_4H_5N_2O$ is 97. Since the molecular weight of caffeine is twice this value, the molecular formula of caffeine is $C_8H_{10}N_4O_2$.

---

**5.5** CHEMICAL
EQUATIONS

Chemical equations are representations of reactions in terms of the symbols and formulas of the elements and compounds involved. The reactants are indicated on the left and the products on the right. An arrow is used instead of the customary equal sign of the algebraic equation; it may be considered as an abbreviation for the word "yields."

The **law of conservation of mass** states that there is no detectable change in mass during the course of an ordinary chemical reaction.[3] This law was first stated formally by Antoine Lavoisier in his work *Traite Elementaire de Chemie* (1789), although the quantitative methods of prior workers assumed such a principle. Insofar as chemical equations are concerned, the law of conservation of mass means that there must be as many atoms of each element, combined or uncombined, indicated on the left of an equation as there are on the right.

Chemical equations report the results of experimentation. One of the goals of chemistry is the discovery and development of principles that make it possible to predict the products of chemical reactions; careful attention will be given to any such generalizations. All too often, however, the products of a particular set of reactants must be memorized, and any prediction is subject to modification if experiment dictates. What may appear reasonable on paper is not necessarily what occurs in the laboratory.

The first step in writing a chemical equation is to ascertain the products of the reaction in question. Carbon disulfide, $CS_2$, reacts with chlorine, $Cl_2$, to produce carbon tetrachloride, $CCl_4$, and disulfur dichloride, $S_2Cl_2$. To represent this, we write

$$CS_2 + Cl_2 \rightarrow CCl_4 + S_2Cl_2$$

This equation is not correct quantitatively and violates the law of conservation of mass. Although one carbon atom and two sulfur atoms are indicated on both the left side and the right side of the equation, two chlorine atoms (one $Cl_2$ molecule) appear on the left and six chlorine atoms appear on the right. The equation can be balanced[4] by indicating that three molecules of chlorine should be used for the reaction. Thus,

$$CS_2 + 3Cl_2 \rightarrow CCl_4 + S_2Cl_2$$

The simplest types of chemical equations are balanced by trial and error as the following examples will illustrate. When steam is passed

---

[3] This is demonstrably untrue for nuclear transformations. In these cases Einstein's postulate of the equivalence of mass and energy (Section 2.9) satisfactorily accounts for the conversion of a small amount of mass into energy. Rigorously, a law of conservation of mass–energy should be stated. The energy changes observed in ordinary chemical reactions may be the result of concomitant mass changes, but such changes in mass are too small to be experimentally detected. Therefore, the law of conservation of mass, as stated, is valid for all reactions except those involving nuclear change (Chapter 21).
[4] Strictly speaking, the expression is not an "equation" until it is balanced.

over hot iron, hydrogen gas and an oxide of iron that has the formula $Fe_3O_4$ are produced. Thus,

$$Fe + H_2O \rightarrow Fe_3O_4 + H_2$$

It is tempting to substitute another oxide of iron, FeO, for $Fe_3O_4$, since this would immediately produce a balanced equation. Such an equation, however, would be without value; experiment indicates that $Fe_3O_4$, not FeO, is a product of the reaction. Balancing an equation is never accomplished by altering the formulas of the products of the reaction. In the equation for the reaction of iron and steam, three atoms of Fe and four molecules of $H_2O$ are needed to provide the iron and oxygen atoms required for the formation of $Fe_3O_4$.

$$3Fe + 4H_2O \rightarrow Fe_3O_4 + H_2$$

The equation is now balanced except for hydrogen; the hydrogen may be balanced as follows:

$$3Fe + 4H_2O \rightarrow Fe_3O_4 + 4H_2$$

In much the same manner we can balance an equation for the complete combustion of ethane ($C_2H_6$) in oxygen. The products of this reaction are carbon dioxide and water.

$$C_2H_6 + O_2 \rightarrow CO_2 + H_2O$$

To balance the two carbon atoms of $C_2H_6$, the production of two molecules of $CO_2$ must be indicated, and the six hydrogen atoms of $C_2H_6$ require that three molecules of $H_2O$ be produced.

$$C_2H_6 + O_2 \rightarrow 2CO_2 + 3H_2O$$

Only the oxygen remains unbalanced; there are seven atoms of oxygen on the right and only two on the left. In order to get seven atoms of oxygen on the left, we would have to take $3\frac{1}{2}$, or $\frac{7}{2}$, molecules of $O_2$.

$$C_2H_6 + \tfrac{7}{2}O_2 \rightarrow 2CO_2 + 3H_2O$$

Customarily, equations are written with whole number coefficients. By multiplying the entire equation by two, we get

$$2C_2H_6 + 7O_2 \rightarrow 4CO_2 + 6H_2O$$

**5.6 PROBLEMS BASED ON CHEMICAL EQUATIONS**

An equation can be interpreted in several different ways. Consider, for example,

$$2H_2 + O_2 \rightarrow 2H_2O$$

On the simplest level this equation shows that hydrogen reacts with oxygen to produce water. On the atomic-molecular level, it states that two molecules of hydrogen react with one molecule of oxygen to produce two molecules of water. In more practical, quantitative terms, it can be read as a statement that 2 mol of hydrogen reacts with 1 mol of oxygen to produce 2 mol of water; in terms of actual weights, 4.0 g of hydrogen reacts with 32.0 g of oxygen to produce 36.0 g of water.

Stoichiometric problems, then, are readily solved by reference to the equation that describes the chemical change in question.

---

*Example 5.10*

Chlorine can be prepared by the reaction

$$MnO_2 + 4HCl \rightarrow MnCl_2 + Cl_2 + 2H_2O$$

(a) How many grams of HCl are required to react with 25.0 g of $MnO_2$? (b) How many grams of $Cl_2$ are produced by the reaction?

*Solution*

(a) The equation gives the mole ratio of $MnO_2$ to HCl:

1 mol $MnO_2$ = 4 mol HCl

This equality is readily converted to the weight relation that is required to solve the problem since the molecular weight of $MnO_2$ is 86.9 and the molecular weight of HCl is 36.5.

86.9 g $MnO_2$ = 4(36.5) g HCl

86.9 g $MnO_2$ = 146.0 g HCl

Therefore,

$$? \text{ g HCl} = 25.0 \text{ g } MnO_2 \left(\frac{146.0 \text{ g HCl}}{86.9 \text{ g } MnO_2}\right)$$
$$= 42.0 \text{ g HCl}$$

(b) To solve this problem, we need a factor that relates the weight of $Cl_2$ produced to the weight of one of the reactants (either $MnO_2$ or HCl may be used for the calculation). If we base our calculation on the quantity of $MnO_2$ consumed in the

reaction, we derive the mole ratio of $Cl_2$ to $MnO_2$ from the chemical equation.

1 mol $Cl_2$ = 1 mol $MnO_2$

The molecular weight of $Cl_2$ is 71.0, and the molecular weight of $MnO_2$ is 86.9. Therefore,

71.0 g $Cl_2$ = 86.9 g $MnO_2$

Hence, our factor is (71.0 g $Cl_2$/86.9 g $MnO_2$), and the solution is

$$? \text{ g } Cl_2 = 25.0 \text{ g } MnO_2 \left(\frac{71.0 \text{ g } Cl_2}{86.9 \text{ g } MnO_2}\right)$$
$$= 20.4 \text{ g } Cl_2$$

If we had based our calculation on the quantity of HCl consumed in the reaction (calculated to be 42.0 g in part a), we would have derived from the equation,

1 mol $Cl_2$ = 4 mol HCl

71.0 g $Cl_2$ = 4(36.5) g HCl

71.0 g $Cl_2$ = 146.0 g HCl

Therefore,

$$? \text{ g } Cl_2 = 42.0 \text{ g HCl} \left(\frac{71.0 \text{ g } Cl_2}{146.0 \text{ g HCl}}\right)$$
$$= 20.4 \text{ g } Cl_2$$

which is the same result as that obtained previously.

---

*Example 5.11*

Iodine(V) oxide is used as a reagent for the determination of carbon monoxide.

$$I_2O_5 + 5CO \rightarrow I_2 + 5CO_2$$

What weight of CO is indicated by the liberation of 0.192 g $I_2$ from excess $I_2O_5$?

**Solution**

The relation by moles of the two substances of interest is obtained from the equation,

$$5 \text{ mol CO} = 1 \text{ mol I}_2$$

Since the molecular weight of CO is 28.0 and the molecular weight of $I_2$ is 253.8,

$$5(28.0) \text{ g CO} = 253.8 \text{ g I}_2$$

$$140.0 \text{ g CO} = 253.8 \text{ g I}_2$$

Hence, the factor we need is $(140.0 \text{ g CO}/253.8 \text{ g I}_2)$.

$$? \text{ g CO} = 0.192 \text{ g I}_2 \left(\frac{140.0 \text{ g CO}}{253.8 \text{ g I}_2}\right)$$

$$= 0.106 \text{ g CO}$$

---

**Example 5.12**

Oxalic acid, $H_2C_2O_4$, is a diprotic acid. Neutralization by an alkali, such as sodium hydroxide (NaOH), yields either a normal salt ($Na_2C_2O_4$, sodium oxalate) or an acid salt ($NaHC_2O_4$, sodium hydrogen oxalate) depending upon the mole ratio of alkali to acid employed. Oxalic acid is poisonous. It occurs in many plants (including the leaves of the rhubarb plant) as the calcium salt ($CaC_2O_4$, a normal salt) or as the potassium acid salt ($KHC_2O_4$). (a) What weight of sodium hydroxide is needed for the complete neutralization of an aqueous solution that contains 1.80 g of $H_2C_2O_4$? (b) What weight of sodium hydroxide should be used to prepare $NaHC_2O_4$ from 2.70 g of $H_2C_2O_4$ in water solution? (c) What are the products, in moles, of a process in which 3.60 g of NaOH in aqueous solution is added to a solution containing 7.20 g of $H_2C_2O_4$?

**Solution**

(a) The molecular weight of NaOH is 40.0, and the molecular weight of $H_2C_2O_4$ is 90.0. The equation for the complete neutralization in water solution (an ionic reaction) is

$$H_2C_2O_4 + 2OH^- \rightarrow C_2O_4^{2-} + 2H_2O$$

From this equation we can see that

$$2 \text{ mol NaOH} = 1 \text{ mol H}_2C_2O_4$$

$$2(40.0) \text{ g NaOH} = 90.0 \text{ g H}_2C_2O_4$$

$$? \text{ g NaOH} = 1.80 \text{ g H}_2C_2O_4 \left(\frac{80.0 \text{ g NaOH}}{90.0 \text{ g H}_2C_2O_4}\right)$$

$$= 1.60 \text{ g NaOH}$$

(b) The ionic equation for the preparation of the acid salt is

$$H_2C_2O_4 + OH^- \rightarrow HC_2O_4^- + H_2O$$

For the reaction

$$1 \text{ mol NaOH} = 1 \text{ mol H}_2C_2O_4$$

$$40.0 \text{ g NaOH} = 90.0 \text{ g H}_2C_2O_4$$

$$? \text{ g NaOH} = 2.70 \text{ g H}_2C_2O_4 \left(\frac{40.0 \text{ g NaOH}}{90.0 \text{ g H}_2C_2O_4}\right)$$

$$= 1.20 \text{ g NaOH}$$

(c) We first determine the number of moles of each reactant used.

$$? \text{ mol NaOH} = 3.60 \text{ g NaOH} \left(\frac{1.00 \text{ mol NaOH}}{40.0 \text{ g NaOH}}\right)$$

$$= 0.0900 \text{ mol NaOH}$$

$$? \text{ mol H}_2C_2O_2 = 7.20 \text{ g H}_2C_2O_4 \left(\frac{1.00 \text{ mol H}_2C_2O_4}{90.0 \text{ g H}_2C_2O_4}\right)$$

$$= 0.0800 \text{ mol H}_2C_2O_4$$

In the course of the addition of the NaOH solution, the acid salt is formed first.

$$H_2C_2O_4 + OH^- \rightarrow HC_2O_4^- + H_2O$$

From this equation we see that for every mole of acid, 1 mol of alkali is used, and 1 mol of acid salt is produced. Since we were given 0.0800 mol of $H_2C_2O_4$, this initial reaction would use 0.0800 mol of NaOH (out of the total of 0.0900 mol NaOH) and produce 0.0800 mol of the acid salt. At this point we assume that we have used all the $H_2C_2O_4$ supplied, have produced 0.0800 mol of the acid salt, and still have 0.0100 mol of NaOH left from the original quantity.

The acid salt can be neutralized by alkali. The equation is

$$HC_2O_4^- + OH^- \rightarrow C_2O_4^{2-} + H_2O$$

This equation shows that for every mole of alkali reacted, 1 mol of acid salt is used, and 1 mol of normal salt is produced. Hence the 0.0100 mol of NaOH that remains from the first step would react with 0.0100 mol of acid salt (reducing the quantity present from 0.0800 mol to 0.0700 mol) and produce 0.0100 mol of normal salt. The final quantities present in solution are 0.0100 mol $Na_2C_2O_4$ and 0.0700 mol $NaHC_2O_4$.

---

*Example 5.13*

What weight of $P_2I_4$ can theoretically be prepared from 5.00 g $P_4O_6$ and 8.00 g $I_2$ according to the following reaction?

$$5P_4O_6 + 8I_2 \rightarrow 4P_2I_4 + 3P_4O_{10}$$

*Solution*

The first step is to determine which reactant limits the quantity of $P_2I_4$ produced. To do this, we first determine the number of moles of each reactant present before reaction.

$$? \text{ mol } P_4O_6 = 5.00 \text{ g } P_4O_6 \left( \frac{1 \text{ mol } P_4O_6}{220.0 \text{ g } P_4O_6} \right)$$
$$= 0.0227 \text{ mol } P_4O_6$$

$$? \text{ mol } I_2 = 8.00 \text{ g } I_2 \left( \frac{1 \text{ mol } I_2}{253.8 \text{ g } I_2} \right)$$
$$= 0.0315 \text{ mol } I_2$$

We select either of these values and determine the number of moles of the other substance required for complete reaction. According to the chemical equation,

$$8 \text{ mol } I_2 = 5 \text{ mol } P_4O_6$$

Therefore,

$$? \text{ mol } I_2 = 0.0227 \text{ mol } P_4O_6 \left( \frac{8 \text{ mol } I_2}{5 \text{ mol } P_4O_6} \right)$$
$$= 0.0363 \text{ mol } I_2$$

Since only 0.0315 mol $I_2$ is supplied, there is insufficient $I_2$ to react with all the $P_4O_6$. The amount of $P_2I_4$ that can be obtained is limited by the amount of $I_2$ used, and the problem is solved on the basis of this quantity of $I_2$.

$$? \text{ g } P_2I_4 = 0.0315 \text{ mol } I_2 \left( \frac{4 \text{ mol } P_2I_4}{8 \text{ mol } I_2} \right) \left( \frac{569.6 \text{ g } P_2I_4}{1 \text{ mol } P_2I_4} \right)$$
$$= 8.97 \text{ g } P_2I_4$$

If we had determined which substance limits the reaction by calculating the number of moles of $P_4O_6$ required for complete reaction of all the $I_2$ supplied (instead of the reverse procedure we employed), we would have noted

$$? \text{ mol } P_4O_6 = 0.0315 \text{ mol } I_2 \left( \frac{5 \text{ mol } P_4O_6}{8 \text{ mol } I_2} \right)$$

$$= 0.0197 \text{ mol } P_4O_6$$

Since 0.0227 mol of $P_4O_6$ was supplied, $P_4O_6$ is in excess of the amount required by the quantity of $I_2$. We again conclude that the amount of product that can be obtained is limited by the amount of $I_2$ available. The problem would be solved in the same way that we previously noted.

*Example 5.14*

If 8.00 g of $P_2I_4$ is obtained from the procedure described in Example 5.13, what is the percent yield?

*Solution*

Frequently, the quantity of product obtained from a reaction is less than the amount calculated on the basis of the equation that describes the change. It may be that part of the reactants do not react, or that part of the reactants react in a way different from that desired (side reactions), or that not all the product is recovered. The *percent yield* relates the amount of product that is actually obtained (8.00 g $P_2I_4$ in this case) to the amount that theory would predict (8.97 g $P_2I_4$, from Example 5.13). Thus,

$$\frac{8.00 \text{ g } P_2I_4}{8.97 \text{ g } P_2I_4} \times 100\% = 89.2\%$$

**5.7 HEAT OF REACTION**

The heat that is liberated or absorbed during a chemical reaction is an important and integral part of the reaction. Reactions that evolve heat are called **exothermic** reactions; those that absorb heat are called **endothermic** reactions. Under constant pressure, a **heat of reaction** (also called the **enthalpy of reaction**) is a manifestation of the difference in **heat content (enthalpy),** $H$, between the products of a reaction and the reactants; therefore, it is given the symbol $\Delta H$. For an exothermic reaction, the products have a lower enthalpy than the reactants; therefore, $\Delta H$ is negative and the system evolves heat. The products of endothermic reactions have a higher enthalpy than the reactants; $\Delta H$ is positive and heat is absorbed.

The enthalpies of chemical substances depend upon temperature and pressure. By convention, $\Delta H$ values are generally reported at 25°C (298°K) and standard atmospheric pressure (Section 6.2). In other words, for exothermic reactions, the heat in excess of that necessary to maintain the temperature of the chemical substances at 25°C is the enthalpy of reaction. For endothermic reactions, the enthalpy of reaction is the heat required to obtain products at 25°C from reactants at the same temperature. If a $\Delta H$ value pertains to a reaction that is run at a temperature other than 298°K, the temperature is usually appended to the symbol as a subscript (e.g., $\Delta H_{300}$).

Thermodynamic values are generally recorded in calories (cal) or kilocalories (kcal), joules (J) or kilojoules (kJ). The **heat capacity** of a substance is the amount of heat required to raise the temperature of the substance by 1°C. Values are usually recorded for 1 mol or for 1 g of the substance considered, and in the latter case the value is called the **specific heat** of the substance. The calorie was originally defined in terms of the specific heat of water, but is now defined by its joule equivalent.

1 cal = 4.184 J (exactly)

The joule, an SI energy unit, is defined as the amount of work done by a force of 1 newton (1 kg m/sec$^2$) acting through a distance of 1 meter. The joule is also a volt–coulomb (Section 10.1). Since it is possible to make electrical measurements with great precision, the joule is a better primary standard than the specific heat of water.

Thermochemical data may be given by writing an equation for the reaction under consideration and listing beside it the $\Delta H$ value for the reaction *as it is written*. The appropriate value of $\Delta H$ is the one required when the equation is read in gram molar quantities. Contrary to usual practice, fractional coefficients may be used to balance the chemical equation; a fractional coefficient simply indicates a fraction of a mole of a substance. Thus,

$$H_2(g) + \tfrac{1}{2}O_2(g) \rightarrow H_2O(l) \qquad \Delta H = -68.3 \text{ kcal}$$

The state of each substance used or produced by the reaction must be indicated in the equation; one of the designations: (g) for gas, (l) for liquid, (s) for solid, or (aq) for "in water solution" is placed after each formula. The need for this convention may be demonstrated by comparing the following reaction with the preceding one:

$$H_2(g) + \tfrac{1}{2}O_2(g) \rightarrow H_2O(g) \qquad \Delta H = -57.8 \text{ kcal}$$

Notice that 10.5 kcal less heat is liberated by the second reaction than by the first reaction because heat is used in converting $H_2O(l)$ to $H_2O(g)$.

When an equation is reversed, the sign of $\Delta H$ is changed; a reaction that is endothermic in one direction is exothermic in the other.

$$H_2(g) + I_2(s) \rightarrow 2HI(g) \qquad \Delta H = +12.40 \text{ kcal}$$

$$2HI(g) \rightarrow H_2(g) + I_2(s) \qquad \Delta H = -12.40 \text{ kcal}$$

Problems involving heat effects (e.g., Example 5.15) are solved in much the same way that simple stoichiometric problems are solved.

*Example 5.15*

The thermite reaction is highly exothermic.

$$2Al(s) + Fe_2O_3(s) \rightarrow 2Fe(s) + Al_2O_3(s)$$
$$\Delta H = -202.6 \text{ kcal}$$

How much heat is liberated when 36.0 g of Al reacts with excess $Fe_2O_3(s)$?

*Solution*

The equation and $\Delta H$ value show that

$$-202.6 \text{ kcal} = 2 \text{ mol Al}$$

$$-202.6 \text{ kcal} = 2(27.0) \text{ g Al}$$

Therefore,

$$? \text{ kcal} = 36.0 \text{ g Al} \left(\frac{-202.6 \text{ kcal}}{54.0 \text{ g Al}}\right) = -135 \text{ kcal}$$

If kilojoules are used, $\Delta H = -848 \text{ kJ}$, and

$$? \text{ kJ} = 36.0 \text{ g Al} \left(\frac{-848 \text{ kJ}}{54.0 \text{ g Al}}\right) = -565 \text{ kJ}$$

A device called a **calorimeter** is used to measure heats of reaction. In the typical calorimeter, a reaction vessel is placed in the center of a larger, well-insulated vessel containing a weighed quantity of water. The reaction is run between known quantities of reactants, and the heat evolved increases the temperature of the water and the calorimeter. The number of calories liberated by the reaction is the product of the rise in temperature and the total heat capacity of the calorimeter and its contents.

The specific heat of water changes slightly with a change in temperature. For our purposes, however, it may be regarded as constant — 1.00 cal/g°C or 4.18 J/g°C. The heat capacity of the calorimeter is determined experimentally, either by running a reaction that evolves a known amount of heat and measuring the increase in temperature or by measuring the temperature increase produced when a measured amount of electrical energy is used to heat the calorimeter that contains a weighed quantity of water.

*Example 5.16*

A bomb-type calorimeter is used to determine the enthalpy of combustion of glucose, $C_6H_{12}O_6$. Glucose, which is the principal sugar in human blood, has a molecular weight of 180.0. A 2.22 g sample of glucose is placed in the bomb, which is then filled with oxygen to a pressure of 25 atm. The bomb is placed in a well-insulated calorimeter vessel that is filled with 1200 g of water. The reaction mixture is ignited electrically, and the reaction causes the temperature of the calorimeter and its contents to increase from 18.00° to 23.19°C. The heat capacity of the calorimeter is 400 cal/°C. Calculate the molar enthalpy of combustion of glucose.

*Solution*

The increase in temperature is

$$23.19°C - 18.00°C = 5.19°C$$

Since 1200 g of $H_2O$ is employed and the specific heat of water is 1.00 cal/g°C, the heat capacity of the water in the calorimeter is

$$C_{H_2O} = (1200 \text{ g})(1.00 \text{ cal/g°C}) = 1200 \text{ cal/°C}$$
$$= 1.20 \text{ kcal/°C}$$

The heat capacity of the calorimeter is

$$C_{cal.} = 400 \text{ cal/}°C = 0.40 \text{ kcal/}°C$$

Therefore, the total heat capacity (of water and calorimeter together) is

$$C_{total} = 1.20 \text{ kcal/}°C + 0.40 \text{ kcal/}°C$$
$$= 1.60 \text{ kcal/}°C$$

Thus, 1.60 kcal is needed to raise the temperature of the assembly by 1°C.

Since the temperature increases 5.19°C, the amount of heat *absorbed* by the calorimeter and water is

$$(1.60 \text{ kcal/}°C)(5.19°C) = 8.30 \text{ kcal}$$

This quantity (8.30 kcal) is also the amount of heat *evolved* by the combustion of 2.22 g of glucose.

For 1 mol of glucose (180 g of glucose),

$$? \text{ kcal} = 180 \text{ g } C_6H_{12}O_6 \left( \frac{-8.30 \text{ kcal}}{2.22 \text{ g } C_6H_{12}O_6} \right)$$
$$= -673 \text{ kcal}$$

The equation for this combustion is

$$C_6H_{12}O_6(s) + 6O_2(g) \rightarrow 6CO_2(g) + 12H_2O(l)$$
$$\Delta H = -673 \text{ kcal}$$

It is just as convenient to use kilojoules as kilocalories in calculations of this type. If the heat capacity of the calorimeter is determined by electrical heating, the electrical energy supplied would probably be measured in joules. The specific heat of water is approximately 4.184 J/g°C. The molar enthalpy of combustion of glucose is therefore −2820 kJ.

No significant increase or decrease in pressure accompanies this reaction. For reactions in which there is an attendant pressure change (e.g., those in which there are more moles of gaseous products than there are moles of gaseous reactants), determination of the enthalpy of reaction requires that a correction be made to the value obtained by the method outlined above. Enthalpy is defined as the heat effect under constant pressure; the calorimeter bomb determines the heat effect under constant volume. (See Section 14.2.)

---

● If SI units are adopted by scientists in this field, dieters will probably count Joules instead of Calories, but the daily allowance will be numerically 4.184 times larger!

The "caloric content" of foods is measured by a method that is similar to the one described in Example 5.16. Glucose, therefore, has a caloric value of 673 kcal/mol, or 3.74 kcal/g. The Calorie of the nutritionist is actually a kilocalorie (the distinction is indicated by the capital C).

**5.8 LAW OF HESS**    The enthalpy of an element or compound depends upon the temperature and pressure. If we wish to compare $\Delta H$ values, the conditions under which they have been measured must be identical. The **standard state** of a substance is the state in which the substance is stable at 1 atm pressure and 25°C (if some other temperature is employed, it is indicated as a subscript on the $\Delta H$ value). The symbol $\Delta H°$ is used to indicate standard enthalpy changes, which apply to reactions involving only materials in their standard states.

Some $\Delta H$ values are given special names. An **enthalpy of combustion** is the $\Delta H$ value for a reaction in which the substance under consideration is completely oxidized by reaction with oxygen; the values are recorded in kcal per mole of the substance that is oxidized (a *reactant*). The following enthalpies of combustion have been measured.

$$CH_4(g) + 2O_2(g) \rightarrow CO_2(g) + 2H_2O(l) \qquad \Delta H° = -212.8 \text{ kcal} \qquad (1)$$

$$H_2(g) + \tfrac{1}{2}O_2(g) \rightarrow H_2O(l) \qquad \Delta H° = -68.3 \text{ kcal} \qquad (2)$$

$$C(graphite) + O_2(g) \rightarrow CO_2(g) \qquad \Delta H° = -94.1 \text{ kcal} \qquad (3)$$

Notice that the standard state of water is $H_2O(l)$, not $H_2O(g)$, and that the standard state of carbon is C (graphite).

The basis of many thermochemical calculations is the **law of constant heat summation,** which was established experimentally by G. H. Hess in 1840. This **law of Hess** states that the change in enthalpy for any chemical reaction is constant, whether the reaction occurs in one step or in several steps. Thus, thermochemical data may be treated algebraically. Consider, for example, calculation of the $\Delta H$ of a reaction in which methane, $CH_4$, is prepared from carbon and hydrogen; it is impossible to measure this enthalpy change directly.

$$C(graphite) + 2H_2(g) \rightarrow CH_4(g)$$

The desired $\Delta H°$ can be obtained from the enthalpies of reaction given in equations (1), (2) and (3). Equation (1) is reversed (and the sign of the $\Delta H°$ value is changed); equation (2) is multiplied by 2 (and the $\Delta H°$ value is multiplied by 2); equation (3) remains as given. Equations (4), (5), and (3) are then added; terms that are common to both sides of the final equation are cancelled in the addition.

$$2H_2O(l) + CO_2(g) \rightarrow CH_4(g) + 2O_2(g) \qquad \Delta H° = +212.8 \text{ kcal} \qquad (4)$$

$$2H_2(g) + O_2(g) \rightarrow 2H_2O(l) \qquad \Delta H° = -136.6 \text{ kcal} \qquad (5)$$

$$\underline{C(graphite) + O_2(g) \rightarrow CO_2(g) \qquad \Delta H° = -94.1 \text{ kcal} \qquad (3)}$$

$$C(graphite) + 2H_2(g) \rightarrow CH_4(g) \qquad \Delta H° = -17.9 \text{ kcal} \qquad (6)$$

The **heat of formation** (or **enthalpy of formation**), $\Delta H_f$, of a compound is the enthalpy change for the reaction in which one mole of a compound is formed from its constituent elements. Enthalpies of formation are expressed in kilocalories per mole of the compound formed (a *product*).

Equation (2) of the preceding discussion is the equation for the enthalpy of formation of $H_2O(l)$, equation (3) gives the enthalpy of formation of $CO_2(g)$, and equation (6) represents the enthalpy of formation of $CH_4(g)$. Notice that equation (2) represents both the enthalpy of combustion of $H_2(g)$ and the enthalpy of formation of $H_2O(l)$. When a compound cannot be made directly from its constituent elements, the law of Hess may be used to calculate its $\Delta H_f$. Notice that the enthalpy of formation of an *element* is zero (the heat effect when an element is prepared from itself).

Enthalpies of formation are convenient to use for thermochemical calculations. Thus the enthalpy of hydrogenation of ethylene, $C_2H_4(g)$,

$$C_2H_4(g) + H_2(g) \rightarrow C_2H_6(g)$$

can be calculated from the enthalpy of formation of ethane, $C_2H_6(g)$, and the enthalpy of formation of ethylene.

$$2C(graphite) + 3H_2(g) \rightarrow C_2H_6(g) \qquad \Delta H_f^\circ = -20.2 \text{ kcal} \qquad (7)$$

$$C_2H_4(g) \rightarrow 2C(graphite) + 2H_2(g) \qquad \Delta H^\circ = -12.5 \text{ kcal} \qquad (8)$$

Equation (8) is the reverse of the equation for the formation of $C_2H_4(g)$. Note that the sign of the $\Delta H$ value had been changed [$\Delta H_f^\circ$ for $C_2H_4(g)$ is $+12.5$ kcal/mol]. Addition of equations (7) and (8) gives

$$C_2H_4(g) + H_2(g) \rightarrow C_2H_6(g) \qquad \Delta H^\circ = -32.7 \text{ kcal} \qquad (9)$$

The enthalpy change of a reaction can be calculated by subtracting the sum of the enthalpies of formation of the reactants from the sum of the enthalpies of formation of the products.

$$\Delta H = \Sigma \, \Delta H_f(\text{products}) - \Sigma \, \Delta H_f(\text{reactants})$$

The preceding problem could have been solved in this way; thus, $\Delta H^\circ$ for equation (9) is

$$\Delta H^\circ = \Delta H_f^\circ(C_2H_6) - \Delta H_f^\circ(C_2H_4)$$
$$= -20.2 \text{ kcal} - (+12.5 \text{ kcal})$$
$$= -32.7 \text{ kcal}$$

When we were using thermochemical equations, we reversed the equation for the enthalpy of formation of $C_2H_4(g)$ and changed the sign of $\Delta H_f^\circ$ and thus indicated the subtraction of this value. Notice that even though our calculation involves no term for $H_2(g)$, the proper amount of $H_2(g)$ is indicated in equation (9).

To calculate the enthalpy of reaction for

$$Fe_2O_3(s) + 3CO(g) \rightarrow 2Fe(s) + 3CO_2(g)$$

the enthalpies of formation of the compounds represented in the equation are treated in the following manner.

$$\Delta H^\circ = 3 \, \Delta H_f^\circ(CO_2) - [\Delta H_f^\circ(Fe_2O_3) + 3 \, \Delta H_f^\circ(CO)]$$
$$= 3(-94.1) - [(-196.5) + 3(-26.4)]$$
$$= -282.3 + 275.7 = -6.6 \text{ kcal}$$

The calculation may be checked by addition of the thermochemical equations for the reactions involved, written in the way indicated in the calculation. Some standard enthalpies of formation are listed in Table 14.1 of Section 14.2.

**5.9 BORN–HABER CYCLE**

The law of Hess is the justification for a method of analysis of thermochemical processes developed independently by Max Born and Fritz Haber in 1919. A **Born–Haber cycle** shows how a reaction or a physical process may be considered to be made up of several component steps. By a relatively simple calculation, the heat effect of one of the steps or of the entire process may be determined. Born–Haber cycles may also be used to analyze reactions or groups of reactions in order to see how they are affected by variations in one of the individual steps.

The calculation of the lattice energy of the sodium chloride crystal will serve as an illustration of a Born–Haber cycle. The enthalpy of formation of NaCl(s) is −98 kcal/mol.

$$Na(s) + \tfrac{1}{2}Cl_2(g) \rightarrow NaCl(s) \qquad \Delta H_f^\circ = -98 \text{ kcal}$$

We can *imagine* one mole of NaCl(s) to be prepared by the following series of steps.

1. Crystalline sodium metal is sublimed into gaseous sodium atoms; 26 kcal of heat is *absorbed* per mole of Na.

$$Na(s) \rightarrow Na(g) \qquad \Delta H_{\text{subl.}} = +26 \text{ kcal}$$

2. *One-half mole* of gaseous $Cl_2$ molecules is dissociated into one *mole* of gaseous Cl atoms; 58 kcal of heat is *absorbed* per *mole* of $Cl_2(g)$. This value is called the bond energy of the Cl—Cl bond or the enthalpy of dissociation of $Cl_2(g)$.

$$\tfrac{1}{2}Cl_2(g) \rightarrow Cl(g) \qquad \tfrac{1}{2}\Delta H_{\text{diss.}} = \tfrac{1}{2}(+58 \text{ kcal}) = +29 \text{ kcal}$$

3. The gaseous sodium atoms are ionized into gaseous sodium ions; the amount of heat *required* can be calculated from the ionization potential of sodium.

$$Na(g) \rightarrow Na^+(g) + e^- \qquad \Delta H_{\text{i.p.}} = +118 \text{ kcal}$$

4. Electrons are added to the gaseous chlorine atoms to produce gaseous chloride ions. The amount of heat *liberated* per mole of Cl(g) is the electron affinity of chlorine.

$$Cl(g) + e^- \rightarrow Cl^-(g) \qquad \Delta H_{\text{e.a.}} = -83 \text{ kcal}$$

5. The gaseous ions condense into one mole of crystalline sodium chloride. This heat effect is defined as the lattice energy.

$$Na^+(g) + Cl^-(g) \rightarrow NaCl(s) \qquad \Delta H_{\text{l.e.}} = ?$$

Figure 5.1 Born–Haber
cycle for NaCl(s).

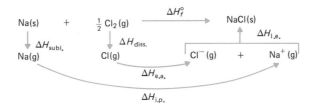

According to the law of Hess, the total $\Delta H$ for all the steps of the hypothetical preparation should equal the value of $\Delta H_f^\circ$ for NaCl(s), −98 kcal/mol. Therefore,

$$\Delta H_f^\circ = \Delta H_{\text{subl.}} + \tfrac{1}{2}\,\Delta H_{\text{diss.}} + \Delta H_{\text{i.p.}} + \Delta H_{\text{e.a.}} + \Delta H_{\text{l.e.}}$$

$$-98 \text{ kcal} = +26 \text{ kcal} + 29 \text{ kcal} + 118 \text{ kcal} - 83 \text{ kcal} + \Delta H_{\text{l.e.}}$$

$$-98 \text{ kcal} = +86 \text{ kcal} + \Delta H_{\text{l.e.}}$$

$$\Delta H_{\text{l.e.}} = -188 \text{ kcal}$$

The Born-Harber cycle for NaCl(s) is summarized in Figure 5.1.

For the NaCl Born–Harber cycle, all the thermal values except the lattice energy can be directly measured, and hence the cycle may be used, as we have used it, to calculate the lattice energy of NaCl. However, the direct determination of electron affinity is difficult; accurate values have been directly obtained for only a few elements. For a given cycle, therefore, both lattice energy and electron affinity may be unknown. In such a case, a lattice energy value calculated from crystal parameters may be employed (Section 7.13) and the cycle used to secure an electron affinity.

Other types of thermochemical cycles have theoretical importance. By means of these cycles, energy effects of processes may be determined that can be evaluated neither by direct experimentation nor by theoretical computation.

## Problems

Thermochemical problems may be solved using either calories or joules as energy units. Thermochemical values expressed in joules appear in brackets.

**5.1** Only one isotope of fluorine occurs in nature, $^{19}_{9}\text{F}$. The atomic weight of fluorine is 19.0. What is the weight (in grams) of a single $^{19}_{9}\text{F}$ atom?

**5.2** If one atom of element X, for which only a single isotope occurs in nature, weighs $2.107 \times 10^{-22}$ g, what is the atomic weight of X?

**5.3** How many moles and how many molecules are present in 10.0

g of each of the following? (a) $H_2$, (b) $F_2$, (c) HF, (d) $F_2O$.

**5.4** How many atoms are present in each of the samples described in Example 5.3?

**5.5** What percentage of wolframite, $FeWO_4$, is tungsten?

**5.6** What percentage of the mineral ilmenite, $FeTiO_3$, is titanium? iron?

**5.7** How many grams of zirconium are theoretically obtainable from 1.00 kilogram of zircon ore that is 73.0% $ZrSiO_4$?

**5.8** How many grams of beryllium are theoretically obtainable from 1.00 kilogram of beryl ore that is 65.0% $Be_3Al_2(SiO_3)_6$?

**5.9** (a) A compound contains 52.9% C and 47.1% O. What is the empirical formula of the compound? (b) The molecular weight of the compound is 68.0. What is the molecular formula of the compound?

**5.10** Ethylene oxide is 54.5% C, 36.4% O, and 9.1% H. What is the empirical formula of ethylene oxide?

**5.11** What is the empirical formula of a compound that is 56.3% P and 43.7% S?

**5.12** What is the empirical formula of a compound that is 68.4% Ce and 31.6% B?

**5.13** What is the empirical formula of a compound that is 1.12% H, 36.0% S, and 62.9% O?

**5.14** What is the empirical formula of a compound that is 4.55% H, 47.0% P, and 48.5% O?

**5.15** What is the empirical formula of a compound that is 30.2% Si, 8.59% O, and 61.2% F?

**5.16** Determine the molecular formulas of the compounds for which the following empirical formulas and molecular weights pertain. (a) $CH_2$, 42.0, (b) $SbO_2$, 307.6, (c) $B_2H_3$, 98.4, (d) SeS, 111.1, (e) SCl, 135.2.

**5.17** Determine the molecular formulas of the compounds for which the following empirical formulas and molecular weights pertain. (a) VOCl, 102.4, (b) $Co(CO)_3$, 571.6, (c) BCl, 185.2, (d) $CN_2H_2$, 126.0, (e) $SiCl_3$, 269.2.

**5.18** A sample of a compound that contains only C and H was burned in oxygen, and 7.92 g of $CO_2$ and 3.24 g of $H_2O$ were obtained. (a) How many moles of C atoms and moles of H atoms did the sample contain? (b) What is the empirical formula of the compound? (c) What was the weight of the sample that was burned?

**5.19** A sample of a compound that contains only C, H, and N was burned in oxygen, and 7.92 g of $CO_2$, 4.86 g of $H_2O$, and 1.26 g of $N_2$ were obtained. (a) How many moles of C atoms, moles of H atoms, and moles of N atoms did the sample contain? (b) What is the empirical formula of the compound? (c) What was the weight of the sample that was burned?

**5.20** Sulfur-containing compounds are an undesirable component of some oils. The amount of sulfur in an oil can be determined by oxidizing the S to the sulfate ion, $SO_4^{2-}$, and precipitating the sulfate ion as barium sulfate, $BaSO_4$, which can be collected, dried, and weighed. From a 5.00 g sample of an oil, 0.262 g of $BaSO_4$ was obtained. What is the percent S in the oil?

**5.21** Write balanced equations for the following reactions.

(a) Tungsten carbide, WC(s), and oxygen (g) yield tungsten(VI) oxide (s) and carbon dioxide (g).

(b) Ammonium dichromate (s) upon heating yields nitrogen (g), water (g), and chromium(III) oxide (s).

(c) Sulfur tetrafluoride (g) and water (l) yield sulfur dioxide (g) and hydrogen fluoride (g).

(d) Gold(I) ions (aq) yield gold(III) ions (aq) and gold (s).

(e) Calcium cyanamide, $CaCN_2(s)$ and water (l) yield calcium carbonate (s) and ammonia (g).

**5.22** Write balanced equations for the following reactions.

(a) Ammonia (g) and fluorine (g) yield dinitrogen tetrafluoride (g) and hydrogen fluoride (g).

(b) Vanadium(V) oxide (s) and hydrogen (g) yield vanadium(III) oxide (s) and water (g).

(c) Nitrogen dioxide (g) and water (l) yield nitric acid (l) and nitrogen oxide (g).

(d) Ethane, $C_2H_6$ (g), and oxygen (g) yield carbon dioxide (g) and water (l).

(e) Boron(III) oxide (s) and carbon (graphite) at high temperatures yield boron carbide, $B_4C(s)$, and carbon monoxide (g).

**5.23** Write balanced equations for the following reactions.

(a) Ammonia (g) and oxygen (g) yield nitrogen (g) and water (g).

(b) Ammonia (g) and oxygen (g) yield nitrogen oxide (g) and water (g) when catalyzed by Pt.

(c) Aluminum sulfide (s) and water (l) yield aluminum hydroxide (s) and dihydrogen sulfide (g).

(d) Silicon tetrachloride (l) and silicon (s) at high temperatures yield disilicon hexachloride (l).

(e) Bismuth (s) and oxygen (g) yield bismuth(III) oxide (s).

**5.24** Element X reacts with $O_2$ to give a compound with the formula $X_2O$. (a) Write the equation for the reaction. (b) If 4.89 g of X reacts with 1.00 g of $O_2$, how many grams of X react with one mole of $O_2$? (c) According to the equation, how many moles of X react with one mole of $O_2$? (d) What is the atomic weight of X?

**5.25** Upon heating, sodium azide $(NaN_3)$ decomposes to elements; this reaction is a laboratory source of pure $N_2(g)$. (a) Write the equation for the reaction. (b) How many moles of $N_2$ are obtained from the decomposition of one mole of $NaN_3$? (c) How many grams of $NaN_3$ are needed to prepare 1.00 g of $N_2$? (d) How many grams of Na are produced when one mole of $N_2$ is prepared? (e) How many grams of Na are produced when 1.00 g of $N_2$ is prepared?

**5.26** Upon heating a 7.29 g sample of $VF_3 \cdot xH_2O$, a hydrate of vanadium(III) fluoride, the water was driven off and 4.86 g of anhydrous $VF_3$ remained. (a) How many grams of water were driven off? How many moles? (b) How many moles of $VF_3$ were present in the anhydrous sample remaining at the end of the experiment? (c) How many moles of water were combined with one mole of $VF_3$ in the hydrate? (d) What is the value of $x$ in the formula?

**5.27** Upon heating a 4.47 g sample of $NiSO_4 \cdot xH_2O$, a hydrate of nickel(II) sulfate, the water was driven off and 2.63 g of anhydrous $NiSO_4$ remained. What is the value of $x$ in the formula $NiSO_4 \cdot xH_2O$?

**5.28** A metal oxide has the formula $XO_3$ and reacts with $H_2$ to give the free metal, X and $H_2O$. (a) Write the equation for the reaction. (b) A 3.31 g sample of $XO_3$ yields 1.24 g of $H_2O$. How many moles of $H_2O$ are produced? How many moles of $XO_3$ are present in the sample? What is the molecular weight of $XO_3$? (c) What is the atomic weight of X?

**5.29** A highly reactive form of silicon can be made by the reaction

$$3CaSi_2 + 2SbCl_3 \rightarrow$$
$$6Si + 2Sb + 3CaCl_2$$

(a) What weight of $CaSi_2$ should be taken to prepare 1.00 g of Si? (b) What weight of $SbCl_3$ would be required for the preparation?

**5.30** Calcium phosphide, $Ca_3P_2$, reacts with water to yield calcium hydroxide and phosphine gas, $PH_3$. (a) Write the equation for the reaction. (b) What weight of calcium phosphide should be used to make 0.750 g of phosphine?

**5.31** Sulfamic acid, $HOSO_2(NH_2)$, is made in the laboratory by a complex reaction that may be summarized as

$$CO(NH_2)_2 + H_2SO_4 + SO_3 \rightarrow$$
$$2HOSO_2(NH_2) + CO_2$$

(a) What weight of sulfamic acid should be secured from 14.0 g of urea, $CO(NH_2)_2$, if an excess of the other reagents is employed? (b) If 39.5 g of sulfamic acid is isolated, what percentage of the theoretical yield is obtained?

**5.32** Antimony(III) chloride can be prepared in the laboratory by the reaction of antimony(III) sulfide and hydrochloric acid; the other product of the reaction is dihydrogen sulfide gas. (a) Write the equation for the reaction. (b) What weight of antimony(III) chloride should be obtained from 5.00 g of antimony(III) sulfide? (c) If 5.00 g of the chloride is isolated, what percentage of the theoretical yield is obtained?

**5.33** Copper(I) chloride may be prepared by reacting an aqueous solution of copper(II) chloride with metallic copper. (a) Write an equation for the reaction. (b) How many grams of copper(I) chloride are obtainable from 2.50 g of copper(II) chloride and 1.27 g of copper?

**5.34** Sodium azide, $NaN_3$, can be prepared by the reaction

$$3NaNH_2 + NaNO_3 \rightarrow$$
$$NaN_3 + 3NaOH + NH_3$$

What weight of $NaN_3$ should be obtained from a mixture of 3.12 g of $NaNH_2$ and 3.40 g of $NaNO_3$?

**5.35** Tartaric acid, $H_2C_4H_4O_6$, is a dibasic acid. The potassium acid salt $KHC_4H_4O_6$ is called *cream of tartar* and is found as a precipitate in wine vats. (a) What weight of NaOH is needed for the complete neutralization of an aqueous solution containing 3.00 g of $H_2C_4H_4O_6$? (b) What weight of NaOH should be used to prepare $NaHC_4H_4O_6$ from 4.50 g of $H_2C_4H_4O_6$ in water solution? (c) What are the products, in moles, of a process in which 2.00 g of NaOH in aqueous solution is added to a water solution containing 6.00 g of $H_2C_4H_4O_6$?

**5.36** A mixture of sodium oxide, $Na_2O$, and barium oxide, BaO, that weighs 6.00 g is dissolved in water, and this solution is then treated with dilute sulfuric acid, $H_2SO_4$. Barium sulfate, $BaSO_4$, precipitates from the solution, but sodium sulfate, $Na_2SO_4$, is soluble and remains in solution. The $BaSO_4$ is collected by filtration and is found to weigh 6.00 g when dried. What percentage of the original sample of mixed oxides is BaO?

**5.37** A 6.00 g sample of a mixture of calcium carbonate, $CaCO_3$, and calcium sulfate, $CaSO_4$, is heated to decompose the carbonate.

$$CaCO_3(s) \rightarrow CaO(s) + CO_2(g)$$

Calcium sulfate is not decomposed by heating. The final weight of the sample is 4.68 g. What percentage of the original mixture is $CaCO_3$?

**5.38** A 3.86 g sample of a mixture of $CaCO_3$ and $NaHCO_3$ is heated and the compounds decompose.

$$CaCO_3(s) \rightarrow CaO(s) + CO_2(g)$$

$$2NaHCO_3(s) \rightarrow$$
$$Na_2CO_3(s) + CO_2(g) + H_2O(g)$$

The decomposition of the sample yielded 1.10 g $CO_2$ and 0.360 g $H_2O$. What percentage of the original mixture was $CaCO_3$?

**5.39** A 1.49 g sample of ethyl alcohol, $C_2H_5OH$, was burned in excess oxygen in a calorimeter that contained 1350 g of water. The heat capacity of the calorimeter was 350 cal/°C. The temperature of the calorimeter and contents increased from 22.22° to 28.45°C. What is the molar enthalpy of combustion of ethyl alcohol?

**5.40** Glycine, $C_2H_5O_2N$, is a simple amino acid that occurs in proteins. A 2.50 g sample of glycine was burned in excess oxygen in a calorimeter that contained 1620 g of water. The temperature of the calorimeter and contents increased from 24.46° to 28.07°C. The heat capacity of the calorimeter was 530 cal/°C. What is the molar enthalpy of combustion of glycine?

**5.41** The enthalpy of combustion of sucrose (commonly called cane sugar), $C_{12}H_{22}O_{11}$, is $-1350.0$ kcal/mol [or $-5648.4$ kJ/mol]. A 2.00 g sample of sucrose was burned in excess oxygen in a calorimeter which contained 1500 g of water. The heat capacity of the calorimeter was 450 cal/°C [or 1883 J/°C]. The initial temperature of the calorimeter and its contents was 18.26°C. What is the final temperature?

**5.42** The enthalpy of combustion of oxalic acid, $H_2C_2O_4$, is $-58.82$ kcal/mol [or $-246.1$ kJ/mol]. A 5.00 g sample of oxalic acid is burned in excess oxygen in a calorimeter containing 1250 g of water. The temperature of the calorimeter and contents increased from 24.00° to 26.04°C. What is the heat capacity of the calorimeter?

**5.43** It is sometimes said that the standard enthalpy of all elements, $H°$, is set equal to zero in determining $\Delta H_f°$ values for compounds. Criticize this statement.

**5.44** In Section 5.8, $\Delta H°$ for the reaction of $Fe_2O_3(s)$ and $CO(g)$ was calculated. Check the calculation by writing and adding thermochemical equations in such a way that the desired equation is obtained.

**5.45** Calculate the enthalpy of combustion of acetylene, $C_2H_2(g)$, to $CO_2(g)$ and $H_2O(l)$. The enthalpies of formation are: $C_2H_2(g)$, +54.19 kcal/mol [or +226.7 kJ/mol]; $CO_2(g)$, −94.05 kcal/mol [or −393.5 kJ/mol]; and $H_2O(l)$, −68.32 kcal/mol [or −285.9 kJ/mol].

**5.46** Calculate the value of $\Delta H°$ for the reaction

$$P_4O_{10}(s) + 6PCl_5(s) \rightarrow$$
$$10\ POCl_3(l)$$

The enthalpies of formation are: $P_4O_{10}(s)$, −720.0 kcal/mol [or −3012 kJ/mol]; $PCl_5(s)$, −110.7 kcal/mol [or −463.2 kJ/mol]; $POCl_3(l)$, −151.0 kcal/mol [or −631.8 kJ/mol].

**5.47** Calculate the value of $\Delta H°$ for the reaction

$$2NiS(s) + 3O_2(g) \rightarrow$$
$$2NiO(s) + 2SO_2(g)$$

The enthalpies of formation are: $NiS(s)$, −17.5 kcal/mol [or −73.2 kJ/mol]; $NiO(s)$, −58.4 kcal/mol [or −244.3 kJ/mol]; $SO_2(g)$, −71.0 kcal/mol [or −297 kJ/mol].

**5.48** Determine the value of $\Delta H°$ for the reaction

$$C_6H_6(l) + 3H_2(g) \rightarrow C_6H_{12}(l)$$

from the enthalpies of combustion of the following reactants. The products of combustion are $CO_2(g)$ and/or $H_2O(l)$. For benzene, $C_6H_6(l)$, $\Delta H° = -781.0$ kcal/mol [or −3268 kJ/mol]; for cyclohexane, $C_6H_{12}(l)$, $\Delta H° = -936.9$ kcal/mol [or −3920 kJ/mol]. For hydrogen, $H_2(g)$, $\Delta H° = -68.3$ kcal/mol [or −286 kJ/mol].

**5.49** The enthalpy of formation of $CO_2(g)$ is −94.1 kcal/mol [or −394 kJ/mol] and the enthalpy of formation of $H_2O(l)$ is −68.3 kcal/mol [or −286 kJ/mol]. Use these values and values given in Problem 5.48 to find the enthalpy of formation of benzene, $C_6H_6(l)$ and that of cyclohexane, $C_6H_{12}(l)$.

**5.50** The enthalpy of formation of $Fe_2O_3(s)$ is −196.5 kcal/mol [or −822.2 kJ/mol] and the enthalpy of combustion of $CO(g)$ is −67.7 kcal/mol [or −283.3 kJ/mol]. Write thermochemical equations to show how these two relations can be used to calculate $\Delta H°$ for

$$Fe_2O_3(s) + 3CO(g) \rightarrow$$
$$2Fe(s) + 3CO_2(g)$$

**5.51** For the following reaction, $\Delta H° = -96.0$ kcal/mol [or −401.7 kJ/mol].

$$CH_4(g) + 4Cl_2(g) \rightarrow$$
$$CCl_4(l) + 4HCl(g)$$

Calculate the enthalpy of formation of $CCl_4(l)$. The enthalpy of formation of $CH_4(g)$ is $-17.9$ kcal/mol [or $-74.9$ kJ/mol] and that of $HCl(g)$ is $-22.1$ kcal/mol [or $-92.5$ kJ/mol].

**5.52** For the following reaction, $\Delta H° = -241.4$ kcal [or $-1010.0$ kJ].

$$2NH_3(g) + 3N_2O(g) \rightarrow$$
$$4N_2(g) + 3H_2O(l)$$

Calculate the enthalpy of formation of $N_2O(g)$. The enthalpy of formation of $NH_3(g)$ is $-11.0$ kcal/mol [or $-46.0$ kJ/mol], and the enthalpy of formation of $H_2O(l)$ is $-68.3$ kcal/mol [or $-285.8$ kJ/mol].

**5.53** Given the following heats of reaction:

$$4NH_3(g) + 5O_2(g) \rightarrow$$
$$4NO(g) + 6H_2O(l)$$
$$\Delta H = -279.4 \text{ kcal}$$
$$[\text{or} -1169 \text{ kJ}]$$

$$4NH_3(g) + 3O_2(g) \rightarrow$$
$$2N_2(g) + 6H_2O(l)$$
$$\Delta H = -365.8 \text{ kcal}$$
$$[\text{or} -1531 \text{ kJ}]$$

calculate the heat of formation of $NO(g)$.

**5.54** Use the following data to calculate the lattice energy of LiF. The enthalpy of formation of LiF is $-139.5$ kcal/mol [or $-583.7$ kJ/mol]. The enthalpy of sublimation of Li is $+37.1$ kcal/mol [or $+155.2$ kJ/mol], and the first ionization energy of Li is $+124.3$ kcal/mol [or $+520.1$ kJ/mol]. The

dissociation energy of $F_2$ is $+37.8$ kcal/mol $F_2$ [or $+158.2$ kJ/mol $F_2$], and the electron affinity of the flourine atom (F) is $-79.6$ kcal/mol F atoms [or $-330.0$ kJ/mol F atoms].

**5.55** Use the following data to calculate the first electron affinity of iodine. The enthalpy of formation of KI is $-78.3$ kcal/mol [or $-327$ kJ/mol], and the lattice energy of KI is $-153.9$ kcal/mol [or $-643.9$ kJ/mol]. The enthalpy of sublimation of K is $+21.5$ kcal/mol [or $+90.0$ kJ/mol], and the first ionization energy of K is $+99.2$ kcal/mol [or $+415$ kJ/mol]. The enthalpy of the reaction

$$I_2(s) \rightarrow 2I(g)$$

is $+51.0$ kcal [or $+213$ kJ].

**5.56** (a) Calculate the lattice energy of CaO from the following data. The enthalpy of formation of CaO is $-151.9$ kcal/mol [or $-635.5$ kJ/mol]. The enthalpy of sublimation of Ca is $+46.0$ kcal/mol [or $+192.5$ kJ/mol]; the first ionization energy of Ca is $+140.9$ kcal/mol [or $589.5$ kJ/mol], and the second ionization energy of Ca is $+273.7$ kcal/mol [or $+1145.2$ kJ/mol]. The dissociation energy of $O_2$ is $+118.3$ kcal/mol $O_2$ [or $+495.0$ kJ/mol $O_2$]; the first electron affinity of the oxygen atom is $-33.9$ kcal/mol O atoms [or $-141.8$ kJ/mol O atoms], and the second electron affinity of the oxygen atom is $+201.8$ kcal/mol O atoms [or $+844.3$ kJ/mol O atoms]. (b) The $Na^+$ ion is about the same size as the $Ca^{2+}$ ion, and the $F^-$ ion is about the same size as the $O^{2-}$ ion. The melting point of CaO is 2580°C and that of NaF is 988°C. What explanation can you offer for the difference?

**5.57** (a) Calculate the lattice energy of $MgCl_2$ from the following data. The enthalpy of formation of $MgCl_2$ is $-153$ kcal/mol [or $-640$ kJ/mol]. The enthalpy of sublimation of Mg is $+36$ kcal/mol [or $+151$ kJ/mol]; the first ionization energy of Mg is $+176$ kcal/mol [or $+736$ kJ/mol], and the second ionization energy is $+345$ kcal/mol [or $+1443$ kJ/mol]. The dissociation energy of $Cl_2$ is $+58$ kcal/mol $Cl_2$ [or $+243$ kJ/mol $Cl_2$], and the first electron affinity of the Cl atom is $-83$ kcal/mol Cl atoms [or $-347$ kJ/mol Cl atoms]. Remember that a mole of $MgCl_2$ contains two moles of $Cl^-$ ions. (b) If Mg formed a $Mg^+$ ion (it does not), the ion would be approximately the same size as the $Na^+$ ion. The lattice energy of the hypothetical MgCl would probably be nearly the same as the lattice energy of NaCl, which is $-188$ kcal/mol [or $-787$ kJ/mol]. Use this value to calculate the enthalpy of formation of "MgCl." Compare the enthalpies of formation and the lattice energies of $MgCl_2$ and "MgCl."

# 6
# Gases

Aristotle (384–322 B.C.), who summarized the theories of earlier Greek philosophers, believed that all terrestial matter was derived from four elements — earth, water, air, and fire. Three of the four elements of Aristotle relate to the three states of matter — solid, liquid, and gas; the fourth relates to energy.

For centuries scientists regarded all gases as forms of air or as mixtures of air with various impurities. Johann van Helmont, the first to perceive the fundamental difference between gases, coined the name *gas* from the Greek word *chaos* in the early seventeenth century. However, not until the time of Antoine Lavoisier (late eighteenth century) was the full importance of this distinction recognized. In 1778 Lavoisier published a correct interpretation of combustion reactions, based on the realization that *air* consists of oxygen *gas* and nitrogen *gas*.

**6.1 THE GASEOUS STATE**    A study of the properties of gases leads to the conclusion that gases consist of widely separated molecules in rapid motion. The most familiar gas, air, is actually a mixture of gases. Experiments show that any two (or more) gases can be used in any proportion to prepare a perfectly homogeneous mixture; no such generalization can be made for liquids. Since the molecules of any gas are thought to be separated by comparatively large distances, it is reasonable that one gas can accommodate molecules of other gases. This molecular model can also be used to explain the compressibility of gases; compression consists of forcing gas molecules closer together.

A gas diffuses throughout any container into which it is introduced. Odorous gases released in a room can soon be detected in all parts of the room. The premise that gas molecules are in constant, rapid motion accounts for the diffusion of gases. Furthermore, gas molecules in their random motion strike the walls of the container, and these myriad impacts explain the fact that gases exert pressure.

**6.2 PRESSURE**    **Pressure** is defined as force per unit area. The pressure of a gas is equal to the force that the gas exerts on the walls of a container divided by the surface area of the container. The SI unit of force is the newton ($1 \text{ N} = 1$ kg m/sec$^2$), and the derived unit of pressure is one newton per square meter ($1 \text{ N/m}^2$), for which the General Conference of Weights and Measures has recently adopted the name *pascal* (Pa). The chemist, however, usually measures gas pressures by relating them to the pressure of the atmosphere.

The atmosphere exerts a pressure on the surface of the earth. The instrument used to measure atmospheric pressure is called a **barometer** and was first devised by Evangelista Torricelli, a pupil of Galileo. A long tube, sealed at one end, is filled with mercury and inverted in an open container of mercury (Figure 6.1). If the tube is sufficiently long, the mercury falls in the tube. However, it will not completely run out; the pressure of the atmosphere on the surface of the mercury in the dish supports the column. Practically no pressure is exerted on the upper surface of the mercury in the column. Since mercury is not very volatile at room temperature, the amount of mercury vapor in the space above the liquid mercury in the tube is small; air has been excluded. A vacuum obtained in this manner is called a Torricellian vacuum. The pressure inside the tube and above the reference level indicated in Figure 6.1 is due only to the weight of the mercury column; this pressure is equal to

Figure 6.1 **Barometer.**

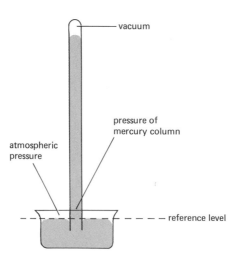

vacuum

pressure of
mercury column

atmospheric
pressure

reference level

the atmospheric pressure outside the tube and above the reference level.

The height of mercury in the tube serves as a measure of the atmospheric pressure. Remember that pressure is force *per unit area.* Whether the tube has a relatively large or small cross-sectional area, a given atmospheric pressure will support the mercury in a tube to the same height.

When normal atmospheric pressure is measured at sea level, the height of a column of mercury is 76.0 cm, or 760 mm, if the mercury is at O °C. The **standard atmosphere (atm)** was originally defined in these terms. The torr (named for Torricelli) is the pressure equivalent of 1 mm of mercury; 1 atm equals 760 torr.

● The standard atmosphere (atm) may be used with SI units for a limited time.

The standard atmosphere is now defined as the pressure equal to 101,325 Pa ($1.01325 \times 10^5$ N/m²); the International Committee of Weights and Measures recommends that the torr not be used. We shall use the atmosphere as the unit of pressure wherever possible instead of the torr. However, in certain instances it is convenient to use mm of Hg, or torr (with subsequent conversion into atmospheres, if needed). For example, the vapor pressures of liquids are usually recorded in mm of mercury, or torr; low pressures, in general, are conveniently expressed in terms of a small pressure unit, such as the torr. In addition, it is convenient to use mm of mercury, or torr, when measuring gas pressures by means of a manometer.

A **manometer,** a device used to measure the pressure of a sample of gas (Figure 6.2), is patterned after the barometer. In Figure 6.2 the right arm of the manometer is open to the atmosphere; thus, atmospheric pressure is exerted on the mercury in this arm. The left arm is connected to a container of gas and allows the gas sample to exert pressure on the mercury. Since the mercury level is lower in the left arm than that in the

Figure 6.2 A type of manometer.

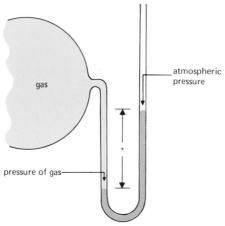

gas

atmospheric pressure

pressure of gas

* difference in height of mercury levels

right, the illustration indicates a situation in which the gas pressure is greater than atmospheric pressure. If the gas sample were under a pressure equal to atmospheric pressure, the mercury would stand at the same level in both sides of the U tube. In the experiment illustrated in Figure 6.2, the difference in height between the two mercury levels (in mm of mercury) must be added to the atmospheric pressure (in mm of mercury, or torr) to obtain the pressure of the gas (in mm of mercury, or torr). The final pressure may be converted into atmospheres by dividing by 760. If the pressure of the gas were less than atmospheric, the mercury in the left arm would stand higher than that in the right arm, and the difference in height would have to be subtracted from atmospheric pressure.

**6.3 BOYLE'S LAW** The quantitative relationship between the volume and pressure of a sample of gas was first summarized by Robert Boyle in 1662. **Boyle's law** states: at constant temperature, the volume of a sample of gas varies inversely with the pressure under which it is measured. If the pressure is doubled, the volume is reduced to half its former value; if the pressure is cut in half, the volume is doubled. Mathematically,

$$V \propto \frac{1}{P}$$

$$V = \frac{k}{P} \quad \text{or} \quad PV = k$$

The value of the constant depends upon the size of the sample and the temperature. The pressure–volume curve for this equation (Figure 6.3)

Figure 6.3 Pressure–
volume curve for an ideal
gas (Boyle's law).

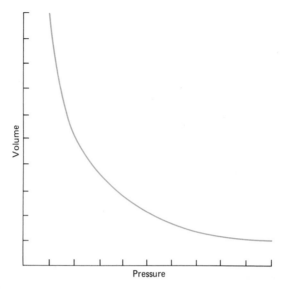

● The SI derived unit of volume is the cubic meter (m³). A liter (also spelled litre) is $10^{-3}$ m³. The liter is not a part of SI, but it may be employed with SI units.

is the first quadrant branch of a hyperbola (negative values for volumes and pressures do not exist). A plot of $V$ versus $1/P$ is a straight line through the origin with a slope equal to the constant, $k$.

Volume is customarily measured in liters. The **liter** is defined as 1 cubic decimeter (1 dm³ = 1000 cm³). Hence, 1 milliliter (1 ml) is 1 cm³, and 1000 ml = 1 liter.

---

*Example 6.1*

A sample of a gas occupies 360 ml under a pressure of 0.750 atm. At constant temperature, what volume will the sample occupy under a pressure of 1.000 atm?

*Solution*

Gas law problems of this type can be solved through the use of correction factors. A correction factor is not equal to unity as the other factors we have employed. For this problem a factor must be used to correct the volume for the change in pressure. Two pressure-correction factors can be derived from the data given: (0.750 atm/1.000 atm) and (1.000 atm/0.750 atm). Which one should be used?

Since the pressure *increases* from 0.750 atm to 1.000 atm, the volume must *decrease* a proportional amount; therefore, the correction factor must be less than 1. Hence,

$$? \text{ ml} = 360 \text{ ml} \left(\frac{0.750 \text{ atm}}{1.000 \text{ atm}}\right) = 270 \text{ ml}$$

---

*Example 6.2*

At 0°C and 5.00 atm pressure, a given sample of a gas occupies 100 liters. If the gas is compressed to 30.0 liters at 0°C, what will be the final pressure?

*Solution*

The volume decreases from 100 to 30.0 liters. Therefore the pressure increases, and the factor must be greater than unity.

$$? \text{ atm} = 5.00 \text{ atm} \left(\frac{100 \text{ liters}}{30.0 \text{ liters}}\right) = 16.7 \text{ atm}$$

**6.4 CHARLES' LAW**  Boyle's law describes the behavior of gas under conditions of constant temperature. Studies to determine the effect of temperature change on the volume and pressure of a gas sample were first undertaken by Jacques Charles (1787); this work was considerably extended by Joseph Gay-Lussac (1802).

Experimental data show that for each Celsius degree rise in temperature, the volume of a gas expands $\frac{1}{273}$ of its value at 0°C if the pressure is held constant. When a 273 ml sample of gas at 0°C is heated to 1°C, the volume of the sample increases by $\frac{1}{273}$ of 273 ml, or 1 ml. At 10°C the volume of the sample would be 283 ml; at −1°C the sample would occupy 272 ml.

Let $V$ equal the volume of a gas sample at temperature $t$ (in °C) and $V_0$ equal the volume of the sample at 0°C. Then

$$V = V_0 + \frac{1}{273}\, tV_0$$

or

$$V = V_0 \left(1 + \frac{t}{273}\right) = V_0 \left(\frac{273 + t}{273}\right)$$

● In the International System the name "kelvin" (K) is given to the unit of thermodynamic temperature. The expression "degrees Kelvin" (°K) should not be used with SI units.

It is convenient to use the absolute temperature scale, on which temperatures are measured in degrees Kelvin (°K). A reading on this scale is obtained by adding 273 to the Celsius value; temperatures on the absolute scale may be denoted by $T$. Thus,

$$T = t + 273$$

Substituting $T$ for $(273 + t)$ in the equation for the dependence of volume on temperature, we get

$$V = V_0 \left(\frac{T}{273}\right) = \left(\frac{V_0}{273}\right) T$$

Since the volume of a given gas sample at 0°C is a constant under constant pressure, this relation can be expressed as

$$V = kT$$

The numerical value of $k$ depends upon the size of the gas sample and the pressure.

**Charles' law** states, therefore, that the volume of a sample of gas varies *directly* with the *absolute* temperature. Volume is *not* directly proportional to the *Celsius* temperature. Thus, doubling the absolute temperature doubles the volume of the gas; doubling the Celsius temperature does not.

The absolute temperature scale was first proposed by William Thomson, Lord Kelvin (1848), and the unit is named in his honor. Any absolute measurement scale must be based on a zero point that represents

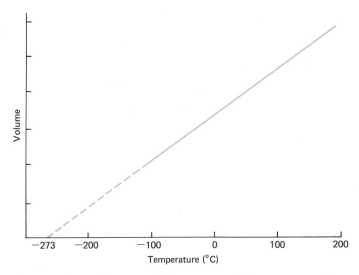

Figure 6.4 Temperature–volume curve for an ideal gas (Charles' law).

the complete absence of the property being measured; on scales of this type negative values are impossible. Thus, 0 cm represents the complete absence of length, and one can say that 10 cm is twice 5 cm because these are absolute measurements. The Celsius temperature scale is not an absolute one; 0°C is the freezing point of water, not the lowest possible temperature. Because of this, negative Celsius temperatures are possible. On the other hand, the Kelvin scale is absolute; 0°K is the lowest possible temperature, and negative Kelvin temperatures are as impossible as negative volumes or negative lengths.

If volume versus temperature is plotted for a sample of gas, a straight line results such as illustrated in Figure 6.4. Since volume is directly

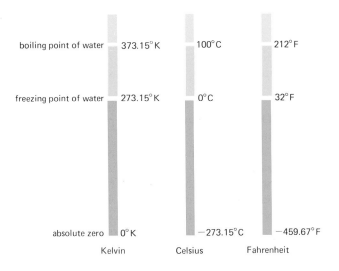

Figure 6.5 Comparison of temperature scales.

proportional to absolute temperature, the volume of the gas *theoretically* should be zero at absolute zero (Section 6.13). Upon cooling, gases liquefy and then solidify before temperatures this low are reached; no substance exists as a gas at a temperature near absolute zero. However, the straight-line curves can be extrapolated to zero volume. The temperature that corresponds to zero volume is −273.15°C; many lines of experimental evidence have confirmed this as absolute zero. The Kelvin degree is the same size as the Celsius degree; however, for the Kelvin scale the zero point is moved to −273.15°C. Conversion to the Kelvin temperature scale can be accomplished, therefore, by adding 273.15 to the Celsius reading. For most problem work this value can be rounded off to 273 without entailing significant error. The Kelvin, Celsius, and Fahrenheit temperature scales are compared in Figure 6.5.

In like manner, the relationship between pressure and temperature can be derived; a temperature–pressure curve is similar to the temperature–volume curve illustrated in Figure 6.4. The pressure of a gas varies directly with absolute temperature when the volume is constant:

$$P = kT$$

It is convenient to define standard reference conditions for work with gases. By convention, **standard temperature and pressure (STP)** are 0°C (273°K) and 1 atm pressure; values for properties of gases are recorded at STP unless otherwise specified.

---

*Example 6.3*

A sample of gas has a volume of 79.5 ml at 45°C. What volume will the sample occupy at 0°C when the pressure is constant?

*Solution*

Since the temperature decreases from 318° to 273°K, the volume must decrease. A factor of less than unity must be employed.

$$? \text{ ml} = 79.5 \text{ ml} \left(\frac{273°\text{K}}{318°\text{K}}\right) = 68.3 \text{ ml}$$

---

*Example 6.4*

A 10.0 liter container is filled with a gas to a pressure of 2.00 atm at 0°C. At what temperature will the pressure inside the container be 2.50 atm?

*Solution*

Since the pressure increases, the temperature must increase. Thus,

$$? °\text{K} = 273°\text{K} \left(\frac{2.50 \text{ atm}}{2.00 \text{ atm}}\right) = 341°\text{K (or } 68°\text{C)}$$

---

**6.5** IDEAL GAS LAW    By combining Boyle's law and Charles' law, it is possible to solve problems in which all three variables — volume, temperature, and pressure — change.

*Example 6.5*

The volume of a sample of a gas is 462 ml at 35°C and 1.15 atm. Calculate the volume of the sample at STP.

*Solution*

A combined gas law problem is best solved by considering each of the changes in conditions separately. First, the temperature decrease from 308°K to 273°K calls for the volume to decrease by a factor of (273°K/308°K). Second, the pressure decreases from 1.15 atm to 1.00 atm. Since volume is inversely proportional to pressure, this change in pressure would increase the volume by a factor of (1.15 atm/1.00 atm).

$$? \text{ ml} = 462 \text{ ml} \left(\frac{273°K}{308°K}\right)\left(\frac{1.15 \text{ atm}}{1.00 \text{ atm}}\right) = 471 \text{ ml}$$

We can derive a mathematical statement for the combined gas law. Since $V \propto 1/P$ and $V \propto T$,

$$V \propto \frac{T}{P}$$

This proportionality can be changed to an equality by the use of a constant.

$$V = k\frac{T}{P}$$

or

$$PV = kT$$

● In SI units $R = 8.314$ J K$^{-1}$ mol$^{-1}$. The joule is a newton meter (1 J = 1 N·m). This value of $R$ is used when $P$ is expressed in pascals (1 Pa = 1 N·m$^{-2}$), $V$ in cubic meters (m$^3$), $n$ in moles (mol), $T$ in kelvins (K).

The value of $k$ in this equation depends upon the size of the gas sample. When one mole of gas is used, the constant is given the designation $R$ and has the numerical value 0.08206 liter atm/°K mol provided the volume is expressed in liters, the pressure in atmospheres, and the temperature in degrees Kelvin. Other values for $R$ are used when other dimensions for these variables are employed. The general form of the equation is

$$PV = nRT$$

where $n$ is the number of moles of the gas under investigation. A gas that follows this equation exactly is called an **ideal gas;** the equation is known as the equation of state for an ideal gas.

*Example 6.6*

What volume will 10.0 g of carbon monoxide occupy at STP? Assume that CO is an ideal gas.

*Solution*

The molecular weight of CO is 28.0, and 10.0 g constitutes 10.0/28.0, or 0.357, mol. The problem is solved by substituting in the equation of state.

$$PV = nRT$$

$$(1 \text{ atm}) V = (0.357 \text{ mol})(0.0821 \text{ liter atm/°K mol})(273°K)$$

$$V = 8.00 \text{ liters}$$

*Example 6.7*
A 500 ml sample of a gas weighs 0.326 g at 100°C and 0.500 atm. What is the molecular weight of the gas?

$$PV = \left(\frac{g}{M}\right) RT$$

The units used for $P$, $V$, and $T$ must correspond to those in which $R$ is expressed. The pressure is 0.500 atm, the volume is 0.500 liter, and the temperature is 373°K.

*Solution*
The number of moles of gas in the sample, $n$, is equal to the weight of the sample, $g$, divided by the molecular weight of the gas, $M$. Therefore,

$$(0.500 \text{ atm})(0.500 \text{ liter}) = \left(\frac{0.326 \text{ g}}{M}\right)(0.0821 \text{ liter atm/°K mol})(373°K)$$

$$M = 39.9 \text{ g/mol}$$

**6.6 KINETIC THEORY**    So far we have examined the laws that correlate observed facts concerning the behavior of gases. The theory that offers a model to explain these laws and observations is known as the **kinetic theory** of gases. This theory was first proposed in substantially the following form by August Krönig (1856) and Rudolf Clausius (1857). However, the theory had its roots in the previous work of many others, notably Daniel Bernoulli (1738) and James Joule (1845–1851); also, later workers, in particular James Clerk-Maxwell (1860) and Ludwig Boltzmann (1871), did much to extend the concept.

The kinetic theory postulates the following for an ideal gas:

1. Gases consist of molecules widely separated in space. The total volume of the molecules is negligible in comparison with the volume of the gas as a whole. Some gases (e.g., the noble gases) are monatomic. In discussions of the kinetic theory of gases, the word molecule is used to designate the smallest particle of any gas.

2. Gas molecules are in constant, rapid, straight-line motion and collide with each other and their container. These collisions are perfectly elastic; energy may be transferred from molecule to molecule, but there is no net decrease in kinetic energy.

3. Gas molecules translate heat energy into kinetic energy. At a given instant, different molecules have different kinetic energies; however, the average kinetic energy of the molecules depends upon the temperature and increases as temperature increases. The molecules of any gas have the same average kinetic energy at a given temperature.

4. No attractive forces exist between the molecules of an ideal gas.

According to the kinetic theory, the gas pressure is caused by molecular collisions with the walls of the container. Therefore, the larger the number of molecules per unit volume — the molecular concentration — the larger is the number of collisions and the higher is the pressure. Reducing the volume of a gas crowds more molecules into a given space and produces a larger molecular concentration and a proportionally higher pressure. Hence, the kinetic theory accounts for Boyle's law.

The average kinetic energy of the molecules of a gas is proportional to the absolute temperature. At absolute zero the molecules are at rest; since the molecular volume of an ideal gas is negligible, the volume of an ideal gas at absolute zero is theoretically zero. As the absolute temperature is increased, the molecules move at increasing speeds. Hence, as the temperature increases, collisions of gas molecules with the walls of the container are more vigorous and more frequent — factors which account for the increased pressures observed when gases are heated. If the pressure is not permitted to change when a gas is heated, its volume must increase so that the number of collisions will be reduced to compensate for the increased intensity of the collisions.

The equation of state for an ideal gas may be derived as follows. Consider a sample of gas that contains $N$ molecules that have a mass of $m$ each. If this sample is enclosed in a cube $l$ cm on a side, the total volume of the gas is $l^3$ cm³. Although the molecules are moving in every possible direction, the derivation is simplified if we assume that one-third of the molecules ($\frac{1}{3} N$) are moving in the direction of the $x$ axis, one-third in the $y$ direction, and one-third in the $z$ direction. For a very large number of molecules, this is a valid simplification since the velocity of each molecule may be divided into an $x$ component, a $y$ component, and a $z$ component.

The pressure of the gas on any wall (surface area $l^2$ cm²) is due to the impacts of the molecules on that wall. The force of each impact can be calculated from the change in momentum per unit time. Consider the shaded wall in Figure 6.6 and take into account only those molecules moving in the direction of the $x$ axis. A molecule moving in this direction will strike this wall every $2 \times l$ cm of its path, since after an impact it must go to the opposite wall (a distance of $l$ cm) and return (a distance of $l$ cm) before the next impact. If the molecule is moving with a velocity of $u$ cm/sec, in one second it will have gone $u$ cm and have made $u/2l$ collisions with the wall under consideration.

Momentum is mass times velocity. Before an impact, the momentum of the molecule is $mu;$ after an impact, the momentum is $-mu$ (the sign is changed because the direction is changed — velocity takes into account speed and direction). Therefore the *change* in momentum equals $2mu$.

In one second a molecule makes $u/2l$ collisions, and the change in

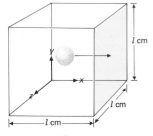

**Figure 6.6** Derivation of the ideal gas laws.

momentum per collision is $2mu$. Therefore, in one second the total change in momentum per molecule is

$$\left(\frac{u}{2l}\right) 2mu = \frac{mu^2}{l}$$

The total change in momentum (force) for all of the molecules striking the wall in one second is

$$\frac{N}{3} \times \frac{mu^2}{l}$$

In this expression, $u^2$ is the average of the squares of all the molecular velocities.

Pressure is force per unit area, and the area of the wall is $l^2$ cm². Therefore, the pressure on the wall under consideration is

$$P = \frac{Nmu^2}{3l} \times \frac{1}{l^2} = \frac{Nmu^2}{3l^3}$$

Since the volume of the cube is $l^3$ cm³, $V = l^3$, and

$$P = \frac{Nmu^2}{3V} \quad \text{or} \quad PV = \tfrac{1}{3}Nmu^2$$

This equation can be written

$$PV = (\tfrac{2}{3}N)(\tfrac{1}{2}mu^2)$$

The kinetic energy of any body is one-half the product of its mass times the square of its speed. The average molecular kinetic energy, therefore, is $\tfrac{1}{2}mu^2$. By substitution,

$$PV = \tfrac{2}{3}N(KE)$$

The average molecular kinetic energy, $KE$, is directly proportional to the absolute temperature, $T$; the number of molecules, $N$, is proportional to the number of moles, $n$. Substitution of these terms requires the inclusion of a constant since these are proportionalities, not equalities. The required constant can be combined with the $\tfrac{2}{3}$ (also a constant); thus,

$$PV = nRT$$

**6.7 AVOGADRO'S PRINCIPLE**   According to the kinetic theory, the molecules of two different gases at the same temperature have the same average kinetic energy. If we use subscripts to relate the measured property to gas A and gas B,

$$KE_A = KE_B$$

and

$$\tfrac{1}{2}m_A u_A^2 = \tfrac{1}{2}m_B u_B^2$$

or

$$m_A u_A^2 = m_B u_B^2$$

If we take equal volumes of these two gases, and if these are collected at the same pressure as well as the same temperature,

$$P_A V_A = P_B V_B$$

From the relationship $PV = \tfrac{1}{3} Nmu^2$,

$$\tfrac{1}{3}N_A m_A u_A^2 = \tfrac{1}{3}N_B m_B u_B^2$$

or

$$N_A m_A u_A^2 = N_B m_B u_B^2$$

But $m_A u_A^2 = m_B u_B^2$; therefore,

$$N_A = N_B$$

We conclude, therefore, that equal volumes of any two gases at the same temperature and pressure contain the same number of molecules. This is a statement of **Avogadro's principle,** which Avogadro derived in a much more empirical manner than described here.

Conversely, equal numbers of molecules of any two gases under the same conditions of temperature and pressure will occupy equal volumes. A mole of any substance contains $6.022 \times 10^{23}$ molecules (Avogadro's number—Section 5.1). According to Avogadro's principle, a mole of any gas should occupy the same volume as a mole of any other gas. At STP this volume, called the **STP molar volume of a gas,** is 22.414 liters (which can be derived by substitution of the appropriate values in $PV = nRT$).

For most calculations the value of the STP molar volume can be rounded off to 22.4 liters. Use of this value simplifies some calculations, although all these calculations can be made by using the equation of state, $PV = nRT$, instead.

---

*Example 6.8*
What is the density of fluorine gas at STP?

*Solution*
The molecular weight of $F_2$ is 38.0. At STP 38.0 g of $F_2$ occupies 22.4 liters. Therefore, the density is 38.0 g/22.4 liters = 1.70 g/liter.

*Example 6.9*

What volume will 10.0 g of carbon dioxide occupy at 27°C and 2.00 atm?

*Solution*

The molecular weight of $CO_2$ is 44.0. Therefore, at STP 22.4 liters of $CO_2$ will weight 44.0 g. To find the volume of 10.0 g of $CO_2$ at STP:

$$? \text{ liters} = 10.0 \text{ g} \left(\frac{22.4 \text{ liters}}{44.0 \text{ g}}\right) = 5.09 \text{ liters}$$

To convert this volume from STP to the conditions specified in the problem:

$$? \text{ liters} = 5.09 \text{ liters} \left(\frac{300°\text{K}}{273°\text{K}}\right)\left(\frac{1.00 \text{ atm}}{2.00 \text{ atm}}\right)$$
$$= 2.80 \text{ liters}$$

This problem can also be solved by use of the equation of state, $PV = nRT$. The number of moles, $n$, is 10.0/44.0, or 0.227.

$$(2.00 \text{ atm}) \, V = (0.227 \text{ mol})(0.0821 \text{ liter atm}/°\text{K mol})(300°\text{K})$$

$$V = 2.80 \text{ liters}$$

*Example 6.10*

(a) Cyclopropane is a gas that is used as a general anesthetic. This gas has a density of 1.56 g/liter at 50°C and 0.984 atm. What is the molecular weight of cyclopropane? (b) The empirical formula of cyclopropane is $CH_2$. What is the molecular formula of the compound?

*Solution*

(a) If we take 1.56 g of the gas, the sample will occupy 1.00 liter under the conditions specified. The volume of the sample at STP would be

$$? \text{ liter} = 1.00 \text{ liter} \left(\frac{273°\text{K}}{323°\text{K}}\right)\left(\frac{0.984 \text{ atm}}{1.00 \text{ atm}}\right)$$
$$= 0.832 \text{ liter}$$

Therefore, 1.56 g of the gas occupies 0.832 liter at STP. The weight of 22.4 liters at STP is the gram molecular weight of the compound.

$$? \text{ g} = 22.4 \text{ liters} \left(\frac{1.56 \text{ g}}{0.832 \text{ liter}}\right) = 42.0 \text{ g}$$

This problem can also be solved by substituting in the equation of state.

$$PV = \left(\frac{\text{g}}{M}\right) RT$$

$$(0.984 \text{ atm})(1.00 \text{ liter}) = \left(\frac{1.56 \text{ g}}{M}\right)(0.0821 \text{ liter atm}/°\text{K mol})(323°\text{K})$$

$$M = 42.0 \text{ g/mol}$$

(b) The formula weight of $CH_2$ is 14.0. Since the molecular weight of the compound is 42.0, the molecular formula is $C_3H_6$.

Avogadro's principle was almost entirely ignored after its presentation in 1811. Stanislao Cannizzaro was the first to see the full significance of this principle, and he was largely responsible for its universal acceptance (1858–1864) after the death of Avogadro.

Cannizzaro was the first to use data on gas densities to assign atomic weights. He based his atomic weight scale on hydrogen and assigned the hydrogen atom a weight of 1, a value which is close to that currently used for hydrogen. Thus, since carbon dioxide gas has a density 22 times greater than that of hydrogen gas, and since the molecular weight of $H_2$ was set equal to 2, $CO_2$ was assigned a molecular weight of 44. Carbon dioxide contains 27.3% carbon; therefore, one molecular weight of carbon dioxide must contain 12 units of carbon.

Similar data for other gaseous compounds containing carbon are summarized in Table 6.1. Notice that the smallest weight of carbon in a **molecular weight of any compound is 12** and that the other weights are multiples of 12. One concludes, therefore, that carbon has an atomic weight of 12, and that the compounds of Table 6.1 contain 1, 2, 2, 1, 1, and 3 atoms of carbon per molecule, respectively.

Table 6.1 Composition of some carbon compounds

| COMPOUND | MOLECULAR WEIGHT | % CARBON | WEIGHT OF CARBON PER MOLE (g) |
|---|---|---|---|
| carbon dioxide | 44 | 27.3 | 12 |
| cyanogen | 52 | 46.2 | 24 |
| ethane | 30 | 80.0 | 24 |
| hydrogen cyanide | 27 | 44.4 | 12 |
| methane | 16 | 75.0 | 12 |
| propane | 44 | 81.8 | 36 |

**6.8** GAY-LUSSAC'S LAW OF COMBINING VOLUMES

In 1808 Gay-Lussac reported the results of his experiments with reacting gases. When measured at constant temperature and pressure, the volumes of gases used or produced in a chemical reaction can be expressed in ratios of small whole numbers. This is a statement of **Gay-Lussac's law of combining volumes.**

For the reaction

$$H_2(g) + Cl_2(g) \rightarrow 2HCl(g)$$

Gay-Lussac found that a given volume of hydrogen reacts with an equivalent volume of chlorine to produce twice this volume of hydrogen chloride. For example, at fixed temperature and pressure, 10 liters of hydrogen will react with 10 liters of chlorine to produce 20 liters of hydrogen chloride. The law is applicable only to gases; the volumes of liquids or solids cannot be treated in a comparable fashion.

The equation shows that equal numbers of hydrogen and chlorine molecules are required for the reaction. According to Avogadro's principle, equal numbers of hydrogen and chlorine molecules are contained in equal volumes of these gases. Since the number of hydrogen chloride molecules produced by the reaction is twice the number of hydrogen molecules used, the volume of hydrogen chloride produced is twice the volume of the hydrogen used.

The total volume of the reacting gases need not equal the volume of the gases produced. This fact is illustrated by another of Gay-Lussac's examples:

$$2CO(g) + O_2(g) \rightarrow 2CO_2(g)$$

For this reaction, two volumes of carbon monoxide and one volume of oxygen are required to produce two volumes of carbon dioxide. The equation for the reaction shows a molecular ratio of carbon monoxide to oxygen to carbon dioxide of 2 to 1 to 2, which is the same as the observed volume ratio. These observations, as the previous ones, are consistent with Avogadro's principle. Thus, if 2 liters of carbon monoxide contain $2x$ molecules, this sample would require 1 liter of oxygen ($x$ molecules) to produce 2 liters of carbon dioxide ($2x$ molecules).

Historically, Gay-Lussac's law of combining volumes preceded Avogadro's principle, and Avogadro derived his statement to explain Gay-Lussac's law.

---

*Example 6.11*

(a) What volume of oxygen is required for the complete combustion of 15.0 liters of ethane, $C_2H_6(g)$, if all gases are measured at 350°C and 1.00 atm?

$$2C_2H_6(g) + 7O_2(g) \rightarrow 4CO_2(g) + 6H_2O(g)$$

(b) What is the total volume of the products of the reaction?

*Solution*

(a) From the equation we see that the volume ratio of $O_2(g)$ to $C_2H_6(g)$ is 7 to 2, or 3.5 to 1.

Therefore, 15.0 liters of $C_2H_6(g)$ would require 3.5(15.0 liters) = 52.5 liters of $O_2(g)$.

(b) The ratio of the volumes of $CO_2(g)$ to $C_2H_6(g)$ indicated by the equation is 4 to 2, or 2 to 1; therefore, 2(15.0 liters) = 30.0 liters of $CO_2(g)$ would be produced. The volume ratio of $H_2O(g)$ to $C_2H_6(g)$ is 6 to 2, or 3 to 1. Hence, 3(15.0 liters) = 45.0 liters of $H_2O(g)$ would be produced. The total volume of the products would be 30.0 liters + 45.0 liters = 75.0 liters.

---

**6.9** WEIGHT-VOLUME RELATIONSHIPS IN REACTIONS

Gay-Lussac's law of combining volumes does not contradict the law of conservation of mass. Listed beneath each formula of the following equation are three equivalent terms derived from the equation. The volumes that are given are measured at STP.

$$2CO(g) \quad + \quad O_2(g) \quad \rightarrow \quad 2CO_2(g)$$

| $2CO(g)$ | $O_2(g)$ | $2CO_2(g)$ |
|---|---|---|
| 2 mol | 1 mol | 2 mol |
| 2(22.4) liters | 22.4 liters | 2(22.4) liters |
| 2(28) g = 56 g | 32 g | 2(44) g = 88 g |

The total weight of the reactants (56 g + 32 g = 88 g) equals the weight of the product (88 g), but the total volume of the reactants (67.2 liters) does not equal the volume of the product (44.8 liters). Each of the three gases involved in the reaction has a different density. In a reaction of this type, in which the volume decreases, the average density increases.

The fact that a gram molecular weight of a gas occupies 22.4 liters at STP can be utilized in stoichiometric calculations.

*Example 6.12*

How many grams of iron are needed to produce 100 liters of hydrogen gas (at STP)?

*Solution*

The equation for the reaction is

$$3Fe(s) + 4H_2O(g) \rightarrow Fe_3O_4(s) + 4H_2(g)$$

According to this equation, 3 mol of iron (167.5 g) produces 4 mol of hydrogen (89.6 liters). Therefore,

$$? \text{ g Fe} = 100 \text{ liters H}_2 \left( \frac{167.5 \text{ g Fe}}{89.6 \text{ liters H}_2} \right)$$

$$= 187 \text{ g Fe}$$

*Example 6.13*

How many liters of carbon monoxide (measured at STP) are needed to reduce 1.00 kg of $Fe_2O_3$?

*Solution*

The equation

$$Fe_2O_3(s) + 3CO(g) \rightarrow 2Fe(s) + 3CO_2(g)$$

shows that 3 mol of CO(g), which occupies 67.2 liters (3 × 22.4 liters), is required to reduce 1 mol of $Fe_2O_3$ (159.6 g). Thus,

$$? \text{ liters CO} = 1000 \text{ g Fe}_2O_3 \left( \frac{67.2 \text{ liters CO}}{159.6 \text{ g Fe}_2O_3} \right)$$

$$= 421 \text{ liters CO}$$

**6.10 DALTON'S LAW OF PARTIAL PRESSURES**

The behavior of a mixture of gases that do not react with one another is frequently of interest. The partial pressure of a component of such a gaseous mixture is the pressure that the component would exert if it were the only gas present in the volume under consideration. **Dalton's law of partial pressures** (1801) states that the total pressure of a mixture of gases is equal to the sum of the partial pressures of the component gases. If the total pressure is $P_{total}$, and the partial pressures are $p_1$, $p_2$, $p_3$, . . . ,

$$P_{total} = p_1 + p_2 + p_3 + \cdots$$

Suppose we have 1 liter of gas A at 0.2 atm pressure and 1 liter of gas B at 0.4 atm pressure. If these two gases are mixed in a 1 *liter container* and the temperature is constant, the pressure of the mixture is 0.6 atm.

According to the kinetic theory, at constant temperature, molecules of gas A have the same average kinetic energy as molecules of gas B. Furthermore, the kinetic theory states that no attractive forces exist between gas molecules that do not react chemically. Hence, the act of mixing two or more gases does not change the average kinetic energy of any of the component gases, and each gas exerts the pressure that it would exert if it were the only gas present in the container.

For a mixture of gas A and gas B, let the pressure of gas A equal $p_A$, the number of moles of gas A equal $n_A$, the number of moles of gas B equal $n_B$, and the total pressure equal $P_{total}$. Then,

$$p_A V = n_A RT$$

or

$$\left(\frac{RT}{V}\right) = \left(\frac{p_A}{n_A}\right)$$

The total number of moles is $n_A + n_B$; therefore,

$$P_{total} V = (n_A + n_B)RT$$

$$P_{total} = (n_A + n_B)\left(\frac{RT}{V}\right)$$

Thus,

$$P_{total} = (n_A + n_B)\left(\frac{p_A}{n_A}\right)$$

Solving for $p_A$ gives

$$p_A = \left(\frac{n_A}{n_A + n_B}\right)P_{total}$$

In like manner,

$$p_B = \left(\frac{n_B}{n_A + n_B}\right)P_{total}$$

The quantity $n_A/(n_A + n_B)$ is called the **mole fraction** of A; it is the ratio of the number of moles of gas A to the total number of moles of gas present. The partial pressure of a component of a gaseous mixture is equal to the mole fraction times the total pressure.

The preparation of gas mixtures at constant pressure is a less common situation. When gases are mixed at constant pressure, the total volume of the mixture is equal to the sum of the volumes of the component gases at that pressure. If 200 ml of oxygen at 1 atm pressure is mixed with 800 ml of nitrogen at 1 atm pressure, the resulting mixture occupies 1000 ml at a pressure of 1 atm. The composition of this mixture approximates that of air. In this final mixture, the oxygen exerts a partial pressure of 0.2 atm, and the nitrogen exerts a partial pressure of 0.8 atm.

In the laboratory it is often convenient to collect a gas over water. In such experiments, water vapor is mixed with the collected gas, and the

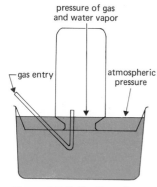

pressure of gas
and water vapor

gas entry

atmospheric
pressure

**Figure 6.7 Collection of gas over water.**

total pressure of the gas sample must be corrected by subtracting the pressure due to the water vapor. In Figure 6.7 the total pressure of the gas sample is equal to the barometric pressure, since the water stands at the same level inside the bottle as outside it. To obtain the pressure of the gas being investigated, it is necessary to subtract the vapor pressure of water at the temperature of the experiment (Table 6.2) from the barometric pressure.

**Table 6.2 Vapor pressure of water**

| TEMPERATURE (°C) | PRESSURE (atm) | (torr) | TEMPERATURE (°C) | PRESSURE (atm) | (torr) |
|---|---|---|---|---|---|
| 0 | 0.0060 | 4.6 | 25 | 0.0313 | 23.8 |
| 1 | 0.0065 | 4.9 | 26 | 0.0332 | 25.2 |
| 2 | 0.0070 | 5.3 | 27 | 0.0352 | 26.7 |
| 3 | 0.0075 | 5.7 | 28 | 0.0373 | 28.3 |
| 4 | 0.0080 | 6.1 | 29 | 0.0395 | 30.0 |
| 5 | 0.0086 | 6.5 | 30 | 0.0419 | 31.8 |
| 6 | 0.0092 | 7.0 | 31 | 0.0443 | 33.7 |
| 7 | 0.0099 | 7.5 | 32 | 0.0470 | 35.7 |
| 8 | 0.0106 | 8.0 | 33 | 0.0496 | 37.7 |
| 9 | 0.0113 | 8.6 | 34 | 0.0525 | 39.9 |
| 10 | 0.0121 | 9.2 | 35 | 0.0555 | 42.2 |
| 11 | 0.0130 | 9.8 | 40 | 0.0728 | 55.3 |
| 12 | 0.0138 | 10.5 | 45 | 0.0946 | 71.9 |
| 13 | 0.0148 | 11.2 | 50 | 0.122 | 92.5 |
| 14 | 0.0158 | 12.0 | 55 | 0.155 | 118.0 |
| 15 | 0.0168 | 12.8 | 60 | 0.197 | 149.4 |
| 16 | 0.0179 | 13.6 | 65 | 0.247 | 187.5 |
| 17 | 0.0191 | 14.5 | 70 | 0.308 | 233.7 |
| 18 | 0.0204 | 15.5 | 75 | 0.380 | 289.1 |
| 19 | 0.0217 | 16.5 | 80 | 0.467 | 355.1 |
| 20 | 0.0231 | 17.5 | 85 | 0.571 | 433.6 |
| 21 | 0.0245 | 18.7 | 90 | 0.692 | 525.8 |
| 22 | 0.0261 | 19.8 | 95 | 0.834 | 633.9 |
| 23 | 0.0277 | 21.1 | 100 | 1.000 | 760.0 |
| 24 | 0.0294 | 22.4 | 105 | 1.192 | 906.1 |

*Example 6.14*
A mixture of 40.0 g of oxygen and 40.0 g of helium has a total pressure of 0.900 atm. What are the partial pressures of oxygen and helium in the mixture?

*Solution*
The molecular weight of $O_2$ is 32.0; 40.0 g of $O_2$ is 40.0/32.0, or 1.25, mol. Helium is a monatomic gas; atomic weight = 4.00. Therefore,

40.0/4.0, or 10.0, mol is present in the mixture. The total number of moles of gases is 11.25. The partial pressure of $O_2$ is

$$? \text{ atm} = \left(\frac{1.25 \text{ mol}}{11.25 \text{ mol}}\right) 0.900 \text{ atm} = 0.100 \text{ atm}$$

The partial pressure of He is

$$? \text{ atm} = \left(\frac{10.00 \text{ mol}}{11.25 \text{ mol}}\right) 0.900 \text{ atm} = 0.800 \text{ atm}$$

*Example 6.15*

A 370 ml sample of oxygen is collected over water at 23°C and a barometric pressure of 0.992 atm. What volume would this sample occupy dry and at STP?

*Solution*

The vapor pressure of water at 23°C is 0.0277 atm. Therefore, the initial pressure of the oxygen is

$$0.992 \text{ atm} - 0.028 \text{ atm} = 0.964 \text{ atm}$$

A pressure change from 0.964 atm to 1.000 atm would decrease the volume by (0.964 atm/1.000 atm). The temperature change from 296°K to 273°K would cause the volume to decrease by a factor of (273°K/296°K). Thus,

$$? \text{ ml} = 370 \text{ ml} \left(\frac{0.964 \text{ atm}}{1.000 \text{ atm}}\right) \left(\frac{273°K}{296°K}\right) = 329 \text{ ml}$$

The problem can also be solved using torr as pressure units. Since 0.992 atm = 754 torr and 0.0277 atm = 21.1 torr, the initial pressure of oxygen is

$$745 \text{ torr} - 21 \text{ torr} = 733 \text{ torr}$$

The conversion factor for the pressure correction is (733 torr/760 torr).

$$? \text{ ml} = 370 \text{ ml} \left(\frac{733 \text{ torr}}{760 \text{ torr}}\right) \left(\frac{273°K}{296°K}\right) = 329 \text{ ml}$$

---

**6.11 MOLECULAR VELOCITIES**

In Section 6.6 we derived the expression

$$PV = \tfrac{1}{3}Nmu^2$$

For one mole of a gas, the number of molecules, $N$, is Avogadro's number, and $N$ times the mass of a single molecule, $m$, is the molecular weight, $M$.

$$PV = \tfrac{1}{3}Mu^2$$

Also for one mole, $PV = RT$; thus,

$$RT = \tfrac{1}{3}Mu^2$$

Rearranging and solving for the molecular velocity, we obtain

$$u = \sqrt{\frac{3RT}{M}}$$

● Since

$$R = 8.3143 \text{ J/K mol}$$
$$1 \text{ J} = 1 \text{ kg m}^2/\text{s}^2$$

and

$$1 \text{ kg} = 10^3 \text{ g}$$

it follows that

$$R = 8.3143 \times 10^3$$
$$\text{g m}^2/\text{s}^2 \text{ K mol}$$

The velocity, $u$, in this equation, as in previous equations, is the root-mean-square velocity. The value of the root-mean-square velocity is obtained by taking the square root of the average of the squares of all of the molecular velocities; it represents the velocity of a molecule that possesses average kinetic energy at the temperature under consideration.

In order to solve the equation for the root-mean-square velocity, $R$ must be expressed in appropriate units. If $u$ is to be obtained in m/sec

**Figure 6.8** Distribution of molecular velocities.

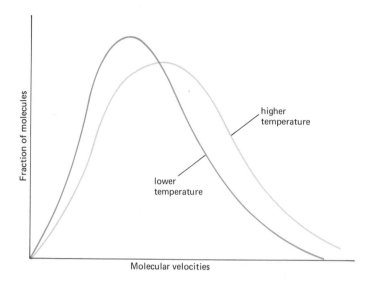

Fraction of molecules

higher temperature

lower temperature

Molecular velocities

and $M$ is expressed in g/mol, the appropriate value of $R$ is $8.3143 \times 10^3$ g m²/sec² °K mol.

The root-mean-square velocity of hydrogen (the lightest gas) at STP is $1.84 \times 10^3$ m/sec; the diffusion of one gas through another gas, however, does not occur at speeds commensurate with this. Although a given molecule travels at a high speed, its direction is continually being changed through collisions with other molecules. At 1 atm pressure and 0°C a hydrogen molecule, on the average, undergoes about $1.4 \times 10^{10}$ collisions in one second. The average distance traveled between collisions is only $1.3 \times 10^{-5}$ cm; this value is called the **mean free path** of hydrogen.

Not all the molecules of a gas possess the same kinetic energy; the velocities of molecules are distributed over a wide range. The nature of this distribution is described by the **Maxwell-Boltzmann distribution law.**

**Figure 6.9** Distribution of molecular energies.

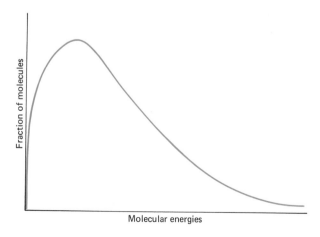

Fraction of molecules

Molecular energies

Figure 6.8 shows the distribution of molecular velocities of a gas at two temperatures. Because of the collisions of molecules and consequent exchange of energy, the velocity of a given molecule is continually changing. However, since a large number of molecules is involved, the distribution of molecular velocities of the total collection is constant. At the higher temperature the curve is broadened and shifted to higher velocities. A typical curve for the distribution of molecular energies is shown in Figure 6.9.

**6.12 GRAHAM'S LAW OF EFFUSION**

Suppose that two gases, A and B, are confined separately in identical containers under the same conditions of temperature and pressure. Molecules of two different gases at the same temperature have the same average kinetic energy; therefore,

$$\tfrac{1}{2}m_A u_A^2 = \tfrac{1}{2}m_B u_B^2$$

or

$$m_A u_A^2 = m_B u_B^2$$

Rearranging this equation, we obtain

$$\frac{u_A^2}{u_B^2} = \frac{m_B}{m_A}$$

Extracting the square root of both sides of the equation, we get

$$\frac{u_A}{u_B} = \sqrt{\frac{m_B}{m_A}}$$

If there is an identical, extremely small orifice in each container, molecules will escape. The rate of escape, $r$, which is called the rate of **molecular effusion,** is equal to the rate at which molecules strike the orifice, and this in turn is roughly proportional to the average molecular velocity. In addition, the ratio of the molecular masses, $m_B/m_A$, is the same as the ratio of the molecular weights, $M_B/M_A$. Thus,

$$\frac{r_A}{r_B} = \sqrt{\frac{M_B}{M_A}}$$

Since the density of a gas, $d$, is proportional to its molecular weight, this relation may also be written

$$\frac{r_A}{r_B} = \sqrt{\frac{d_B}{d_A}}$$

Thomas Graham experimentally derived this **law of effusion** in his studies of 1828–1833. For molecular effusion, the orifice must be small in comparison to the mean free path of the molecules. If high pressures and large orifices are employed, collisions of the molecules while passing through the orifice cause the gas to escape in the form of a jet that does not obey Graham's law.

It is not surprising that the lighter of two molecules with the same kinetic energy will move more rapidly than the heavier one. Graham's law confirms this and gives the quantitative relationship. The molecular weights of oxygen and hydrogen are 32 and 2, respectively; hence, hydrogen molecules will effuse 4 times faster than oxygen molecules under the same conditions ($\sqrt{\frac{32}{2}} = 4$).

This principle has been used for the separation of isotopes. Naturally occurring uranium consists of 0.7% $^{235}U$ and 99.3% $^{238}U$; therefore, the uranium hexafluoride produced by the reaction of uranium and fluorine consists of a mixture of $^{235}UF_6$ and $^{238}UF_6$. When this mixture is heated above its boiling point (56°C) and passed at low pressure through a porous barrier, the lighter $^{235}UF_6$ effuses 1.004 times faster than $^{238}UF_6$ ($\sqrt{352/349} = 1.004$). The emerging gas, therefore, has a higher $^{235}UF_6$ content than the original mixture. This procedure must be repeated thousands of times to effect a significant separation.

**6.13 DEVIATIONS FROM THE IDEAL GAS LAWS**

The gas laws describe the behavior of an ideal or perfect gas—a gas defined by the kinetic theory. Under ordinary conditions of temperature and pressure, real gases follow the behavior expressed by these laws rather closely; this is not true, however, at high pressures and low temperatures.

The kinetic theory assumes that gas molecules are points in space and that their volume is negligible; hence, at 0°K, the temperature at which molecular motion ceases, the volume of an ideal gas is zero. Real gases, of course, do not have zero molecular volumes; although the space between the molecules can be reduced by increasing the pressure, the molecular volume cannot be compressed. At pressures up to a few atmospheres, the gas molecules are widely separated, and the free space between the molecules is large in comparison to the molecular volume. Under such conditions, neglecting the volume of the molecules does not introduce a large error. However, at high pressures the molecules of a gas are relatively close together, and the molecular volume constitutes a significant fraction of the total volume. Under these conditions, the volume of a real gas is appreciably larger than that predicted for an ideal gas.

The kinetic theory also assumes that there are no attractive forces between gas molecules. This assumption is belied by the fact that all gases can be liquefied. At high temperatures, gas molecules move so rapidly that the attractive forces between molecules are effectively overcome. On the other hand, at low temperatures, the forces of attraction pull the molecules together so that the volume observed is less than that predicted by the ideal gas law.

It is interesting to note that these two effects work in opposite directions — one increases the volume and the other decreases the volume. Thus, under moderate conditions, the ideal gas law is followed fairly closely by real gases. However, under more extreme conditions, the net deviation depends upon which effect is dominant.

Johannes van der Waals in 1873 modified the equation of state for an ideal gas to take into account these two effects. The **van der Waals equation** is

$$\left(P + \frac{n^2 a}{V^2}\right)(V - nb) = nRT$$

The numerical values of the constants $a$ and $b$ are determined by experiment; these values are specific to the gas under consideration (see Table 6.3).

molecular volume, $\frac{4}{3}\pi r^3$

excluded volume for two molecules, $8\left(\frac{4}{3}\pi r^3\right)$

Figure 6.10 van der Waals correction, $b$, for excluded volume.

Table 6.3 van der Waals constants

|  | $a$ (liter² atm/mol²) | $b$ (liter/mol) |
|---|---|---|
| $H_2$ | 0.244 | 0.0266 |
| He | 0.0341 | 0.0237 |
| $N_2$ | 1.39 | 0.0391 |
| $O_2$ | 1.36 | 0.0318 |
| $Cl_2$ | 6.49 | 0.0562 |
| $NH_3$ | 4.17 | 0.0371 |
| CO | 1.49 | 0.0399 |
| $CO_2$ | 3.59 | 0.0427 |

The constant $b$, multiplied by $n$, is subtracted from the total volume of the gas to correct for that portion of the volume that is not compressible because of the intrinsic volume of the gas molecules. If the molecules are assumed to be spherical and have a radius $r$, the excluded volume per molecule is not merely the volume of the molecule, $\frac{4}{3}\pi r^3$. Since the closest approach of *two* molecules is $2r$ (Figure 6.10), the excluded volume for *two* molecules is $\frac{4}{3}\pi(2r)^3$ which is $8(\frac{4}{3}\pi r^3)$. For one molecule this volume is $4(\frac{4}{3}\pi r^3)$, which is four times the molecular volume. Hence, for a mole of molecules, $N$ molecules,

$$b = 4N(\tfrac{4}{3}\pi r^3)$$

where $b$ is given in cm³/mole and $r$ is given in cm.

The pressure term of the equation is corrected by a factor $n^2 a/V^2$ to take into account the intermolecular attractions (Section 4.6). These attractive forces draw the gas molecules together. Hence, this effect augments pressure in reducing the volume of the gas, and the correction term is added to $P$.

The term $(n/V)$ represents a concentration (mol/liter). If $x$ molecules are confined in a liter, there are $(x - 1)$ ways for a given molecule to collide or interact with another molecule since it cannot collide with itself. This factor applies to all the molecules; therefore, a total of $\frac{1}{2}x(x - 1)$ possible interactions exists for the entire collection of molecules. The fraction $\frac{1}{2}$ is added so that a given interaction is not counted twice — once for each of the molecules entering into it. If a large number of molecules are present, $(x - 1)$ is approximately equal to $x$, and the proportionality is $\frac{1}{2}x^2$ to a good approximation. Hence, the number of interactions between gas molecules is proportional to the *square* of the concentration. The van der Waals constant $a$ may be regarded as a proportionality constant (incorporating the $\frac{1}{2}$), and the correction term is $n^2a/V^2$.

When the volume of a mole of gas is large and pressure is low, the effect of subtracting $nb$ from the volume is negligible, as is the magnitude of the correction term $n^2a/V^2$. Under such conditions, gases follow the ideal gas law.

## 6.14 LIQUEFACTION OF GASES

Liquefaction of a gas occurs under conditions that permit the intermolecular attractive forces to bind the gas molecules together in the liquid form. If the pressure is high, the molecules are close together, and the effect of the attractive forces is appreciable. The attractive forces are opposed by the motion of the gas molecules; thus, liquefaction is favored by low temperatures where the average kinetic energy of the molecules is low. The behavior of a gas deviates more and more from ideality as the temperature is lowered and the pressure is raised. At extremes of these conditions, gases liquefy.

The higher the temperature of a gas, the more difficult it is to liquefy and the higher the pressure that must be employed (Table 6.4). For each gas there is a temperature above which it is impossible to liquefy the gas no matter how high the applied pressure. This temperature is called the **critical temperature** of the gas under consideration. The **critical pressure** is the minimum pressure needed to liquefy a gas at its critical temperature. The critical constants of some common gases are listed in Table 6.5.

The critical temperature of a gas gives an indication of the strength of the intermolecular attractive forces of that gas. A substance with weak attractive forces would have a low critical temperature; above this temperature, the molecular motion is too violent to permit the relatively weak forces to hold the molecules in the liquid state. The substances of Table 6.5 are listed in order of increasing critical temperature; the magnitude of the intermolecular attractive forces (related to the $a$ of Table 6.3)

**Table 6.4** Pressures needed to liquefy carbon dioxide (vapor pressures of liquid $CO_2$)

| TEMPERATURE (°C) | PRESSURE (atm) |
|---|---|
| −50 | 6.7 |
| −30 | 14.1 |
| −10 | 26.1 |
| 10 | 44.4 |
| 20 | 56.5 |
| 30 | 71.2 |
| 31 | 72.8 |

Table 6.5 Critical point data

| GAS | CRITICAL TEMPERATURE (°K) | CRITICAL PRESSURE (atm) |
|---|---|---|
| He | 5.3 | 2.26 |
| $H_2$ | 33.3 | 12.8 |
| $N_2$ | 126.1 | 33.5 |
| CO | 134.0 | 35.0 |
| $O_2$ | 154.4 | 49.7 |
| $CH_4$ | 190.2 | 45.6 |
| $CO_2$ | 304.2 | 72.8 |
| $NH_3$ | 405.6 | 111.5 |
| $H_2O$ | 647.2 | 217.7 |

● The SI unit of energy is named in honor of James Joule, and the unit of thermodynamic temperature is named in honor of Lord Kelvin.

increases in this same order. Helium, which has weak attractive forces, can exist as a liquid below 5.3°K only; the strong attractive forces of water permit it to be liquefied up to a temperature of 647.2°K. The critical constants have been used to evaluate the constants of the van der Waals equation.

The data of Table 6.5 show that it is necessary to cool many gases below room temperature (ca. 295°K) before these substances can be liquefied. Commercial liquefaction procedures make use of the **Joule-Thomson effect** to cool gases. When a compressed gas is allowed to expand to a lower pressure (e.g., through an orifice) the gas cools. In the expansion, work is done against the intermolecular attractive forces. The energy used in performing this work must be taken from the kinetic energy of the gas molecules themselves; hence, the temperature of the gas decreases. This effect was studied by James Joule and William Thomson (Lord Kelvin) in the years 1852–1862. The liquefaction of air is accomplished by first allowing cooled, compressed air to expand. The temperature of the air falls to a lower level. This cooled air is used to precool entering compressed air, and the expansion of this compressed air results in the attainment of even lower temperatures. The cooled, expanded air is recycled through the compression chamber. Eventually the cooling and compression produce liquid air.

## Problems

**6.1** The space above the liquid mercury in a column of a barometer is occupied only by mercury vapor. The vapor pressure of mercury at 25°C is $2.05 \times 10^{-6}$ atm. Assuming mercury vapor to be an ideal monatomic gas, calculate the number of atoms of mercury per milliliter at this temperature and pressure.

**6.2** If a barometer were constructed using water instead of mercury, to what height, in mm, would the water rise at 0°C and a pressure of 1.00 atm? The vapor pressure of water at 0°C is 0.00605 atm. Assume that the density of water at 0°C is 1.00 g/cm³.

**6.3** If a barometer were constructed using ethyl alcohol instead of mercury, to what height, in mm, would the ethyl alcohol rise at 0°C and a pressure of 1.00 atm? The vapor pressure of ethyl alcohol at 0°C is 0.0161 atm, and the density of ethyl alcohol at 0°C is 0.806 g/cm³.

**6.4** For each of the following pairs of variables, which apply to measurements made on a sample of an ideal gas, draw a rough graph to show how one quantity varies with the other. (a) $P$ vs. $V$, $T$ constant; (b) $T$ vs. $V$, $P$ constant; (c) $P$ vs. $T$, $V$ constant; (d) $PV$ vs. $V$, $T$ constant.

**6.5** The volume of a sample of gas is 3.00 liters at 3.00 atm. If the temperature is held constant, what is the volume of the sample at (a) 1.00 atm, (b) 0.30 atm, (c) 10.00 atm?

**6.6** A container is filled with a gas to a pressure of 3.00 atm at 0°C. (a) What pressure will develop inside the sealed container if it is warmed to 100°C? (b) At what temperature would the pressure be 2.00 atm? (c) At what temperature would the pressure be 4.00 atm?

**6.7** The volume of gas in a gas thermometer is 100.0 ml at 0°C.

What is the temperature when the volume is 125.0 ml? Assume that the pressure remains constant.

**6.8** A sample of gas occupies 300 ml at 30°C and 1.30 atm. What volume will the sample occupy at STP?

**6.9** A sample of gas occupies 3.36 liters at STP. What volume will it occupy at 100°C and 0.500 atm?

**6.10** The volume of a sample of gas is 250 ml at STP. What is the temperature when the volume of the sample is 200 ml and the pressure is 2.50 atm?

**6.11** The volume of a sample of gas is 2.50 liters at 27°C and 1.10 atm. What is the pressure of the gas at 127°C if the volume is 3.50 liters?

**6.12** What volume would 2.10 g of $CH_4(g)$ occupy at 77°C and 0.750 atm?

**6.13** What is the density of $PH_3(g)$, in grams per liter, at 22°C and 0.950 atm?

**6.14** What is the total pressure of a mixture containing 33.0 g of $CO(g)$ and 36.0 g of $CH_4(g)$ confined to a volume of 42.0 liters at 0°C?

**6.15** If the temperature is held constant at 0°C, at what pressure will the density of $He(g)$ be 1.00 g/liter?

**6.16** If the pressure is held constant at 1.00 atm, at what temperature will the density of $O_2(g)$ be 1.00 g/liter?

**6.17** A 1.500 g sample of a liquid was vaporized at 150°C. The vapor occupied 450 ml under a pressure of 0.965 atm. What is the molecular weight of the liquid?

**6.18** A gas has a density of 2.04 g/liter at 32°C and a pressure of 0.850 atm. What is the molecular weight of the gas?

**6.19** What volume would 1 billion ($10^9$) molecules of $H_2(g)$ occupy at STP?

**6.20** Calculate the number of molecules in 1.00 liter of $H_2(g)$ at 100°C and 1.00 atm.

**6.21** A given volume of helium weighs twice as much as the same volume of hydrogen, yet a helium-filled balloon can lift 92.5% of the weight that can be lifted by a balloon of the same volume filled with hydrogen. Explain. Note that air is a mixture of gases with an average molecular weight of 28.6.

**6.22** Hydrogen cyanide, a highly poisonous compound, is commercially prepared by the following reaction run at a high temperature in the presence of a catalyst.

$$2CH_4(g) + 3O_2(g) + 2NH_3(g) \rightarrow$$
$$2HCN(g) + 6H_2O(g)$$

How many liters of $CH_4(g)$, $O_2(g)$ and $NH_3(g)$ are required and how many liters of $H_2O(g)$ are produced in the preparation of 15.0 liters of HCN(g)? Assume that all gas volumes are measured under the same conditions of temperature and pressure.

**6.23** Cyanogen, $(CN)_2(g)$, is a flammable, extremely poisonous gas. It can be prepared by a catalyzed, gas-phase reaction between HCN(g) and $NO_2(g)$; the products of the reaction are $(CN)_2(g)$, NO(g), and $H_2O(g)$. Assume that all gas volumes are measured under the

same conditions of temperature and pressure.

(a) Write the equation for the reaction. (b) How many liters of $(CN)_2(g)$ can be prepared by the reaction of 10.0 liters of HCN(g) with 10.0 liters of $NO_2(g)$? (c) How many liters of NO(g) and $H_2O(g)$ are prepared in the reaction described in part (b)?

**6.24** How many liters of HCN(g) can be prepared from 100.0 liters of $CH_4(g)$, 100.0 liters of $O_2(g)$, and 100.0 liters of $NH_3(g)$ in a reaction such as the one described by the equation given in Problem 6.22? Assume that the volumes are measured at constant conditions of temperature and pressure.

**6.25** Ammonia, $NH_3(g)$, reacts with oxygen at 850°C in the presence of a Pt catalyst to yield NO(g) and $H_2O(g)$. Write an equation for the reaction. What volume of NO(g) can be obtained from 10.0 liters of $NH_3(g)$ and 10.0 liters of $O_2(g)$? The volumes of all gases are measured under the same conditions.

**6.26** In the absence of a catalyst, ammonia, $NH_3(g)$, reacts with oxygen gas to yield nitrogen gas and water vapor. Write an equation for the reaction. What volume of $N_2(g)$ can be obtained from 10.0 liters of $NH_3(g)$ and 10.0 liters of $O_2(g)$? The volumes of all gases are measured under the same conditions.

**6.27** At temperature above 50°C, nitrogen oxide, NO(g), decomposes to yield dinitrogen oxide, $N_2O(g)$, and nitrogen dioxide, $NO_2(g)$. Write an equation for the reaction. What total volume of $N_2O(g)$ and $NO_2(g)$ together would result from the decomposition of 20.0 liters of $NO(g)$ at 150°C and 1.00 atm? Assume that all gas volumes are measured under the same conditions. What are the partial pressures of $N_2O(g)$ and $NO_2(g)$ in this gas mixture?

**6.28** Chlorine dioxide, $ClO_2(g)$, is an extremely reactive gas that sometimes decomposes explosively. The danger of working with $ClO_2(g)$ can be minimized by diluting it with another gas, such as $CO_2(g)$, so that the partial pressure of $ClO_2(g)$ in the mixture is relatively low. Chlorine dioxide can be made by the following reaction.

$$2NaClO_3(aq) + SO_2(g) + H_2SO_4(aq) \rightarrow$$
$$2ClO_2(g) + 2NaHSO_4(aq)$$

What volume of $SO_2(g)$, measured at STP, is required to prepare 2.00 liters of $ClO_2(g)$, measured at 0°C and a partial pressure of $ClO_2(g)$ of 0.0500 atm?

**6.29** Calcium hydride, $CaH_2(s)$, reacts with water to yield hydrogen gas and calcium hydroxide. Write the equation for the reaction. What volume of $H_2(g)$, measured at STP, would be obtained from the reaction of 1.000 g of $CaH_2$ with excess $H_2O$?

**6.30** Calcium metal, $Ca(s)$, reacts with water to yield hydrogen gas and calcium hydroxide. Write the equation for the reaction. What volume of $H_2(g)$, measured at STP, would be obtained from the reaction of 1.000 g of $Ca$ with excess $H_2O$? Compare your answer to that of Problem 6.29.

**6.31** Aluminum carbide, $Al_4C_3(s)$, reacts with water to yield aluminum hydroxide, $Al(OH)_3(s)$, and methane, $CH_4(g)$. Write the equation for the reaction. How many grams of $Al_4C_3$ would be needed to prepare 1.000 liter of $CH_4$ (measured at STP)?

**6.32** Write the equation for the reaction of lanthanum carbide, $La_2(C_2)_3(s)$, with water to yield lanthanum hydroxide, $La(OH)_3(s)$, and acetylene, $C_2H_2(g)$. How many grams of $La_2(C_2)_3$ would be needed to prepare 1.000 liter of $C_2H_2$ (measured at STP)?

**6.33** The complete combustion of 5.60 liters of a gaseous compound that contains only C and H requires 28.0 liters of $O_2(g)$ and produces 16.8 liters of $CO_2(g)$ and 18.0 g of $H_2O(l)$. All gas measurements are made at STP. (a) Calculate the number of moles of each substance involved in the reaction. (b) Use your answers from part (a) to derive the whole-number coefficients of the balanced chemical equation for the reaction. (c) Determine a formula for the hydrocarbon and write the equation for the reaction.

**6.34** The complete combustion of a 0.350 g sample of a compound that contains only carbon and hydrogen produced 560 ml of $CO_2(g)$, measured at STP, and 0.450 g of $H_2O$. (a) What is the empirical formula of the compound? (b) If the original sample of the compound occupied a volume of 140 ml at STP, what is the molecular formula of the compound?

**6.35** Both $CaH_2(s)$ and $LiH(s)$ react with water to produce $H_2(g)$

and the corresponding hydroxide (either $Ca(OH)_2$ or $LiOH$). A 0.980 g sample of a mixture of $CaH_2$ and $LiH$ produces 1.344 liters of $H_2$ (measured at STP). What percentage of the mixture is $CaH_2$?

**6.36** Upon heating, sodium azide, $NaN_3(s)$, decomposes into its constituent elements. Write the equation for the reaction. A 0.400 g sample of $NaN_3$ is decomposed. What volume of $N_2(g)$, measured at 25°C and 0.980 atm, is obtained?

**6.37** A sample of magnesium nitride, $Mg_3N_2(s)$, is reacted with water and 758 ml of wet $NH_3(g)$ is collected at 22°C and a total pressure of 0.985 atm.

$$Mg_3N_2(s) + 6H_2O(1) \rightarrow$$
$$3Mg(OH)_2(s) + 2NH_3(g)$$

Use the equation of state to determine the number of moles of *dry* $NH_3(g)$ produced. The vapor pressure of water at 22°C is 0.0261 atm. How many grams did the sample of $Mg_3N_2$ weigh?

**6.38** A mixture of 7.70 g of $CO(g)$ and 7.70 g of $CO_2(g)$ exerts a pressure of 0.770 atm. What is the partial pressure of each gas?

**6.39** The partial pressures of $He(g)$ and $CO_2(g)$ in a mixture of the two gases are 0.900 atm and 0.300 atm respectively. (a) What is the mole fraction of $He(g)$ in the mixture? of $CO_2(g)$? (b) If the mixture occupies 44.8 liters at STP, what is the total number of moles of gas present? (c) What is the composition of the mixture by weight? (d) What is the weight percent of $He(g)$ in the mixture?

**6.40** The blood of deep sea divers becomes saturated with $N_2$ and $O_2$ of the air under the comparatively high pressures characteristic of the depths at which divers work. If this pressure is relieved too rapidly (e.g., by too rapid an ascent to the surface), the $N_2$ comes out of solution rapidly and forms bubbles in the circulatory system of the afflicted divers. This condition, known as the "bends," may be fatal. One solution to this problem involves the use of an artificial atmosphere of He and $O_2$ in place of air (He is not very soluble in blood). If the helium–oxygen gas mixture is 21.0% $O_2$ by volume (which is the percentage of $O_2$ found in air), what are the partial pressures of He and $O_2$ in the mixture at 1.000 atm? What are the mole fractions of He and $O_2$ in the mixture? What is the percentage of $O_2$ by weight in the mixture? How many grams of He should be mixed with 100.0 g $O_2$ to prepare the mixture?

**6.41** A sample of $N_2(g)$, collected over water at 25°C and a pressure of 1.050 atm, occupies 150 ml. What volume would the $N_2$ occupy if dry and at STP? The vapor pressure of water at 25°C is 0.031 atm.

**6.42** A 1.000 liter of $N_2(g)$ is collected over water at 20°C and a total pressure of 0.990 atm. The vapor pressure of water is 0.0231 atm at 20°C. (a) How many grams of water does the sample contain? (b) How many grams of $N_2$ are present in the sample?

**6.43** A sample of a gas, collected over water at 35°C, occupies a volume of 1.000 liter. The wet gas exerts a pressure of 1.000 atm. When dried, the sample of gas occupies 1.000 liter and exerts a pressure of 1.000 atm at 53°C. What is the vapor pressure of water at 35°C?

**6.44** Calculate the root-mean-square velocity of the oxygen molecule at 100°K and at 300°K.

**6.45** At what temperature would the root-mean-square velocity of the oxygen molecule equal the root-mean-square velocity of the hydrogen molecule at 100°K?

**6.46** Given 1.00 mol of $H_2(g)$ and 1.00 mol of $O_2(g)$ measured under the same conditions, compare quantitatively the $H_2$ sample to the $O_2$ sample in terms of each of the following properties: (a) number of molecules, (b) mass of a single molecule, (c) root-mean-square velocity of a molecule, (d) change in momentum for a single molecular collision with the wall of the container, (e) total number of collisions with the wall per unit time, (f) total change in momentum for all molecular collisions with the wall per unit time.

**6.47** Use Graham's law to calculate the molecular weight of an unknown gas if a given volume of this gas effuses through an apparatus in 189 sec, and the same volume of $N_2$, under the same conditions, effuses through the apparatus in 250 sec.

**6.48** At 27°C and 1 atm, the density of $CO_2(g)$ is 1.79 g/liter. The rate of effusion of $CO_2(g)$ through an apparatus is 7.50 ml/sec. What is the density of an unknown gas if it effuses at a rate of 6.90 ml/sec through the same apparatus and under the same conditions? (b) What is the approximate molecular weight of the gas?

**6.49** Calculate the pressure exerted by 1.00 mol of $CO_2$ gas confined to a volume of 1.00 liter at 0°C by (a) the ideal gas law and (b) the van der Waals equation.

**6.50** Calculate (to four significant figures) the pressure exerted by 1.000 mol of $CO_2$ gas confined to a volume of 10.00 liters at 0°C by (a) the ideal gas law and (b) the van der Waals equation. (c) Compare the results to each other and to those obtained in Problem 6.49.

**6.51** Which substance of those listed in Table 6.3 would you expect to have the highest critical temperature? Why?

**6.52** On the basis of the value of the van der Waals constant $b$ recorded in Table 6.3, determine the volume of a molecule of $NH_3$. At STP what percentage of the total volume of nitrogen gas is molecular volume?

**6.53** (a) If the radius of the xenon atom is 2.21 Å, what is the value of the van der Waals constant $b$, in liters/mol? (b) Calculate a value for the van der Waals constant $a$ for Xe(g) given your value from part (a) and the fact that 1.000 mol of Xe(g) confined to a volume of 1.000 liter at 27°C exerts a pressure of 23.45 atm.

**6.54** Use the values of the van der Waals constants $a$ and $b$ that are given in Table 6.3 to compare strength of the intermolecular attractive forces in $H_2(g)$ and He(g). Note that the normal boiling point of $H_2(g)$ is 20.4°K and that of He(g) is 4.2°K (the lowest of any known substance).

**6.55** The value of the van der Waals constant $b$ for Ar(g) is 0.0322 liter/mol. On the basis of this value, what is the radius of the argon atom? At STP what percentage of the total volume of argon gas is molecular volume?

# 7

# Liquids and Solids

The kinetic energies of gas molecules decrease when the temperature is lowered. Consequently the intermolecular attractive forces cause the gas molecules to condense into a liquid when the gas has been cooled sufficiently. The molecules are closer together and the attractive forces exert a greater influence in a liquid than in a gas. Molecular motion, therefore, is more restricted in the liquid state than in the gaseous state.

Additional cooling causes the kinetic energies of the molecules to decrease further and ultimately produces a solid. In a crystalline solid the molecules assume positions in a crystal lattice, and the motion of the molecules is restricted to vibration about these fixed points.

The comparatively *high* kinetic energies of gas molecules cause the intermolecular attractive forces to assume a role that can be minimized in the development of a satisfactory theory of gases. The comparatively *low* kinetic energies of molecules (or ions) in crystals are easily overcome by the attractive forces to produce highly ordered, crystalline structures that have been well characterized by diffraction techniques. Our understanding of the intermediate state, the liquid state, however, is not so complete as that of the other two states.

**7.1 THE LIQUID STATE**  The liquid state is intermediate in character between the complete molecular randomness that characterizes gases and the orderly arrangement of molecules typical of crystalline solids. In liquids, the molecules move slowly enough that the intermolecular forces of attraction can hold them together in a definite volume; however, the molecular motion is too rapid for the attractive forces to fix the molecules into the definite positions of a crystal lattice. A liquid, therefore, retains its volume but not its shape; liquids flow to assume the shape of the container.

The space between the molecules of liquids is reduced almost to a minimum by the intermolecular attractions; a change in pressure has almost no effect on the volume of a liquid. An increase in temperature, however, increases the volume of most liquids slightly and consequently decreases the liquid density. As the temperature of a liquid is increased, the kinetic energy of the molecules increases, and this increased molecular motion works against the attractive forces. This expansion is, however, much less than that observed for gases in which the effect of the attractive forces is negligible.

Two miscible (mutually soluble) liquids will diffuse into each other when placed together. If a liquid is carefully overlaid with a second, less dense liquid, a sharp boundary line is observed between the two. This boundary gradually becomes less distinct and, in time, disappears completely as the molecules of the two liquids mix. The diffusion of liquids, however, is a much slower process than the diffusion of gases. Since the molecules of liquids are relatively close together, the mean free path of liquid molecules is relatively short; liquid molecules suffer many more collisions per unit time than do gas molecules.

Any liquid exhibits resistance to flow, a property known as **viscosity.** One way of determining the viscosity of a liquid is to measure the time that it takes for a definite amount of the liquid to pass through a tube of small diameter under a given pressure. Resistance to flow is largely due to the attractions between molecules, and the measurement of the viscosity of a liquid gives a simple estimate of the strength of these attractions. Allowance should be made, however, for factors such as molecular weight and the structure of the molecules. Liquids with large, irregularly shaped molecules are generally more viscous than liquids with small, spherical molecules. In general, as the temperature of a liquid is increased, the cohesive forces are less able to cope with the increasing molecular motion, and the viscosity decreases. On the other hand, increasing the pressure generally increases the viscosity of a given liquid.

Another property of liquids related to the intermolecular forces of attraction is surface tension. A molecule in the center of a liquid is attracted equally in all directions by surrounding molecules. Molecules on the surface of a liquid, however, are attracted only toward the interior of the liquid (Figure 7.1). The surface molecules, therefore, are pulled inward, and the surface area of a liquid tends to be minimized. This behavior accounts for the spherical shape of liquid drops. **Surface tension** is a measure of this inward force on the surface of a liquid, the force which must be overcome to expand the surface area. The surface tension of a liquid decreases with increasing temperature since the increased molecular agitation tends to decrease the effect of the intermolecular cohesive forces.

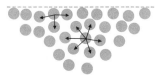

**Figure 7.1** Schematic diagram indicating the unbalanced intermolecular forces on the surface molecules of a liquid.

## 7.2 EVAPORATION

The kinetic energy of the molecules of a liquid follows a Maxwell–Boltzmann distribution similar to the distribution of kinetic energy among gas molecules (Figure 7.2). The kinetic energy of a given molecule of a liquid is continually changing as the molecule collides with other molecules, but at any given instant, some of the molecules of the total collection have relatively high energies and some have relatively low energies. The molecules with kinetic energies sufficiently high to overcome the attractive forces of surrounding molecules can escape from the liquid and enter the gas phase if they are close to the surface and are moving in the right direction; they use part of their energy to work against the attractive forces when they escape.

In time, the loss of a number of high-energy molecules causes the

**Figure 7.2** Distribution of kinetic energy among molecules of a liquid.

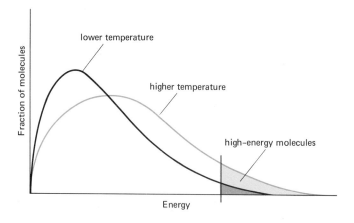

Fraction of molecules

lower temperature

higher temperature

high-energy molecules

Energy

average kinetic energy of the molecules remaining in the liquid to decrease, and the temperature of the liquid falls proportionately. When liquids evaporate from an open container at room temperature, heat flows into the liquid from the surroundings to maintain the temperature of the liquid. Thus, the supply of high-energy molecules is replenished, and the process continues until all of the liquid has evaporated. The total quantity of heat required to vaporize a mole of liquid at a given temperature is called the **molar heat of vaporization** of that liquid. For example,

$$H_2O(l) \rightarrow H_2O(g) \qquad \Delta H = +9.7 \text{ kcal}$$

The transfer of heat from the surroundings explains why a swimmer emerging from the water becomes chilled as the water evaporates from his skin. Likewise, the regulation of body temperature is, in part, accomplished by the evaporation of perspiration from the skin. Various cooling devices have made use of this principle. A water cooler of the Middle East consists of a jar of unglazed pottery filled with water. The water saturates the clay of the pottery and evaporates from the outer surface of the jar, thus cooling the water remaining in the jar.

The rate of evaporation increases as the temperature of a liquid is raised. When the temperature is increased, the average kinetic energy of the molecules increases, and the number of molecules with energies high enough for them to escape into the vapor phase constitutes an increased fraction of the total number of molecules (Figure 7.2).

**7.3 VAPOR PRESSURE**  When an evaporating liquid is confined in a closed container, the vapor molecules cannot escape from the vicinity of the liquid, and in the course of their random motion, some of the vapor molecules return to the liquid.

$$H_2O(l) \rightleftharpoons H_2O(g)$$

The rate of return depends upon the concentration of molecules in the vapor; the more vapor molecules per unit volume, the greater the chance that some of them will strike the liquid and be recaptured. Thus, a situation develops in which the vaporization of the liquid is opposed by the condensation of the vapor. Eventually the system reaches a point where the rate of condensation equals the rate of vaporization.

This condition, in which the rates of two opposite tendencies are equal, is called a state of **equilibrium.** At equilibrium the concentration of molecules in the vapor state is constant because molecules leave the vapor through condensation at the same rate that molecules add to the vapor through vaporization. Similarly, the quantity of liquid is a constant.

It is important to note that equilibrium does not imply a static situation; in any system, the numbers of molecules present in both phases are constant because the two opposing changes are taking place at equal rates and not because vaporization and condensation have ceased. In a closed container, equilibrium is the only situation that can prevail for any substantial length of time. If the rate of vaporization were greater than the rate of condensation for an extended period, all the substance would eventually become vapor. The reverse situation is equally improbable; all the material would end as liquid.

The pressure of vapor in equilibrium with a liquid at a given temperature is called the **vapor pressure.** If the temperature of a liquid–vapor system is constant, the vapor pressure will be constant. The pressure of the vapor depends upon the average kinetic energy of the vapor molecules (which is constant since the temperature is constant) and upon the concentration of vapor molecules. The concentration of vapor molecules, as well as their average kinetic energy, is fixed by temperature since the vapor concentration must be such that the rate of condensation is equal to the rate of vaporization, and the rate of vaporization is dictated by the temperature.

For any sample of a liquid at a given temperature, then, the equilibrium concentration of vapor is constant, regardless of the size of the sample or the dimensions of the container. The absolute quantity of vapor may not be the same from system to system, but the vapor concentration is. If the volume of vapor is changed, the consequent alteration of the vapor concentration causes the rate of condensation to increase or decrease until the concentration is reestablished at its former (equi-

**Figure 7.3** Vapor pressure curves for water, ethyl alcohol, and ethyl ether.

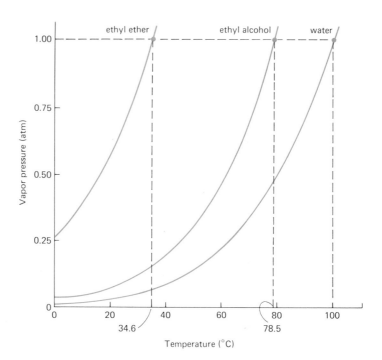

librium) value. Hence, vapor pressure is fixed by temperature alone; volume changes do not alter it. A list of the vapor pressures of water at various temperatures appears in Table 6.2.

As the temperature is increased, the vapor pressure of a liquid increases. We have noted that an increase in temperature causes the rate of vaporization to increase; equilibrium is possible only if the rate of condensation increases proportionately, and this will occur only when the concentration of vapor molecules becomes adjusted to a higher level. This factor and the augmented kinetic energy of the vapor molecules accompanying a rise in temperature account for the observed increase in vapor pressure.

Figure 7.3 shows the temperature–vapor pressure curves for ethyl ether, ethyl alcohol, and water. The curve for each substance could be extended to the critical temperature of that substance; at this point the vapor pressure would equal the critical pressure. Above the critical temperature only one phase—the gas phase—can exist.

Liquids with weak intermolecular attractive forces have relatively high vapor pressures. At 20°C ethyl ether has a vapor pressure of 0.582 atm (or 422 torr), and water has a vapor pressure of 0.0230 atm (or 17.5 torr). These data indicate that the forces of attraction are stronger in water than in ethyl ether.

**7.4 BOILING POINT**  The temperature at which the vapor pressure of a liquid equals the atmospheric pressure is called the **boiling point** of the liquid. At this temperature, vapor produced in the interior of a liquid results in the bubble formation and turbulence characteristic of boiling. Bubble formation is impossible at temperatures below the boiling point; the atmospheric pressure on the surface of the liquid prevents the formation of bubbles with internal pressures that are less than atmospheric.

The temperature of a boiling liquid remains constant until all the liquid has been vaporized. In an open container the maximum vapor pressure that can be attained by any liquid is the atmospheric pressure. We have noted in the preceding section that the vapor pressure of a liquid is determined by temperature alone; hence, if the vapor pressure is fixed, the temperature is fixed. Heat must be added to a boiling liquid to maintain the temperature, because in the boiling process the high-energy molecules are lost by the liquid. If the rate of heat addition is increased above the minimum needed to maintain the temperature of the boiling liquid, the rate of boiling increases, but the temperature of the liquid does not rise.

The boiling point of a liquid changes with changes in external pressure. Water, for example, will boil at 98.6°C at a pressure of 0.950 atm (or 722 torr) and at 101.4°C at a pressure of 1.05 atm (or 798 torr). Only at a pressure of 1.00 atm (or 760 torr) will water boil at 100°C. The **normal**

**boiling point** of a liquid is defined as the temperature at which the vapor pressure of the liquid equals 1 atm. Boiling points given in reference books are understood to be normal boiling points.

The normal boiling points of ethyl ether (34.6°C), ethyl alcohol (78.5°C), and water are indicated on the vapor pressure curves of Figure 7.3. The boiling point of a liquid can be read from its vapor pressure curve by finding the temperature at which the vapor pressure of the liquid equals the prevailing pressure.

The fluctuations in atmospheric pressure at any one geographic location cause a maximum variation of about 2°C in the boiling point of water. The variations from place to place, however, can be greater than this. The average barometric pressure at sea level is 1.000 atm; at higher elevations, average barometric pressures are less. Thus, at an elevation of 5000 feet above sea level, the average barometric pressure is 0.836 atm; at this pressure, water boils at 95.1°C. Water boils at 90.1°C at 0.695 atm, which is the average atmospheric pressure at 10,000 feet above sea level.

If a liquid has a high normal boiling point or decomposes when heated, it can be made to boil at low temperatures by reducing the pressure. This procedure is followed in vacuum distillation; water can be made to boil at 10°C, which is considerably below room temperature, by adjusting the pressure to 0.0121 atm (see Table 6.2 and Figure 7.3). Many food products are concentrated by removing unwanted water under reduced pressure; in these procedures, the product is not subjected to temperatures that bring about decomposition or discoloration.

**7.5 HEAT OF VAPORIZATION**

The **molar heat of vaporization,** $\Delta H_v$, is the quantity of energy that must be supplied to vaporize a mole of a liquid at a specified temperature. Heats of vaporization are usually recorded at the normal boiling point in kcal/mol (Table 7.1).

Table 7.1 Heats of vaporization of liquids at their normal boiling points

| LIQUID | $t_b$ NORMAL BOILING POINT (°C) | $\Delta H_v$ HEAT OF VAPORIZATION (kcal/mol) | (kJ/mol) | $\Delta S_v = \Delta H_v/T_b$ ENTROPY OF VAPORIZATION (cal/°K mol) | (J/°K mol) |
|---|---|---|---|---|---|
| water | 100.0 | 9.72 | 40.7 | 26.0 | 109 |
| benzene | 80.1 | 7.35 | 30.8 | 20.8 | 87 |
| ethyl alcohol | 78.5 | 9.22 | 38.6 | 26.2 | 110 |
| carbon tetrachloride | 76.7 | 7.17 | 30.0 | 20.5 | 86 |
| chloroform | 61.3 | 7.02 | 29.4 | 21.0 | 88 |
| carbon disulfide | 46.3 | 6.40 | 26.8 | 20.0 | 84 |
| ethyl ether | 34.6 | 6.21 | 26.0 | 20.2 | 84 |

In Section 7.2 a kinetic picture of the vaporization process was presented, and the heat of vaporization was interpreted from this viewpoint. A slightly different interpretation based on a consideration of the heat content (enthalpy, $H$) of the vapor and of the liquid follows. The temperature of a vapor is the same as the temperature of the liquid with which it is in equilibrium; hence, the molecules of each phase have the same average kinetic energy. The phases, however, differ in total internal energy, which includes potential energy as well as kinetic energy. The molecules of the liquid are held together by cohesive forces; the molecules of the vapor are essentially free. When the liquid is converted into a gas, energy must be supplied to separate the molecules. Thus, the energy of the gas phase is higher than that of the liquid phase by the amount of this difference.

The difference in heat content of the phases ($\Delta H$) takes into account another factor. The volume of a gas is considerably larger than the volume of the liquid from which it is derived (e.g., about 1700 ml of steam is produced by the vaporization of 1 ml of water at 100°C); energy must be supplied to do the work of pushing back the atmosphere to make room for the vapor. The heat of vaporization includes both the energy required to overcome the intermolecular cohesive forces and the energy needed to expand the vapor.

When a mole of vapor is condensed to a liquid, the difference in heat content of the phases is released rather than absorbed. In this instance the heat effect is called the **molar heat of condensation;** it has a negative sign but is numerically equal to the molar heat of vaporization at the same temperature.

The heat of vaporization of a given liquid decreases as the temperature increases and equals zero at the critical temperature of the substance. This parallels an increase in the fraction of high-energy molecules; at the critical temperature all the molecules have sufficient energy to vaporize.

Over a relatively narrow temperature range, the heat of vaporization can be considered to be constant. Under such conditions the vapor pressure of a liquid, $p$ (in atm), is related to the temperature at which it is measured, $T$ (in °K), by the equation

$$\log p = -\frac{\Delta H_v}{2.303 R T} + C$$

where $\Delta H_v$ is the molar heat of vaporization (in cal/mol, or J/mol), $R$ is the gas constant (1.987 cal/°K mol, or 8.314 J/°K mol), and $C$ is a constant characteristic of the liquid under study. The vapor pressure–temperature curves shown in Figure 7.3 can be mathematically described by fitting appropriate values into this equation.

If we wish to compare the vapor pressure of a given liquid, $p_1$, at one temperature, $T_1$, to the vapor pressure of the same liquid, $p_2$, at a second temperature, $T_2$, we can derive a very useful equation.

At $T_2$:   $\log p_2 = -\dfrac{\Delta H_v}{2.303R}\left(\dfrac{1}{T_2}\right) + C$

At $T_1$:   $\log p_1 = -\dfrac{\Delta H_v}{2.303\,R}\left(\dfrac{1}{T_1}\right) + C$

By subtracting the second equation from the first, we get

$$\log p_2 - \log p_1 = -\dfrac{\Delta H_v}{2.303R}\left(\dfrac{1}{T_2} - \dfrac{1}{T_1}\right)$$

which can be rearranged to

$$\log\left(\dfrac{p_2}{p_1}\right) = \dfrac{\Delta H_v}{2.303R}\left(\dfrac{T_2 - T_1}{T_1 T_2}\right)$$

This equation, known as the **Clausius–Clapeyron equation,** was first proposed by Benoît Clapeyron in 1834 and later derived from thermodynamic theory of Rudolf Clausius.

---

**Example 7.1**

The normal boiling point of chloroform is 334°K. At 328°K the vapor pressure of chloroform is 0.823 atm. What is the heat of vaporization for this temperature range?

*Solution*

If we set $T_2 = 334°K$, then $p_2 = 1.000$ atm; $T_1 = 328°K$, and $p_1 = 0.823$ atm.

$$\log\left(\dfrac{p_2}{p_1}\right) = \dfrac{\Delta H_v}{2.303R}\left(\dfrac{T_2 - T_1}{T_1 T_2}\right)$$

$$\log\left(\dfrac{1.000\ \text{atm}}{0.823\ \text{atm}}\right) = \dfrac{\Delta H_v}{(2.303)(1.987\ \text{cal/°K mol})}\left[\dfrac{334°K - 328°K}{(328°K)(334°K)}\right]$$

$$\Delta H_v = 7070\ \text{cal/mol}$$
$$= 7.07\ \text{kcal/mol}$$

---

**Example 7.2**

The vapor pressure of carbon disulfide at 301°K is 0.526 atm. What is the vapor pressure of $CS_2$ at 273°K? The heat of vaporization over this temperature range may be taken as 6.60 kcal/mol.

*Solution*

We set $T_2 = 301°K$, $p_2 = 0.526$ atm, and $T_1 = 273°K$. Thus,

$$\log\left(\dfrac{p_2}{p_1}\right) = \dfrac{\Delta H_v}{2.303R}\left(\dfrac{T_2 - T_1}{T_1 T_2}\right)$$

$$\log\left(\dfrac{0.526\ \text{atm}}{p_1}\right) = \dfrac{6,600\ \text{cal/mol}}{(2.303)(1.987\ \text{cal/°K mol})}\left[\dfrac{301°K - 273°K}{(273°K)(301°K)}\right]$$

$$= 0.491$$

$$\dfrac{0.526\ \text{atm}}{p_1} = 3.10$$

$$p_1 = 0.170\ \text{atm}$$

In general, the higher the heat of vaporization of a substance, the stronger are the intermolecular forces of attraction. In 1884 Frederick Trouton discovered that for many liquids the molar heat of vaporization (at the normal boiling point) divided by the normal boiling point (on the Kelvin scale) is a constant: 21 cal/°K mol (**Trouton's rule**). Thus,

$$\frac{\Delta H_v}{T_b} = 21 \text{ cal/°K mol}$$

● Trouton's rule is expressed in SI units as

$$\frac{\Delta H_v}{T_b} = 88 \text{ J/K mol}$$

This rule is only approximate, as the values listed in Table 7.1 attest.

The value ($\Delta H_v/T_b$) of Trouton's rule is called the **molar entropy of vaporization** and is given the symbol $\Delta S_v$. The **entropy**, $S$, of a system is a measure of the system's randomness, or disorder (Section 14.4). A positive change in entropy, $\Delta S$, indicates an increase in disorder; a negative value of $\Delta S$ indicates that the system has become more ordered.

The change from liquid to vapor involves an increase in molecular randomness. The extent of this increase is about the same for all nonpolar liquids; therefore, the values of $\Delta S_v$ are about the same, 21 cal/°K mol. In water and ethyl alcohol, the intermolecular forces of attraction are unusually strong (Section 8.9); hence, $\Delta S_v$ values for these liquids are comparatively large (Table 7.1), since they represent changes for liquids that are more highly ordered than usual.

The entropy of vaporization of a given liquid, like the heat of vaporization, decreases as the temperature is increased. With increasing temperature the molecular randomness of the liquid phase increases and approaches that of the gas phase; at the critical temperature they are equal, only one phase exists, $\Delta H_v = 0$, and $\Delta S_v = 0$.

**7.6 THE FREEZING POINT**

As a liquid is cooled, the molecules move more and more slowly. Eventually a temperature is reached at which some of the molecules have sufficiently low kinetic energies to allow the intermolecular attractions to hold them in a crystal lattice; the substance then starts to freeze. Gradually the low-energy molecules assume positions in the crystal lattice; the molecules remaining in the liquid have a higher temperature because of the loss of these low-energy molecules. Heat must be removed from the liquid to maintain the temperature.

The **normal freezing point** of a liquid is the temperature at which solid and liquid are in equilibrium under a total pressure of 1 atm. At the freezing point the temperature of the solid–liquid system remains constant until all of the liquid is frozen. The quantity of heat that must be removed to freeze a mole of a substance at the freezing point is called the **molar heat of crystallization.** This quantity represents the difference in the heat content ($\Delta H$) between the liquid and the solid.

At times the molecules of a liquid, as they are cooled, continue the random motion characteristic of the liquid state at temperatures below the freezing point; such liquids are referred to as **undercooled** or **supercooled.** These systems can usually be caused to revert to the freezing temperature and the stable solid-liquid equilibrium by scratching the interior walls of the container with a stirring rod or by adding a seed crystal around which crystallization can occur. The crystallization process supplies heat, and the temperature is brought back to the freezing point until normal crystallization is complete.

Some supercooled liquids can exist for long periods, or even permanently, in this state. When these liquids are cooled, molecules solidify in a random arrangement typical of the liquid state rather than in an orderly geometric pattern of a crystal. Substances of this type have relatively high viscosities and generally possess complex molecular forms for which crystallization would be difficult. They are frequently called amorphous solids, vitreous materials, or glasses; examples include glass, tar, and certain plastics. Some materials (e.g., glass) can be obtained in the crystalline form by careful cooling and change very slowly to the crystalline form from the supercooled state. Amorphous solids have no definite freezing or melting point; rather, these transitions take place over a temperature range. They exhibit conchoidal fracture (the fragments have curved, shell-like surfaces) rather than the cleavage along definite planes at definite angles shown by crystalline materials.

When a crystalline substance is heated, the temperature at which solid–liquid equilibrium is attained under air at 1 atm pressure is called the **melting point;** it is, of course, the same temperature as the freezing point of the substance. The quantity of heat that must be *added* to melt a mole of the material at the melting point is called the **molar heat of fusion** and is numerically equal to the heat of crystallization.

The molar heats of fusion and **molar entropies of fusion** of several substances are listed in Table 7.2. The transition from solid to liquid represents an increase in disorder (and consequently in entropy) in the same way that the transition from liquid to gas does. Entropies of fusion, however, are generally of a lower order than entropies of vaporization

Table 7.2 Heats of fusion of solids at the melting point

| SOLID | $t_f$ MELTING POINT (°C) | $\Delta H_f$ HEAT OF FUSION (kcal/mol) | (kJ/mol) | $\Delta S_f = \Delta H_f/T_f$ ENTROPY OF FUSION (cal/°K mol) | (J/°K mol) |
|---|---|---|---|---|---|
| water | 0.0 | 1.44 | 6.02 | 5.26 | 22.0 |
| benzene | 5.5 | 2.35 | 9.83 | 8.44 | 35.3 |
| ethyl alcohol | −117.2 | 1.10 | 4.60 | 7.10 | 29.7 |
| carbon tetrachloride | −22.9 | 0.60 | 2.51 | 2.40 | 10.0 |
| chloroform | −63.5 | 2.20 | 9.20 | 10.50 | 43.9 |
| carbon disulfide | −112.1 | 1.05 | 4.39 | 6.52 | 27.3 |

because molecular disorder is greater in the gaseous state than in the liquid or solid states. The variation in $\Delta S_f$ values from substance to substance is indicative of the differences in the lattice energies of the crystalline solids; $\Delta S_f$ values are not nearly so constant as $\Delta S_v$ values.

**7.7 VAPOR PRESSURE OF A SOLID**

Molecules in a crystal vibrate about their lattice positions. A distribution of kinetic energy exists among these molecules similar to that for liquids and gases but on a lower level. Energy is transmitted from molecule to molecule within a crystal; the energy of any one molecule, therefore, is not constant. High-energy molecules on the surface of the crystal can overcome the attractive forces of the crystal and escape into the vapor phase. If the crystal is in a closed container, equilibrium is eventually reached in which the rate of the molecules leaving the solid equals the rate of the vapor molecules returning to the crystal. The vapor pressure of a solid at a given temperature is a measure of the number of molecules in the vapor at equilibrium.

Every solid has a vapor pressure, although some pressures are very low. The magnitude of the vapor pressure is inversely proportional to the strength of the attractive forces; thus, ionic crystals have very low vapor pressures.

Since the ability of molecules to escape the intermolecular forces of attraction depends upon their kinetic energies, the vapor pressure of a solid increases as the temperature increases. The temperature–vapor pressure curve for ice is illustrated in Figure 7.4. This curve intersects the vapor pressure curve for water at the freezing point. In the absence of air, the normal freezing point of water (1 atm total pressure) is 0.0025°C. *In air*, however, and under a total pressure of 1 atm, the freezing point of water is 0.0000°C, which is the commonly reported value. The difference in freezing point is caused by the presence of dissolved air in the water (Section 9.9). The vapor pressures plotted in Figure 7.4 are the partial pressures of $H_2O$ in air with the total pressure equal to 1 atm. Freezing points are usually determined in air; however, in any event, any change in freezing point of a given substance caused by the presence of air is generally very small.

At the freezing point the vapor pressures of solid and liquid are equal. If this were not true, one form would gradually be converted into the other. For example, if the vapor pressure of the liquid were higher than that of the solid, the vapor concentration would be higher than the equilibrium value for the solid–vapor system.

solid $\rightleftharpoons$ vapor

In this circumstance the rate of crystallization would increase, the amount of solid would increase, and the concentration of vapor would decrease.

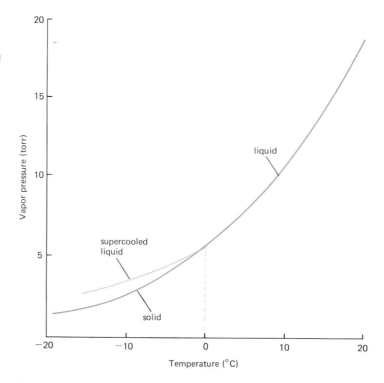

**Figure 7.4** Vapor pressure curves for ice and water near the freezing point. Vapor pressures are partial pressures of $H_2O$ under air at a total pressure of 1.00 atm.

This decrease in the concentration of vapor would cause more of the liquid to vaporize in order to reestablish the equilibrium vapor pressure of the liquid.

$$\text{liquid} \rightleftharpoons \text{vapor}$$

The process would continue until all the liquid was converted into solid. At a given temperature, the stable form is the one that has the lower vapor pressure.

**7.8 PHASE DIAGRAMS**  The temperature–pressure phase diagram for water conveniently illustrates the conditions under which water can exist as solid, liquid, or vapor, as well as the conditions that bring about changes in the state of water. Figure 7.5 is a schematic representation of the water system; it is not drawn to scale, and some of its features are exaggerated to give prominence to important details. Every substance has a unique phase diagram that describes only systems in equilibrium; this diagram must be derived from experimental observations.

The diagram of Figure 7.5 relates to what is called a **one-component system;** that is, it pertains to the behavior of water in the absence of any

**Figure 7.5** Phase diagram for water (not drawn to scale).

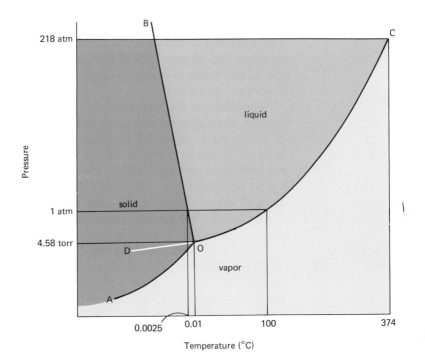

other substance. Thus, the total pressure of any system described by the diagram cannot be due in any part to the pressure of a gas other than water vapor. The vapor pressure curves plotted in Figure 7.4 (measured in air under a constant total pressure of 1 atm) therefore deviate slightly, but only very slightly, from the vapor pressure curves of Figure 7.5 (for which the vapor pressure of water is the *total* pressure). The easiest way to interpret the phase diagram for water is to visualize the total pressure acting on a system in mechanical terms, for example, as a piston acting on the material comprising the system contained in a cylinder.

In Figure 7.5 curve OC is a vapor pressure curve for liquid and terminates at the critical point, C. Any point on this line describes a set of temperature and pressure conditions under which liquid and vapor can exist in equilibrium. The extension DO is the curve for supercooled liquid; systems between liquid and vapor described by points on this line are **metastable.** (The term metastable is applied to systems that are not in the most stable state possible at the temperature in question.) Curve AO is a vapor pressure curve for solid and represents a set of points that describe the possible temperature and pressure conditions for solid–vapor equilibria. The line BO, the melting point curve, represents conditions for equilibria between solid and liquid.

These curves intersect at point O, the **triple point.** Solid, liquid, and vapor can exist together in equilibrium under the conditions represented by this point: 0.01°C (273.16°K) and a pressure of 0.00603 atm (or 4.58 torr) due to the vapor. When a point described by the temperature-pressure coordinates falls in one of the regions labeled solid, liquid, or vapor, a situation is indicated in which only one phase can exist under the conditions described.

The slope of the melting point (or freezing point) curve, BO, shows that the freezing point decreases as the pressure is increased. A variation of this type is observed for only a few substances such as gallium, bismuth, and water; it indicates an unusual situation in which the liquid expands upon freezing. At 0°C a mole of water occupies 18.00 cm³, and a mole of ice occupies 19.63 cm³. Thus, there is an expansion when one mole of liquid water freezes into ice. An increase in pressure on the system would oppose this expansion and the freezing process. Hence, the freezing point of water is lowered as the total pressure is increased. In Figure 7.5 the slope of the line BO is exaggerated.

Phase changes brought about by temperature changes at constant pressure may be read from a phase diagram by interpreting a horizontal line drawn at the reference pressure (like the line drawn at 1.00 atm in Figure 7.5). The point where this line intersects curve BO indicates the

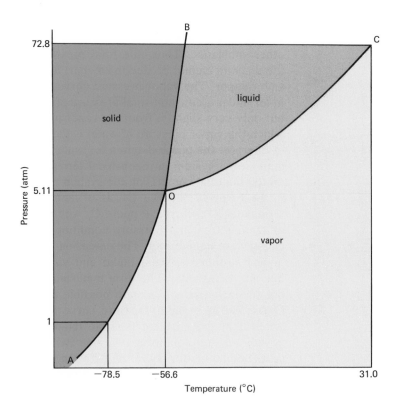

**Figure 7.6** Phase diagram for carbon dioxide (not drawn to scale).

normal melting point (or freezing point), and the point where the 1.00 atm line intersects curve CO represents the normal boiling point. Beyond this point, only vapor exists.

In like manner, phase changes brought about by pressure changes at constant temperature may be read from a vertical line drawn at the reference temperature. If the pressure is increased, for example, at 0.0025°C (Figure 7.5), the point where the vertical line crosses AO is the pressure where vapor changes to solid, and the point where the vertical line crosses BO represents the pressure where solid changes to liquid. Above this point, only liquid exists.

For materials that contract upon freezing (i.e., the solid phase is more dense than the liquid phase), the freezing-point curve inclines in the opposite direction, and the freezing point increases as the pressure is increased. This behavior is characteristic of most substances, and the freezing-point curves of most phase diagrams slant to the right as is seen in the phase diagram for carbon dioxide in Figure 7.6.

The process in which a solid goes directly into a vapor without going through the liquid state is known as **sublimation;** this process is reversible. The phase diagram for carbon dioxide is typical for substances that sublime at ordinary pressures rather than melt and then boil. The triple point of the carbon dioxide system is −55.6°C at a pressure of 5.11 atm. Liquid carbon dioxide exists only at pressures greater than 5.11 atm. When solid carbon dioxide (dry ice) is heated at 1 atm pressure, it is converted directly into gas at −78.5°C; the **molar heat of sublimation** is the heat that must be added to a mole of solid to convert it directly into a gas.

**7.9 CRYSTALS**     The orderly arrangement of atoms, ions, or molecules in a crystalline solid was discussed in Section 4.8. The constituent particles of a crystal are arranged in a repeating three-dimensional pattern called a **crystal (or space) lattice.** The smallest section of a crystal lattice that can be used to describe the lattice is the **unit cell.** A crystal can be reproduced, in theory, by stacking its unit cells in three dimensions.

Crystal lattices may be grouped according to symmetry into six **crystal systems** (Figure 7.7). A crystal system may be described by the dimensions of a unit cell along its three axes ($a$, $b$, $c$) and the sizes of the three angles between the axes ($\alpha$, $\beta$, $\gamma$). The hexagonal system is sometimes divided into two symmetrically related systems: hexagonal and rhombohedral. Crystals have the same symmetry as their constituent unit cells.

In illustrations of crystal lattices, points are customarily used to indicate the centers of the ions, atoms, or molecules. Six point lattices can be drawn by placing points at the corners of the unit cells shown in

**Figure 7.7** Crystal systems.

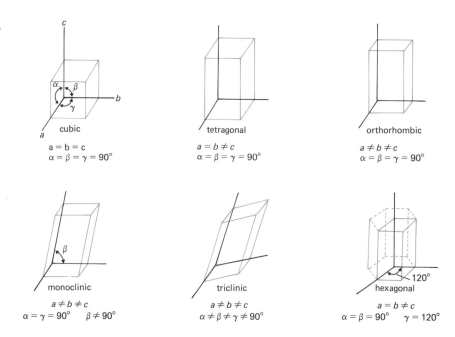

cubic
a = b = c
α = β = γ = 90°

tetragonal
a = b ≠ c
α = β = γ = 90°

orthorhombic
a ≠ b ≠ c
α = β = γ = 90°

monoclinic
a ≠ b ≠ c
α = γ = 90°    β ≠ 90°

triclinic
a ≠ b ≠ c
α ≠ β ≠ γ ≠ 90°

hexagonal
a = b ≠ c
α = β = 90°    γ = 120°

Figure 7.7. However, it is possible to have points at positions other than the corners. Thus, three cubic lattices are known (Figure 7.8): simple cubic, body-centered cubic, and face-centered cubic. Altogether, there are 14 possible lattices in which each point is surrounded in an identical manner by other points.

In counting the number of atoms per unit cell, one must keep in mind that atoms on corners or faces are shared with adjoining cells. Eight unit cells share each corner atom, and two unit cells share each face-centered

**Figure 7.8** Cubic lattices.

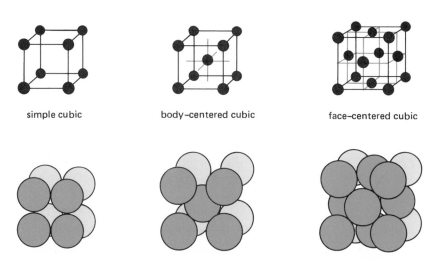

simple cubic

body–centered cubic

face–centered cubic

**Figure 7.9** In cubic crystals, (a) a corner atom is shared by eight unit cells, and (b) a face-centered atom is shared by two unit cells.

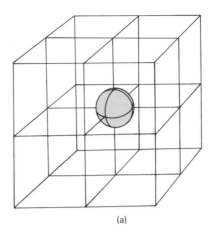

(a)

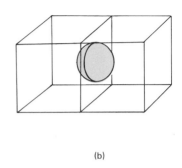

(b)

atom (Figure 7.9). The simple cubic unit cell, therefore, contains the equivalent of only one atom (8 corners at $\frac{1}{8}$ each). The body-centered unit cell contains two atoms (8 corners at $\frac{1}{8}$ each and one unshared atom in the center). The face-centered unit cell contains the equivalent of four atoms (8 corners at $\frac{1}{8}$ each and 6 face-centered atoms at $\frac{1}{2}$ each).

---

**Example 7.3**

Nickel crystallizes in a face-centered cubic crystal; the edge of the unit cell is 3.52 Å. The atomic weight of nickel is 58.7, and its density is 8.94 g/cm³. From these data calculate Avogadro's number.

**Solution**

There are four nickel atoms in each unit cell. If we let $N$ = Avogadro's number, the weight of one atom is $(58.7/N)$ g. The weight of one unit cell is $4(58.7/N)$ g.

The volume of one unit cell is $(3.52 \times 10^{-8})^3$ cm³. The density of nickel is

$$\frac{4(58.7/N) \text{ g}}{(3.52 \times 10^{-8})^3 \text{ cm}^3} = 8.94 \text{ g/cm}^3$$

Therefore,

$$N = 6.02 \times 10^{23}$$

An alternative solution is

$$? \text{ atoms} = 58.7 \text{ g Ni} \left(\frac{1 \text{ cm}^3}{8.94 \text{ g Ni}}\right)\left(\frac{4 \text{ atoms}}{(3.52 \times 10^{-8})^3 \text{ cm}^3}\right)$$

$$= 6.02 \times 10^{23}$$

---

**7.10 X-RAY DIFFRACTION OF CRYSTALS**

Much of what is known about the internal structure of crystals has been learned from X-ray diffraction experiments. Two X-ray waves that are in phase reinforce each other and produce a wave that is stronger than either of the original waves. The resultant wave has a greater amplitude than either of the primary waves, but the wavelength, $\lambda$, remains the same. Two waves that are completely out of phase cancel each other; the resultant has negligible intensity (see Figure 4.11).

Figure 7.10 Derivation of
the Bragg equation.

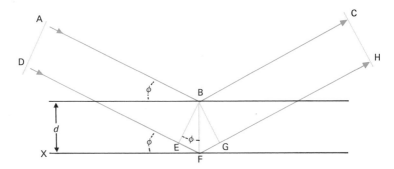

Figure 7.10 illustrates the determination of crystal spacings by use of X rays of a single wavelength. The rays impinge upon parallel planes of the crystal at an angle $\phi$; the angle of reflection equals the angle of incidence. Some of the rays are reflected from the upper plane, some from the second plane, and some from lower planes. A strong reflected beam will result only if all the rays are in phase. In the illustration the ray *DFH* travels farther than the ray *ABC* by an amount equal to *EF* + *FG*. These rays will be in phase at *CH* only if this difference equals a whole number of wavelengths. Thus,

$$EF + FG = n\lambda$$

where $n$ is a simple integer.

The line *BE* is drawn perpendicular to *DF*. Angle *BEF*, therefore, equals 90°. Since the sum of the angles of any triangle equals 180°, the sum of the other two angles (*EBF* and *EFB*) of the triangle *BEF* must also equal 90°.

$$\angle EBF + \angle EFB = 90°$$

Angle *XFB* is a right angle, and angle *XFE* is $\phi$; therefore, angle *EFB* is equal to 90° minus $\phi$. Consequently,

$$\angle EBF + (90° - \phi) = 90°$$

and angle *EBF* equals $\phi$.

The sine of this angle, $\phi$, is equal to *EF/BF*, and since *BF* is equal to $d$ (the distance between the planes of the crystal),

$$\sin \phi = \frac{EF}{d}$$

or

$$EF = d \sin \phi$$

Likewise,

$$FG = d \sin \phi$$

therefore,

$$EF + FG = 2d \sin \phi$$

or

$$n\lambda = 2d \sin \phi$$

This equation was derived by William Henry Bragg and his son William Lawrence Bragg in 1913.

The Bragg equation can be rearranged:

$$\sin \phi = \frac{n\lambda}{2d}$$

Thus with X rays of a definite wavelength, reflections at various angles will be observed for a given set of planes with a spacing equal to $d$. These reflections correspond to $n = 1, 2, 3$, and so on, and are spoken of as first order, second order, third order, and so on. With each successive order, the angle $\phi$ increases, and the intensity of the reflected beam weakens.

Figure 7.11 is a schematic representation of an X-ray spectrometer. An X-ray beam defined by a slit system impinges upon a crystal that is mounted on a turntable. A detector (photographic plate, ionization chamber, or Geiger counter) is positioned as shown in the figure. As the crystal is rotated, strong signals flash out as angles are passed that satisfy the Bragg equation. Any set of regularly positioned planes that contain

● In scientific literature crystal distances are customarily recorded in ångstrom units. This unit, however, is to be used for a limited period of time. The appropriate SI unit is the nanometer.

$$1 \text{ Å} = 0.1 \text{ nm}$$

Figure 7.11 X-ray diffraction of crystals (schematic).

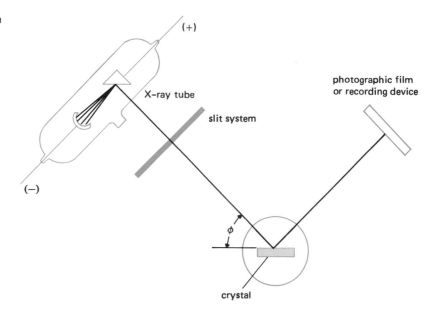

atoms can give rise to reflections — not only those that form the faces of the unit cells. Thus, the value of $d$ is not necessarily the edge of the unit cell, although the two are always mathematically related.

---

*Example 7.4*

The diffraction of a crystal of barium with X radiation of wavelength 2.29 Å gives a first-order reflection at 27° 8′. What is the distance between the diffracted planes?

*Solution*

Substitution into the Bragg equation gives

$$n\lambda = 2d \sin \phi$$

$$1(2.29 \text{ Å}) = 2d(0.456)$$

$$d = 2.51 \text{ Å}$$

---

**7.11 CRYSTAL STRUCTURE OF METALS**

In the overwhelming majority of cases, metal crystals belong to one of three classifications: body-centered cubic (Figure 7.8), face-centered cubic (Figure 7.8), and hexagonal close packed (Figure 7.12). The geometric arrangement of atoms in the face-centered cubic and the hexagonal close-packed crystals is such that each atom has a coordination number of 12 (each is adjoined by 12 other atoms). If the atoms are viewed as spheres, there is a minimum of empty space in these two types of crystals (about 26%), and both of the crystal lattices are called close-packed structures. The body-centered cubic arrangement is slightly more open than either of the close-packed arrangements (about 32% empty space); each atom of a body-centered cubic crystal has a co-ordination number of 8.

The difference between the two closed-packed structures may be derived from a consideration of Figure 7.13. The shaded circles of the diagram represent the first layer of spheres which are placed as close together as possible. The second layer of spheres (open circles of Figure 7.13) are placed in the hollows formed by adjacent spheres of the first layer. The first two layers of both the face-centered cubic and the hexagonal close-packed arrangements are the same; the difference arises in the third and subsequent layers.

In the hexagonal close-packed arrangement, the spheres of the third layer are placed so that they are directly over those of the first layer; the sequence of layers may be represented as *ababab.* . . . In the face-

**Figure 7.12** Hexagonal close-packed structure.

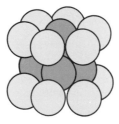

**Figure 7.13** Schematic representation of the first two layers of the close-packed arrangements.

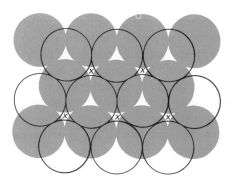

centered cubic structure, however, the spheres of the third layer are placed over the holes (marked *x* in Figure 7.13) formed by the arrangements of the first two layers. The spheres of the fourth layer of the face-centered cubic structure are placed so that they are directly over those of the first layer, and the sequence of layers is *abcabc*. . . .

The holes that the *second* layer of spheres cap are called **tetrahedral holes** because each is formed by four spheres. The *x* holes are **octahedral holes,** and each is surrounded by six spheres. In an extended three-dimensional model of either close-packed structure, there are the same number of octahedral holes as spheres and twice as many tetrahedral holes as spheres. The importance of these holes is discussed in Section 7.12.

The crystal structure of metals are summarized in Table 7.3. The

**Table 7.3** Crystal structures of metals

| Li | Be |    |    |    |    |    |    |    |    |    |    |    |    |
|----|----|----|----|----|----|----|----|----|----|----|----|----|----|
| Na | Mg |    |    |    |    |    |    |    |    |    |    |    | Al |
| K  | Ca | Sc | Ti | V  | Cr |    | Fe | Co | Ni | Cu | Zn |    |    |
| Rb | Sr | Y  | Zr | Nb | Mo | Tc | Ru | Rh | Pd | Ag | Cd |    |    |
| Cs | Ba | La | Hf | Ta | W  | Re | Os | Ir | Pt | Au |    | Tl | Pb |

| Ce | Pr | Nd |   | Eu | Gd | Tb | Dy | Ho | Er | Tm | Yb | Lu |
|----|----|----|---|----|----|----|----|----|----|----|----|----|

☐ hexagonal close packed

☐ face-centered cubic

☐ body-centered cubic

division between the three structures (hexagonal close packed, face-centered cubic, and body-centered cubic) is about even. The close packing of most metallic crystals helps to explain the relatively high densities of metals. The structures of a few metals (for example, manganese and mercury) do not fall into any of the three categories, and the symbols for these metals do not appear in the table. Some metals exhibit **crystal allotropy;** that is, different crystal forms of the same metal are stable under different conditions. For example, a modification of calcium exists in each of the three structures. The form of each metal that is indicated in Table 7.3 is the one that is stable under ordinary conditions. In addition to many metals, the noble gases also crystallize in the face-centered cubic lattice.

**7.12 IONIC CRYSTALS**

Ionic lattices are more complicated than metallic lattices. An arrangement must accommodate ions of opposite charge and different size in the proper stoichiometric ratio and in such a way that electrostatic attractions outweigh electrostatic repulsions.

Since the anion is larger than the cation in most ionic compounds, it is advantageous to view an ionic crystal as being composed of a lattice of anions with cations placed in lattice holes. Three types of lattice holes are illustrated in Figure 7.14. By imagining that the lattice points of the diagrams expand into spheres that touch, one can get an idea of the actual appearance of these holes.

If cations were placed in the spots marked $y$ in the diagrams of Figure 7.14, the number of anions around each cation (and presumed to be touching it) would vary according to the type of hole occupied: eight for cubic, six for octahedral, and four for tetrahedral. Clearly, the cubic hole is best in terms of crystal stability brought about by plus-minus attractions.

However, another factor must be considered: the radius of the cation, $r^+$, in comparison to the radius of the anion, $r^-$. If a large cation were to be placed in a cubic hole, the anions of the lattice could be forced apart,

**Figure 7.14** Types of holes in crystal lattices and radius ratio criteria for cation occupancy.

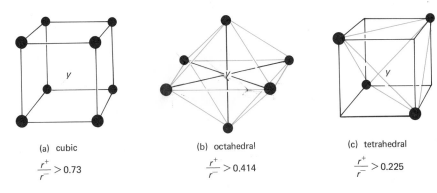

(a) cubic

$$\frac{r^+}{r^-} > 0.73$$

(b) octahedral

$$\frac{r^+}{r^-} > 0.414$$

(c) tetrahedral

$$\frac{r^+}{r^-} > 0.225$$

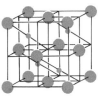

Figure 7.15 Crystal structures of ionic compounds of the MX type. Gold colored circles represent cations.

cesium chloride
CsCl

sodium chloride
NaCl

zinc blende
ZnS

but if a small cation were to be so situated, the electrostatic repulsions of the anions would prevent them from being squeezed together (Figure 3.6). In this latter case, the cation could not touch each anion surrounding it, and the structure would not be stable because the magnitude of the plus-minus attractions would be reduced. In order for a cation to occupy a cubic hole, the radius of the cation must be at least 0.73 times as large as the radius of the anion (radius ratio, $r^+/r^-$, greater than 0.73). The value 0.73 is derived from a calculation based on solid geometry.

If a cation is too small for a cubic hole, the next best bet is the slightly smaller octahedral hole (coordination number, 6). In order to maintain cation–anion contact, the radius of a cation filling an octahedral hole must be at least 0.414 times as large as the anion radius.

The tetrahedral hole is the smallest of the three, but its coordination number is only 4. Cations that are small in relation to the size of the anions of the lattice fill tetrahedral holes; the critical radius ratio is only 0.225.

The three most common crystal types for compounds of formula MX are shown in Figure 7.15. In each diagram there are as many cations shown belonging to a unit cell as there are anions (1 each in CsCl, 4 each in both NaCl and ZnS, taking into account sharing by adjacent unit cells). Therefore, a 1 : 1 stoichiometric ratio is represented in each case.

Compare the diagrams of Figure 7.14 with those of Figure 7.15. In the **cesium chloride** type crystal, the chloride ions form a simple cubic lattice with cesium ions filling cubic holes. In the **sodium chloride** crystal the chloride ions assume a face-centered cubic lattice, and the sodium ions fill octahedral holes. The sulfide ions of the **zinc blende** structure also form a face-centered cubic lattice, but the zinc ions fill alternate tetrahedral holes. Since there are twice as many tetrahedral holes as spheres in a close-packed structure (Section 7.11), the zinc ions of ZnS fill only half of the total number of these holes.

Predictions based on radius ratio alone are sometimes not borne out because the hard sphere model of ions that we have used fails to take ion distortion into account (Section 3.8). Many ionic compounds occur in the three structures we have discussed; examples are given in Table 7.4.

Table 7.4 Crystal structures of some ionic compounds

| STRUCTURE | EXAMPLES |
|---|---|
| cesium chloride | CsCl, CsBr, CsI, TlCl, TlBr, TlI, $NH_4Cl$, $NH_4Br$ |
| sodium chloride | halides of $Li^+$, $Na^+$, $K^+$, $Rb^+$<br>oxides and sulfides of $Mg^{2+}$, $Ca^{2+}$, $Sr^{2+}$, $Ba^{2+}$, $Mn^{2+}$, $Ni^{2+}$<br>AgF, AgCl, AgBr, $NH_4I$ |
| zinc blende | sulfides of $Be^{2+}$, $Zn^{2+}$, $Cd^{2+}$, $Hg^{2+}$<br>CuCl, CuBr, CuI, AgI, ZnO |
| fluorite | fluorides of $Ca^{2+}$, $Sr^{2+}$, $Ba^{2+}$, $Cd^{2+}$, $Pb^{2+}$<br>$BaCl_2$, $SrCl_2$, $ZrO_2$, $ThO_2$, $UO_2$ |
| antifluorite | oxides and sulfides of $Li^+$, $Na^+$, $K^+$, $Rb^+$ |
| rutile | fluorides of $Mg^{2+}$, $Ni^{2+}$, $Mn^{2+}$, $Zn^{2+}$, $Fe^{2+}$<br>oxides of $Ti^{4+}$, $Mn^{4+}$, $Sn^{4+}$, $Te^{4+}$ |

Analysis of the crystal forms of compounds of formula $MX_2$ or $M_2X$ proceeds along similar lines. The antifluorite and rutile structures are shown in Figure 7.16. In the **antifluorite** form, cations occupy all the tetrahedral holes in a face-centered cubic anion lattice. The **fluorite** structure (named for the mineral fluorite, $CaF_2$) is not shown in Figure 7.16; however, it is similar to the antifluorite structure except that cation and anion positions are exchanged. The **rutile** unit cell is not cubic and is difficult to describe. Notice, however, that the titanium ion in the center of the unit cell fills an octahedral hole formed by oxide ions.

**7.13 LATTICE ENERGY**

The lattice energy of an ionic compound may be calculated if the crystal geometry and distances between the ions in the crystals are known. An expression for the electrostatic interactions of the ions may be derived by treating the charges of the ions as though they were located entirely at points in the centers of the ions. The expression for the potential energy of interaction of point charges—that is, the energy associated with bring-

Figure 7.16 Crystal structures of ionic compounds of the $MX_2$ type. Gold colored circles represent cations.

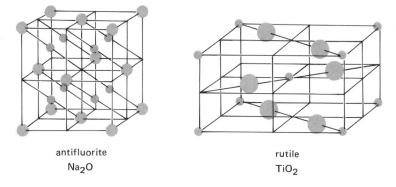

antifluorite
$Na_2O$

rutile
$TiO_2$

ing charges together from an infinite distance to a separation distance, $d$ — may be derived from Coulomb's law (Section 2.11).

$$PE = \frac{q_1 q_2}{d} \tag{1}$$

If the charges are of like sign (both positive or both negative), they will repel each other, and the potential energy will be positive; energy is absorbed by the process. On the other hand, if the charges have unlike signs, they will attract each other, and the potential energy will be negative; energy is evolved by the process.

In a sodium chloride crystal the distance between neighboring $Na^+$ and $Cl^-$ ions is 2.8 Å, which we shall designate as $r$. The $Na^+$ and $Cl^-$ ions have charges of $+e$ and $-e$, respectively, where $e$ is the unit electrical charge ($4.803 \times 10^{-10}$ esu). Calculation of the electrostatic potential of a single sodium ion proceeds by determining the contributions to the potential energy from interactions with all the surrounding ions of the crystal.

As a greatly simplified example, we shall consider the potential energy of the sodium ion (dark circle) in the center of the extremely small crystal fragment illustrated in Figure 7.17. The fragment is a cube $2r$ on a side and may be considered to be made up of eight smaller cubes each $r$ on a side. Notice that the diagonal of a face of one of the smaller cubes is $r\sqrt{2}$ and that the cube diagonal of the smaller cube is $r\sqrt{3}$. These facts may be confirmed by use of the Pythagorean theorem.

**Figure 7.17** Fragment of NaCl crystal.

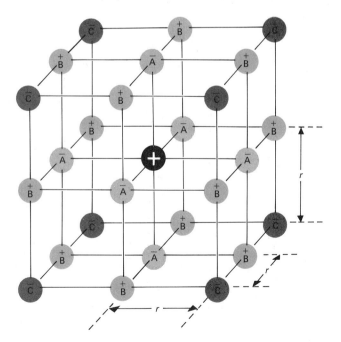

The nearest neighbors of the reference ion are the six $Cl^-$ ions that are labeled A in the diagram; these ions are at a distance of $r$ from the $Na^+$ ion. The potential energy contributed by six of these ions is

$$PE = 6\frac{(+e)(-e)}{r} = -\frac{e^2}{r}(6) \tag{2}$$

The next nearest neighbors of the central $Na^+$ ion are the twelve $Na^+$ ions that are labeled B in Figure 7.17. Each of these ions is situated at a distance $r\sqrt{2}$ from the central $Na^+$ ion (along face diagonals of the small cubes). Since the charge on each of these ions is the same as that of the reference ion, a positive potential energy—a repulsion—is indicated. The total contribution from this group of ions is

$$PE = 12\frac{(+e)(+e)}{r\sqrt{2}} = -\frac{e^2}{r}\left(-\frac{12}{\sqrt{2}}\right) \tag{3}$$

The eight $Cl^-$ ions that are labeled C in Figure 7.17 are at a distance of $r\sqrt{3}$ from the central $Na^+$ ion (along cube diagonals of the small cubes). Their contribution is

$$PE = 8\frac{(+e)(-e)}{r\sqrt{3}} = -\frac{e^2}{r}\left(\frac{8}{\sqrt{3}}\right) \tag{4}$$

If we continued in this way for a crystal of a more realistic size, we would find that there were six $Na^+$ ions at $2r$, 24 $Cl^-$ ions at $r\sqrt{5}$, 24 $Na^+$ ions at $r\sqrt{6}$, and so on. However, we can illustrate the process by summing expressions (2), (3), and (4) as long as we keep in mind that many more terms would be required for a complete treatment. Thus,

$$PE = -\frac{e^2}{r}\left(6 - \frac{12}{\sqrt{2}} + \frac{8}{\sqrt{3}} - \ldots\right) \tag{5}$$

The sum of the terms of the infinite series in parentheses is called the **Madelung constant** and is given the symbol $A$. Thus, for the reference ion (a sodium ion)

$$PE = -\frac{e^2}{r}A \tag{6}$$

The expression for the chloride ion is identical.

A mole of sodium chloride contains $N$ $Na^+$ ions and $N$ $Cl^-$ ions, where $N$ is Avogadro's number. The total electrostatic energy for the formation of a mole of NaCl, however, is not $2N$ times expression (6), since this procedure would count each interaction twice—once for the $Na^+$ ion and again for the $Cl^-$ ion. Hence, the electrostatic, potential energy per mole is one-half of $2N$, or $N$, times equation (6)

$$PE = -\frac{e^2}{r}NA \tag{7}$$

Notice that energy is evolved (minus sign) in the formation of a crystal from its constituent ions.

Equation (7) is a general expression for any type of crystal in which both ions bear unit charges. Differences in crystal geometry are reflected in the terms of the series used to derive the Madelung constant; the value of the constant, therefore, depends upon the crystal structure. For crystals of the NaCl type, $A$ is 1.748; for zinc blende crystals, $A$ is 1.638; for CsCl type crystals, $A$ is 1.763.

For crystals formed between ions that do not both have unit charges, equation (7) becomes

$$PE = - \frac{(Z_c)(Z_a)e^2}{r} NA \qquad (8)$$

where $Z_c$ and $Z_a$ are the number of unit charges on the cation and anion, respectively. These numbers do not include signs. The value of $A$ for the fluorite structure is 5.039 and that for the rutile structure is 4.816.

Equations (7) and (8) give values for lattice energies that are about 10% too high. Ions in crystals are not charges at fixed points in a lattice. The electron clouds of ions interact, and there are repulsive forces (even between ions of unlike sign) caused by this electronic interaction. A more exact calculation of lattice energy includes a correction for these repulsions — the magnitude of the correction factor being determined by measurement of the extent to which the crystal resists compression. More highly refined calculations include more corrections (for example, one that takes van der Waals forces into account). The importance of lattice energy calculations in relation to the Born–Haber cycle has been discussed in Section 5.9.

**7.14 DEFECT STRUCTURES**

Few crystals are perfect; many have some type of lattice defect. **Dislocations** are crystal imperfections that occur where planes of atoms are misaligned. For example, one type of dislocation is caused by the insertion (perpendicular to a face of the crystal) of an extra plane of atoms part way through a crystal. Atoms within the part of the crystal containing the extra plane are compressed.

**Point defects** are caused by missing or misplaced ions. One type of defect consists of a cation that has been moved from its proper position (thus creating a vacancy) to a place between regular lattice sites (an **interstitial** position). Another type of point defect consists of a pair of vacancies — one cation and one anion; the ions for these lattice positions are missing from the structure completely. These two point defects do not alter the stoichiometry of the crystal.

Certain crystals are imperfect because their compositions are not stoichiometric. Thus, samples of iron(II) oxide, FeO, usually contain more oxygen atoms than iron atoms, whereas zinc oxide, ZnO, usually has more zinc atoms than oxygen atoms. These departures from stoichiometry are usually quite small, on the order of 0.1%. There are several causes for **nonstoichiometry,** and in each the electrical neutrality of the crystal is preserved. Extra metal *atoms* or nonmetal *atoms* may be included in interstitial positions between the *ions* of the crystal (such as extra Zn atoms in ZnO). In addition, metal *atoms* or nonmetal *atoms* may assume regular lattice positions in place of *ions;* in these cases, ionic vacancies exist in the crystal lattice so that the whole lattice is electrically neutral. For example, in FeO oxygen atoms assume positions normally occupied by oxide ions, and there are missing $Fe^{2+}$ ions so that a 1 to 1 cation–anion ratio is maintained. It is postulated that in some crystals of this type (e.g., KCl), electrons from the "extra" metal atoms occupy the holes created by the anion vacancies. Nonstoichiometric substances are sometimes referred to as **berthollides,** named for Claude Louis Berthollet, who believed that the compositions of compounds varied continuously within limits. Joseph Proust, who upheld the law of definite proportions, engaged Berthollet in an argument of eight years' duration (1799–1807) concerning the composition of compounds.

The presence of **impurities** frequently accounts for crystal defects. For example, a $Mg^{2+}$ ion may occur in a lattice position in a NaCl crystal in place of a $Na^+$ ion; electrical neutrality requires that another lattice position be vacant, since the charge of the $Mg^{2+}$ ion is twice that of the $Na^+$ ion.

The addition of impurities to certain crystals produces conduction effects. Pure silicon and germanium function as intrinsic semiconductors (Section 4.7, Figure 4.31). Both these elements crystallize in the diamond-type network lattice in which each atom is bonded to four other atoms (Figure 4.33). At room temperature the conductivity of either silicon or germanium is extremely low, since the electrons are fixed by the bonding of the crystal. At higher temperatures, however, the crystal bonding begins to break down; electrons are freed and are able to move through the structure, thereby causing the conductivity to increase.

The addition of small traces of certain impurities to either silicon or germanium enhances the conductivity of these materials and produces what are called **extrinsic** (or **impurity**) **semiconductors.** For example, if boron is added to pure silicon at the rate of one B atom per million Si atoms, the conductivity is increased by a factor of approximately one hundred thousand (from $4 \times 10^{-6}$/ohm cm to 0.8/ohm cm at room temperature). Each Si atom has four valence electrons that are used in the bonding of the network lattice. A boron atom, however, has only three valence electrons. Hence, a B atom that assumes a position of a Si atom in the crystal can form only three of the four bonds required for a perfect lattice, and an electron vacancy (or hole) is introduced. An electron from a nearby bond can move into this vacancy, thus completing the four bonds on the B atom but, at the same time, leaving a vacancy at the original site of the electron. In this way electrons move through the structure, and the holes move in a direction opposite to that of the conduction electrons.

This type of extrinsic semiconductor, in which holes enable electron motion, is called a **p-type** semiconductor; the p stands for positive. The term is somewhat misleading, however, since the crystal, which is formed entirely from neutral atoms, is electrically neutral. The structure is electron deficient only with regard to the covalent bonding requirements of the lattice; it never has an excess of positive charge.

In addition to boron, other III A elements (Al, Ga, or In) can be added in small amounts to pure silicon or germanium to produce p-type semiconductors. Impurities that owe their effect to their ability to accept electrons by means of the holes they produce in the crystal lattice are known as **acceptor impurities.**

The energy-level band diagram for p-type semiconductors is shown in Figure 7.18. Electron vacancies are responsible for the addition of *empty* impurity levels, just above the filled valence band, to a diagram that otherwise would be characteristic of an intrinsic semiconductor. Promotion of electrons from the valence band to the unoccupied impurity levels requires the addition of comparatively little energy, and conduction takes place in this manner.

Traces of group V A elements (P, As, Sb, or Bi) when added to silicon or germanium produce a second type of extrinsic semiconductor known as an **n-type;** the n stands for negative. In this instance, each impurity atom has five valence electrons — one more than required by the bonding of the host crystal. The extra electron can move through the structure and function as a conduction electron. Impurities of this type are known as **donor impurities,** since they provide conduction electrons. The n-type semiconductor is negative only in the peculiar sense that more electrons are present than are required by the bonding scheme of the crystal; the substance is electrically neutral.

The energy-level diagram for the n-type semiconductor is shown in Figure 7.18. *Filled* impurity levels are introduced just below the empty conduction band by the addition of donor impurities. The electrons from these levels are easily excited into the conduction band by heating, and

**Figure 7.18** Energy-level band diagrams for extrinsic semiconductors.

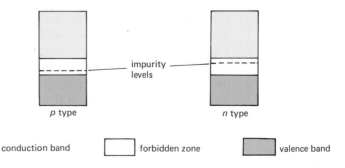

p type          impurity levels          n type

conduction band     forbidden zone     valence band

conduction occurs by this means. Semiconductors find use in photocells and transistors.

## Problems

Thermochemical problems may be solved using either calories or joules as energy units. Thermochemical values expressed in joules appear in brackets.

**7.1** Briefly explain how and why each of the following gives an indication of the strength of the intermolecular forces of attraction: (a) critical temperature, (b) surface tension, (c) viscosity, (d) vapor pressure, (e) heat of vaporization, (f) heat of fusion, (g) normal boiling point.

**7.2** Compare and contrast: (a) enthalpy, entropy; (b) boiling point, normal boiling point; (c) freezing point of water, triple point of water; (d) crystalline solid, vitreous solid; (e) $p$-type semiconductor, $n$-type semiconductor.

**7.3** The vapor pressure of $n$-heptane is 0.775 atm at 90°C and 0.275 atm at 60°C. Use the Clausius–Clapeyron equation to calculate a value for the molar heat of vaporization for this temperature range.

**7.4** The normal boiling point of chlorobenzene is 132°C, and the heat of vaporization of the compound is 8730 cal/mol [or 36,500 J/mol]. Use the Clausius–Clapeyron equation to determine the temperature at which the compound will boil under a pressure of 0.500 atm.

**7.5** The vapor pressure of cyclo-hexane at 25°C is 0.130 atm. Use the Clausius–Clapeyron equation to derive the normal boiling point of cyclohexane. Assume that $\Delta H_v$ is 7.60 kcal/mol [or 31.8 kJ/mol] over the temperature range of interest.

**7.6** The vapor pressure of cyclohexane is 0.527 atm at 61°C. Assume that $\Delta H_v$ is 7.60 kcal/mol [or 31.8 kJ/mol] over the temperature range of interest and calculate the vapor pressure of cyclohexane at 50°C.

**7.7** The normal boiling point of hydrogen, $H_2$, is 20.38°K and that of deuterium ($D_2$, which is $^2_1H_2$) is 23.59°K. Discuss the reason for the difference.

**7.8** The normal boiling point of HBr is −67°C. Use Trouton's rule to estimate the molar heat of vaporization of HBr.

**7.9** Calculate the molar entropy of vaporization, $\Delta S_v$, for hydrazine, $H_2NNH_2$. The heat of vaporization of hydrazine is 10.0 kcal/mol [or 41.8 kJ/mol] at the normal boiling point, 113.5°C. How well does this value compare with the one predicted by Trouton's rule? Can you explain the difference?

**7.10** A hydride of silicon has a

normal boiling point of 53°C and a heat of vaporization at the boiling point of 73.5 cal/g [or 308 J/g]. Use Trouton's rule to estimate the molecular weight of the compound.

**7.11** The heat of vaporization of tin(IV) chloride, $SnCl_4$, at the normal boiling point is 31.0 cal/g [or 130 J/g]. Use Trouton's rule to estimate the normal boiling point of $SnCl_4$.

**7.12** Explain why the entropy of vaporization of a substance is considerably larger than the entropy of fusion of that substance (Tables 7.1 and 7.2).

**7.13** Use the following data to draw a rough phase diagram for hydrogen. Normal melting point, 14.01°K; normal boiling point, 20.38°K; triple point, 13.95°K, $7 \times 10^{-2}$ atm; critical point, 33.3°K, 12.8 atm; vapor pressure of solid at 10°K, $1 \times 10^{-3}$ atm.

**7.14** Figure 7.5 is the phase diagram for water. Describe the phase changes that occur, and the approximate pressures at which they occur, when the pressure on a $H_2O$ system is gradually increased (a) at a constant temperature of −1°C, (b) at a constant temperature of 50°C, (c) at a constant temperature of −50°C.

**7.15** Refer to Figure 7.5 and describe the phase changes that occur, and the approximate temperatures at which they occur, when water is heated from −10°C to 110°C (a) under a pressure of $1 \times 10^{-3}$ atm, (b) under a pressure of 0.5 atm, (c) under a pressure of 1.1 atm.

**7.16** Refer to Figure 7.6 and describe the phase changes that occur and the approximate pressures at which they occur when the pressure on a $CO_2$ system is gradually increased (a) at a constant temperature of −60°C, (b) at a constant temperature of 0°C.

**7.17** Refer to Figure 7.6 and describe the phase changes that occur, and the approximate temperatures at which they occur, when carbon dioxide is heated (a) at a constant pressure of 2.0 atm, (b) at a constant pressure of 6.0 atm.

**7.18** Briefly explain how and why the slope of the melting-point curve of a phase diagram for a substance depends upon the relative densities of the solid and liquid forms of the substance.

**7.19** Sublimation is sometimes used to purify solids. The impure material is heated and the pure crystalline product condenses on a cold surface. Is it possible to purify ice by sublimation? What conditions would have to be employed?

**7.20** Xenon crystallizes in a face-centered cubic lattice, and the edge of the unit cell is 6.20 Å. What is the density of crystalline Xe?

**7.21** Silver crystallizes in a cubic system, and the edge of the unit cell is 4.08 Å. The density of silver is 10.6 g/cm³. How many atoms of Ag are contained in one unit cell of Ag? What type of cubic unit cell does Ag form?

**7.22** An element crystallizes in a body-centered cubic lattice, and the edge of the unit cell is 3.05 Å.

The density of the crystalline solid is 5.96 g/cm³. What is the atomic weight of the element?

**7.23** Sodium crystallizes in a body-centered cubic lattice, and the edge of the unit cell is 4.30 Å. Calculate the dimensions of a cube that would contain 23.0 g of Na (1 mol).

**7.24** Calcium crystallizes in a face-centered cubic lattice. The density of Ca is 1.55 g/cm³. What is the length of an edge of the unit cell?

**7.25** Indium crystallizes in a face-centered tetragonal lattice (see Figure 7.7). The base of the unit cell is 4.58 Å on a side and the height of the unit cell is 4.94 Å. What is the density of indium?

**7.26** How many atoms "belong" to the cell of the hexagonal close-packed structure shown in Figure 7.12?

**7.27** In the diffraction of a crystal using X rays with a wavelength equal to 0.710 Å, a first-order reflection was obtained at an angle of 12°. What is the distance between the diffracted planes? The sine of 12° is 0.208.

**7.28** In the X-ray diffraction of a set of crystal planes for which $d$ is 1.80 Å, a first-order reflection was obtained at an angle of 22°. What is the wavelength of the X rays that were employed? The sine of 22° is 0.375.

**7.29** In the diffraction of a crystal using X rays with a wavelength of 1.54 Å, a first-order reflection is found at an angle of 11°. What is the wavelength of X rays that show the same reflection at an angle of 13°55'? The sine of 11° is 0.191; the sine of 13°55' is 0.241.

**7.30** At what angle would a first-order reflection be observed in the X-ray diffraction of a set of crystal planes for which $d$ is 2.04 Å if the X rays used have a wavelength of 1.54 Å? At what angle would a second-order reflection from this same set of planes be observed?

**7.31** How many ions of each type are shown in the unit cell of the cesium chloride crystal depicted in Figure 7.15? The density of TlCl, which crystallizes in a cesium chloride lattice, is 7.00 g/cm³. What is the length of the edge of the unit cell shown in the figure? What is the shortest distance, $r$, between a $Tl^+$ ion and a $Cl^-$ ion?

**7.32** How many ions of each type are shown in the unit cell of zinc sulfide depicted in Figure 7.15? The density of BeS, which crystallizes in a zinc blende lattice, is 2.36 g/cm³. What is the length of the edge of the unit cell shown in the figure? What is the shortest distance, $r$, between a $Be^{2+}$ ion and a $S^{2-}$ ion?

**7.33** (a) How many $Na^+$ ions and $Cl^-$ ions belong to the cell lattice of the NaCl crystal shown in Figure 7.15? (b) Calcium sulfide, CaS, crystallizes in the sodium chloride lattice. The shortest distance between a $Ca^{2+}$ ion and a $S^{2-}$ ion is 2.83 Å. What is the length of the edge of the cell similar to that shown in Figure 7.15? What is the density, in g/cm³, of CaS?

**7.34** Consider the cubic hole depicted in Figure 7.14. Assume that a cation (radius, $r^+$) is inserted in the position marked $y$, and that this cation touches all of the anions (radius, $r^-$) of the structure. Also assume contact between any two anions that are adjacent along an edge of the cube. By calculating the body diagonal of the cube, determine the value of the radius ratio, $r^+/r^-$, that is consistent with the criteria given.

**7.35** Consider the octahedral hole depicted in Figure 7.14. Assume that a cation (radius, $r^+$) is inserted in the position marked $y$, and that this cation touches all of the anions (radius, $r^-$) of the structure. Also assume contact between any two anions that are adjacent along an edge of the octahedron. By calculating the diagonal of the square formed by the four anions in the center of the structure, determine the value of the radius ratio, $r^+/r^-$, that is consistent with the criteria given.

**7.36** Consider the tetrahedral hole depicted in Figure 7.14. The anions (radius, $r^-$) along the diagonal of a face of the cube shown touch one another; therefore, the diagonal of a face of the cube is $2r^-$. (a) What is the length of an edge of the cube shown in terms of $r^-$? (b) What is the length of the body diagonal of the cube in terms of $r^-$? (c) Assume that a cation (radius, $r^+$) is placed in the position marked $y$ in the figure. The body diagonal of the cube is equal to $(2r^+ + 2r^-)$. Calculate the radius ratio, $r^+/r^-$, for the tetrahedral hole.

**7.37** In a given crystal the distance between the centers of a cation and a neighboring anion is approximately equal to the sum of the ionic radii of the two ions. Some ionic radii are: $Na^+$, 0.95 Å; $K^+$, 1.33 Å; $Ag^+$, 1.13 Å; $Mg^{2+}$, 0.65 Å; $Sr^{2+}$, 1.13 Å; $Mn^{2+}$, 0.80 Å; $F^-$, 1.36 Å; $I^-$, 2.16 Å; $O^{2-}$, 1.40 Å; $S^{2-}$, 1.84 Å. Arrange the following (each of which crystallizes in a sodium chloride type lattice) in decreasing order of lattice energy (most negative value first): AgF, KF, MgO, NaI, MgS, and MnO.

**7.38** The distance between a $K^+$ ion and a $F^-$ ion, $r$, in a KF crystal is 2.67 Å, and KF crystallizes in a sodium chloride lattice (Madelung constant, 1.75). (a) Calculate the electrostatic potential energy per mole of crystalline KF. The value of $e$ is $4.80 \times 10^{-10}$ esu, or $4.80 \times 10^{-10}$ $g^{1/2}cm^{3/2}/sec$; 1 J = $10^7$ g $cm^2/sec^2$; 1 cal = 4.18 J. (b) The lattice energy of KF is $-194$ kcal/mol [or $-812$ kJ/mol]. Compare your result to this value.

**7.39** The distance between a $Cs^+$ ion and a $I^-$ ion, $r$, in a CsI crystal is 3.96 Å, and CsI crystallizes in a cesium chloride lattice (Madelung constant, 1.76). (a) Calculate the electrostatic potential energy per mole of crystalline CsI. The value of $e$ is $4.80 \times 10^{-10}$ esu, or $4.80 \times 10^{-10}$ $g^{1/2}cm^{3/2}/sec$; 1 J = $10^7$ g $cm^2/sec^2$; 1 cal = 4.18 J. (b) The lattice energy of CsI is $-142$ kcal/mol [or $-594$ kJ/mol]. Compare your result to this value.

**7.40** List and describe the types of lattice defects found in crystals.

# 8

# Oxygen and Hydrogen

Water was one of Aristotle's four elements (earth, water, air, and fire). To Johann van Helmont (1579–1644), water had a special significance. In his famous tree experiment, van Helmont planted a willow sapling that he had weighed in a vessel that contained a weighed quantity of earth. For five years he watered the tree with distilled water and took precautions so that dust would not add to the weight of the earth in the vessel. At the end of that time the weight of the earth had not changed, but the weight of the tree had increased greatly. He concluded that water was an element but that earth was not, since earth (the new growth of the willow tree) had evidently been produced from water. It is ironic that the man who first recognized that carbon dioxide is a distinct gas did not know the role played by that gas in the growth of plants.

In the late eighteenth century both Joseph Priestly and Henry Cavendish showed that water is produced by the reaction of oxygen and hydrogen (for which they used other names); their explanations of the phenomenon, however, were incorrect. The modern interpretation of the reaction was published in 1784 by Lavoisier, who recognized that water is a compound composed of two elements; Lavoisier named the elements *oxygen* and *hydrogen*.

*on Earth*

**8.1 OCCURRENCE AND PREPARATION OF OXYGEN**

Oxygen is the most abundant element (Table 8.1). Free oxygen makes up about 23.2% by weight of the atmosphere and about 21.0% by volume. Water is approximately 89% oxygen. Most minerals contain oxygen in the combined state. Silica, $SiO_2$, is a common ingredient of many minerals and the chief constituent of sand. Silicon is second to oxygen in the order of natural abundance because of the widespread occurrence of silica. Other oxygen-containing minerals are oxides, sulfates, and carbonates. Oxygen is a constituent of the compounds that make up plant and animal matter; the human body is more than 60% oxygen.

Table 8.1 Abundance of the elements (earth's crust, bodies of water, and atmosphere)

| RANK | ELEMENT | PERCENT BY WEIGHT |
|------|---------|-------------------|
| 1 | O | 49.2 |
| 2 | Si | 25.7 |
| 3 | Al | 7.5 |
| 4 | Fe | 4.7 |
| 5 | Ca | 3.4 |
| 6 | Na | 2.6 |
| 7 | K | 2.4 |
| 8 | Mg | 1.9 |
| 9 | H | 0.9 |
| 10 | Ti | 0.6 |
| 11 | Cl | 0.2 |
| 12 | P | 0.1 |
| 13 | Mn | 0.1 |
| 14 | C | 0.09 |
| 15 | S | 0.05 |
| 16 | Ba | 0.05 |
|  | all others | 0.51 |

Oxygen may be prepared by the thermal decomposition of certain oxygen-containing compounds. The oxides of metals of low reactivity (e.g., Hg, Ag, and Au) decompose on heating to give oxygen and the free metals.

$$2HgO(s) \rightarrow 2Hg(l) + O_2(g)$$

Some higher oxides, upon heating, release only a part of their oxygen and go to lower oxides.

$$2PbO_2(s) \rightarrow 2PbO(s) + O_2(g)$$

$$3MnO_2(s) \rightarrow Mn_3O_4(s) + O_2(g)$$

Oxygen and the oxide ion are produced when the peroxide ion, $O_2^{2-}$, is heated.

$$2\left[:\ddot{O}:\ddot{O}:\right]^{2-} \rightarrow 2\left[:\ddot{O}:\right]^{2-} + O_2$$

Thus,

$$2Na_2O_2(s) \rightarrow 2Na_2O(s) + O_2(g)$$

$$2BaO_2(s) \rightarrow 2BaO(s) + O_2(g)$$

(Note that $PbO_2$ and $MnO_2$ are not peroxides.) If sodium peroxide is added to water, oxygen and a solution of sodium hydroxide result.

$$2O_2^{2-}(aq) + 2H_2O \rightarrow 4OH^-(aq) + O_2(g)$$

Certain other compounds release all, or part, of their oxygen upon heating. Nitrates of the I A metals form nitrites.

$$2NaNO_3(l) \rightarrow 2NaNO_2(l) + O_2(g)$$

Potassium chlorate loses all its oxygen; a catalyst ($MnO_2$) is generally used to lower the temperature required for this decomposition.

$$2KClO_3(s) \rightarrow 2KCl(s) + 3O_2(g)$$

Small amounts of pure oxygen are produced commercially by the electrolysis of water.

$$2H_2O \xrightarrow{\text{electrolysis}} 2H_2(g) + O_2(g)$$

Most of the oxygen produced industrially, however, is obtained from the liquefaction of air. The principal components of air are nitrogen (78% by volume) and oxygen (21% by volume). In the fractionation of liquid air, nitrogen (boiling point, $-196°C$) is boiled away from the oxygen (boiling point, $-183°C$).

**8.2 PROPERTIES OF OXYGEN**

Oxygen is a colorless, odorless, and tasteless gas under ordinary conditions. The normal boiling point of oxygen is $-182.9°C$, and the normal melting point is $-218.4°C$; solid and liquid oxygen are pale blue in color. The critical temperature of oxygen is $-118°C$, and the critical pressure is 49.7 atm.

At STP the density of oxygen gas is 1.429 g/liter. Approximately 31 ml of the gas will dissolve in 1 liter of water under ordinary conditions; evidently this small amount is enough to sustain fish and aquatic plant life under water.

There are three naturally occurring isotopes of oxygen: $^{16}_8O$ (99.759%),

$^{18}_{8}O$ (0.204%), and $^{17}_{8}O$ (0.037%). The isotopes $^{14}O$, $^{15}O$, $^{19}O$, and $^{20}O$ are artificial and unstable.

The reactions of oxygen are often more sluggish than would be predicted from the fact that oxygen is a very electronegative element (3.5), second in this property only to fluorine (4.0). The reason for this slowness is that the bond energy of oxygen is high (118 kcal/mol); therefore, reactions that require the oxygen-to-oxygen bond to be broken occur only at high temperatures. Many of these reactions are relatively highly exothermic and produce sufficient heat to sustain themselves after having once been initiated by external heating.

Whether self-sustaining or not, most oxygen reactions occur at temperatures considerably higher than room temperature. The most facile reactions of oxygen are those in which the oxygen-to-oxygen bond of the $O_2$ molecule is not completely broken. Examples of this type of reaction are the reactions of oxygen in which peroxides are prepared. Oxygen forms three different anions: the superoxide, peroxide, and oxide ions. Molecular orbital diagrams for oxygen, the superoxide ion, and the peroxide ion are given in Figure 8.1. The **superoxide ion,** $O_2^-$, can be considered to arise from the addition of one electron to the $\pi^*2p$ orbital

**Figure 8.1** Molecular orbital energy-level diagrams for oxygen, the superoxide ion, and the peroxide ion.

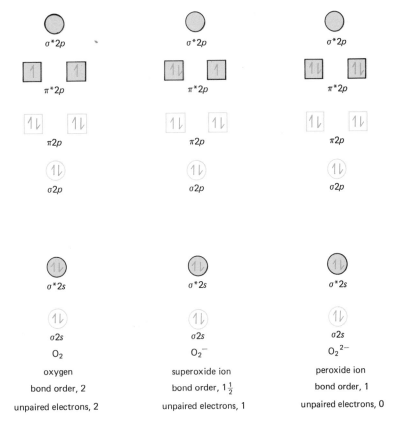

of the $O_2$ molecule, which reduces the number of unpaired electrons to 1 and the bond order to $1\frac{1}{2}$. The **peroxide ion,** $O_2^{2-}$, contains two more electrons (in the $\pi^*2p$ orbitals) than the $O_2$ molecule; hence, the bond order is reduced to 1, and the ion is diamagnetic. The **oxide ion,** $O^{2-}$, is isoelectronic with neon and is diamagnetic.

All metals except a few of the less-reactive metals (e.g., Ag and Au) react with oxygen. Oxides of all metals are known but some must be made indirectly. The most reactive metals of group I A (and those with largest atomic radii) — Cs, Rb, and K — react with oxygen to produce superoxides. For example,

$$Cs(s) + O_2(g) \rightarrow CsO_2(s)$$

Sodium peroxide is produced by the reaction of sodium with oxygen.

$$2Na(s) + O_2(g) \rightarrow Na_2O_2(s)$$

Lithium metal forms a normal oxide with $O_2$ rather than a peroxide or a superoxide because the small $Li^+$ ion cannot form a stable lattice with the larger $O_2^{2-}$ or $O_2^-$ ions.

$$4Li(s) + O_2(g) \rightarrow 2Li_2O(s)$$

Generally, oxides form at much higher temperatures than either peroxides or superoxides. The normal oxides of potassium, rubidium, and cesium can be obtained by heating oxygen with an excess of the metal.

With the exception of barium (which reacts with oxygen to yield barium peroxide), the remaining metals generally produce normal oxides in their reactions with oxygen.

$$2Mg(s) + O_2(g) \rightarrow 2MgO(s)$$

$$4Al(s) + 3O_2(g) \rightarrow 2Al_2O_3(s)$$

Analogous reactions can be written for the preparation of CaO, CuO, ZnO, PbO, and other oxides. The reaction of mercury and oxygen is reversible.

$$2Hg(l) + O_2(g) \rightleftharpoons 2HgO(s)$$

For metals that have more than one electrovalence number, the oxide produced generally depends upon the quantity of oxygen, the quantity of the metal, and the reaction conditions. Thus, the reaction of iron and oxygen can be made to yield FeO (low pressure of oxygen, temperature above 600°C), $Fe_3O_4$ (finely divided iron, heated in air at 500°C), or $Fe_2O_3$ (iron heated in air at temperatures above 500°C). Hydrated $Fe_2O_3$ is iron rust.

Except for the noble gases and the group VII A elements, all nonmetals react with oxygen. Oxides of the halogens and those of the heavier members of the noble gas family (Section 11.1) have been prepared by indirect means. The reaction of oxygen with hydrogen produces water. The product of the reaction of carbon with oxygen depends upon the proportion of carbon to oxygen employed.

$$2C(s) + O_2(g) \rightarrow 2CO(g)$$

$$C(s) + O_2(g) \rightarrow CO_2(g)$$

In like manner, the product of the reaction of phosphorus and oxygen depends upon whether phosphorus is reacted in a limited oxygen supply ($P_4O_6$) or in excess oxygen ($P_4O_{10}$). Sulfur reacts to produce $SO_2$.

$$S(s) + O_2(g) \rightarrow SO_2(g)$$

The reaction of nitrogen with oxygen requires extremely high temperatures. The following reaction occurs in a high-energy electric arc.

$$N_2(g) + O_2(g) \rightarrow 2NO(g)$$

Additional important oxides of sulfur (e.g., $SO_3$) and nitrogen (e.g., $NO_2$ and $N_2O_5$) are prepared by means other than the direct combination of the elements. Lower oxides can be reacted with oxygen to produce higher oxides. For example,

$$2Cu_2O(s) + O_2(g) \rightarrow 4CuO(s)$$

$$2CO(g) + O_2(g) \rightarrow 2CO_2(g)$$

Most reactions of compounds with oxygen yield the same products that would be obtained if the individual elements that compose the compounds were reacted directly. Thus,

$$2H_2S(g) + 3O_2(g) \rightarrow 2H_2O(g) + 2SO_2(g)$$

$$CS_2(l) + 3O_2(g) \rightarrow CO_2(g) + 2SO_2(g)$$

$$2C_2H_2(g) + 5O_2(g) \rightarrow 4CO_2(g) + 2H_2O(g)$$

$$C_2H_6O(l) + 3O_2(g) \rightarrow 2CO_2(g) + 3H_2O(g)$$

The reaction of zinc sulfide with oxygen illustrates a metallurgical process known as roasting; many sulfide ores are subjected to this procedure (Section 18.3).

$$2ZnS(s) + 3O_2(g) \rightarrow 2ZnO(s) + 2SO_2(g)$$

The products of the reaction of a hydrocarbon with oxygen depend upon the amount of oxygen supplied. Thus, when natural gas (methane, $CH_4$) is burned in air, $H_2O(g)$, $C(s)$, $CO(g)$, and $CO_2(g)$ are produced by the oxidation.

In addition to water, hydrogen and oxygen form a compound called hydrogen peroxide, $H_2O_2$, which is a colorless liquid that boils at 150.2°C and freezes at $-0.41$°C. Hydrogen peroxide can be made by treating peroxides with acids.

$$BaO_2(s) + 2H^+(aq) + SO_4^{2-}(aq) \rightarrow H_2O_2(aq) + BaSO_4(s)$$

In this preparation the barium sulfate, which is insoluble, can be removed by filtration.

The peroxide linkage (—O—O—) exists in covalent compounds and ions in addition to $H_2O_2$ and the $O_2^{2-}$ ion. For example, the peroxydisulfate ion, written $S_2O_8^{2-}$ or $[O_3SOOSO_3]^{2-}$, contains this linkage (Section 11.12). The $S_2O_8^{2-}$ ion is produced by the electrolysis of sulfuric acid under suitable conditions; the reaction of this ion with water serves as a commercial preparation of hydrogen peroxide.

$$S_2O_8^{2-}(aq) + 2H_2O \rightarrow H_2O_2(aq) + 2HSO_4^{-}(aq)$$

Hydrogen peroxide is a weak, diprotic acid in water solution. One or two hydrogens can be neutralized by sodium hydroxide to produce either sodium hydroperoxide ($NaHO_2$) or sodium peroxide ($Na_2O_2$). In the laboratory, $H_2O_2$ is used as an oxidizing or reducing agent (Section 8.5).

**8.3  ACIDIC AND ALKALINE OXIDES**

Many oxides of metals and nonmetals react with water to produce alkalies and acids (section 3.11). Thus, the oxides of group I A metals and those of calcium, strontium, and barium dissolve in water to produce hydroxides. All these oxides are ionic; when one of them dissolves in water, the oxide ion reacts with the water.

$$O^{2-}(aq) + H_2O \rightarrow 2OH^-(aq)$$

Hydroxides of other metals exist, but they are prepared by other methods, since the corresponding oxides, as well as the hydroxides, are insoluble in water. When heated, most hydroxides are converted to oxides.

$$Mg(OH)_2(s) \rightarrow MgO(s) + H_2O(g)$$

Many covalent, nonmetal oxides react with water to produce acids.

$$SO_3 + H_2O \rightarrow H_2SO_4$$

$$P_4O_{10} + 6H_2O \rightarrow 4H_3PO_4$$

$$Cl_2O + H_2O \rightarrow 2HOCl$$

$$N_2O_5 + H_2O \rightarrow 2HNO_3$$

$$SO_2 + H_2O \rightarrow H_2SO_3$$

Most of the polyatomic anions listed in Table 3.9 are derived from acids that can be prepared in this manner. Binary acids, of course, cannot be derived from oxides.

The oxides of metals, therefore, are frequently called **alkaline oxides,** and the oxides of the nonmetals are called **acidic oxides.** The oxides themselves can be made to undergo neutralization reactions. Thus, barium sulfate can be prepared by the action of sulfuric acid on either barium hydroxide or barium oxide.

$$2H^+(aq) + SO_4^{2-}(aq) + Ba^{2+}(aq) + 2OH^-(aq) \rightarrow BaSO_4(s) + 2H_2O$$

$$2H^+(aq) + SO_4^{2-}(aq) + BaO(s) \rightarrow BaSO_4(s) + H_2O$$

The insoluble $Fe_2O_3$ will react with acids, even though it will not react with water to produce a hydroxide.

$$6H^+(aq) + Fe_2O_3(s) \rightarrow 2Fe^{3+}(aq) + 3H_2O$$

Oxides of nonmetals will neutralize alkalies. The same products are obtained from the reaction of an acidic oxide as are obtained from the reaction of the corresponding acid.

$$H_2SO_3(aq) + 2OH^-(aq) \rightarrow SO_3^{2-}(aq) + 2H_2O$$

$$SO_2(g) + 2OH^-(aq) \rightarrow SO_3^{2-}(aq) + H_2O$$

Some oxides react with both acids and alkalies. Oxides of this type are called **amphoteric** and are generally derived from elements in the center of the periodic table, on the borderline between the metals and the nonmetals.

$$ZnO(s) + 2OH^-(aq) + H_2O \rightarrow Zn(OH)_4^{2-}(aq)$$

zincate ion

$$ZnO(s) + 2H^+(aq) \rightarrow Zn^{2+}(aq) + H_2O$$

$$Al_2O_3(s) + 2OH^-(aq) + 3H_2O \rightarrow 2Al(OH)_4^-(aq)$$

aluminate ion

$$Al_2O_3(s) + 6H^+(aq) \rightarrow 2Al^{3+}(aq) + 3H_2O$$

**8.4 OZONE**  The existence of an element in more than one form in the same physical state is called **allotropy,** and the forms are called **allotropes.** A number of elements exhibit allotropy, for example, carbon, sulfur, and phosphorus. Oxygen exists in a triatomic form, ozone, in addition to the common diatomic modification.

The ozone molecule is diamagnetic and has an angular structure. Both oxygen-to-oxygen bonds have the same length (1.27 Å), which is intermediate between the double-bond distance (1.10 Å) and the single-bond distance (1.48 A). The molecule may be represented as a resonance hybrid:

Ozone is a pale blue gas with a characteristic odor; predictably, its density is $1\frac{1}{2}$ times that of $O_2$. The normal boiling point of ozone is $-112°C$, and the normal melting point is $-250°C$. It is slightly more soluble in water than is $O_2$.

Ozone is produced by passing a silent electric discharge through oxygen gas. The reaction proceeds through the dissociation of an $O_2$ molecule into oxygen atoms and the combination of an O atom with a second $O_2$ molecule.

$$\tfrac{1}{2}O_2 \rightarrow O \qquad \Delta H = +59 \text{ kcal}$$

$$O + O_2 \rightarrow O_3 \qquad \Delta H = -25 \text{ kcal}$$

The energy released in the second step, in which a new bond is formed, is not sufficient to compensate for the energy required by the first step, in which a bond is broken. Hence, the overall reaction for the preparation of ozone is endothermic.

$$\tfrac{3}{2}O_2 \rightarrow O_3 \qquad \Delta H = +34 \text{ kcal}$$

Ozone is highly reactive; it is explosive at temperatures above $300°C$ or in the presence of substances that catalyze its decomposition. Ozone will react with many substances at temperatures that are not high enough to produce reaction with $O_2$. The higher reactivity of $O_3$ in comparison to $O_2$ is consistent with the higher energy content of $O_3$.

**8.5 OXIDATION–REDUCTION REACTIONS**

The term oxidation was originally applied to reactions in which substances combined with oxygen, and reduction was defined as the removal of oxygen from an oxygen-containing compound. The meanings of the terms have gradually been broadened. Today oxidation and reduction are defined on the basis of change in oxidation number (Section 3.10).

**Oxidation** is the process in which an atom undergoes an algebraic increase in oxidation number and **reduction** is the process in which an atom undergoes an algebraic decrease in oxidation number. On this basis, oxidation–reduction is involved in the reaction

$$\overset{0}{S} + \overset{0}{O_2} \rightarrow \overset{4+}{S}\overset{2-}{O_2}$$

whereas oxidation–reduction is not involved in the reaction

$$\overset{4+}{S}\overset{2-}{O_2} + \overset{1+}{H_2}\overset{2-}{O} \rightarrow \overset{1+}{H_2}\overset{4+}{S}\overset{2-}{O_3}$$

The oxidation number of each type of atom is written above its symbol. In the first reaction, sulfur is oxidized and oxygen is reduced.

It is apparent, from the manner in which oxidation numbers are assigned, that neither oxidation nor reduction can occur alone; furthermore, each process must occur to the same extent in a given reaction. According to the rules by which oxidation numbers are assigned, the sum of the oxidation numbers of any chemical species equals the charge on that species (zero if the entire compound is considered). In any reaction the sum of the charges of the reactant species (zero for reactions involving only molecules) must equal the sum of the charges of the product species (again zero for reactions involving only molecules). Charge conservation demands, therefore, that any oxidation (increase in oxidation number) be accompanied by a commensurate reduction (decrease in oxidation number).

Since one substance cannot be reduced unless another is simultaneously oxidized, the material that is reduced is responsible for the oxidation; therefore, it is called the **oxidizing agent** or **oxidant.** Because of the interdependence of the two processes, the converse is also true: the material that is itself oxidized is called the **reducing agent** or **reductant.** Therefore,

$$\overset{0}{S} + \overset{0}{O_2} \rightarrow \overset{4+\ 2-}{S\ O_2}$$

$$\underset{\substack{\text{oxidized} \\ \text{reducing agent}}}{\phantom{S}} \qquad \underset{\substack{\text{reduced} \\ \text{oxidizing agent}}}{\phantom{O_2}}$$

Equations for oxidation–reduction reactions are usually more difficult to balance than those for reactions that do not entail oxidation and reduction (such as neutralization reactions). It is advantageous to balance oxidation–reduction equations systematically; there are two methods that are commonly used to balance them: the oxidation-number method and the ion–electron method. Both of these methods will be discussed. For clarity, the physical state of the atoms, molecules, and ions will not be indicated in the examples that follow. In addition, the symbol $H^+$, instead of $H^+(aq)$ or $H_3O^+$, will be employed; it should be kept in mind, however, that all ions are hydrated in aqueous solution.

There are three steps in the **oxidation-number method** of balancing oxidation-reduction equations. An example of oxidation–reduction is the reaction of nitric acid with hydrogen sulfide; the unbalanced expression for this change is

$$HNO_3 + H_2S \rightarrow NO + S + H_2O$$

1. The oxidation numbers of the atoms in the equation are determined in order to identify those undergoing oxidation or reduction. Thus,

$$H\ \overset{5+}{N}\ O_3 + H_2\overset{2-}{S} \rightarrow \overset{2+}{N}\ O + \overset{0}{S} + H_2O$$

2. Coefficients are placed before the appropriate formulas so that the total decrease in oxidation number equals the total increase in oxidation number. In this example, nitrogen is reduced (5+ to 2+, a decrease

of 3) and sulfur is oxidized (2— to 0, an increase of 2). Thus, it is necessary that 2 molecules of $HNO_3$ and 2 molecules of NO, as well as 3 molecules of $H_2S$ and 3 atoms of S, be indicated. In this way the total increase in oxidation number will be 6 and will equal the total decrease in oxidation number of 6.

$$2HNO_3 + 3H_2S \rightarrow 2NO + 3S + H_2O$$

3. Balancing is completed by inspection; this method takes care of only those materials that undergo oxidation-number change. In this example there are now 8 hydrogen atoms on the left; therefore, $4H_2O$ molecules must be indicated on the right.

$$2HNO_3 + 3H_2S \rightarrow 2NO + 3S + 4H_2O$$

The final, balanced equation should be checked to ensure that there are as many atoms of each element on the left as there are on the right.

Nitric acid is frequently employed as an oxidizing agent, and it functions in this capacity in this reaction (the nitrogen of $HNO_3$ is reduced). The sulfur of $H_2S$ is oxidized; therefore, $H_2S$ acts as a reducing agent.

The oxidation-number method may be used to balance **net ionic equations,** in which only those ions and molecules that actually take part in the reaction are shown. For example, the $K^+$ ion does not take part in the reaction of potassium chlorate, $KClO_3$, with iodine and is not shown in the equation. The steps in balancing are:

$$1.\ H_2O + \overset{0}{I_2} + \overset{5+}{ClO_3^-} \rightarrow \overset{5+}{IO_3^-} + \overset{1-}{Cl^-} + H^+$$

2. *Each* iodine atom undergoes an increase of 5 (from 0 to 5+), but there are *two* iodine atoms in $I_2$. The increase in oxidation number is therefore 10. Chlorine undergoes a decrease of 6 (from 5+ to 1−). The lowest common multiple of 6 and 10 is 30. Therefore, $3I_2$ molecules must be indicated (a total increase of 30) and $5ClO_3^-$ ions are needed (a total decrease of 30). The coefficients of the products, $IO_3^-$ and $Cl^-$, follow from this assignment.

$$H_2O + 3I_2 + 5ClO_3^- \rightarrow 6IO_3^- + 5Cl^- + H^+$$

3. If $H_2O$ is ignored, there are now 15 oxygen atoms on the left and 18 oxygen atoms on the right. To make up 3 oxygen atoms on the left, we must indicate $3H_2O$ molecules. It then follows that the coefficient of $H^+$ must be 6 to balance the hydrogens of the $H_2O$ molecules.

$$3H_2O + 3I_2 + 5ClO_3^- \rightarrow 6IO_3^- + 5Cl^- + 6H^+$$

An ionic equation must indicate charge balance as well as mass balance. Since the algebraic sum of the charges on the left (5−) equals that on the right (5−), the equation is balanced.

Reactions in which electrons are transferred are clearly examples of oxidation–reduction reactions. In the reaction of sodium and chlorine, a sodium atom loses its valence electron to a chlorine atom.

$$\overset{0}{2Na} + \overset{0}{Cl_2} \rightarrow \overset{1+}{2Na^+} + \overset{1-}{2Cl^-}$$

For simple ions the oxidation number is the same as the charge on the ion. It follows, then, that electron loss represents a type of oxidation, and electron gain represents a type of reduction. This equation can be divided into two **partial equations** that represent **half reactions.**

Oxidation: $\quad 2Na \rightarrow 2Na^+ + 2e^-$

Reduction: $\quad 2e^- + Cl_2 \rightarrow 2Cl^-$

The **ion–electron method** of balancing oxidation–reduction equations employs partial equations. Only those ions that are involved in the reaction are shown; un-ionized (or slightly ionized) species and insoluble substances that take part in the reaction are written in molecular form. The equation for the reaction between dichromate ion and chloride ion in acid solution will be used to illustrate the steps of the method.

1. Two skeleton partial equations for the half reactions are written, and the central element of each partial equation is balanced.

$$Cr_2O_7^{2-} \rightarrow 2Cr^{3+}$$

$$2Cl^- \rightarrow Cl_2$$

2. The hydrogen and oxygen atoms are then balanced. Since this reaction occurs in acid solution, $H^+$ and $H_2O$ can be added where needed. For each oxygen atom that is needed, one $H_2O$ molecule is added to the side of the partial equation that is deficient. The hydrogen is then brought into balance by the addition of $2H^+$ to the opposite side. Thus, 7 oxygens must be added to the right side of the first partial equation; the second partial equation is already in material balance.

$$14H^+ + Cr_2O_7^{2-} \rightarrow 2Cr^{3+} + 7H_2O$$

$$2Cl^- \rightarrow Cl_2$$

3. The next step is to balance the partial equations electrically. In the first partial equation, the net charge is 12+ on the left side of the equation (14+ and 2−) and 6+ on the right side. Six electrons must be added to the left. The second equation is balanced electrically by the addition of 2 electrons to the right.

$$6e^- + 14H^+ + Cr_2O_7^{2-} \rightarrow 2Cr^{3+} + 7H_2O$$

$$2Cl^- \rightarrow Cl_2 + 2e^-$$

4. The number of electrons lost must equal the number of electrons gained. Therefore, the oxidation equation is multiplied through by 3.

$$6e^- + 14H^+ + Cr_2O_7^{2-} \rightarrow 2Cr^{3+} + 7H_2O$$

$$6Cl^- \rightarrow 3Cl_2 + 6e^-$$

5. Addition of the two partial equations gives the final equation.

$$14H^+ + Cr_2O_7^{2-} + 6Cl^- \rightarrow 2Cr^{3+} + 3Cl_2 + 7H_2O$$

These steps are illustrated for the reaction in which $As_4O_6$ reacts with $MnO_4^-$ to produce $H_3AsO_4$ and $Mn^{2+}$.

1. $MnO_4^- \rightarrow Mn^{2+}$

$$As_4O_6 \rightarrow 4H_3AsO_4$$

2. The first partial equation can be brought into material balance by the addition of $4H_2O$ to the right side and $8H^+$ to the left side. In the second partial equation, $10H_2O$ must be added to the left side to make up the needed 10 oxygens. If we stopped at this point, there would be 20 hydrogen atoms on the left and 12 on the right; therefore, $8H^+$ must be added to the right.

$$8H^+ + MnO_4^- \rightarrow Mn^{2+} + 4H_2O$$

$$10H_2O + As_4O_6 \rightarrow 4H_3AsO_4 + 8H^+$$

3. To balance the net charges, electrons are added.

$$5e^- + 8H^+ + MnO_4^- \rightarrow Mn^{2+} + 4H_2O$$

$$10H_2O + As_4O_6 \rightarrow 4H_3AsO_4 + 8H^+ + 8e^-$$

4. The first partial equation must be multiplied through by 8 and the second by 5 so that the same number of electrons are lost in the oxidation partial equation as are gained in the reduction partial equation.

$$40e^- + 64H^+ + 8MnO_4^- \rightarrow 8Mn^{2+} + 32H_2O$$

$$50H_2O + 5As_4O_6 \rightarrow 20H_3AsO_4 + 40H^+ + 40e^-$$

5. When these two partials are added, water molecules and hydrogen ions must be canceled as well as electrons. It is poor form to leave an equation with $64H^+$ on the left and $40H^+$ on the right.

$$24H^+ + 18H_2O + 5As_4O_6 + 8MnO_4^- \rightarrow 20H_3AsO_4 + 8Mn^{2+}$$

Equations for reactions that take place in alkaline solution are balanced in a manner slightly different from those that occur in acidic solution. All the steps are the same except the second one; $H^+$ cannot be used to balance equations for alkaline reactions. As an illustration, consider the reaction between $MnO_4^-$ and $N_2H_4$ that takes place in alkaline solution.

1. $MnO_4^- \rightarrow MnO_2$

   $N_2H_4 \rightarrow N_2$

2. For reactions occurring in alkaline solution, $OH^-$ and $H_2O$ are used to balance oxygen and hydrogen. For each oxygen that is needed, $2OH^-$ ions are added to the side of the partial equation that is deficient, and one $H_2O$ molecule is added to the opposite side. For each hydrogen that is needed, one $H_2O$ molecule is added to the side that is deficient, and one $OH^-$ ion is added to the opposite side.

   In the first partial equation, the right side is deficient by 2 oxygen atoms. We add, therefore, $4OH^-$ to the right side and $2H_2O$ to the left.

   $$2H_2O + MnO_4^- \rightarrow MnO_2 + 4OH^-$$

   In order to bring the second partial equation into material balance, we must add four hydrogen atoms to the right side. For *each* hydrogen atom needed, we add one $H_2O$ to the side deficient in hydrogen and one $OH^-$ to the opposite side. In the present case we add $4H_2O$ to the right side and $4OH^-$ to the left to make up the four hydrogen atoms needed on the right.

   $$4OH^- + N_2H_4 \rightarrow N_2 + 4H_2O$$

3. Electrons are added to effect charge balances.

   $$3e^- + 2H_2O + MnO_4^- \rightarrow MnO_2 + 4OH^-$$

   $$4OH^- + N_2H_4 \rightarrow N_2 + 4H_2O + 4e^-$$

4. The lowest common multiple of 3 and 4 is 12. Therefore, the first partial equation is multiplied through by 4 and the second by 3 so that the number of electrons gained equals the number lost.

   $$12e^- + 8H_2O + 4MnO_4^- \rightarrow 4MnO_2 + 16OH^-$$

   $$12OH^- + 3N_2H_4 \rightarrow 3N_2 + 12H_2O + 12e^-$$

5. Addition of these partial equations, with cancellation of $OH^-$ ions and $H_2O$ molecules as well as electrons, gives the final equation.

   $$4MnO_4^- + 3N_2H_4 \rightarrow 4MnO_2 + 3N_2 + 4H_2O + 4OH^-$$

As a final example, consider the following skeleton equation for a reaction in alkaline solution.

$$Br_2 \rightarrow BrO_3^- + Br^-$$

In this reaction the same substance, $Br_2$, is both oxidized and reduced. Such reactions are called **disproportionations** or **auto-oxidation–reduction reactions.**

1. $$Br_2 \rightarrow 2BrO_3^-$$
$$Br_2 \rightarrow 2Br^-$$

2. $$12OH^- + Br_2 \rightarrow 2BrO_3^- + 6H_2O$$
$$Br_2 \rightarrow 2Br^-$$

3. $$12OH^- + Br_2 \rightarrow 2BrO_3^- + 6H_2O + 10e^-$$
$$2e^- + Br_2 \rightarrow 2Br^-$$

4. $$12OH^- + Br_2 \rightarrow 2BrO_3^- + 6H_2O + 10e^-$$
$$10e^- + 5Br_2 \rightarrow 10Br^-$$

5. $$12OH^- + 6Br_2 \rightarrow 2BrO_3^- + 10Br^- + 6H_2O$$

When the ion–electron method is applied to a disproportionation reaction, the coefficients of the resulting equation usually are divisible by some common number since one reactant was used in both partial equations. The coefficients of this equation are all divisible by 2 and should be reduced to the lowest possible terms.

$$6OH^- + 3Br_2 \rightarrow BrO_3^- + 5Br^- + 3H_2O$$

Most oxidation–reduction equations may be balanced by the ion-electron method, which is especially convenient for electrochemical reactions and reactions of ions in water solution. However, several misconceptions that can arise must be pointed out. Half reactions cannot occur alone, and partial equations do not represent complete chemical changes. Even in electrochemical cells, where the two half reactions take place at different electrodes, the two half reactions always occur simultaneously.

Whereas the partial equations probably represent an overall, if not detailed, view of the way an oxidation–reduction reaction occurs in an electrochemical cell, the same reaction in a beaker may not take place in this way at all. The method should *not* be interpreted as necessarily giving the correct mechanism by which a reaction occurs. It is, at times, difficult to recognize whether a given reaction is a legitimate example of an electron-exchange reaction. The reaction

$$\overset{4+}{S}O_3^{2-} + \overset{5+}{Cl}O_3^- \rightarrow \overset{6+}{S}O_4^{2-} + \overset{3+}{Cl}O_2^-$$

looks like an electron-exchange reaction, can be made to take place in an electrochemical cell, and can be balanced by the ion–electron method. However, this reaction has been shown to proceed by direct oxygen exchange (from $ClO_3^-$ to $SO_3^{2-}$) and *not* by electron exchange.

**8.6 EQUIVALENT**
**WEIGHTS**

It is often convenient to use equivalent weights for stoichiometric calculations. The definition of equivalent weight depends upon the type of reaction being considered; however, the definition is always framed so that one equivalent weight of a given reactant will react with exactly one equivalent weight of another. Therefore, unlike mole ratios, there is always a simple 1:1 stoichiometric ratio between numbers of equivalents entering into a reaction.

Two important types of reactions for which equivalent weights are defined are neutralization reactions and oxidation–reduction reactions. For neutralization reactions, equivalent weights are based on the fact that one proton reacts with one hydroxide ion.

$$H^+(aq) + OH^-(aq) \rightarrow H_2O$$

Thus, the weight of an acid that supplies one mole of $H^+$ is the gram equivalent weight of that acid. A gram equivalent weight of a hydroxide is the weight that will supply one mole of $OH^-$. It is obvious, then, that one gram equivalent of an acid will completely neutralize one gram equivalent of an alkali. Table 8.2 lists the equivalent weights of three acids and three hydroxides; these values can be derived by dividing the molecular weight of the compound by the number of replaceable hydrogens or hydroxides per formula unit. In some instances polyprotic acids undergo reactions in which not all the hydrogens are neutralized; for these reactions, the equivalent weight of the acid must be calculated on the basis of the number of hydrogens that are neutralized.

Table 8.2 Equivalent weights of some acids and bases (complete neutralization)

| COMPOUND | MOLECULAR WEIGHT | EQUIVALENT WEIGHT |
|---|---|---|
| $HNO_3$ | 63.01 | 63.01 |
| $H_2SO_4$ | 98.07 | 49.04 |
| $H_3PO_4$ | 98.00 | 32.67 |
| NaOH | 40.00 | 40.00 |
| $Ca(OH)_2$ | 74.09 | 37.05 |
| $Al(OH)_3$ | 78.00 | 26.00 |

*Example 8.1*

Using equivalent weights, calculate the weight of NaOH required to produce $Na_2HAsO_4$ from 10.0 g of $H_3AsO_4$.

*Solution*

The equivalent weight of NaOH is the same as the molecular weight, 40.0. The equivalent weight of $H_3AsO_4$ is the molecular weight, 141.9, divided by 2, or 71.0, since only two hydrogens are replaced in this neutralization reaction.

$$? \text{ g NaOH} = 10.0 \text{ g H}_3\text{AsO}_4 \left( \frac{40.0 \text{ g NaOH}}{71.0 \text{ g H}_3\text{AsO}_4} \right)$$

$$= 5.63 \text{ g NaOH}$$

In an oxidation–reduction reaction, the total increase in oxidation number of one species must equal the total decrease in the oxidation number of another species. Therefore, a system of equivalent weights for oxidizing and reducing agents is based on oxidation number changes. The equivalent weight of a compound that functions as an oxidizing or reducing agent is defined as the molecular weight of the compound divided by the total decrease or increase in oxidation number of all the atoms in one formula unit of the compound under consideration. Obviously, one gram equivalent weight of an oxidizing agent requires one gram equivalent weight of a reducing agent for complete reaction.

Thus, in the reaction

$$\overset{0}{I_2} + 10 \, H \overset{5+}{N} O_3 \rightarrow 2 \, H \overset{5+}{I} O_3 + 10 \, \overset{4+}{N} O_2 + 4H_2O$$

each iodine atom undergoes an increase in oxidation number of 5, making a total increase of 10 per molecule of $I_2$. The molecular weight of $I_2$ is 253.81, and in this reaction, therefore, the equivalent weight of iodine is 253.81/10, or 25.381. In this reaction each nitrogen atom of $HNO_3$ undergoes a decrease in oxidization number of 1; therefore, the equivalent weight of $HNO_3$ is the same as its molecular weight, 63.01.

The stoichiometric weight ratio of reductant ($I_2$) to oxidant ($HNO_3$) derived from equivalent weights is the same as that derived from the balanced chemical equation. Thus, the balanced chemical equation for the reaction of $I_2$ and $HNO_3$ shows 1 mol of $I_2$, which is 10 equivalents, reacting with 10 mol of $HNO_3$, which is also 10 equivalents.

It is also possible to use half equations to derive equivalent weights; the increase or decrease in oxidation number is the same as the number of electrons indicated as being lost or gained in the partial equations.

The equivalent weight of an oxidizing agent or a reducing agent is not invariant; for different reactions a given compound may have different equivalent weights. When potassium permanganate serves as an oxidizing agent in acid solution, the partial equation for the reduction of permanganate ion is

$$5e^- + 8H^+ + \overset{7+}{Mn}O_4^- \rightarrow Mn^{2+} + H_2O$$

The equivalent weight of $KMnO_4$ for this reaction is the molecular weight (158.04) divided by the change in the oxidation number of Mn (5), which is 31.61.

The partial equation for the use of $KMnO_4$ as an oxidizing agent in alkaline solution is

$$3e^- + 2H_2O + \overset{7+}{Mn}O_4^- \rightarrow \overset{4+}{Mn}O_2 + 4OH^-$$

The equivalent weight of $KMnO_4$ for this reaction is 158.04/3, or 52.68.

Some analytical determinations employ solutions that contain a known concentration of a given oxidant or reductant. The concentrations of these solutions and the mathematical treatment of these analyses are usually based on equivalent weights. The preparation and use of these solutions are described in later chapters.

---

*Example 8.2*
What weight of $KMnO_4$ is needed to oxidize 15.0 g of $As_4O_6$ in acidic solution? The products of the reaction are $Mn^{2+}$ and $H_3AsO_4$.

is transformed into $H_3AsO_4$, *each* As atom undergoes an increase in oxidation number of 2 (from 3+ to 5+). The total change for 4As atoms is 8, and the equivalent weight of $As_4O_6$ is 395.6/8, or 49.5.

*Solution*
The equivalent weight of $KMnO_4$ (to $Mn^{2+}$) is 31.6 (see preceding discussion). When $As_4O_6$

$$? \text{ g KMnO}_4 = 15.0 \text{ g As}_4\text{O}_6 \left( \frac{31.6 \text{ g KMnO}_4}{49.5 \text{ g As}_4\text{O}_6} \right)$$
$$= 9.58 \text{ g KMnO}_4$$

---

**8.7 OCCURRENCE AND PREPARATION OF HYDROGEN**

In the crust, bodies of water, and atmosphere of the earth, hydrogen constitutes less than 1% of the weight of the whole and ranks ninth of all elements in order of abundance (Table 8.1). The most important naturally occurring compound of hydrogen is water, which is the principal source of the element. Hydrogen forms a series of compounds with carbon known as the hydrocarbons; these occur in coal, natural gas, and petroleum. A few minerals, such as clay and certain hydrates, contain hydrogen. Compounds of hydrogen (with carbon, oxygen, and occasionally other elements) constitute the principal part of all plant and animal matter. Free hydrogen occurs in nature only in negligible amounts (e.g., as a component of volcanic gases).

The very reactive metals (Ca, Sr, Ba, and the group I A metals) react with water at room temperature to produce hydrogen and solutions of hydroxides.

$$2Na(s) + 2H_2O \rightarrow 2Na^+(aq) + 2OH^-(aq) + H_2(g)$$

$$Ca(s) + 2H_2O \rightarrow Ca^{2+}(aq) + 2OH^-(aq) + H_2(g)$$

At high temperatures some additional metals, which have slightly lower reactivity, will displace hydrogen from water (steam); in these cases, metal oxides (the anhydrides of hydroxides) are obtained instead of metal hydroxides.

$$Mg(s) + H_2O(g) \rightarrow MgO(s) + H_2(g)$$

The reaction of iron and steam serves as a commercial source of hydrogen.

$$3Fe(s) + 4H_2O(g) \rightarrow Fe_3O_4(s) + 4H_2(g)$$

A still longer list of metals (including Mg, Al, Mn, Zn, Cr, Fe, Cd, Co, Ni, Sn, and Pb) will react with aqueous solutions of acids to produce hydrogen. This method is a convenient laboratory preparation, but it is too expensive to be of commercial importance. The very reactive metals react too violently to be useful.

$$Zn(s) + 2H^+(aq) \rightarrow Zn^{2+}(aq) + H_2(g)$$

$$Fe(s) + 2H^+(aq) \rightarrow Fe^{2+}(aq) + H_2(g)$$

$$2Al(s) + 6H^+(aq) \rightarrow 2Al^{3+}(aq) + 3H_2(g)$$

The anions of certain acids (e.g., nitric acid) also react with the metals; these acids cannot be used to prepare pure hydrogen.

The reactions of metals and acids are examples of a type of oxidation–reduction reaction known as a **displacement reaction;** the metals are said to displace hydrogen from its compounds, the acids. Displacement reactions in which metals participate are often electron-exchange reactions. For example, in the reaction

$$Mg(s) + Cu^{2+}(aq) \rightarrow Mg^{2+}(aq) + Cu(s)$$

the more reactive metal, magnesium, loses its valence electrons to the ion of the less reactive metal, copper. Certain electrochemical determinations provide a measure of the tendency of a given metal toward electron loss and the tendency of a given metal ion toward electron gain; these **electrode potentials** are a topic of Chapter 10.

Coke and steam react at high temperatures (1000°C) to produce a gaseous mixture known as **water gas.**

$$C(s) + H_2O(g) \rightarrow CO(g) + H_2(g)$$

This mixture is used as a fuel. Since it is difficult to separate hydrogen from carbon monoxide, the water gas is subjected to a second step if hydrogen alone is desired. Water gas, mixed with additional steam, is passed over a catalyst at a temperature of about 500°C; this converts the carbon monoxide into carbon dioxide.

$$CO(g) + H_2O(g) \rightarrow CO_2(g) + H_2(g)$$

The carbon dioxide and hydrogen are separated by taking advantage of the solubility of carbon dioxide in cold water under pressure.

Hydrocarbons react with steam in the presence of suitable catalysts to give mixtures of carbon monoxide and hydrogen.

$$CH_4(g) + H_2O(g) \rightarrow CO(g) + 3H_2(g)$$

The production of hydrogen and water gas from these reactions is industrially important.

Hydrogen is obtained from the catalytic decomposition of hydrocarbons at high temperatures.

$$CH_4(g) \rightarrow C(s) + 2H_2(g)$$

In gasoline refining, petroleum hydrocarbons are "cracked" into compounds of lower molecular weight; hydrogen is a by-product.

Very pure, but relatively expensive, hydrogen is commercially prepared by the electrolysis of water that contains a small amount of sulfuric acid or sodium hydroxide.

$$2H_2O \xrightarrow{\text{electrolysis}} 2H_2(g) + O_2(g)$$

Hydrogen is a by-product (as is chlorine) in the industrial preparation of sodium hydroxide by the electrolysis of concentrated aqueous solutions of sodium chloride.

$$2Na^+(aq) + 2Cl^-(aq) + 2H_2O \xrightarrow{\text{electrolysis}}$$
$$2Na^+(aq) + 2OH^-(aq) + H_2(g) + Cl_2(g)$$

Certain metals and nonmetals displace hydrogen from alkaline solutions.

$$Zn(s) + 2OH^-(aq) + 2H_2O \rightarrow H_2(g) + Zn(OH)_4^{2-}(aq)$$
zincate ion

$$2Al(s) + 2OH^-(aq) + 6H_2O \rightarrow 3H_2(g) + 2Al(OH)_4^-(aq)$$
aluminate ion

$$Si(s) + 2OH^-(aq) + H_2O \rightarrow 2H_2(g) + SiO_3^{2-}(aq)$$
silicate ion

## 8.8 PROPERTIES OF HYDROGEN

Hydrogen is a colorless, odorless, tasteless gas. It has the lowest density of any chemical substance; at STP one liter of hydrogen weighs 0.0899 g. The two atoms of hydrogen of the $H_2$ molecule are joined by a single covalent bond, giving each atom a stable helium electronic configuration. The molecule is nonpolar; the weak nature of the intermolecular forces of attraction is indicated by the low normal boiling point ($-252.7°C$), the normal melting point ($-259.1°C$) and the critical temperature ($-240°C$ at a critical pressure of 12.8 atm). Hydrogen is virtually insoluble in water; approximately 2 ml of hydrogen will dissolve in 1 liter of water at room temperature and atmospheric pressure.

There are three isotopes of hydrogen. The most abundant isotope, $_1^1H$, constitutes 99.985% of naturally occurring hydrogen; deuterium, $_1^2H$ (also indicated as $_1^2D$) constitutes 0.015%; and the radioactive tritium, $_1^3H$ (also indicated as $_1^3T$) occurs only in trace amounts.

The bond energy of the H—H bond is 103 kcal/mol. In the course of most reactions this bond must be broken so that the H atoms can form new bonds with other atoms. Since this bond energy is relatively high, most reactions of hydrogen take place at elevated temperatures.

Hydrogen reacts with the metals of group I A and the heavier metals of group II A (Ba, Sr, and Ca) to form **saltlike hydrides.** The hydride ion ($H^-$) of these compounds is isoelectronic with helium and achieves this stable configuration by the addition of an electron from the metal. Since the electron affinity of hydrogen is low ($-0.7$ eV), hydrogen reacts in this manner only with the most reactive metals.

$$2Na(s) + H_2(g) \rightarrow 2NaH(s)$$

$$Ca(s) + H_2(g) \rightarrow CaH_2(s)$$

The saltlike hydrides react with water to form hydrogen.

$$H^- + H_2O \rightarrow H_2(g) + OH^-(aq)$$

With certain other metals, such as platinum, palladium, and nickel, hydrogen forms **interstitial hydrides.** Many of these hydrides are non-stoichiometric; their apparent formulas depend upon the conditions under which they are prepared. Palladium can absorb up to 900 times its own volume of hydrogen. The interstitial hydrides resemble the metals from which they are derived; the crystal structures of the metals do not change (or change only slightly) as hydrogen is added. Whether these substances should be regarded as compounds is debatable; the name, interstitial hydrides, arises from the view that the hydrogen is only ab-sorbed into the interstices of the metallic crystal.

However, magnetic studies indicate that electron pairing of some sort is involved in the formation of these hydrides. Many of the metals have unpaired electrons and, hence, are paramagnetic. A gradual loss of paramagnetism is observed as the hydrides form from the metal. Since the electrons of molecular hydrogen are paired, these studies imply that the hydrogen is incorporated into the crystal in atomic form. This would account for the catalytic activity of platinum, palladium, nickel, and other metals for many reactions of hydrogen; if the H—H bond is broken or even only weakened, the absorbed hydrogen should be much more reactive than ordinary $H_2$ gas.

**Complex hydrides** of boron and aluminum are important and useful compounds for chemical syntheses. Important examples are sodium borohydride and lithium aluminum hydride.

$$Na^+ \begin{bmatrix} H \\ H:\overset{\cdot\cdot}{B}:H \\ H \end{bmatrix}^- \qquad Li^+ \begin{bmatrix} H \\ H:\overset{\cdot\cdot}{Al}:H \\ H \end{bmatrix}^-$$

Since these compounds, as well as the saltlike hydrides, react with water to liberate hydrogen, they are prepared by reactions in ether.

$$4LiH + AlCl_3 \rightarrow Li[AlH_4] + 3LiCl$$

Hydrogen forms covalent compounds with most of the nonmetals. It reacts with the halogens to form colorless, polar covalent gases.

$$H_2(g) + Cl_2(g) \rightarrow 2HCl(g)$$

Reactions with fluorine or chlorine will take place at room temperature or below, but the reactions with the less reactive bromine or iodine require higher temperatures (400°C to 600°C).

The reaction of hydrogen and oxygen is highly exothermic and is the basis of the high temperature (ca. 2800°C) produced by the oxyhydrogen torch. The reaction of hydrogen with sulfur is more difficult and requires high temperatures.

$$H_2(g) + S(g) \rightarrow H_2S(g)$$

Hydrogen reacts with nitrogen at high pressures (300 to 1000 atm), at high temperatures (400°C to 600°C), and in the presence of a catalyst to produce ammonia (Haber process).

$$3H_2(g) + N_2(g) \rightarrow 2NH_3(g)$$

Hydrogen does not react readily with carbon, but at high temperatures, hydrogen can be made catalytically to react with finely divided carbon (Bergius process); hydrocarbons are produced in these reactions.

$$2H_2(g) + C(g) \rightarrow CH_4(g)$$

The ionization potential of hydrogen is relatively high (13.60 eV). In none of the compounds formed between hydrogen and the nonmetals does hydrogen exist as positive ions; the bonding in all these compounds is polar covalent. Because of its large charge-to-radius ratio, the unassociated proton ($H^+$) does not exist in ordinary chemical systems; in pure acids the hydrogen atom is covalently bonded to the rest of the molecule, and in aqueous solutions the proton is hydrated.

There is evidence that the proton bonds to a water molecule by means of a pair of electrons of the oxygen atom to form an ion, $H_3O^+$, called the **hydronium ion.** The three hydrogen atoms of the $H_3O^+$ ion are equivalent, and the ion is in the form of a trigonal pyramid with the oxygen atom at the apex. The hydronium ion is known to occur in a few crystalline hydrates of acids such as $H_3O^+$, $ClO_4^-$; $H_3O^+$, $Cl^-$; and $H_3O^+$, $NO_3^-$.

However, there is evidence to indicate that the $H_3O^+$ ion is associated with three additional water molecules (by means of hydrogen bonds, Section 8.9) to form an ion with the formula $H_9O_4^+$. Other types of experimental evidence tend to support the idea that several types of hydrated species exist simultaneously in solution. We shall represent the hydrated proton as $H^+(aq)$ unless it is especially convenient to employ the formula $H_3O^+$.

Hydrogen reacts with many metal oxides to produce water and the free metal.

$$CuO(s) + H_2(g) \rightarrow Cu(s) + H_2O(g)$$

$$WO_3(s) + 3H_2(g) \rightarrow W(s) + 3H_2O(g)$$

$$FeO(s) + H_2(g) \rightarrow Fe(s) + H_2O(g)$$

Some of these reactions are employed in the metallurgy of oxide ores; for example, in the commercial production of tungsten metal, $WO_3$ is reduced to the free metal by hydrogen.

Carbon monoxide and hydrogen react at high temperatures and high pressures in the presence of a catalyst to produce methyl alcohol.

$$CO(g) + 2H_2(g) \rightarrow CH_3OH(g)$$

**8.9 THE HYDROGEN BOND**

The intermolecular attractions of certain hydrogen-containing compounds are unusually strong. These attractions occur in compounds in which hydrogen is covalently bonded to highly electronegative elements of small atomic size. In these compounds the electronegative element exerts such a strong attraction on the bonding electrons that the hydrogen is left with a significant $\delta^+$ charge. In fact, the hydrogen is almost an exposed proton since this element has no screening electrons. The hydrogen of one molecule and a pair of unshared electrons on the electronegative atom of another molecule are mutually attracted and form what is called a **hydrogen bond.** Because of its small size, each hydrogen atom is capable of forming only one hydrogen bond. The association of HF, $H_2O$, and $NH_3$ by hydrogen bonds (indicated by dotted lines) can be roughly diagramed as follows:

$$H—F \cdots H—F \cdots \qquad H—\underset{\underset{H}{|}}{O} \cdots H—\underset{\underset{H}{|}}{O} \cdots \qquad H—\underset{\underset{H}{|}}{\overset{\overset{H}{|}}{N}} \cdots H—\underset{\underset{H}{|}}{\overset{\overset{H}{|}}{N}} \cdots$$

Unusual properties are characteristic of compounds in which hydrogen bonding occurs. In Figure 8.2 the normal boiling points of the hydrogen compounds of the elements of groups IV A, V A, VI A, and VII A are plotted. The series $CH_4$, $SiH_4$, $GeH_4$, and $SnH_4$ illustrates the expected trend in boiling point for compounds in which the only intermolecular forces are London forces; the boiling point increases as the

**Figure 8.2** Boiling points of the hydrogen compounds of the elements of groups IV A, V A, VI A, and VII A.

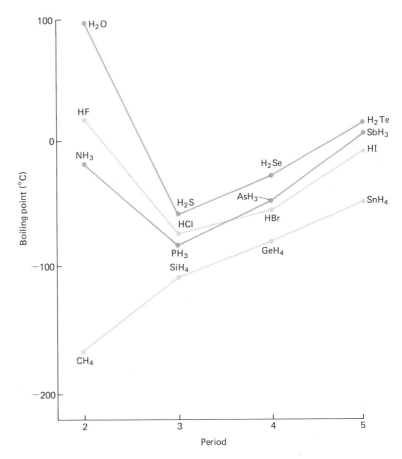

molecular size increases. The hydrogen compounds of the IV A elements are nonpolar molecules; the central atom of each molecule has no unshared electron pair.

In each compound of the other three series of Figure 8.2, however, London forces are aided by dipole–dipole forces in holding the molecules together (Section 4.6). Nevertheless, the boiling point of the first member of each series (HF, $H_2O$, and $NH_3$) is unusually high in comparison to those of the other members of the series. In each of these three compounds, hydrogen bonding increases the difficulty of separating the molecules from the liquid state. Significant hydrogen bonding is not found in any of the other compounds for which a boiling point is plotted in Figure 8.2. In addition to high boiling points, compounds that are associated by hydrogen bonding have abnormally high melting points, heats of vaporization, heats of fusion, and viscosities.

Hydrogen bonding occurs not only between the identical molecules of some pure compounds but also between the different molecules that comprise certain solutions. There are two requirements for strong hydrogen bonding. First, the molecule that supplies the proton for the formation of the hydrogen bond (the *proton donor*) must be strongly polar so that the hydrogen atom will have a relatively high positive charge. The strength of the hydrogen bonds N—H $\cdots$ N < O—H $\cdots$ O < F—H $\cdots$ F parallels the increasing electronegativity of the atom bonded to hydrogen, N < O < F. Hence, the high positive charge on the hydrogen atom attracts the electron pair from another molecule strongly, and the small size of the hydrogen atom permits the second molecule to approach closely.

The second requirement for the formation of a strong hydrogen bond is that the atom (of the *proton acceptor*) that supplies the electron pair for the hydrogen bond must be relatively small. Really effective hydrogen bonds are formed only by fluorine, oxygen, and nitrogen compounds. Chlorine compounds form weak hydrogen bonds, as evidenced by the slight displacement of the boiling point of HCl (Figure 8.2). Chlorine has approximately the same electronegativity as nitrogen; however, a chlorine atom is larger than a nitrogen atom, and the electron cloud of the chlorine atom is, therefore, more diffuse than that of the nitrogen atom.

An examination of Figure 8.2 will show that hydrogen bonding has a greater effect on the boiling point of water than on the boiling point of hydrogen fluoride. This effect is observed even though the O—H $\cdots$ O bond is only about two-thirds as strong as the F—H $\cdots$ F bond. There are, *on the average,* twice as many hydrogen bonds *per molecule* in $H_2O$ as there are *per molecule* in HF. The oxygen atom of each water molecule has two hydrogen atoms and two unshared electron pairs. The fluorine atom of the hydrogen fluoride molecule has three free electron pairs that can bond with hydrogen atoms but only one hydrogen atom with which it can form a hydrogen bond.

Other properties of water are affected to an unusual degree by hydrogen bonding. The tetrahedral arrangement of the hydrogen atoms and the unshared electron pairs of oxygen in water (Section 4.3) cause the hydrogen bonds of the ice crystal to be arranged in this manner and leads to the open structure of the ice crystal. Ice, therefore, has a comparatively low density. In water at the freezing point, the molecules are arranged more closely together; water, therefore, has a higher density than ice—an unusual situation. It should be noted that $H_2O$ molecules are associated by hydrogen bonds in the liquid state but not to the same extent, nor in the rigid manner, as they are associated in ice.

Hydrogen bonding also accounts for the unexpectedly high solubilities of some compounds containing oxygen, nitrogen, and fluorine in certain hydrogen-containing solvents, notably water. Thus, ammonia and methanol dissolve in water through the formation of hydrogen bonds.

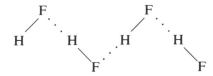

In addition, certain oxygen-containing anions (e.g., sulfate, $SO_4^{2-}$) dissolve in water through hydrogen-bond formation.

Rather than the somewhat random arrangement of some polar molecules with the usual dipole–dipole attractions, there is some evidence that the hydrogen bond is directional. It is assumed, at least in certain cases, that the hydrogen atom of one molecule approaches the second molecule in the direction of the hybrid orbitals of the unshared electron pair (Section 4.2). This directional character accounts for the tetrahedral structure observed in the ice crystal as well as the zigzag arrangement of HF molecules in solid hydrogen fluoride.

Hydrogen bonding plays an important role in determining the structures and properties of molecules of living systems. For example, the

**Figure 8.3** Schematic representations of the structure of DNA. (a) Dots represent hydrogen bonding between units of the chains of the double helix. (b) The dark spheres represent carbon atoms of five-carbon sugar units, the white spheres represent the H and O atoms of sugar and phosphate groups, and the colored spheres represent C and N atoms of units that are nitrogen-containing ring structures.

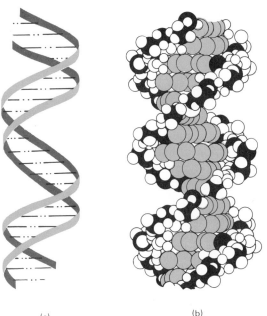

(a)

(b)

fibrous proteins (such as are found in silk, hair, and muscle) consist of high molecular-weight, long-chain protein molecules, which are organized in bundles held together by hydrogen bonding. In deoxyribonucleic acid (DNA), two long spiral chains form a double helix, and the hydrogen bonding that occurs between subunits of these chains holds the structure together (Figure 8.3). One theory proposed to account for the replication of DNA molecules assumes that the chains first come apart through the breaking of the hydrogen bonds. Each chain of the pair then forms a new complementary chain from simpler molecules that exist in the cell. In this respect, the parent chain acts as a template for the formation of the daughter chain; the geometry of the parent, including the location of active sites for hydrogen bonding (which is especially important), determines the molecular structure of the daughter. The process ends with two double helices where only one had been before.

## Problems

Thermochemical problems may be solved using either calories or joules as energy units. Thermochemical values expressed in joules appear in brackets.

**8.1** Write chemical equations for the preparation of oxygen from (a) $HgO$, (b) $BaO_2$, (c) $Na_2O_2$ and $H_2O$, (d) $NaNO_3$, (e) $KClO_3$, (f) $H_2O$.

**8.2** Write chemical equations for the preparation of hydrogen from (a) Na and $H_2O$, (b) Mg and steam, (c) Fe and steam, (d) Zn and $H^+(aq)$, (e) Zn and $OH^-(aq)$, (f) C and steam.

**8.3** Write a chemical equation for the reaction of each of the following with oxygen: (a) Cs, (b) Na, (c) Li, (d) C, (e) S, (f) $Cu_2O$, (g) $P_4$.

**8.4** Write equations for the complete combustion in $O_2$ of (a) $C_8H_{18}$, (b) $C_4H_{10}S$, (c) $C_2H_5OH$, (d) PbS.

**8.5** Write a chemical equation for the reaction of each of the following with hydrogen: (a) Na, (b) $Cu_2O$, (c) $Cl_2$, (d) CO, (e) $WO_3$, (f) $N_2$.

**8.6** Write a chemical equation for the reaction of each of the following with water: (a) $CaH_2$, (b) $Li[AlH_4]$, in which the hydroxides of Li and Al are two of the products, (c) $SO_3$, (d) CaO, (e) $Na_2O$, (f) $N_2O_3$.

**8.7** What are the anhydrides of (a) KOH, (b) $Ca(OH)_2$, (c) $Al(OH)_3$, (d) $H_2CO_3$, (e) $H_3PO_3$, (f) $HClO_4$, (g) $H_3BO_3$, (h) $HNO_2$, (i) HOCl.

**8.8** Complete and balance the following equations. Assume complete neutralization for the acid–alkali reactions.

(a) $Sc_2O_3(s) + H^+(aq) \rightarrow$

(b) $Ca(OH)_2(s) + CO_2(g) \rightarrow$

(c) $CaO(s) + SiO_2(s) \xrightarrow{\text{heat}}$

(d) $SO_2(g) + OH^-(aq) \rightarrow$

(e) $Ag_2O(s) + HCl(g) \rightarrow$

**8.9** Describe the structure and properties of (a) salt-like hydrides,

(b) interstitial hydrides, (c) complex hydrides, (d) covalent hydrides.

**8.10** Starting with elements only, write a series of reactions for the preparation of $Na_2SO_3$.

**8.11** For each of the following reactions, identify the substance oxidized, the substance reduced, the oxidizing agent, and the reducing agent.

(a) $Zn + Cl_2 \rightarrow ZnCl_2$

(b) $2ReCl_5 + SbCl_3 \rightarrow$
$\quad\quad 2ReCl_4 + SbCl_5$

(c) $Mg + CuCl_2 \rightarrow MgCl_2 + Cu$

(d) $2NO + O_2 \rightarrow 2NO_2$

(e) $WO_3 + 3H_2 \rightarrow W + 3H_2O$

**8.12** For each of the following reactions, identify the substance oxidized, the substance reduced, the oxidizing agent, and the reducing agent.

(a) $Cl_2 + 2NaBr \rightarrow 2NaCl + Br_2$

(b) $Zn + 2HCl \rightarrow ZnCl_2 + H_2$

(c) $Fe_2O_3 + Al \rightarrow Al_2O_3 + 2Fe$

(d) $OF_2 + H_2O \rightarrow O_2 + 2HF$

(e) $2HgO \rightarrow 2Hg + O_2$

**8.13** Balance the following by the change-in-oxidation-number method.

(a) $Sb + H^+ + NO_3^- \rightarrow$
$\quad\quad Sb_4O_6 + NO + H_2O$

(b) $NaI + H_2SO_4 \rightarrow$
$\quad\quad H_2S + I_2 + Na_2SO_4 + H_2O$

(c) $IO_3^- + H_2O + SO_2 \rightarrow$
$\quad\quad I_2 + SO_4^{2-} + H^+$

(d) $NF_3 + AlCl_3 \rightarrow N_2 + Cl_2 + AlF_3$

(e) $As_4O_6 + Cl_2 + H_2O \rightarrow$
$\quad\quad H_3AsO_4 + HCl$

**8.14** Balance the following by the change-in-oxidation-number method.

(a) $Fe^{2+} + H^+ + ClO_3^- \rightarrow$
$\quad\quad Fe^{3+} + Cl^- + H_2O$

(b) $Pt + H^+ + NO_3^- + Cl^- \rightarrow$
$\quad\quad PtCl_6^{2-} + NO + H_2O$

(c) $Cu + H^+ + SO_4^{2-} \rightarrow$
$\quad\quad Cu^{2+} + SO_2 + H_2O$

(d) $Pb + PbO_2 + H^+ + SO_4^{2-} \rightarrow$
$\quad\quad PbSO_4 + H_2O$

(e) $MnO_2 + HI \rightarrow MnI_2 + I_2 + H_2O$

**8.15** Complete and balance the following equations by the ion-electron method. All reactions occur in acid solution.

(a) $ClO_3^- + I^- \rightarrow Cl^- + I_2$

(b) $Zn + NO_3^- \rightarrow Zn^{2+} + NH_4^+$

(c) $H_3AsO_3 + BrO_3^- \rightarrow$
$\quad\quad H_3AsO_4 + Br^-$

(d) $H_2SeO_3 + H_2S \rightarrow Se + HSO_4^-$

(e) $ReO_2 + Cl_2 \rightarrow HReO_4 + Cl^-$

**8.16** Complete and balance the following equations by the ion-electron method. All reactions occur in acid solution.

(a) $Fe^{2+} + Cr_2O_7^{2-} \rightarrow Fe^{3+} + Cr^{3+}$

(b) $HNO_2 + MnO_4^- \rightarrow NO_3^- + Mn^{2+}$

(c) $As_2S_3 + ClO_3^- \rightarrow$
$\quad\quad H_3AsO_4 + S + Cl^-$

(d) $IO_3^- + N_2H_4 \rightarrow I^- + N_2$

(e) $Cu + NO_3^- \rightarrow Cu^{2+} + NO$

**8.17** Complete and balance the following equations by the ion-electron method. All reactions occur in acid solution.

(a) $P_4 + HOCl \rightarrow H_3PO_4 + Cl^-$

(b) $XeO_3 + I^- \rightarrow Xe + I_3^-$

(c) $UO^{2+} + Cr_2O_7^{2-} \rightarrow$
$\quad\quad UO_2^{2+} + Cr^{3+}$

(d) $H_2C_2O_4 + BrO_3^- \rightarrow CO_2 + Br^-$

(e) $Te + NO_3^- \rightarrow TeO_2 + NO$

**8.18** Complete and balance the following equations by the ion-electron method. All reactions occur in acid solution.

(a) $Zn + AsO_4^{3-} \rightarrow Zn^{2+} + AsH_3$

(b) $H_5IO_6 + I^- \rightarrow I_2$

(c) $MnO_4^- + Cl^- \rightarrow Mn^{3+} + Cl_2$

(d) $Co(NO_2)_6^{3-} + MnO_4^- \rightarrow$
$\qquad Co^{2+} + NO_3^- + Mn^{2+}$

(e) $CNS^- + IO_3^- + Cl^- \rightarrow$
$\qquad CN^- + SO_4^{2-} + ICl_2^-$

**8.19** Complete and balance the following equations by the ion–electron method. All reactions occur in alkaline solution.

(a) $S^{2-} + I_2 \rightarrow SO_4^{2-} + I^-$

(b) $CN^- + MnO_4^- \rightarrow$
$\qquad CNO^- + MnO_2$

(c) $Au + CN^- + O_2 \rightarrow$
$\qquad Au(CN)_2^- + OH^-$

(d) $Si + OH^- \rightarrow SiO_3^{2-} + H_2$

(e) $Cr(OH)_3 + BrO^- \rightarrow$
$\qquad CrO_4^{2-} + Br^-$

**8.20** Complete and balance the following equations by the ion–electron method. All reactions occur in alkaline solution.

(a) $Al + H_2O \rightarrow Al(OH)_4^- + H_2$

(b) $S_2O_3^{2-} + OCl^- \rightarrow SO_4^{2-} + Cl^-$

(c) $I_2 + Cl_2 \rightarrow H_3IO_6^{2-} + Cl^-$

(d) $Bi(OH)_3 + Sn(OH)_4^{2-} \rightarrow$
$\qquad Bi + Sn(OH)_6^{2-}$

(e) $NiO_2 + Fe \rightarrow$
$\qquad Ni(OH)_2 + Fe(OH)_3$

**8.21** Complete and balance the following equations by the ion–electron method. All reactions occur in alkaline solution.

(a) $HClO_2 \rightarrow ClO_2 + Cl^-$

(b) $MnO_4^- + I^- \rightarrow MnO_4^{2-} + IO_4^-$

(c) $P_4 \rightarrow HPO_3^{2-} + PH_3$

(d) $SbH_3 + H_2O \rightarrow Sb(OH)_4^- + H_2$

(e) $CO(NH_2)_2 + OBr^- \rightarrow$
$\qquad CO_2 + N_2 + Br^-$

**8.22** Complete and balance the following equations by the ion–electron method. All reactions occur in alkaline solution.

(a) $Mn(OH)_2 + O_2 \rightarrow Mn(OH)_3$

(b) $Cl_2 \rightarrow ClO_3^- + Cl^-$

(c) $HXeO_4^- \rightarrow XeO_6^{4-} + Xe + O_2$

(d) $As + OH^- \rightarrow AsO_3^{3-} + H_2$

(e) $S_2O_4^{2-} + O_2 \rightarrow SO_3^{2-} + OH^-$

**8.23** Because the oxygen of $H_2O_2$ can be either oxidized (to $O_2$) or reduced (to $H_2O$), hydrogen peroxide can function as a reducing agent or as an oxidizing agent. Write and balance equations (by the ion–electron method) to show the following reactions of $H_2O_2$. (a) the oxidization of PbS to $PbSO_4$ in acid solution, (b) the oxidation of $Cr(OH)_3$ to $CrO_4^{2-}$ in alkaline solution, (c) the reduction of $MnO_4^-$ to $Mn^{2+}$ in acid solution, (d) the reduction of $Ag_2O$ to Ag in alkaline solution.

**8.24** What fraction of a mole is one equivalent weight of each of the following substances? (a) $As_4O_6$ in a reaction in which $H_3AsO_4$ is produced, (b) $N_2H_4$ in a reaction in which $N_2$ is produced, (c) $H_2S_2O_8$ in a reaction in which $SO_4^{2-}$ is produced, (d) $H_2SO_4$ in a reaction in which $NaHSO_4$ is produced, (e) $H_2SO_4$ in a reaction in which $H_2S$ is produced.

**8.25** What fraction of a mole is one equivalent weight of each of the following substances? (a) $KIO_3$ in a reaction in which $I_2$ is produced, (b) $KIO_3$ in a reaction in which ICl is produced, (c) $HIO_3$ in a reaction in which $KIO_3$ is produced, (d) $HIO_3$ in a reaction in which $H_5IO_6$ is produced, (e) $H_5IO_6$ in a reaction in which $Na_2H_3IO_6$ is produced.

**8.26** A 2.70 g sample of oxalic acid requires 2.40 g of NaOH for complete neutralization. What is the equivalent weight of oxalic acid?

**8.27** The reaction of $I_2O_5$ and CO

yields $I_2$ and $CO_2$. (a) What are the equivalent weights of $I_2O_5$ and CO for this reaction? (b) What weight of CO is present in a gas sample that requires 1.00 g of $I_2O_5$ for complete reaction?

**8.28** The equivalent weight of a diprotic acid is 104.0 for the production of the acid salt and 52.0 for the production of the normal salt. If 1.000 g of $Ca(OH)_2$ is reacted with 2.807 g of the acid to produce a pure salt, has the normal salt or the acid salt been prepared?

**8.29** (a) What is the equivalent weight of $KBrO_3$ (in a transformation producing $Br^-$), of Se (in a transformation producing $H_2SeO_3$), and of Se (in a transformation producing $H_2SeO_4$)? (b) In a reaction between $KBrO_3$ and Se, 1.02 g of $KBrO_3$ reacts with 0.725 g of Se. What is the formula of the selenium-containing product of the reaction?

**8.30** (a) What is the equivalent weight of $H_2O_2$ for a reaction in which $H_2O$ is produced? (b) What is the equivalent weight of $N_2H_4$ for a reaction in which $N_2$ is produced? (c) How many grams of $H_2O_2$ are needed to react with 4.00 g of $N_2H_4$?

**8.31** (a) What is the equivalent weight of $K_2Cr_2O_7$ for a reaction in which $Cr^{3+}$ is produced? (b) $K_2Cr_2O_7$ oxidizes $Fe^{2+}$ to $Fe^{3+}$. How many grams of $Fe^{2+}$ will react with 1.000 g of $K_2Cr_2O_7$?

**8.32** Ethylene glycol, $H_4C_2(OH)_2$, which is used as an antifreeze, is oxidized to formaldehyde, $H_2CO$, by periodic acid, $HIO_4$. The periodic acid is reduced to the $IO_3^-$ ion. (a) Write the partial equations for the half reactions and deter-

mine the equivalent weights of $H_4C_2(OH)_2$ and $HIO_4$. (b) How many grams of ethylene glycol were present in a sample that required 7.68 g of $HIO_4$ for complete reaction?

**8.33** The reaction of sodium tellurate, $Na_2TeO_4$, with hydrogen iodide, HI, yields Te and $I_2$. (a) What are the equivalent weights of $Na_2TeO_4$ and HI for this reaction? (b) What quantity of HI is needed to react with 5.00 g of $Na_2TeO_4$?

**8.34** The effect of hydrogen bonding on the properties of the following hydrides falls in the order: $H_2O > HF > NH_3$. Explain this observation.

**8.35** Look at Problem 7.9 in the last chapter. Are you now better able to offer an explanation for the high molar entropy of vaporization of hydrazine, $N_2H_4$?

**8.36** The compound $KHF_2$ can be prepared from the reaction of KF and HF in water solution. Explain the structure of the $HF_2^-$ ion.

**8.37** Diagram the structures of the following molecules and explain how the water solubility of each compound is enhanced by hydrogen bonding. (a) $NH_3$ (b) $H_2N—OH$ (c) $H_3COH$ (d) $H_2CO$

**8.38** Although there are exceptions, most acid salts are more soluble in water than the corresponding normal salts. Offer an explanation for this generalization.

**8.39** Offer an explanation as to why chloroform, $HCCl_3$, and acetone,

$$CH_3—\overset{\overset{\textstyle O}{\|}}{C}—CH_3,$$ mixtures have higher boiling points than either pure component.

**8.40** (a) Calculate the molar entropy of vaporization at the boiling point for formic acid, $HCO_2H$. The normal boiling point of formic acid is 101°C and the heat of vaporization is 120 cal/g [or 502 J/g]. How does your answer compare to the Trouton's rule value? (b) The formic acid molecule consists of a central C atom to which a double-bonded O atom, an —O—H group, and a H atom are bonded. Formic acid vapor at temperatures near the boiling point consists of double molecules, or dimers, that are joined by hydrogen bonds. Diagram the formic acid dimer. How does the existence of these dimers in the vapor explain the result obtained in part (a)?

# 9

# Solutions

Solutions are homogeneous mixtures (Section 1.1). They are usually classified according to their physical state; gaseous, liquid, and solid solutions can be prepared. Dalton's law of partial pressures describes the behavior of gaseous solutions, of which air is the most common example. Certain alloys are solid solutions; coinage silver is copper dissolved in silver, and brass is a solid solution of zinc in copper. Not all alloys are solid solutions, however; some are heterogeneous mixtures, and some are intermetallic compounds. Liquid solutions are the most common and are probably the most important to the chemist.

An understanding of the laws governing the behavior of gaseous solutions (as well as pure gases) is mainly the product of scientific investigations of the seventeenth and eighteenth centuries. The three theories upon which our understanding of liquid solutions has largely been built were all proposed within a brief period of time; they are Raoult's law of ideal solutions (1878–87), van't Hoff's theory of dilute solutions (1886), and Arrhenius's theory of ionization (1887). In the late nineteenth century significant work was also done on the nature of solid solutions; the phase rule of J. Willard Gibbs (developed in 1873–78) is central to these investigations.

## 9.1 NATURE OF SOLUTIONS

The component of a solution that is present in greatest quantity is usually called the **solvent,** and the other components are called **solutes.** This terminology is loose and arbitrary. It is sometimes convenient to designate a component as the solvent even though it is present in only small amount; at other times, the assignment of the terms solute and solvent has little significance (e.g., in characterizing gaseous solutions).

Certain pairs of substances are miscible in all proportions. Complete miscibility is characteristic of the components of all gaseous solutions and some pairs of components of liquid and solid solutions. However, for most materials there is a limit on the amount of the substance that will dissolve in a given solvent. The **solubility** of a substance in a particular solvent at a specified temperature is the maximum amount of the solute that will dissolve in a definite amount of the solvent and produce a stable system.

For a given solution, the amount of solute dissolved in a unit volume of solution (or a unit amount of solvent) is the **concentration** of the solute. Solutions containing a relatively low concentration of solute are called **dilute solutions;** those of relatively high concentrations are called **concentrated solutions.**

If an excess of solute (more than will normally dissolve) is added to a quantity of a liquid solvent, an equilibrium is established between the pure solute and the dissolved solute.

$$\text{solute}_{\text{pure}} \rightleftharpoons \text{solute}_{\text{dissolved}}$$

The pure solute may be a solid, liquid, or gas. At equilibrium in such a system, the rate at which the pure solute dissolves equals the rate at which the dissolved solute comes out of solution. The concentration of the dissolved solute, therefore, is a constant. A solution of this type is called a **saturated solution,** and its concentration is the solubility of the solute in question.

That such dynamic equilibria exist has been shown experimentally. If small crystals of a solid solute are placed in contact with a saturated solution of the solute, the crystals are observed to change in size and shape. Throughout this experiment, however, the concentration of the saturated solution does not change, nor does the quantity of excess solute decrease or increase.

An **unsaturated solution** has a lower concentration of solute than a saturated solution. On the other hand, it is sometimes possible to prepare a **supersaturated solution,** one in which the concentration of solute is higher than that of a saturated solution. A supersaturated solution, however, is metastable, and if a very small amount of pure solute is added to it, the solute that is in excess of that needed to saturate the solution will precipitate.

Reactions that are run in solution are usually rapid since: the reactants are in a small state of subdivision (solutes exist in solution as molecules, atoms, or ions), the attractions between the particles (molecules, atoms, or ions) of the pure solute have been at least partially overcome, and the particles are free to move about through the solution.

**9.2** THE SOLUTION PROCESS  London forces are the only intermolecular forces between nonpolar covalent molecules. On the other hand, the intermolecular attractions between polar covalent molecules are due to dipole–dipole forces as well as to London forces. In substances in which there is hydrogen bonding (an exceptionally strong type of dipole–dipole attraction), the intermolecular forces are unusually strong.

Nonpolar substances and polar substances are generally immiscible; for example, carbon tetrachloride (a nonpolar substance) is insoluble in water (a polar substance). The attraction of one water molecule for another water molecule is much greater than any attraction between a carbon tetrachloride molecule and a water molecule. Hence, carbon tetrachloride molecules are "squeezed out," and these two substances form a two-liquid-layer system.

Iodine is a nonpolar material and is soluble in carbon tetrachloride. The attractions between $I_2$ molecules in solid iodine are approximately of the same type and magnitude as those between $CCl_4$ molecules in pure carbon tetrachloride. Hence, significant iodine–carbon tetrachloride attractions are possible, and iodine molecules can mix with carbon tetrachloride molecules. The resulting solution is a random molecular mixture.

Methyl alcohol, $CH_3OH$, like water, consists of polar molecules that are highly associated. In both pure liquids the molecules are attracted to one another through hydrogen bonding.

Methyl alcohol and water are miscible in all proportions. In solutions of methyl alcohol in water, $CH_3OH$ and $H_2O$ molecules are associated through hydrogen bonding.

$$\underset{\mathsf{H}}{\mathsf{H-O}} \cdots \underset{\underset{\mathsf{H}}{\overset{|}{\mathsf{H-C-H}}}}{\mathsf{H-O}} \cdots \underset{\mathsf{H}}{\mathsf{H-O}}$$

Methyl alcohol does not dissolve in nonpolar solvents. The strong intermolecular attractions of pure methyl alcohol are not overcome unless the solvent molecules can form attractions of equal, or almost equal, strength with the methyl alcohol molecules.

In general, polar materials dissolve only in polar solvents, and nonpolar substances are soluble in nonpolar solvents. This is the first rule of solubility: "like dissolves like." Network crystals (e.g., the diamond), in which the atoms comprising the crystal are held together by covalent bonds, are insoluble in all liquids. This crystalline structure is far too stable to be broken down by a solution process; any potential solute–solvent attractions cannot approach the strength of the covalent bonding of the crystal.

Polar liquids (water, in particular) can function as solvents for many ionic compounds. The ions of the solute are electrostatically attracted by the polar solvent molecules—negative ions by the positive poles of the solvent molecules, positive ions by the negative poles of the solvent

**Figure 9.1** Solution of an ionic crystal in water.

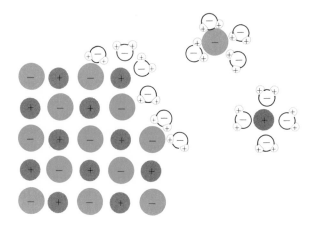

molecules. These ion–dipole attractions can be relatively strong. Figure 9.1 diagrams the solution of an ionic crystal in water. The ions in the center of the crystal are attracted equally in all directions by the oppositely charged ions of the crystal. The electrostatic attractions on the ions of the surface of the crystal, however, are unbalanced. Water molecules are attracted to these surface ions, the positive ends of the water molecules to the anions and the negative ends of the water molecules to the cations. The ion–dipole attractions thus formed allow the ions to escape from the crystal and drift into the liquid phase. The dissolved ions are **hydrated** and move through the solution surrounded by a sheath of water molecules; all ions are hydrated in water solution.

A number of covalent compounds that contain metals and nonmetals (e.g., aluminum chloride) ionize in water solution. True cations with charges of 3+ or higher (such as $Al^{3+}$), however, seldom exist in pure compounds. In addition, other species with relatively high ratios of charge to size (e.g., the very small $Be^{2+}$ ion) form pure compounds that are largely covalent. The same factor that is principally responsible for the covalent character of these compounds (a high ratio of ionic charge to radius) also leads to the formation of very stable hydrated ions.

$$BeCl_2(s) + 4H_2O \rightarrow Be(H_2O)_4^{2+}(aq) + 2Cl^-(aq)$$

**9.3 HYDRATES** The nature of the interactions between water molecules and ions in aqueous solution is a subject of extensive study. Negative ions in aqueous solution are hydrated by means of ion–dipole attractions or, where possible, by hydrogen bonding. Unless the positive ion itself contains hydrogen atoms (as, e.g., the ammonium ion, $NH_4^+$), hydrogen bonding between a positive ion and a water molecule is impossible.

Nevertheless, the attractions between water molecules and most metal cations are strong ones. At one time coordinate covalent bonds were thought to bind these particles together into **complex ions,** such a bond being formed by the use of a vacant orbital of the metal cation and a pair of unshared electrons from the oxygen of a water molecule. This valence bond approach, especially when modified to take into account hybridized orbitals (Section 4.2), remains a useful one, but the ligand field theory and the molecular orbital theory (Section 19.5) offer interpretations that are more in agreement with the observed properties of these hydrated ions.

Most metal cations are hydrated by a definite number of water molecules; the most common number is six, although four is not rare. For sixfold coordination, the water molecules are arranged at the corners of a regular octahedron as illustrated in Figure 9.2 for the ion $[Cr(H_2O)_6]^{3+}$. Certain other molecules and anions (called **ligands**) form complex ions

**Figure 9.2 The hydrated Cr³⁺ ion.**

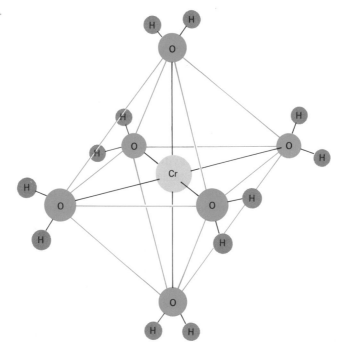

with metal cations (see Chapter 19). The number of ligands coordinated around the central ion is called the **coordination number** of that ion. In aqueous solution, hydrated complex cations are undoubtedly further hydrated by additional water molecules that are hydrogen bonded to the coordinated water molecules; however, the water molecules of the outer layers are much more loosely held than those of the inner coordination sphere.

Frequently, hydrated ions persist in the crystalline materials that are obtained by the evaporation of aqueous solutions of salts. Thus, the solid compound $FeCl_3 \cdot 6H_2O$ is actually $[Fe(H_2O)_6]Cl_3$, and $[Fe(H_2O)_6]^{3+}$ ions can be identified in the crystal. Other examples of solid hydrates in which the water of hydration is coordinated to the cation are $[Co(H_2O)_6]Cl_2$, $[Cr(H_2O)_6]Cl_3$, and $[Be(H_2O)_4]Cl_2$.

Solid hydrates in which water molecules are coordinated to anions by hydrogen bonding are known but are not common. For example, in the compound $ZnSO_4 \cdot 7H_2O$, six water molecules are coordinated around the zinc ion, and the seventh is hydrogen bonded to the sulfate anion. Other examples include the analogous compounds $NiSO_4 \cdot 7H_2O$ and $CoSO_4 \cdot 7H_2O$, as well as $CuSO_4 \cdot 5H_2O$ which may be represented $[Cu(H_2O)_4]SO_4 \cdot H_2O$.

There are two other ways in which water can occur in a crystalline hydrate in addition to association with cation or anion. Water molecules may assume positions in the crystal lattice without being associated with a specific ion (e.g., in $BaCl_2 \cdot 2H_2O$). Removal of the water from $BaCl_2 \cdot 2H_2O$ destroys the crystal lattice. Water molecules may also occur between the layers or in the interstices of a crystal lattice; the hydrates of the zeolites, a group of silicate minerals, incorporate water in this manner. Removal of zeolitic water from a hydrate, however, changes the crystal very little, if at all.

**9.4 HEAT OF SOLUTION**

When a solute dissolves in a solvent, energy is absorbed or evolved, the actual quantity of energy per mole of solute depending upon the concentration of the final solution. Thus, when a **heat of solution** is recorded, the number of moles of solvent used to dissolve one mole of solute in the determination of the heat effect must be indicated in some way. If this distinction is not made, it may be assumed that the heat of solution relates to a process in which one mole of solute is dissolved in an infinitely large quantity of solvent (heat of solution at infinite dilution); the heat of solution per mole of a given solute is virtually a constant when very dilute solutions are prepared.

The heat effect observed when a solution is prepared is the net result of the energy *required* to break apart certain chemical bonds or attractions (solute–solute and solvent–solvent) and the energy *released* by the formation of new ones (solute–solvent). Thus, if an ionic solid (MX) dissolves in water, the heat of solution is a manifestation of the energy required to break apart the crystal lattice (the lattice energy; $\Delta H$, positive):

$$MX(s) \rightarrow M^+(g) + X^-(g)$$

and the energy liberated when the ions are hydrated (the heat of hydration; $\Delta H$, negative):

$$M^+(g) + X^-(g) \rightarrow M^+(aq) + X^-(aq)$$

The heat of hydration is itself the sum of two heat effects—the energy needed to break the hydrogen bonds between some of the solvent molecules and the energy released when these water molecules hydrate the ions of the solute. However, it is difficult to investigate these latter two effects separately.

The overall solution process is exothermic ($\Delta H$, negative) if more energy is liberated by the hydration process than is required to break apart the crystal lattice (see data for AgF in Table 9.1). On the other hand, heat is absorbed ($\Delta H$, positive) when the solute dissolves in water if the lattice energy cannot be supplied by the hydration energy (see data for KCl in Table 9.1). The data of Table 9.1 apply to the preparation of very dilute solutions.

Table 9.1 Calculation of heat of solution[a]

| SUBSTANCE | LATTICE ENERGY (kcal/mol) | HEAT OF HYDRATION (kcal/mol) | HEAT OF SOLUTION (kcal/mol) |
|---|---|---|---|
| KCl | +167.6 | −163.5 | +4.1 |
| AgF | +217.7 | −222.6 | −4.9 |

[a] Infinite dilution, 298°K. Heat of hydration pertains to the hydration of gaseous ions.

The magnitude of the hydration energy of a salt depends upon the concentration of charge (ratio of charge to surface area) of its ions. In general, a high concentration of charge results when an ion has a small size, a high charge, or a combination of these two factors. Ions with high concentrations of charge attract water molecules strongly, and this leads to large heats of hydration. This is generally not reflected in the heat of solution, however, because the same factor—high ionic charge concentration—also leads to very stable crystal structures and consequently large lattice energies.

Similar considerations apply to the solution of nonionic materials. The lattice energies of molecular crystals are not so large as those of ionic crystals since the forces holding molecular crystals together are not so strong as those holding ionic crystals together; however, solvation energies for these nonionic materials are also low. For molecular substances that dissolve in nonpolar solvents without ionization and without appreciable solute–solvent interaction, the heat of solution is endothermic and has about the same magnitude as the heat of fusion of the solute.

Gases generally dissolve in liquids with the evolution of heat. Since no energy is required to separate the molecules of a gas, the predominant heat effect of such a solution process is the solvation of the gas molecules; therefore, the process is exothermic. Calculation of the heat of solution of hydrogen chloride gas in water, however, requires that the heat evolved by the formation of hydrated ions be reduced by the energy required for the ionization of the HCl molecules.

An entropy effect (Sections 7.5 and 14.4) occurs when a solution is prepared. In some cases the entropy increases ($\Delta S$, positive); the change from pure substances to solution represents an increase in randomness, or disorder. In other instances (e.g., the dissolution of a gas in a liquid), the entropy, or disorder, of the system decreases when the solution is prepared ($\Delta S$, negative); that is, the solution is more highly ordered (less disordered) than the pure components.

Entropy may be regarded as a probability function. A disordered condition is far more statistically probable than an ordered one. The probability that a bridge hand dealt from a well-shuffled deck of cards will contain a random mixture of suits is higher than the probability that it will contain thirteen cards of the same suit. In any natural process there is a tendency for disorder, or randomness, to increase.

However, there is also a tendency for the energy of a system to decrease in the course of a natural process; that is, the system tends to go from a more energetic state to a less energetic state. Both of these factors (the tendency toward minimum energy and the tendency toward maximum entropy) must be considered in deciding whether a given process is spontaneous. It is convenient to use a function called the **Gibbs free energy**, $G$, for this purpose.[1]

At constant temperature the change in Gibbs free energy, $\Delta G$, for a given process is

$$\Delta G = \Delta H - T \Delta S$$

A system is more ordered at low temperatures than at high temperatures; as the temperature increases, the randomness increases. The probability factor, $T \Delta S$, includes both temperature and change in entropy. Gibbs free energy and enthalpy are usually measured in cal or J; the factor $T \Delta S$ has comparable dimensions (°K times cal/°K or °K times J/°K). A favorable $\Delta H$ is negative (minimum energy), whereas a favorable $T \Delta S$ is positive (maximum entropy). Since the factor $T \Delta S$ appears with a negative sign in the equation for $\Delta G$, $\Delta G$ values for spontaneous processes are negative.

We are now in a position to understand why endothermic solution processes occur. Why should KCl dissolve in water when more energy is required to break apart the crystal lattice than is supplied by the hydration of the ions? The increase in entropy for this process compensates for the unfavorable enthalpy change.

In the preparation of aqueous solutions of other ionic compounds (e.g., the preparation of dilute solutions of AgF), the entropy decreases; dissolution occurs because a favorable enthalpy change overshadows an unfavorable entropy change. In this latter case, water molecules must hydrate the ions in a highly ordered manner to produce a solution that represents a decrease in randomness, or an increase in order (Figures 9.1 and 9.2). Entropy effects are difficult to predict qualitatively.

The classification of ionic substances according to their solubility in water is difficult. Nothing is completely "insoluble" in water; the degree of solubility varies greatly from one "soluble" substance to another. Furthermore, some substances (such as $Al_2S_3$) react with water to yield other compounds [$Al(OH)_3$ and $H_2S$ in this instance].

Nevertheless, a solubility classification scheme is useful even though

---

[1] This function was named in honor of J. Willard Gibbs, who developed the application of thermodynamic principles to chemistry.

it must be regarded as approximate. The rules given in Table 9.2 apply to common ionic substances. Compounds that dissolve to the extent of at least 10 g/liter are listed as *soluble;* those that fail to dissolve to the extent of 1 g/liter are classed as *insoluble;* those compounds with solubilities that are intermediate between these limits are listed as *slightly soluble* (and marked with a star in the table). These standards are common but arbitrary. The solubility rules can be used to write equations for reactions in which precipitates are formed (Example 9.1)

**Table 9.2** Solubilities of some ionic compounds in water[a]

MAINLY WATER-SOLUBLE

| | |
|---|---|
| $NO_3^-$ | All nitrates are soluble. |
| $C_2H_3O_2^-$ | All acetates are soluble. |
| $ClO_3^-$ | All chlorates are soluble. |
| $Cl^-$ | All chlorides are soluble except $AgCl$, $Hg_2Cl_2$, $CuCl$, and $PbCl_2$*. |
| $Br^-$ | All bromides are soluble except $AgBr$, $Hg_2Br_2$, $CuBr$, $PbBr_2$*, and $HgBr_2$*. |
| $I^-$ | All iodides are soluble except $AgI$, $Hg_2I_2$, $CuI$, $PbI_2$, and $HgI_2$. |
| $SO_4^{2-}$ | All sulfates are soluble except $CaSO_4$*, $SrSO_4$, $BaSO_4$, $PbSO_4$, $Hg_2SO_4$, and $Ag_2SO_4$*. |

MAINLY WATER-INSOLUBLE

| | |
|---|---|
| $S^{2-}$ | All sulfides are insoluble except those of the I A and II A elements and $(NH_4)_2S$. |
| $CO_3^{2-}$ | All carbonates are insoluble except those of the I A elements and $(NH_4)_2CO_3$. |
| $PO_4^{3-}$ | All phosphates are insoluble except those of the I A elements and $(NH_4)_3PO_4$. |
| $OH^-$ | All hydroxides are insoluble except those of the I A elements, $Ba(OH)_2$, $Sr(OH)_2$*, and $Ca(OH)_2$*. |

[a] The following cations are considered: those of the IA and IIA families, $NH_4^+$, $Al^{3+}$, $Cd^{2+}$, $Co^{2+}$, $Cr^{3+}$, $Cu^+$, $Cu^{2+}$, $Fe^{2+}$, $Fe^{3+}$, $Hg_2^{2+}$, $Hg^{2+}$, $Mn^{2+}$, $Ni^{2+}$, $Pb^{2+}$, $Sn^{2+}$, and $Zn^{2+}$.
  Soluble compounds dissolve to the extent of at least 10 g/liter. Slightly soluble compounds (marked with an *) dissolve to the extent of from 1 g/liter to 10 g/liter.

---

*Example 9.1*

Use the solubility rules to write balanced chemical equations for the reactions that occur when aqueous solutions of the following are mixed. (a) $AgNO_3$ and $NH_4Cl$, (b) $Na_2SO_4$ and $CuCl_2$, (c) $ZnSO_4$ and $Ba(OH)_2$.

*Solution*

(a) Both $AgNO_3$ and $NH_4Cl$ are soluble, and hence we can assume that $Ag^+$, $NO_3^-$, $NH_4^+$, and $Cl^-$ ions are present in solution initially. The interaction of these ions could conceivably produce $AgCl$ and/or $NH_4NO_3$. The solubility rules list AgCl as insoluble, but $NH_4NO_3$ is soluble (since all nitrates are soluble). Therefore, we write

$$Ag^+(aq) + NO_3^-(aq) + NH_4^+(aq) + Cl^-(aq) \rightarrow$$
$$AgCl(s) + NH_4^+(aq) + NO_3^-(aq)$$

An examination of this equation shows that the $NH_4^+$ and $NO_3^-$ ions are not involved in the reaction. An equation showing only those ions that actually take part in the reaction (such as those used in the ion–electron method) is called a *net ionic equation*. In the present case the net ionic equation is

$$Ag^+(aq) + Cl^-(aq) \rightarrow AgCl(s)$$

(b) Both $Na_2SO_4$ and $CuCl_2$ are soluble. Furthermore, the potential products, NaCl and $CuSO_4$, are also soluble. No reaction would occur.

$$2Na^+(aq) + SO_4^{2-}(aq) + Cu^{2+}(aq) + 2 Cl^-(aq) \rightarrow$$
$$\text{N.R.}$$

(c) Both $ZnSO_4$ and $Ba(OH)_2$ are soluble, and both of the possible products, $Zn(OH)_2$ and $BaSO_4$, are insoluble. Therefore, a mixture of precipitates is obtained.

$$Zn^{2+}(aq) + SO_4^{2-}(aq) + Ba^{2+}(aq) + 2OH^-(aq) \rightarrow$$
$$Zn(OH)_2(s) + BaSO_4(s)$$

**9.5 EFFECT OF TEMPERATURE AND PRESSURE ON SOLUBILITY**

The effect of a temperature change on the solubility of a substance depends upon whether heat is absorbed or evolved *when a saturated solution is prepared*. Suppose that a small quantity of solute dissolves in a nearly saturated solution with the absorption of heat. The crystallization of a small quantity of this solute from the saturated solution would evolve a commensurate quantity of heat. We can represent the equilibrium between excess solid solute and saturated solution as

$$\text{energy} + \text{solute} + H_2O \rightleftharpoons \text{saturated solution}$$

The effect of adding heat to such a system can be predicted by means of a principle first proposed by Henri Le Chatelier in 1888. **Le Chatelier's principle** states that the application of a stress to a system in equilibrium results in the system reacting in a way that counteracts the stress and establishes a new equilibrium state. If the temperature of the system previously described is increased (energy added), the equilibrium point will shift to the right (the direction in which energy is absorbed), and more solute will dissolve. If the system is cooled, the point of equilibrium will shift to the left (the direction in which energy is evolved), and solute will precipitate out of solution. The solubility of materials that dissolve in nearly saturated solutions with the absorption of heat increases with increasing temperature; this is true of the solubility of most ionic solutes in water.

When a small amount of an ionic solute dissolves in a solution that is nearly saturated, the heat evolved by the hydration of ions is usually less than the heat evolved when the same amount of the solute dissolves in pure water. In a nearly saturated solution, such a large proportion of water molecules may be already involved in the hydration of ions that the proportion remaining may not be sufficient to hydrate the added ions completely. The heat of solution is evidence of the deficit or surplus that results from the amount of heat supplied by the hydration· process and the amount of heat required to break the lattice structure; the heat of hydration and heat of solution are usually close in magnitude.

Therefore, even though a small amount of a solute may dissolve in pure water with the evolution of heat, the dissolution of a small amount of the same solute in a nearly saturated solution may proceed with the absorption of heat. The data of Table 9.1 pertain to the preparation of very dilute solutions. Many such preparations are exothermic; however, the large majority of ionic solutes dissolve in nearly saturated solutions with the absorption of heat.

When a substance dissolves with the evolution of heat,

$$\text{solute} + H_2O \rightleftharpoons \text{saturated solution} + \text{energy}$$

Le Chatelier's principle indicates that the solubility of the solute increases when the temperature is lowered, and the solubility decreases when the temperature is raised. A few ionic compounds (such as $Li_2CO_3$ and $Na_2SO_4$) behave in this fashion. In addition, the solubility of all gases decreases as the temperature is raised; warming a soft drink causes bubbles of carbon dioxide gas to come out of solution.

The solubility differential with respect to change in temperature depends upon the magnitude of the heat of solution. The solubilities of substances with small heats of solution do not change much with changes in temperature.

Changes in pressure ordinarily have little effect upon the solubility of solid and liquid solutes. However, increasing or decreasing the pressure on a solution that contains a dissolved gas has a definite effect. William Henry in 1803 discovered that the amount of a gas that dissolves in a given quantity of a liquid at constant temperature is directly proportional to the partial pressure of the gas above the solution. **Henry's law** is valid only for dilute solutions and relatively low pressures. Gases that are extremely soluble generally react chemically with the solvent (e.g., hydrogen chloride gas in water reacts to produce hydrochloric acid); these solutions do not follow Henry's law.

**9.6 CONCENTRATION OF SOLUTIONS**

The concentration of solute in a solution may be expressed in several different ways.

1. The **weight percentage** of solute in a solution is 100 times the weight of solute divided by the *total* weight of the solution. Thus, 100 g

of a 10% aqueous solution of sulfuric acid contains 10 g of $H_2SO_4$ and 90 g of water. Notice that a solution of this concentration would contain 11 g of $H_2SO_4$ per 100 g of water. Volume percentages are not commonly employed; figures recorded as percentages should be understood to be based on weight unless specific notation to the contrary is made.

2. The **mole fraction,** $X$, of a component of a solution is the ratio of the number of moles of that component to the total number of moles of all components present in the solution (Section 6.10, Example 6.14). A 100 g sample of a 10% $H_2SO_4$ solution contains $10/98 = 0.10$ mol of $H_2SO_4$ and $90/18 = 5.0$ mol of $H_2O$. The mole fraction of $H_2SO_4$ is, therefore,

$$X_{H_2SO_4} = \frac{0.10}{5.0 + 0.10} = 0.020$$

Figure 9.3 Volumetric flask.

Since the mole fractions of all the components of a solution taken together must equal 1.00, $X_{H_2O} = 0.98$.

3. The **molarity,** $M$, of a solution is the number of moles of solute per liter of solution. Thus, a $6M$ solution of sulfuric acid is prepared by taking 6 mol of $H_2SO_4$ ($6 \times 98.08 = 588.48$ g) and adding the acid, with careful mixing, to sufficient water to make exactly 1 liter of solution.

Notice that the definition is based on the total *volume* of the solution. When a liquid solution is prepared, the volume of the solution rarely equals the sum of the volumes of the pure components. Usually the final volume of the solution is larger or smaller than the total of the volumes of the constituents. Hence, it is not practical to attempt to predict the amount of solvent that should be employed to prepare a given solution; rather, molar solutions (as well as others that are based on total volume) are generally prepared by the use of the volumetric flasks (Figure 9.3). In the preparation of a solution, the correct amount of solute is placed in the flask, and water is then added, with careful and constant mixing, until the solution fills the flask to the calibration mark on the neck.

It is a simple matter to calculate the quantity of solute present in a given sample of a solution when the concentration of the solution is expressed in terms of molarity. Concentrations defined on the basis of the total volume of solution have this decided advantage. Thus, 1 liter of a $3M$ solution contains 3 mol of solute, 500 ml contains 1.5 mol, 250 ml contains 0.75 mol, and 100 ml contains 0.3 mol. A disadvantage of basing concentrations on volume of solution, however, is that such concentrations change slightly with temperature changes because of expansion or contraction of liquid solutions. For exact work, therefore, a solution should be prepared at the temperature at which it is to be used, and a volumetric flask calibrated for this temperature should be employed.

*Example 9.2*

What weight of a concentrated nitric acid solution that is 70.0% $HNO_3$ should be used to prepare 250 ml of 2.00$M$ $HNO_3$? If the density of the concentrated nitric acid solution is 1.42 g/ml, what volume of this acid should be used?

*Solution*

The molecular weight of $HNO_3$ is 63.0; therefore, $2(63.0) = 126.0$ g

of $HNO_3$ should be contained in 1000 ml of 2.00$M$ solution. In 100 g of concentrated nitric acid (70.0%), there would be 70.0 g $HNO_3$. Therefore,

$$? \text{ g concd. } HNO_3 = 250 \text{ ml soln.} \left( \frac{126.0 \text{ g } HNO_3}{1000 \text{ ml soln.}} \right) \left( \frac{100.0 \text{ g concd. } HNO_3}{70.0 \text{ g } HNO_3} \right)$$

$$= 45.0 \text{ g concd. } HNO_3$$

To convert this value into ml of concentrated acid,

$$? \text{ ml concd. } HNO_3 = 45.0 \text{ g concd. } HNO_3 \left( \frac{1.00 \text{ ml concd. } HNO_3}{1.42 \text{ g concd. } HNO_3} \right)$$

$$= 31.7 \text{ ml concd. } HNO_3$$

*Example 9.3*

What is the molarity of a 20.0% $H_2SO_4$ solution if the density of the solution is 1.14 g/ml?

*Solution*

$$? \text{ mol } H_2SO_4 = 1000 \text{ ml soln.} \left( \frac{1.14 \text{ g soln.}}{1.00 \text{ ml soln.}} \right) \left( \frac{20.0 \text{ g } H_2SO_4}{100 \text{ g soln.}} \right) \left( \frac{1 \text{ mol } H_2SO_4}{98.08 \text{ g } H_2SO_4} \right)$$

$$= 2.32 \text{ mol } H_2SO_4$$

The solution is therefore 2.32$M$.

4. The **normality,** $N$, of a solution is the number of gram equivalent weights of solute per liter of solution. The equivalent weight of a substance depends upon the reaction that the substance undergoes. Hence, the normality of a solution depends upon the intended use of the solution. If sulfuric acid is to be used in a neutralization reaction, the equivalent weight of $H_2SO_4$ is one-half the molecular weight, or we can say that 1 mol of $H_2SO_4$ is 2 equivalents (Section 8.6). Hence, a 1$M$ solution of $H_2SO_4$ is 2$N$.

For solutions of $KMnO_4$ that are to be employed in oxidation–reduction reactions in which the equivalent weight of $KMnO_4$ is $\frac{1}{5}$ the molecular weight (Section 8.6), the normality of a given solution is five times the molarity since 1 mol of $KMnO_4$ is 5 equivalents. On the other hand, the normality of an NaOH solution that is to be used in a neutralization reaction is the same as its molarity since the equivalent weight of NaOH is the same as the molecular weight. The normality of a solution is always a simple whole number multiple (including 1) of the molarity of the solution.

Since the normality of a solution is based upon its total volume, the same volumetric technique is employed in the preparation of normal solutions as is used in the preparation of molar solutions. Similarly, the normality of a solution varies slightly with temperature.

---

*Example 9.4*

What weight of $K_2Cr_2O_7$ should be used to prepare 500.0 ml of 0.1000*N* solution if the dichromate ion functions as an oxidizing agent?

$$6e^- + 14H^+ + Cr_2O_7^{2-} \rightarrow 2Cr^{3+} + 7H_2O$$

What is the molarity of this solution?

*Solution*

The molecular weight of $K_2Cr_2O_7$ is 294.19, and according to the partial equation, the equivalent weight of $K_2Cr_2O_7$ is $\frac{1}{6}$ of the molecular weight, or 49.03. Thus,

$$? \text{ g } K_2Cr_2O_7 = 500 \text{ ml soln.} \left( \frac{0.1(49.03) \text{ g } K_2Cr_2O_7}{1000 \text{ ml soln.}} \right)$$
$$= 2.452 \text{ g } K_2Cr_2O_7$$

Since 1 equivalent of $K_2Cr_2O_7$ is $\frac{1}{6}$ of a mole, this solution is $\frac{1}{6}(0.1)M$, or 0.01667*M*.

---

5. The **molality,** *m,* of a solution is the number of moles of solute per 1000 g of solvent. Thus, a 1*m* aqueous solution of sulfuric acid can be prepared by adding 1 mol (98.08 g) of $H_2SO_4$ to 1000 g of water. The final volume of a molal solution is of no importance, and 1*m* aqueous solutions of different solutes, each containing 1000 g of water, will have different volumes. However, these solutions all will have the same mole fractions of solute and solvent (see Example 9.5).

The molality of a given solution does not vary with temperature since the solution is prepared on the basis of the masses of the components; mass does not vary with temperature changes. The *molality* of a *very dilute* aqueous solution is approximately the same as the *molarity* of the solution since 1000 g of water occupies approximately 1000 ml.

---

*Example 9.5*

What are the mole fractions of solute and solvent in a 1.0000*m* aqueous solution?

*Solution*

The molecular weight of $H_2O$ is 18.015. Hence,

$$? \text{ mol } H_2O = 1000 \text{ g } H_2O \left( \frac{1 \text{ mol } H_2O}{18.015 \text{ g } H_2O} \right)$$
$$= 55.51 \text{ mol } H_2O$$

The total number of moles in a solution containing 1 mol of solute and 1000 g of $H_2O$ is, therefore, $1 + 55.51 = 56.51$. Hence,

$$X_{H_2O} = \frac{55.51}{56.51} = 0.9823$$

$$X_{solute} = \frac{1.000}{56.51} = 0.0177$$

This last result could have been obtained by subtracting $X_{H_2O}$ from 1.0000.

*Example 9.6*

What is the molality of a $0.5000M$ aqueous solution of sucrose ($C_{12}H_{22}O_{11}$) if the density of the solution is 1.0638 g/ml?

*Solution*

The molecular weight of $C_{12}H_{22}O_{11}$ is 342.30. One liter of the solution weighs 1063.8 g and contains $342.30/2 = 171.15$ g $C_{12}H_{22}O_{11}$ and $(1063.8 - 171.2) = 892.6$ g $H_2O$.

$$? \text{ mol } C_{12}H_{22}O_{11} = 1000 \text{ g } H_2O \left( \frac{0.5 \text{ mol } C_{12}H_{22}O_{11}}{892.6 \text{ g } H_2O} \right)$$

$$= 0.5601 \text{ mol } C_{12}H_{22}O_{11}$$

The solution is therefore $0.5601m$.

---

**9.7 VOLUMETRIC ANALYSIS**

The concentration of a given solution is constant from sample to sample no matter how much of the solution is taken, but the actual quantity of solute in a given sample is a different matter. Molar and normal concentrations state the amount of solute present in a liter of the solution. To determine the number of moles or the number of equivalents in a sample of a molar or normal solution, one must multiply the concentration of the solution by the volume, in liters, of the sample. Thus, a $6M$ solution contains, by definition, 6 mol of solute in 1 liter of solution; 500 ml of this solution (0.5 liter) would contain 0.5(6), or 3, mol of solute; 100 ml of this solution (0.1 liter) would contain 0.1(6), or 0.6, mol of solute.

For a sample of a solution of volume $V$ (*in liters*) and of molarity $M$ or normality $N$:

$V \times M = $ number of moles of solute in sample
$V \times N = $ number of equivalents of solute in sample

Calculations for reactions that are run between substances in solution are simplified if the concentrations of the solutions are expressed in normalities. One equivalent weight of a substance (1) will react with exactly 1 equivalent weight of another (2). Hence, $n$ equivalents of 1 will react with exactly $n$ equivalents of 2.

$$n_1 = n_2$$

But the number of equivalents of 1 in a sample of a solution of 1 is equal to the volume of the sample, in liters, times the normality of the solution.

$$n_1 = V_1 N_1 \quad \text{(volume in liters)}$$

Likewise,

$$n_2 = V_2 N_2 \quad \text{(volume in liters)}$$

We can therefore derive the relationship

$$V_1 N_1 = V_2 N_2$$

Since a volume term appears on both sides of the equality, any unit can be used to express $V_1$ and $V_2$ provided that both volumes are expressed in the same unit. This equation may be employed for calculations involving reactions between samples of two solutions.

Under some circumstances, another equation is useful. The number of equivalents in a sample is equal to the weight of the sample ($g$) divided by the equivalent weight ($eq$) of the substance. Thus,

$$n_2 = \frac{g_2}{eq_2}$$

We can use this expression together with $n_1 = V_1 N_1$ to derive

$$V_1 N_1 = \frac{g_2}{eq_2} \quad \text{(volume in liters)}$$

If the volume is expressed in milliliters,

$$V_1 N_1 = \frac{1000 g_2}{eq_2} \quad \text{(volume in milliliters)}$$

This expression is convenient for calculations for reactions between a weighed substance (2) and another (1) in solution. If sample 2 is impure, the actual weight of the reactive part of the sample is equal to the fractional purity ($p_2$) times the total weight of the sample ($g_2$); thus,

$$V_1 N_1 = \frac{1000 p_2 g_2}{eq_2}$$

Figure 9.4 **Buret.**

Solutions of known concentration are called **standard solutions,** and analytical procedures that make use of such solutions are called **volumetric analyses.** For example, a standard solution of an alkali may be used to determine the concentration of an acid solution in a process known as a **titration.** The standard solution of alkali is placed in a **buret** (Figure 9.4), which is a graduated tube fitted with a stopcock at the lower end to permit the solution to be withdrawn in controlled amounts. A carefully measured volume of the unknown acid solution is placed in a flask and a few drops of a colored substance known as an **indicator** are added. The standard alkali solution is gradually added to the flask from the buret until the indicator changes color; throughout the addition the contents of the flask are kept well mixed by swirling. At the **end point,** shown by the color change of the indicator, we assume that all the acid has been neutralized, and equivalent amounts of acid and alkali have been used. The volume of the standard alkali used is read from the buret, and the concentration of the acid is calculated as illustrated in Example 9.7.

**Example 9.7**

What is the normality of an acid solution if 50.00 ml of the solution requires 37.52 ml of 0.1429 $N$ alkali for neutralization?

*Solution*

$$V_1 N_1 = V_2 N_2$$

$$(50.00 \text{ ml})x = (37.52 \text{ ml})(0.1429 N)$$

$$x = 0.1702 N$$

**Example 9.8**

A 0.2676 g sample of a pure, solid acid dissolved in 100 ml of water requires 35.68 ml of 0.1190 $N$ alkali for neutralization. What is the equivalent weight of the acid? What is the molecular weight of the acid if it is diprotic?

*Solution*

$$V_1 N_1 = \frac{g_2}{eq_2} \quad \text{(volume in liters)}$$

$$(0.03568)(0.1190) = \frac{0.2676}{x}$$

$$x = 63.03$$

Since the acid is diprotic, the equivalent weight is $\frac{1}{2}$ the molecular weight, and the molecular weight must be 126.06.

**Example 9.9**

A 0.4308 g sample of iron ore is dissolved in acid and the iron converted into the iron(II) state. This solution is titrated against permanganate; the titration employs 32.31 ml of 0.1248 $N$ KMnO$_4$ which oxidizes Fe$^{2+}$ to Fe$^{3+}$. What is the percentage of iron in the ore?

*Solution*

Since the oxidation number of iron increases by 1 unit, the equivalent weight of iron is the same as the atomic weight, 55.85.

$$V_1 N_1 = \frac{p_2 g_2}{eq_2} \quad \text{(volume in liters)}$$

$$(0.03231)(0.1248) = \frac{x(0.4308)}{55.85}$$

$$x = 0.5228$$

The ore is 52.28% iron.

**9.8 VAPOR PRESSURE OF SOLUTIONS**

The vapor pressure of any solution ($P_{total}$) is the sum of the partial pressure of the components ($p_A$, $p_B$, and so on). Thus, for a binary solution

$$P_{total} = p_A + p_B$$

For **ideal solutions** the partial pressure of each component can be calculated from the mole fraction of the component present in the solution ($X_A$) and the vapor pressure of the pure component at the temperature in question ($p_A^\circ$). Thus,

$$p_A = X_A p_A^\circ \qquad p_B = X_B p_B^\circ$$

$$P_{total} = X_A p_A^\circ + X_B p_B^\circ$$

**Figure 9.5** Typical total and partial pressure curves for solutions that follow Raoult's law.

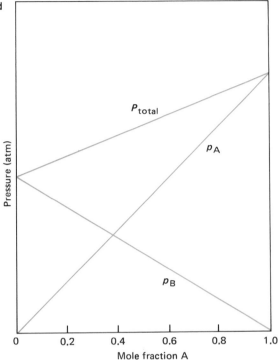

In other words, ideal solutions have properties that are the weighted averages of the properties of the components. This relationship was discovered by François Raoult in 1886 and is called **Raoult's law.**

Figure 9.5 illustrates partial pressure curves for components A and B of ideal binary solutions as well as the total pressure curve, which is the sum of the two partial pressure curves. A and B represent liquids that are miscible in all proportions. In the figure, pressures are plotted versus mole fraction of A; since the mole fractions of A and B must add up to 1.0, the mole fraction of B is easily derived from the scale of the x axis.

Not all solutions are ideal. The vapor pressures of some solutions show negative deviations from Raoult's law (Figure 9.6); other nonideal solutions show positive deviations (Figure 9.7). The colored lines of Figures 9.6 and 9.7 are the partial pressure and total pressure curves calculated by means of Raoult's law.

In ideal solutions the intermolecular attractions between the components (A–B attractions) are of the same magnitude as the intermolecular attractions found in the pure components (A–A attractions and B–B attractions). Consequently, no heat effect is observed when an

**Figure 9.6** Typical total and partial vapor pressure curves for solutions that show negative deviations from Raoult's law. (Colored lines are based on Raoult's law.)

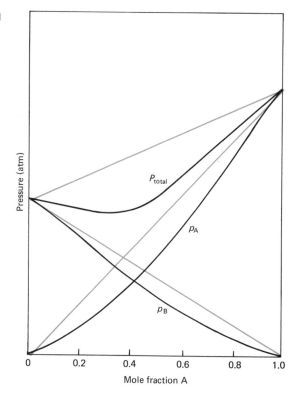

**Figure 9.6** Typical total and partial vapor pressure curves for solutions that show negative deviations from Raoult's law. (Colored lines are based on Raoult's law.)

ideal solution is prepared. Negative deviations are observed when the intermolecular attractions of the solution (A–B) are stronger than those of either of the pure components (A–A and B–B); hence, the escaping tendencies of the components are reduced when they are mixed. Such solutions generally have exothermic heats of solution. Positive deviations result when the pure components have stronger intermolecular attractions (A–A and B–B) than the mixture (A–B); heat is absorbed when these solutions are prepared.

Even though a pair of components does not form ideal solutions over the entire range of concentrations, the behavior of the *solvent* of most *dilute* solutions approaches ideality. In Figures 9.6 and 9.7 the extreme right portion of the $p_A$ curve (where $X_A$ approaches 1.0) and the extreme left portion of the $p_B$ curve (where $X_B$ approaches 1.0) are close to the curves calculated by means of Raoult's law.

Hence, dilute solutions of *nonvolatile* molecular solutes generally follow Raoult's law. When the solute has a very low vapor pressure, the vapor pressure of the solution is almost entirely due to the solvent. If component A is the solvent and B is the solute,

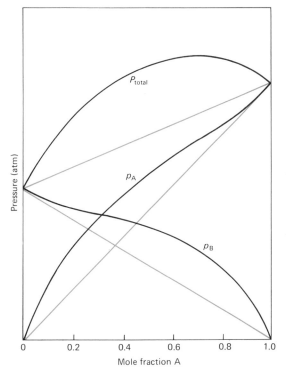

**Figure 9.7** Typical total and partial vapor pressure curves for solutions that show positive deviations from Raoult's law. (Colored lines are based on Raoult's law.)

$$P_{total} = X_A p_A^\circ$$

or

$$P_{total} = (1 - X_B) p_A^\circ$$

Therefore, the vapor pressure of a solution prepared from 1 mol of a nonvolatile, nonionizing solute and 99 mol of solvent would be 99% of the vapor pressure of the pure solvent at the same temperature. The escape of solvent molecules to the vapor phase would be reduced because only 99% of the surface molecules in the solution are solvent molecules; 1% of the surface molecules are, by definition, nonvolatile.

Raoult's law is only strictly applicable to this type of solution when the solution is dilute and solvent–solute interactions are unimportant. Under this condition, the particular nonvolatile solute does not matter. It is the number of dissolved solute particles per given quantity of solvent that determines the vapor–pressure lowering—not their size, shape, or weight. The properties of solutions containing nonvolatile solutes that are in ionic form are considered in Section 9.12.

### Example 9.10

If heptane and octane form ideal solutions, what is the vapor pressure at 40°C of a solution containing 1.0 mol of heptane and 4.0 mol of octane? At 40°C the vapor pressure of heptane is 0.121 atm, and the vapor pressure of octane is 0.041 atm.

*Solution*

$$X_{\text{heptane}} = \tfrac{1}{5} \quad \text{and} \quad X_{\text{octane}} = \tfrac{4}{5}$$

Hence,

$$P_{\text{total}} = \tfrac{1}{5}(0.121 \text{ atm}) + \tfrac{4}{5}(0.041 \text{ atm})$$
$$= 0.057 \text{ atm}$$

### Example 9.11

Assuming ideality, calculate the vapor pressure of a 1.00$m$ solution of a nonvolatile molecular solute in water at 50°C. The vapor pressure of water at 50°C is 0.122 atm.

*Solution*

The mole fraction of water in a 1.00$m$ solution is 0.982 (from Example 9.5). Hence, the vapor pressure of a 1.00$m$ solution at 50°C is

$$0.982(0.122 \text{ atm}) = 0.120 \text{ atm}$$

### Example 9.12

A solution is prepared from 2.00 g of a nonvolatile solute and 90.0 g of water. The vapor pressure of the solution at 60°C is 0.194 atm. According to Raoult's law, what is the approximate molecular weight of the solute? The vapor pressure of pure water at 60°C is 0.196 atm.

*Solution*

$$P_{\text{total}} = X_A p_A^{\circ}$$
$$0.194 \text{ atm} = X_A(0.196 \text{ atm})$$
$$X_A = 0.990$$

The amount of water used in the preparation of the solution (90.0 g) is 5.00 mol. Therefore,

$$X_A = \frac{n_A}{n_A + n_B}$$

$$0.990 = \frac{5.00}{5.00 + n_B}$$

$$n_B = 0.0505 \text{ mol}$$

Thus, the 2.00 g of solute used in the solution is 0.0505 mol.

$$? \text{ g} = 1.00 \text{ mol} \left(\frac{2.00 \text{ g}}{0.0505 \text{ mol}}\right) = 39.6 \text{ g}$$

---

**9.9 BOILING POINT AND FREEZING POINT OF SOLUTIONS**

The effect of a dissolved nonvolatile solute on the vapor pressure of a liquid is difficult to measure, but the consequences of this vapor pressure depression – elevation of the boiling point and depression of the freezing point – are relatively easily determined with considerable accuracy.

The boiling point of a liquid is defined as the temperature at which the vapor pressure of the liquid is equal to the prevailing atmospheric pressure; boiling points measured under 1 atmosphere pressure are termed normal boiling points (Section 7.4.) Since the addition of a nonvolatile solute decreases the vapor pressure of a liquid, a solution will not boil at the normal boiling point of the solvent. It is necessary to increase the temperature above this point in order to attain a vapor pressure over the solution of 1 atmosphere. Hence, the boiling point of a solution contain-

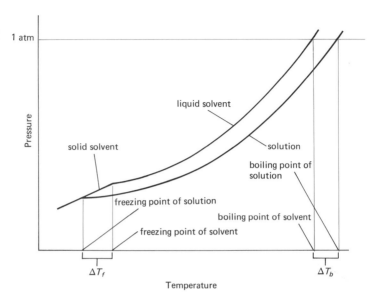

**Figure 9.8** Vapor pressure curves of a pure solvent and a solution of a nonvolatile solute (not drawn to scale).

ing a nonvolatile molecular solute is higher than that of the pure solvent, and the elevation is proportional to the concentration of solute in the solution.

This effect is illustrated by the vapor pressure curves plotted in Figure 9.8. The extent to which the vapor pressure curve of the solution lies below the vapor pressure curve of the solvent is proportional to the mole fraction of solute in the solution. The elevation of the boiling point, $\Delta T_b$, reflects this displacement of the vapor pressure curve; for a given solvent the elevation is a constant for all solutions of the same concentration.

Concentrations are customarily expressed in molalities, rather than mole fractions, for problems involving boiling-point elevations. For example, the boiling point of a $1\,m$ aqueous solution is $0.512°C$ higher than the boiling point of water. Table 9.3 lists molal boiling-point eleva-

**Table 9.3** Molal boiling-point elevation and freezing-point depression constants

| SOLVENT | BOILING POINT (°C) | $K_b$ (°/m) | FREEZING POINT (°C) | $K_f$ (°/m) |
|---|---|---|---|---|
| acetic acid | 118.1 | 3.07 | 16.6 | −3.90 |
| benzene | 80.1 | 2.53 | 5.5 | −5.12 |
| camphor | − | − | 179. | −39.7 |
| carbon tetrachloride | 76.8 | 5.02 | −22.8 | −29.8 |
| chloroform | 61.2 | 3.63 | −63.5 | −4.68 |
| ethyl alcohol | 78.4 | 1.22 | −114.6 | −1.99 |
| naphthalene | − | − | 80.2 | −6.80 |
| water | 100.0 | 0.512 | 0.0 | −1.86 |

tion constants for several solvents. The boiling point of a $0.5m$ solution would be expected to be elevated by an amount equal to $\frac{1}{2}$ the molal constant. Thus, the boiling-point elevation, $\Delta T_b$, of a solution can be calculated by multiplying the molal boiling-point elevation constant of the solvent, $K_b$, by the molality of the solution, $m$.

$$\Delta T_b = mK_b$$

In reality, this relationship is only approximate; a more exact statement would require that the concentration be expressed in mole fraction of solute, not in molality. However, molalities of **dilute** solutions are proportional (at least with sufficient accuracy) to mole fractions of solute, and since Raoult's law only describes the behavior of most real solutions satisfactorily if they are dilute, the use of molalities in these calculations is justified.

The mole fraction of solute in a solution, $X_B$, is

$$X_B = \frac{n_B}{n_A + n_B}$$

where $n_B$ is the number of moles of solute, and $n_A$ is the number of moles of solvent. If the solution is dilute, the number of moles of solute is negligible in comparison with the number of moles of solvent, and the denominator may be simplified by dropping the term $n_B$.

$$X_B \approx \frac{n_B}{n_A}$$

The molality of the solution, $m$, is the number of moles of solute per 1000 g of solvent. If the molecular weight of the solvent is $M_A$, $n_A = 1000/M_A$, and

$$X_B \approx \frac{m}{1000/M_A} \approx \frac{M_A}{1000}m$$

Hence, there is a direct relation between molality and mole fraction of solute for dilute solutions.

At the freezing point the vapor pressure of solid and liquid are equal (Section 7.7). In Figure 9.8 the vapor pressure curves of the liquid solvent and the solid solvent intersect at the freezing point of the solvent. At this temperature, however, the vapor pressure of the solution is lower than the equilibrium vapor pressure of the pure solvent. The vapor pressure curve of the solution intersects the vapor pressure curve of the solid solvent at a lower temperature, and hence the freezing point of the solution is lower than that of the pure solvent. As in the case of boiling-point elevations, freezing-point depressions depend upon the concentration of the solution and the solvent employed. Molal freezing-point depression constants for some solvents are listed in Table 9.3. The freezing-point depression, $\Delta T_f$, of a solution can be calculated from the molality of the solution and the constant for the solvent $K_f$.

$$\Delta T_f = mK_f$$

This statement assumes that the solute does not form a solid solution with the solvent; if this is not the case, the relationship is not valid.

The following problems illustrate the use of these relations — for boiling-point elevation and freezing-point depression — in calculations of the boiling points and freezing points of solutions and, more importantly, the molecular weights of solutes.

---

**Example 9.13**

What is the boiling point and freezing point of a solution prepared by dissolving 2.40 g of biphenyl ($C_{12}H_{10}$) in 75.0 g of benzene? The molecular weight of biphenyl is 154.

*Solution*

The molality of the solution is

$$? \text{ mol } C_{12}H_{10} = 1000 \text{ g benzene} \left(\frac{2.40 \text{ g } C_{12}H_{10}}{75.0 \text{ g benzene}}\right)\left(\frac{1 \text{ mol } C_{12}H_{10}}{154 \text{ g } C_{12}H_{10}}\right)$$

$$= 0.208 \text{ mol } C_{12}H_{10}$$

The boiling-point elevation is

$$? \,°C = 0.208m \; C_{12}H_{10}\left(\frac{2.53\,°C}{1\,m\,C_{12}H_{10}}\right) = 0.526\,°C$$

The boiling point of the solution is

$$80.1\,°C + 0.5\,°C = 80.6\,°C$$

The freezing-point depression is

$$? \,°C = 0.208m \; C_{12}H_{10}\left(\frac{-5.12\,°C}{1\,m\,C_{12}H_{10}}\right) = -1.06\,°C$$

The freezing point of the solution is

$$5.5\,°C - 1.1\,°C = 4.4\,°C$$

---

**Example 9.14**

A solution prepared from 0.300 g of an unknown nonvolatile solute in 30.0 g carbon tetrachloride has a boiling point 0.392°C higher than that of pure $CCl_4$. What is the molecular weight of the solute?

*Solution*

$K_b$ for $CCl_4$ is 5.02°C, and the molality of the solution is therefore

$$?m = 0.392\,°C\left(\frac{1\,m}{5.02\,°C}\right) = 0.0781\,m$$

Thus, 1000 g of $CCl_4$ would contain 0.0781 mol of solute. The molecular weight of the solute is therefore

$$? \text{ g} = 1 \text{ mol solute} \left(\frac{1000 \text{ g } CCl_4}{0.0781 \text{ mol solute}}\right)\left(\frac{0.300 \text{ g solute}}{30.0 \text{ g } CCl_4}\right)$$

$$= 128 \text{ g}$$

**9.10 OSMOSIS**  The properties of solutions that depend principally upon the concentration of dissolved particles, rather than upon the nature of these particles, are called **colligative properties.** For solutions of nonvolatile solutes these properties include: vapor–pressure lowering, freezing-point depression, boiling-point elevation, and osmotic pressure. The last of these, osmotic pressure, is the topic of this section.

A membrane, such as cellophane or parchment, that permits some molecules, but not all, to pass through it is called a **semipermeable membrane.** Figure 9.9 shows a membrane that is permeable to water but not to sucrose (cane sugar) placed between pure water and a sugar solution. Water molecules, but not sugar molecules, can go through the membrane in either direction. However, there are more water molecules per unit volume on the left side (the side containing pure water) than on the right; therefore, the rate of passage through the membrane from left to right exceeds the rate in the opposite direction.

As a result, the number of water molecules on the right increases, the sugar solution becomes more dilute, and the height of the solution in the right arm of the U tube increases. This process is called **osmosis.** The difference in height between the levels of the liquids in the two arms of the U tube is a measure of the **osmotic pressure.**

The increased hydrostatic pressure on the right tends to force water molecules through the membrane from right to left, so that ultimately the rate of passage to the left equals the rate to the right. The final condition, therefore, is an equilibrium state with equal rates of passage of water molecules through the membrane in both directions. If an increased pressure that is in excess of the equilibrium, osmotic pressure is applied to the solution in the right arm, it is possible to force water in a direction contrary to that normally observed. This process, called **reverse osmosis,** is used to secure pure water from salt water.

Figure 9.9 Osmosis.

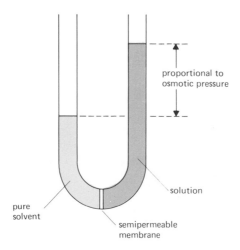

proportional to
osmotic pressure

solution

pure
solvent

semipermeable
membrane

Similarities exist between the behavior of water molecules in osmosis and the behavior of gas molecules in diffusion. In both processes molecules diffuse from regions of high concentrations to regions of low concentration. In 1887 Jacobus van't Hoff discovered the relation

$$\pi V = nRT$$

where $\pi$ is the osmotic pressure (in atm), $n$ is the number of moles of solute dissolved in volume $V$ (in liters), $T$ is the absolute temperature, and $R$ is the gas constant (0.08206 liter atm/°K mol). The similarity between this equation and the equation of state for an ideal gas is striking. The equation may be written in the form

$$\pi = \left(\frac{n}{V}\right) RT$$

$$\pi = MRT$$

where $M$ is the molarity of the solution. Osmosis plays an important role in plant and animal physiological processes; the passage of substances through the semipermeable walls of a living cell, the action of the kidneys, and the rise of sap in trees are prime examples.

**9.11 DISTILLATION** A solution of a nonvolatile solute can be separated into its components by **simple distillation.** This procedure consists of boiling away the volatile solvent from the solute. The solvent is collected by condensing the vapor; the solute is the residue that remains after the distillation.

A solution of two volatile components that follows Raoult's law (Figure 9.5) can be separated into its components by a process known as **fractional distillation.** According to Raoult's law, each component contributes to the vapor pressure of the solution in proportion to its mole fraction times its vapor pressure in the pure state. Let us assume that at the temperature of the experiment component A has a vapor pressure of 0.200 atm and component B has a vapor pressure of 0.400 atm. For a solution with a mole fraction of B equal to 0.60, the total vapor pressure is

$$
\begin{aligned}
P_{total} &= X_A p_A^\circ + X_B p_B^\circ \\
&= 0.40(0.200 \text{ atm}) + 0.60(0.400 \text{ atm}) \\
&= 0.080 \text{ atm} + 0.240 \text{ atm} \\
&= 0.320 \text{ atm}
\end{aligned}
$$

The composition of the *vapor* in equilibrium with this solution can be calculated by comparing the partial pressure of each component with the total vapor pressure of the solution (Section 6.10). Thus, in the vapor

$$X_{A,\ vapor} = \frac{0.080 \text{ atm}}{0.320 \text{ atm}} = 0.25 \qquad X_{B,\ vapor} = \frac{0.240 \text{ atm}}{0.320 \text{ atm}} = 0.75$$

Hence, a solution of concentration $X_B = 0.60$ would be in equilibrium with a vapor of concentration $X_{B, \text{ vapor}} = 0.75$. For ideal solutions the vapor is always richer than the liquid in the more volatile component (which in this instance is B—it has the higher vapor pressure).

In a distillation of a solution of A and B, the vapor that comes off and condenses is richer in B than the liquid remaining behind. The actual compositions of vapor and liquid change as the distillation proceeds, but at any given time this generalization is true. By collecting the condensed vapor in several fractions and subjecting these fractions to repeated distillations, the components of the original mixture can eventually be obtained in substantially pure form.

For systems that deviate from Raoult's law, the situation is somewhat different. If the deviation is positive (Figure 9.7), there is a maximum in the total vapor pressure curve corresponding to a solution, of definite composition, that has a vapor pressure higher than either of the pure components. This means that such a solution, called a **minimum boiling azeotrope,** will boil at a lower temperature than either of the two pure components. Ethyl alcohol and water form a minimum boiling azeotrope that contains 4.0 weight percent water and has a normal boiling point of 78.17°C (ethyl alcohol and water boil at 78.3°C and 100°C, respectively).

If a system shows a negative deviation from Raoult's law (Figure 9.6), there will be a minimum in the $P_{\text{total}}$ curve. The solution that has a concentration corresponding to this minimum will have a vapor pressure, at any given temperature, lower than either pure component. Thus, this solution boils at a temperature *higher* than either pure component and is called a **maximum boiling azeotrope.** Hydrochloric acid and water form such an azeotrope containing 20.22 weight percent HCl and boiling at 108.6°C (pure HCl has a boiling point of −80°C).

The vapor in equilibrium with a maximum or minimum boiling azeotrope has the same concentration as the liquid; hence, azeotropes, like pure substances, distill without change. Fractional distillation of a solution containing two components that form an azeotrope will eventually produce one pure component and the azeotrope but not both pure components.

**9.12 SOLUTIONS OF ELECTROLYTES**

Solutes may be classified according to the ability of their aqueous solutions to conduct electricity (Section 10.2). The electrical conductivity of a solution depends upon the presence of ions; pure water, itself, is very slightly ionized and is a poor conductor (Section 16.2). An **electrolyte** is a solute of a solution that is a better electrical conductor than the parent solvent alone; in solution a solute of this type exists (in whole or in part) as ions. Covalent solutes that are exclusively molecular in solution and hence do nothing to enhance the conductivity of the solvent are called **nonelectrolytes.**

Electrolytes may be further divided into two groups: **strong electrolytes** and **weak electrolytes.** The conductivity of a $1m$ solution of a weak electrolyte is considerably lower than the conductivity of a $1m$ solution of a strong electrolyte. Weak electrolytes are polar covalent materials that are incompletely dissociated into ions in solution; strong electrolytes are essentially completely ionic in solution.

In 1887 Svanté Arrhenius proposed his "chemical theory of electrolytes." In addition to data from electrical conductivity experiments, a principal source of evidence advanced by Arrhenius in support of his theory was derived from studies of the deviations from Raoult's law exhibited by solutions of electrolytes. The freezing points of *very dilute* solutions of electrolytes are lowered more than those of ideal solutions of nonelectrolytes with corresponding concentrations. The freezing-point depression for a dilute solution of a salt such as NaCl or $AgNO_3$ is approximately twice that of a solution of a nonelectrolyte with a similar molal concentration; the depression for a dilute solution of a salt such as $K_2SO_4$ or $BaCl_2$ is approximately three times as great as "expected" (see Table 9.4).

Table 9.4 Observed freezing-point depressions for some aqueous solutions compared to calculated depressions[a]

| | CONCENTRATION OF SOLUTION | | |
| SOLUTE | $0.001m$ | $0.01m$ | $0.1m$ |
| --- | --- | --- | --- |
| nonelectrolyte | 0.00186°C | 0.0186°C | 0.186°C |
| sucrose | 0.00186 | 0.0186 | 0.188 |
| 2 ions/formula | 0.00372 | 0.0372 | 0.372 |
| NaCl | 0.00366 | 0.0360 | 0.348 |
| 3 ions/formula | 0.00558 | 0.0558 | 0.558 |
| $K_2SO_4$ | 0.00528 | 0.0501 | 0.432 |
| 4 ions/formula | 0.00744 | 0.0744 | 0.744 |
| $K_3[Fe(CN)_6]$ | 0.00710 | 0.0626 | 0.530 |

[a] Calculated on the assumptions that $K_f$ for water is 1.86°C/molal over the entire range of concentrations, the salts are 100% ionic in solution, and the ions act independently of one another in their effect on the freezing point of the solution.

Arrhenius explained these results on the supposition that electrolytes are dissociated into ions in solution—NaCl and $AgNO_3$ form two ions ($Na^+$, $Cl^-$ and $Ag^+$, $NO_3^-$) per "molecule," whereas $K_2SO_4$ and $BaCl_2$ form three ions ($2K^+$, $SO_4^{2+}$ and $Ba^{2+}$, $2Cl^-$) per "molecule." Today we know that these salts are ionic even when pure; "molecules" of these materials do not exist. Substances of this type are, therefore, completely ionic in solution and are strong electrolytes. Similar deviations are observed for boiling-point elevations.

Theoretically the net effect on the freezing-point or boiling-point depends only on the number of nonvolatile particles in solution and not on their nature. The question, then, is not why the deviations for these salt solutions are larger than those expected for solutions of nondissociating solutes; it is, rather, why the deviations are not *exactly* twice the "expected" in the case of solutions of salts containing two ions per formula unit or three times the "expected" in the case of solutions of salts containing three ions per formula unit. This will be the topic of Section 9.13.

The hydroxides of the group I A metals are ionic crystalline materials, are water soluble, and are strong electrolytes; $Ba(OH)_2$, $Sr(OH)_2$, and $Ca(OH)_2$ are only moderately soluble in water,[2] but they too are strong electrolytes.

We have previously noted that certain covalent substances are ionized in water solution—a process more in harmony with Arrhenius' original views. Perchloric acid is completely ionic in water solution; it is virtually impossible to detect undissociated $HClO_4$ molecules.

$$H_2O + HClO_4 \rightarrow H_3O^+ + ClO_4^-$$

Other acids, though covalent when pure, are strong electrolytes in aqueous solution; the common ones that may be considered ionic in solutions of $1M$ concentration or less are $HClO_4$, $HCl$, $HBr$, $HI$, $HNO_3$, and $H_2SO_4$.[3]

Some covalent materials may be ionized to a lesser extent in water and are classed as weak electrolytes—for example, acetic acid, $HC_2H_3O_2$.

$$H_2O + HC_2H_3O_2 \rightleftharpoons H_3O^+ + C_2H_3O_2^-$$

A $0.01m$ solution of acetic acid is approximately 4% ionic at room temperature; at any given time, 96% of the acetic acid present in the solution is in molecular form. This process is reversible and therefore a double arrow is shown in the preceding equation. In the solution, molecules are constantly dissociating into ions at the same rate that ions, upon interionic contact, are re-forming molecules. Thus, this reversible system is in a state of chemical equilibrium; such equilibria are the topic of Chapter 16. Most acids are weak electrolytes.

---

[2] There is some evidence that in nearly saturated solutions of these compounds dissociation may be incomplete. Thus, fairly large concentrations of the ion $CaOH^+$, as well as $Ca^{2+}$ and $OH^-$, have been identified in such solutions, presumably from the dissociations: $Ca(OH)_2 \rightarrow CaOH^+ + OH^-$ and $CaOH^+ \rightleftharpoons Ca^{2+} + OH^-$.
[3] The first ionization only: $H_2O + H_2SO_4 \rightarrow H_3O^+ + HSO_4^-$.

Certain covalent compounds, of which ammonia is the most common example, react with water to produce hydroxide ions.

$$NH_3 + H_2O \rightleftharpoons NH_4^+ + OH^-$$

Thus, it is possible to have alkaline solutions of weak electrolytes.

Most salts are ionic compounds; these substances, if soluble, are strong electrolytes in aqueous solution. A number of covalent compounds containing metals and nonmetals (e.g., aluminum chloride, Section 9.2) are ionized almost completely in water solution and are classed as strong electrolytes. A small number of covalent compounds of metals, however, dissolve in water but are not ionized completely; they are weak electrolytes. Thus, mercuric chloride solutions contain $HgCl_2$ molecules in addition to $HgCl^+$, $Cl^-$, low concentrations of $Hg^{2+}$, and other more complicated species.

**9.13 INTERIONIC ATTRACTIONS IN SOLUTION**

The **van't Hoff factor,** $i$, may be defined as the ratio of the observed freezing-point depression for a solution ($\Delta T_f$) to the depression calculated from the molality of the solution and the molal freezing-point constant for the solvent ($mK_f$).

$$i = \frac{\Delta T_f}{mK_f}$$

Since a similar relation is used for boiling-point elevations,

$$i = \frac{\Delta T_b}{mK_b}$$

we shall use unspecified $\Delta T$ and $K$ in the discussion that follows.

If we rearrange the preceding expressions to

$$\Delta T = imK$$

the significance of the $i$ factor, which was named for Jacobus van't Hoff who first employed it, becomes clear. The expression of Section 9.9, $\Delta T = mK$, was derived on the assumption that the solute *does not dissociate* ($i = 1$).

When dissociation of the solute occurs, it is necessary to correct the molality of the solution for freezing-point and boiling-point calculations. Thus if AB is a strong electrolyte and therefore dissociates completely in solution, in a $0.001m$ solution of AB there would be 0.001 mol of $A^+$ and 0.001 mol of $B^-$ contained in 1000 g of water. On the assumption that each ion acts independently in its effect on the boiling point and freezing point, the effective concentration of this hypothetical solution would be $0.002m$. Hence, in this simplified example, $i = 2$, $m = 0.001$, and

$$\Delta T = 2(0.001 \ m)K$$

Inspection of the $i$ values recorded in Table 9.5 (which were derived from freezing-point determinations) reveals that they do not exactly equal the number of ions per formula unit for each of the strong electrolytes listed. Thus for $0.001m$ solutions, the $i$ value of NaCl is 1.97 (not 2); that of $K_2SO_4$, 2.84 (not 3); and that of $K_3[Fe(CN)_6]$, 3.82 (not 4). Furthermore, the $i$ value changes with concentration of the solution and approaches the value expected for complete dissociation as the solution becomes more and more dilute.

Table 9.5 Van't Hoff factor, $i$, for various strong electrolytes in solution[a]

| ELECTROLYTE | CONCENTRATION OF SOLUTION | | |
| --- | --- | --- | --- |
| | $0.001m$ | $0.01m$ | $0.1m$ |
| NaCl | 1.97 | 1.94 | 1.87 |
| $MgSO_4$ | 1.82 | 1.53 | 1.21 |
| $K_2SO_4$ | 2.84 | 2.69 | 2.32 |
| $K_3[Fe(CN)_6]$ | 3.82 | 3.36 | 2.85 |

[a] From freezing-point determinations.

In 1923 Peter Debye and Erich Hückel proposed a quantitative theory to explain the behavior of dilute solutions of electrolytes.[4] According to the **Debye-Hückel theory,** interionic attractions occur in solutions of electrolytes so that the ions are not completely independent of one another as in the case for uncharged molecules in solution. (As has been noted in Section 9.8, solute–solute interactions cause deviations from Raoult's law.) Thus, the electrical forces that operate between the oppositely charged ions reduce the effectiveness of these ions.

The $i$ values in Table 9.5 show that the interionic attractions in solutions of $MgSO_4$ produce a stronger effect than those in solutions of NaCl of corresponding concentration even though both solutes contain two moles of ions per mole of compound; thus for $0.001m$ NaCl, $i = 1.97$, whereas for $0.001m$ $MgSO_4$, $i = 1.82$. Both the ions of magnesium sulfate are doubly charged ($Mg^{2+}$, $SO_4^{2-}$), whereas the ions of sodium chloride are singly charged ($Na^+$, $Cl^-$); hence, the interionic attractions are stronger in solutions of magnesium sulfate.

The effect of dilution upon $i$ values can also be explained on the basis of these interionic attractions. As a solution is diluted, the ions spread farther and farther apart, their influence upon one another diminishes accordingly, and the $i$ factor approaches its limiting value. Hence, in

[4] It should be noted that the Debye-Hückel theory accounts only for the behavior of dilute solutions of electrolytes. Concentrated solutions are imperfectly understood at present. Among the factors that have been postulated to explain the anomalous behavior of concentrated solutions are the lack of sufficient water to hydrate the ions and the presence of ion pairs and ion triplets (ions closely associated with one another).

*very* dilute solutions the Debye-Huckel effect is negligible, and the ions essentially behave independently.

Interionic attractions, therefore, make a solution behave as though its ion concentrations were less than they actually are. The **activity** of an ion, *a*, is defined as the "effective concentration" of that ion; it is related to the actual concentration of the ion, *c*, by an **activity coefficient, γ.**

$$a = \gamma c$$

The activity coefficient is always less than 1 and approaches unity as the concentration of the solution decreases. The calculation of activity coefficients from experimental data is complicated. It is important to recognize that the activity of a given ion depends upon the concentrations of all ions present in the solution whether they are all derived from the same compound or not, since all ions can enter into interionic attractions.

Some characteristics of solutes are summarized in Table 9.6.

Table 9.6 Characteristics of solutes

| SOLUTE | FORM OF SOLUTE IN $1\,m$ SOLUTION | VAN'T HOFF FACTOR FOR $\Delta T_f$; DILUTE SOLUTIONS OF MOLALITY, $m$ $\Delta T_f = iK_f m$ | EXAMPLES |
|---|---|---|---|
| nonelectrolytes | molecules | $i = 1$ | $C_{12}H_{22}O_{11}$ (sucrose) $CON_2H_4$ (urea) $C_3H_5(OH)_3$ (glycerol) |
| strong electrolytes | ions | $i \approx n^a$ | NaCl KOH |
| weak electrolytes | molecules and ions | $1 < i < n^a$ | $HC_2H_3O_2$ $NH_3$ $HgCl_2$ |

$^a$ $n$ = number of moles of ions per mole of solute.

## Problems

Thermochemical problems may be solved using either calories or joules as energy units. Thermochemical values expressed in joules appear in brackets.

**9.1** Define the following terms: heat of solution, heat of hydration, ideal solution, standard solution, activity of ions, Le Chatelier's principle, minimum-boiling azeotrope.

**9.2** Describe three ways in which water occurs in crystalline hydrates.

**9.3** Why is the heat of solution of $I_2$ in $CCl_4$ about the same as the heat of fusion of pure $I_2$? Why cannot a similar statement be made concerning the dissolution of ionic substances in water?

**9.4** Why is $Br_2$ more soluble in $CCl_4$ than $I_2$?

**9.5** The following data pertain to the preparation of very dilute aqueous solutions at 298°K. Heats of solution: $CuSO_4$, $-17.5$ kcal/mol [or $-73.2$ kJ/mol]; $Mg(NO_3)_2$, $-20.4$ kcal/mol [or $-85.4$ kJ/mol]; NaCl, $+0.93$ kcal/mol [or $+3.89$ kJ/mol]. Entropy changes ($\Delta S$): $CuSO_4$, $-46.6$ cal/°K mol [or $-195.0$ J/°K mol]; $Mg(NO_3)_2$, $+2.6$ cal/°K mol [or $+10.9$ J/°K mol]; NaCl, $+10.3$ cal/°K mol [or $+43.1$ J/°K mol]. Calculate the change in Gibbs free energy for the preparation of each solution and discuss the results.

**9.6** The heat of solution of NaCl at 298°K for the preparation of a very dilute aqueous solution is $+0.93$ kcal/mol [or $+3.9$ kJ/mol]. The lattice energy of NaCl is $-188$ kcal/mol [or $-787$ kJ/mol]. What is the heat of hydration of the ions of this compound? Describe the process(es) to which this value pertains.

**9.7** Which of each of the following pairs is the more soluble in water? (a) $CH_3OH$ or $CH_3CH_3$, (b) $CCl_4$ or NaCl, (c) $CH_3F$ or $CH_3Cl$.

**9.8** (a) The molecules in pure ethanol, $CH_3CH_2OH$, are extensively associated through hydrogen bonding; however, those in pure acetone, $CH_3-\overset{\displaystyle O}{\overset{\|}{C}}-CH_3$, and in pure chloroform, $CHCl_3$, are not. Explain why this is so. (b) On the basis of your answer to part (a), explain why the vapor pressures of chloroform and acetone solutions exhibit negative deviations from Raoult's law, whereas the vapor pressures of chloroform and ethanol solutions exhibit positive deviations.

**9.9** Use the solubility rules given in Table 9.2 to write balanced ionic equations for the reactions that occur when aqueous solutions of the following pairs are mixed. (a) sodium carbonate and strontium acetate, (b) tin(II) chloride and ammonium sulfate, (c) sodium sulfide and cadmium nitrate, (d) magnesium chloride and barium hydroxide, (e) iron(III) bromide and potassium phosphate.

**9.10** Use the solubility rules given in Table 9.2 to write balanced ionic equations for the reactions that occur when aqueous solutions of the following pairs are mixed. (a) lithium chlorate and aluminum chloride, (b) zinc sulfate and lead(II) nitrate, (c) strontium hydroxide and nickel(II) sulfate, (d) manganese(II) chloride and cobalt(II) sulfate, (e) copper(II) bromide and mercury(I) acetate.

**9.11** For aqueous solutions of $CO_2(g)$ at 10°C, the value of $K$ in Henry's law, $p = KX$, is $1.04 \times 10^3$ atm. In the equation, $p$ is the partial pressure of the gas over a saturated solution, and $X$ is the mole fraction of the gas in the solution. How many moles of $CO_2(g)$ will dissolve in 1.00 mol of $H_2O$ if the pressure of $CO_2(g)$ over the solution is (a) 1.00 atm, (b) 0.500 atm?

**9.12** Use the data of Problem 9.11 to determine the partial pressure of $CO_2(g)$ at 10°C over a saturated solution of $CO_2$ containing 0.130 mol of $CO_2$ and 100 mol of $H_2O$.

**9.13** For aqueous solutions of $N_2(g)$ at 20°C, the value of $K$ in Henry's law, $p = KX$, is $6.41 \times 10^4$ atm. In the equation, $p$ is the partial pressure of the gas over a saturated solution, and $X$ is the mole fraction of the dissolved gas in the solution. (a) How many grams of $N_2(g)$ dissolve in 100 g of water at 20°C if the pressure of $N_2$ over the solution is 1.00 atm? (b) What is the percent, by weight, of $N_2$ in the solution?

**9.14** For aqueous solutions of $He(g)$ at 20°C, the value of $K$ in Henry's law, $p = KX$, is $1.38 \times 10^5$ atm. In the equation, $p$ is the partial pressure of the gas over a saturated solution, and $X$ is the mole fraction of the dissolved gas in the solution. (a) How many grams of $He(g)$ dissolve in 100 g of water at 20°C if the pressure of He over the solution is 1.00 atm? (b) What is the percent, by weight, of He in the solution? (c) Compare your result to that of Problem 9.13 in the light of comments made in Problem 6.40.

**9.15** (a) Air is approximately 80.0% $N_2$ and 20.0% $O_2$ by volume. What are the partial pressures of $N_2$ and $O_2$ in air when the total pressure is 1.00 atm? (b) Henry's law may be stated as $p = KX$, where $p$ is the partial pressure of the gas over a saturated solution, $X$ is the mole fraction of the dissolved gas, and $K$ is a constant. At 0°C the values of $K$ for $N_2$ and $O_2$ are $5.38 \times 10^4$ atm and $2.51 \times 10^4$ atm, respectively. What are the mole fractions of $N_2$ and $O_2$ in a saturated aqueous solution at 0°C under air at a pressure of 1.00 atm? (c) What is the freezing point of the solution?

**9.16** The volume of an aqueous solution changes as the temperature changes. Thus, a concentration of a given solution determined at one temperature may not be correct for that same solution at another temperature. Which methods of expressing concentration yield results that are temperature-independent, and which yield results that are temperature-dependent?

**9.17** A given aqueous solution has a density of $x$ g/ml and is $y$% solute by weight. Derive a mathematical equation relating the molarity of the solution, $M$, to the molality of the solution $m$.

**9.18** (a) What is the molarity of a 0.2 $N$ solution of $H_3PO_4$ (calculated on the basis of neutralization to the $HPO_4^{2-}$ ion)? (b) What is the normality of a 0.1 $M$ $Ba(OH)_2$ solution (based on complete neutralization of the alkali)? (c) What is the molarity of 0.03 $N$ $K_2Cr_2O_7$ solution (calculated on the basis of a reaction in which

$Cr^{3+}$ is produced)? (d) What is the normality of a 0.05 $M$ $KIO_3$ solution for a reaction in which $ICl_2^-$ is produced?

**9.19** Concentrated hydrochloric acid is 37.0% HCl by weight and has a density of 1.18 g/ml. What is the molarity of concentrated HCl?

**9.20** Concentrated $H_2SO_4$ is 96.0% $H_2SO_4$ and has a density of 1.84 g/ml. How many grams of concentrated $H_2SO_4$ should be used to prepare 500 ml of 0.300 $M$ $H_2SO_4$? How many milliliters of concentrated $H_2SO_4$ should be used to prepare this solution?

**9.21** Concentrated $H_3PO_4$ is 85.0% $H_3PO_4$ and has a density of 1.70 g/ml. How many grams of concentrated $H_3PO_4$ should be used to prepare 250 ml of 0.400 $M$ $H_3PO_4$? How many milliliters of concentrated $H_3PO_4$ should be used to prepare this solution?

**9.22** How many grams of $CaCl_2$ should be used to prepare 750 ml of 0.150 $M$ $CaCl_2$ solution if the solid $CaCl_2$ is 96.0% pure?

**9.23** A 10.0% solution of $AgNO_3$ has a density of 1.09 g/ml. What is (a) the molarity and (b) the molality of this solution?

**9.24** A 25.0 ml sample of concentrated HI solution is diluted to a final volume of 500 ml. The concentrated HI solution is 47.0% HI and has a density of 1.50 g/ml. What is the molarity of the final solution?

**9.25** A quantity of Na is added to water and 140 ml of $H_2(g)$, measured dry and at STP, and 250 ml of a solution of NaOH are produced. Write the equation for the

reaction. What is the molarity of the NaOH solution?

**9.26** How many milliliters of 0.150 $N$ NaOH would be required to neutralize 35.0 ml of 0.13 $N$ $H_2SO_4$?

**9.27** A 15.0 ml sample of an acid requires 37.3 ml of 0.303 $N$ NaOH for neutralization. What is the normality of the acid?

**9.28** An antacid tablet containing magnesium hydroxide requires 30.5 ml of 0.450 $N$ HCl for neutralization. How many grams of $Mg(OH)_2$ did the tablet contain?

**9.29** Sodium bicarbonate, $NaHCO_3$, and aluminum hydroxide, $Al(OH)_3$, are two commonly used antacids. The $NaHCO_3$ reacts with stomach acid (HCl) to produce NaCl, $CO_2$, and $H_2O$. Aluminum hydroxide reacts to form the normal salt. (a) What volume of 0.800 $N$ HCl is required to neutralize 1.00 g of $NaHCO_3$? (b) What volume of 0.800 $N$ HCl is required to neutralize 1.00 g of $Al(OH)_3$?

**9.30** Citric acid, which may be obtained from lemon juice, has the empirical formula $C_6H_8O_7$. A 0.250 g sample of pure citric acid requires 37.2 ml of 0.105 $N$ NaOH for complete neutralization. (a) What is the equivalent weight of citric acid? (b) How many acidic hydrogens per molecule does citric acid have?

**9.31** Sodium carbonate reacts with acids:

$$Na_2CO_3 + 2HCl \rightarrow 2NaCl + H_2O + CO_2$$

(a) What is the equivalent weight of $Na_2CO_3$ for this reaction?

(b) If a 0.283 g sample of impure $Na_2CO_3$ reacts with exactly 32.1 ml of 0.123 $N$ HCl, what is the weight percent of $Na_2CO_3$ in the material?

**9.32** A solution contains both NaOH and $Na_2CO_3$ (see Problem 9.31). A 15.0 ml portion of the solution requires 33.3 ml of 0.200 $N$ HCl for complete neutralization. A solution of $BaCl_2$ is added to another 15.0 ml portion of the $NaOH$–$Na_2CO_3$ solution, and $BaCO_3$ precipitates. The titration of the resulting solution with 0.200 $N$ HCl requires 13.3 ml of the acid. In this titration only enough acid is added to neutralize the solution; the solid $BaCO_3$ does not react. What are the normal and molar concentrations of NaOH and $Na_2CO_3$ in the original solution?

**9.33** A 25.0 ml sample of vinegar, which contains acetic acid $(HC_2H_3O_2)$, requires 37.5 ml of 0.460 $N$ NaOH for neutralization. The density of the vinegar is 1.00 g/ml. What is the concentration of acetic acid in the vinegar in terms of (a) normality and (b) weight percent?

**9.34** A 0.250 g sample of impure oxalic acid, $H_2C_2O_4$, requires 30.1 ml of 0.120 $N$ NaOH for complete neutralization. What is the percentage of $H_2C_2O_4$ in the material?

**9.35** Pure sodium carbonate is used to standardize a solution of HCl (see Problem 9.31). A 0.200 g sample of $Na_2CO_3$ requires 41.2 ml of the acid for complete neutralization. What is the normality of the acid?

**9.36** A 5.00 g sample of hemo-globin is treated in such a way as to produce small water-soluble molecules and ions by destroying the hemoglobin molecule. The iron in the aqueous solution that results from this procedure is reduced to $Fe^{2+}$ and titrated against standard $KMnO_4$ solution. In the titration, $Fe^{2+}$ is oxidized to $Fe^{3+}$ and $MnO_4^-$ is reduced to $Mn^{2+}$. The sample requires 30.5 ml of 0.0100 $N$ $KMnO_4$. (a) What is the percent iron in hemoglobin? (b) The molecular weight of hemoglobin is 65,600. How many iron atoms are there in one hemoglobin molecule?

**9.37** Iron wire containing 99.8% Fe is used to standardize a solution of $K_2Cr_2O_7$. A 0.185 g sample of iron wire is dissolved in acid and converted to the $Fe^{2+}$ ion. This solution requires 29.3 ml of the dichromate solution for complete reaction. In the reaction, $Fe^{2+}$ is oxidized to $Fe^{3+}$ and $Cr_2O_7^{2-}$ is reduced to $Cr^{3+}$. (a) What is normality of the $K_2Cr_2O_7$ solution? (b) What is the molarity of the $K_2Cr_2O_7$ solution?

**9.38** Hydrazine $(N_2H_4)$ reacts with the $BrO_3^-$ ion in acid solution to produce $N_2(g)$ and $Br^-$. A 0.132 g sample of impure hydrazine requires 38.3 ml of 0.103 $N$ $KBrO_3$ for complete reaction. What percent of the sample is hydrazine?

**9.39** A 0.502 g sample of impure $BaO_2$ was dissolved in acid to produce an amount of $H_2O_2$ equivalent to the amount of $BaO_2$ present in the sample. The resulting solution was titrated against standard $KMnO_4$. In the titration the $H_2O_2$ was oxidized to $O_2$ and the $MnO_4^-$ was reduced to $Mn^{2+}$. The titration required 33.3 ml of 0.130 $N$ $KMnO_4$ solution. What percent of the material is $BaO_2$?

**9.40** Benzene, $C_6H_6$, and toluene, $C_7H_8$, form ideal solutions. At 60°C the vapor pressure of pure benzene is 0.507 atm and the vapor pressure of pure toluene is 0.184 atm. (a) What is the vapor pressure of 60°C of a solution containing 6.5 g of benzene and 23.0 g of toluene? (b) What is the composition of the vapor (mole fraction of benzene) in equilibrium with this solution at 60°C?

**9.41** (a) Use the data given in Problem 9.40 to calculate the mole fraction of benzene in a solution that has a vapor pressure of 0.350 atm at 60°C. (b) What is the mole fraction of benzene in the vapor that is in equilibrium with this solution?

**9.42** Liquids A and B form ideal solutions. The vapor pressure of pure B is 0.650 atm at the boiling point of a solution prepared from 0.200 mol of B and 0.600 mol of A. (a) What is the vapor pressure of pure A at this temperature? (b) What is the mole fraction of A in the vapor that is in equilibrium with this solution when the solution first begins to boil?

**9.43** A solution containing 1 mol of chloroform and 4 mol of acetone has a vapor pressure of 0.400 atm at 35°C. At this temperature the vapor pressure of pure chloroform is 0.359 atm and that of acetone is 0.453 atm. (a) What would the vapor pressure of the solution be if chloroform and acetone formed ideal solutions? (b) Does the vapor pressure of the solution show a positive or negative deviation from that predicted by Raoult's law? (c) Is heat evolved or absorbed when this solution is prepared? (d) Do chloroform and acetone form a minimum- or maximum-boiling azeotrope?

**9.44** A solution containing 1 mol of carbon disulfide and 4 mol of acetone has a vapor pressure of 0.750 atm at 35°C. At this temperature the vapor pressure of pure carbon disulfide is 0.674 atm and that of acetone is 0.453 atm. (a) What would the vapor pressure of the solution be if it were an ideal solution? (b) Does the vapor pressure of the solution show a positive or negative deviation from that predicted by Raoult's law? (c) Is heat evolved or absorbed when this solution is prepared? (d) Do carbon disulfide and acetone form a minimum- or maximum-boiling azeotrope?

**9.45** A solution containing 20.0 g of a nonvolatile solute in exactly 1.00 mol of a volatile solvent has a vapor pressure of 0.500 atm at 20°C. A second mole of solvent is added to the mixture, and the resulting solution has a vapor pressure of 0.550 atm at 20°C. (a) What is the molecular weight of the solute? (b) What is the vapor pressure of the pure solvent at 20°C?

**9.46** Which of the following measurements will give the best value for the molecular weight of a substance, such as a protein, that has a high molecular weight: vapor–pressure lowering, freezing-point depression, boiling-point elevation, or osmotic pressure? Why?

**9.47** Twice the percent, by volume, of ethyl alcohol ($C_2H_5OH$) in an alcoholic beverage is defined as the *proof* of the liquor. Gin is 90 proof. Assume that the density of ethyl alcohol is 0.810 g/ml and that of water is 1.00 g/ml. What is the freezing point of gin?

**9.48** The antifreeze commonly used in car radiators is ethylene glycol, $C_2H_4(OH)_2$. How many grams of ethylene glycol must be added to 1000 g of water to produce a solution that will freeze at $-15°C$?

**9.49** Lauryl alcohol is obtained from coconut oil and is used to make detergents. A solution of 5.00 g of lauryl alcohol in 100.0 g of benzene boils at 80.78°C. The normal boiling point of benzene is 80.10°C and $K_b$ for benzene is $+2.53°C/m$. What is the molecular weight of lauryl alcohol?

**9.50** A solution containing 22.0 g of ascorbic acid (vitamin C) in 100 g of water freezes at $-2.33°C$. What is the molecular weight of ascorbic acid?

**9.51** A solution containing 9.66 g of a nonvolatile solute in 300 g of $CCl_4$ boils at 78.6°C. What is the molecular weight of the solute?

**9.52** What is the freezing point of a solution containing 1.00 g of naphthalene, $C_{10}H_8$, in 250 g of camphor?

**9.53** What is the osmotic pressure of an aqueous solution containing 1.00 g of urea, $CON_2H_4$, in 100 ml of solution at 20°C?

**9.54** An aqueous solution containing 1.00 g/liter of the ethoxylate of lauryl alcohol, which is a detergent, has an osmotic pressure of 0.0448 atm at 20°C. What is the molecular weight of the compound?

**9.55** A solution containing 1.00 g/liter of insulin exhibits an osmotic pressure of $6.68 \times 10^{-3}$ atm at 20°C. What is the approximate molecular weight of insulin?

**9.56** The osmotic pressure of blood at body temperature (37°C) is the same as the osmotic pressure of a 0.160 $M$ solution of NaCl. For electrolytes the van't Hoff equation is $\pi = iMRT$, and the value of $i$ for a NaCl solution of this concentration is 1.85. (a) What is the osmotic pressure of blood at 37°C? (b) What is the total concentration of solute in blood? Assume that the dissolved substances do not ionize.

**9.57** One of the problems of producing fresh water from salt water by reverse osmosis is developing semipermeable membranes strong enough to withstand the high pressure required. Sea water contains dissolved salts (3.50% by weight). Assume that the solute is entirely NaCl (over 90% is), that the density of sea water is 1.03 g/ml, and that the $i$ factor for a NaCl solution of this concentration is 1.83. For solutions of electrolytes, $\pi = iMRT$. What is the osmotic pressure of sea water at 20°C? What pressure would have to be applied to effect reverse osmosis?

**9.58** A solution prepared by adding a given weight of a solute to 20.0 g of benzene freezes at a temperature 0.384°C below the freezing point of pure benzene. The freezing point of a solution prepared from the same weight of the solute and 20.00 g of water freezes at −0.4185°C. Assume that the solute is undissociated in benzene solution but is completely dissociated into ions in water solution. How many ions result from the dissociation of one molecule in water solution?

**9.59** A solution containing 3.81 g of $MgCl_2$ in 400 g of water freezes at −0.497°C. What is the van't Hoff factor, $i$, for the freezing point of this solution?

**9.60** A solution is prepared from 3.00 g of NaOH and 75.0 g of water. If the van't Hoff factor, $i$, for this solution is 1.83, at what temperature will the solution freeze?

**9.61** What is the freezing point of a 0.125 $m$ aqueous solution of a weak acid, HX, if the acid is 4.00% ionized? Ignore interionic attractions.

**9.62** A 0.0200 $m$ solution of a weak acid, HY, in water freezes at −0.0385°C. Ignore interionic attractions and calculate the percentage ionization of the weak acid.

# 10

# Electrochemistry

All chemical reactions are fundamentally electrical in nature since electrons are involved (in various ways) in all types of chemical bonding. Electrochemistry, however, is primarily the study of oxidation–reduction phenomena.

The relations between chemical change and electrical energy have theoretical as well as practical importance. Chemical reactions can be used to produce electrical energy (in cells that are called either **voltaic** or **galvanic cells**); electrical energy can be used to bring about chemical transformations (in **electrolytic cells**). In addition, the study of electrochemical processes leads to an understanding, as well as to the systemization, of oxidation–reduction phenomena that take place outside cells.

For example, data from electrochemical studies are used to determine the most favorable conditions for a given oxidation–reduction reaction, whether a proposed chemical transformation will occur, and the magnitude and nature of the energy effect that accompanies a given chemical change.

## 10.1 METALLIC CONDUCTION

An electric current is the flow of electric charge. In metals this charge is carried by electrons, and electrical conduction of this type is called **metallic conduction.** The current results from the application of an electric force supplied by a battery or some other source of electrical energy; a complete circuit is necessary to produce a current.

Metallic crystals have been described in terms of mobile electron clouds permeating relatively fixed lattices of positive metal ions (Section 4.7). When electrons are forced into one end of a metal wire, the impressed electrons displace other electrons of the cloud at the point of entry. The displaced electrons, in turn, assume new positions by pushing neighboring electrons ahead, and this effect is transmitted down the length of the wire until electrons are forced out of the wire at the opposite end. The current source may be regarded as an electron pump, for it serves to force electrons into one end of the circuit and drain them off from the other end. At any position in the wire, electrical neutrality is preserved, since the rate of electrons in equals the rate of electrons out.

The analogy between the flow of electricity and the flow of a liquid is an old one; in earlier times electricity was described in terms of a current of "electric fluid." Conventions of long standing, which may be traced back to Benjamin Franklin (1747) and which were adopted before the electron was identified, ascribe a positive charge to this current. We shall interpret electrical circuits in terms of the movement of electrons. Remember, however, that conventional electric current is arbitrarily described as positive and as flowing in the opposite direction.

● The standard for the ampere (A), the SI base unit of electric current, is described inside the back cover of this book. The coulomb (C) is an SI derived unit that is used to measure quantity of electric charge.

$$1 \text{ C} = 1 \text{ A} \cdot \text{s}$$

Electric current is measured in **amperes.** Quantity of electric charge is measured in **coulombs;** the coulomb is defined as the quantity of electricity carried in 1 second by a current of 1 ampere. Therefore,

1 amp = 1 coulomb/sec

and

1 coulomb = 1 amp sec

● The volt (V) is the SI derived unit of electric potential difference.

$1 \text{ V} = 1 \text{ J/A} \cdot \text{s}$
$\quad = 1 \text{ m}^2 \cdot \text{kg} \cdot \text{s}^{-3} \cdot \text{A}^{-1}$

The SI derived unit of electric resistance is the ohm ($\Omega$).

$1 \Omega = 1 \text{ V/A}$
$\quad = 1 \text{ m}^2 \cdot \text{kg} \cdot \text{s}^{-3} \cdot \text{A}^{-2}$

The current is forced through the circuit by an electrical potential difference, which is measured in **volts.** It takes 1 joule of work to move 1 coulomb from a lower to a higher potential when the potential difference is 1 volt. One volt, therefore, equals 1 joule/coulomb, and 1 volt coulomb is a unit of energy and equals 1 joule.

The higher the potential difference between two points in a given wire, the more current the wire will carry between those two points. Georg Ohm in 1826 expressed the quantitative relation between potential difference, $\mathscr{E}$, in volts, and current, $I$, in amperes, as

$$I = \mathscr{E}/R \quad \text{or} \quad \mathscr{E} = IR$$

where the proportionality constant, $R$, of **Ohm's law** is called the resistance. Resistance is measured in **ohms;** 1 volt is required to force a current of 1 amp through a resistance of 1 ohm.

Resistance to the flow of electricity in metals is probably caused by the vibration of the metal ions about their lattice positions; these vibrations interfere with the motion of the electrons and retard the current. As the temperature is increased, the thermal motion of the metal ions is increased; hence, the resistance of metals increases, and the metals become poorer conductors.

**10.2 ELECTROLYTIC CONDUCTION**

**Electrolytic conduction,** in which the charge is carried by ions, will not occur unless the ions of the electrolyte are free to move. Hence, electrolytic conduction is exhibited principally by molten salts and by aqueous solutions of electrolytes. Furthermore, a sustained current through an electrolytic conductor requires that chemical change accompany the movement of ions.

These principles of electrolytic conduction are best illustrated by reference to an electrolytic cell such as the one diagramed in Figure 10.1 for the electrolysis of molten NaCl between inert electrodes.[1] The current source pumps electrons into the left-hand electrode, which therefore may be considered to be negatively charged; electrons are drained from the right-hand, positive electrode. In the electric field thus produced, sodium ions (cations) are attracted toward the negative pole (cathode), and chloride ions (anions) are attracted toward the positive pole (anode). Electric charge in electrolytic conduction is carried by cations moving toward the **cathode** and anions moving in the opposite direction, toward the **anode.**

For a complete circuit, electrode reactions must accompany the movement of ions. At the cathode some chemical species (not necessarily the charge carrier) must accept electrons and be reduced; at the anode, electrons must be removed from some chemical species, which as a consequence is oxidized. The conventions relating to the terms anode and cathode are summarized in Table 10.1.

In the diagramed cell, sodium ions are reduced at the cathode

$$\text{Na}^+ + e^- \rightarrow \text{Na}$$

Inert electrodes are not involved in electrode reactions.

Table 10.1 Electrode conventions

|  | CATHODE | ANODE |
|---|---|---|
| ions attracted | cations | anions |
| direction of electron movement | into cell | out of cell |
| half reaction | reduction | oxidation |
| sign |  |  |
|   electrolysis cell | negative | positive |
|   galvanic cell | positive | negative |

and chloride ions are oxidized at the anode

$$2Cl^- \rightarrow Cl_2 + 2e^-$$

Proper addition of these two partial equations gives the reaction for the entire cell

$$2NaCl(l) \xrightarrow{\text{electrolysis}} 2Na(l) + Cl_2(g)$$

In the actual operation of the commercial cell used to produce metallic sodium, calcium chloride is added to lower the melting point of sodium chloride, and the cell is operated at a temperature of approximately 600°C. At this temperature sodium metal is a liquid.

We can trace the flow of negative charge through the circuit of Figure 10.1 as follows. Electrons leave the current source and are pumped into the cathode where they are picked up by and reduce sodium ions that have been attracted to this negative electrode. Chloride ions move away from the cathode toward the anode and thus carry negative charge in this direction. At the anode, electrons are removed from the chloride ions, thus oxidizing them to chlorine gas; these electrons are pumped out of the cell by the current source. In this manner the circuit is completed.

Electrolytic conduction, then, rests on the mobility of ions, and any-

Figure 10.1 Electrolysis of molten sodium chloride.

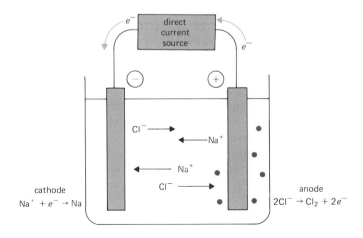

thing that inhibits the motion of these ions causes resistance to the current. Factors that influence the electrical conductivity of solutions of electrolytes include interionic attractions, solvation of ions, and viscosity of the solvent; these factors rest on solute–solute attractions, solute–solvent attractions, and solvent–solvent attractions, respectively. The average kinetic energy of the solute ions increases as the temperature is raised, and therefore, the resistance of electrolytic conductors generally decreases as the temperature is raised (i.e., conduction increases). Furthermore, the effect of each of the three previously mentioned factors decreases as the temperature is increased.

At all times electrical neutrality is preserved throughout all parts of the electrolytic liquid—there are as many cations per unit volume as there are anions.

**10.3 ELECTROLYSIS**   The electrolysis of molten sodium chloride serves as a commercial source of sodium metal and chlorine gas; analogous procedures are used to prepare other very active metals (such as potassium and calcium). When certain aqueous solutions are electrolyzed, however, water is involved in the electrode reactions rather than the ions derived from the solute. Hence, the current-carrying ions are not necessarily discharged at the electrodes.

In the electrolysis of aqueous sodium sulfate, sodium ions move toward the cathode and sulfate ions move toward the anode (Figure 10.2). Both these ions are difficult to discharge. When this electrolysis is conducted between inert electrodes, hydrogen gas is evolved at the cathode, and the solution surrounding the electrode becomes alkaline. Reduction occurs at the cathode, but rather than the reduction of the sodium ion

$$e^- + Na^+ \rightarrow Na$$

**Figure 10.2** Electrolysis of aqueous sodium sulfate.

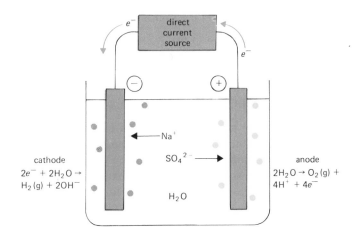

the *net change* that occurs is the reduction of water

$$2e^- + 2H_2O \rightarrow H_2(g) + 2OH^-$$

Water is an extremely weak electrolyte; pure water is approximately $2 \times 10^{-7}\%$ ionized at 25°C

$$2H_2O \rightleftharpoons H_3O^+ + OH^-$$

or, more briefly,

$$H_2O \rightleftharpoons H^+ + OH^-$$

The exact mechanism of the cathode reaction in the electrolysis of aqueous $Na_2SO_4$ is not known. It may be that the hydrogen ions from water are discharged and that the reaction proceeds as follows:

$$H_2O \rightleftharpoons H^+ + OH^-$$

$$2e^- + 2H^+ \rightarrow H_2(g)$$

Multiplication of the first equation by 2 followed by addition of the equations gives the net change

$$2e^- + 2H_2O \rightarrow H_2(g) + 2OH^-$$

In general, water is reduced at the cathode (producing hydrogen gas and hydroxide ions) whenever the cation of the solute is difficult to reduce.

Oxidation occurs at the anode, and in the electrolysis of aqueous $Na_2SO_4$, the anions ($SO_4^{2-}$) that migrate toward the anode are difficult to oxidize.

$$2SO_4^{2-} \rightarrow S_2O_8^{2-} + 2e^-$$

Therefore, the oxidation of water occurs preferentially. The mode of this reaction may be

$$H_2O \rightleftharpoons H^+ + OH^-$$

$$4OH^- \rightarrow O_2(g) + 2H_2O + 4e^-$$

Multiplying the first equation by 4 and adding the equations, we get the net change

$$2H_2O \rightarrow O_2(g) + 4H^+ + 4e^-$$

At the anode the evolution of oxygen gas is observed, and the solution surrounding the pole becomes acidic. In general, water is oxidized at the anode (producing oxygen gas and hydrogen ions) whenever the anion of the solute is difficult to oxidize.

The complete reaction for the electrolysis of aqueous $Na_2SO_4$ may be obtained by adding the cathode and anode reactions

$$2[2e^- + 2H_2O \rightarrow H_2(g) + 2OH^-]$$
$$2H_2O \rightarrow O_2(g) + 4H^+ + 4e^-$$
$$\overline{6H_2O \rightarrow 2H_2(g) + O_2(g) + 4H^+ + 4OH^-}$$

If the solution is mixed, the hydrogen and hydroxide ions that are produced neutralize one another, and the net change

$$2H_2O \xrightarrow{\text{electrolysis}} 2H_2(g) + O_2(g)$$

is merely the electrolysis of water. In the course of the electrolysis, the hydrogen ions migrate away from the anode, where they are produced, toward the cathode. In like manner, the hydroxide ions move toward the anode. These ions neutralize one another in the solution between the two electrodes.

The electrolysis of an aqueous solution of NaCl between inert electrodes serves as an example of a process in which the anion of the electrolyte is discharged, but the cation is not.

anode: $\qquad 2Cl^- \rightarrow Cl_2(g) + 2e^-$

cathode: $\quad 2e^- + 2H_2O \rightarrow H_2(g) + 2OH^-$
$$\overline{2H_2O + 2Cl^- \rightarrow H_2(g) + Cl_2(g) + 2OH^-}$$

Since the sodium ion remains unchanged in the solution, the reaction may be indicated

$$2H_2O + 2(Na^+, Cl^-) \xrightarrow{\text{electrolysis}} H_2(g) + Cl_2(g) + 2(Na^+, OH^-)$$

This process is a commercial source of hydrogen gas, chlorine gas, and, by evaporation of the solution left after electrolysis, sodium hydroxide.

In the electrolysis of a solution of $CuSO_4$ between inert electrodes (see right portion of Figure 10.4, given subsequently), the current is carried by the $Cu^{2+}$ and $SO_4^{2-}$ ions. The current-carrying cations are discharged, but the anions are not.

anode: $\qquad 2H_2O \rightarrow O_2(g) + 4H^+ + 4e^-$

cathode: $\quad 2e^- + Cu^{2+} \rightarrow Cu(s)$
$$\overline{2Cu^{2+} + 2H_2O \rightarrow O_2(g) + 2Cu(s) + 4H^+}$$

It is, of course, possible to have both ions of the solute discharged during the electrolysis of an aqueous solution. An example is the electrolysis of $CuCl_2$ between inert electrodes.

anode: $\qquad 2Cl^- \rightarrow Cl_2(g) + 2e^-$

cathode: $\quad 2e^- + Cu^{2+} \rightarrow Cu(s)$
$$\overline{Cu^{2+} + 2Cl^- \rightarrow Cu(s) + Cl_2(g)}$$

**Figure 10.3** Electrolysis of aqueous cupric sulfate between copper electrodes.

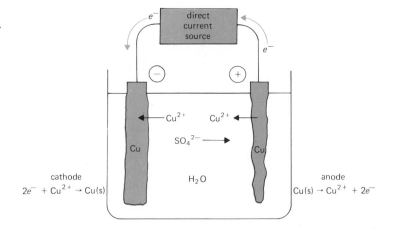

**Figure 10.3** Electrolysis of aqueous cupric sulfate between copper electrodes.

It is also possible to have the electrode itself enter into an electrode reaction. If aqueous $CuSO_4$ is electrolyzed between copper electrodes (Figure 10.3), $Cu^{2+}$ ions are reduced at the cathode

$$2e^- + Cu^{2+} \rightarrow Cu$$

but of the *three* possible anode oxidations

$$2SO_4^{2-} \rightarrow 2S_2O_8^{2-} + 2e^-$$

$$2H_2O \rightarrow O_2(g) + 4H^+ + 4e^-$$

$$Cu(s) \rightarrow Cu^{2+} + 2e^-$$

the oxidation of the copper metal of the electrode is observed to occur. Hence, at the anode, copper from the electrode goes into solution as $Cu^{2+}$ ions, and at the cathode, $Cu^{2+}$ ions plate out as $Cu(s)$ on the electrode. This process is used to refine copper. Impure copper is used as the anode of an electrolytic cell, and a solution of $CuSO_4$ is electrolyzed; pure copper plates out on the cathode. Active electrodes are also used in electroplating processes; in silver plating, silver anodes are employed.

**10.4 FARADAY'S LAWS**  The quantitative relationships between electricity and chemical change were first described by Michael Faraday. In 1832 he showed that the weight of a chemical substance liberated at an electrode is directly proportional to the amount of current passed through the cell. In 1833 he stated that the weights of different substances produced by a given amount of current are proportional to the equivalent weights of the substances.

**Faraday's laws** are readily interpreted by reference to the electrolysis of molten sodium chloride. The change at the cathode requires one electron for every sodium ion reduced.

$$Na^+ + e^- \rightarrow Na$$

If Avogadro's number of electrons (1 mol) are consumed at this electrode, 22.9898 g of sodium metal (1 mol) is produced. The corresponding quantity of electricity is called the **faraday** (F) and has been found to equal 96,487 coulombs (for ordinary problem work this value is customarily rounded off to 96,500 coulombs). If 2 F of electricity (2 mol of electrons) are used, 2 mol of sodium are produced; 0.5 F would liberate 0.5 mol of sodium. Since the charge on the sodium ion is 1+, the equivalent weight of sodium is the same as the atomic weight.

In the same time that electrons equivalent to 1 F are added to the cathode, that same number of electrons are removed from the anode.

$$2Cl^- \rightarrow Cl_2 + 2e^-$$

The removal of 1 mol of electrons from this anode would result in the discharge of 35.453 g of chloride ion (1 mol), thus producing 0.5 mol of chlorine gas (1 gram equivalent). If 2 F of electricity flow through the cell, 2 mol of chloride ions are discharged.

One faraday, then, will liberate 1 gram equivalent of an element or a compound. Rather than use equivalent weights, however, it is possible to interpret electrode reactions in terms of moles and faradays. Thus, the anode oxidation of the hydroxide ion,

**Figure 10.4** Silver coulometer in series with a cell for electrolysis.

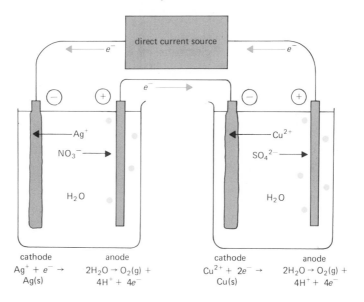

cathode
$Ag^+ + e^- \rightarrow$
$Ag(s)$

anode
$2H_2O \rightarrow O_2(g) +$
$4H^+ + 4e^-$

cathode
$Cu^{2+} + 2e^- \rightarrow$
$Cu(s)$

anode
$2H_2O \rightarrow O_2(g) +$
$4H^+ + 4e^-$

$$4OH^- \rightarrow O_2(g) + 2H_2O + 4e^-$$

may be read: four hydroxide ions produce one oxygen molecule, two water molecules, and four electrons, *or* 4 mol of hydroxide ion produces 1 mol of oxygen and 2 mol of water when 4 faradays of electricity is passed through the cell.

In Figure 10.4 two electrolytic cells are set up in series; electricity passes through one cell first and then through the other before returning to the current source. If silver nitrate is electrolyzed in one of the cells, the cathode reaction is

$$Ag^+ + e^- \rightarrow Ag(s)$$

and metallic silver is plated out on the electrode used for the electrolysis. By weighing the electrode before and after the electrolysis, one can determine the quantity of silver plated out and hence the number of coulombs that have passed through the cell. One faraday would plate out 107.868 g of silver; therefore, 1 coulomb is equivalent to

$$107.868/96,487 = 1.1180 \times 10^{-3} \text{ g of silver}$$

The same number of coulombs pass through both cells in a given time when these cells are arranged in series. Therefore, the number of coulombs used in an electrolysis may be determined through the addition, in series, of this **silver coulometer** to the circuit of the experimental cell.

---

*Example 10.1*
The charge on a single electron is $1.6021 \times 10^{-19}$ coulomb. Calculate Avogadro's number from the fact that 1 F = 96,487 coulombs.

*Solution*

$$? \text{ electrons} = 9.6487 \times 10^4 \text{ coulombs} \left( \frac{1 \text{ electron}}{1.6021 \times 10^{-19} \text{ coulomb}} \right)$$

$$= 6.0225 \times 10^{23} \text{ electrons}$$

---

*Example 10.2*
In the electrolysis of $CuSO_4$, what weight of copper is plated out on the cathode by a current of 0.750 amp in 10.0 min?

*Solution*
The number of faradays used may be calculated as follows:

$$? \text{ F} = 10.0 \text{ min} \left( \frac{60 \text{ sec}}{1 \text{ min}} \right) \left( \frac{0.75 \text{ coulomb}}{1 \text{ sec}} \right) \left( \frac{1 \text{ F}}{96,500 \text{ coulombs}} \right)$$

$$= 0.00466 \text{ F}$$

The cathode reaction is $Cu^{2+}$ (aq) $+ 2e^- \rightarrow Cu(s)$, and therefore 2 F plate out 63.5 g Cu(s)

$$? \text{ g Cu} = 0.00466 \text{ F} \left( \frac{63.5 \text{ g Cu(s)}}{2F} \right) = 0.148 \text{ g Cu(s)}$$

*Example 10.3*
(a) What volume of $O_2(g)$ at STP is liberated at the anode in the electrolysis of $CuSO_4$ described in Example 10.2? (b) If 100 ml of $1.00M$ $CuSO_4$ is employed in the cell, what is the $H^+(aq)$ concentration at the end of the electrolysis? Assume that there is no volume change for the solution during the experiment and that the anode reaction is $2H_2O \rightarrow 4H^+(aq) + O_2(g) + 4e^-$.

*Solution*
(a) Four faradays produce 22.4 liters of $O_2(g)$ at STP.

$$? \text{ liters } O_2(g) = 0.00466 \text{ F} \left( \frac{22.4 \text{ liters } O_2(g)}{4F} \right)$$
$$= 0.0261 \text{ liter } O_2(g)$$

(b) Four faradays also produce 4 mol of $H^+(aq)$

$$? \text{ mol } H^+(aq) = 0.00466 \text{ F} \left( \frac{1 \text{ mol } H^+(aq)}{1F} \right)$$
$$= 0.00466 \text{ mol } H^+(aq)$$

The small contribution of $H^+(aq)$ from the ionization of water may be ignored, and we may assume that there are 0.00466 mol $H^+(aq)$ in 100 ml of solution.

$$? \text{ mol } H^+(aq) = 1000 \text{ ml solution} \left( \frac{0.00466 \text{ mol } H^+(aq)}{100 \text{ ml solution}} \right)$$
$$= 0.0466 \text{ mol } H^+(aq)$$

The solution is therefore $0.0466M$ in hydrogen ion.

*Example 10.4*
(a) What weight of copper is plated out in the electrolysis of $CuSO_4$ in the same time that it takes to deposit 1.00 g of Ag in a silver coulometer that is arranged in series with the $CuSO_4$ cell? (b) If a current of 1.00 amp is used, how many minutes are required to plate out this quantity of copper?

*Solution*
(a) From the electrode reactions we see that 2 F deposit 63.5 g Cu and 1 F deposits 107.9 g Ag.

$$? \text{ g Cu} = 1.00 \text{ g Ag} \left( \frac{1F}{107.9 \text{ g Ag}} \right) \left( \frac{63.5 \text{ g Cu}}{2F} \right) = 0.294 \text{ g Cu}$$

(b)

$$? \text{ min} = 1.00 \text{ g Ag} \left( \frac{96,500 \text{ coulombs}}{107.9 \text{ g Ag}} \right) \left( \frac{1 \text{ sec}}{1 \text{ coulomb}} \right) \left( \frac{1 \text{ min}}{60 \text{ sec}} \right)$$
$$= 14.9 \text{ min}$$

**10.5  ELECTROLYTIC CONDUCTANCE**    A material with a high electrical resistance is a poor electrical conductor; the ability of a substance to conduct electrical current is inversely related to the substance's electrical resistance. **Conductance** is defined as the reciprocal of resistance and has the units of reciprocal ohms (1/ohm). From Ohm's law, $\mathscr{E} = IR$,

$$R = \frac{\mathscr{E}}{I}$$

and

$$\text{conductance} = \frac{1}{R} = \frac{I}{\mathscr{E}}$$

● The International System includes the siemens (S) as a derived unit to be used for measurement of conductance.

$$1 \text{ S} = 1/\Omega = \text{A}/\text{V}$$

Therefore, the conductance of a given solution is equal to the current ($I$) when the potential difference is 1 V ($\mathscr{E} = 1$ V). Conductance measurements are made using a rapidly alternating current (in which the direction of the current reverses at regular time intervals) to minimize the effect of electrode reactions, which would change the composition of the solution.

The conductance of a given solution varies with the cell dimensions — directly with the area of the electrodes ($A$) and inversely with the distance between them ($d$). Thus,

$$\text{conductance} = \frac{1}{R} = \frac{\kappa A}{d}$$

The proportionality constant, $\kappa$, is the **specific conductance** — the conductance when $A$ is 1 cm$^2$ and $d$ is 1 cm; $\kappa$ has the units 1/ohm cm.

It is not necessary to construct a cell of these exact dimensions in order to determine specific conductances. By measuring the resistance of a solution of known $\kappa$ in an experimental cell, one can calculate the cell constant, $k$,

$$\kappa = \frac{k}{R}$$

and by means of this constant convert resistances measured in that cell into specific conductances. Since resistance (and hence conductance) varies with temperature, conductance measurements must be made at constant temperature.

The charges of solute ions are critical in determining how much current a solution may carry; therefore, comparisons of conductance data must be made between values for solutions of corresponding *normality*. The equivalent weight used in determining the normality is obtained by dividing the molecular weight of the solute by the total positive charge represented in the formula of the compound.

**Equivalent conductance,** $\Lambda$, is defined as the conductance, between electrodes 1 cm apart, of a volume of solution that contains 1 gram equivalent of solute; hence, it represents the conductance of a collection of cations and anions equal to Avogadro's number of positive charges and Avogadro's number of negative charges. Since $\kappa$ is the conductance of 1 cm$^3$ of solution, $1000\kappa$ is the conductance of 1 liter of solution. Hence,

$$\Lambda = \frac{1000\,\kappa}{N}$$

where $N$, the normality of the solution, is the number of gram equivalents per liter; $\Lambda$ therefore has the units cm$^2$/ohm equivalent.

*Example 10.5*

A conductance cell filled with 0.0200N KCl has a resistance of 163.3 ohms at 25°C. When filled with 0.0500N AgNO$_3$, it had a resistance of 78.5 ohms. The specific conductance of 0.0200N KCl is 2.768 × 10$^{-3}$/ohm cm. (a) What is the specific conductance of 0.500N AgNO$_3$? (b) What is the equivalent conductance of this solution?

*Solution*

(a) The cell constant may be determined by

$$\kappa = \frac{k}{R}$$

$$k = \kappa R = (2.768 \times 10^{-3}/\text{ohm cm})(163.3 \text{ ohm})$$
$$= 0.4520/\text{cm}$$

For 0.0500N AgNO$_3$,

$$\kappa = \frac{0.4520/\text{cm}}{78.5 \text{ ohm}} = 5.76 \times 10^{-3}/\text{ohm cm}$$

(b)

$$\Lambda = \frac{1000 \, \kappa}{N}$$
$$= \frac{(1000 \text{ cm}^3/\text{liter})(5.76 \times 10^{-3}/\text{ohm cm})}{5.00 \times 10^{-2} \text{ equivalent/liter}}$$
$$= 115 \text{ cm}^2/\text{ohm equivalent}$$

An examination of the data of Table 10.2 shows that the equivalent conductance of solutions of a given electrolyte, for example, NaCl, increases with decreasing concentration and approaches a limiting value, $\Lambda_0$, at infinite dilution. Owing to interionic attraction, any ion has a diffuse atmosphere of oppositely charged ions surrounding it; its motion in an electric field is therefore retarded. At infinite dilution the ions are far enough apart to act independently of one another, and theoretically this ion-drag effect is eliminated.

**Table 10.2** Equivalent conductances, $\Lambda$, of various electrolytes at 25°C (in cm$^2$/ohm equivalent)

| ELECTROLYTE | CONCENTRATION (equivalents/liter) | | | |
|---|---|---|---|---|
| | 0.000 | 0.001 | 0.010 | 0.100 |
| NaCl | 126.5 | 123.7 | 118.5 | 106.7 |
| KCl | 149.9 | 147.0 | 141.3 | 129.0 |
| BaCl$_2$ | 140.0 | 134.3 | 123.9 | 105.2 |
| CuSO$_4$ | 133.0 | 115.2 | 83.3 | 50.5 |

As would be expected, the magnitude of the effect caused by interionic attractions depends upon the size of the charges of the solute ions. The equivalent conductance of 0.1N NaCl is about 84% of the $\Lambda_0$ value, the equivalent conductance of 0.1N BaCl$_2$ is about 75% of the $\Lambda_0$ value, and the equivalent conductance of 0.1N CuSO$_4$ is about 38% of the $\Lambda_0$ value.

Equivalent conductance measurements provide a method for the determination of the **degree of dissociation** ($\alpha$) of a weak electrolyte in solution. In a solution of a weak electrolyte such as acetic acid

$$H_2O + HC_2H_3O_2 \rightleftharpoons H_3O^+ + C_2H_3O_2^-$$

the fraction of the compound in ionic form increases as the solution is diluted; weak electrolytes are theoretically 100% ionic at infinite dilution. At ordinary concentrations the concentrations of ions in solutions of weak electrolytes are so small that interionic-attraction effects may be ignored. The undissociated compound, since it is uncharged, does not contribute to the conductance, and the equivalent conductance of a solution is attributed entirely to the fraction of the compound in ionic form. Since $\Lambda_0$ is the equivalent conductance that represents complete dissociation,

$$\Lambda = \alpha \Lambda_0$$

or

$$\alpha = \frac{\Lambda}{\Lambda_0}$$

The equivalent conductance at infinite dilution, $\Lambda_0$, of a strong electrolyte may be obtained by extrapolation of data obtained for solutions of various concentration (see Table 10.2). However, this procedure cannot be used to secure $\Lambda_0$ values for weak electrolytes since equivalent conductance values for solutions of ordinary concentrations are low and rise very rapidly as the concentration approaches zero (i.e., at high dilutions). In order to get $\Lambda_0$ values for weak electrolytes, the assumption is made that $\Lambda_0$ values for compounds represent the sum of the equivalent conductances of the constituent ions; this procedure is illustrated in Example 10.6.

---

**Example 10.6**

Given the following $\Lambda_0$ values at 25°C: HCl, 426.2 cm²/ohm equivalent; $NaC_2H_3O_2$, 91.0 cm²/ohm equivalent; NaCl, 126.5 cm²/ohm equivalent. What is $\Lambda_0$ for acetic acid, $HC_2H_3O_2$, at 25°C?

**Solution**

$$\Lambda_0(HC_2H_3O_2) = \Lambda_0(HCl) + \Lambda_0(NaC_2H_3O_2) - \Lambda_0(NaCl)$$
$$= (426.2 + 91.0 - 126.5) \text{ cm}^2/\text{ohm equivalent}$$
$$= 390.7 \text{ cm}^2/\text{ohm equivalent}$$

---

*Example 10.7*
At 25°C the equivalent conductance of 0.100$N$ $HC_2H_3O_2$ is 5.2 cm²/ohm equivalent; $\Lambda_0$ for acetic acid is 390.7 cm²/ohm equivalent. What is the degree of dissociation of 0.100$N$ $HC_2H_3O_2$ at 25°C?

*Solution*

$$\alpha = \frac{\Lambda}{\Lambda_0}$$

$$\alpha = \frac{5.2 \text{ cm}^2/\text{ohm equivalent}}{390.7 \text{ cm}^2/\text{ohm equivalent}}$$

$$\alpha = 0.013$$

At 25°C, 0.100$N$ acetic acid is 1.3% ionized.

**10.6 VOLTAIC CELLS**

A cell that is used as a source of electrical energy is called a voltaic cell or a galvanic cell after Alessandro Volta (1800) or Luigi Galvani (1780), who first experimented with the conversion of chemical energy into electrical energy.

The reaction between metallic zinc and copper(II) ions in solution is illustrative of a spontaneous change in which electrons are transferred.

$$Zn(s) + Cu^{2+}(aq) \rightarrow Zn^{2+}(aq) + Cu(s)$$

The exact mechanism by which electron transfer occurs is not known; however, we may represent the above reaction as a combination of two half reactions.

$$Zn(s) \rightarrow Zn^{2+}(aq) + 2e^-$$

$$2e^- + Cu^{2+}(aq) \rightarrow Cu(s)$$

In a voltaic cell these half reactions are made to occur at different electrodes so that the transfer of electrons takes place through the external electrical circuit rather than directly between zinc metal and copper(II) ions.

The cell diagramed in Figure 10.5 is designed to make use of this reaction to produce an electric current. The half cell on the left contains a zinc metal electrode and $ZnSO_4$ solution; the half cell on the right consists of a copper metal electrode in a solution of $CuSO_4$. The half cells are separated by a porous partition that prevents the mechanical mixing of the solutions but permits the passage of ions under the influence of the flow of electricity.

When the zinc and copper electrodes are joined by a wire, electrons flow from the zinc electrode to the copper electrode. At the zinc electrode the zinc metal is *oxidized* to zinc ions; this electrode is the anode, and the electrons that are the product of the oxidation *leave* the cell from this pole (Table 10.1). The electrons travel the external circuit to the copper

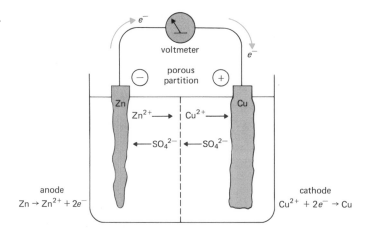

Figure 10.5 The Daniell Cell.

electrode where they are used in the reduction of copper(II) ions to metallic copper; the copper thus produced plates out on the electrode. The copper electrode is the cathode; here, the electrons *enter* the cell and *reduction* occurs.

Since electrons are produced at the zinc electrode, this anode is designated as the negative pole. Electrons travel from the negative pole to the positive pole in the external circuit of any voltaic cell when the cell is operating. The cathode, where electrons are used in the electrode reaction, is therefore the positive pole. Within the cell the movement of ions complete the electric circuit. At first glance, it is surprising that anions, which are negatively charged, should travel toward an anode that is the negative electrode. Conversely, cations, which carry a positive charge, travel toward the cathode, which is the positive pole.

Careful consideration of the electrode reactions provides the answer to this apparent anomaly. At the anode, zinc ions are being produced and electrons left behind in the metal. At all times the electrical neutrality of the solution is maintained; in the solution surrounding the electrode there must be as much negative charge from anions as there is positive charge from cations. Hence, $SO_4^{2-}$ ions move toward the anode to neutralize the effect of the $Zn^{2+}$ ions that are being produced. At the same time, zinc ions move away from the anode toward the cathode. At the cathode, electrons are being used to reduce $Cu^{2+}$ ions to copper metal. While the $Cu^{2+}$ ions are being discharged, more $Cu^{2+}$ ions move into the region surrounding the cathode to take the place of the ions being removed. If this did not occur, a surplus of $SO_4^{2-}$ ions would build up around the cathode.

The porous partition is added to prevent mechanical mixing of the solutions of the half cells. If $Cu^{2+}$ ions came into contact with the zinc metal electrode, electrons would be transferred directly rather than through the circuit. In the normal operation of the cell, this "short circuit" does not occur because the $Cu^{2+}$ ions move in a direction away from the zinc electrode.

Actually this cell would work if a solution of an electrolyte other than $ZnSO_4$ were used in the anode compartment, and if a metal other than copper were used for the cathode. However, the substitutes must be chosen so that the electrolyte in the anode compartment does not react with the zinc electrode and the cathode does not react with $Cu^{2+}$ ions.

**10.7 ELECTROMOTIVE FORCE**

If $1M$ $ZnSO_4$ and $1M$ $CuSO_4$ solutions are employed in the Daniell cell, the cell may be represented by the notation

$$Zn(s)|Zn^{2+}(1M)|Cu^{2+}(1M)|Cu(s)$$

in which the vertical lines represent phase boundaries. By convention, the substance forming the anode is listed first and is followed by the other materials comprising the cell listed in the order that one would encounter them leading from the anode to the cathode, the composition of which is given last.

Electric current is produced by a voltaic cell as a result of the **electromotive force** (emf) of the cell, which is measured in volts. The greater the tendency for the cell reaction to occur, the higher the emf of the cell. The emf of a given cell, however, also depends upon the concentrations of the substances used to make the cell.

A **standard emf,** $\mathscr{E}°$, pertains to the electromotive force of a cell, at 25°C, in which all reactants and products are present in their standard states. The standard state of a solid or a liquid is, of course, the pure solid or pure liquid itself. The standard state of a gas or a substance in solution is a defined state of *ideal unit activity*—that is, corrections are applied for deviations from ideality caused by intermolecular and interionic attractions (Section 9.13). For our discussion we shall make the assumption that the activity of ions may be represented by their molar concentrations and the activity of gases by their pressures in atmospheres. Hence, according to this approximation, a standard cell would contain ions at $1M$ concentrations and gases at 1 atm pressures. In the cell notations that follow, concentrations will be indicated only if they deviate from standard.

If the emf of a cell is to be used as a reliable measure of the tendency for the cell reaction to occur, the voltage must be the maximum value obtainable for the particular cell under consideration. If there is an appreciable flow of electricity during measurement, the voltage measured, $\mathscr{E}$, will be reduced by an amount, $IR$, equal to the product of the amount of current, $I$, and the resistance of the cell, $R$. In addition, when the cell delivers current, the electrode reactions produce concentration changes that reduce the voltage.

The emf of a cell, therefore, must be measured with no appreciable flow of electricity through the cell; this measurement is accomplished by the use of a potentiometer. The circuit of a potentiometer includes a current source of variable voltage and a means of measuring this voltage. The cell being studied is connected to the potentiometer circuit in such a way that the emf of the cell is opposed by the emf of the potentiometer current source.

If the emf of the cell is larger than that of the potentiometer, electrons will flow in the normal direction for a spontaneously discharging cell of that type. On the other hand, if the emf of the potentiometer current source is larger than that of the cell, electrons will flow in the opposite direction, thus causing the cell reaction to be reversed. When the two emf's are exactly balanced, no electrons flow; this voltage is the **reversible emf** of the cell. The emf of a standard Daniell cell is 1.10 V.

Faraday's laws apply to the cell reactions of voltaic, as well as electrolytic, cells. One precaution must be observed, however. Electricity is generated by the simultaneous oxidation and reduction half reactions that occur at the anode and cathode, respectively; both must occur if the cell is to deliver current. Two faradays of electricity will be produced, therefore, by the oxidation of 1 mol of zinc at the anode *together with* the reduction of 1 mol of $Cu^{2+}$ ions at the cathode. The partial equations

anode: $\qquad\qquad Zn \rightarrow Zn^{2+} + 2e^-$

cathode: $\quad 2e^- + Cu^{2+} \rightarrow Cu$

when read in terms of moles, represent the flow of two times Avogadro's number of electrons or the production of 2 F of electricity.

The quantity of electrical energy, in joules, produced by a cell is the product of the quantity of electricity delivered, in coulombs, and the emf of the cell, in volts (see Section 10.1). The electrical energy *produced* by the reaction between 1 mol of zinc metal and 1 mol of copper(II) ions may be calculated as follows:

$$-2(96,500 \text{ coulombs})(1.10 \text{ V}) = -212,000 \text{ J}$$

or since 1 cal = 4.184 J,

$$\frac{-212,000 \text{ J}}{4.18 \text{ J/cal}} = -50,700 \text{ cal}$$

The $\mathscr{E}^\circ$ used in the preceding calculation is the reversible emf of the standard Daniell cell and hence the maximum voltage for this cell. Therefore, the value secured (−50,700 cal) is the maximum work that can be obtained from the operation of this type of cell. The maximum *net* work[2]

---

[2] Some reactions proceed with an increase in volume, and the system must do work to expand against the atmosphere in order to maintain a constant pressure. The energy for this pressure–volume work is not available for any other purpose; it must be expended in this

that can be obtained from a chemical reaction conducted at a constant temperature and pressure is a measure of the *decrease* in the Gibbs free energy (Sections 9.4 and 14.5) of the system. Hence,

$$\Delta G = -nF\mathscr{E}$$

where $n$ is the number of moles of electrons transferred in the reaction (or the number of faradays produced), $F$ is the value of the faraday in appropriate units, and $\mathscr{E}$ is the emf in volts. If $F$ is expressed as 96,487 coulombs, $\Delta G$ is obtained in joules; if the value 23,061 cal/V is used for $F$, $\Delta G$ is obtained in calories. A change in free energy derived from a standard emf, $\mathscr{E}°$, is given the symbol $\Delta G°$.

The free energy change of a reaction is a measure of the tendency of the reaction to occur. If work must be done on a system to bring about a change, the change is not spontaneous. At constant temperature and pressure a spontaneous change is one from which net work can be obtained. Hence, for any spontaneous reaction the free energy of the system decreases; $\Delta G$ is negative. Since $\Delta G = -nF\mathscr{E}$, only if $\mathscr{E}$ is positive will the cell reaction be spontaneous and serve as a source of electrical energy.

**10.8 ELECTRODE POTENTIALS**

In the same way that a cell reaction may be regarded as the sum of two half reactions, the emf of a cell may be thought of as the sum of two half-cell potentials. However, it is impossible to determine the absolute value of the potential of a single half cell. A relative scale has been established by assigning a value of zero to the voltage of a standard reference half cell and expressing all half-cell potentials relative to this reference electrode.

The reference half cell used is the **standard hydrogen electrode,** which consists of hydrogen gas, at 1 atm pressure, bubbling over a platinum electrode (coated with finely divided platinum to increase its surface) that is immersed in an acid solution containing $H^+(aq)$ at unit activity. In Figure 10.6 a standard hydrogen electrode is shown connected by means of a salt bridge to a standard $Cu^{2+}/Cu$ electrode. A **salt bridge** is a tube filled with a concentrated solution of a salt (usually $KCl$), which conducts the current between the half cells but prevents the mixing of the solutions of the half cells. The cell of Figure 10.6 may be diagramed as

$$Pt|H_2|H^+\|Cu^{2+}|Cu$$

A double bar indicates a salt bridge. The hydrogen electrode is the anode, the copper electrode is the cathode, the emf of the cell is 0.34 V.

---

way if the reaction is to occur at constant pressure. Pressure–volume work is not included in the potentiometric measurement of the electrical work of any cell. Net work (or available work) is work other than pressure–volume work.

Figure 10.6 Standard hydrogen electrode and a Cu²⁺/Cu
electrode.

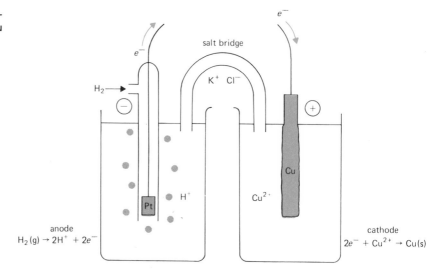

The cell emf is considered to be the sum of the half-cell potential for the oxidation half reaction (which we shall give the symbol $\mathscr{E}_{ox}$) and the half-cell potential for the reduction half reaction (which we shall indicate as $\mathscr{E}$ without subscript). For the cell of Figure 10.6,

anode:  $\qquad$ $H_2 \rightarrow 2H^+ + 2e^-$ $\qquad$ $\mathscr{E}^\circ_{ox} = 0.00$ V

cathode:  $\quad 2e^- + Cu^{2+} \rightarrow Cu$ $\qquad$ $\mathscr{E}^\circ = +0.34$ V

Since the hydrogen electrode is arbitrarily assigned a potential of zero, the entire cell emf is ascribed to the standard Cu²⁺/Cu electrode. The value +0.34 V is called the **standard electrode potential** of the Cu²⁺/Cu electrode; notice that electrode potentials are given for reduction half reactions.

There are several interpretations of the meaning of the positive sign given to this electrode potential. (1) The copper electrode is the positive electrode (cathode). (2) Copper(II) ions are more active electron acceptors than H⁺(aq) ions (greater tendency for reduction). (3) The reduction of Cu²⁺ ions is the spontaneous direction of change for the Cu²⁺/Cu electrode of this cell.

If a cell is constructed from a standard hydrogen electrode and a standard Zn²⁺/Zn electrode, the zinc electrode is the anode, and the emf of the cell is 0.76 V. Thus.

anode:  $\qquad$ $Zn \rightarrow Zn^{2+} + 2e^-$ $\qquad$ $\mathscr{E}^\circ_{ox} = +0.76$ V

cathode:  $\quad 2e^- + 2H^+ \rightarrow H_2$ $\qquad$ $\mathscr{E}^\circ = 0.00$ V

The value +0.76 V is sometimes called an **oxidation potential**, since it corresponds to an oxidation half reaction. An electrode potential, however, is a **reduction potential.** To obtain the electrode potential of the $Zn^{2+}/Zn$ couple, we must change the sign of the oxidation potential so that the potential corresponds to the reverse half reaction, a reduction.

$$2e^- + Zn^{2+} \rightarrow Zn \qquad \mathscr{E}^\circ = -0.76 \text{ V}$$

The negative sign of this electrode potential may be interpreted in several ways. (1) The zinc electrode is the negative electrode (anode). (2) Zinc ions are less active electron acceptors than $H^+(aq)$ ions (lesser tendency for reduction). (3) Oxidation of Zn (*not* reduction of $Zn^{2+}$) is the spontaneous direction of change for the $Zn^{2+}/Zn$ electrode of this cell.

It is not necessary to use a cell containing a standard hydrogen electrode to obtain a standard electrode potential. For example, the standard potential of the $Ni^{2+}/Ni$ electrode may be determined from the cell

$$Ni \,|\, Ni^{2+} \,\|\, Cu^{2+} \,|\, Cu$$

The emf of this cell is 0.59 V, and the nickel electrode functions as the anode.

$$Ni + Cu^{2+} \rightarrow Ni^{2+} + Cu \qquad \mathscr{E}^\circ = +0.59 \text{ V}$$

The standard electrode potential of the $Cu^{2+}/Cu$ electrode has been determined.

$$2e^- + Cu^{2+} \rightarrow Cu \qquad \mathscr{E}^\circ = +0.34 \text{ V}$$

If we subtract the $Cu^{2+}/Cu$ half reaction from the cell reaction and subtract the half-cell potential from the cell emf, we obtain

$$Ni \rightarrow Ni^{2+} + 2e^- \qquad \mathscr{E}^\circ_{ox} = +0.25 \text{ V}$$

The electrode potential is therefore

$$2e^- + Ni^{2+} \rightarrow Ni \qquad \mathscr{E}^\circ = -0.25 \text{ V}$$

Standard electrode potentials are listed in Table 10.3, and a more complete list is found in the appendix.[3] The table is constructed with the most positive electrode potential (greatest tendency for reduction) at the bottom. Hence, if a pair of electrodes is combined to make a voltaic cell, the reduction half reaction (cathode) of the cell will be that listed for the electrode that stands lower in the table, and the oxidation half reaction (anode) will be the *reverse* of that shown for the electrode that stands higher in the table.

---

[3] The definition of the standard electrode potential used agrees with that suggested by the International Union of Pure and Applied Chemistry (IUPAC) in 1953. Prior to this time, the common American practice was to tabulate oxidation potentials rather than electrode potentials.

Table 10.3 Standard electrode potentials at 25°C[a]

| HALF REACTION | $\mathscr{E}°$ (volts) |
|---|---|
| $Li^+ + e^- \rightleftharpoons Li$ | −3.045 |
| $K^+ + e^- \rightleftharpoons K$ | −2.925 |
| $Ba^{2+} + 2e^- \rightleftharpoons Ba$ | −2.906 |
| $Ca^{2+} + 2e^- \rightleftharpoons Ca$ | −2.866 |
| $Na^+ + e^- \rightleftharpoons Na$ | −2.714 |
| $Mg^{2+} + 2e^- \rightleftharpoons Mg$ | −2.363 |
| $Al^{3+} + 3e^- \rightleftharpoons Al$ | −1.662 |
| $2H_2O + 2e^- \rightleftharpoons H_2 + 2OH^-$ | −0.82806 |
| $Zn^{2+} + 2e^- \rightleftharpoons Zn$ | −0.7628 |
| $Cr^{3+} + 3e^- \rightleftharpoons Cr$ | −0.744 |
| $Fe^{2+} + 2e^- \rightleftharpoons Fe$ | −0.4402 |
| $Cd^{2+} + 2e^- \rightleftharpoons Cd$ | −0.4029 |
| $Ni^{2+} + 2e^- \rightleftharpoons Ni$ | −0.250 |
| $Sn^{2+} + 2e^- \rightleftharpoons Sn$ | −0.136 |
| $Pb^{2+} + 2e^- \rightleftharpoons Pb$ | −0.126 |
| $2H^+ + 2e^- \rightleftharpoons H_2$ | 0 |
| $Cu^{2+} + 2e^- \rightleftharpoons Cu$ | +0.337 |
| $Cu^+ + e^- \rightleftharpoons Cu$ | +0.521 |
| $I_2 + 2e^- \rightleftharpoons 2I^-$ | +0.5355 |
| $Fe^{3+} + e^- \rightleftharpoons Fe^{2+}$ | +0.771 |
| $Ag^+ + e^- \rightleftharpoons Ag$ | +0.7991 |
| $Br_2 + 2e^- \rightleftharpoons 2Br^-$ | +1.0652 |
| $O_2 + 4H^+ + 4e^- \rightleftharpoons 2H_2O$ | +1.229 |
| $Cr_2O_7^{2-} + 14H^+ + 6e^- \rightleftharpoons 2Cr^{3+} + 7H_2O$ | +1.33 |
| $Cl_2 + 2e^- \rightleftharpoons 2Cl^-$ | +1.3595 |
| $MnO_4^- + 8H^+ + 5e^- \rightleftharpoons Mn^{2+} + 4H_2O$ | +1.51 |
| $F_2 + 2e^- \rightleftharpoons 2F^-$ | +2.87 |

[a] Data from A. J. de Bethune and N. A. Swendeman Loud, "Table of Electrode Potentials and Temperature Coefficients," pp. 414–424 in *Encyclopedia of Electrochemistry* (C. A. Hampel, editor), Van Nostrand Reinhold, New York, 1964, and from A. J. de Bethune and N. A. Swendeman Loud, *Standard Aqueous Electrode Potentials and Temperature Coefficients*, 19 pp., C. A. Hampel, publisher, Skokie, Illinois, 1964.

For example, consider a cell constructed from standard $Ni^{2+}/Ni$ and $Ag^+/Ag$ electrodes. The table entries for these electrodes are

$$Ni^{2+} + 2e^- \rightleftharpoons Ni \qquad \mathscr{E}° = -0.250 \text{ V}$$

$$Ag^+ + e^- \rightleftharpoons Ag \qquad \mathscr{E}° = +0.799 \text{ V}$$

Of the two ions, the $Ag^+$ ion shows the greater tendency for reduction. The $Ag^+/Ag$ electrode is therefore the cathode, and the $Ni^{2+}/Ni$ electrode is the anode. The half reaction that takes place at an anode is an oxidation, and the half-cell potential is an oxidation potential.

anode: $\qquad Ni \rightarrow Ni^{2+} + 2e^- \qquad \mathscr{E}°_{ox} = +0.250 \text{ V}$

cathode: $\quad 2e^- + 2Ag^+ \rightarrow 2Ag \qquad \mathscr{E}° = +0.799 \text{ V}$

The cell reaction and cell emf may be obtained by addition.

$$Ni + 2Ag^+ \rightarrow Ni^{2+} + 2Ag \qquad \mathscr{E}° = +1.049 \text{ V}$$

Notice that the half reaction for the reduction of $Ag^+$ must be multiplied by 2 before the addition so that the electrons lost and gained in the half reactions will cancel. The $\mathscr{E}°$ for the $Ag^+/Ag$ electrode, however, is *not* multiplied by 2. The magnitude of an electrode potential depends upon the temperature and the concentrations of materials used in the construction of the half cell; these variables are fixed for *standard* electrode potentials. Indication of the stoichiometry of the cell reaction does not imply that a concentration change has been made.

Actually the half reactions implied by the half-cell potentials are

$$2\cancel{H}^+ + Ni \rightarrow \cancel{H_2} + Ni^{2+}$$
$$\cancel{H_2} + 2Ag^+ \rightarrow 2\cancel{H}^+ + 2Ag$$
$$\overline{Ni + 2Ag^+ \rightarrow Ni^{2+} + 2Ag}$$

Notice, however, that in the addition of these half reactions, the $H_2$ molecules and $H^+$ ions cancel.

An alternative way to use the table of electrode potentials to calculate a cell emf consists of subtracting the electrode potential of the anode (value nearer the top of the table column) from the electrode potential of the cathode (value nearer the bottom of the table column). This mathematical procedure amounts to the same thing as changing the sign of the electrode potential of the anode (to correspond to an oxidation) and adding the value to the electrode potential of the cathode.

By convention, the *anode* of a cell is indicated on the *left* in a cell notation. Thus, the $Cd^{2+}/Cd$ electrode is the anode of the cell

$$Cd \,|\, Cd^{2+} \,\|\, Ag^+ \,|\, Ag$$

and the cell reaction indicated is

$$Cd + 2Ag^+ \rightarrow Cd^{2+} + 2Ag$$

The standard emf of the cell is

$$\mathscr{E}° = \mathscr{E}°_{\text{right}} - \mathscr{E}°_{\text{left}}$$

Substitution of $\mathscr{E}°$ values from the table gives

$$\mathscr{E}° = (+0.799) - (-0.403)$$
$$= +1.202 \text{ V}$$

The positive sign of the cell emf indicates that the cell is properly diagramed and the reaction proceeds as indicated.

If we had mistakenly diagramed the cell as

$$Ag \,|\, Ag^+ \,\|\, Cd^{2+} \,|\, Cd$$

which corresponds to the reaction

$$2Ag + Cd^{2+} \rightarrow 2Ag^+ + Cd$$

we would have obtained the result

$$\begin{aligned}\mathscr{E}^\circ &= \mathscr{E}^\circ_{\text{right}} - \mathscr{E}^\circ_{\text{left}}\\ &= (-0.403) - (+0.799)\\ &= -1.202 \text{ V}\end{aligned}$$

The negative sign of the cell emf means that the cell will not function in the manner indicated by the cell notation. Rather, the cell reaction would be the reverse of that shown, and the diagram should be reversed.

Electrode potentials are also useful for the evaluation of oxidation–reduction reactions that take place outside of electrochemical cells. Either of the two ways described for the analysis of a cell may be used to evaluate an oxidation–reduction reaction. The half reactions of Table 10.3 are listed as reductions, from left to right, with the oxidized form of the element, molecule, or ion accepting electrons to produce the reduced form.

$$\text{(oxidized form)} + ne^- \rightleftharpoons \text{(reduced form)}$$

Because of the way in which the table is constructed (greatest tendency for reduction at the bottom), any *reduced form,* at unit activity, should be able to reduce the *oxidized form,* at unit activity, of a half reaction listed below it in the table. From the positions in the table of the following half reactions

$$2e^- + I_2 \rightleftharpoons 2I^- \qquad \mathscr{E}^\circ = +0.536 \text{ V}$$

$$2e^- + Cl_2 \rightleftharpoons 2Cl^- \qquad \mathscr{E}^\circ = +1.360 \text{ V}$$

we would correctly predict that the reduced form, $I^-$, should be able to reduce the oxidized form, $Cl_2$.

$$Cl_2(g) + 2I^-(aq) \rightarrow 2Cl^-(aq) + I_2(s) \qquad \mathscr{E}^\circ = +0.824$$

The species with the greatest tendency toward oxidation (best reducing agent) is the reduced form of the half reaction at the top of the table (Li in Table 10.3); the oxidized form of the half reaction at the bottom of the table ($F_2$ in Table 10.3) has the greatest tendency to be reduced (best oxidizing agent).

There are several factors that must be kept in mind when using a table of electrode potentials to predict the course of a chemical reaction. Because $\mathscr{E}$ changes with changes in concentration, many presumably unfavored reactions can be made to occur by altering the concentrations of the reacting species. In addition, some theoretically favored reactions proceed at such a slow rate that they are of no practical consequence.

Correct use of the table also demands that all pertinent half reactions of a given element be considered before making a prediction. On the basis of the half reactions

$$3e^- + Fe^{3+} \rightleftharpoons Fe \qquad \mathscr{E}^\circ = -0.036 \text{ V}$$

$$2e^- + 2H^+ \rightleftharpoons H_2 \qquad \mathscr{E}^\circ = 0.000 \text{ V}$$

one might predict that the products of the reaction of iron with $H^+$ would be hydrogen gas and $Fe^{3+}$ ions ($\mathscr{E}^\circ$ for the complete reaction, $+0.036$ V). However, the oxidation state iron(II) lies between metallic iron and the oxidation state iron(III). Once an iron atom has lost two electrons and becomes a $Fe^{2+}$ ion, further oxidation is opposed, as may be seen from the reverse of the following

$$e^- + Fe^{3+} \rightleftharpoons Fe^{2+} \qquad \mathscr{E}^\circ = +0.771 \text{ V}$$

Thus, the reaction yields $Fe^{2+}$ ions only. This fact could have been predicted by an examination of the half reaction

$$2e^- + Fe^{2+} \rightleftharpoons Fe \qquad \mathscr{E}^\circ = -0.440 \text{ V}$$

The $\mathscr{E}_{ox}$ for the production of $Fe^{2+}$ ions from the reaction of iron metal and $H^+$ ions ($+0.440$ V) is greater than that for the production of $Fe^{3+}$ ions ($+0.036$V), and hence the former is favored.

We may summarize the electrode potentials for iron and its ions as follows:

$$Fe^{3+} \xrightarrow{\;+0.771V\;} Fe^{2+} \xrightarrow{\;-0.440V\;} Fe$$
$$\underset{-0.036V}{\underline{\hspace{8cm}}}$$

The preceding predictions are immediately evident from this diagram, if we remember that oxidation is the reverse of the relation corresponding to an electrode potential.

Occasionally an oxidation state of an element is unstable toward disproportionation. The electrode potentials for copper and its ions may be summarized as follows:

$$Cu^{2+} \xrightarrow{\;+0.153V\;} Cu^+ \xrightarrow{\;+0.521V\;} Cu$$
$$\underset{+0.337V}{\underline{\hspace{8cm}}}$$

From this we see that the $Cu^+$ ion is not a very stable one. In water, $Cu^+$ ions disproportionate to copper metal and $Cu^{2+}$ ions.

$$2Cu^+(aq) \rightarrow Cu(s) + Cu^{2+}(aq)$$

The $\mathscr{E}^\circ$ for this reaction is $+0.521 - 0.153 = +0.368$ V. Species instable toward disproportionation may be readily recognized by the fact that the electrode potential for the reduction to the next lower oxidation state is more positive than the electrode potential for the couple with the next higher oxidation state. An inspection of the diagram for iron and its ions shows that the $Fe^{2+}$ ion is stable toward such disproportionation.

**10.9 EFFECT OF CONCENTRATION ON CELL POTENTIALS**

The potential of a cell in which the constituents are present at concentrations other than standard may be calculated by means of the **Nernst equation,** named for Walther Nernst, who developed it in 1889.

$$\mathscr{E} = \mathscr{E}^\circ - \frac{RT}{nF} \ln Q$$

where $\mathscr{E}^\circ$ is the standard cell emf, $R$ is the gas constant (1.987 cal/°K mol), $T$ is the absolute temperature, $n$ is the number of moles of electrons transferred in the reaction (or the number of faradays involved), $F$ is the faraday constant (23,060 cal/V mol), and $\ln Q$ is the natural logarithm of the reaction quotient.

The reaction quotient, $Q$, is a fraction derived from the activities of the dissolved substances and gases employed in the cell; the activity of each substance is raised to a power equal to the coefficient of that substance in the balanced chemical equation. The *numerator* of $Q$ is the product of the activity terms for the materials on the *right* of the chemical equation; the *denominator* of $Q$ is the product of the activity terms for the substances on the *left* of the chemical equation. Since the activity of a pure solid is assumed to be unity at all times, the activity term for a solid is always equal to 1.

For a hypothetical cell reaction,

$$w\text{W} + x\text{X} \rightarrow y\text{Y} + z\text{Z}$$

where the lower case letters represent the coefficients of the balanced chemical equation,

$$Q = \frac{(a_Y)^y (a_Z)^z}{(a_W)^w (a_X)^x}$$

For our work we shall make the assumptions that the activity of a substance in solution is given by the molar concentration of the substance and the activity of a gas is equal to the partial pressure of the gas in atmospheres; these assumptions lead to $\mathscr{E}$ values that are approximately correct.

If we substitute the values of the constants into the Nernst equation and multiply by 2.303 to convert the natural logarithm into a logarithm to the base 10, for $T = 298.2°K$ (which is 25°C),

$$\mathscr{E} = \mathscr{E}^\circ - \frac{0.05916}{n} \log Q$$

When the activities of all substances are unity (standard states), $Q = 1$, $\log Q = 0$, and $\mathscr{E} = \mathscr{E}°$. The Nernst equation may be used to determine the emf of a cell constructed from nonstandard electrodes or to calculate the electrode potential of a half cell in which all species are not present at unit activity.

---

### Example 10.8

What is the electrode potential of a $Zn^{2+}/Zn$ electrode in which the concentration of $Zn^{2+}$ ions is $0.1M$?

$$\mathscr{E} = \mathscr{E}° - \frac{0.0592}{2} \log \left( \frac{1}{[Zn^{2+}]} \right)$$

$\mathscr{E}°$ for the $Zn^{2+}/Zn$ electrode is $-0.76$ V.

#### Solution

The partial equation

$$2e^- + Zn^{2+} \rightarrow Zn$$

shows 2 electrons gained. If the symbol $[Zn^{2+}]$ is used to designate the molar concentration of $Zn^{2+}$ ions,

$$\mathscr{E} = -0.76 - \frac{0.0592}{2} \log \left( \frac{1}{0.1} \right)$$

$$\mathscr{E} = -0.76 - 0.0296(1) = -0.79 \text{ V}$$

---

### Example 10.9

(a) What is the potential for the cell

$$Ni \,|\, Ni^{2+}(0.01M) \,\|\, Cl^-(0.2M) \,|\, Cl_2(1 \text{ atm}) \,|\, Pt$$

(b) What is $\Delta G$ for the cell reaction?

#### Solution

(a) Two faradays of electricity are involved in the cell reaction

$$Ni + Cl_2 \rightarrow Ni^{2+} + 2Cl^-$$

and therefore $n = 2$. The standard cell emf is

$$\mathscr{E}° = \mathscr{E}°(Cl_2/Cl^-) - \mathscr{E}°(Ni^{2+}/Ni)$$
$$= (+1.36) - (-0.25) = +1.61 \text{ V}$$

Therefore,

$$\mathscr{E} = \mathscr{E}° - \frac{0.0592}{2} \log \left( \frac{[Cl^-]^2 [Ni^{2+}]}{[Cl_2]} \right)$$

$$\mathscr{E} = +1.61 - \frac{0.0592}{2} \log \left( \frac{(0.2)^2(0.01)}{(1)} \right)$$

$$\mathscr{E} = +1.61 - 0.0296 \log (0.0004)$$

$$\mathscr{E} = +1.61 + 0.10 = +1.71 \text{ V}$$

(b)

$$\Delta G = -nF\mathscr{E}$$

$$\Delta G = -2(23,100 \text{ cal/V})(1.71 \text{ V})$$
$$= -79,000 \text{ cal} = -7.90 \text{ kcal}$$

---

### Example 10.10

What is the $\mathscr{E}$ of the cell

$$Sn \,|\, Sn^{2+}(1.0M) \,\|\, Pb^{2+}(0.0010M) \,|\, Pb$$

#### Solution

The following may be obtained from a table of standard electrode potentials

$$Sn \rightarrow Sn^{2+} + 2e^- \qquad \mathscr{E}°_{ox} = +0.136 \text{ V}$$

$$2e^- + Pb^{2+} \rightarrow Pb \qquad \mathscr{E}° = -0.126 \text{ V}$$

Thus the reaction in a standard cell is

$$Sn + Pb^{2+} \rightarrow Sn^{2+} + Pb \qquad \mathscr{E}° = +0.010 \text{ V}$$

For the cell as diagramed in the problem,

$$\mathscr{E} = \mathscr{E}° - \frac{0.0592}{2} \log \left( \frac{[Sn^{2+}]}{[Pb^{2+}]} \right)$$

$$\mathscr{E} = +0.010 - \frac{0.0592}{2} \log \left( \frac{1.0}{0.0010} \right)$$

$$\mathscr{E} = +0.010 - 0.0296(3)$$
$$= +0.010 - 0.089 = -0.079 \text{ V}$$

This result means that the cell will not function in the manner implied by the diagram.

Instead, it would operate in the reverse order and in a direction opposite to that of the standard cell. The cell is properly diagramed

$$Pb \mid Pb^{2+}(0.0010M) \| Sn^{2+}(1.0M) \mid Sn$$

and the reaction of the cell is

$$Pb + Sn^{2+} \rightarrow Pb^{2+} + Sn \qquad \mathscr{E} = +0.079 \text{ V}$$

From this example we see that concentration effects can sometimes reverse the direction that a reaction is expected to take.

---

Notice that the results of the preceding examples are qualitatively in agreement with predictions based on Le Chatelier's principle (Section 9.5). Increasing the concentrations of reactants and decreasing the concentrations of products would be expected to increase the driving force of the reaction and lead to a more positive $\mathscr{E}$. On the other hand, decreasing the concentrations of reactants and increasing the concentrations of products would be expected to retard a reaction and result in a more negative $\mathscr{E}$.

**10.10** CONCENTRATION CELLS

Since an electrode potential depends upon the concentration of the ions used in the electrode, a cell may be constructed from two electrodes composed of the same materials but differing in concentration of ions. For example,

$$Cu \mid Cu^{2+}(0.010M) \| Cu^{2+}(0.10M) \mid Cu$$

From the partial equation,

$$2e^- + Cu^{2+} \rightleftharpoons Cu \qquad \mathscr{E}° = +0.34 \text{ V}$$

and Le Chatelier's principle, we can predict that *increasing* the concentration of $Cu^{2+}$ ions would drive the reaction to the right and raise the reduction potential, whereas *decreasing* the concentration of $Cu^{2+}$ ions would drive the reaction to the left and raise the oxidation potential (or lower the reduction potential). Hence, of the two electrodes in the cell diagramed, the left electrode would have the stronger tendency for oxidation and the right would have the stronger tendency for reduction. The "reaction" of the cell is

$$Cu + Cu^{2+}(0.10M) \rightarrow Cu^{2+}(0.010M) + Cu$$

$\mathscr{E}°$ for this cell is zero since the same electrode is involved in each half cell. Then

$$\mathscr{E} = 0.00 - \frac{0.0592}{2} \log \left( \frac{0.010}{0.10} \right)$$

$$\mathscr{E} = -0.0296(-1) = +0.0296 \text{ V}$$

**10.11 ELECTRODE POTENTIALS AND ELECTROLYSIS**

Electrode potentials are determined under reversible conditions. The measured voltage of a spontaneously discharging voltaic cell will not equal the calculated electromotive force for the cell since in the production of an electric current, the cell is operating in an irreversible manner. The cell emf calculated by means of electrode potentials is the maximum voltage that the cell can develop and is determined by the application of a measured external voltage of a magnitude that will just stop the flow of electrons. If an external voltage infinitesimally greater than that which the cell develops were applied, the cell would run in the reverse direction; electrolysis would occur. Electrode potentials, therefore, give the minimum voltage necessary for electrolysis.

In practice, however, it is found that a higher voltage than calculated must be applied to effect electrolysis. Part of the increased voltage is required to overcome the electrical resistance of the cell, and the remainder is needed to overcome two additional effects. The first is called **concentration polarization** and is caused by the changes in the concentration of the electrolyte around the electrodes as the electrolysis proceeds and ions are removed or produced. The concentration gradient set up within the cell is, in effect, a concentration cell and produces a back emf opposing the applied voltage. Concentration polarization can be reduced by stirring the electrolyte during the course of the electrolysis.

The other effect that makes the voltage required for an electrolysis higher than the reversible emf is called **overvoltage.** Overvoltage is thought to be caused by a slow rate of reaction at the electrodes; excess applied voltage is required to make the electrolysis proceed at an appreciable rate. Overvoltages for the deposition of metals are low, but those required for the liberation of hydrogen gas or oxygen gas are appreciable, vary with the material used for the electrode, and may amount to a volt or more.

Electrolysis is an important procedure for the quantitative analysis of ores, alloys, and other metal-containing compounds and mixtures. As the voltage is increased in the electrolysis of an aqueous solution, ions or molecules are discharged at the *cathode* in order of decreasing value of electrode potential—the most positive value (nearest to the bottom of the table) first. In addition, the reduction of water is possible. The $\mathscr{E}°$ value for this reduction is $-0.828$ V, but in neutral aqueous solutions the concentration of $OH^-$ is $10^{-7}M$, not $1M$. The $\mathscr{E}$ value for neutral aqueous solutions, obtained by means of the Nernst equation, is

$$2e^- + 2H_2O \rightleftharpoons H_2 + 2OH^- \qquad \mathscr{E} = -0.414 \text{ V}$$

One might predict that it would be impossible to reduce any aqueous species with an electrode potential more negative than $-0.414$ V, the reduction potential for water. However, because of the high hydrogen overvoltage, it is possible to plate out, on the cathode, metals that have electrode potentials more negative than the one for the reduction of water—down to, and including, zinc ($\mathscr{E}° = -0.763$ V).

At the *anode* the species present are oxidized in inverse order of electrode potential—most negative value (uppermost on the table) first. The half reactions are the reverse of those listed in the table. In the electrolysis of aqueous solutions, the oxidation of water is possible. For neutral aqueous solutions (concentration of $H^+$, $10^{-7}M$), this oxidation is the reverse of

$$4e^- + 4H^+ + O_2 \rightleftharpoons 2H_2O \qquad \mathscr{E} = +0.815 \text{ V}$$

For the liberation of oxygen, however, high overvoltages are required. Thus, the oxidation of chloride ion to chlorine gas takes place in the electrolysis of aqueous solutions of chlorides.

$$2e^- + Cl_2(g) \rightleftharpoons 2Cl^- \qquad \mathscr{E}° = +1.36 \text{ V}$$

The overvoltage of chlorine is less than the overvoltage of oxygen.

The products of electrolysis vary with the concentration of ions in the solution, since the half-cell emf's are dependent upon concentrations. For example, the electrolysis of *dilute* solutions of chlorides yields oxygen at the anode rather than chlorine. Furthermore, after the primary electrode reactions in which electrons are transferred, secondary reactions may occur. Thus, if chlorine is liberated in an alkaline solution, $ClO^-$ or $ClO_3^-$ ions are formed from the reaction of the chlorine with the hydroxide ions; if chlorine is liberated on a silver anode, AgCl is formed.

---

*Example 10.11*

A solution is $1M$ in each of the following: $AgNO_3$, $Cu(NO_3)_2$, $Fe(NO_3)_2$, and $NaNO_3$. In the electrolysis of this solution, between inert electrodes, the voltage is gradually increased. What would be the order in which the materials would plate out on the cathode, and at what voltage would each be discharged? Assume that the overvoltage of oxygen is 0.600 V, the overvoltage of hydrogen is 0.300 V, and the solution is neutral. Ignore concentration polarization and cell resistance.

*Solution*

The only species at the anode (which is inert) are the nitrate ion and water. Since it is impossible to oxidize the nitrate ion, water will be oxidized:

$$2H_2O \rightarrow O_2(g) + 4H^+ + 4e^- \qquad \mathscr{E}_{ox} = -0.815 \text{ V}$$

If the overvoltage of oxygen is taken into account, the half-cell potential for the oxidation is $-1.415$ V.

The following reductions should be considered as possible cathode half reactions:

$$e^- + Na^+ \rightarrow Na \qquad \mathscr{E}° = -2.714 \text{ V}$$

$$2e^- + Fe^{2+} \rightarrow Fe \qquad \mathscr{E}° = -0.440 \text{ V}$$

$$2e^- + Cu^{2+} \rightarrow Cu \qquad \mathscr{E}° = +0.337 \text{ V}$$

$$e^- + Ag^+ \rightarrow Ag \qquad \mathscr{E}° = +0.799 \text{ V}$$

In addition, the reduction of water

$$2e^- + 2H_2O \rightarrow H_2(g) + 2OH^-$$

$$\mathscr{E} = -0.414 \text{ V}$$

must be considered. If the overvoltage of hydrogen is taken into account, the reduction potential for water is $-0.714$ V.

The silver will plate out first. The voltage at which this will occur will be equal to the oxidation potential of water ($-1.415$ V) plus the reduction potential of Ag($+0.799$ V), or $-0.616$ V. (The values are negative since the voltage is applied.) After the silver has been removed, the copper will plate out at a voltage of $-1.088$ V, followed by the iron at a voltage of $-1.855$ V. The reduction of water (half-cell potential = $-0.714$ V) will occur before the reduction of sodium ions (half-cell potential = $-2.714$ V). Hence, at a voltage of 2.129 V, hydrogen will be evolved at the cathode. Sodium ions can be reduced only in the absence of water, that is, by the electrolysis of molten sodium salts.

**10.12 SOME COMMERCIAL VOLTAIC CELLS**

Several voltaic cells are of commercial importance. The **dry cell** (Figure 10.7) consists of a zinc metal container (which serves as the anode) that is filled with a moist paste of ammonium chloride and zinc chloride and contains a graphite electrode (the cathode) surrounded by manganese dioxide. The electrode reactions are complex, but they may be approximately represented by

anode: $\qquad\qquad$ Zn $\rightarrow$ Zn$^{2+}$ + 2$e^-$

cathode: $\quad 2e^- + 2MnO_2 + 2NH_4^+ \rightarrow Mn_2O_3 + H_2O + 2NH_3$

The dry cell generates a voltage of approximately 1.25 to 1.50 V.

A newer type of dry cell which has found use in small electrical devices (such as hearing aids) consists of a zinc container as the anode, a carbon rod as the cathode, and moist mercury(II) oxide mixed with potas-

Figure 10.7 The dry cell.

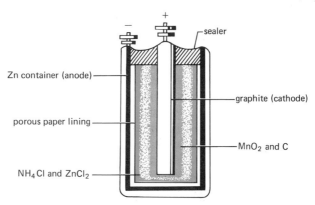

Zn container (anode)

porous paper lining

NH$_4$Cl and ZnCl$_2$

sealer

graphite (cathode)

MnO$_2$ and C

sium hydroxide as the electrolyte. A lining of porous paper keeps the electrolyte separated from the zinc anode. The cell has a potential of approximately 1.35 V.

anode: $\quad\quad\quad\quad\quad$ $Zn + 2OH^- \rightarrow Zn(OH)_2 + 2e^-$

cathode: $\quad$ $2e^- + HgO + H_2O \rightarrow Hg + 2OH^-$

The **lead storage cell** consists of a lead anode and a grid of lead packed with lead dioxide as the cathode. The electrolyte is sulfuric acid, and the half-cell reactions are

anode: $\quad\quad\quad\quad\quad\quad$ $Pb(s) + SO_4^{2-} \rightarrow PbSO_4(s) + 2e^-$

cathode: $\quad$ $2e^- + PbO_2(s) + SO_4^{2-} + 4H^+ \rightarrow PbSO_4(s) + 2H_2O$

In practice, the current obtainable from a lead storage cell is increased by constructing the cell from a number of cathode plates joined together and arranged alternately with a number of anode plates, which are also joined together. The potential difference of one cell is approximately 2 V. A storage battery consists of three or six such cells joined in series to produce a 6- or a 12-volt battery.

The electrode reactions of the storage battery can be reversed by the application of an external current source and in this manner the battery can be recharged. Since sulfuric acid is consumed as the storage battery delivers current, the state of charge of the battery can be determined by measuring the density of the battery electrolyte.

The **nickel–cadmium storage cell** has a longer life than the lead storage cell but is more expensive to manufacture.

anode: $\quad\quad\quad\quad\quad$ $Cd(s) + 2OH^- \rightarrow Cd(OH)_2 + 2e^-$

cathode: $\quad$ $2e^- + NiO_2(s) + 2H_2O \rightarrow Ni(OH)_2 + 2OH^-$

The potential of each cell of a nickel–cadmium battery is approximately 1.4 V, and the battery is rechargeable.

**10.13 FUEL CELLS** Electrical cells that are designed to convert the energy from the combustion of fuels such as hydrogen, carbon monoxide, or methane directly into electrical energy are called fuel cells. Since in theory 100% of the free energy released by a combustion ($\Delta G$) should be obtainable from an efficient fuel cell, extensive research into their development is currently being undertaken. Although approximately only 60 to 70% efficiency has been realized as yet, present fuel cells are about twice as efficient as processes in which the heat of combustion is used to generate electricity by mechanical means.

In a typical fuel cell, hydrogen and oxygen are bubbled through porous carbon electrodes into concentrated aqueous sodium hydroxide or potassium hydroxide. Catalysts are incorporated in the electrodes.

$$C \mid H_2(g) \mid OH^- \mid O_2(g) \mid C$$

The gaseous materials are consumed and are continuously supplied. The electrode reactions are

anode: $\qquad$ $2H_2(g) + 4OH^- \rightarrow 4H_2O + 4e^-$

cathode: $\quad 4e^- + O_2(g) + 2H_2O \rightarrow 4OH^-$

The complete cell reaction is

$$2H_2(g) + O_2(g) \rightarrow 2H_2O(1)$$

The cell is maintained at an elevated temperature, and the water produced by the cell reaction evaporates as it is formed.

Research in the design of fuel cells is currently being directed toward lowering the cost of the processes. Another problem involves the development of electrode catalysts that will cause the half-cell reactions to occur more rapidly.

## Problems

**10.1** Given the following: 1 F = 96,490 coulombs, 1 cal = 4.184 J, 1 volt coulomb = 1 J, Avogadro's number = $6.022 \times 10^{23}$. Determine (a) the charge on a single electron in coulombs, (b) the number of calories equal to 1 electron volt, (c) the number of joules equal to 1 electron volt, (d) the value of the electron volt in kcal/mol, (e) the value of the electron volt in kJ/mol, (f) the value of the faraday in cal/V, (g) the value of the faraday in J/V.

**10.2** What is the effect of an increase in temperature on the conductivity of metals and on the conductivity of solutions of electrolytes? Discuss the causes of resistance to the current in these two types of conductors.

**10.3** Sketch a cell for the electrolysis of $CdCl_2$ between inert electrodes. On the sketch indicate: (a) the signs of the electrodes, (b) the cathode and anode, (c) the directions in which the ions move, (d) the direction in which the electrons move, (e) the electrode reactions.

**10.4** Sketch a voltaic cell in which the reaction is

$$Cd + Cl_2 \rightarrow Cd^{2+} + 2Cl^-$$

On the sketch indicate: (a) the signs of the electrodes, (b) the cathode and anode, (c) the directions in which the ions move, (d) the direction in which the electrons move, (e) the electrode reactions, (f) the cell voltage.

**10.5** A $Co^{2+}$ solution is electrolyzed using a current of 0.400 amp. What weight of Co is plated out in 30.0 minutes?

**10.6** An acidic solution containing $BiO^+$ ions is electrolyzed using a current of 0.750 amp. What weight of Bi metal is plated out in 45.0 minutes? Write the equation for the reaction at the cathode.

**10.7** How many minutes will it take to plate out 10.0 g of Cu metal from a $CuSO_4$ solution using a current of 3.00 amp?

**10.8** (a) If 0.690 g of Ag is deposited on the cathode of a silver coulometer, how many coulombs have passed through the circuit? (b) If the process takes 20.0 minutes, what was the rate of the current?

**10.9** In the Hall process, aluminum is produced by the electrolysis of molten $Al_2O_3$. The electrode reactions are

anode:    $C + 2O^{2-} \rightarrow CO_2 + 4e^-$

cathode:    $3e^- + Al^{3+} \rightarrow Al$

In the process the carbon of which the anode is composed is gradually consumed by the anode reaction. What is the weight loss of the anode when 1.00 kg of aluminum is prepared?

**10.10** How long would it take to produce enough aluminum by the Hall process (see Problem 10.9) to make a case of aluminum softdrink cans (24 cans) if each can uses 5.00 g of Al, a current of 50,000 amp is employed, and the efficiency of the cell is 90.0%?

**10.11** A nuclear power station operates 100,000 amp below capacity for six hours each day. A scientist proposes generating hydrogen with this excess current by electrolyzing water so that the hydrogen can be burned during peak consumption hours and produce extra power when it is needed. What volume tank must be constructed to hold the hydrogen (measured at STP)?

**10.12** In the electrolysis of anhydrous $MgCl_2$, how many liters of $Cl_2(g)$, measured at STP, are produced in the same time that it takes to produce 1.00 kg of Mg?

**10.13** (a) In the electrolysis of aqueous NaCl, what volume of $Cl_2(g)$ is produced in the same time that it takes to liberate 3.00 liters of $H_2(g)$? Assume that the volumes of both gases are measured at STP. (b) The volume of the solution used in the electrolysis is 300 ml. What is the molarity of NaOH after the process described in (a)?

**10.14** What is the equivalent weight of a metal if a current of 0.300 amp will cause 0.129 g of the metal to plate out of a solution undergoing electrolysis in 40.0 minutes?

**10.15** If 300 ml of $0.300M$ $NiCl_2$ solution is electrolyzed using a current of 3.00 amp for 40.0 minutes, what are the final concentrations of $Ni^{2+}$ ions and $Cl^-$ ions? Assume that the volume of the solution does not change during the course of the electrolysis.

**10.16** How many hours will it take to plate out all of the copper in 300 ml of a $0.300M$ $Cu^{2+}$ solution using a current of 0.750 amp?

**10.17** The equivalent conductance of a 0.100$N$ solution of $CuSO_4$ is 50.6 cm²/ohm equivalent at 25°C. A cell with electrodes that are 2.00 cm³ in surface area and 0.600 cm apart is filled with 0.100$N$ $CuSO_4$. What is the current, in amperes, when the potential difference between the electrodes is 3.00 V?

**10.18** A conductance cell is filled with 0.100$N$ KCl (specific conductance, $\kappa$, $1.29 \times 10^{-2}$/ohm cm) at 25°C and the measured resistance is 93.6 ohms. When the cell is filled with 0.0100$M$ $CaCl_2$, the measured resistance is 522.2 ohms at 25°C. What are the specific and equivalent conductances of the 0.0100$M$ $CaCl_2$?

**10.19** Barium sulfate is only slightly soluble in water, and a saturated solution of $BaSO_4$ may be regarded as infinitely dilute. The equivalent conductance of $BaSO_4$ at infinite dilution, $\Lambda_0$, (estimated from the equivalent conductances of solutions of other sulfates and other barium salts) is 143 cm²/ohm equivalent. The specific conductance, $\kappa$, of a saturated solution of $BaSO_4$ is $2.85 \times 10^{-6}$/ohm cm after correction for the conductance of pure water. What is the solubility of $BaSO_4$ in moles per liter?

**10.20** Given the following $\Lambda_0$ values at 25°C (in cm²/ohm equivalent): $KNO_3$, 145.0; KCl, 149.9; HCl, 426.2; $BaCl_2$, 140.0; LiCl, 115.0. Calculate the *equivalent* conductance at infinite dilution of (a) $HNO_3$, (b) $LiNO_3$, (c) $Ba(NO_3)_2$.

**10.21** (a) Given the following $\Lambda_0$ values at 25°C (in cm²/ohm equivalent): $MgCl_2$, 129.4; NaOH, 247.8; NaCl, 126.5. What is the equivalent conductance at infinite dilution of $Mg(OH)_2$? (b) Magnesium hydroxide is only slightly soluble in water, and a saturated solution may be regarded as infinitely dilute. At 25°C the specific conductance, $\kappa$, of a saturated solution is $6.57 \times 10^{-5}$/ohm cm after correction for the conductance of pure water. What is the solubility of $Mg(OH)_2$ in moles per liter at 25°C?

**10.22** At 25°C the specific conductance, $\kappa$, of a 0.0100$N$ solution of propanoic acid, $HC_3H_5O_2$, is $1.41 \times 10^{-4}$/ohm cm. The equivalent conductance at infinite dilution of propanoic acid at 25°C is 385.6 cm²/ohm equivalent. Determine the degree of dissociation, $\alpha$, of propanoic acid in a 0.0100$N$ solution.

**10.23** (a) What is $\mathscr{E}°$ for the cell:

$$Cd\,|\,Cd^{2+}\,\|\,Cu^{2+}\,|\,Cu?$$

(b) Write the equation for the cell reaction. (c) Which electrode is positive?

**10.24** (a) Give the notation for a cell that utilizes the reaction

$$Cl_2(g) + 2I^-(aq) \rightarrow 2Cl^-(aq) + I_2(s)$$

(b) What is $\mathscr{E}°$ for the cell? (c) Which electrode is the cathode?

**10.25** For the cell:

$$Sc\,|\,Sc^{3+}\,\|\,Cu^{2+}\,|\,Cu,$$

$\mathscr{E}°$ is +2.414 V. Use the emf of the cell and the $\mathscr{E}°$ for the $Cu^{2+}/Cu$

couple to calculate $\mathscr{E}°$ for the $Sc^{3+}/Sc$ half reaction.

**10.26** For the cell:

$$Mg \,|\, Mg^{2+} \,\|\, Lu^{3+} \,|\, Lu$$

$\mathscr{E}°$ is +0.108 V. Use the emf of the cell and the $\mathscr{E}°$ for the $Mg^{2+}/Mg$ couple to calculate $\mathscr{E}°$ for the $Lu^{3+}/Lu$ half reaction.

**10.27** From the table of standard electrode potentials (reactions in acid solution) that appears in the appendix, select a suitable substance for each of the following transformations (assume that all soluble substances are present in $1M$ concentrations): (a) an oxidizing agent capable of oxidizing Ni to $Ni^{2+}$ but not Pb to $Pb^{2+}$, (b) an oxidizing agent capable of oxidizing $H_2S$ to S but not $H_2SO_3$ to $SO_4^{2-}$, (c) a reducing agent capable of reducing $Ce^{4+}$ to $Ce^{3+}$ but not $PbO_2$ to $Pb^{2+}$, (d) a reducing agent capable of reducing $Cr^{3+}$ to $Cr^{2+}$ but not $Cr^{3+}$ to Cr.

**10.28** Use $\mathscr{E}°$ values to predict whether or not the following skeleton equations represent reactions that will occur in acid solution with all soluble substances present in $1M$ concentrations. All reactants and products other than $H^+$ and $H_2O$ are shown below. Complete and balance the equation for each reaction that is predicted to occur.
(a) $Au^+ \rightarrow Au + Au^{3+}$
(b) $Fe + Fe^{3+} \rightarrow Fe^{2+}$
(c) $Co^{2+} \rightarrow Co + Co^{3+}$
(d) $Cr_2O_7^{2-} + Cr \rightarrow Cr^{3+}$

**10.29** Predict whether or not each of the following reactions will occur spontaneously in acid solution and write a balanced chemical

equation for each reaction that is predicted to occur. Assume that soluble reactants and products are present in $1M$ concentrations. (a) Oxidation of $Co^{2+}$ to $Co^{3+}$ by $Br_2$ (reduced to $Br^-$), (b) reduction of $Br_2$ to $Br^-$ by $H_2O_2$ (oxidized to $O_2$), (c) oxidation of $Mn^{2+}$ to $MnO_4^-$ by $Ce^{4+}$ (reduced to $Ce^{3+}$), (d) oxidation of $Cl^-$ to $Cl_2$ by $NO_3^-$ (reduced to NO), (e) reduction of $Mn^{2+}$ to Mn by Zn (oxidized to $Zn^{2+}$).

**10.30** Given the following standard electrode potential diagram (acid solution):

(a) Is the $In^+$ ion stable toward disproportionation in water solution? (b) Which ion is produced when In metal reacts with $H^+(aq)$? (c) The $\mathscr{E}°$ for $Cl_2/Cl^-$ is +1.36 V. Will In react with $Cl_2$? What is the product? (d) Write balanced chemical equations for all reactions.

**10.31** Given the following standard electrode potential diagram (acid solution):

(a) Is the $Tl^+$ ion stable toward disproportionation in water solution? (b) Which ion is produced when Tl metal reacts with $H^+(aq)$? (c) The $\mathscr{E}°$ for $Cl_2/Cl^-$ is +1.36 V. Will Tl react with $Cl_2$? What is the product? (d) Write balanced chemical equations for all reactions.

**10.32** Given the following standard electrode potential diagram (acid solution):

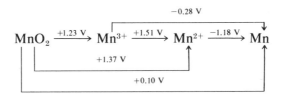

(a) Write a balanced chemical equation for the disproportionation of any species shown that disproportionates. (b) Write the equation for the reaction of Mn metal with $H^+(aq)$.

**10.33** The following reactions occur at 25°C with all soluble substances present in $1M$ concentrations:

$$Zn + 2Eu^{3+} \rightarrow Zn^{2+} + 2Eu^{2+}$$

$$Zr + 2Zn^{2+} \rightarrow Zr^{4+} + 2Zn$$

$$4Sc + 3Zr^{4+} \rightarrow 4Sc^{3+} + 3Zr$$

From this information alone, predict whether the following reactions will occur under similar conditions.
(a) $Sc + 3Eu^{3+} \rightarrow Sc^{3+} + 3Eu^{2+}$
(b) $4Eu^{2+} + Zr^{4+} \rightarrow 4Eu^{3+} + Zr$
(c) $2Sc + 3Zn^{2+} \rightarrow 2Sc^{3+} + 3Zn$

**10.34** Given the following:

$$PbSO_4 + 2e^- \rightleftharpoons Pb + SO_4^{2-} \qquad \mathscr{E}° = -0.359 \text{ V}$$

$$Pb^{2+} + 2e^- \rightleftharpoons Pb \qquad \mathscr{E}° = -0.126 \text{ V}$$

(a) Write the notation for a cell utilizing these half reactions. (b) Write the equation for the cell reaction. (c) Calculate $\mathscr{E}°$ for the cell. (d) Determine $\Delta G°$.

**10.35** Given the following:

$$AgI + e^- \rightleftharpoons Ag + I^- \qquad \mathscr{E}° = -0.152 \text{ V}$$

$$Ag^+ + e^- \rightleftharpoons Ag \qquad \mathscr{E}° = +0.799 \text{ V}$$

(a) Write the notation for a cell utilizing these half reactions. (b) Write the equation for the cell reaction. (c) Calculate $\mathscr{E}°$ for the cell. (d) Determine $\Delta G°$.

**10.36** (a) Diagram a cell for which the cell reaction is

$$H^+ + OH^- \rightarrow H_2O$$

(b) Calculate $\mathscr{E}°$ for the cell and $\Delta G°$ for the reaction.

**10.37** (a) Diagram a cell for which the cell reaction in acid solution is

$$2H_2 + O_2 \rightarrow 2H_2O$$

(b) Calculate $\mathscr{E}°$ for the cell and $\Delta G°$ for the reaction.

**10.38** (a) For the reaction (at 25°C)

$$Pb(s) + Cl_2(g) \rightarrow PbCl_2(s)$$

$\Delta G° = -75.04$ kcal. Calculate $\mathscr{E}°$ for a cell utilizing this reaction. (b) A half cell can be prepared in which a Pb electrode is in contact with $PbCl_2(s)$, which in turn is in contact with a solution containing $Cl^-$ ions.

$$2e^- + PbCl_2(s) \rightleftharpoons Pb(s) + 2Cl^-(aq)$$

Determine $\mathscr{E}°$ for this half reaction by designing a cell for the reaction given in part (a) and by using the cell emf calculated in part (a). (c) Diagram a cell and calculate $\mathscr{E}°$ for the cell using the reaction:

$$Pb^{2+} + 2Cl^- \rightarrow PbCl_2(s)$$

**10.39** Why does the measurement of the density of the electrolyte in a lead storage battery give an indication of the state of charge of the battery?

**10.40** What is the concentration of $Ni^{2+}$ in the cell:

$$Ni \mid Ni^{2+}(?M) \parallel Cu^{2+}(0.750M) \mid Cu$$

if the emf of the cell is 0.601 V?

**10.41** (a) Calculate the emf of the cell:

$$Ga \mid Ga^{3+}(0.0300M) \parallel H^+(3.00M) \mid H_2(0.100 \text{ atm}) \mid Pt$$

(b) Write an equation for the cell reaction. (c) Which electrode is negative?

**10.42** (a) According to the $\mathscr{E}°$ value, should the following reaction proceed spontaneously at 25°C with all soluble substances present at $1.00M$ concentration?

$$MnO_2(s) + 4H^+(aq) + 2Cl^-(aq) \rightarrow Mn^{2+}(aq) + 2H_2O + Cl_2(g)$$

(b) Would the reaction be spontaneous if the concentrations of $H^+$ and $Cl^-$ were each increased to $10.0M$?

**10.43** For a half reaction of the form

$$M^{2+} + 2e^- \rightleftharpoons M$$

what would be the effect on the electrode potential if (a) the concentration of $M^{2+}$ were doubled, (b) the concentration of $M^{2+}$ were cut in half?

**10.44** For the half reaction

$$Cr_2O_7^{2-} + 14H^+ + 6e^- \rightleftharpoons 2Cr^{3+} + 7H_2O$$

$\mathscr{E}°$ is +1.33 V. What would the potential be if the concentration of $H^+(aq)$ were reduced to $0.100M$?

**10.45** Given the following standard cell:

Co | $Co^{2+}$ ‖ $Ni^{2+}$ | Ni

(a) Calculate the value of $\mathscr{E}°$ for the cell. (b) When the cell operates, the concentration of $Co^{2+}$ increases and the concentration of $Ni^{2+}$ decreases. What are the concentrations of the ions when $\mathscr{E}$ is zero?

**10.46** (a) Determine the emf, at 25°C, of the cell

Pt | $H_2$(g, 1.00 atm) | $H^+(0.0250M)$ ‖ $H^+(5.00M)$ | $H_2$(g, 1.00 atm) | Pt

(b) Write an equation for the "cell reaction." (c) Determine $\mathscr{E}$ for the cell if the pressure of $H_2$(g) is changed to 2.00 atm in the anode half cell and to 0.100 atm in the cathode half cell.

**10.47** (a) Determine the emf of a concentration cell prepared from two $Ga^{3+}/Ga$ half cells, one in which the concentration of $Ga^{3+}$ is $2.00M$ and a second in which the concentration of $Ga^{3+}$ is $0.300M$. (b) Write an equation for the "cell reaction." (c) Which electrode is negative?

# 11

# The Nonmetals, Part I

In his book *Tráite Elémentaire de Chimie* (*Elementary Treatise on Chemistry*) published in 1789, Lavoisier proposed the modern definition of a chemical element and correctly identified 23 elements—although he incorrectly included light, heat, and several simple compounds in his list. During the next forty years, 31 elements were discovered—principally as the result of the analysis of minerals. By the end of 1830, 54 elements were known.

The search for new elements was given fresh stimulation in 1859 by Bunsen and Kirchhoff, who introduced the spectroscope as a means of identifying elements. As a result, approximately 60 elements were known by the time that Mendeleev and Lothar Meyer published their versions of the periodic table (in 1869 and 1870, respectively). Mendeleev left blanks for undiscovered elements in his table, and he predicted the properties of these unknown substances. Several elements were discovered as the result of the efforts of chemists to fill in these blanks.

The existence of the noble gases was unforeseen by Mendeleev. After the fortuitous discovery of argon by Rayleigh in 1892, five of the noble gases were isolated and identified in the years 1892–98. By 1925, 88 naturally-occurring elements were known. Elements that do not occur in nature (or occur in extremely small amounts) have been made by nuclear processes. Work of this type (from 1940 to the present) has expanded the number of known elements to its present value, 105.

In this chapter we shall survey the chemistry of the group 0, group VII A, and group VI A nonmetals.

## THE NOBLE GASES

The outstanding characteristic of the noble gases is their low order of chemical reactivity. Until 1962 no true compounds of these elements were known, and hence they were called the "inert gases." Since 1962 approximately 25 compounds of the heavier elements of this group have been prepared.

## 11.1 PROPERTIES OF THE NOBLE GASES

The properties of the noble gases reflect their very stable electronic configurations. The atoms have no tendency to combine with each other to form molecules; each element occurs as a colorless, monatomic gas. Each noble gas has the highest first ionization potential of any element in its period (Table 11.1). The low melting points and boiling points of these elements are evidence of the weak nature of the London forces of attraction that operate between the atoms.

With increasing atomic number, the atomic size increases, and the outer electrons become slightly less tightly held. Therefore, the ionization potential decreases regularly from helium to radon. This factor (i.e.,

Table 11.1 Some properties of the noble gases

| GAS | MELTING POINT (°C) | BOILING POINT (°C) | IONIZATION POTENTIAL (eV) | ABUNDANCE IN ATMOSPHERE (VOLUME %)[a] |
|-----|-----|-----|-----|-----|
| He | —[b] | −268.9 | 24.6 | $5 \times 10^{-4}$ |
| Ne | −248.6 | −245.9 | 21.6 | $2 \times 10^{-3}$ |
| Ar | −189.3 | −185.8 | 15.8 | 0.94 |
| Kr | −157 | −152.9 | 14.0 | $1 \times 10^{-4}$ |
| Xe | −112 | −107.1 | 12.1 | $9 \times 10^{-6}$ |
| Rn | −71 | −61.8 | 10.8 | trace |

[a] Dry air at sea level.
[b] −272.2°C at 26 atm pressure.

increasing size of the electron cloud) also accounts for the increasing strength of London forces and consequently the increasing boiling point and melting point from helium to radon.

All the noble gases occur in the atmosphere (Table 11.1), and Ne, Ar, Kr, and Xe are by-products of the fractionation of liquid air. Certain natural gas deposits (located principally in Kansas) contain a higher percentage of helium than is found in air; these deposits constitute the major commercial source of helium. Isotopes of radon are produced in nature by the decay of certain radioactive elements. All the isotopes of radon are themselves radioactive; $^{222}_{86}$Rn is produced by the radioactive decay of $^{226}_{88}$Ra and has a half-life of 3.82 days, the longest of any radon isotope.

Argon is used to fill electric light bulbs; the gas does not react with the hot filament but rather conducts heat away from it, thus prolonging its life. Argon is also used as an inert atmosphere in welding and high-temperature metallurgical processes; the gas protects the hot metals from air oxidation. Helium is used in lighter-than-air craft (its density is about 14% of the density of air) and in low temperature work (it has the lowest boiling point of any known substance). Neon signs are made from discharge tubes containing neon gas at a low pressure. Radon has been used as a source of $\alpha$ particles in cancer therapy.

The first chemical reaction of a noble gas to be observed, the reaction of xenon with platinum hexafluoride ($PtF_6$), was reported by Neil Bartlett in 1962. Platinum hexafluoride is a powerful oxidizing agent; it reacts with oxygen to give $[O_2^+][PtF_6^-]$. Since the first ionization potential of *molecular* oxygen (12.2 eV)

$$O_2 \rightarrow O_2^+ + e^-$$

is close to the first ionization potential of xenon (12.1 eV), Bartlett reasoned that xenon should react with $PtF_6$. Experiment verified this prediction, and a red crystalline solid, consisting of $[Xe^+][PtF_6^-]$ and other compounds, is produced by the reaction.

The best characterized noble-gas compounds are the xenon fluorides: $XeF_2$, $XeF_4$, and $XeF_6$. Each of these compounds may be prepared by

**Figure 11.1** Structures of $XeF_4$, $XeOF_4$, and $XeO_6^{4-}$.

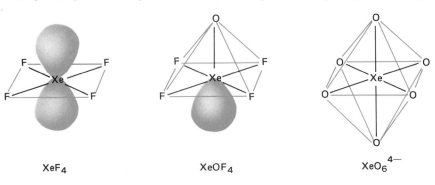

XeF₄          XeOF₄          XeO₆⁴⁻

Table 11.2 Properties of some compounds of xenon

| OXIDATION STATE | COMPOUND | FORM | MELTING POINT (°C) |
|---|---|---|---|
| 2+ | $XeF_2$ | colorless crystals | 129 |
| 4+ | $XeF_4$ | colorless crystals | 117 |
| 6+ | $XeF_6$ | colorless crystals | 50 |
|  | $XeOF_4$ | colorless liquid | −46 |
|  | $XeO_3$ | colorless crystals | − |
| 8+ | $XeO_4$ | colorless gas | − |
|  | $Na_4XeO_6 \cdot 8H_2O$ | colorless crystals | − |

direct reaction of Xe and $F_2$; the choice of reaction conditions, particularly the proportion of Xe to $F_2$ used, determines which fluoride is obtained. The xenon fluorides are colorless, crystalline solids (Table 11.2).

Oxygen-containing compounds of xenon are produced by reactions of the xenon fluorides with water. Partial hydrolysis of $XeF_6$ yields $XeOF_4$, a colorless liquid.

$$XeF_6(s) + H_2O \rightarrow XeOF_4(l) + 2HF(g)$$

Complete hydrolysis of $XeF_6$ or hydrolysis of $XeOF_4$ produces a solution that yields solid $XeO_3$ upon evaporation.

$$XeF_6(s) + 3H_2O \rightarrow XeO_3(aq) + 6HF(aq)$$

$$XeOF_4(l) + 2H_2O \rightarrow XeO_3(aq) + 4HF(aq)$$

The order of the group 0 elements according to decreasing reactivity (increasing ionization potential, Table 11.1) should be: Rn > Xe > Kr > Ar > Ne > He. Thus, radon should be the most reactive noble gas. There is evidence that Rn reacts with fluorine; however, the radioactive disintegration of Rn isotopes makes the chemistry of Rn difficult to assess. Krypton is not so reactive as Xe, but a few compounds of Kr have been prepared. The most important of these compounds is $KrF_2$, which is made by subjecting a mixture of Kr and $F_2$ to an electric discharge. No compounds of He, Ne, or Ar have been prepared as yet.

The structures of most xenon compounds have been determined and may be interpreted by a valence bond approach (Figures 11.1 and 11.2).

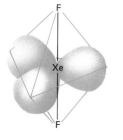

XeF₂

Figure 11.2 Structure of $XeF_2$.

## ■ THE HALOGENS

The elements of group VII A, fluorine, chlorine, bromine, iodine, and astatine, are called the **halogens.** The name "halogen" is derived from Greek and means "salt former;" these elements, with the exception of astatine, occur extensively in nature in the form of halide salts. Astatine probably occurs in nature, in extremely small amounts, as a short-lived intermediate of natural radioactive decay processes. However, most of our meager information about the chemistry of astatine comes from the study of the small amounts of a radioactive isotope of this element prepared by nuclear transmutation reactions.

**11.2 GROUP PROPERTIES**

The electronic configurations of the halogens are listed in Table 11.3. Each halogen atom has one electron less than the noble gas that follows it in the periodic classification; there is a marked tendency on the part of a halogen atom to attain a noble-gas configuration by the formation of a uninegative ion or a single covalent bond. Positive oxidation states exist for all the elements except fluorine.

Table 11.3 Electronic configurations of the halogens

| ELEMENT | $Z$ | $1s$ | $2s$ | $2p$ | $3s$ | $3p$ | $3d$ | $4s$ | $4p$ | $4d$ | $4f$ | $5s$ | $5p$ | $5d$ | $6s$ | $6p$ |
|---------|-----|------|------|------|------|------|------|------|------|------|------|------|------|------|------|------|
| F | 9 | 2 | 2 | 5 | | | | | | | | | | | | |
| Cl | 17 | 2 | 2 | 6 | 2 | 5 | | | | | | | | | | |
| Br | 35 | 2 | 2 | 6 | 2 | 6 | 10 | 2 | 5 | | | | | | | |
| I | 53 | 2 | 2 | 6 | 2 | 6 | 10 | 2 | 6 | 10 | | 2 | 5 | | | |
| At | 85 | 2 | 2 | 6 | 2 | 6 | 10 | 2 | 6 | 10 | 14 | 2 | 6 | 10 | 2 | 5 |

Some properties of the halogens are summarized in Table 11.4; the symbol X stands for any halogen. Most of the properties listed increase or decrease in a regular fashion within the halogen series arranged by atomic number. The halogen of each period has a high ionization potential, second only to that of the noble gas of the period, and within the group, ionization potential decreases with increasing atomic radius.

Under ordinary conditions the halogens exist as diatomic molecules with a single covalent bond joining the atoms of a molecule. The molecules are held together in the solid and liquid state by London forces. Of all the halogen molecules, $I_2$ is the largest, has the most electrons, and is

Table 11.4 Some properties of the halogens

| PROPERTY | $F_2$ | $Cl_2$ | $Br_2$ | $I_2$ |
|----------|-------|--------|--------|-------|
| color | pale yellow | yellow-green | red-brown | violet-black |
| melting point (°C) | −218 | −101 | −7 | +113 |
| boiling point (°C) | −188 | −35 | +59 | +183 |
| atomic radius (Å) | 0.72 | 0.99 | 1.14 | 1.33 |
| ionic radius, $X^-$ (Å) | 1.36 | 1.81 | 1.95 | 2.16 |
| first ionization potential (eV) | 17.4 | 13.0 | 11.8 | 10.4 |
| electron affinity (eV) | −3.45 | −3.61 | −3.36 | −3.06 |
| electronegativity | 4.0 | 3.0 | 2.8 | 2.5 |
| bond energy (kcal/mol) | 37 | 58 | 46 | 36 |
| standard electrode potential (V) $2e^- + X_2 \rightleftharpoons 2X^-$ | +2.87 | +1.36 | +1.07 | +0.54 |

the most polarizable; it is not surprising, therefore, that the intermolecular attractions of $I_2$ are the strongest and that iodine has the highest melting point and boiling point. At ordinary temperatures and pressures, $I_2$ is a solid, $Br_2$ is a liquid, and $Cl_2$ and $F_2$ are gases.

Each halogen is the most reactive nonmetal of its period, and fluorine is the most reactive of all the nonmetals. Fluorine has the highest electronegativity of any element and is one of the strongest oxidizing agents known. The electronegativity of the halogens decreases in the order $F > Cl > Br > I$, and the oxidizing power of the halogens decreases in the same order.

The values for electron affinity and bond energy (listed in Table 11.4) do not change in a regular order from $F_2$ to $I_2$. One might expect that the values for fluorine would be the largest in view of the trends in electronegativities and electrode potentials. The reasons for the comparatively small electron affinity of fluorine and for the relatively low F—F bond energy are not completely understood; it is thought that the repulsion of the small, highly dense electron cloud of the fluorine atom is responsible for these effects.

Recall that electrode potentials refer to processes that occur in water solution. On this basis, the superior oxidizing ability of fluorine in comparison to the other halogens can be explained (in spite of fluorine's relatively small electron affinity) by consideration of the several steps involved in the conversion of the free halogens to halide ions in water solution (Table 11.5).

Table 11.5 Enthalpies of some halogen reactions (kcal)

| REACTION | | | $F_2$ | $Cl_2$ | $Br_2$ | $I_2$ |
|---|---|---|---|---|---|---|
| $X_2$(standard state) | $\rightarrow$ | $2X$(g) | +37 | +58 | +53 | +51 |
| $2X$(g) $+ 2e^-$(g) | $\rightarrow$ | $2X^-$(g) | −159 | −166 | −155 | −141 |
| $2X^-$(g) | $\rightarrow$ | $2X^-$(aq) | −246 | −178 | −162 | −144 |
| $X_2$(standard state) $+ 2e^-$(g) | $\rightarrow$ | $2X^-$(aq) | −368 | −286 | −264 | −234 |

In the first step, heat is *absorbed* in the dissociation of the halogen molecules into gaseous halogen atoms. The enthalpy of dissociation of iodine includes the energy needed to convert $I_2$(s) to $I_2$(g) as well as the bond energy of $I_2$, and the enthalpy of dissociation of bromine includes the heat needed to convert $Br_2$(l) to $Br_2$(g). In the second step, heat is *evolved* (the electron affinity) by the formation of the gaseous halide ion. In the third step, heat is *evolved* by the hydration of the ions. The fluoride ion, which is the smallest halide ion and has the highest charge density, has the highest enthalpy of hydration.

For each of the halogens, the algebraic sum of the enthalpies listed in Table 11.5 gives a measure of the vigor of the formation of the halide ion in water solution. It should be noted that the final equation of Table 11.5 involves gaseous electrons; however, the enthalpy values relating to this equation are adequate for the purpose of comparison. It is obvious that the high enthalpy of hydration of the fluoride ion (and to a lesser extent the low enthalpy of dissociation of the fluorine molecule) more than compensate for the difference in electron affinity. Therefore in water solution, fluorine is the most easily reduced halogen and consequently the strongest oxidizing agent.

In anhydrous reactions, fluorine is again the most reactive halogen; this is shown by the enthalpies of formation of the halides. The values listed in Table 11.6 are the heat effects when one mole of each of the compounds listed is prepared from the elements in their standard states (25°C). In each series more heat is liberated in the preparation of the fluoride than in the production of any other halide, and the values decline in the expected order. The relatively low energy of the F—F bond (energy that is absorbed in the preparation) coupled with the relatively high energies of the short, strong bonds that fluorine forms with most other nonmetals (energy evolved) is believed to explain this effect for the covalent halides. Crystalline fluorides have high lattice energies because the $F^-$ ion is relatively small and has a high charge density.

Table 11.6 Enthalpies of formation of some halides (kcal/mol)

|        | $X = F$  | $X = Cl$ | $X = Br$ | $X = I$ |
|--------|---------|---------|---------|--------|
| NaX(s) | −136.0  | −98.2   | −86.3   | −69.5  |
| KX(s)  | −134.5  | −104.2  | −94.2   | −78.9  |
| HX(g)  | −64.2   | −22.1   | −8.7    | +6.2   |

The relative oxidizing ability of the halogens may be observed in displacement reactions. Thus, fluorine can displace chlorine, bromine, and iodine from their salts; chlorine can displace bromine and iodine from their salts; and bromine can displace iodine from iodides.

$$F_2(g) + 2NaCl(s) \rightarrow 2NaF(s) + Cl_2(g)$$

$$Cl_2(g) + 2Br^-(aq) \rightarrow 2Cl^-(aq) + Br_2(l)$$

$$Br_2(l) + 2I^-(aq) \rightarrow 2Br^-(aq) + I_2(s)$$

Since fluorine actively oxidizes water (producing $O_2$), displacement reactions involving $F_2$ cannot be run in water solution.

**11.3 PREPARATION OF THE HALOGENS**

The principal natural sources of the halogens are listed in Table 11.7. Fluorine must be prepared by an electrochemical process because no suitable chemical agent is sufficiently powerful to oxidize fluoride ion to fluorine. Furthermore, since the oxidation of water is easier to accom-

Table 11.7 Occurrence of the halogens

| ELEMENT | PERCENT OF EARTH'S CRUST | OCCURRENCE |
|---|---|---|
| fluorine | 0.1 | $CaF_2$ (fluorspar), $Na_3AlF_6$ (cryolite) |
| chlorine | 0.2 | $Cl^-$ (sea water and underground brines) $NaCl$ (rock salt) |
| bromine | 0.001 | $Br^-$ (sea water, underground brines, solid salt beds) |
| iodine | 0.001 | $I^-$ (oil-well brines, sea water) $NaIO_3$, $NaIO_4$ (impurities in Chilean saltpeter, $NaNO_3$) |

plish than the oxidation of the fluoride ion, the electrolysis must be carried out under anhydrous conditions. In actual practice, a solution of potassium fluoride in anhydrous hydrogen fluoride is electrolyzed. Pure HF does not conduct electric current; the KF reacts with HF to produce ions ($K^+$ and $HF_2^-$) that can act as charge carriers. The $HF_2^-$ ion is formed from a $F^-$ ion hydrogen-bonded to a HF molecule ($F—H\cdots F^-$); this hydrogen bond is so strong that the H atom is exactly midway between the two F atoms. The hydrogen fluoride used in the electrolysis is commercially derived from fluorospar, $CaF_2$ (Section 11.5).

$$2KHF_2(l) \xrightarrow[\text{heat}]{\text{electrolysis}} H_2(g) + F_2(g) + 2KF(l)$$

The principal industrial source of chlorine is the electrolysis of aqueous sodium chloride, from which sodium hydroxide and hydrogen are also products.

$$2Na^+(aq) + 2Cl^-(aq) + 2H_2O \xrightarrow{\text{electrolysis}}$$
$$H_2(g) + Cl_2(g) + 2Na^+(aq) + 2OH^-(aq)$$

Chlorine is also obtained, as a by-product, from the industrial processes in which the reactive metals Na, Ca, and Mg are prepared. In each of these processes an anhydrous molten chloride is electrolyzed and $Cl_2$ gas is produced at the anode. For example,

$$2NaCl(l) \xrightarrow{\text{electrolysis}} 2Na(l) + Cl_2(g)$$

Bromine is commercially prepared by the oxidation of the bromide ion of salt brines or sea water with chlorine as the oxidizing agent.

$$Cl_2(g) + 2Br^-(aq) \rightarrow 2Cl^-(aq) + Br_2(l)$$

The liberated $Br_2$ is removed from the solution by a stream of air and, in subsequent steps, collected from the air and purified.

In the United States the principal commercial source of iodine is the iodide ion found in oil-well brines; free iodine is obtained by chlorine displacement.

$$Cl_2(g) + 2I^-(aq) \rightarrow 2Cl^-(aq) + I_2(s)$$

In addition, iodine is commercially obtained from the iodate impurity found in Chilean nitrates; sodium bisulfite is used to reduce the iodate ion.

$$2IO_3^-(aq) + 5HSO_3^-(aq) \rightarrow I_2(s) + 5SO_4^{2-}(aq) + 3H^+(aq) + H_2O$$

With the exception of fluorine (which must be prepared electrochemically), the free halogens are usually prepared in the laboratory by the action of oxidizing agents on aqueous solutions of the hydrogen halides or on solutions containing the sodium halides and sulfuric acid.

From a table of standard electrode potentials we can get an approximate idea of what oxidizing agents will satisfactorily oxidize a given halide ion. Thus, any couple with a standard electrode potential more positive than $+1.36$ should oxidize the chloride ion as well as the bromide ion ($\mathscr{E}° = +1.07$ V) and the iodide ion ($\mathscr{E}° = +0.54$ V). Recall, however, that standard electrode potentials are listed for half reactions at 25°C with all materials in their standard states. Thus, even though the standard electrode potential of $MnO_2 \rightarrow Mn^{2+}$ is only $+1.23$ V, $MnO_2$ is capable of oxidizing the chloride ion if concentrated HCl is used (rather than HCl at unit activity) and if the reaction is heated. In actual practice, $KMnO_4$, $K_2Cr_2O_7$, $PbO_2$, and $MnO_2$ are frequently used to prepare the free halogens from halide ions.

**Table 11.8** Some reactions of the halogens ($X_2 = F_2$, $Cl_2$, $Br_2$, or $I_2$)

| GENERAL REACTION | REMARKS |
| --- | --- |
| $nX_2 + 2M \rightarrow 2MX_n$ | $F_2$, $Cl_2$ with practically all metals; $Br_2$, $I_2$ with all except noble metals |
| $X_2 + H_2 \rightarrow 2HX$ | |
| $3X_2 + 2P \rightarrow 2PX_3$ | with excess P; similar reactions with As, Sb, and Bi |
| $5X_2 + 2P \rightarrow 2PX_5$ | with excess $X_2$, but not with $I_2$; $SbF_5$, $SbCl_5$, $AsF_5$, $AsCl_5$, and $BiF_5$ may be similarly prepared |
| $X_2 + 2S \rightarrow S_2X_2$ | with $Cl_2$, $Br_2$ |
| $X_2 + H_2O \rightarrow H^+ + X^- + HOX$ | not with $F_2$ |
| $2X_2 + 2H_2O \rightarrow 4H^+ + 4X^- + O_2$ | $F_2$ rapidly; $Cl_2$, $Br_2$ slowly in sunlight |
| $X_2 + H_2S \rightarrow 2HX + S$ | |
| $X_2 + CO \rightarrow COX_2$ | $Cl_2$, $Br_2$ |
| $X_2 + SO_2 \rightarrow SO_2X_2$ | $F_2$, $Cl_2$ |
| $X_2 + 2X'^- \rightarrow X_2' + 2X^-$ | $F_2 > Cl_2 > Br_2 > I_2$ |
| $X_2 + X_2' \rightarrow 2XX'$ | formation of the interhalogen compounds (all except IF) |

$$MnO_2(s) + 4H^+(aq) + 2Cl^-(aq) \rightarrow Mn^{2+}(aq) + Cl_2(g) + 2H_2O$$

$$2MnO_4^-(aq) + 16H^+(aq) + 10Br^-(aq) \rightarrow 2Mn^{2+}(aq) + 5Br_2(l) + 8H_2O$$

$$Cr_2O_7^{2-}(aq) + 14H^+(aq) + 6I^-(aq) \rightarrow 2Cr^{3+}(aq) + 3I_2(s) + 7H_2O.$$

Some of the reactions of the halogens are summarized in Table 11.8.

**11.4 THE INTER-HALOGEN COMPOUNDS**

The halogens react with each other to produce a number of interhalogen compounds. All the compounds with the formula $XX'$ (such as BrCl) are known except IF. Four $XX'_3$ compounds have been prepared ($ClF_3$, $BrF_3$, $ICl_3$, and $IF_3$), and three $XX'_5$ compounds are known ($ClF_5$, $BrF_5$, and $IF_5$). The only $XX'_7$ compound that has been made is $IF_7$.

With the exception of $ICl_3$, all the molecules for which $n$ of the formula $XX'_n$ is greater than 1 are halogen fluorides in which fluorine atoms (the smallest and most electronegative of all the halogen atoms) surround a Cl, Br, or I atom. The stability of these compounds increases as the size of the central atom increases. Hence, neither $BrF_7$ nor $ClF_7$ has been prepared, although $IF_7$ is known; $ClF_5$ readily decomposes into $ClF_3$ and $F_2$, whereas $BrF_5$ and $IF_5$ are stable at temperatures above 400°C.

The structures of the higher interhalogen compounds have received considerable attention because the bonding of the central atom in each of these molecules violates the octet principle (Figure 11.3). The X atom of the $XX'_3$ compounds utilizes $dsp^3$ hybrid orbitals for bonding, and the molecules are T-shaped. The $XX'_5$ molecules are square pyramidal and the central atoms use $d^2sp^3$ hybrid orbitals. The unshared electron pairs of the central atoms of these two types of molecules introduce some distortion. The $IF_7$ molecule has a pentagonal bipyramidal structure and is bonded through $d^3sp^3$ hybrid orbitals of I.

**11.5 THE HYDROGEN HALIDES**

Each of the hydrogen halides may be prepared by the direct reaction of hydrogen with the corresponding free halogen; the vigor of the reaction decreases markedly from fluorine to iodine.

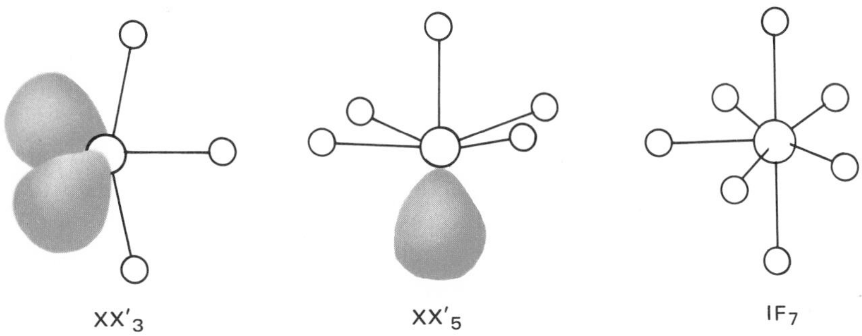

**Figure 11.3** Structures of $XX'_3$, $XX'_5$, and $IF_7$. The unshared pairs in $XX'_3$ and $XX'_5$ molecules are responsible for some distortion from the regular T-shaped and square pyramidal structures.

XX'₃          XX'₅          IF₇

Both HF and HCl are made industrially by the action of warm concentrated sulfuric acid on the corresponding naturally occurring halide, $CaF_2$ and NaCl.

$$CaF_2(s) + H_2SO_4(l) \rightarrow CaSO_4(s) + 2HF(g)$$

$$NaCl(s) + H_2SO_4(l) \rightarrow NaHSO_4(s) + HCl(g)$$

All hydrogen halides are colorless gases at room temperature; sulfuric acid, on the other hand, is a high-boiling liquid. Thus, the foregoing reactions are examples of a general method for the preparation of a volatile acid from its salts by means of a nonvolatile acid. At higher temperatures (ca. 500°C), further reaction occurs between $NaHSO_4$ and NaCl.

$$NaCl(s) + NaHSO_4(l) \rightarrow HCl(g) + Na_2SO_4(s)$$

Hydrogen bromide and hydrogen iodide cannot be made by the action of concentrated sulfuric acid on bromides and iodides because hot, concentrated sulfuric acid oxidizes these anions to the free halogens. The bromide and iodide ions are easier to oxidize than the fluoride and chloride ions.

$$2NaBr(s) + 2H_2SO_4(l) \rightarrow Br_2(g) + SO_2(g) + Na_2SO_4(s) + 2H_2O(g)$$

Since the iodide ion is a stronger reducing agent (more easily oxidized) than the bromide ion, S and $H_2S$, as well as $SO_2$, are obtained as reduction products from the reaction of NaI with hot concentrated sulfuric acid.

Pure HBr or HI can be obtained by the action of phosphoric acid on NaBr or NaI; phosphoric acid is an essentially nonvolatile acid and is a poor oxidizing agent.

$$NaBr(s) + H_3PO_4(l) \rightarrow HBr(g) + NaH_2PO_4(s)$$

$$NaI(s) + H_3PO_4(l) \rightarrow HI(g) + NaH_2PO_4(s)$$

The hydrogen halides may be prepared by the reaction of water on the appropriate phosphorus trihalide.

$$PX_3 + 3H_2O \rightarrow 3HX(g) + H_3PO_3(aq)$$

Convenient laboratory preparations of HBr and HI have been developed in which red phosphorus, bromine or iodine, and a limited amount of water are employed and in which no attempt is made to isolate the phosphorus trihalide intermediate.

Hydrogen fluoride molecules associate with each other through hydrogen bonding. The vapor consists of aggregates up to $(HF)_6$ at temperatures near the boiling point (19.4°C) but is less highly associated at higher temperatures. Gaseous HCl, HBr, and HI consist of single molecules. Liquid HF and solid HF are more highly hydrogen bonded than gaseous HF, and the boiling point and melting point of HF are abnormally high in comparison to those of the other hydrogen halides (Figure 8.2).

All hydrogen halides are very soluble in water; water solutions are called hydrohalic acids. For example, aqueous HI is called hydroiodic acid. The H—F bond is stronger than any other H—X bond; HF is a weak acid in water solution, whereas HCl, HBr, and HI are completely dissociated.

$$HF(aq) \rightleftharpoons H^+(aq) + F^-(aq)$$

The $F^-$ ions from this dissociation are largely associated with HF molecules.

$$F^-(aq) + HF(aq) \rightleftharpoons HF_2^-(aq)$$

Concentrated HF solutions are more strongly ionic than dilute solutions and contain high concentrations of ions of the type $HF_2^-$, $H_2F_3^-$, and higher.

Hydrofluoric acid reacts with silica, $SiO_2$, and glass, which is made from silica.

$$SiO_2(s) + 6HF(aq) \rightarrow 2H^+(aq) + SiF_6^{2-}(aq) + 2H_2O$$

When warmed, the reaction is

$$SiO_2(g) + 4HF(aq) \rightarrow SiF_4(g) + 2H_2O(g)$$

For this reason, hydrofluoric acid must be stored in wax or plastic containers instead of glass bottles.

The fluorosilicate ion, $SiF_6^{2-}$, is an example of a large group of complex ions formed by the halide ions. Halo complexes are formed by most metals (with the notable exceptions of the group I A, group II A, and lanthanide metals) and with some nonmetals (e.g., $BF_4^-$). The formulas of these complex ions are most commonly of the types $(MX_4)^{n-4}$ and $(MX_6)^{n-6}$, where $n$ is the oxidation number of the central atom of the complex.

## 11.6 THE METAL HALIDES

Metal halides can be prepared by direct interaction of the elements, by the reactions of the hydrogen halides with hydroxides or oxides, and by the reactions of the hydrogen halides with carbonates.

$$K_2CO_3(s) + 2HF(l) \rightarrow 2KF(s) + CO_2(g) + H_2O$$

The character of the bonding in metal halides varies widely as do the physical properties of these compounds. A metal that has a low ionization energy generally forms halides that are highly ionic and consequently have high melting and boiling points. On the other hand, metals that have comparatively high ionization energies react, particularly with halogens of low electron affinity, to form halides in which the bonding has a high degree of covalent character; such compounds have comparatively low melting points and boiling points. In general, the high-melting halides are, in the fused state, better conductors of electricity than are the low-melting compounds.

In general, the halides of the I A metals, the II A metals (with the exception of Be), and most of the inner-transition metals are largely ionic; halides of the remaining metals are covalent to a varying degree. Within a series of halides of metals of the same period, covalent character increases from compound to compound as the oxidation number of the metal increases and the size of the cation undergoes a concomitant decrease. The equivalent conductances of the molten chlorides of some fourth-period metals are plotted in Figure 11.4. The "cations" of the compounds are isoelectronic. Potassium chloride is a completely ionic solid, and molten KCl has the highest conductance of the four compounds; $TiCl_4$ is a covalent liquid and is nonconducting.

If a metal exhibits more than one oxidation state, the halides of the highest oxidation state are the most covalent. Thus, the second member of each of the following pairs is a highly covalent, volatile liquid (melting points are given in parentheses): $SnCl_2$ (246°C), $SnCl_4$ (−33°C); $PbCl_2$ (501°C), $PbCl_4$ (−15°C); $SbCl_3$ (73.4°C), $SbCl_5$ (2.8°C).

Since fluorine is the most electronegative halogen, fluorides are the most ionic of the halides; ionic character decreases in the order: fluoride > chloride > bromide > iodide. The trend in the melting points of the sodium halides is typical:

| NaF | NaCl | NaBr | NaI |
|------|------|------|------|
| 993°C | 801°C | 755°C | 651°C |

The halides of aluminum are an excellent example of the relationship between ionic or covalent character and the size of the halide ion.

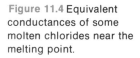

Figure 11.4 Equivalent conductances of some molten chlorides near the melting point.

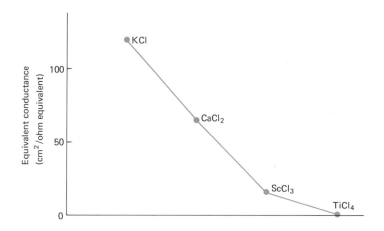

Aluminum fluoride is an ionic substance. Aluminum chloride is semi-covalent and crystallizes in a layer lattice in which electrically neutral layers are held together by London forces. Aluminum bromide and aluminum iodide are essentially covalent; the crystals consist of $Al_2Br_6$ and $Al_2I_6$ molecules, respectively.

The water solubility of fluorides is considerably different from that of the chlorides, bromides, and iodides. The fluorides of lithium, the group II A metals, and the lanthanides are only slightly soluble, whereas the other halides of these metals are relatively soluble.

Most chlorides, bromides, and iodides are soluble in water. The cations that form slightly soluble compounds with these halide ions include silver(I), mercury(I), lead(II), copper(I), and thallium(I).

The insolubility of the silver salts of $Cl^-$, $Br^-$, and $I^-$ is the basis of a common test for these halide ions; AgCl is white, AgBr is cream, and AgI is yellow. The silver halide precipitates may be formed by the addition of a solution of silver nitrate to a solution containing the appropriate halide ion. Silver iodide is insoluble in excess ammonia; however, AgCl readily dissolves to form the $Ag(NH_3)_2^+$ complex ion and AgBr dissolves with difficulty. Silver fluoride is soluble; generally, $MgF_2$ or $CaF_2$ precipitates are used to confirm the presence of the fluoride ion in a solution.

If the iodide ion in an aqueous solution is oxidized to iodine (the usual procedure employs chlorine), the $I_2$ can be extracted by carbon tetrachloride, which forms a two-liquid-layer system with water. The solution of $I_2$ in $CCl_4$ is violet colored. In the corresponding test for the bromide ion, the $Br_2$–$CCl_4$ solution is brown.

## 11.7 OXYACIDS OF THE HALOGENS

The oxyacids of the halogens are listed in Table 11.9. The only oxyacid of fluorine that has been prepared is hypofluorous acid, HOF, a thermally unstable compound. The acids of chlorine and their salts are the most important of these compounds.

Table 11.9 Oxyhalogen acids

| OXIDIZATION STATE OF THE HALOGEN | FORMULA OF ACID | | | | NAME OF ACID | NAME OF ANION DERIVED FROM ACID |
|---|---|---|---|---|---|---|
| 1+ | HOF | HOCl | HOBr | HOI | hypohalous acid | hypohalite ion |
| 3+ | — | $HClO_2$ | — | — | halous acid | halite ion |
| 5+ | — | $HClO_3^a$ | $HBrO_3^a$ | $HIO_3^a$ | halic acid | halate ion |
| 7+ | — | $HClO_4^a$ | $HBrO_4^a$ | $\begin{cases} HIO_4 \\ H_4I_2O_9 \\ H_5IO_6 \end{cases}$ | perhalic acid | perhalate ion |

[a] Strong acids.

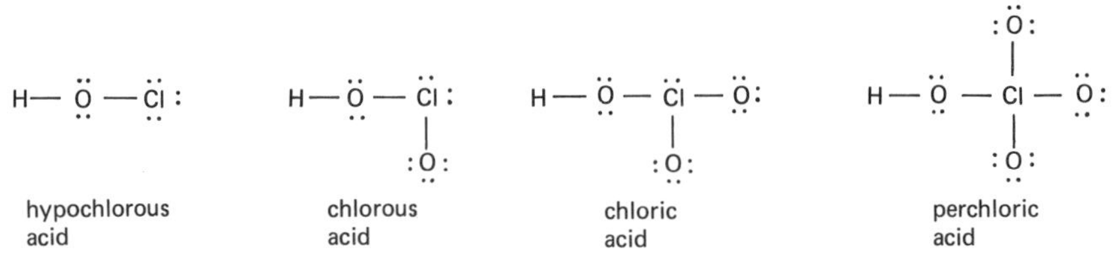

**Figure 11.5** Lewis formulas for the oxychlorine acids.

$$H— \overset{..}{\underset{..}{O}} — \overset{..}{\underset{..}{Cl}} :$$

$$H— \overset{..}{\underset{..}{O}} — \overset{..}{\underset{\underset{:O:}{|}}{Cl}} :$$

$$H— \overset{..}{\underset{..}{O}} — \overset{..}{\underset{\underset{:O:}{|}}{Cl}} — \overset{..}{\underset{..}{O}} :$$

$$H— \overset{..}{\underset{..}{O}} — \overset{\overset{:O:}{|}}{\underset{\underset{:O:}{|}}{Cl}} — \overset{..}{\underset{..}{O}} :$$

hypochlorous acid      chlorous acid      chloric acid      perchloric acid

Lewis formulas for the oxychlorine acids are shown in Figure 11.5; removal of $H^+$ from each of these structures gives the electronic formula of the corresponding anion. However, Lewis structures, in which each Cl and O atom has a valence-electron octet, do not show that the Cl—O bonds in these compounds can have a considerable amount of double-bond character. For example, the experimentally observed Cl—O bond length in the $ClO_4^-$ ion (1.46 Å) is appreciably shorter than the bond length predicted from atomic radii on the basis of single-bond formation (about 1.70 Å).

In these compounds the Cl atom is assumed to use $sp^3$ hybrid orbitals for $\sigma$ bonding with unshared electron pairs occupying some of the tetrahedral positions; the observed structures are consistent with this view (Figure 11.6). The double-bond character of the bonds, which does not appreciably distort the geometry of the molecules, is assumed to arise from the overlap of unfilled $3d$ orbitals of Cl with filled $2p$ orbitals of O and is called *$p\pi$–$d\pi$* **bonding.**

Such $p\pi$–$d\pi$ bonding is particularly important in compounds in which a third-period nonmetal (Si, P, S, or Cl) is bonded to O, N, or F. Since the second-period nonmetals have no $d$ orbitals in their valence levels, $p\pi$–$d\pi$ bonding does not occur in compounds of these elements (although double bonds may form by use of the $s$ and $p$ orbitals of these elements).

The oxybromine and oxyiodine anions have configurations similar to the analogous oxychlorine anions shown in Figure 11.6. The $H_5IO_6$ molecule is octahedral with five OH groups and one O atom in the six positions surrounding the central I atom. Presumably this molecule is bonded by means of $d^2sp^3$ hybrid orbitals of I.

The standard electrode potentials for $Cl_2$, $Br_2$, $I_2$, and the compounds of these elements in acidic and alkaline solution are summarized in

**Figure 11.6** Structures of the oxychlorine anions.

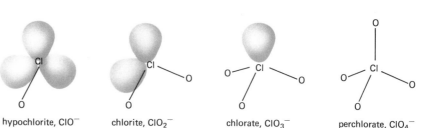

hypochlorite, $ClO^-$      chlorite, $ClO_2^-$      chlorate, $ClO_3^-$      perchlorate, $ClO_4^-$

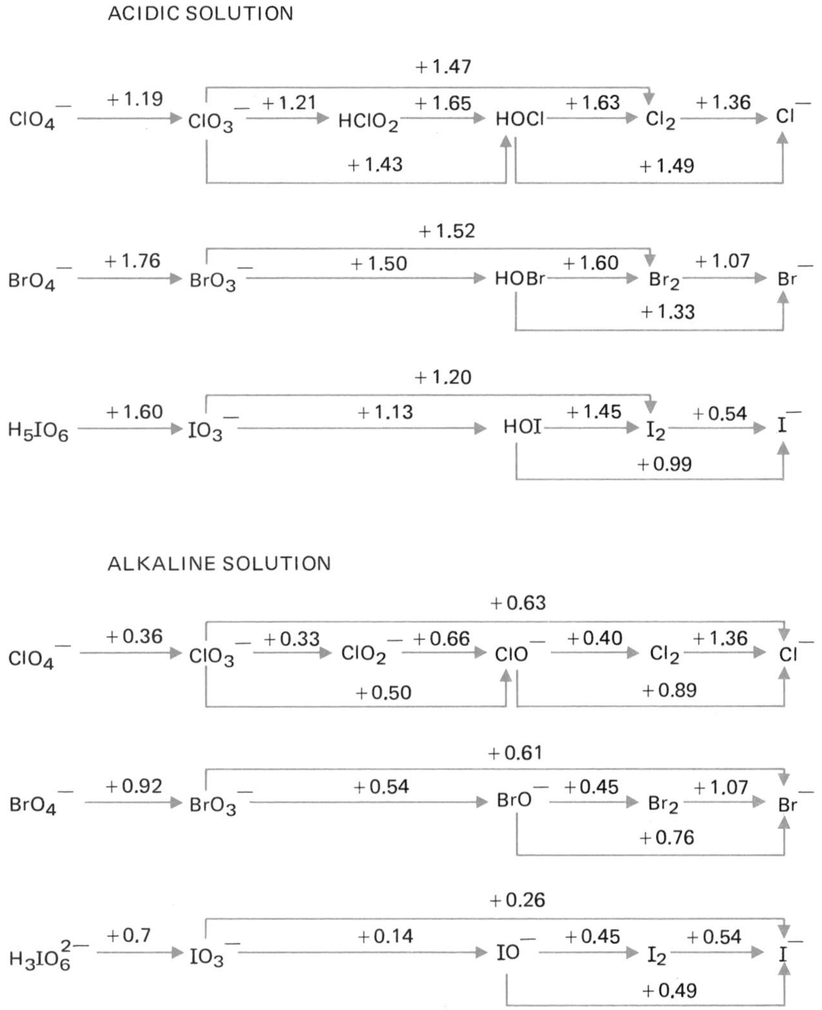

Figure 11.7 Standard electrode potential diagrams for chlorine, bromine, iodine, and their compounds.

Figure 11.7. The number shown over each arrow is the $\mathscr{E}°$, in volts, for the reduction of the species on the left to that on the right; an oxidation potential for a transformation from right to left may, of course, be obtained by changing the sign of $\mathscr{E}°$.

Perchloric acid ($HClO_4$), perbromic acid ($HBrO_4$), and the halic acids ($HClO_3$, $HBrO_3$, and $HIO_3$) are strong acids, but the remaining oxyacids are incompletely dissociated in water solution and exist in solution largely in molecular form. Hence, molecular formulas are shown in Figure 11.7 for the weak acids in acidic solution. In general, acid strength increases with increasing oxygen content.

Much of the chemistry of these compounds is effectively correlated by means of these electrode potential diagrams. Remember, however, that standard electrode potentials refer to reductions that take place at 25°C with all substances present in their standard states. Concentration changes and temperature changes alter $\mathscr{E}°$ values. Furthermore, a cell emf tells nothing about the speed of the transformation to which it applies; some reactions for which the cell emf's are positive occur so slowly that they are of no practical importance.

One of the outstanding characteristics of the oxyhalogen compounds is their ability to function as oxidizing agents; all the $\mathscr{E}°$ values are positive. The $\mathscr{E}°$ values also show that, in general, these compounds are stronger oxidizing agents in acidic solution than in alkaline solution.

Many of these oxyhalogen compounds are unstable toward disproportionation. Such materials are readily identified from the diagrams of Figure 11.7. A given substance is unstable toward disproportionation if the potential for a reduction to a lower oxidation state is more positive than the potential for a transformation from a higher oxidation state to the substance being studied. For example, in alkaline solution

$$ClO^- \xrightarrow{+0.40} Cl_2 \xrightarrow{+1.36} Cl^-$$

Since $+1.36$ is more positive than $+0.40$, $Cl_2$ will disproportionate.

$$
\begin{array}{ll}
4OH^- + Cl_2 \rightarrow 2ClO^- + 2H_2O + 2e^- & \mathscr{E}°_{ox} = -0.40 \text{ V} \\
2e^- + Cl_2 \rightarrow 2Cl^- & \mathscr{E}° = +1.36 \text{ V} \\
\hline
2OH^- + Cl_2 \rightarrow ClO^- + Cl^- + H_2O & \mathscr{E}° = +0.96 \text{ V}
\end{array}
$$

The $\mathscr{E}°$ values indicate that HOCl, HOBr, HOI, $HClO_2$, and $ClO_3^-$ should disproportionate in acidic solution. In alkaline solution all species should disproportionate except the halide ions, $ClO_4^-$, $BrO_4^-$, $BrO_3^-$, and $H_3IO_6^{2-}$.

In acidic solution any reduction that has a potential more positive than $+1.23$ V theoretically can bring about the oxidation of water to $O_2$, since

$$2H_2O \rightarrow O_2 + 4H^+ + 4e^- \qquad \mathscr{E}°_{ox} = -1.23 \text{ V}$$

In alkaline solution any reduction that has a potential more positive than $+0.40$ V is theoretically capable of bringing about the oxidation:

$$4OH^- \rightarrow O_2 + 2H_2O + 4e^- \qquad \mathscr{E}°_{ox} = -0.40 \text{ V}$$

Both these oxidations, however, occur slowly, so that it is possible to observe all the oxyhalogen compounds in solution.

The hypohalous acids (HOX) are the weakest halogen oxyacids. They exist in solution but cannot be prepared in pure form. Each of the three halogens is slightly soluble in water and reacts to produce a low concentration of the corresponding hypohalous acid.

$$X_2 + H_2O \rightarrow H^+(aq) + X^-(aq) + HOX(aq)$$

All standard potentials for these reactions are negative, and the reactions proceed to a very limited extent (see Example 10.10). Thus, at 25°C saturated solutions of the halogens contain the following HOX concentrations: HOCl, $3 \times 10^{-2}$ $M$; HOBr, $1 \times 10^{-3}$ $M$; HOI, $6 \times 10^{-6}$ $M$.

The $X_2$–$H_2O$ reactions may be driven to the right, and the yields of HOX increased, by the addition of $Ag_2O$ or $HgO$ to the reaction mixture (Section 13.10). These oxides precipitate the $X^-$ ion and remove $H^+(aq)$.

$$2X_2 + Ag_2O(s) + H_2O \rightarrow 2AgX(s) + 2HOX(aq)$$

$$2X_2 + 4HgO(s) + H_2O \rightarrow HgX_2 \cdot 3HgO(s) + 2HOX(aq)$$

Each of the three halogens dissolves in alkaline solution to produce a hypohalite ion and a halide ion.

$$X_2 + 2OH^-(aq) \rightarrow X^-(aq) + XO^-(aq) + H_2O$$

The emf for each of these reactions is positive, and each reaction is rapid. However, the hypohalite ions disproportionate in alkaline solution to the halate ions and halide ions. Fortunately the disproportionation reactions of $ClO^-$ and $BrO^-$ are slow at temperatures around 0°C, so that these ions can be prepared by this method. The hypoiodite ion, however, disproportionates rapidly even at low temperatures.

Solutions of sodium hypochlorite, used commercially in cotton bleaching (e.g., Clorox), are prepared by electrolyzing cold sodium chloride solutions such as the ones used in the preparation of chlorine (Section 11.3). In this process, however, the products of the electrolysis are not kept separate; rather, the electrolyte is vigorously mixed so that the chlorine produced at the anode reacts with the hydroxide ion produced at the cathode.

anode: $\qquad\qquad 2Cl^- \rightarrow Cl_2 + 2e^-$

cathode: $\quad 2e^- + 2H_2O \rightarrow 2OH^- + H_2$

$\qquad\qquad Cl_2 + 2OH^- \rightarrow OCl^- + Cl^- + H_2O$

The overall equation for the entire process is

$$Cl^- + H_2O \xrightarrow[\text{cold}]{\text{electrolysis}} OCl^- + H_2$$

The hypohalous acids and hypohalites are good oxidizing agents, particularly in acidic solution. The compounds decompose not only by disproportionation but also by the liberation of $O_2$ from the solutions, since all the relevant $\mathscr{E}^\circ$ values for acidic half reactions are greater than $+1.23$ V and those for alkaline half reactions are greater than $+0.40$ V.

The reactions in which $O_2$ is liberated are slow, however, and are catalyzed by metal salts.

The only halous acid known is chlorous acid ($HClO_2$). This compound cannot be isolated in pure form, and even in aqueous solution it decomposes rapidly. Chlorous acid is a weak acid, but it is stronger than hypochlorous acid. The disproportionation of $HOCl$ in acidic solution does not yield $HClO_2$. The electrode potentials indicate, however, that it should be possible to make the chlorite ion by the disproportionation of $ClO^-$ in alkaline solution; however, the disproportionation of $ClO^-$ into $ClO_3^-$ and $Cl^-$ is more favorable, and $ClO_2^-$ cannot be made from $ClO^-$.

Chlorites are comparatively stable in alkaline solution and may be prepared by passing chlorine dioxide gas (a very reactive odd-electron molecule) into a solution of an alkali.

$$2ClO_2(g) + 2OH^-(aq) \rightarrow ClO_2^-(aq) + ClO_3^-(aq) + H_2O$$

A better preparation, in which no chlorate ion is produced, is the reaction of $ClO_2$ with an alkaline solution of sodium peroxide (which forms the hydroperoxide ion, $HO_2^-$, in water).

$$2ClO_2(g) + HO_2^-(aq) + OH^-(aq) \rightarrow 2ClO_2^-(aq) + O_2(g) + H_2O$$

Solid chlorites are dangerous chemicals; they detonate when heated and are explosive in contact with combustible material.

We have mentioned the disproportionation reactions of the hypohalite ions.

$$3XO^- \rightarrow XO_3^- + 2X^-$$

Thus if a free halogen is added to a hot, concentrated solution of alkali, the corresponding halide and halate ions are produced rather than the halide and hypohalite ions.

$$3X_2 + 6OH^-(aq) \rightarrow 5X^-(aq) + XO_3^-(aq) + 3H_2O$$

The chlorates are commercially prepared by the electrolysis of hot, concentrated solutions of chlorides (instead of the cold solutions used for the electrochemical preparation of the hypochlorites). The electrolyte is stirred vigorously so that the chlorine produced at the anode reacts with the hydroxide ion that is a product of the reduction at the cathode.

anode: $\qquad\qquad 2Cl^- \rightarrow Cl_2 + 2e^-$

cathode: $\quad 2e^- + 2H_2O \rightarrow 2OH^- + H_2$

$$3Cl_2 + 6OH^- \rightarrow 5Cl^- + ClO_3^- + 3H_2O$$

If the three foregoing equations are added after the first two equations have each been multiplied through by 3, the equation for the overall process is obtained.

$$Cl^- + 3H_2O \xrightarrow[\text{hot}]{\text{electrolysis}} ClO_3^- + 3H_2$$

The chlorate crystallizes from the concentrated solution employed as the electrolyte of the cell. Although chlorates are generally water soluble, they are much less soluble than the corresponding chlorides.

Solutions of a halic acid can be prepared by adding sulfuric acid to a solution of the barium salt of the acid.

$$Ba^{2+}(aq) + 2XO_3^-(aq) + 2H^+(aq) + SO_4^{2-}(aq) \rightarrow$$
$$BaSO_4(s) + 2H^+(aq) + 2XO_3^-(aq)$$

Pure $HBrO_3$ or $HClO_3$ cannot be isolated from aqueous solutions because of decomposition. Iodic acid, $HIO_3$, however, can be obtained as a white solid; this acid is generally prepared by oxidizing iodine with concentrated nitric acid. All halic acids are strong. The halates, as well as the halic acids, are strong oxidizing agents. The reactions of chlorates with easily oxidized materials may be explosive.

Chlorine dioxide, which is used in the preparation of chlorites, may be prepared by the reduction of a chlorate in aqueous solution using sulfur dioxide gas as the reducing agent.

$$2ClO_3^-(aq) + SO_2(g) \rightarrow ClO_2(g) + SO_4^{2-}(aq)$$

Chlorates decompose upon heating in a variety of ways. At high temperatures, and particularly in the presence of a catalyst, chlorates decompose into chlorides and oxygen (Section 8.1).

$$2KClO_3(s) \xrightarrow[\text{MnO}_2]{\text{heat}} 2KCl(s) + 3O_2(g)$$

At more moderate temperatures, in the absence of a catalyst, the decomposition yields perchlorates and chlorides.

$$4KClO_3(s) \xrightarrow{\text{heat}} 3KClO_4(s) + KCl(s)$$

Perchlorate salts are made by the controlled thermal decomposition of a chlorate or by the electrolysis of a cold solution of a chlorate. The free acid, a clear hygroscopic liquid, may be prepared by distilling a mixture of a perchlorate salt with concentrated sulfuric acid. A number of crystalline hydrates of perchloric acid are known. The compound $HClO_4 \cdot H_2O$ is of interest; the lattice positions of the crystal are occupied by $H_3O^+$ and $ClO_4^-$ ions, and the solid is isomorphous (of the same crystalline structure) with $NH_4ClO_4$. Perchloric acid is a strong acid and a strong oxidizing agent. Concentrated $HClO_4$ may react violently when heated with organic substances; the reactions are frequently explosive.

Perbromates are prepared by the oxidation of bromates in alkaline solution using $F_2$ as the oxidizing agent. Several periodic acids have been prepared; the most common one is $H_5IO_6$. Periodates are made by the oxidation of iodates, usually by means of $Cl_2$ in alkaline solution. Salts of the anions $IO_4^-$, $I_2O_9^{4-}$, $IO_6^{5-}$, and $IO_3^{3-}$ have been obtained.

■ **SULFUR, SELENIUM, AND TELLURIUM**

Group VI A includes oxygen, sulfur, selenium, tellurium, and polonium. The chemistry of oxygen, the most important and most abundant element of the group, has been discussed in Chapter 8. Polonium is produced by the radioactive disintegration of radium; all isotopes of polonium are radioactive. The most abundant isotope, $^{210}Po$, has a half-life of only 138.7 days and not much is definitely known about the chemistry of this element.

**11.8 GROUP PROPERTIES**

The electronic configurations of the group VI A elements are listed in Table 11.10; each element is two electrons short of a noble-gas structure. Hence, these elements attain a noble-gas electronic configuration in the formation of ionic compounds by accepting two electrons per atom.

$$2Na^+ \quad :\overset{\cdot\cdot}{\underset{\cdot\cdot}{S}}:^{2-}$$

The elements also acquire noble-gas configurations through covalent-bond formation.

$$H:\overset{\cdot\cdot}{\underset{\cdot\cdot}{Se}}:$$
$$\overset{\cdot\cdot}{H}$$

Table 11.10 Electronic configurations of the group VI A elements

| ELEMENT | $Z$ | $1s$ | $2s$ | $2p$ | $3s$ | $3p$ | $3d$ | $4s$ | $4p$ | $4d$ | $4f$ | $5s$ | $5p$ | $5d$ | $6s$ | $6p$ |
|---------|-----|------|------|------|------|------|------|------|------|------|------|------|------|------|------|------|
| O | 8 | 2 | 2 | 4 | | | | | | | | | | | | |
| S | 16 | 2 | 2 | 6 | 2 | 4 | | | | | | | | | | |
| Se | 34 | 2 | 2 | 6 | 2 | 6 | 10 | 2 | 4 | | | | | | | |
| Te | 52 | 2 | 2 | 6 | 2 | 6 | 10 | 2 | 6 | 10 | | 2 | 4 | | | |
| Po | 84 | 2 | 2 | 6 | 2 | 6 | 10 | 2 | 6 | 10 | 14 | 2 | 6 | 10 | 2 | 4 |

Certain properties of the group VI A elements are summarized in Table 11.11. Each member of the group is a less active nonmetal than the halogen of its period. The electronegatives of the elements decrease, in the expected manner, with increasing atomic number. Oxygen is the second most electronegative element (fluorine is first); sulfur is about as electronegative as iodine. Thus, the oxides of most metals are ionic, whereas the sulfides, selenides, and tellurides of only the most active metals (such as the I A and II A metals) are truly ionic compounds.

Table 11.11 Some properties of the group VI A elements

| PROPERTY | OXYGEN | SULFUR | SELENIUM | TELLURIUM |
|---|---|---|---|---|
| color | colorless | yellow | red to black | silver-white |
| molecular formula | $O_2$ | $S_8$ rings | $Se_8$ rings $(Se)_n$ chains | $(Te)_n$ chains |
| melting point (°C) | −218.4 | 119 | 217 | 452 |
| boiling point (°C) | −182.9 | 444.6 | 688 | 1390 |
| atomic radius (Å) | 0.74 | 1.04 | 1.17 | 1.37 |
| ionic radius (2 − ion) (Å) | 1.40 | 1.84 | 1.98 | 2.21 |
| first ionization potential (eV) | 13.6 | 10.4 | 9.8 | 9.0 |
| electronegativity | 3.5 | 2.5 | 2.4 | 2.1 |
| bond energy (single bonds) (kcal/mol) | 33 | 51 | 44 | 33 |
| $\mathscr{E}°$ for reduction of element to $H_2X$ in acidic solution (V) | +1.23 | +0.14 | −0.40 | −0.72 |

The group VI A elements are predominantly nonmetallic in chemical behavior; however, metallic characteristics appear in the heavier members of the group. The trend in increasing metallic character parallels, as expected, increasing atomic number, increasing atomic radius, and decreasing ionization potential. Polonium is the most metallic member of the group; it appears to be capable of forming a $Po^{2+}$ ion that exists in aqueous solution, but the 2− state of polonium (e.g., in $H_2Po$) is unstable. Whereas tellurium is essentially nonmetallic in character, unstable salts of tellurium with anions of strong acids have been reported. The ordinary form of tellurium is metallic; selenium exists in both metallic and nonmetallic crystalline modifications.

Sulfur, selenium, and tellurium exist in positive oxidation states in compounds in which they are combined with more electronegative elements (such as oxygen and the halogens); oxygen is considered to have a positive oxidation number only in the few compounds that it forms with fluorine. For sulfur, selenium, and tellurium, the oxidation states of 4+ and 6+ are particularly important.

The electrode potentials listed in Table 11.11 give an idea of the strength of the group VI A elements as oxidizing agents. Oxygen is a strong oxidizing agent, but there is a striking decrease in this property from oxygen to tellurium; in fact, $H_2Te$ and $H_2Se$ are better *reducing* agents than hydrogen. Compare the $\mathscr{E}°$ values listed in Table 11.11 with those given for the halogens in Table 11.4.

**11.9 THE ELEMENTS**     All members of this group exist in more than one allotropic modification. For a given element the difference may be in molecular complexity ($O_2$ and $O_3$ for oxygen, Section 8.4), in crystalline form, or in both.

**Figure 11.8** Structure of the $S_8$ molecule.

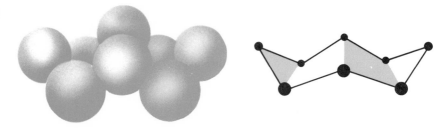

The most important solid modifications of sulfur belong to the rhombic and monoclinic crystal systems (Section 7.9). The crystals of both allotropes are built from $S_8$ molecules. These molecules are in the form of puckered, eight-membered rings of S atoms in which the S atoms are bonded to each other by single covalent bonds and the S—S—S bond angle is about 105° (Figure 11.8). In chemical equations, therefore, elementary sulfur should be indicated by the formula $S_8$. However, the usual practice is to designate sulfur by the symbol S; this practice leads to less complicated equations that are nevertheless stoichiometrically valid.

The phase diagram for sulfur is shown in Figure 11.9. The rhombic form of sulfur is the stable modification under ordinary conditions of temperature and pressure. Any point that falls within the triangle ABC of the diagram describes a set of temperature and pressure conditions under which monoclinic sulfur is the stable modification. If rhombic sulfur is heated under its own vapor pressure (lower curve of Figure 11.9), it slowly undergoes a transition to monoclinic sulfur at 95.5°C, and the monoclinic sulfur, upon continued heating, melts at 119°C. The transition from one solid form to another is slow; rapid heating of the rhombic form causes it to melt without going through the monoclinic form at 114°C (lines AD, DC, and DB of Figure 11.9). There are three triple points (A, B, and C) and one metastable triple point (between rhombic, liquid, and gas at point D).

Liquid sulfur undergoes a series of changes as its temperature is increased. At the melting point liquid sulfur is a light yellow, mobile liquid consisting principally of $S_8$ molecules. Upon continued heating the sulfur changes into a red-brown, highly viscous material. The viscosity reaches a maximum between 160°C and 200°C; upon further heating (up to the boiling point at 444.6°C), the viscosity of the liquid decreases. It is believed that the viscosity effect is caused by the dissociation of the $S_8$ rings and the formation of long chains of S atoms; at temperatures approaching the boiling point, it appears that these chains break into fragments. If sulfur is heated to approximately 200°C and then poured into cold water, a red-brown rubbery mass called plastic sulfur is obtained. It is assumed that plastic sulfur consists mainly of long chains of sulfur atoms; X-ray analysis of plastic sulfur has shown that it has the molecu-

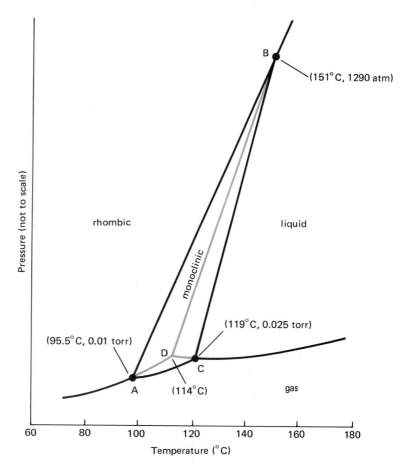

**Figure 11.9** Phase diagram for sulfur.

lar structure characteristic of fibers. At room temperature, plastic sulfur, which is a supercooled liquid, slowly crystallizes and the $S_8$ rings re-form. Sulfur vapor has been shown to consist of $S_8$, $S_6$, $S_4$, and $S_2$ molecules; $S_2$ is paramagnetic, like $O_2$.

The stable modification of selenium at room temperature is a gray, metallic, hexagonal form, the crystals of which are constructed of zigzag chains of selenium atoms. It is this form that is used in photoelectric cells; the normally low electrical conductivity of hexagonal selenium is increased about 200 times by exposure of the element to light. There are two monoclinic forms of selenium, both red; one of these has been shown to be made up of $Se_8$ rings that are similar to $S_8$ rings. In addition, amorphous forms of selenium have been described.

The common form of tellurium consists of silver-white, metallic hexagonal crystals built from zigzag chains of tellurium atoms. A black, amorphous form of tellurium exists.

**Table 11.12** Occurrence of sulfur, selenium, and tellurium

| ELEMENT | PERCENT OF EARTH'S CRUST | OCCURRENCE |
|---|---|---|
| sulfur | 0.05 | native<br>$FeS_2$ (pyrite), PbS (galena), HgS (cinnabar), ZnS (sphalerite), $Cu_2S$ (chalcocite), $CuFeS_2$ (chalcopyrite) $CaSO_4 \cdot 2 \; H_2O$ (gypsum), $BaSO_4$ (barite), $MgSO_4 \cdot 7 \; H_2O$ (epsomite) |
| selenium | $9 \times 10^{-6}$ | small amounts of Se in some S deposits<br>rare minerals: $Cu_2Se$, PbSe, $Ag_2Se$<br>low concentrations in sulfide ores of Cu, Fe, Pb, Ni |
| tellurium | $2 \times 10^{-7}$ | small amounts of Te in some S deposits<br>rare minerals: $AuTe_2$, PbTe, $Ag_2Te$, $Au_2Te$, $Cu_2Te$<br>low concentrations in sulfide ores of Cu, Fe |

The two modifications of polonium that have been reported belong to the cubic and rhombohedral systems.

The principal forms in which sulfur, selenium, and tellurium occur in nature are listed in Table 11.12. Sulfur is obtained from large underground beds of the free element by the **Frasch process** (Figure 11.10). The sulfur is melted underground by water that is heated to approximately 170°C under pressure and forced down to the deposits. A froth of sulfur, air,

**Figure 11.10** The Frasch process.

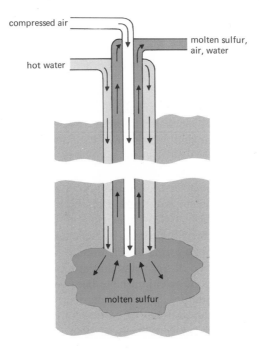

compressed air

molten sulfur, air, water

hot water

molten sulfur

and water is forced to the surface by hot, compressed air. The sulfur thus obtained is about 99.5% pure.

The principal commercial source of both selenium and tellurium is the anode sludge obtained from the electrolytic refining of copper (Section 18.5).

Some reactions of sulfur, selenium, and tellurium are summarized in Table 11.13.

**Table 11.13.** Some reactions of sulfur, selenium, and tellurium

| REACTION OF SULFUR | REMARKS |
|---|---|
| $nS + mM \rightarrow M_mS_n$ | Se, Te react similarly with many metals (not noble metals) |
| $nS + S^{2-} \rightarrow S_{n+1}^{2-}$ | for S and Te, $n = 1$ to 5; for Se, $n = 1$ to 4 |
| $S + H_2 \rightarrow H_2S$ | $S > Se > Te$; elevated temperatures; compounds are better prepared by actions of dilute HCl on sulfides, selenides, or tellurides |
| $S + O_2 \rightarrow SO_2$ | $S > Se > Te$; dioxides of Se and Te are easier to prepare with a mixture of $O_2 + NO_2$ |
| $S + 3F_2 \rightarrow SF_6$ | S, Se, Te with excess $F_2$ |
| $S + 2F_2 \rightarrow SF_4$ | S, Se; $TeF_4$ is made indirectly $(TeF_6 + Te)$ |
| $2S + X_2 \rightarrow S_2X_2$ | S, Se; $X_2 = Cl_2$ or $Br_2$ |
| $S + 2X_2 \rightarrow SX_4$ | S, Se, Te with excess $Cl_2$; Se, Te with excess $Br_2$; Te with excess $I_2$ |
| $S_2Cl_2 + Cl_2 \rightarrow 2SCl_2$ | $SCl_2$ only, $SBr_2$ unknown; $SeCl_2$, $SeBr_2$ (only in vapor state); $TeCl_2$, $TeBr_2$ are made by thermal decomposition of higher halides |
| $S + 4HNO_3 \rightarrow SO_2 + 4NO_2 + 2H_2O$ | hot, concentrated nitric acid; S yields mixtures of $SO_2$ and $SO_4^{2-}$; Se yields $H_2SeO_3$ ($SeO_2 \cdot H_2O$); Te yields $2TeO_2 \cdot HNO_3$ |

## 11.10 HYDROGEN COMPOUNDS AND DERIVATIVES

The hydrogen compounds of sulfur, selenium, and tellurium can be prepared by the direct combination of the elements at elevated temperatures. However, direct combination is not a satisfactory laboratory source of the compounds; in addition to the inconvenience of this method of preparation, $H_2S$, $H_2Se$, and $H_2Te$ are unstable at high temperatures, and the products are contaminated by starting materials.

The hydrogen compounds are readily obtained by the action of dilute acid on sulfides, selenides, and tellurides; for example,

$$FeS(s) + 2H^+(aq) \rightarrow Fe^{2+}(aq) + H_2S(g)$$

The reaction of thioacetamide with water is a convenient laboratory source of $H_2S$.

$$\text{thioacetamide (aq)} + H_2O \rightarrow \text{acetamide (aq)} + H_2S(g)$$

Hydrogen sulfide, hydrogen selenide, and hydrogen telluride are colorless, unpleasant-smelling, highly poisonous gases; they are composed of angular molecules, similar to the water molecule (Figure 4.8).

Hydrogen sulfide reacts with oxygen to yield water and either $SO_2$ or free S depending upon the amount of oxygen employed. The combustion of $H_2Se$ or $H_2Te$ in oxygen produces water and either Se or Te.

The compounds $H_2S$, $H_2Se$, and $H_2Te$ are all moderately soluble in water and are weak acids in aqueous solution. The trend in acid strength parallels that of the hydrogen halides — the elements of highest atomic number form the strongest acids. Thus, $H_2Te$ is the strongest acid, and $H_2S$ is the weakest acid of the group. The acids are diprotic and ionize in two steps.

$$H_2S(aq) \rightleftharpoons H^+(aq) + HS^-(aq)$$

$$HS^-(aq) \rightleftharpoons H^+(aq) + S^{2-}(aq)$$

The aqueous hydrogen sulfide system is discussed in Section 16.8.

The sulfides of the group I A and group II A metals are water soluble. The sulfides of the remaining metals are either very slightly soluble or decompose in the presence of water to form slightly soluble hydroxides.

$$Al_2S_3(s) + 6H_2O \rightarrow 2Al(OH)_3(s) + 3H_2S(g)$$

Precipitation of the slightly soluble sulfides, under varying conditions, is extensively used in analytical procedures for the separation and identification of cations in solution (Section 17.3). An analytical test for the sulfide ion consists of generating $H_2S$ gas by the addition of acid to the sulfide; the $H_2S$ may be identified by its odor or by the formation of insoluble, black PbS on a filter paper wet with a solution of a soluble $Pb^{2+}$ salt.

$$Pb^{2+}(aq) + H_2S(g) \rightarrow PbS(s) + 2H^+(aq)$$

Sulfur dissolves in solutions of soluble sulfides and forms a mixture of polysulfide anions.

$$S^{2-}(aq) + nS(s) \rightarrow S^{2-}_{n+1}(aq)$$

Polyselenide and polytelluride ions can be prepared by analogous reactions. For sulfur, ions varying in complexity from $S_2^{2-}$ to $S_6^{2-}$ have been prepared; $Se_5^{2-}$ and $Te_6^{2-}$ are the highest polyselenides and polytellurides known. The atoms of a polysulfide ion are joined into chains by single covalent bonds.

$$\left[ \begin{array}{c} \ddot{\underset{\cdot\cdot}{S}} \qquad \ddot{\underset{\cdot\cdot}{S}} \\ \diagdown \qquad \diagup \qquad \diagdown \\ \underset{\cdot\cdot}{\ddot{S}} \qquad\quad \underset{\cdot\cdot}{\ddot{S}} \end{array} \right]^{2-}$$

The structure of the disulfide ion, $S_2^{2-}$, is similar to that of the peroxide ion (Section 8.2). The mineral pyrite, $FeS_2$, is iron(II) disulfide.

At room temperature, polysulfides decompose in acid solution to yield mainly $H_2S$ and free S; however, careful treatment of a polysulfide solution with concentrated HCl at $-15°C$ yields $H_2S_2$, $H_2S_3$, and small quantities of higher homologs. The hydrogen polysulfides are unstable, yellow oils.

## 11.11 THE 4+ OXIDATION STATE

The important compounds in which sulfur appears in a 4+ oxidation state are sulfur dioxide ($SO_2$), sulfurous acid ($H_2SO_3$), and the salts of sulfurous acid — the sulfites. Both selenium and tellurium form compounds analogous to those of sulfur.

Sulfur dioxide is commercially obtained by burning sulfur

$$S(s) + O_2(g) \rightarrow SO_2(g)$$

or by roasting sulfide ores (such as ZnS, PbS, $Cu_2S$, and $FeS_2$) in air (Section 18.3)

$$2ZnS(s) + 3O_2(g) \rightarrow 2ZnO(s) + 2SO_2(g)$$

Sulfur dioxide is a colorless gas. It has a sharp, irritating odor and is poisonous. The molecules of $SO_2$ are angular, and the structure of the compound may be represented as a resonance hybrid (Section 4.1).

The S—O bonds have additional double-bond character from $p\pi$–$d\pi$ bonding.

The polar nature of the $SO_2$ molecule is reflected in the ease with which sulfur dioxide may be liquefied. Sulfur dioxide liquefies at $-10°C$ (the normal boiling point) under a pressure of 1 atm, and at 20°C a pressure of about 3 atm will liquefy the gas. It is this property of $SO_2$ that makes the compound useful as a refrigerant.

Sulfur dioxide is moderately soluble in water, producing solutions of sulfurous acid, $H_2SO_3$. The acid is not very stable, and pure $H_2SO_3$ cannot be isolated. The extent of the reaction between dissolved $SO_2$ molecules and water is not known; probably both $H_2SO_3$ and $SO_2$ molecules exist in the solution in equilibrium.

$$SO_2(aq) + H_2O \rightleftharpoons H_2SO_3(aq)$$

Sulfurous acid is a weak, diprotic acid.

$$H_2SO_3(aq) \rightleftharpoons H^+(aq) + HSO_3^-(aq)$$

or

$$SO_2(aq) + H_2O \rightleftharpoons H^+(aq) + HSO_3^-(aq)$$

and

$$HSO_3^-(aq) \rightleftharpoons H^+(aq) + SO_3^{2-}(aq)$$

Therefore, sulfurous acid forms two series of salts: normal salts (e.g., $Na_2SO_3$, sodium sulfite) and acid salts (e.g., $NaHSO_3$, sodium bisulfite or sodium hydrogen sulfite).

Sulfites are often prepared by bubbling $SO_2$ gas through a solution of a hydroxide.

$$2OH^- + SO_2(g) \rightarrow SO_3^{2-}(aq) + H_2O$$

If the addition is continued, acid sulfites are produced.

$$H_2O + SO_3^{2-}(aq) + SO_2(g) \rightarrow 2HSO_3^-(aq)$$

The salts of sulfurous acid can be isolated from solution.

Since sulfurous acid is unstable, the addition of an acid to a sulfite or to an acid sulfite liberates $SO_2$ gas; this is a convenient way to prepare the gas in the laboratory.

$$SO_3^{2-}(aq) + 2H^+(aq) \rightarrow SO_2(g) + H_2O$$

Sulfur dioxide, sulfurous acid, and sulfites can function as mild oxidizing agents; however, reactions in which these compounds react as reducing agents (and are oxidized to the sulfate ion, $SO_4^{2-}$) are more numerous and more important. Many substances (such as potassium permanganate, potassium dichromate, chlorine, and bromine) oxidize sulfite to sulfate. In fact, sulfites are usually contaminated by traces of sulfates because of oxidation by the oxygen of the air.

$$2SO_3^{2-} + O_2(g) \rightarrow 2SO_4^{2-}$$

Sulfites may be identified by the production of $SO_2$ gas upon acidification and by oxidation to the sulfate ion,

$$6H^+(aq) + 5SO_3^{2-}(aq) + 2MnO_4^-(aq) \rightarrow$$
$$5SO_4^{2-}(aq) + 2Mn^{2+}(aq) + 3H_2O$$

followed by the precipitation of the sulfate as the insoluble barium salt.

$$Ba^{2+}(aq) + SO_4^{2-}(aq) \rightarrow BaSO_4(s)$$

The dioxides of selenium and tellurium may be prepared by direct combination of the elements with oxygen; however, $SeO_2$ and $TeO_2$ are usually made by heating the product obtained from the oxidation of selenium or tellurium by concentrated nitric acid (see Table 11.13). Both $SeO_2$ and $TeO_2$ are white solids.

Selenous acid, $H_2SeO_3$, is formed when the very soluble $SeO_2$ is dissolved in water; it is a weak, diprotic acid and may be obtained in pure form by the evaporation of a solution of $SeO_2$. Tellurium dioxide is only slightly soluble in water, and pure $H_2TeO_3$ has never been prepared. Consequently, only very dilute solutions of tellurous acid have been studied.

Tellerium dioxide, as well as selenium dioxide, will dissolve in aqueous solutions of hydroxides to produce tellurites and selenites. If an excess of the dioxide is employed, the corresponding acid salt is obtained.

The selenium and tellurium compounds of the 4+ oxidation state are better oxidizing agents and poorer reducing agents than the corresponding sulfur compounds: $SeO_2$ is a good oxidizing agent and is employed in certain organic syntheses.

**11.12 THE 6+ OXIDATION STATE**

Sulfur trioxide is produced when sulfur dioxide reacts with atmospheric oxygen. Since the reaction is very slow at ordinary temperatures, the commercial preparation is conducted at elevated temperatures (400° to 700°C) and in the presence of a catalyst (such as divanadium pentoxide or spongy platinum).

$$2SO_2(g) + O_2(g) \rightarrow 2SO_3(g)$$

Sulfur trioxide is a volatile material (boiling point, 44.8°C); in the gas phase it consists of single molecules that are planar triangular in form with O—S—O bond angles of 120°. The electronic structure of the molecule may be represented as a resonance hybrid:

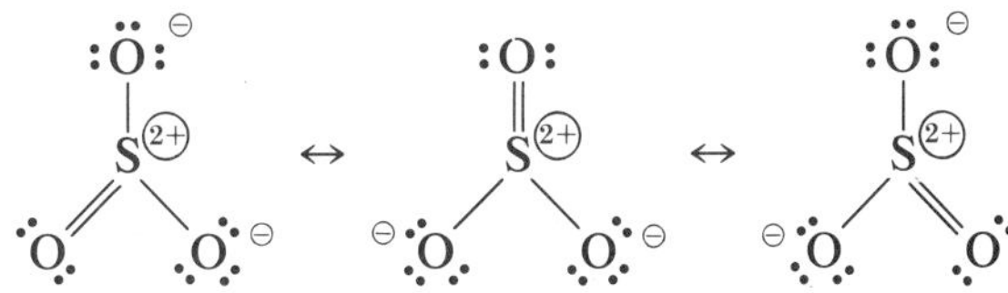

Additional $\pi$ bonding, in excess of that represented by the resonance forms, arises from the overlap of filled $2p$ orbitals of O atoms with empty $3d$ orbitals of S. The compound exists in at least three solid modifications that are formed by the condensation of $SO_3$ units into larger molecules called polymers. The commercial product, which is a variable mixture of two forms of the compound, is a colorless solid that melts at approximately 40°C.

Sulfur trioxide is an extremely reactive substance and a strong oxidizing agent. It is the anhydride of sulfuric acid, $H_2SO_4$, and reacts vigorously with water to produce the acid and with metallic oxides to produce sulfates (Section 8.3).

Sulfuric acid is an important industrial chemical. Most sulfuric acid is made by the **contact process** in which $SO_2$ is catalytically oxidized to $SO_3$ in the manner previously described. The $SO_3$ vapor is bubbled through $H_2SO_4$ and pyrosulfuric acid ($H_2S_2O_7$) is formed.

$$SO_3(g) + H_2SO_4(l) \rightarrow H_2S_2O_7(l)$$

Water is then added to the pyrosulfuric acid to make sulfuric acid of the desired concentration.

$$H_2S_2O_7(l) + H_2O \rightarrow 2H_2SO_4(l)$$

This procedure, in which pyrosulfuric acid is formed, is easier to control than the direct reaction of $SO_3$ with water.

The **lead chamber process** is an older method for the manufacture of sulfuric acid. Sulfur dioxide, oxygen, water vapor, and oxides of nitrogen are mixed in a lead-lined chamber. The reactions are complex and are not completely understood, but it is recognized that nitrosylsulfuric acid, $HOSO_2(ONO)$, is formed as an intermediate and that this compound then reacts with water to form sulfuric acid. We may summarize the process by the simplified equations

$$2SO_2(g) + NO(g) + NO_2(g) + O_2(g) + H_2O(g) \rightarrow 2HOSO_2(ONO)(s)$$

$$2HOSO_2(ONO)(s) + H_2O(g) \rightarrow 2H_2SO_4(l) + NO(g) + NO_2(g)$$

The oxides of nitrogen are regenerated and are reused. The sulfuric acid produced by the lead chamber process has a concentration of only 60 to 78% and is less pure than that made by the contact process.

Sulfuric acid is a colorless, oily liquid that freezes at 10.4°C and begins to boil at approximately 290°C with decomposition into water and sulfur trioxide. The electronic structure of sulfuric acid may be represented as

$$\text{H}-\overset{\displaystyle \ddot{\text{O}}:}{\underset{\displaystyle \ddot{\text{O}}:}{\overset{|}{\underset{|}{\ddot{\text{O}}-\text{S}-\ddot{\text{O}}}}}-\text{H}$$

However, the S—O bonds have a considerable degree of double-bond character from $p\pi$–$d\pi$ bonding, and the electronic formula

$$\text{H}-\overset{..}{\underset{..}{\text{O}}}-\overset{\overset{\text{:O:}}{\|}}{\underset{\underset{\text{:O:}}{\|}}{\text{S}}}-\overset{..}{\underset{..}{\text{O}}}-\text{H}$$

is sometimes used. The $H_2SO_4$ molecule, as well as the $SO_4^{2-}$ ion, is tetrahedral.

When concentrated sulfuric acid is added to water, a great deal of heat is evolved; this heat effect may occur because of the formation of the hydronium ion, $H_3O^+$. Sulfuric acid has a strong affinity for water and forms a series of hydrates (such as $H_2SO_4 \cdot H_2O$, $H_2SO_4 \cdot 2H_2O$, and $H_2SO_4 \cdot 4H_2O$). Thus, sulfuric acid is used as a drying agent; gases that do not react with $H_2SO_4$ may be dried by being bubbled through the acid. The dehydrating power of sulfuric acid is also seen in the charring action of the acid on carbohydrates.

$$C_{12}H_{22}O_{11}(s) \xrightarrow{\text{H}_2\text{SO}_4} 12C(s) + 11H_2O(g)$$
$$\text{sucrose}$$

In aqueous solution, sulfuric acid ionizes in two steps.

$$H_2SO_4 \rightarrow H^+(aq) + HSO_4^-(aq)$$

$$HSO_4^-(aq) \rightleftharpoons H^+(aq) + SO_4^{2-}(aq)$$

Sulfuric acid is a strong electrolyte as far as the first dissociation is concerned; however, the second dissociation is not complete. The acid forms two series of salts: normal salts (such as sodium sulfate, $Na_2SO_4$) and acid salts (such as sodium bisulfate, $NaHSO_4$). Most sulfates are soluble in water. However, barium sulfate ($BaSO_4$), strontium sulfate ($SrSO_4$), lead sulfate ($PbSO_4$), and mercury(I) sulfate ($Hg_2SO_4$) are practically insoluble; calcium sulfate ($CaSO_4$) and silver sulfate ($Ag_2SO_4$) are but slightly soluble. The formation of white, insoluble barium sulfate, the least soluble of the substances listed, is commonly used as a laboratory test for the sulfate ion.

Since sulfuric acid has a relatively high boiling point (or decomposition temperature), it is used to secure more volatile acids from their salts. This use is illustrated by the preparations of HF and HCl (Section 11.5), as well as by the preparation of $HNO_3$ (Section 12.7).

Sulfuric acid at unit concentration and 25°C is not a particularly good oxidizing agent (note the relatively low standard electrode potential in Table 11.14); however, hot, concentrated sulfuric acid is a moderately effective oxidizing agent. The oxidizing ability of hot, concentrated $H_2SO_4$ on bromides and iodides has already been noted (Section 11.5). This reagent will also oxidize many nonmetals.

$$C(s) + 2H_2SO_4(l) \rightarrow CO_2(g) + 2SO_2(g) + 2H_2O(g)$$

Most metals are oxidized by hot, concentrated sulfuric acid, including those metals of relatively low reactivity that are not oxidized by hydronium ion. For example, copper will not displace hydrogen from aqueous acids; copper metal is, however, oxidized by hot, concentrated sulfuric acid, although hydrogen gas is not a product of the reaction.

$$Cu(s) + 2H_2SO_4(l) \rightarrow CuSO_4(s) + SO_2(g) + 2H_2O(g)$$

**Table 11.14** Standard electrode potentials of sulfuric, selenic, and telluric acids

| REACTION | STANDARD ELECTRODE POTENTIAL |
|---|---|
| $2e^- + 4H^+ + SO_4^{2-} \rightleftharpoons H_2SO_3 + H_2O$ | $\mathscr{E}° = +0.17$ V |
| $2e^- + 4H^+ + SeO_4^{2-} \rightleftharpoons H_2SeO_3 + H_2O$ | $\mathscr{E}° = +1.15$ V |
| $2e^- + 2H^+ + H_6TeO_6 \rightleftharpoons TeO_2 + 4H_2O$ | $\mathscr{E}° = +1.02$ V |

Selenic acid, $H_2SeO_4$, is prepared by the oxidization of selenous acid, $H_2SeO_3$; selenates may be prepared by the oxidation of the corresponding selenites. Selenium trioxide, the acid anhydride of selenic acid, is not very stable and decomposes to $SeO_2$ and $O_2$ on warming. Low yields of $SeO_3$ mixed with $SeO_2$ are produced when an electric discharge is passed through selenium vapor and oxygen; the trioxide may also be prepared by the reaction of potassium selenate, $K_2SeO_4$, and $SO_3$.

Selenic acid is very similar to sulfuric acid. In aqueous solution the first dissociation of $H_2SeO_4$ is strong and the second dissociation is weak. The acid forms normal and acid salts. Like the sulfate ion, the selenate ion is tetrahedral.

Telluric acid is prepared by the action of vigorous oxidizing agents on elementary tellurium. Unlike $H_2SO_4$ and $H_2SeO_4$, the formula of telluric acid is $H_6TeO_6$, which may be regarded as a hydrated form of the nonexistent $H_2TeO_4$. No compound of formula $H_2TeO_4$ has ever been prepared, although salts of an acid corresponding to this formula exist. Tellurium, like iodine, is a large atom and can accommodate six oxygen atoms, and telluric acid, like $IO_6^{5-}$, is octahedral. Telluric acid is not very similar to sulfuric acid, and $H_6TeO_6$ functions in aqueous solution as a very weak diprotic acid. If $H_6TeO_6$ is heated to approximately 350°C, water is driven off and solid tellurium trioxide results. This compound is not very water soluble but reacts with alkalies to produce tellurates.

Both selenic and telluric acids are stronger oxidizing agents than sulfuric acid (Table 11.14), although the reactions of $H_2SeO_4$ and $H_6TeO_6$ often proceed slowly.

There are other acids of sulfur in which S is in a 6+ oxidation state. Pyrosulfuric acid (also called disulfuric acid)

$$\begin{array}{ccc}
& O & & O & \\
& \| & & \| & \\
H-O- & S & -O- & S & -O-H \\
& \| & & \| & \\
& O & & O &
\end{array}$$

has been mentioned as the product of the reaction of $SO_3$ with $H_2SO_4$ in a 1:1 molar ratio. The pyrosulfate ion has been shown to have a structure in which two $SO_4$ tetrahedra are joined by an O atom common to both tetrahedra: $[O_3SOSO_3]^{2-}$. Pyrosulfuric acid is a stronger oxidizing agent and a stronger dehydrating agent than sulfuric acid.

A peroxy acid is an acid that contains a peroxide group (—O—O—) somewhere in the molecule. Two peroxy acids of sulfur exist: peroxymonosulfuric acid ($H_2SO_5$) and peroxydisulfuric acid ($H_2S_2O_8$).

$$\begin{array}{cccccc}
& O & & & O & & O \\
& \| & & & \| & & \| \\
H-O-O- & S & -O-H & \qquad H-O- & S & -O-O- & S & -O-H \\
& \| & & & \| & & \| \\
& O & & & O & & O
\end{array}$$

The structure of the peroxydisulfate ion consists of two complete $SO_4$ tetrahedra joined by an O—O bond.

Peroxydisulfuric acid is prepared by the electrolysis of moderately concentrated solutions of sulfuric acid (50 to 70%) at temperatures below room temperature (5 to 10°C). Potassium and ammonium salts of this acid are prepared by the electrolysis of the corresponding acid sulfates. The anode reaction in these electrolyses may be represented by the partial equation

$$2HSO_4^- \rightarrow S_2O_8^{2-} + 2H^+ + 2e^-$$

The reaction of peroxydisulfuric acid with water yields peroxymonosulfuric acid.

$$H_2S_2O_8 + H_2O \rightarrow H_2SO_5 + H_2SO_4$$

Upon further hydrolysis, $H_2SO_5$ is decomposed into hydrogen peroxide and $H_2SO_4$.

$$H_2SO_5 + H_2O \rightarrow H_2SO_4 + H_2O_2$$

The acids may also be prepared by the reaction of one mole of hydrogen peroxide with one or two moles of chlorosulfonic acid, $HOSO_2Cl$ (the product of the reaction of $SO_3$ and HCl).

Both peroxymonosulfuric acid and peroxydisulfuric acid are low melting solids. The first ionization of peroxydisulfuric acid is strong. The peroxydisulfate ion is one of the strongest oxidizing agents known.

$$2e^- + S_2O_8^{2-} \rightleftharpoons 2SO_4^{2-} \qquad E° = +2.01 \text{ V}$$

The oxidations, however, are slow and are usually catalyzed by $Ag^+$ ions. In both peroxy acids, sulfur is assumed to be in its highest oxidation state (6+). However, the oxygen atoms of the peroxide grouping are each assigned an oxidation number of 1−, and in the course of an oxidation, it is the peroxide oxygens that change oxidation state (from 1− to 2−).

There is another significant group of sulfur-containing anions. The most important representative of the group is the thiosulfate ion, $S_2O_3^{2-}$.

$$\left[ \begin{array}{c} O \\ | \\ O-S-O \\ | \\ S \end{array} \right]^{2-}$$

This tetrahedral ion may be regarded as a sulfate ion in which one oxygen atom has been replaced by a sulfur atom. In fact, the prefix "thio-" is used to name any species that may be considered to be derived from another compound by replacing an oxygen atom by a sulfur atom; the prefix is simply appended to the base name of the unaltered compound. Thus, $CNO^-$ is the cyanate ion and $CNS^-$ is the thiocyanate ion; $CO(NH_2)_2$ is urea and $CS(NH_2)_2$ is thiourea.

Thiosulfates may be prepared by the reaction of sulfur with sulfites in aqueous solution.

$$SO_3^{2-}(aq) + S(s) \rightarrow S_2O_3^{2-}(aq)$$

The corresponding acid does not exist. Upon acidification, thiosulfates decompose to elementary sulfur and $SO_2$ gas.

$$S_2O_3^{2-}(aq) + 2H^+(aq) \rightarrow S(s) + SO_2(g) + H_2O$$

The two sulfur atoms of the thiosulfate ion are not equivalent. This has been shown by the reactions of a compound derived from a sulfite and radioactive sulfur, $^{35}_{16}S$. When a compound thus prepared is decomposed by acidification, all the activity ends in the elementary sulfur.

$$SO_3^{2-}(aq) + {}^{35}S(s) \rightarrow {}^{35}SSO_3^{2-}(aq)$$

$${}^{35}SSO_3^{2-}(aq) + 2H^+(aq) \rightarrow {}^{35}S(s) + SO_2(g) + H_2O$$

Hence, the central sulfur atom is generally assigned an oxidation number of 6+ (as in the sulfate ion), and the coordinated sulfur is usually assigned an oxidation number of 2− (corresponding to the oxidation number of the oxygen it replaces); the average oxidation state of sulfur in the ion is 2+.

The thiosulfate ion is readily oxidized to the tetrathionate ion, $S_4O_6^{2-}$. The tetrathionate ion may be regarded as an analog of the peroxydisulfate ion, $S_2O_8^{2-}$, in which the peroxide group (—O—O—) is replaced by a disulfide group (—S—S—). Other thionate ions exist − for example, the

Figure 11.11 Electrode potential diagrams for sulfur and its compounds.

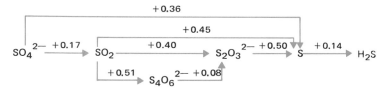

ACIDIC SOLUTION

ALKALINE SOLUTION

dithionate ion, $S_2O_6^{2-}$, and the trithionate ion, $S_3O_6^{2-}$; the latter is structurally similar to the pyrosulfate ion, $S_2O_7^{2-}$, with a sulfur atom replacing the central oxygen atom.

**11.13 SULFUR ELECTRODE POTENTIAL DIAGRAMS**

Electrode potential diagrams for sulfur and its compounds appear in Figure 11.11. On the basis of the $\mathscr{E}°$ values, several substances shown in the diagrams should be able to disproportionate. In alkaline solution (but not in acid), sulfur itself forms $S^{2-}$ and $S_2O_3^{2-}$. The thiosulfate ion is stable in alkaline solution but disproportionates to S and $SO_2$ in acid. The electrode potentials also show that both $SO_2$, in acid, and $SO_3^{2-}$, in base, are unstable toward disproportionation. The latter two reactions, however, are slow under ordinary conditions. None of the materials shown in the diagrams is capable of liberating $O_2$ from solution.

At standard concentrations and in acidic solution, the oxysulfur compounds shown in the diagram are only moderately strong oxidizing agents. In alkaline solution the oxysulfur compounds are poor oxidants. In fact, all the ions with the exception of $SO_4^{2-}$ are easily oxidized in base and can function as reducing agents. The thiosulfate ion is easily oxidized to the tetrathionate ion $S_4O_6^{2-}$; $\mathscr{E}_0°$ for this transformation is only +0.08 V.

**Problems**

**11.1** For each of the following, state the type of hybrid orbitals employed by Xe and draw diagrams to show the geometric configuration. (a) $XeF_2$, (b) $XeF_4$, (c) $XeO_3$, (d) $XeOF_4$, (e) $XeO_4$, (f) $XeO_6^{4-}$.

**11.2** Write equations to show how

to prepare (a) $XeF_2$, (b) $XeF_4$, (c) $XeF_6$, (d) $XeOF_4$, (e) $XeO_3$.

**11.3** What reasons can you give for the fact that the only binary noble-gas compounds known are the fluorides and oxides of Kr, Xe, and Rn?

**11.4** (a) Use the ion-electron method to write an equation for the oxidation, in acid solution, of $Mn^{2+}$ to $MnO_4^-$ by $XeO_3$, which is reduced to Xe(g). (b) In a reaction between $XeO_3$ and $Mn^{2+}$ that occurs in 200 ml of solution, the Xe(g) produced occupied a volume of 448 ml at STP. What is the normality and molarity of the $MnO_4^-$ solution resulting from the reaction? What weight of $XeO_3$ was consumed by the reaction?

**11.5** What reason can you give for the fact that the boiling point of helium is lower than the boiling point of hydrogen?

**11.6** Write chemical equations to show how $Cl_2(g)$ may be prepared from $Cl^-(aq)$ by use of (a) $MnO_2(s)$, (b) $PbO_2(s)$, (c) $MnO_4^-(aq)$, (d) $Cr_2O_7^{2-}(aq)$. Analogous equations may be written for reactions in which bromine and iodine are prepared from bromide ions and iodide ions. However, fluorine cannot be prepared from fluoride ions by use of these oxidizing agents. Why not?

**11.7** Write chemical equations to show how to prepare the following: (a) $F_2$ from $CaF_2$, (b) $Cl_2$ from NaCl, (c) $Br_2$ from sea water, (d) $I_2$ from $NaIO_3$, (e) HBr from $PBr_3$.

**11.8** Write chemical equations for the reactions of $Cl_2$ with (a) $H_2$, (b) Zn, (c) P, (d) S, (e) $H_2S$, (f) CO, (g) $SO_2$, (h) $I^-(aq)$, (i) cold $H_2O$.

**11.9** Explain why fluorine is a more reactive nonmetal than chlorine despite the fact that the electron affinity of chlorine is $-3.6$ eV, whereas that of fluorine is only $-3.5$ eV.

**11.10** Write equations for the reaction of HF with (a) $SiO_2$, (b) $Na_2CO_3$, (c) KF, (d) CaO.

**11.11** Write equations to show why concentrated HF(aq) is more strongly ionic than dilute HF(aq).

**11.12** Since HCl(g) can be prepared from NaCl and $H_2SO_4$, why is it that the reaction of NaBr and $H_2SO_4$ cannot be used to prepare HBr(g)?

**11.13** From each of the following pairs of compounds, select the compound that would have the higher conductivity in the molten state. Explain the basis of your prediction in each case. (a) $FeCl_2$ or $FeCl_3$, (b) RbCl or $SrCl_2$, (c) $BeF_2$ or $BeCl_2$.

**11.14** Discuss the bonding and structure of the interhalogen compounds $XX_3'$, $XX_5'$, and $XX_7'$.

**11.15** Write equations for the electrolysis of (a) dry, molten NaCl, (b) cold NaCl solution, (c) cold NaCl solution with the electrolyte stirred, (d) hot, concentrated NaCl solution with the electrolyte stirred, (e) cold $NaClO_3$ solution.

**11.16** Describe the changes in sulfur that occur as the temperature is increased.

**11.17** Describe the Frasch process for mining elementary sulfur.

**11.18** Describe the phases that exist at each of the three stable triple points and at the one metastable triple point for sulfur (Figure 11.9).

**11.19** Write equations for the reactions of sulfur with (a) $O_2$, (b) $S^{2-}$(aq), (c) $SO_3^{2-}$(aq), (d) Fe, (e) $F_2$, (f) $Cl_2$, (g) $HNO_3$.

**11.20** Compare the lead-chamber and contact processes for the manufacture of sulfuric acid. Which process yields the purer product?

**11.21** Write equations for the reactions of $H_2SO_4$ with (a) $C_{12}H_{22}O_{11}$, (b) $NaNO_3$, (c) Cu, (d) Zn, (e) ZnS, (f) $Fe_2O_3$.

**11.22** Write a sequence of equations representing reactions that could be used to synthesize each of the following and that start with elementary S. (a) $H_2S$ (not by direct union of the elements), (b) $H_2SO_3$, (c) $Na_2S_2O_3$, (d) $NaHSO_4$, (e) $H_2S_2O_7$.

**11.23** Write equations for the reaction of sulfur dioxide with (a) $O_2$ (Pt catalyst), (b) $Cl_2$(g), (c) $H_2O$, (d) $ClO_3^-$(aq), (e) $OH^-$(aq), (f) $SO_3^{2-}$(aq) and $H_2O$.

**11.24** Draw Lewis structures for and describe the geometric structure of (a) $S_3^{2-}$, (b) $SO_3^{2-}$, (c) $SO_4^{2-}$, (d) $S_2O_3^{2-}$, (e) $S_4O_6^{2-}$, (f) $H_2S_2O_7$, (g) $H_2S_2O_8$.

**11.25** Write an equation for the reaction of water with (a) $CH_3C(NH_2)S$, (b) $Al_2S_3$, (c) $SeO_3$, (d) $TeO_3$, (e) $H_2S_2O_7$, (f) $H_2S_2O_8$.

**11.26** For each of the following acids, an oxide exists that reacts with water to produce the acid. What are these acid anhydrides? (a) $H_2SO_3$, (b) $H_2SeO_4$, (c) $H_6TeO_6$, (d) HOCl, (e) $HClO_4$, (f) $HIO_3$, (g) $H_5IO_6$.

**11.27** Write equations for the reaction of $O_2$(g) with (a) $H_2S$, (b) $H_2Te$, (c) PbS, (d) $Na_2SO_3$, (e) $PtF_6$.

**11.28** Write equations for the reaction of HCl(aq) with (a) $Na_2SO_3$, (b) $Na_2S$, (c) $Na_2S_2O_3$.

**11.29** Discuss $p\pi$–$d\pi$ bonding. What is the origin of the phenomenon? What is its effect on structures in which it occurs?

**11.30** Explain why (a) $H_6SO_6$ does not exist but $H_6TeO_6$ does, (b) $HCl_2^-$ is not known but $HF_2^-$ is, (c) $OF_4$ cannot be prepared but $SF_4$ can be.

**11.31** What is the difference in meaning between the prefixes *per-* and *peroxy-* as applied in the naming of acids?

**11.32** Describe a laboratory test for (a) $F^-$(aq), (b) $Cl^-$(aq), (c) $Br^-$(aq), (d) $I^-$(aq), (e) $S^{2-}$(aq), (f) $SO_3^{2-}$(aq), (g) $SO_4^{2-}$(aq), (h) $S_2O_3^{2-}$(aq).

# 12

# The Nonmetals, Part II

The earliest known books on chemical topics were written in Alexandria, Egypt, during the Hellenistic age. The name *chemistry* is derived from the word *chemeia* (probably of Egyptian origin) that the Greeks used to mean *the Egyptian art*. When the Arabs conquered the centers of Hellenistic civilization, the "Egyptian art" passed into their hands. The word *alchemy* was coined by the addition of the Arabic definite article *al-* to the root *chemeia*.

Alchemy (as well as algebra) was introduced into western culture when the Europeans made contact with the Arab world in the middle ages. The principal aim of the European alchemists was the transmutation of base metals into gold. Their work is important to the development of chemistry largely because they accumulated a sizable body of data that proved to be of value to later chemists. Four of the elements that are discussed in this chapter—phosphorus, arsenic, antimony, and bismuth— were discovered by alchemists. In his book *The Sceptical Chymist* (published in 1661) Boyle dropped the prefix *al-* and scoffed at the theories of the alchemists.

The group V A elements, as well as carbon and silicon of group IV A and boron of group III A, are discussed in this chapter.

## ■ THE GROUP V A ELEMENTS

The elements of group V A—nitrogen, phosphorus, arsenic, antimony, and bismuth—collectively show a wider range of properties than is exhibited by either the elements of group VI A or the elements of group VII A.

## 12.1 GROUP PROPERTIES

Within any group of the periodic classification, metallic character increases (and nonmetallic character decreases) with increasing atomic number, atomic weight, and atomic size. This trend is particularly striking in group V A. Thus the electronegativities and first ionization potentials of the elements in the group decrease from values typical of a nonmetal (N) to those characteristic of a metal (Bi); these values are listed in Table 12.1. Nitrogen and phosphorus are generally regarded as nonmetals, arsenic and antimony as semimetals or metalloids, and bismuth as a metal. Bismuth, however, is not strongly metallic; its electrical and heat conductivities are low.

The electronic configurations of the elements are listed in Table 12.2. Each element has three electrons less than the noble gas of its period, and the formation of trinegative ions might be expected. Nitrogen forms the nitride ion, $N^{3-}$, in combination with certain reactive metals; phos-

Table 12.1 Some properties of the group V A elements

| PROPERTY | NITROGEN | PHOSPHORUS | ARSENIC | ANTIMONY | BISMUTH |
|---|---|---|---|---|---|
| color | colorless | white, red, black | gray metallic, yellow | gray metallic, yellow | gray metallic |
| molecular formula | $N_2$ | $P_4$ (white) $P_n$ (black) | $As_n$ (metallic) $As_4$ (yellow) | $Sb_n$ (metallic) $Sb_4$ (yellow) | $Bi_n$ |
| melting point (°C) | −210 | 44.1 (white) | 814 (36 atm) (metallic) | 630.5 (metallic) | 271 |
| boiling point (°C) | −195.8 | 280 | 633 (sublimes) | 1325 | 1560 |
| atomic radius (Å) | 0.74 | 1.10 | 1.21 | 1.41 | 1.52 |
| ionic radius (Å) | 1.4($N^{3-}$) | 1.85($P^{3-}$) | | 0.92($Sb^{3-}$) | 1.08($Bi^{3-}$) |
| first ionization potential (eV) | 14.5 | 11.0 | 10.0 | 8.6 | 8.0 |
| electronegativity | 3.0 | 2.1 | 2.0 | 1.9 | 1.9 |

phorus forms the phosphide ion, $P^{3-}$, less readily. The remaining elements of the group, however, have lower electronegativities and are more metallic than nitrogen and phosphorus; hence, arsenic, antimony, and bismuth have no tendency to form comparable anions.

Table 12.2 Electronic configurations of the group V A elements

| ELEMENT | Z | 1s | 2s | 2p | 3s | 3p | 3d | 4s | 4p | 4d | 4f | 5s | 5p | 5d | 6s | 6p |
|---------|---|----|----|----|----|----|----|----|----|----|----|----|----|----|----|----|
| N | 7 | 2 | 2 | 3 | | | | | | | | | | | | |
| P | 15 | 2 | 2 | 6 | 2 | 3 | | | | | | | | | | |
| As | 33 | 2 | 2 | 6 | 2 | 6 | 10 | 2 | 3 | | | | | | | |
| Sb | 51 | 2 | 2 | 6 | 2 | 6 | 10 | 2 | 6 | 10 | | 2 | 3 | | | |
| Bi | 83 | 2 | 2 | 6 | 2 | 6 | 10 | 2 | 6 | 10 | 14 | 2 | 6 | 10 | 2 | 3 |

The loss of electrons and consequent formation of cations, which is characteristic of metals, is observed for the heavier members of the group. High ionization potentials prohibit the loss of all five valence electrons by any element; therefore, 5+ ions do not exist, and the 5+ oxidation state is only attained through covalent bonding. In addition, most of the compounds in which the group V A elements appear in the 3+ oxidation state are covalent. However, antimony and bismuth can form $d^{10}s^2$ ions, $Sb^{3+}$ and $Bi^{3+}$, through loss of the $p$ electrons of their valence levels. The compounds $Sb_2(SO_4)_3$, $BiF_3$, and $Bi(ClO_4)_3 \cdot 5H_2O$ are ionic. The 3+ ions of antimony and bismuth react with water to form antimonyl and bismuthyl ions ($SbO^+$ and $BiO^+$), as well as hydrated forms of these ions (e.g., $Bi(OH)_2^+$).

$$Bi^{3+}(aq) + H_2O \rightleftharpoons BiO^+(aq) + 2H^+(aq)$$

Nitrogen, phosphorus, and arsenic do not form simple cations.

The oxides of the group V A elements become less acidic and more basic as the metallic character of the element increases. Thus $N_2O_3$, $P_4O_6$, and $As_4O_6$ are acidic oxides; they dissolve in water to form acids, and they dissolve in solutions of alkalies to form salts of these acids. The compound $Sb_4O_6$ is amphoteric; it will dissolve in hydrochloric acid as well as in sodium hydroxide. The comparable oxide of bismuth is strictly basic; $Bi_2O_3$ is not soluble in alkalies, but the compound will dissolve in acids to produce bismuth salts.

All the oxides in which the elements exhibit a 5+ oxidation state are acidic, but the acidity declines markedly from $N_2O_5$ to $Bi_2O_5$. In addition, the stability of the 5+ oxidation state decreases with increasing atomic number; $Bi_2O_5$ is extremely unstable and has never been prepared in a pure state.

Many of the properties of nitrogen are anomalous in comparison to those of the other V A elements; this departure is characteristic of the first members of the groups of the periodic classification. Nitrogen is the smallest and the most electronegative atom of group V A; however, free nitrogen, $N_2$, is not very reactive (Section 12.2). In fact, phosphorus is more reactive toward oxygen than is nitrogen; phosphorus is the only member of group V A that does not occur in nature as an uncombined element.

Nitrogen has no $d$ orbitals in its valence level ($n = 2$), and the maximum number of covalent bonds formed by nitrogen is four (e.g., in $NH_4^+$). In the valence levels of the other V A elements, there are empty $d$ orbitals which may be utilized in covalent bond formation. Hence, P, As, Sb, and Bi form as many as six covalent bonds in such species as $PCl_5$, $PCl_6^-$, $AsF_5$, $SbCl_6^-$, and $BiCl_5^{2-}$.

For the group as a whole, the 3−, 3+, and 5+ oxidation states are most common; the importance and stability of the 5+ and 3− states decline from the lighter to the heavier elements. Nitrogen, however, appears in every oxidation state from 3− to 5+. Nitrogen also has a tendency toward the formation of multiple bonds (e.g., in the cyanide ion, $C{\equiv}N^-$). The other V A elements do not form $\pi$ bonds with $p$ orbitals, but some multiple bond character can arise in the compounds of these elements (particularly those of P) from $p\pi$–$d\pi$ bonding.

## 12.2 THE ELEMENTS

Nitrogen constitutes about 78% by volume of the earth's atmosphere and is produced commercially by the fractional distillation of liquid air. Very pure nitrogen is conveniently prepared by heating an aqueous solution saturated with ammonium chloride and sodium nitrite,

$$NH_4^+(aq) + NO_2^-(aq) \rightarrow N_2(g) + 2H_2O$$

or by heating either sodium azide or barium azide,

$$2NaN_3(s) \rightarrow 2Na(l) + 3N_2(g)$$

Free nitrogen is surprisingly unreactive, partly because of the great strength of the bonding in the $N_2$ molecule.

$$:N{\equiv}N:$$

According to the molecular orbital theory, two $\pi$ bonds and one $\sigma$ bond join the atoms of a $N_2$ molecule, and the bond order is 3 (Section 4.4). The energy required to dissociate molecular $N_2$ into atoms is very high (225 kcal/mol).

Nitrogen is a constituent element of all plant and animal protein (Section 20.10). The cells of living systems cannot assimilate the nitrogen of the air to use in the synthesis of proteins. However, nitrogen of the air is converted by several natural **nitrogen-fixation processes** into compounds that can be used by plants. During storms, lightning flashes cause some nitrogen and oxygen of the air to form nitrogen oxide, which by a series of reactions involving additional oxygen and water is converted into nitric acid (Section 12.7) and ultimately into nitrates in the soil. Certain soil bacteria, as well as nitrogen-fixing bacteria in the root nodules of leguminous plants (such as peas, beans, and alfalfa) fix atmospheric nitrogen into compounds that the plant can assimilate. The Haber process (Section 12.4) for the manufacture of ammonia from nitrogen is the principal commercial nitrogen-fixation process; a large proportion of the ammonia produced is used to make fertilizers.

Phosphorus is industrially prepared by heating a mixture of phosphate rock, sand, and coke in an electric furnace.

$$2Ca_3(PO_4)_2(s) + 6SiO_2(s) \rightarrow 6CaSiO_3(l) + P_4O_{10}(g)$$

$$P_4O_{10}(g) + 10C(s) \rightarrow P_4(g) + 10CO(g)$$

The calcium silicate is withdrawn as a molten slag from the bottom of the furnace, and the product gases are passed through water, which condenses the phosphorus vapor into a white solid.

Arsenic, antimony, and bismuth are obtained by carbon reduction of their oxides at elevated temperatures.

$$As_4O_6(s) + 6C(s) \rightarrow As_4(g) + 6CO(g)$$

An important industrial source of the oxides is the flue dust obtained from the processes used in the production of certain metals, notably copper and lead. In addition, the oxides are obtained by roasting the sulfide ores of the elements in air; for example,

$$2Sb_2S_3(s) + 9O_2(g) \rightarrow Sb_4O_6(g) + 6SO_2(g)$$

Table 12.3 Occurrence of the group V A elements

| ELEMENT | PERCENT OF EARTH'S CRUST | OCCURRENCE |
|---|---|---|
| nitrogen | 0.0046 (0.03 including atmosphere) | $N_2$ (atmosphere), $NaNO_3$ (Chilean saltpeter) |
| phosphorus | 0.12 | $Ca_3(PO_4)_2$ (phosphate rock), $Ca_5(PO_4)_3F$ and $Ca_5(PO_4)_3Cl$ (apatite) |
| arsenic | $5 \times 10^{-4}$ | FeAsS(arsenopyrite), $As_4S_4$ (realgar), $As_2S_3$ (orpiment), $As_4O_6$ (arsenolite); native As; in ores of Cu, Pb, Co, Ni, Zn, Sn, Ag, and Au |
| antimony | $5 \times 10^{-5}$ | $Sb_2S_3$ (stibnite), $Sb_4O_6$ (senarmontite); native Sb; in ores of Cu, Pb, Ag, and Hg |
| bismuth | $1 \times 10^{-5}$ | $Bi_2S_3$ (bismuthinite), $Bi_2O_3$ (bismite); native Bi; in ores of Cu, Pb, Sn, Co, Ni, Ag, and Au |

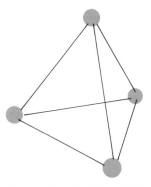

Figure 12.1 Structure of the $P_4$ molecule.

Although arsenic, antimony, and bismuth all occur as native ores, only the deposits of native bismuth are sufficiently large to be of commercial importance. The principal natural sources of the group V A elements are listed in Table 12.3.

Phosphorus, arsenic, and antimony occur in allotropic modifications. There are three important forms of phosphorus: white, red, and black. White phosphorus, a waxy solid, is obtained by condensing phosphorus vapor. Crystals of white phosphorus are formed from $P_4$ molecules (Figure 12.1) in which each phosphorus atom has an unshared pair of electrons and completes its octet by forming single covalent bonds with the other three phosphorus atoms of the molecule.

White phosphorus is soluble in a number of nonpolar solvents (e.g., benzene, carbon disulfide, and ethyl ether). In such solutions, in liquid white phosphorus, and in phosphorus vapor, the element exists as $P_4$ molecules. At temperatures above 800°C a slight dissociation of the $P_4$ molecules of the vapor into $P_2$ molecules is observed; these latter molecules are assumed to have a structure similar to that of the $N_2$ molecule. White phosphorus is the most reactive form of the element and is stored under water to protect it from atmospheric oxygen.

Red phosphorus may be prepared by heating white phosphorus to about 250°C in the absence of air. It is a polymeric material in which many phosphorus atoms are joined in a network, but the details of the structure of red phosphorus are not known. Red phosphorus is not soluble in common solvents and is considerably less reactive than the white variety. It does not react with oxygen at room temperature.

Black phosphorus, a less common allotrope, is made by subjecting the element to very high pressures or by a slow crystallization of liquid white phosphorus in the presence of mercury as a catalyst and a seed of black phosphorus. Crystalline black phosphorus consists of layers of phosphorus atoms covalently joined into a network (Figure 12.2). The distance between phosphorus atoms of adjacent layers is much greater than the distance between bonded phosphorus atoms of the same layer; it is assumed that the layers are held together by comparatively weak

Figure 12.2 Structure of a layer of the black phosphorus crystal.

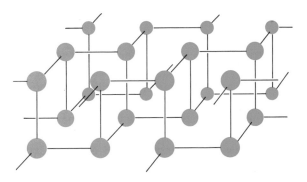

London forces. Hence, black phosphorus is a flaky material much like graphite (which also has a layer-type crystal, Section 12.10), and like graphite, black phosphorus is an electrical conductor. Black phosphorus is the least soluble and least reactive form of the element.

Arsenic and antimony exist in soft, yellow, nonmetallic modifications which are thought to be formed from tetrahedral $As_4$ and $Sb_4$ molecules analogous to the $P_4$ molecules of white phosphorus. These yellow forms may be obtained by the rapid condensation of vapors and are soluble in carbon disulfide. They are unstable and are readily converted into stable, gray, metallic modifications.

Bismuth commonly occurs as a light gray metal with a reddish cast; the element does not exist in a yellow modification. The metallic modifications of arsenic, antimony, and bismuth are comparatively soft and brittle and have a metallic luster. Their crystalline structures are similar to the structure of black phosphorus, and they are electrical conductors.

## 12.3 NITRIDES AND PHOSPHIDES

Elementary nitrogen reacts with a number of metals at high temperatures. It forms ionic compounds that contain the $N^{3-}$ ion with lithium, the group II A metals, cadmium, and zinc. These **ionic nitrides** are high-melting, white, crystalline solids; they readily react with water to yield ammonia and hydroxides.

$$Li_3N(s) + 3H_2O \rightarrow 3Li^+(aq) + 3OH^-(aq) + NH_3(g)$$

$$Ca_3N_2(s) + 6H_2O \rightarrow 3Ca^{2+}(aq) + 6OH^-(aq) + 2NH_3(g)$$

**Interstitial nitrides** may be made at elevated temperatures from many transition metals, which are in powdered form, and nitrogen or ammonia. These substances, such as $VN$, $Fe_4N$, $W_2N$, and $TiN$, contain nitrogen atoms in the interstices of the metal lattices; they frequently deviate slightly from exact stoichiometry. Interstitial nitrides have extremely high melting points, resemble metals in appearance, are good electrical conductors, are very hard, and are chemically inert.

**Covalent nitrides** include such compounds as $S_4N_4$, $P_3N_5$, $Si_3N_4$, $Sn_3N_4$, $BN$, and $AlN$. Some of these compounds are molecular in form; others are materials in which a large number of the constituent atoms are covalently joined into a crystal lattice. Aluminum nitride and boron nitride, which are representative of the latter type of compound, may be prepared by the reaction of the elements at high temperatures; $AlN$ and $BN$ react with water ($BN$ only at red heat) to give ammonia and $Al_2O_3$ or $B_2O_3$. Since one B atom and one N atom may be considered to be equivalent to two C atoms, $BN$ may be considered to be isoelectronic with carbon; the compound is known in two crystalline modifications — one resembling graphite and another extremely hard form resembling the diamond.

Many metals unite with white phosphorus directly to form phosphides; these compounds have not been so well characterized as the nitrides. The group II A metals form phosphides of formula $M_3P_2$; lithium forms $Li_3P$, and sodium forms $Na_3P$. In addition, the I A metals form compounds with formulas corresponding to $M_2P_5$. All these compounds readily hydrolyze to form phosphine, $PH_3$. For example,

$$Ca_3P_2(s) + 6H_2O \rightarrow 3Ca^{2+}(aq) + 6OH^-(aq) + 2PH_3(g)$$

The phosphides of the III A elements (such as BP, AlP, GaP, and so on) form covalent network crystals similar to silicon, and like silicon, these substances are semiconductors. Many phosphides of the transition metals are known (e.g., FeP, $Fe_2P$, $Co_2P_5$, RuP, and $OsP_2$); these materials are gray-black, semimetallic substances that are electrical conductors and are insoluble in water.

The reactions of metals with arsenic, antimony, and, to a lesser extent, bismuth yield arsenides, stibnides, and bismuthides. These compounds become progressively more difficult to prepare as the atomic number of the group V A element increases.

**12.4 HYDROGEN COMPOUNDS**

The group V A elements all form hydrogen compounds, the most important of which is ammonia, $NH_3$. Large quantities of ammonia are commercially prepared by the direct union of the elements (**Haber process**).

$$N_2(g) + 3H_2(g) \rightleftharpoons 2NH_3(g)$$

Ammonia is the only hydrogen compound of the V A elements that can be prepared directly. The reaction is conducted under high pressures (from 100 to 1000 atm), at 400° to 550°C, and in the presence of a catalyst. One catalyst, so employed, consists of finely divided iron and $Fe_3O_4$ containing small amounts of $K_2O$ and $Al_2O_3$.

Smaller quantities of ammonia are produced as a by-product in the manufacture of coke by the destructive distillation of coal. Ammonia was formerly produced commercially by the reaction of calcium cyanamide, CaNCN, with steam under pressure.

$$CaNCN(s) + 3H_2O(g) \rightarrow CaCO_3(s) + 2NH_3(g)$$

However, the Haber process has largely displaced this method as a commercial source of ammonia, and calcium cyanamide is produced chiefly as a fertilizer and as a raw material in the manufacture of certain nitrogen-containing organic compounds. Calcium cyanamide is produced in a two-step process. Calcium carbide, $CaC_2$, is made by the reaction of CaO and coke in an electric furnace

$$CaO(s) + 3C(s) \rightarrow CaC_2(s) + CO(g)$$

and the calcium carbide is reacted with relatively pure nitrogen at approximately 1000°C to produce calcium cyanamide.

$$CaC_2(s) + N_2(g) \rightarrow CaNCN(s) + C(s)$$

In the laboratory, ammonia is conveniently prepared by the hydrolysis of nitrides (Section 12.3) or by heating an ammonium salt with a strong alkali [such as NaOH or Ca(OH)$_2$] either dry or in solution.

$$NH_4^+(aq) + OH^-(aq) \rightarrow NH_3(g) + H_2O$$

The ammonia molecule,

$$H-\overset{\cdot\cdot}{N}-H \atop | \atop H$$

is trigonal pyramidal with the nitrogen atom at the apex (Section 4.2); this compound is associated through hydrogen bonding in the liquid and solid states (Section 8.9).

Aqueous solutions of ammonia are alkaline (Sections 15.3 and 16.1).

$$NH_3(aq) + H_2O \rightleftharpoons NH_4^+(aq) + OH^-(aq)$$

In solution or as a dry gas, ammonia reacts with acids to produce ammonium salts.

$$NH_3(g) + HCl(g) \rightarrow NH_4Cl(s)$$

The ammonium ion is tetrahedral.

Nitrogen is formed when ammonia is burned in pure oxygen.

$$4NH_3(g) + 3O_2(g) \rightarrow 2N_2(g) + 6H_2O(g)$$

However, when a mixture of ammonia and air is passed over platinum gauze at 1000°C, nitric oxide, NO, is produced.

$$4NH_3(g) + 5O_2(g) \rightarrow 4NO(g) + 6H_2O(g)$$

This catalyzed oxidation of $NH_3$ is a part of the Ostwald process for the manufacture of nitric acid (see Section 12.7).

Hydrazine, $N_2H_4$, may be considered to be derived from $NH_3$ by the replacement of a H atom by a —$NH_2$ group.

$$H-\overset{\cdot\cdot}{N}-\overset{\cdot\cdot}{N}-H \atop | \quad | \atop H \quad H$$

The compound, which is a liquid, may be prepared by oxidizing $NH_3$ with NaOCl. Hydrazine is less basic than $NH_3$ but does form cations in which one proton or two protons are bonded to the free electron pairs of the molecule. For example,

$$\left[ \begin{array}{c} \text{H} \\ | \\ \text{H}-\overset{..}{\text{N}}-\text{N}-\text{H} \\ | \quad | \\ \text{H} \quad \text{H} \end{array} \right]^{+} \text{Cl}^{-} \qquad \left[ \begin{array}{c} \text{H} \quad \text{H} \\ | \quad | \\ \text{H}-\text{N}-\text{N}-\text{H} \\ | \quad | \\ \text{H} \quad \text{H} \end{array} \right]^{2+} \text{2Cl}^{-}$$

Hydrazine is a strong reducing agent and has found some use in rocket fuels.

Hydroxylamine, $NH_2OH$, is another compound which, like hydrazine, may be considered to be derived from the $NH_3$ molecule.

$$\text{H}-\overset{..}{\text{N}}-\text{O}-\text{H}$$
$$| \atop \text{H}$$

Like hydrazine, hydroxylamine is a weaker base than $NH_3$, but salts containing the $[NH_3OH]^+$ ion may be prepared.

Hydrazoic acid, $HN_3$, is another hydrogen compound of nitrogen. The structure of hydrazoic acid may be represented as a resonance hybrid.

$$\text{H}-\overset{..}{\text{N}}=\overset{\oplus}{\text{N}}=\overset{..}{\text{N}}\overset{\ominus}{:} \quad \leftrightarrow \quad \text{H}-\overset{..}{\underset{..}{\text{N}}}-\overset{\ominus}{\text{N}}\equiv\overset{\oplus}{\text{N}}:$$

The two N—N bond distances of the molecule are not the same; the distance from the central N to the N bearing the H atom (1.24 Å) is longer than the other (1.13 Å). Hydrazoic acid may be made by reacting hydrazine (which forms the $[N_2H_5]^+$ ion in acidic solution) with nitrous acid ($HNO_2$).

$$N_2H_5^+(aq) + HNO_2(aq) \rightarrow HN_3(aq) + H^+(aq) + 2H_2O$$

The free acid, a low-boiling liquid, may be obtained by distillation of the water solution. Hydrazoic acid is a weak acid. The heavy metal salts of the acid, such as lead azide $[Pb(N_3)_2]$, explode upon being struck and are used in detonation caps.

Phosphine ($PH_3$) is a very poisonous, colorless gas that is prepared by the hydrolysis of phosphides or by the reaction of white phosphorus with concentrated solutions of alkalies.

$$P_4(s) + 3OH^-(aq) + 3H_2O \rightarrow PH_3(g) + \quad 3H_2PO_2^-(aq)$$
$$\text{hypophosphite ion}$$

The $PH_3$ molecule is pyramidal, similar to the $NH_3$ molecule; however, unlike $NH_3$, the compound is not associated by hydrogen bonding in the liquid state.

Phosphine is much less basic than ammonia. Phosphonium compounds, such as $PH_4I(s)$ which can be made from dry $PH_3(g)$ and $HI(g)$, are unstable. They decompose at relatively low temperatures, or in aqueous solution, to yield the component gases.

Arsine ($AsH_3$), stibine ($SbH_3$), and bismuthine ($BiH_3$) are extremely poisonous gases that may be produced by the hydrolysis of arsenides, stibnides, and bismuthides (e.g., $Na_3As$, $Zn_3Sb_2$, and $Mg_3Bi_2$). The yields of the hydrogen compounds become poorer with increasing molecular weight; very poor yields of bismuthine are obtained by this method. The stability of the hydrogen compounds declines in the series from $NH_3$ to $BiH_3$; bismuthine is very unstable and decomposes to the elements at room temperature. Arsine and stibine may be similarly decomposed by warming.

Compounds of these V A elements in higher oxidation states may be reduced by reactive metals in sulfuric acid solution to prepare the corresponding hydrogen compounds.

$$H_3AsO_3(aq) + 6H^+(aq) + 3Zn(s) \rightarrow AsH_3(g) + 3Zn^{2+}(aq) + 3H_2O$$

In the Marsh test for arsenic or antimony, the material to be tested is subjected to a metal reduction such as the one described by the preceding equation. If As or Sb is present in the sample, either $AsH_3$ gas or $SbH_3$ gas is produced. The gas issuing from the reaction flask is passed through a heated tube in which it thermally decomposes and deposits, in the case of a positive test, either an As or an Sb mirror on the tube.

Arsine, stibine, and bismuthine have no basic properties and do not form salts with acids.

**12.5 HALOGEN COMPOUNDS**

The most important halides of the V A elements are the trihalides (e.g., $NF_3$) and the pentahalides (e.g., $PF_5$). All four binary trihalides of each of the V A elements have been made, but the tribromide and triiodide of nitrogen can be isolated only in the form of ammoniates ($NBr_3·6NH_3$ and $NI_3·xNH_3$). The nitrogen trihalides are prepared by the halogenation of ammonia gas ($NF_3$, $NBr_3$), of an ammonium salt in acidic solution ($NCl_3$, $NBr_3$), or of concentrated aqueous ammonia ($NI_3$). Each of the trihalides of P, As, Sb, and Bi is prepared by direct halogenation of the V A element using a stoichiometric excess of the V A element to prevent pentahalide formation.

Bismuth trifluoride is an ionic compound, but the other trihalides are covalent. In the gaseous state the covalent trihalides exist as trigonal pyramidal molecules; this configuration is expected for $sp^3$ hybridization with an unshared electron pair occupying one of the hybrid orbitals (Figure 12.3). This molecular form persists in the liquid state and in all of the solids except $AsI_3$, $SbI_3$, and $BiI_3$, which crystallize in covalent layer lattices.

Nitrogen trifluoride is a very stable colorless gas, whereas the other trihalides of nitrogen are explosively unstable. The trihalides undergo hydrolysis:

**Figure 12.3** Molecular structure of the covalent trihalides of group V A elements.

$$NCl_3(l) + 3H_2O \rightarrow NH_3(g) + 3HOCl(aq)$$

$$PCl_3(l) + 3H_2O \rightarrow H_3PO_3(aq) + 3H^+(aq) + 3Cl^-(aq)$$

$$AsCl_3(l) + 3H_2O \rightarrow H_3AsO_3(aq) + 3H^+(aq) + 3Cl^-(aq)$$

$$SbCl_3(s) + \quad H_2O \rightarrow SbOCl(s) + 2H^+(aq) + 2Cl^-(aq)$$

$$BiCl_3(s) + \quad H_2O \rightarrow BiOCl(s) + 2H^+(aq) + 2Cl^-(aq)$$

Nitrogen is more electronegative than Cl, and in the hydrolysis of $NCl_3$, N and Cl appear in 3− and 1+ oxidation states, respectively. In each of the other hydrolysis reactions, the V A element appears in a 3+ oxidation state and Cl in a 1− state. The metallic character of the V A elements increases with increasing atomic number, and Sb and Bi occur as oxo cations ($SbO^+$ and $BiO^+$) in the hydrolysis products of $SbCl_3$ and $BiCl_3$.

The pentahalide series is not so complete as the trihalide series. Since N has no *d* orbitals in its valence level, N can form no more than four covalent bonds; hence, pentahalides of N do not exist. All pentahalides of phosphorus are known with the exception of the pentaiodide; presumably there is not sufficient room around a phosphorus atom to accommodate five large iodine atoms. In addition, $AsF_5$, $SbF_5$, $BiF_5$, and $SbCl_5$ have been prepared.

The pentahalides may be prepared by direct reaction of the elements using an excess of the halogen and by the reaction of the halogen with the trihalide.

$$PCl_3(l) + Cl_2(g) \rightleftharpoons PCl_5(s)$$

The preceding reaction is reversible. In the gas phase the pentahalides dissociate in varying degrees.

The pentahalides are trigonal bipyramidal molecules ($dsp^3$ hybridization) in the gaseous and liquid state (Figure 12.4). The crystal lattice of $SbCl_5$ consists of such molecules; however, solid $PCl_5$ and $PBr_5$ form ionic lattices composed of $PCl_4^+$ and $PCl_6^-$ and $PBr_4^+$ and $Br^-$, respectively. Apparently it is impossible to pack six bromine atoms around a phosphorus atom since $PBr_6^-$ does not form. The cations are tetrahedral ($sp^3$) and the $PCl_6^-$ ion is octahedral ($d^2sp^3$); see Figure 12.5.

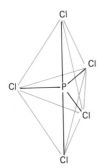

**Figure 12.4** Structure of the gaseous $PCl_5$ molecule.

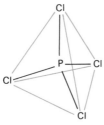 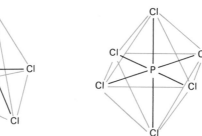

**Figure 12.5** Structures of $PCl_4^+$ and $PCl_6^-$.

The phosphorus pentahalides undergo hydrolysis in two steps; for example,

$$PCl_5(s) + H_2O \rightarrow POCl_3(l) + 2H^+(aq) + 2Cl^-(aq)$$

$$POCl_3(l) + 3H_2O \rightarrow H_3PO_4(aq) + 3H^+(aq) + 3Cl^-(aq)$$

The phosphoryl halides, $POX_3$ (X = F, Cl, or Br) can be prepared by the hydrolysis of the appropriate pentahalide in a limited amount of water or by the reaction of the trihalide with oxygen.

$$2PCl_3(l) + O_2(g) \rightarrow 2POCl_3(l)$$

Molecules of the phosphoryl halides have a $PX_3$ grouping arranged as a trigonal pyramid (Figure 12.3) with an oxygen atom bonded to the phosphorus atom, thus forming a distorted tetrahedron.

A number of mixed trihalides (e.g., $NF_2Cl$, $PFBr_2$, and $SbBrI_2$) and mixed pentahalides (e.g., $PCl_2F_3$, $PClF_4$, and $SbCl_3F_2$) have been prepared. In addition, halides are known that conform to the general formula $E_2X_4$: $N_2F_4$, $P_2Cl_4$, $P_2I_4$, and $As_2I_4$. These compounds have molecular structures similar to the structure of hydrazine, $N_2H_4$.

**Isomers** are substances that have the same molecular formula but differ in the way the constituent atoms are arranged into molecules. Dinitrogen difluoride exists in two isomeric forms.

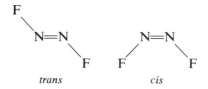

*trans*          *cis*

The double bond, composed of a $\sigma$ bond and a $\pi$ bond, between the two nitrogen atoms prevents free rotation about the nitrogen–nitrogen axis. In the *cis* isomer both fluorine atoms are on the same side of the double-bonded nitrogen atoms, whereas in the *trans* isomer the fluorine atoms are on opposite sides; both molecules are planar.

**12.6 SULFIDES**  The sulfides of the V A elements are listed in Table 12.4; other sulfides have been reported, but their existence is questioned. Tetranitrogen tetrasulfide, $N_4S_4$, is obtained by the reaction of anhydrous ammonia with either disulfur dichloride, $S_2Cl_2$, or sulfur dichloride, $SCl_2$; it is a yellow-orange crystalline solid.

With the exception of $N_4S_4$, all the sulfides listed in Table 12.4 may be prepared by the direct reaction of the elements. Each of the four phosphorus sulfides is a yellow crystalline material; $P_4S_3$ is used in the manufacture of strike-anywhere matches.

Table 12.4 Sulfides of the V A elements

| NITROGEN | PHOSPHORUS | ARSENIC | ANTIMONY | BISMUTH |
|----------|-----------|---------|----------|---------|
| $N_4S_4$ | $P_4S_3$ | $As_4S_3$ | $Sb_2S_3$ | $Bi_2S_3$ |
|  | $P_4S_5$ | $As_4S_4$ | $Sb_2S_5$ |  |
|  | $P_4S_7$ | $As_2S_3$ |  |  |
|  | $P_4S_{10}$ | $As_2S_5$ |  |  |

The sulfides of bismuth and antimony, as well as $As_2S_3$ and $As_2S_5$, may be precipitated from acidic solutions of appropriate compounds of these V A elements by hydrogen sulfide; $As_2S_3$ and $As_2S_5$ are yellow, $Sb_2S_3$ is orange-red to black, $Sb_2S_5$ is orange, and $Bi_2S_3$ is dark brown. The arsenic and antimony sulfides are dissolved by excess sulfide ion to form thio anions; the 3+ compounds form the thioarsenite, $AsS_3^{3-}$, and thioantimonite, $SbS_3^{3-}$, ions, and the 5+ compounds form the thioarsenate, $AsS_4^{3-}$, and the thioantimonate, $SbS_4^{3-}$, ions. Bismuth sulfide, $Bi_2S_3$, is not dissolved by sulfide ion. The formation of thio anions is used in qualitative analysis to effect separations of sulfide precipitates (Section 17.4).

## 12.7 OXIDES AND OXYACIDS OF NITROGEN

Oxides are known for every oxidation state of nitrogen from 1+ to 5+. Dinitrogen oxide (also called nitrous oxide), $N_2O$, is prepared by gently heating molten ammonium nitrate.

$$NH_4NO_3(l) \rightarrow N_2O(g) + 2H_2O(g)$$

It is a colorless gas and is relatively unreactive. However, at temperatures around 500°C, dinitrogen oxide decomposes to nitrogen and oxygen; hence, $N_2O$ supports combustion. Molecules of $N_2O$ are linear, and the electronic structure of the compound may be represented as a resonance hybrid.

$$\overset{\ominus}{:}\ddot{N}=\overset{\oplus}{N}=\ddot{O}: \leftrightarrow :N\equiv\overset{\oplus}{N}-\overset{\ominus}{\underset{\cdot\cdot}{\ddot{O}}}:$$

Dinitrogen oxide is commonly called "laughing gas" because of the effect it produces when breathed in small amounts. The gas is used as a general anesthetic, and because of its solubility in cream, it is the gas used to charge whipped cream aerosol cans.

Nitrogen oxide (also called nitric oxide), NO, may be prepared by the direct reaction of the elements at high temperatures.

$$N_2(g) + O_2(g) \rightarrow 2NO(g)$$

The reaction is endothermic ($\Delta H = +21.6$ kcal/mol), but even at 3000°C the yield of NO is only approximately 4%. In a successful preparation the hot gases from the reaction must be rapidly cooled to prevent the decomposition of NO into nitrogen and oxygen. By this reaction, atmospheric nitrogen is fixed during lightning storms; this reaction also serves as the basis of the **arc process** of nitrogen fixation in which an electric arc is used to provide the high temperatures necessary for the direct combination of nitrogen and oxygen. As a commercial source of NO, the arc process has been supplanted by the catalytic oxidation of ammonia from the Haber process (Section 12.4).

The NO molecule contains an odd number of electrons, which means that one electron must be unpaired; for this reason, NO is paramagnetic. The electronic structure of the molecule may be represented by the resonance forms

$$:\dot{N}\!=\!\ddot{O}: \leftrightarrow :\overset{\ominus}{\ddot{N}}\!=\!\overset{\oplus}{\dot{O}}:$$

However, the structure is best described by molecular orbital theory, which assigns a bond order of $2\frac{1}{2}$ to the molecule and indicates that the odd electron is in a $\pi^*$ orbital (Section 4.4). The loss of an electron from NO produces the nitrosonium ion, $NO^+$. Since the electron is lost from an antibonding orbital, the $NO^+$ ion has a bond order of 3, and the bond distance in $NO^+$ (1.06 Å) is shorter than the bond distance in the NO molecule (1.14 Å). Ionic compounds of $NO^+$ are known (e.g., $NO^+$ $HSO_4^-$, $NO^+$ $ClO_4^-$, and $NO^+$ $BF_4^-$).

Whereas odd-electron molecules are generally very reactive and highly colored, nitric oxide is only moderately reactive and is a colorless gas (condensing to a blue liquid and blue solid at low temperatures). In addition, NO shows little tendency to associate into $N_2O_2$ molecules by electron pairing. Nitrogen oxide reacts instantly with oxygen at room temperature to form nitrogen dioxide.

$$2NO(g) + O_2(g) \rightarrow 2NO_2(g)$$

Dinitrogen trioxide, $N_2O_3$, forms as a blue liquid when an equimolar mixture of nitric oxide and nitrogen dioxide is cooled to $-20°C$.

$$NO(g) + NO_2(g) \rightarrow N_2O_3(l)$$

The compound is unstable under ordinary conditions and decomposes into NO and $NO_2$. Both NO and $NO_2$ are odd-electron molecules; $N_2O_3$ is formed by electron pairing, and the $N_2O_3$ molecule is thought to contain a N—N bond. Dinitrogen trioxide is the anhydride of nitrous acid, $HNO_2$, and dissolves in aqueous alkali to produce the nitrite ion, $NO_2^-$.

Nitrogen dioxide, $NO_2$, and dinitrogen tetroxide, $N_2O_4$, exist in equilibrium.

$$2NO_2 \rightleftharpoons N_2O_4$$

Nitrogen dioxide consists of odd-electron molecules, is paramagnetic, and is brown in color; the dimer, in which the electrons are paired, is diamagnetic and colorless. In the solid state the oxide is colorless and consists of pure $N_2O_4$. The liquid is yellow in color and consists of a dilute solution of $NO_2$ in $N_2O_4$. As the temperature is raised, the gas contains more and more $NO_2$ and becomes deeper and deeper brown in color. At 135°C approximately 99% of the mixture is $NO_2$.

Nitrogen dioxide molecules are angular.

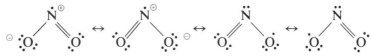

The structure of $N_2O_4$ is thought to be planar with two $NO_2$ units joined by a N—N bond.

Nitrogen dioxide is produced by the reaction of nitric oxide with oxygen. In the laboratory the compound is conveniently prepared by heating lead nitrate.

$$2Pb(NO_3)_2(s) \rightarrow 2PbO(s) + 4NO_2(g) + O_2(g)$$

Dinitrogen pentoxide, $N_2O_5$, is the acid anhydride of nitric acid; $N_2O_5$ may be prepared from this acid by dehydration using phosphorus (V) oxide.

$$4HNO_3(g) + P_4O_{10}(s) \rightarrow 4HPO_3(s) + N_2O_5(g)$$

The compound is a colorless, crystalline material that sublimes at 32.5°C. The vapor consists of $N_2O_5$ molecules which are thought to be planar with the two nitrogen atoms joined through an oxygen atom $(O_2NONO_2)$.

The electronic structure of the molecule may be represented as a resonance hybrid of

and other equivalent structures that have different arrangements of the double bonds. The compound is unstable in the vapor state and decomposes according to the equation

$$2N_2O_5(g) \rightarrow 4NO_2(g) + O_2(g)$$

Crystals of $N_2O_5$ are composed of nitronium, $NO_2^+$, and nitrate, $NO_3^-$, ions; the compound is dissociated into these two ions in solutions in anhydrous sulfuric acid, nitric acid, and phosphoric acid. The nitrate ion is triangular planar. The nitronium ion is linear; it is isoelectronic with $CO_2$ and may be considered as a nitrogen dioxide molecule minus the odd electron. The ion is probably a reaction intermediate in certain reactions of nitric acid in the presence of sulfuric acid (nitrations); ionic nitronium compounds have been prepared (e.g., $NO_2^+ClO_4^-$, $NO_2^+BF_4^-$, $NO_2^+PF_6^-$).

The most important oxyacid of nitrogen is nitric acid, $HNO_3$, in which nitrogen exhibits an oxidation number of 5+. Commercially, nitric acid is produced by the **Ostwald process.** Nitric oxide from the catalytic oxidation of ammonia is reacted with oxygen to form nitrogen dioxide. This gas, together with excess oxygen, is passed into a tower where it reacts with warm water.

$$3NO_2(g) + H_2O \rightarrow 2H^+(aq) + 2NO_3^-(aq) + NO(g)$$

The excess oxygen converts the NO into $NO_2$; this $NO_2$ then reacts with water as before. In this cyclic manner the nitric oxide is eventually completely converted into nitric acid. The product of the Ostwald process is about 70% $HNO_3$ and is known as concentrated nitric acid; more concentrated solutions may be prepared from it by distillation.

Pure nitric acid is a colorless liquid that boils at 83°C. It may be prepared in the laboratory by heating sodium nitrate with concentrated sulfuric acid.

$$NaNO_3(s) + H_2SO_4(l) \rightarrow NaHSO_4(s) + HNO_3(g)$$

This preparation (from Chilean saltpeter) is a minor commercial source of the acid.

The $HNO_3$ molecule is planar and may be represented as a resonance hybrid.

Nitric acid is a strong acid and is almost completely dissociated in aqueous solution. Most salts of nitric acid, which are called nitrates, are very soluble in water. The nitrate ion is triangular planar.

Nitric acid is a powerful oxidizing agent; it oxidizes most nonmetals (generally to oxides or oxyacids of their highest oxidation state) and all metals with the exception of a few of the noble metals. Many unreactive metals, such as silver and copper, that do not react to yield hydrogen with nonoxidizing acids, such as HCl, dissolve in nitric acid.

In nitric acid oxidations, hydrogen is almost never obtained; instead, a variety of nitrogen-containing compounds, in which nitrogen is in a lower oxidation state, is produced (Table 12.5). The product to which $HNO_3$ is reduced depends upon the concentration of the acid, the temperature, and the nature of the material being oxidized. Generally, a mixture of products is obtained, but the principal product in many cases is NO when dilute $HNO_3$ is employed and the nitrogen(IV) oxides when concentrated $HNO_3$ is used.

Dilute:

$$3Cu(s) + 8H^+(aq) + 2NO_3^-(aq) \rightarrow 3Cu^{2+}(aq) + 2NO(g) + 4H_2O$$

Concentrated:

$$Cu(s) + 4H^+(aq) + 2NO_3^-(aq) \rightarrow Cu^{2+}(aq) + 2NO_2(g) + 2H_2O$$

In some instances, however, strong reducing agents are known to produce almost pure compounds of nitrogen in lower oxidation states; for example, the reaction of zinc and dilute nitric acid yields $NH_3$ as the reduction product of $HNO_3$.

Table 12.5 Standard electrode potentials for reductions of the nitrate ion

| HALF REACTION | STANDARD ELECTRODE POTENTIAL |
|---|---|
| $e^- + 2H^+ + NO_3^- \rightleftharpoons NO_2 + H_2O$ | $\mathscr{E}° = +0.80$ V |
| $8e^- + 10H^+ + NO_3^- \rightleftharpoons NH_4^+ + 3H_2O$ | $\mathscr{E}° = +0.88$ V |
| $2e^- + 3H^+ + NO_3^- \rightleftharpoons HNO_2 + H_2O$ | $\mathscr{E}° = +0.94$ V |
| $3e^- + 4H^+ + NO_3^- \rightleftharpoons NO + 2H_2O$ | $\mathscr{E}° = +0.96$ V |
| $8e^- + 10H^+ + 2NO_3^- \rightleftharpoons N_2O + 5H_2O$ | $\mathscr{E}° = +1.12$ V |
| $10e^- - 6H^+ + 2NO_3^- \rightleftharpoons N_2 + 6H_2O$ | $\mathscr{E}° = +1.25$ V |

The half reactions for the reduction of the nitrate ion in acid solution (Table 12.5) show the $\mathscr{E}°$ values to be strongly dependent upon the $H^+(aq)$ concentration. This concentration dependence is experimentally observed; below a concentration of $2M$, nitric acid has little more oxidizing power than solutions of HCl of corresponding concentration.

Pure nitrous acid, $HNO_2$, has never been isolated, but a solution of this acid may be prepared by adding an equimolar mixture of NO and $NO_2$ to water.

$$NO(g) + NO_2(g) + H_2O \rightleftharpoons 2HNO_2(aq)$$

The reaction is exothermic and reversible; when warmed, nitrous acid decomposes, and the $NO_2$ thus produced reacts with water to give nitric acid. The overall reaction may be indicated

$$3HNO_2(aq) \xrightarrow{\text{heat}} H^+(aq) + NO_3^-(aq) + H_2O + 2NO(g)$$

Consequently, solutions of the acid are usually prepared by adding a strong acid to a cold aqueous solution of a nitrite.

$$H^+(aq) + NO_2^-(aq) \rightarrow HNO_2(aq)$$

The acid is weak and may function as an oxidizing agent or a reducing agent. The electronic structure may be represented as

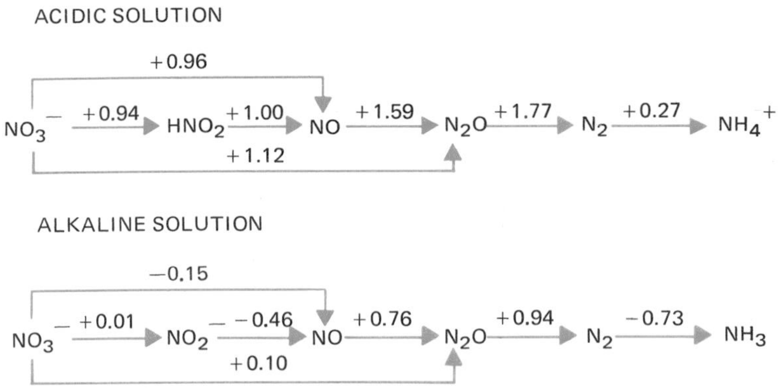

ACIDIC SOLUTION

$$NO_3^- \xrightarrow{+0.94} HNO_2 \xrightarrow{+1.00} NO \xrightarrow{+1.59} N_2O \xrightarrow{+1.77} N_2 \xrightarrow{+0.27} NH_4^+$$

with $+0.96$ spanning $NO_3^-$ to $NO$ and $+1.12$ spanning $NO_3^-$ to $NO$

ALKALINE SOLUTION

$$NO_3^- \xrightarrow{+0.01} NO_2^- \xrightarrow{-0.46} NO \xrightarrow{+0.76} N_2O \xrightarrow{+0.94} N_2 \xrightarrow{-0.73} NH_3$$

with $-0.15$ spanning $NO_3^-$ to $NO$ and $+0.10$ spanning $NO_3^-$ to $NO$

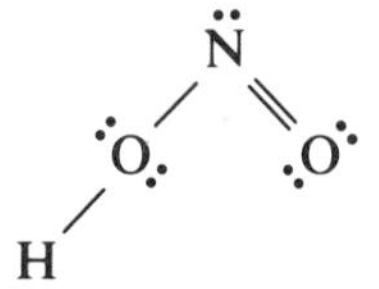

Nitrites are prepared by the addition of NO and $NO_2$ to solutions of alkalies. The nitrites of the I A metals are formed when the nitrates are heated; they may also be prepared by heating the nitrate with a reducing agent, such as lead, iron, or coke.

$$NaNO_3(l) + C(s) \rightarrow NaNO_2(l) + CO(g)$$

The nitrite ion is angular.

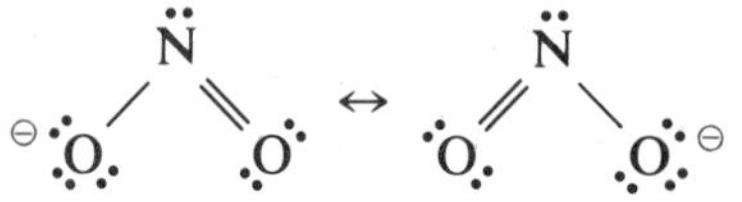

Electrode potential diagrams for $N_2$ and some of its compounds appear in Figure 12.6. Except for $NH_4^+$ and $NH_3$ (which cannot function as oxidizing agents), the compounds of nitrogen shown in the diagrams are much stronger oxidizing agents in acidic solution than in alkaline solution. In fact, the $NO_3^-$ ion is a very weak oxidant in alkaline solution. Conversely, a given compound of nitrogen is easier to oxidize in alkaline solution than in acidic solution with the exception of $NO_3^-$ which cannot be oxidized. The potentials also indicate that $HNO_2$ is unstable toward disproportionation to NO and $NO_3^-$ in acid, whereas in alkaline solution the $NO_2^-$ ion is stable toward such disproportionation.

**12.8 OXIDES AND OXYACIDS OF PHOSPHORUS, ARSENIC, ANTIMONY, AND BISMUTH**

When white phosphorus is burned in a limited supply of air, the chief product is phosphorus(III) oxide (also called phosphor*ous* oxide), $P_4O_6$. It is a colorless material that melts at 23.8°C. The combustion of phosphorus in excess oxygen yields phosphorus(V) oxide (also called phosphor*ic* oxide), $P_4O_{10}$, a white powder that sublimes at approximately

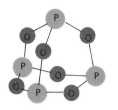

$P_4O_6$

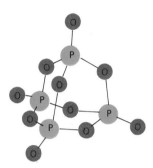

$P_4O_{10}$

**Figure 12.7** Structures of $P_4O_6$ and $P_4O_{10}$.

$PO_4{}^{3-}$

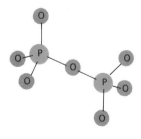

$P_2O_7{}^{4-}$

**Figure 12.8** Structures of $PO_4^{3-}$ and $P_2O_7^{4-}$.

360°C. The structures of these molecules are based on $P_4$ tetrahedra (Figure 12.7). Before the true molecular structures of these compounds were known, empirical formulas were used; consequently, the compounds are often designated as $P_2O_3$, phosphorus trioxide, and $P_2O_5$, phosphorus pentoxide. Phosphorus(V) oxide exists in three crystalline modifications, two of which are polymeric. When $P_4O_6$ is heated above 210°C, red phosphorus and a third oxide, $(PO_2)_n$, are formed; $(PO_2)_n$ is a polymeric material and is relatively unimportant.

Phosphorus(V) oxide has a great affinity for water and is a very effective drying agent. Many different phosphoric acids may be prepared by the addition of water to $P_4O_{10}$. The most important acid of phosphorus in the 5+ state, orthophosphoric acid (simply called phosphoric acid) results from the complete hydration of the oxide.

$$P_4O_{10} + 6H_2O \rightarrow 4H_3PO_4$$

Phosphoric acid is obtained commercially by this means or by treating phosphate rock with sulfuric acid.

$$Ca_3(PO_4)_2(s) + 3H_2SO_4(l) \rightarrow 2H_3PO_4(l) + 3CaSO_4(s)$$

The compound is a colorless solid but is generally sold as an 85% solution. The electronic structure of $H_3PO_4$ may be represented as

since the P—O bond has double-bond character from $p\pi$–$d\pi$ bonding. The $H_3PO_4$ molecule and the ions derived from it are tetrahedral (Figure 12.8).

Phosphoric acid is a weak, triprotic acid without effective oxidizing power.

$$H_3PO_4(aq) \rightleftharpoons H^+(aq) + H_2PO_4^-(aq)$$

$$H_2PO_4^-(aq) \rightleftharpoons H^+(aq) + HPO_4^{2-}(aq)$$

$$HPO_4^{2-}(aq) \rightleftharpoons H^+(aq) + PO_4^{3-}(aq)$$

Hence, three series of salts may be derived from $H_3PO_4$. The product of a given neutralization depends upon the stoichiometric ratio of $H_3PO_4$ to alkali.

Phosphates are important ingredients of commercial fertilizers. Phosphate rock is too insoluble in water to be used for this purpose; however, the more soluble calcium dihydrogen phosphate may be obtained by treatment of phosphate rock with an acid.

$$Ca_3(PO_4)_2(s) + 2H_2SO_4(l) \rightarrow Ca(H_2PO_4)_2(s) + 2CaSO_4(s)$$

The mixture of $Ca(H_2PO_4)_2$ and $CaSO_4$ is called **superphosphate fertilizer.** A higher yield of the dihydrogen phosphate is obtained if phosphoric acid is employed in the reaction instead of sulfuric acid.

$$Ca_3(PO_4)_2(s) + 4H_3PO_4(l) \rightarrow 3Ca(H_2PO_4)_2(s)$$

Since nitrates are also important constituents of fertilizers, the mixture obtained by treatment of phosphate rock with nitric acid is a highly effective fertilizer.

$$Ca_3(PO_4)_2(s) + 4HNO_3(l) \rightarrow Ca(H_2PO_4)_2(s) + 2Ca(NO_3)_2(s)$$

Condensed phosphoric acids have more than one P atom per molecule. The members of one group of condensed phosphoric acids, the **polyphosphoric acids,** conform to the general formula $H_{n+2}P_nO_{3n+1}$. Examples are

$H_4P_2O_7$, diphosphoric acid          $H_5P_3O_{10}$, triphosphoric acid

The polyphosphoric acids and polyphosphates have chain structures based on $PO_4$ tetrahedra which are joined through O atoms that are common to adjacent tetrahedra (Figure 12.8).

The **metaphosphoric acids** constitute another group of condensed phosphoric acids. These compounds have the general formula $H_nP_nO_{3n}$. Some of the metaphosphoric acids are cyclic. For example,

$H_3P_3O_9$, trimetaphosphoric acid          $H_4P_4O_{12}$, tetrametaphosphoric acid

In addition, there are high-molecular-weight, long-chain metaphosphoric acids which are always obtained as complex mixtures that are assigned the formula $(HPO_3)_n$. The molecules of these mixtures are based on long

$$\begin{array}{c} \text{O} \\ \| \\ \text{chains of } -\text{P}-\text{O}- \text{ units joined in such a way that each P atom is tetra-} \\ | \\ \text{OH} \end{array}$$

hedrally bonded to four O atoms, but the complete structures are very complicated and involve phosphorus–oxygen units that link two chains together.

The condensed phosphoric acids may be obtained by the controlled addition of water to $P_4O_{10}$. For example,

$$P_4O_{10}(s) + 4H_2O \rightarrow 2H_4P_2O_7(s)$$
$$\text{diphosphoric acid}$$

$$P_4O_{10}(s) + 2H_2O \xrightarrow{0°C} H_4P_4O_{12}(s)$$
$$\text{tetrametaphosphoric acid}$$

The dehydration of $H_3PO_4$ by heating also yields condensed acids. For example,

$$2H_3PO_4(l) \xrightarrow{215°C} H_4P_2O_7(l) + H_2O(g)$$
$$\text{diphosphoric acid}$$

$$nH_3PO_4(l) \xrightarrow{325°C} (HPO_3)_n(l) + nH_2O(g)$$
$$\text{metaphosphoric}$$
$$\text{acid mixture}$$

On standing in water, all condensed phosphoric acids revert to $H_3PO_4$.

Phosphor*ous* acid, $H_3PO_3$, may be prepared by adding $P_4O_6$ to cold water,

$$P_4O_6(s) + 6H_2O \rightarrow 4H_3PO_3(aq)$$

or by the hydrolysis of $PCl_3$ (Section 12.5). Condensed phosphorous acids also exist. Even though the $H_3PO_3$ molecule contains three hydrogen atoms, phosphorous acid is a weak, *diprotic* acid and is probably better formulated as $H_2(HPO_3)$.

$$H_2(HPO_3)(aq) \rightleftharpoons H^+(aq) + H(HPO_3)^-(aq)$$

$$H(HPO_3)^-(aq) \rightleftharpoons H^+(aq) + HPO_3^{2-}(aq)$$

The sodium salts $NaH_2PO_3$ and $Na_2HPO_3$ are known, but it is impossible to prepare $Na_3PO_3$.

Solutions of salts of hypophosphorous acid, $H_3PO_2$, are prepared by boiling white phosphorus with solutions of alkalies (Section 12.4). The acid, which is a colorless crystalline material, may be obtained by treating a solution of barium hypophosphite with sulfuric acid, removing the precipitated barium sulfate by filtration, and evaporating the solution. Hypophosphorous acid is a weak monoprotic acid; hence the formula of the compound is sometimes written $H(H_2PO_2)$.

$$H(H_2PO_2)(aq) \rightleftharpoons H^+(aq) + H_2PO_2^-(aq)$$

The numbers of protons released by orthophosphoric, phosphorous, and hypophosphorous acids may be explained by the electronic structures of these compounds.

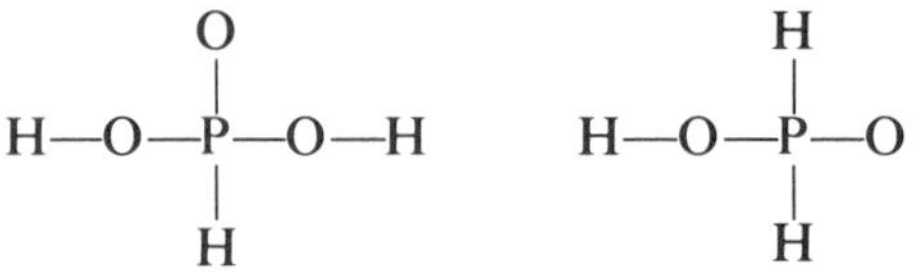

$$
\begin{array}{cc}
\quad\ \text{O} & \quad\ \text{H} \\
\quad\ | & \quad\ | \\
\text{H—O—P—O—H} & \text{H—O—P—O} \\
\quad\ | & \quad\ | \\
\quad\ \text{H} & \quad\ \text{H} \\
\text{phosphorous acid} & \text{hypophosphorous acid}
\end{array}
$$

Only the hydrogen atoms bonded to oxygen atoms are acidic; those bonded to phosphorus atoms do not dissociate as $H^+(aq)$ in water. In each of these molecules the P—O bonds have $p\pi$-$d\pi$ double-bond character.

Electrode potential diagrams for phosphorus and some of its compounds appear in Figure 12.9. The most striking feature of the diagrams is that all the potentials are negative—in contrast to the potentials of the nitrogen diagram. Thus none of the substances is a good oxidizing agent, particularly in alkaline solution. Rather, $H_3PO_3$, $H_3PO_2$, and the salts of these acids have strong reducing properties; the acids are readily oxidized to $H_3PO_4$ and the anions are readily oxidized to $PO_4^{3-}$. With the exception of $HPO_3^{2-}$ in alkaline solution, all species of intermediate oxidation state disproportionate; $PH_3$ is one of the products of each disproportionation.

When arsenic, antimony, or bismuth is heated in air, a 3+ oxide is formed: $As_4O_6$, $Sb_4O_6$, or $Bi_2O_3$. Aqueous solutions of $As_4O_6$ are acidic and thought to contain arsenious acid ($H_3AsO_3$); the other oxides are practically insoluble in water.

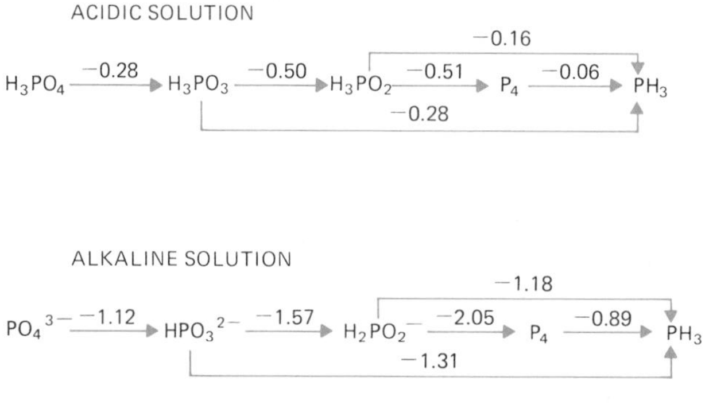

**Figure 12.9** Electrode potential diagrams for phosphorus and some of its compounds.

ACIDIC SOLUTION

$$
\text{H}_3\text{PO}_4 \xrightarrow{-0.28} \text{H}_3\text{PO}_3 \xrightarrow{-0.50} \text{H}_3\text{PO}_2 \xrightarrow[{-0.28}]{\overset{-0.16}{\phantom{xxxx}}\ -0.51} \text{P}_4 \xrightarrow{-0.06} \text{PH}_3
$$

ALKALINE SOLUTION

$$
\text{PO}_4{}^{3-} \xrightarrow{-1.12} \text{HPO}_3{}^{2-} \xrightarrow{-1.57} \text{H}_2\text{PO}_2{}^- \xrightarrow[{-1.31}]{\overset{-1.18}{\phantom{xxxx}}\ -2.05} \text{P}_4 \xrightarrow{-0.89} \text{PH}_3
$$

The increasing metallic character in the series As, Sb, Bi is apparent in the decreasing acidic character of these oxides. Both $As_4O_6$ and $Sb_4O_6$ dissolve in aqueous alkalies to form arsenites (salts of $AsO_3^{3-}$ and other forms) or antimonites (salts of $SbO_2^-$); $Bi_2O_3$, however, will not dissolve in aqueous alkalies, thus displaying a reactivity typical of metallic oxides. On the other hand, $Bi_2O_3$ and $Sb_4O_6$ (which is amphoteric) will dissolve in acids to give salts of the $BiO^+$ or $SbO^+$ ions [salts of $Bi^{3+}$ or $Sb^{3+}$ form if the concentration of $H^+(aq)$ is high enough]. Bismuth(III) hydroxide, the only true hydroxide of the group, precipitates when $OH^-(aq)$ is added to a solution of a $Bi^{3+}$ salt.

The structures of the 5+ oxides of arsenic and antimony are not known, and empirical formulas ($As_2O_5$ and $Sb_2O_5$) are employed. These oxides are prepared by the action of concentrated $HNO_3$ on the elements or the 3+ oxides, followed by the dehydration of the products of these reactions ($2H_3AsO_4 \cdot H_2O$ and $Sb_2O_5 \cdot xH_2O$). The 5+ oxides are exclusively acidic. Orthoarsenic acid ($H_3AsO_4$, a triprotic acid) forms when $As_2O_5$ is dissolved in water. Arsenates (salts of $AsO_4^{3-}$) or antimonates (salts of $Sb(OH)_6^-$) can be prepared by dissolving $As_2O_5$ or $Sb_2O_5$ in aqueous alkalies. Sodium antimonate, $NaSb(OH)_6$, is one of the least soluble sodium salts, and its formation is often used as a test for $Na^+$.

Bismuth(V) oxide has never been prepared in the pure state; it is unstable and readily loses oxygen. The red-brown product obtained by the action of strong oxidizing agents (such as $Cl_2$, $OCl^-$, and $S_2O_8^{2-}$) on a suspension of $Bi_2O_3$ in an alkaline solution is thought to be impure $Bi_2O_5$.

■ **CARBON AND SILICON**

Carbon, silicon, germanium, tin, and lead comprise group IV A. The compounds of carbon are more numerous than the compounds of any other element with the exception of hydrogen. In fact, approximately ten compounds that contain carbon are known for every compound that does not. The chemistry of the compounds of carbon (most of which also contain hydrogen) is the subject of **organic chemistry** (Chapter 20.).

**12.9 GROUP PROPERTIES**

The transition from nonmetallic character to metallic character with increasing atomic number that is exhibited by the elements of group V A is also evident in the chemistry of the IV A elements. Carbon is strictly a nonmetal (although graphite is an electrical conductor). Silicon is essentially a nonmetal in its chemical behavior; however, its electrical and physical properties are those of a semimetal. Germanium is a semimetal; its properties are more metallic than nonmetallic. Tin and lead are truly metallic, although some vestiges of nonmetallic character remain (e.g., the oxides and hydroxides of tin and lead are amphoteric).

Carbon exists in network crystals with the atoms of the crystal held together by covalent bonds (Section 12.10). A large amount of energy is required to rupture some or all of these bonds in fusion or vaporization; hence, carbon has the highest melting point and boiling point of the family (Table 12.6). The heaviest member of the family, lead, exists in a typical metallic lattice. The crystalline forms of the intervening members show a transition between the two extremes displayed by carbon and lead, and this accounts for the trend in melting points and boiling points of the elements (Table 12.6).

Table 12.6 Some properties of the group IV A elements

| PROPERTY | CARBON | SILICON | GERMANIUM | TIN | LEAD |
|---|---|---|---|---|---|
| melting point (°C) | 3570 | 1420 | 959 | 232 | 327 |
| boiling point (°C) | 4200 | 2355 | 2700 | 2360 | 1755 |
| atomic radius (Å) | 0.77 | 1.17 | 1.22 | 1.41 | 1.54 |
| ionization potential (eV) | | | | | |
| first | 11.3 | 8.1 | 8.1 | 7.3 | 7.4 |
| second | 24.4 | 16.3 | 15.9 | 14.5 | 15.0 |
| third | 47.9 | 33.5 | 34.1 | 30.5 | 32.0 |
| fourth | 64.5 | 45.1 | 45.5 | 39.4 | 42.1 |
| electronegativity | 2.5 | 1.8 | 1.8 | 1.8 | 1.8 |

The crystalline forms of silicon and germanium are similar to the diamond; however, the bonds are not so strong as those in the diamond, and silicon and germanium are semiconductors (the diamond is not). Silicon and germanium may be used to prepare impurity semiconductors (Section 7.14) that are employed in transistors. Although one modification of tin has a diamond-type structure, the principal form of tin is metallic.

The electronic configurations of the elements are listed in Table 12.7. With the possible exception of carbon, the assumption of a noble-gas configuration through the formation of a $4-$ ion by electron gain is not observed; the electronegativities of the group IV A elements are generally low (Table 12.6).

Table 12.7 Electronic configurations of the group IV A elements

| ELEMENT | $Z$ | $1s$ | $2s$ | $2p$ | $3s$ | $3p$ | $3d$ | $4s$ | $4p$ | $4d$ | $4f$ | $5s$ | $5p$ | $5d$ | $6s$ | $6p$ |
|---|---|---|---|---|---|---|---|---|---|---|---|---|---|---|---|---|
| C | 6 | 2 | 2 | 2 | | | | | | | | | | | | |
| Si | 14 | 2 | 2 | 6 | 2 | 2 | | | | | | | | | | |
| Ge | 32 | 2 | 2 | 6 | 2 | 6 | 10 | 2 | 2 | | | | | | | |
| Sn | 50 | 2 | 2 | 6 | 2 | 6 | 10 | 2 | 6 | 10 | | 2 | 2 | | | |
| Pb | 82 | 2 | 2 | 6 | 2 | 6 | 10 | 2 | 6 | 10 | 14 | 2 | 6 | 10 | 2 | 2 |

The ionization potentials of the elements (Table 12.6) show that the energy required for the removal of all four valence electrons from any given element is extremely high. Consequently, simple 4+ ions of group IV A elements are unknown. Germanium, tin, and lead appear to be able to form $d^{10}s^2$ ions by the loss of two electrons. However, only some of the compounds of $Pb^{2+}$ (such as $PbF_2$ and $PbCl_2$) are ionic. The compounds of $Ge^{2+}$ and $Sn^{2+}$ are predominantly covalent, and those of $Ge^{2+}$ are relatively unstable toward disproportionation to the 4+ and 0 states.

In the majority of their compounds, the IV A elements are covalently bonded. Through the formation of four covalent bonds per atom, an element can attain the electronic configuration of the noble gas of its period. These bonds may be regarded as having been formed by the use of $sp^3$ hybrid orbitals; compounds of the type $AB_4$ are tetrahedral. All the carbon family elements can form such compounds, but only a few compounds of this type are known for lead ($PbF_4$, $PbCl_4$, and $PbH_4$) and they, with the exception of $PbF_4$, are thermally unstable.

In the case of carbon, the formation of four covalent bonds saturates the valence level. However, the other members of the group have empty $d$ orbitals available in their valence levels and can form species in which the atom of the IV A element exhibits a covalence greater than four. The ions $SiF_6^{2-}$, $GeCl_6^{2-}$, $SnBr_6^{2-}$, $Sn(OH)_6^{2-}$, $Pb(OH)_6^{2-}$, and $PbCl_6^{4-}$ utilize $d^2sp^3$ hybrid bonding orbitals; they are octahedral.

The most important way in which carbon differs from the remaining elements of group IV A (as well as from all other elements) is the pronounced ability of carbon to form compounds in which many carbon atoms are bonded to each other in chains or rings. This property, called **catenation,** is exhibited by other elements near carbon in the periodic classification (such as boron, nitrogen, phosphorus, sulfur, oxygen, silicon, germanium, and tin) but to a much lesser extent than carbon; this property of carbon accounts for the large number of organic compounds.

In group IV A the tendency for self-linkage diminishes markedly with increasing atomic number. The hydrides, which have the general formula $E_nH_{2n+2}$ (where E is a group IV A element), illustrate this trend. There appears to be no limit to the number of carbon atoms that can bond together to form chains, and a very large number of hydrocarbons are known. For the other IV A elements, the most complex hydrides that have been prepared are $Si_6H_{14}$, $Ge_9H_{20}$, $Sn_2H_6$, and $PbH_4$. The carbon–carbon single bond energy (82 kcal/mol) is much greater than that of the silicon–silicon bond (43 kcal/mol), the germanium–germanium bond (40 kcal/mol), or the tin–tin bond (37 kcal/mol).

In addition, carbon chains are much more stable than the chains of any other element. The reason for this difference in stability may be derived, in part, from a comparison of some bond energies:

| C—C | 82 kcal/mol | Si—Si | 43 kcal/mol |
|------|-------------|-------|-------------|
| C—O | 84 kcal/mol | Si—O | 89 kcal/mol |
| C—H | 99 kcal/mol | Si—H | 81 kcal/mol |
| C—Cl | 78 kcal/mol | Si—Cl | 86 kcal/mol |
| C—F | 105 kcal/mol | Si—F | 128 kcal/mol |

The C—C bond energy is approximately of the same order of magnitude as the energies of the bond between carbon and other elements. The Si—Si bond, however, is weaker than the bonds that silicon forms with other elements.

Another characteristic of carbon is its pronounced ability to form multiple bonds with itself and with other nonmetals. Groupings such as

$$\overset{\diagdown}{\diagup}C=C\overset{\diagup}{\diagdown}, \quad -C{\equiv}C-, \quad -C{\equiv}N, \quad \overset{\diagdown}{\diagup}C=O, \quad \text{and} \quad \overset{\diagdown}{\diagup}C=S$$

are frequently encountered. No other IV A element uses $p$ orbitals to form $\pi$ bonds.

Only the truly nonmetallic members of the family, carbon and silicon, are treated in the sections that follow.

## 12.10 THE ELEMENTS

Figure 12.10 Arrangement of atoms in a diamond crystal.

Carbon constitutes approximately 0.03% of the earth's crust; in addition, the atmosphere contains 0.03% $CO_2$ by volume; carbon is also an important constituent of all plant and animal matter. The allotropes of carbon, diamond and graphite, as well as impure forms of the element such as coal (which also contains combined carbon), occur in nature. In combined form the element occurs in compounds with hydrogen (which are called **hydrocarbons** and are found in natural gas and petroleum), in the atmosphere as $CO_2$, and in carbonate minerals such as limestone ($CaCO_3$), dolomite ($CaCO_3 \cdot MgCO_3$), siderite ($FeCO_3$) witherite ($BaCO_3$), and malachite ($CuCO_3 \cdot Cu(OH)_2$).

In diamond each carbon atom is bonded through $sp^3$ hybrid orbitals to four other carbon atoms arranged tetrahedrally (Figure 12.10). Strong bonds hold this network crystal together; furthermore, all valence electrons of each carbon atom are paired in bonding orbitals—the valence level of each carbon atom can hold no more than eight electrons. Thus the diamond is extremely hard, high melting, stable, and a nonconductor of electricity.

Whereas the diamond is a colorless, transparent material with a high refractivity, graphite is a soft, black solid with a slight metallic luster. The graphite crystal is composed of layers formed from hexagonal rings

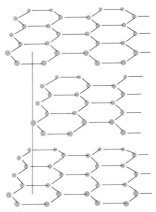

**Figure 12.11** Arrangement of atoms in a graphite crystal.

of carbon atoms (Figure 12.11). The layers are held together by relatively weak London forces; the distance from carbon atom to carbon atom in adjacent planes is 3.35 Å as compared to a distance of 1.415 Å between bonded carbon atoms of a plane. Since it is easy for the layers to slide over one another, graphite is soft and has a slippery feel; it is less dense than diamond.

The nature of the bonding in the layers of the graphite crystal accounts for some of the properties of this substance. Each carbon atom is bonded to three other carbon atoms, and all the bonds are perfectly equivalent. The C—C bond distance in graphite (1.415 Å) compared to that in the diamond (1.54 Å) suggests that a degree of multiple bonding exists in the former; graphite may be represented as a resonance hybrid (Figure 12.12) in which each bond is a $1\frac{1}{3}$ bond.

Each C atom in graphite forms $\sigma$ bonds with three other C atoms through the use of $sp^2$ hybrid orbitals. Therefore, the structure is planar with the three $\sigma$ bonds of each atom directed to the corners of equilateral triangles. This bonding accounts for three of the four valence electrons of each C atom; the fourth, a $p$ electron, is not involved in $\sigma$ bond formation (Figure 12.13a).

If only two adjacent atoms in a molecule had this electron arrangement, the additional $p$ electrons would pair to form a localized $\pi$ bond, and thus the atoms would be joined by a double bond. However, in graphite *each* C atom has an additional $p$ electron, and the resonance forms depict the possibility of forming conventional double bonds in three ways. Since each $p$ orbital overlaps with more than one other $p$ orbital, an extended $\pi$ bonding system that encompasses the entire structure forms (Figure 12.13b). The electrons in this $\pi$ bonding system are not localized between two atoms but are free to move throughout the entire layer. Hence, graphite has a metallic luster and is an electrical conductor. The conductivity is fairly large in a direction parallel to the layers but is small in a direction perpendicular to the planes of the crystal.

Silicon, which constitutes approximately 28% of the earth's crust, is the second most abundant element (oxygen is first). The element does not occur free in nature; rather, it is found as silicon dioxide (sometimes called silica) and in an enormous variety of silicate minerals.

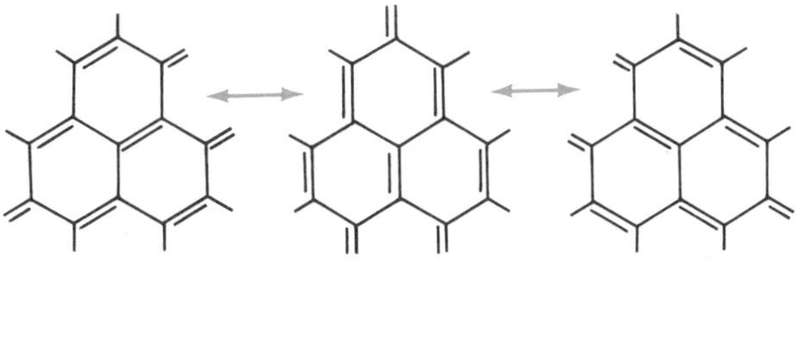

**Figure 12.12** Resonance forms for a fragment of a graphite layer.

Figure 12.13 Schematic representation of the formation of the multicenter bonding system of a graphite fragment. ($\sigma$ bonds are shown as solid lines.)

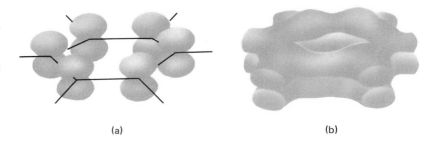

(a)                                    (b)

Silicon is prepared by the reduction of silicon dioxide by coke at high temperatures in an electric furnace.

$$SiO_2(l) + 2C(s) \rightarrow Si(l) + 2CO(g)$$

If a larger quantity of carbon is employed, silicon carbide (which is called "Carborundum"), SiC, is produced rather than silicon. The only known modification of silicon has a structure similar to diamond. Crystalline silicon is a gray, lustrous solid that is a semiconductor. The bonds in silicon are not so strong as those in diamond, and the bonding electrons are not so firmly localized. Evidently, in silicon, electrons may be thermally excited to a conductance band that is energetically close to the valence band (Section 4.7).

Very pure silicon, which is used in transistors, is prepared by a series of steps. First, impure silicon is reacted with chlorine to produce $SiCl_4$. The tetrachloride, a volatile liquid, is purified by fractional distillation and then reduced by hydrogen to elementary silicon. This product is further purified by zone refining (see Figure 18.6). In this process a short section of one end of a silicon rod is melted, and this melted zone is caused to move slowly along the rod to the other end by the movement of the heater. Pure silicon crystallizes from the melt, and the impurities are swept along in the melted zone to one end of the rod which is subsequently sawed off and discarded.

**12.11  CARBIDES AND SILICIDES**

A large number of carbides and silicides are known—the carbides have been much more extensively studied than the silicides. Carbides may be made by heating the appropriate metal or its oxide with carbon, carbon monoxide, or a hydrocarbon. As is the case with the hydrides and nitrides, carbides may be classified as **saltlike, interstitial,** or **covalent.**

The saltlike carbides may be divided into two types. The I A and II A metals, as well as $Cu^+$, $Ag^+$, $Au^+$, $Zn^{2+}$, and $Cd^{2+}$, form carbides that are believed to contain the acetylide ion, $C_2^{2-}$, which has the structure

$$[:C{\equiv}C:]^{2-}$$

Upon hydrolysis, acetylides yield acetylene, $C_2H_2$,

$$H-C\equiv C-H$$

$$CaC_2(s) + 2H_2O \rightarrow Ca(OH)_2(aq) + C_2H_2(g)$$

Beryllium carbide, $Be_2C$, and aluminum carbide, $Al_4C_3$, contain the methanide ion, $C^{4-}$, and yield methane, $CH_4$, upon hydrolysis.

$$Al_4C_3(s) + 12H_2O \rightarrow 4Al(OH)_3(s) + 3CH_4(g)$$

Interstitial carbides are formed by transition elements and consist of metallic lattices with carbon atoms in the interstices. These materials, such as $TiC$, $TaC$, $W_2C$, $VC$, and $Mo_2C$, are in general very hard, high melting, and chemically inert. They resemble metals in appearance and electrical conductivity.

For the formation of interstitial carbides, the interstices of the metallic crystal must be sufficiently large to accommodate carbon atoms, and this requires that the metals have a radius greater than 1.3 Å. Some transition metals that have radii less than 1.3 Å form carbides intermediate in character between the saltlike carbides and the interstitial carbides. The metal lattices of these substances, such as $Fe_3C$, $Cr_3C_2$, $Co_3C$, and $Ni_3C$, are distorted; these materials are readily hydrolyzed.

Silicon carbide, produced by the reaction of $SiO_2$ and C, is a covalent carbide. It is hard, stable, and chemically inert. The silicon carbide crystal consists of a three-dimensional network formed from tetrahedra of alternating carbon and silicon atoms. Boron carbide, $B_4C$, which is even harder than silicon carbide, is made by the reduction of $B_2O_3$ by carbon in an electric furnace.

Silicon dissolves in almost all molten metals, and in many of these instances, definite compounds, called silicides, are produced. The compounds $Mg_2Si$, $CaSi_2$, $Li_3Si$, $FeSi$, $Fe_2Si$, $Ni_2Si$, $NiSi$, $VSi_2$, and $CoSi_3$ are among those known. Although probably none of the silicides are truly ionic, certain of them hydrolyze to produce hydrogen–silicon compounds, called **silicon hydrides** or **silanes.**

The silanes are compounds of general formula $Si_nH_{2n+2}$; compounds where $n$ equals 1 to 6 are well characterized. The hydrolysis of $Mg_2Si$ by dilute acids gives a mixture of silanes; for example,

$$Mg_2Si(s) + 4H^+(aq) \rightarrow 2Mg^{2+}(aq) + SiH_4(g)$$

The silanes structurally resemble the hydrocarbons that conform to the formula $C_nH_{2n+2}$, which are called the **alkanes** (Figure 12.14). Although there is presumably no limit to the number of carbon atoms that can join together to form alkanes, the number of silanes appears to be limited because of the comparative weakness of the Si—Si bond. In addi-

Figure 12.14 Arrangement of atoms in (a) CH₄ and SiH₄ and (b) C₂H₆ and Si₂H₆. Open circles, C or Si; dark circles, H.

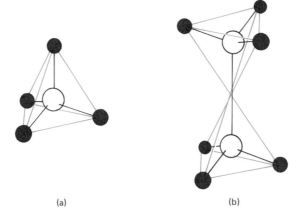

(a)                         (b)

tion, the alkanes are much less reactive than the silanes, which are spontaneously flammable in air.

$$2Si_2H_6(g) + 7O_2(g) \rightarrow 4SiO_2(s) + 6H_2O(g)$$

The C—C bonds of the alkanes are all $\sigma$ bonds, but there are hydrocarbons, such as acetylene, which contain multiple bonds between carbon atoms. Silicon hydrides that contain multiple bonds are unknown. The hydrocarbons are discussed in Chapter 20.

**12.12  OXIDES AND OXYACIDS**  Three oxides of carbon are well characterized: $C_3O_2$ (carbon suboxide), CO (carbon monoxide), and $CO_2$ (carbon dioxide). Carbon monoxide is formed by the combustion of carbon in a limited supply of oxygen at high temperatures (ca. 1000°C).

$$2C(s) + O_2(g) \rightarrow 2CO(g)$$

It is also produced by the reaction of $CO_2$ and C at high temperatures and is a constituent of water gas (Section 8.7).

Carbon monoxide is isoelectronic with nitrogen (Section 4.4)

$$:C\equiv O:$$

and has two $\pi$ bonds and one $\sigma$ bond joining the atoms. Certain transition metals and transition metal salts react with CO to produce metal carbonyls in which the CO molecule uses the unshared electron pair of the C atom for bond formation. Nickel carbonyl, $Ni(CO)_4$, is formed by the reaction of finely divided Ni and CO. The $Ni(CO)_4$ molecule is diamagnetic, has a tetrahedral arrangement of four CO groups around a central Ni atom, and may be assumed to be bonded through $sp^3$ hybrid orbitals of Ni. In like manner, the Fe atom in $Fe(CO)_5$ exhibits $dsp^3$ hybridization (the molecule is trigonal bipyramidal and diamagnetic), and the Cr atom in $Cr(CO)_6$ displays $d^2sp^3$ hybridization (the molecule is octahedral and diamagnetic).

Since each CO molecule contributes a pair of electrons to the bonding, a metal with an unpaired electron might be expected to form a carbonyl that would have an unpaired electron. In these cases, however, metal–metal bonds are usually formed, and a binuclear carbonyl results. For example, in a hypothetical $Mn(CO)_5$ formed by using $d^2sp^3$ hybrid orbitals of Mn, five of the hybrid orbitals would be occupied by the five electron pairs from the CO molecules, and the sixth would contain an unpaired electron from the Mn atom. The odd electron is paired by the formation of a Mn—Mn bond between two $Mn(CO)_5$ units, and the formula of the compound is $Mn_2(CO)_{10}$. A large number of carbonyl compounds are known.

Carbon monoxide burns in air.

$$2CO(g) + O_2(g) \rightarrow 2CO_2(g)$$

Since this reaction is highly exothermic ($\Delta H = -67.6$ kcal/mol), carbon monoxide can be used as a fuel. The compound reacts with the halogens in sunlight to produce the carbonyl halides ($COX_2$) and with sulfur vapor at high temperatures to form carbonyl sulfide (COS).

Carbon monoxide is used as a reducing agent in metallurgical processes; at high temperatures it reacts with many metal oxides to yield the free metal and $CO_2$.

$$FeO(s) + CO(g) \rightarrow Fe(l) + CO_2(g)$$

The catalyzed reactions of carbon monoxide and hydrogen are commercially important for the production of hydrocarbons and methanol. Carbon monoxide is a poisonous gas because in the lungs it combines with the hemoglobin of the blood, thus preventing the hemoglobin from combining with oxygen.

Carbon dioxide is formed by the complete combustion of carbon or compounds of carbon (principally the hydrocarbons). It is also produced by the reaction of carbonates or hydrogen carbonates with acids,

$$CaCO_3(s) + 2H^+(aq) \rightarrow Ca^{2+}(aq) + CO_2(g) + H_2O$$

and by heating carbonates,

$$CaCO_3(s) \rightarrow CaO(s) + CO_2(g)$$

The molecule is linear and nonpolar. The electronic structure

$$:\overset{..}{O}=C=\overset{..}{O}:$$

is only an approximation because the bonds of $CO_2$ are shorter and stronger than carbon-to-oxygen double bonds of other molecules. A treatment of the bonding has been proposed that incorporates delocalized

$\pi$ molecular orbitals that extend over all three atoms, together with the $\sigma$ bonds that join the O atoms to the C atom.

Plants convert carbon dioxide and water vapor that are present in air into carbohydrates — compounds of the general formula $C_x(H_2O)_y$. The energy for this process, which is called **photosynthesis,** is supplied by sunlight; this process is catalyzed by the green-coloring matter of plants, **chlorophyll.**

Carbon dioxide is moderately soluble in water; it is the acid anhydride of carbonic acid, $H_2CO_3$. The acid, however, has never been obtained in the pure state; solutions of $CO_2$ in water consist mainly of dissolved $CO_2$ molecules — less than 1% of the dissolved material is in the form of $H_2CO_3$ molecules. Carbonic acid is a weak, diprotic acid, and equations for the ionizations are best written

$$CO_2(aq) + H_2O \rightleftharpoons H^+(aq) + HCO_3^-(aq)$$

$$HCO_3^-(aq) \rightleftharpoons H^+(aq) + CO_3^{2-}(aq)$$

Two series of salts are formed: the normal carbonates, such as $Na_2CO_3$ and $CaCO_3$, and the hydrogen carbonates, such as $NaHCO_3$ and $Ca(HCO_3)_2$. The structure of the carbonate ion has been discussed in Section 4.5 (Figure 4.24).

Sodium carbonate is an important industrial chemical, and more than 90% of the compound is produced commercially by the **Solvay process.** Ammonia and carbon dioxide, under pressure, are passed into a saturated solution of sodium chloride. The ammonia and carbon dioxide react.

$$NH_3(g) + CO_2(g) + H_2O \rightleftharpoons NH_4^+(aq) + HCO_3^-(aq)$$

Sodium bicarbonate precipitates from the solution because of the high concentration of sodium ion, and ammonium chloride is left behind.

$$Na^+(aq) + HCO_3^-(aq) \rightarrow NaHCO_3(s)$$

The sodium bicarbonate is removed by filtration and washed; heating decomposes it into the normal carbonate.

$$2NaHCO_3(s) \rightarrow Na_2CO_3(s) + CO_2(g) + H_2O(g)$$

The ammonia is recovered, for reuse, from the mother liquor by heating this solution with quicklime, CaO.

$$2NH_4^+(aq) + CaO(s) \rightarrow 2NH_3(g) + Ca^{2+}(aq) + H_2O$$

The calcium oxide is obtained from limestone, $CaCO_3$, by heating; the $CO_2$ produced as a by-product of the preparation of CaO, as well as that produced by heating $NaHCO_3$, is used in the initial step of the process.

Although it is postulated that silicon monoxide exists in the vapor state, the only well-characterized oxide of silicon is $SiO_2$. In contrast to the oxides of carbon, which are volatile molecular species held together by London forces in the solid state, $SiO_2$ forms very stable, nonvolatile, three-dimensional network crystals (melting point, $\sim 1700°C$). One of the three crystal modifications of $SiO_2$ has a lattice that may be considered to be derived from the diamond lattice with silicon atoms replacing carbon atoms and an oxygen atom midway between each pair of silicon atoms.

The chemistry of silicon is dominated by compounds that contain the Si—O linkage. Silicon dioxide is the product of the reaction of the elements; it is produced by reaction of the spontaneously flammable silicon hydrides with air. Hydrous $SiO_2$ is the product of the hydrolysis of many silicon compounds. Silicon dioxide occurs in several forms in nature; among them are sand, flint, agate, jasper, onyx, and quartz.

Silicon dioxide is an acidic oxide; however, no acids of silicon have ever been isolated. The oxide does not react directly with water; acidification of a water solution of a soluble silicate yields only hydrous $SiO_2$. Silicates may be made by heating metal oxides or metal carbonates with $SiO_2$. Certain silicates of the I A metals (with a molar ratio of silicon dioxide to metal oxide of not more than 2 to 1) are water soluble.

A large number of silicates, of various types, occur in nature. The basic structure of all silicates is a tetrahedral $SiO_4$ unit (Figure 12.15). The simple ion $SiO_4^{4-}$ occurs in certain minerals, for example, zircon ($ZrSiO_4$). More complicated ions are known in which two or more $SiO_4$ tetrahedra are joined through common oxygen atoms. Silicon has four valence electrons and forms covalent bonds with four oxygen atoms in forming an $SiO_4$ unit; each oxygen of the unit then completes its octet by accepting an electron–which accounts for the 4– charge of the ion. When two tetrahedra are joined, the bridge oxygen atom (the atom that is shared by both tetrahedra) completes its octet by covalent bonding to *two* silicon atoms. Hence the ion formed by joining two $SiO_4$ tetrahedra has a charge of 6– and the formula $Si_2O_7^{6-}$; the mineral thortveitite ($Sc_2Si_2O_7$) contains this ion. Cyclic silicate ions are known in which two oxygen atoms of each $SiO_4$ unit act as bridge atoms; bentonite ($BaTiSi_3O_9$) and beryl ($Be_3Al_2Si_6O_{18}$) contain such ions.

The sharing of two oxygen atoms per $SiO_4$ tetrahedron can lead to polymeric anion chains of the type $(SiO_3^{2-})_n$ found in the pyroxenes (e.g., spodumene, $LiAl(SiO_3)_2$) and the double-chain type $(Si_4O_{11}^{6-})_n$ found in the amphiboles (e.g., the asbestos minerals). These anion chains extend throughout the crystal; consequently, minerals of these types have a fibrous nature.

If three oxygen atoms of each $SiO_4$ tetrahedron are used as bridge atoms, a sheetlike anion results. The anion $(Si_2O_5^{2-})_n$ occurs in talc, $Mg_3(Si_2O_5)(OH)_2$; because of the layer structure, this material feels slippery. Occasionally, aluminum atoms take the places of some of the silicon atoms in certain anions. The hypothetical $AlO_4$ tetrahedron would

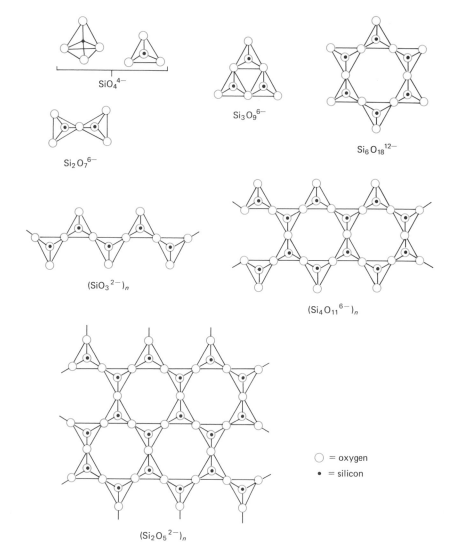

**Figure 12.15** Schematic representation of the arrangement of atoms in the silicate ions.

$SiO_4^{4-}$

$Si_2O_7^{6-}$

$Si_3O_9^{6-}$

$Si_6O_{18}^{12-}$

$(SiO_3^{2-})_n$

$(Si_4O_{11}^{6-})_n$

$(Si_2O_5^{2-})_n$

◯ = oxygen

• = silicon

have a 5− charge; consequently, such substitutions increase the negative charge of the anion. Muscovite, $KAl_3Si_3O_{10}(OH)_2$ contains a sheetlike aluminosilicate anion with one-fourth of the silicon atoms of the $(Si_2O_5^{2-})_n$ structure replaced by aluminum atoms.

If all four oxygen atoms of each $SiO_4$ tetrahedron are used as bridge atoms, $SiO_2$ results. However, if some of the silicon atoms of this three-dimensional network are replaced by aluminum atoms, an anion results. Framework aluminosilicates, such as the feldspars and the zeolites, are of this type.

Glass is a mixture of silicates made by fusing $SiO_2$ with metal oxides and carbonates. Common soda-lime glass is made from $Na_2CO_3$, $CaCO_3$, and $SiO_2$. Special glasses may be made by the addition of other acidic and basic oxides (such as $Al_2O_3$, $B_2O_3$, PbO, and $K_2O$). Cement is a complex aluminosilicate mixture made from limestone ($CaCO_3$) and clay ($H_4Al_2Si_2O_9$).

**12.13 SULFUR COMPOUNDS AND HALOGEN COMPOUNDS**

Carbon disulfide, $CS_2$, is a volatile liquid prepared by heating carbon and sulfur together in an electric furnace. The electronic structure of the molecule is analogous to $CO_2$. Silicon disulfide $SiS_2$, prepared by the reaction of the elements at high temperatures, is a colorless, crystalline material that hydrolyzes to $SiO_2$ and $H_2S$. Solid $SiS_2$ has a chainlike structure

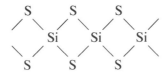

Carbon tetrachloride is made by heating chlorine with carbon disulfide

$$CS_2(g) + 3Cl_2(g) \rightarrow CCl_4(g) + S_2Cl_2(g)$$

Carbon tetrachloride, which is a liquid under ordinary conditions, is a good solvent for many nonpolar materials. Carbon tetrafluoride, $CF_4$, which may be obtained by the fluorination of almost any carbon-containing compound, is a very stable gas. The mixed halide $CCl_2F_2$, which is called "Freon," is a very stable, odorless, nontoxic gas that is used as a refrigerant. Carbon tetrabromide, $CBr_4$, and carbon tetraiodide, $CI_4$, are solids; they are thermally unstable, presumably because of the difficulty of carbon in accommodating four large atoms around itself.

The silicon tetrahalides, $SiX_4$, are formed by the reactions of the elements; $SiF_4$ is a gas; $SiCl_4$ and $SiBr_4$ are liquids; and $SiI_4$ is a solid. Unlike the tetrahalides of carbon, the silicon tetrahalides readily hydrolyze

$$SiX_4 + (n + 2)H_2O \rightarrow SiO_2 \cdot nH_2O(s) + 4H^+(aq) + 4X^-(aq)$$

The difference in chemical reactivity of the tetrahalides of carbon and silicon is attributed to the availability of unoccupied $d$ orbitals in silicon which make it possible for a water molecule to attach itself and thus facilitate hydrolysis.

With $SiF_4$ a secondary reaction follows the one previously given. The hydrofluoric acid formed in the initial reaction reacts with $SiF_4$ to form fluosilicic acid, a strong acid.

$$SiF_4(g) + 2HF(aq) \rightarrow 2H^+(aq) + SiF_6^{2-}(aq)$$

The $SiF_6^{2-}$ ion is octahedral and uses $d^2sp^3$ hybrid bonding orbitals of silicon.

Catenated halides of carbon and silicon are known (e.g., $C_2Cl_6$ and $Si_2Cl_6$).

**12.14 CARBON–NITROGEN COMPOUNDS**

There are many compounds in which carbon is bonded to nitrogen. Sodium cyanide is commercially prepared by the reaction of sodium amide with carbon at red heat. Sodium amide is a product of the iron-catalyzed reaction of Na metal and liquid ammonia.

$$2Na(s) + 2NH_3(l) \rightarrow 2Na^+ + 2NH_2^- + H_2(g)$$

$$NaNH_2(l) + C(s) \rightarrow NaCN(l) + H_2(g)$$

The cyanide ion is isoelectronic with $N_2$ and CO (Section 4.4)

$$[:C\equiv N:]^-$$

and it has a strong tendency to form covalent complexes with metal cations. Covalent complexes such as $Ag(CN)_2^-$, $Cd(CN)_4^{2-}$, $Ni(CN)_4^{2-}$, $Hg(CN)_4^{2-}$, $Fe(CN)_6^{4-}$, and $Cr(CN)_6^{3-}$ are known to exist.

Treatment of a cyanide with dilute sulfuric acid produces hydrogen cyanide, HCN. The compound is a low-boiling liquid (boiling point, 26.5°C); it readily dissolves in water to produce solutions of hydrocyanic acid, a weak acid. Hydrogen cyanide and the cyanides are extremely poisonous.

The thermal decomposition of certain heavy metal cyanides, such as AgCN, $Hg(CN)_2$, and AuCN, yields the metal and cyanogen, $C_2N_2$, or $(CN)_2$. The molecule is linear.

$$:N\equiv C-C\equiv N:$$

This very poisonous, colorless gas is sometimes called a pseudohalogen because it resembles the halogens in its chemical reactions. For example,

$$Cl_2(g) + 2OH^-(aq) \rightarrow Cl^-(aq) + OCl^-(aq) + H_2O$$

$$(CN)_2(g) + 2OH^-(aq) \rightarrow CN^-(aq) + OCN^-(aq) + H_2O$$

Cyanides resemble chlorides in their water solubility.

The ion $OCN^-$, produced by the disproportionation reaction of cyanogen in alkaline solution, is the cyanate ion; the corresponding acid, cyanic acid (HOCN), is not stable. An analogous ion, the thiocyanate ion, $SCN^-$, is prepared by fusing I A cyanides with sulfur.

**■ BORON**

Of the group III A elements—boron, aluminum, gallium, indium, and thallium—boron alone is a nonmetal.

**12.15 GROUP PROPERTIES** The electronic configurations of the elements are listed in Table 12.8, and properties of the elements appear in Table 12.9. Boron is a much smaller atom than the others of the group; this difference accounts for the sharp distinction in properties between the nonmetallic boron and the other group members, which are metallic. The atomic sizes of Ga, In, and Tl are influenced by the electronic inner-building of elements that immediately precede them in the periodic classification (particularly in the case of Tl which follows the lanthanides); hence, atomic radius does not rapidly and regularly increase with increasing atomic number for these elements. The comparatively small sizes of the elements account for the relatively high ionization potentials which do not decline in the expected manner as the group is descended.

Table 12.8 Electronic configurations of the group III A elements

| ELEMENT | $Z$ | $1s$ | $2s$ | $2p$ | $3s$ | $3p$ | $3d$ | $4s$ | $4p$ | $4d$ | $4f$ | $5s$ | $5p$ | $5d$ | $6s$ | $6p$ |
|---|---|---|---|---|---|---|---|---|---|---|---|---|---|---|---|---|
| B | 5 | 2 | 2 | 1 | | | | | | | | | | | | |
| Al | 13 | 2 | 2 | 6 | 2 | 1 | | | | | | | | | | |
| Ga | 31 | 2 | 2 | 6 | 2 | 6 | 10 | 2 | 1 | | | | | | | |
| In | 49 | 2 | 2 | 6 | 2 | 6 | 10 | 2 | 6 | 10 | | 2 | 1 | | | |
| Tl | 81 | 2 | 2 | 6 | 2 | 6 | 10 | 2 | 6 | 10 | 14 | 2 | 6 | 10 | 2 | 1 |

Table 12.9 Some properties of the group III A elements

| PROPERTY | BORON | ALUMINUM | GALLIUM | INDIUM | THALLIUM |
|---|---|---|---|---|---|
| melting point (°C) | 2300 | 659 | 30 | 155 | 304 |
| boiling point (°C) | 2550 | 2500 | 2070 | 2100 | 1457 |
| atomic radius (Å) | 0.80 | 1.25 | 1.25 | 1.50 | 1.55 |
| ionic radius, $M^{3+}$ (Å) | 0.20 | 0.52 | 0.62 | 0.81 | 0.95 |
| ionization potential (eV) | | | | | |
| first | 8.3 | 6.0 | 6.0 | 5.8 | 6.1 |
| second | 25.1 | 18.8 | 20.4 | 18.8 | 20.3 |
| third | 37.9 | 28.4 | 30.6 | 27.9 | 29.7 |
| heat of hydration, $M^{3+}$ (kcal/mol) | – | −1121 | −1124 | −994 | −984 |
| electrode potential, $\mathscr{E}°$, $(M^{3+}/M)$(V) | – | −1.67 | −0.52 | −0.34 | +0.72 |

None of the elements has the slightest tendency to form simple anions. Rather, as would seem reasonable from the electronic configurations of the elements, the most important oxidation state is 3+. This relatively high charge coupled with comparatively small ionic radii leads to species with significant polarizing abilities.

Hence, the compounds of the elements in the 3+ state are predominantly covalent; this covalent character also follows from the comparatively high values of the first three ionization potentials of the elements. With the exception of boron, which is exclusively nonmetallic in its chemistry, the III A elements exist as 3+ ions in water solution; these ions are strongly hydrated, however, and the heats of hydration are high.

The electrode potentials are indicative of a higher degree of metallic reactivity than the ionization potentials of the gaseous ions might lead us to expect. This is because of the high heats of hydration that result from the high charge concentration of the small ions. The high charge density also accounts for the marked tendency of the elements to form complex ions.

The oxides and hydroxides (and also hydrous oxides) exhibit the usual trend in decreasing acidic character with increasing atomic number. Thus $B_2O_3$ and $B(OH)_3$ (usually written $H_3BO_3$) are acidic, the compounds of aluminum and gallium are amphoteric, and $In_2O_3$ and $Tl_2O_3$ have no acidic character at all.

The $ns^2np^1$ configuration of the valence level would lead us to anticipate the existence of 1+, $d^{10}s^2$ ions for the larger members of the group. This is indeed the case for thallium. The 1+ state of this element is considerably more stabie than the 3+ state; TlOH is a strong electrolyte. Compounds of indium and gallium in which these elements are in a 1+ state are known, but this state is less important and less stable than the 3+ state.

Boron has a slight tendency for catenation (e.g., the atoms of $B_2Cl_4$ are arranged $Cl_2BBCl_2$), but no other member of the family displays this characteristic.

**12.16  THE ELEMENT**  Boron (which composes $3 \times 10^{-4}\%$ of the earth's crust) does not occur free in nature; the principal ores of boron are borates such as kernite ($Na_2B_4O_7 \cdot 4H_2O$), borax ($Na_2B_4O_7 \cdot 10H_2O$), colemanite ($Ca_2B_6O_{11} \cdot 5H_2O$), and ulexite ($NaCaB_5O_9 \cdot 8H_2O$). A somewhat impure, dark-brown, amorphous form of the element is prepared by heating boric oxide, $B_2O_3$, with powdered magnesium. Pure, crystalline boron may be obtained by reducing $BBr_3$ by hydrogen on a hot tungsten filament (approximately 1500°C).

$$2BBr_3(g) + 3H_2(g) \rightarrow 2B(s) + 6HBr(g)$$

The crystals, which condense on the filament, are black with a metallic luster.

The structures of three types of boron crystals have been determined, and at least one other allotrope is known to exist. Groups of 12 B atoms arranged in regular icosahedra (solid figures, each with 20 equilateral-triangular faces) occur in each of these crystalline modifications (Figure 12.16). In the $\alpha$-rhombohedral form, each B atom participates not only in the bonding of its icosahedron but also in bonding that icosahedron to

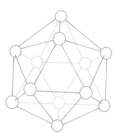

Figure 12.16 Arrangement of atoms in a $B_{12}$ icosahedron.

**Figure 12.17** Formation of a three-center B—B—B bond.

others. Ordinary electron-pair bonds cannot account for the large number of boron–boron interactions of the structure since a boron atom has only three electrons and four orbitals in its valence level.

In metallic crystals there are too few valence electrons to form covalent bonds between neighboring metal atoms of the crystal lattice; the electrons belong to the crystal as a whole and serve to bind many nuclei together. However, boron is a nonmetal, and the properties of crystalline boron (low electrical conductivity, extreme hardness, brittleness, and high melting point) are more typical of a covalent, network type of crystal than of a metallic crystal.

The bonding in elementary boron has been interpreted as involving **three-center bonds** as well as the more common two-center bonds. In a three-center bond, three atoms are held together by an electron pair in a single molecular orbital. The molecular orbital is assumed to arise from the overlap of three atomic orbitals, one from each of the three bonded atoms. Several types of three-center bonds have been postulated, including the BBB bond found in crystalline boron (Figure 12.17) and a BHB bond found in the boron hydrides (Section 12.17).

The bonding of α-rhombohedral boron may be described by assuming that a number of three-center BBB bonds, together with two-center BB bonds, are used to hold each icosahedron together as well as to bond it to neighboring icosahedra. Such a description is not without its defects, however, since more than 70 equivalent resonance forms can be drawn for the structure that includes the proper number of bonds of each type.

**12.17 COMPOUNDS OF BORON**

At very high temperatures (about 2000°C) boron reacts with many metals to form borides. These substances are very hard, are chemically stable, and have metallic conductivity. In the crystals of some metallic borides, the boron atoms are interstitial; in others, chains, octahedra, or layers of boron atoms are present. Magnesium boride, $MgB_2$, unlike the other borides, is readily hydrolyzed to produce a mixture of boron hydrides.

Boron reacts with ammonia or nitrogen at elevated temperatures to produce boron nitride, BN. This material is isoelectronic with carbon and has a crystal structure similar to graphite but with alternating boron and nitrogen atoms. At very high temperatures and pressures this modification of BN is converted to another form that has a diamond-type lattice and is harder than diamond.

At high temperatures, boron reacts with the halogens to yield the trihalides. $BF_3$ and $BCl_3$ are gases, $BBr_3$ is a liquid, and $BI_3$ is a solid. The molecules of the trihalides are triangular planar. Presumably, $sp^2$ hybrid orbitals of boron are used for bonding.

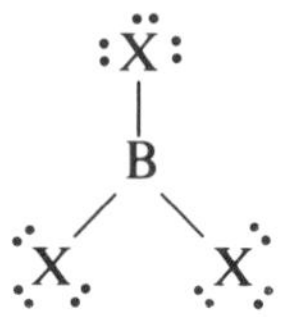

Since the boron atom does not have an octet of electrons in each of these molecules, the trihalides react as electron acceptors (Lewis acids, see Section 15.7).

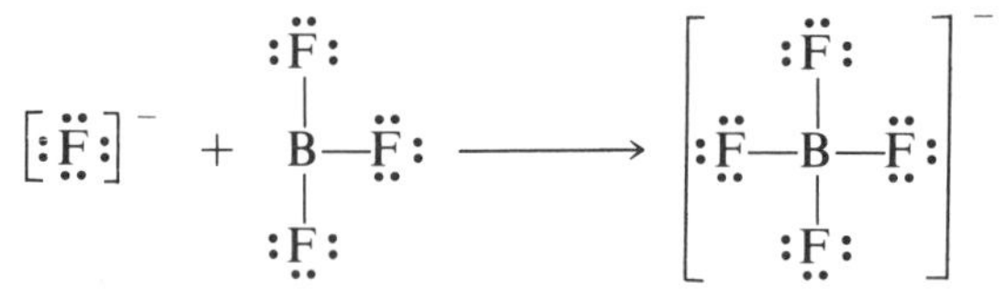

$$NH_3 + BF_3 \rightarrow H_3N:BF_3$$
$$H_2O + BF_3 \rightarrow H_2O:BF_3$$

The fluoborate ion, $BF_4^-$, is tetrahedral and is bonded by $sp^3$ hybrid bonds.

Many borates occur in nature. Some may be prepared by fusion of metallic oxides with $B_2O_3$ or boric acid; hydrated borates may be obtained by crystallization of the solution resulting from the neutralization of boric acid with aqueous alkali. The borate anions may be considered to be built up from triangular $BO_3$ units, with or without $BO_4$ tetrahedra, by the sharing of oxygen atoms between units in much the same way as the silicates are constructed. Borax, $Na_2B_4O_5(OH)_4 \cdot 8H_2O$, which can be prepared by neutralizing boric acid with NaOH in aqueous solution, contains the $B_4O_5(OH)_4^{2-}$ ion.

The acidification of an aqueous solution of any borate precipitates orthoboric acid (also called boric acid), $H_3BO_3$ or $B(OH)_3$, as soft, white crystals. This acid, which may also be prepared by the complete hydration of $B_2O_3$, has a layer-type crystal lattice in which there is extensive hydrogen bonding between $B(OH)_3$ molecules (Figure 12.18). The crystal structure of the material accounts for its cleavage into sheets and its slippery feel. Boric acid and the borates are toxic.

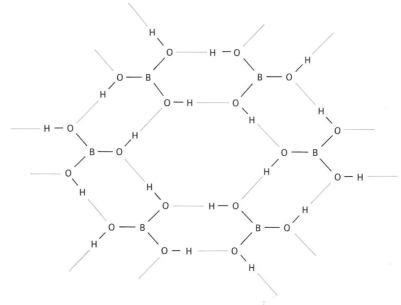

Boric acid is a very weak, **monobasic** acid in solution. Its ionization in fairly dilute solution may be represented as

$$B(OH)_3(aq) + H_2O \rightleftharpoons H^+(aq) + B(OH)_4^-(aq)$$

or

$$(H_2O)B(OH)_3(aq) \rightleftharpoons H^+(aq) + B(OH)_4^-(aq)$$

The $B(OH)_4^-$ ion is tetrahedral. In more concentrated solutions, polymeric anions, such as $B_3O_3(OH)_4^-$, exist.

When boric acid is heated to about 175°C, metaboric acid, $HBO_2$, results. The meta- acid reverts to the more highly hydrated ortho- form in water. Although salts of many hypothetical boric acids are known, these are the only two acids that exist in the pure condition.

Metaboric acid may be dehydrated by heating into boric oxide, $B_2O_3$. The oxide may also be prepared by heating boron in air. Crystals of $B_2O_3$ consist of interconnected spiral chains of $BO_4$ tetrahedra. The material, however, is usually obtained as a glass.

Boron forms two series of hydrides: $B_nH_{n+4}$, where $n = 2, 5, 6, 8, 10$, or 18, and $B_nH_{n+6}$, where $n = 4, 5, 6, 9$, or 10. A few compounds that do not conform to either of these formulas have been reported. The simplest hydride of boron is diborane, $B_2H_6$. The molecule $BH_3$ cannot be isolated but is thought to be a transitory intermediate in some of the reactions of the boron hydrides.

Diborane, which is a gas under ordinary conditions, can be prepared by the reaction of $BF_3$ and lithium hydride in ether.

$$6LiH + 8BF_3 \rightarrow 6LiBF_4 + B_2H_6$$

A mixture of higher boranes is obtained by heating diborane.

The compounds of formula $B_nH_{n+4}$ are thermally more stable (with regard to decomposition to the elements) than the second group, $B_nH_{n+6}$. The compounds $B_2H_6$, $B_5H_9$, and $B_5H_{11}$ are spontaneously flammable in air; all boranes hydrolyze, at various rates, to yield boric acid and hydrogen.

The molecular structure of diborane is shown in Figure 12.19. In the molecule, each B atom has two H atoms (called terminal hydrogen atoms) bonded to it by regular two-center, electron-pair bonds, and the resulting $BH_2$ fragments are joined together by two hydrogen bridges (three-center BHB bonds). The two B atoms and the four terminal H atoms are co-planar, and the bridge H atoms lie above and below this plane.

Each B atom may be considered to use $sp^3$ hybrid orbitals for the formation of the molecule. Two of the $sp^3$ orbitals of each B atom are used to form ordinary two-center bonds with terminal H atoms, and the other two hybrid orbitals are used to form three-center BHB bonds. Each three-center bond consists of a pair of electrons in a bonding molecular orbital formed by the overlap of the $1s$ orbital of the H atom with orbitals of the two B atoms (Figure 12.19). Of the twelve valence electrons in the molecule (six from the two B atoms and six from the six H atoms), four pairs are utilized in the two-center bonds to the four terminal H atoms, and two pairs are used in the two BHB three-center bonds. The molecule is diamagnetic. The bond distance between a B atom and a terminal H is shorter than the distance between a B atom and a bridge H, thus indicating a stronger bond in the former case.

The structures of the higher boranes involve BBB as well as BHB three-center bonds. Several of the higher boranes have skeletons of boron atoms that are fragments of the $B_{12}$ icosahedron found in crystalline boron.

Boron forms a series of borohydride ions. The most important one is $BH_4^-$, a tetrahedral ion that may be prepared by the reaction of lithium hydride and $B_2H_6$ in ether.

**Figure 12.19** (a) Structure of diborane. (b) Formation of bonding orbitals. (Shapes of orbitals simplified.)

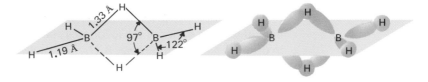

(a)                    (b)

$$2LiH + B_2H_6 \rightarrow 2LiBH_4$$

Lithium borohydride, and the alkali borohydrides in general, are ionic materials that are important reducing agents. Other borohydride ions are known, such as $B_3H_8^-$, $B_{10}H_{13}^-$, and $B_{12}H_{12}^{2-}$. The boron skeleton of the $B_{12}H_{12}^{2-}$ ion is a $B_{12}$ icosahedron.

## Problems

**12.1** Discuss how the properties of the elements of group V A and their compounds change with increasing atomic number.

**12.2** Write all the equations that you can for the reactions of elementary nitrogen. Explain the low chemical reactivity of $N_2$.

**12.3** Why is it advantageous for a farmer to rotate the crops sown in a given field?

**12.4** In 1892 Lord Rayleigh observed that the density of nitrogen prepared from air (by the removal of $H_2O$, $CO_2$, and $O_2$) was different from the density of nitrogen prepared from $NH_4Cl$ and $NaNO_2$. On the basis of this observation, Rayleigh and William Ramsey isolated argon (1894). Assume that air is 1.00% Ar, 78.00% $N_2$, and 21.00% $O_2$ (by volume). How should the densities of the two "nitrogen" samples compare?

**12.5** The normal boiling point of the compound ethylene diamine, $H_2NCH_2CH_2NH_2$, is 117°C and that of propyl amine, $CH_3CH_2CH_2NH_2$, is 49°C. The molecules, however, are similar in size and molecular weight. What reason can you give for the difference in boiling point?

**12.6** Discuss the comparative re-

activities, solubilities, and electrical conductivities of the three allotropic forms of phosphorus in terms of molecular structure.

**12.7** Write chemical equations for the separate reactions of each of the following with $H_2O$, with aqueous NaOH, and with aqueous HCl: (a) $P_4O_6$, (b) $Sb_4O_6$, and $Bi_2O_3$.

**12.8** What are the characteristics of the three types of nitrides?

**12.9** Boron nitride, BN, exists in two crystalline modifications—one resembling diamond and the other graphite. Describe, using drawings, the structures of these four substances.

**12.10** Discuss the preparation, structure, and properties of ammonia. Why is the boiling point of $NH_3$ high in comparison to the boiling points of $PH_3$, $AsH_3$, and $SbH_3$?

**12.11** Write equations for the following reactions of ammonia.

(a) $NH_3(aq) + Ag^+(aq) \rightarrow$

(b) $NH_3(aq) + H^+(aq) \rightarrow$

(c) $NH_3(aq) + H_2O + CO_2(aq) \rightarrow$

(d) $NH_3(g) + O_2(g) \xrightarrow{heat}$

(e) $NH_3(g) + O_2(g) \xrightarrow[heat]{Pt}$

(f) $NH_3(g) + HCl(g) \rightarrow$

(g) $NH_3(g) + V(s) \xrightarrow{\text{heat}}$

(h) $NH_3(l) + Na(s) \rightarrow$

**12.12** Write a series of equations for the preparation of nitric acid starting with $N_2$ as the source of nitrogen.

**12.13** Write an equation for the reaction of $HNO_3$ with (a) Cu, (b) Zn, (c) $P_4O_{10}$, (d) $NH_3$, (e) $Ca(OH)_2$.

**12.14** (a) List the oxides of nitrogen. (b) Write a chemical equation for the preparation of each of them. (c) Diagram the electronic structure of each oxide.

**12.15** State the products of the thermal decomposition of (a) $NH_4NO_3$, (b) $NH_4NO_2$, (c) $Pb(NO_3)_2$, (d) $NaNO_3$, (e) $NaN_3$.

**12.16** Write equations for the following oxidation–reduction reactions: (a) the reaction of $NH_3$ with $OCl^-$ that produces $N_2H_4$ and $Cl^-$; (b) the reaction of $NO_2$ with $H_2$ in HCl solution that produces $[NH_3OH]Cl$; (c) the electrolysis of $NH_4F$ in anhydrous HF that produces $NF_3$ and $H_2$.

**12.17** Draw electronic structures for orthophosphoric, phosphorous, and hypophosphorous acids. Tell how these structures explain the number of $H^+(aq)$ dissociated per mole when each of these compounds is dissolved in water.

**12.18** Write equations for the preparations of all acids that may be obtained by the hydrolysis of $P_4O_{10}$.

**12.19** Write an equation for the reaction of each of the following with water: (a) $Li_3N$, (b) AlN, (c) $Ca_3P_2$, (d) CaNCN, (e) $NCl_3$, (f) $PCl_3$, (g) $PCl_5$, (h) $H_4P_2O_7$, (i) $NO_2$, (j) $CaC_2$, (k) $Be_2C$, (l) $SiCl_4$, (m) $BF_3$.

**12.20** What is the acid anhydride of each of the following? (a) $H_2N_2O_2$, (b) $HNO_2$, (c) $HNO_3$, (d) $H_4P_2O_7$, (e) $H_3P_3O_9$, (f) $H_3PO_3$, (g) $H_3AsO_4$, (h) $H_2CO_3$, (i) $B(OH)_3$.

**12.21** Diagram the resonance forms of (a) $NO_2^-$, (b) $NO_2$, (c) NO, (d) $NO_3^-$, (e) $HNO_3$, (f) $HN_3$, (g) graphite, (h) $CO_3^{2-}$.

**12.22** Write equations for the reactions of HCl with (a) $H_2NNH_2$, (b) $H_2NOH$, (c) $NaNO_2$, (d) $CaCO_3$.

**12.23** Write equations for the reactions of $O_2$ with (a) As, (b) $P_4$, (c) $PCl_3$, (d) NO, (e) $Si_2H_6$, (f) $B_2H_6$, (g) $Sb_2S_3$.

**12.24** Write equations for the reactions of water with (a) $Na_3As$, (b) $As_2O_5$, (c) $Mg_3Bi_2$, (d) $As_4O_6$, (e) $SbCl_3$, (f) $PH_4I$, (g) $MgB_2$.

**12.25** Draw Lewis structures for (a) $P_2H_4$, (b) $C_2N_2$, (c) $H_2C_2$, (d) CO, (e) $C_2H_6$, (f) cis- $N_2F_2$, (g) trans- $N_2F_2$, (h) $H_3NBF_3$.

**12.26** Describe the shape of each of the following and give the type of hybrid orbitals employed by the central atom in the bonding. (a) $NH_3$, (b) $NH_2^-$, (c) $PCl_4^+$, (d) $PCl_6^-$, (e) $BF_4^-$, (f) $SbCl_5$, (g) $Sb(OH)_6^-$.

**12.27** Discuss the ways in which the chemistry of carbon differs from the chemistry of the other elements of group IV A.

**12.28** What explanation may be offered for the fact that carbon has a greater tendency toward catenation than any other element?

**12.29** For each of the following carbonyls, all of which are diamagnetic, describe the geometric configuration of the molecule and the type of hybrid orbitals employed by the metal atom. (a) $Co_2(CO)_8$, (b) $Mo(CO)_6$, (c) $Re_2(CO)_{10}$, (d) $Ru(CO)_5$.

**12.30** Write a series of equations that lead to the preparation of $Na_2CO_3$ and that employ only $N_2$, $H_2O$, $NaCl$, $CaCO_3$, and substances derived from this list.

**12.31** Explain the cause of the difference in reactivity toward water of $CF_4$ and $SiF_4$.

**12.32** Write an equation for the reaction that occurs when each of the following is heated with carbon. (a) $P_4O_{10}$, (b) $As_4O_6$, (c) $CaO$, (d) $NaNO_3$, (e) $NaNH_2$, (f) $SiO_2$.

**12.33** Write an equation for the reaction that occurs when each of the following is heated. (a) $NaHCO_3$, (b) $CaCO_3$, (c) $AgCN$, (d) $H_3PO_4$, (e) $Ca_3(PO_4)_2$ and $SiO_2$, (f) $CaCO_3$ and $SiO_2$.

**12.34** Describe the crystal structure of (a) diamond, (b) graphite, (c) $SiC$, (d) black phosphorus, (e) $Bi$, (f) $H_3BO_3$.

**12.35** Describe the shape of the following molecules: (a) $P_4$, (b) $POCl_3$, (c) $P_4O_6$, (d) $P_4O_{10}$, (e) $B_2H_6$.

**12.36** Describe the commercial processes used in the manufacture of (a) $P_4$, (b) $NH_3$, (c) $B$, (d) $H_3PO_4$, (e) $HNO_3$, (f) superphos-

phate fertilizer, (g) $SiC$, (h) $As_4$.

**12.37** Write equations for the reactions of aqueous $NaOH$ with (a) $NH_4Cl$, (b) $P_4$, (c) $As_4O_6$, (d) $Sb_2O_5$, (e) $H_3BO_3$, (f) $Si$.

**12.38** Write an equation to show how orthoboric acid ionizes in water solution.

**12.39** Complete and balance the following equations.

(a) $B_2O_3 + Mg \xrightarrow{\text{heat}}$

(b) $BBr_3 + H_2 \xrightarrow{\text{heat}}$

(c) $B + N_2 \xrightarrow{\text{heat}}$

(d) $B + Mg \xrightarrow{\text{heat}}$

(e) $BF_3 + F^- \rightarrow$

(f) $B_2O_3 + H_2O \rightarrow$

(g) $B(OH)_3 + Na^+ + OH^- \rightarrow$

(h) $B(OH)_3 \xrightarrow[\text{heating}]{\text{mild}}$

(i) $B(OH)_3 \xrightarrow[\text{heating}]{\text{strong}}$

(j) $LiH + B_2H_6 \rightarrow$

**12.40** In a molecule of $B_4H_{10}$, the four boron atoms form a puckered square and are joined by B—H—B three-center bonds. Two of the boron atoms, which are diagonally across the square from one another, are joined by an electron-pair bond, and the remaining hydrogens are bonded to boron atoms by electron-pair bonds. Diagram the structure of the molecule.

**12.41** What is a three-center bond?

# 13

# Chemical Kinetics and Chemical Equilibrium

**Chemical kinetics** is the study of the rates of chemical reactions, the means by which reaction rates may be controlled, and the ways in which reactions proceed on the atomic-molecular level.

Reactions are frequently classed as **homogeneous** or **heterogeneous**. A homogeneous reaction occurs in a single phase; for example,

$$H^+(aq) + OH^-(aq) \rightarrow H_2O(l)$$

$$2NO(g) + O_2(g) \rightarrow 2NO_2(g)$$

Heterogeneous reactions take place on a phase boundary; for example,

$$Zn(s) + 2H^+(aq) \rightarrow Zn^{2+}(aq) + H_2(g)$$

$$2Mg(s) + O_2(g) \rightarrow 2MgO(s)$$

Many chemical reactions are reversible. For example, the reaction

$$N_2(g) + 3H_2(g) \rightleftharpoons 2NH_3(g)$$

may be made to give a good yield of ammonia through a judicious choice of conditions; under other reaction conditions virtually complete dissociation of ammonia is observed.

**13.1 MOLECULAR COLLISIONS AND REACTION RATES**

Consider a hypothetical homogeneous gas phase reaction between two diatomic molecules $A_2$ and $B_2$.

$$A_2(g) + B_2(g) \rightleftharpoons 2AB(g)$$

Not every reaction proceeds in a manner described by the net chemical equation; many reactions occur through a series of steps (called the **reaction mechanism**), and the net equation represents the sum of the equations for the steps. In the case of the reaction between $A_2$ and $B_2$, however, let us assume that the reaction proceeds, as depicted by the equation, through the collisions of $A_2$ and $B_2$ molecules (Figure 13.1).

**Figure 13.1** Collision between an $A_2$ molecule and a $B_2$ molecule resulting in a reaction.

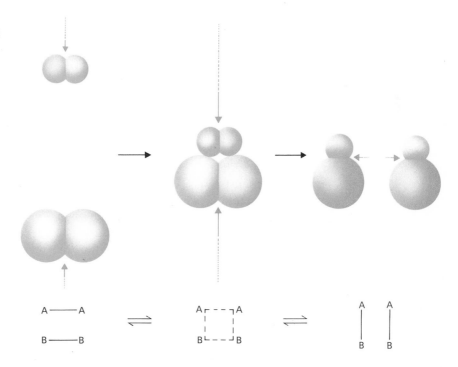

**Figure 13.2** Collision between an $A_2$ molecule and a $B_2$ molecule producing no reaction.

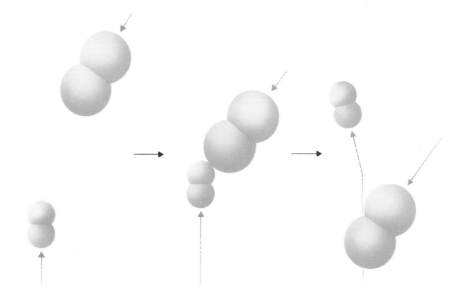

Not every collision between molecules results in a reaction. Comparison of the calculated total number of collisions per unit time with the observed rate of a reaction shows that generally only a small fraction of the total number of collisions is effective.

There are two reasons why a given collision may not be effective. First, the molecules may be improperly aligned (Figure 13.2). Second, the impact of the collision may be so gentle that the molecules rebound unchanged. The negatively charged electron clouds surrounding molecules produce a force of repulsion that acts between molecules; this repulsion causes molecules with low energies to rebound when they collide. In a collision between high-energy molecules, however, the molecules possess sufficient energy to overcome the forces of repulsion, and the rearrangement typical of the chemical reaction may then follow. For an effective collision the sum of the energies of the colliding molecules must equal or exceed some minimum value.

Therefore, when the rate of a given reaction is increased through an alteration in an experimental condition, this increase is brought about by either an increase in the total number of collisions per unit time, an increase in the fraction of the total number of collisions that are effective, or a combination of both factors.

**13.2 TEMPERATURE AND REACTION RATE**

The rates of virtually all chemical reactions increase when the temperature is raised; this effect is observed for endothermic as well as for exothermic reactions. The more rapid molecular motion resulting from an increase in temperature brings about a larger number of molecular collisions per unit time. However, raising the temperature from 25°C to 35°C causes the average speed of the molecules to increase by about 2%, whereas the same temperature rise brings about an increase in the reaction rate of approximately 200 to 300%. Obviously, increasing the temperature must increase the fraction of molecular collisions that are effective, and this factor must far outweigh the concomitant increase in the total number of collisions per unit time.

By an examination of Figure 13.3 we can understand why proportionately more molecular collisions result in reactions at a higher temperature than at a lower temperature. Two molecular energy distribution curves are shown — one for a temperature, $t_1$, and another for a higher temperature, $t_2$. The minimum energy required for reaction is indicated on the diagram; the number of molecules at $t_1$ with energies equal to or greater than this minimum energy is proportional to the area, $a$, under the curve for $t_1$. The curve for temperature $t_2$ is shifted only slightly in the direction of higher energy; however, at $t_2$ the number of molecules possessing sufficient energy to react successfully upon collision is greatly increased and is proportional to the area $a + b$. Thus a temperature rise produces an increase in reaction rate principally because the proportion of effective collisions is increased. The increase in the total number of collisions per unit time is only a minor factor. The influence of temperature on reaction rates is analyzed mathematically in Section 13.6.

**Figure 13.3** Molecular energy distributions at temperatures $t_1$ and $t_2$.

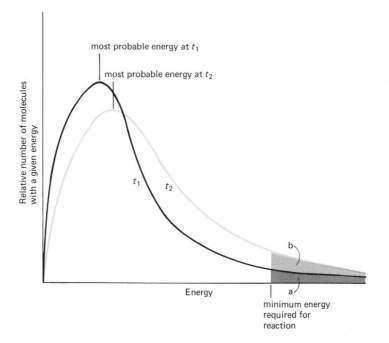

In a successful collision between reacting molecules, the molecules form a transitory intermediate called an **activated complex.** Let us again consider the reaction between $A_2$ and $B_2$. In a gentle collision the $A_2$ and $B_2$ molecules are repelled by the molecular electron clouds and never get close enough for A—B bonds to form. However, a collision between high-energy molecules results in the formation of a short-lived molecule, $A_2B_2$, which is the activated complex for the reaction; the atoms of the complex assume the square configuration diagramed in Figure 13.1. The $A_2B_2$ complex may split to form two AB molecules or may decompose to re-form $A_2$ and $B_2$ molecules. In fact, this same activated complex is produced by the collision of two high-energy AB molecules and is the intermediate in the reverse reaction – the formation of $A_2$ and $B_2$ from AB.

$$A_2 + B_2 \rightleftharpoons \begin{matrix} A \cdots A \\ \vdots \quad \vdots \\ B \cdots B \end{matrix} \rightleftharpoons 2AB$$

The bonding arrangement of the activated complex is not so stable as that of either the reactants or the products; the activated complex represents a state of relatively high potential energy. A potential diagram for the hypothetical reaction

$$A_2 + B_2 \rightleftharpoons A_2B_2 \rightleftharpoons 2AB$$

is shown in Figure 13.4; potential energy is plotted against a reaction coordinate that may be interpreted as showing the progress of the reaction. The difference between the potenital energy of the reactants,

Figure 13.4 Potential energy diagram for the hypothetical reaction $A_2 + B_2 \rightleftharpoons A_2B_2 \rightleftharpoons$ 2AB.

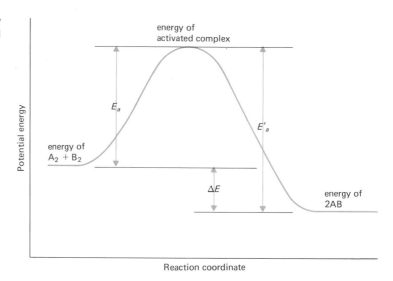

$A_2 + B_2$, and the potential energy of the activated complex is called the **energy of activation** and is given the symbol $E_a$; typical values of $E_a$ range from 15 to 60 kcal/mol.

The energy of activation constitutes a potential energy barrier between the reactants and products. Even though the energy of the reactants, $A_2 + B_2$, is higher than that of the products and the overall reaction is exothermic, the system must climb a potential energy hill before it can coast down to a state of lower potential energy. In a successful collision part of the kinetic energy of the fast-moving $A_2$ and $B_2$ molecules is used to provide the energy of activation and thus produce the high-energy molecular arrangement of the activated complex. If the activated complex re-forms $A_2$ and $B_2$ molecules, the energy of activation is released in the form of the kinetic energy of the molecules. If the activated complex splits into the product, two molecules of AB, the energy indicated as $E_a'$ on the diagram is released as kinetic energy. The overall difference between the energy released and the energy absorbed is the $\Delta E$ (or $\Delta H$, depending on conditions) for the reaction.

$$\Delta E = E_a - E_a'$$

As shown in Figure 13.4, $E_a' > E_a$; therefore $\Delta E$ is negative, and the forward reaction is exothermic.

The reaction diagramed in Figure 13.4 is reversible; $E_a'$ is the energy of activation of the reverse reaction. If the diagram is read from right to left, $E_a'$ is absorbed and $E_a$ evolved.

$$\Delta E = E_a' - E_a$$

The reverse reaction is endothermic; $\Delta E$ is positive since $E_a' > E_a$.

If molecules of $A_2$ and $B_2$ with relatively low kinetic energies approach each other, sufficient energy is lacking to produce the arrangement of the activated complex by forcing the molecules together in opposition to the repulsive forces. In this instance the molecules possess only sufficient energy between them to get part way up the hill; then, repelling each other, they coast back down the hill and fly apart unchanged. Since only high-energy molecules can get over the hill, it is easy to understand the role of temperature in influencing reaction rate through changes in the fraction of high-energy molecules.

**13.3 CATALYSTS**   A **catalyst** is a substance that increases the rate of a chemical reaction without being used up in the reaction; the catalyst may be recovered unchanged at the conclusion of the process. For example, the decomposition of $KClO_3$ can be greatly accelerated by the addition of a small quantity of $MnO_2$. In the equation for the change, the catalyst is indicated over the arrow since its use does not affect the overall stoichiometry of the reaction.

$$2KClO_3 \xrightarrow{\text{MnO}_2} 2KCl + 3O_2$$

Catalysts cannot cause thermodynamically impossible reactions to occur. Furthermore, it is not the mere *presence* of a catalyst (presumably acting as a cheering section) that causes the effect on the reaction rate. In a catalyzed reaction the catalyst is actually consumed in one step and regenerated in a subsequent step; thus it is used over and over again without undergoing any permanent change.

A catalyst, then, functions by opening an alternative path by which a reaction can proceed; the mechanism of the catalyzed reaction is different from that of the uncatalyzed reaction. The catalyzed path has a lower energy of activation than the uncatalyzed path (Figure 13.5), a fact that accounts for the more rapid reaction rate. When a catalyst is used, proportionately more molecules possess the energy required for a successful collision (Figure 13.6); thus, of the total number of collisions per unit time, the fraction that results in reaction is increased.

Two additional observations may be derived from an examination of Figure 13.5. First, $\Delta E$ for the catalyzed reaction is the same as $\Delta E$ for the uncatalyzed reaction. Second, the energy of activation for the reverse reaction, $E_a'$, is lowered by the use of the catalyst to the same extent that the energy of activation of the forward reaction, $E_a$, is lowered. This means that a catalyst has the same effect on the reverse reaction that it

**Figure 13.5** Potential energy diagrams for catalyzed and uncatalyzed reactions.

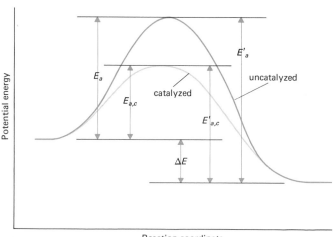

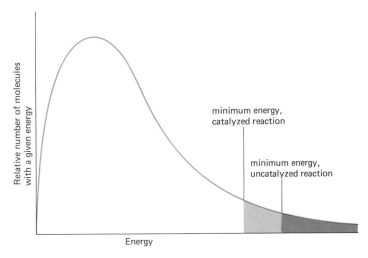

**Figure 13.6** Molecular energy distribution showing the effect of a catalyst on the number of molecules possessing sufficient energy to react.

has on the forward reaction. If a catalyst doubles the speed of a forward reaction, the same catalyst will double the speed of the reverse reaction.

Many industrial processes depend upon catalytic procedures, but even more important to man are the natural catalysts known as **enzymes.** These extremely complicated substances catalyze life processes such as digestion and cell synthesis. The large number of complex chemical reactions that occur in the body, and are necessary for life, can occur at the relatively low temperature of the body because of the action of enzymes. Thousands of enzymes are known to exist, each serving a specific function. Research into the structure and action of enzymes offers great promise for the advancement of knowledge of the causes of disease and the mechanism of growth.

In **homogeneous** catalysis the substance serving as a catalyst is present in the same phase as the reactants; in **heterogeneous,** or **surface,** catalysis the reactants and catalyst comprise two separate phases, and the reaction occurs on a catalytic surface. An example of homogeneous catalysis in the gas phase is the effect of chlorine on the decomposition of dinitrogen oxide.

Gaseous $N_2O$ is a relatively inert substance at room temperature; however, at temperatures in the neighborhood of 1000°K it decomposes according to the equation

$$2N_2O(g) \rightarrow 2N_2(g) + O_2(g)$$

Kinetic studies show that the reaction proceeds by means of collisions between two molecules of $N_2O$. The reaction is catalyzed by traces of chlorine gas.

A proposed mechanism for the catalyzed path follows. At the temperature of the experiment, and particularly in the presence of light, some chlorine molecules are dissociated into chlorine atoms.

$$Cl_2(g) \rightarrow 2Cl(g)$$

These chlorine atoms readily react with $N_2O$

$$N_2O(g) + Cl(g) \rightarrow N_2(g) + ClO(g)$$

and the decomposition of ClO is rapid

$$2ClO(g) \rightarrow Cl_2(g) + O_2(g)$$

In this last step the catalyst ($Cl_2$) is returned to its original form; the final products of the catalyzed reaction ($N_2$ and $O_2$) are the same as those of the uncatalyzed reaction.

Another example of homogeneous catalysis is the decomposition of ozone in the presence of $N_2O_5$. Dinitrogen pentoxide decomposes readily into oxygen and lower oxides of nitrogen. For example,

$$2N_2O_5(g) \rightarrow 2N_2O_4(g) + O_2(g)$$

Ozone reacts rapidly with $N_2O_4$ to produce oxygen and regenerate the catalyst, $N_2O_5$.

$$O_3(g) + N_2O_4(g) \rightarrow O_2(g) + N_2O_5(g)$$

The proper addition of these two equations gives the net change

$$2O_3(g) \xrightarrow{N_2O_5} 3O_2(g)$$

It must be emphasized that these mechanisms are merely plausible hypotheses. Undoubtedly, the actual mechanism of the $N_2O_5$-catalyzed decomposition of ozone is much more complicated and may involve steps such as

$$N_2O_4(g) \rightarrow 2NO_2(g)$$

$$NO_2(g) + O_3(g) \rightarrow NO_3(g) + O_2(g)$$

$$NO_2(g) + NO_3(g) \rightarrow N_2O_5(g)$$

Homogeneous catalysis also occurs in aqueous solution; many reactions are catalyzed by acids and bases. The decomposition of hydrogen peroxide

$$2H_2O_2(aq) \rightarrow 2H_2O(l) + O_2(g)$$

is catalyzed by the presence of iodide ion. A postulated mechanism for the catalyzed reaction is

$$H_2O_2(aq) + I^-(aq) \rightarrow H_2O(l) + IO^-(aq)$$

$$H_2O_2(aq) + IO^-(aq) \rightarrow H_2O(l) + O_2(g) + I^-(aq)$$

Heterogeneous catalysis generally proceeds through **chemical adsorption** (or **chemisorption**) of the reactants on the surface of the catalyst. Adsorption is a process in which molecules adhere to the surface of a solid; charcoal, for example, is used in gas masks as an adsorbent for noxious gases. In ordinary **physical adsorption** the molecules are held to the surface of the adsorbent by van der Waals forces; thus the adsorbed gas molecules are affected to about the same extent that they would be if the gas were liquefied.

In chemisorption the adsorbed molecules are held to the catalytic surface by bonds comparable to those of chemical compounds. In the process of forming bonds with the adsorbent, chemisorbed molecules undergo internal electron rearrangements. The bonding within certain molecules is stretched and weakened, and the bonds of some molecules are even broken. For example, hydrogen is adsorbed in atomic form on the surface of platinum. Thus the chemisorbed layer of molecules functions as an activated complex in a surface-catalyzed reaction.

At the present time chemisorption and surface catalysis are far from completely understood; we can only make reasonable postulates for the mechanisms of specific reactions. A suggested mechanism for the decomposition of $N_2O$ on gold is diagramed in Figure 13.7. Molecules of $N_2O$ are chemisorbed on the surface of the gold, and the following reactions take place on this surface.

$$N_2O \rightarrow N_2(g) + O$$

$$O + O \rightarrow O_2(g)$$

**Figure 13.7** Proposed mode of decomposition of $N_2O$ on Au.

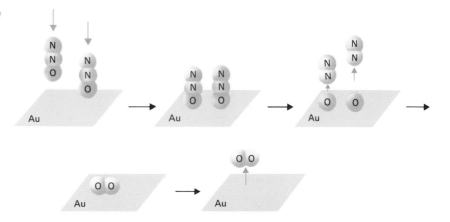

The release of the gaseous products from the surface is called **desorption.**

Some empirical observations have great utility in the selection of suitable catalysts for particular reactions; for example, most reactions of gaseous hydrogen are catalyzed by platinum. However, the fundamental theory of catalytic processes has not been developed to the point where it can be used to select or design a catalyst for a specific use. It appears that the electronic structure of the solid, as well as the geometric arrangement and spacing of the atoms on the surface of the solid, is a critical consideration.

It has been proposed that lattice defects or irregularities in the surface of the catalyst are active sites for catalysis; this hypothesis has been used to explain the action of **promoters,** which are materials that enhance the activity of catalysts. For example, in the synthesis of ammonia

$$N_2(g) + 3H_2(g) \xrightarrow{\text{Fe}} 2NH_3(g)$$

an iron catalyst is made more effective by incorporating traces of potassium or vanadium.

Catalytic **poisons** are substances that inhibit the activity of catalysts. For example, small amounts of arsenic destroy the power of platinum to catalyze the preparation of sulfur trioxide from sulfur dioxide.

$$2SO_2(g) + O_2(g) \xrightarrow{\text{Pt}} 2SO_3(g)$$

Presumably, platinum arsenide forms on the surface of the platinum and destroys its catalytic activity.

Catalysts are generally highly specific in their activity. In some cases a given substance will catalyze the synthesis of one set of products from certain reagents, whereas another substance will catalyze the synthesis of completely different products from the same reactants. In these instances, of course, both reactions are thermodynamically feasible. Carbon monoxide and hydrogen can be made to yield a wide variety of products depending upon the catalyst employed and the conditions of the reaction.

If a cobalt or nickel catalyst is used, CO and $H_2$ produce mixtures of hydrocarbons. For example,

$$CO(g) + 3H_2(g) \xrightarrow{\text{Ni}} CH_4(g) + H_2O(g)$$

On the other hand, methanol is the product of the reaction of CO and $H_2$ when a mixture of zinc and chromic oxides is employed as a catalyst.

$$CO(g) + 2H_2(g) \xrightarrow{\text{ZnO/Cr}_2\text{O}_3} CH_3OH(g)$$

**13.4 HETEROGENEOUS REACTIONS**

Heterogeneous reactions occur on a surface or phase boundary; hence, the rate of such a reaction can be greatly increased if the surface is increased. In a reaction between a solid and a gas, or a solid and a liquid, the total number of collisions per unit time can be increased by pulverizing the solid. A finely divided combustible material may burn with such rapidity as to produce an explosion. The rate of a heterogeneous process may be increased by agitation of the reacting mixture, since agitation keeps the reactive surface continuously exposed. Thus, sugar dissolves more quickly in coffee if the mixture is stirred.

For this same reason, surface catalysts are usually prepared in a way that will produce a large surface area for a given amount of solid. Solids used as catalysts are generally porous and extremely finely divided.

**13.5 CONCENTRATIONS AND REACTION RATES**

Reaction rate is generally dependent upon the concentrations of the reacting substances. When the concentration of a reactant is increased, more molecules are crowded into a given volume, and the total number of molecular collisions per unit time is increased; this results in an increase in the rate of the reaction. If all other conditions are held constant, there is no change in the proportion of collisions that are effective.

Many different techniques are employed to study reaction rates — the appropriate one depending upon the reaction under consideration. In addition to chemical analysis, physical methods are sometimes used to follow the progress of chemical reactions. These physical methods are applicable only when some measurable change takes place in the course of the reaction; for example, physical methods have been devised to study reactions in which there is a change in one of the following: total pressure, color (appearance or disappearance of a colored substance), acidity or alkalinity, conductivity, volume, or viscosity.

Since the concentrations of the reactants decrease as the reaction proceeds, the rate of a chemical reaction decreases with time as well. In Figure 13.8 concentrations are plotted against time for the hypothetical reaction

$$A_2 + B_2 \rightarrow 2AB$$

The rate of the reaction, in terms of the change in concentration of $A_2$, $B_2$, or $AB$ per unit time is given by the slope of the appropriate curve. The rate of this reaction may be expressed in terms of the rate of disappearance of $A_2$ and $B_2$ or the rate of appearance of $AB$; because of the stoichiometry of the reaction, these rates are not equal. The rate of appearance of $AB$ is twice that of the rate of disappearance of $A_2$ or $B_2$.

Concentrations are generally expressed in moles per liter (molarity for substances in solution); in mathematical expressions the concentration of a substance is usually represented by its formula enclosed in brackets.

**Figure 13.8** Curves showing
changes in concentrations
of materials with time for
the reaction $A_2 + B_2 \rightarrow 2AB$.

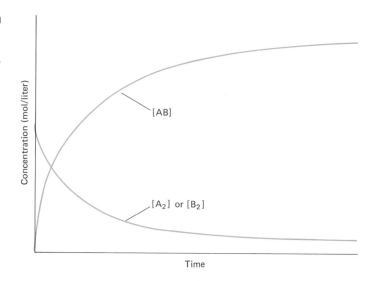

**Figure 13.8** Curves showing changes in concentrations of materials with time for the reaction $A_2 + B_2 \rightarrow 2AB$.

Since the pressure of a gas is a measure of its concentration, expressions are frequently given in terms of the partial pressures of reacting gases; the partial pressure of nitrogen is represented by the symbol $p_{N_2}$.

For a chemical reaction a mathematical expression (a **rate equation**) relating the concentrations of reactants to the reaction rate may usually be determined experimentally. The chemical equation for a reaction stoichiometrically describes only the initial reactants and final products; frequently the chemical change itself occurs by way of a mechanism consisting of several steps. Each of these steps proceeds at its own rate—some rapid, some slow—and each is dependent upon the concentrations of the initial reactants and/or the reaction intermediates. Therefore, the rate expression for the chemical reaction may not generally be derived from an inspection of the balanced chemical equation for the overall chemical change but must be determined experimentally.

However, if experimental results indicate that a reaction occurs by a simple one-step mechanism, a rate expression can be derived from the chemical equation; such reactions are *not* numerous. For example, we have assumed that the reaction

$$A_2(g) + B_2(g) \rightarrow 2AB(g)$$

proceeds in one step through collisions between $A_2$ molecules and $B_2$ molecules. If the concentration of $A_2$ is doubled in a given system, the number of $A_2$ molecules is doubled. In this instance the number of $A_2$–$B_2$ collisions per second is doubled, and the reaction rate is doubled. If the concentration of $B_2$ is doubled and the concentration of $A_2$ is constant, the same effect on the rate is observed. Doubling both the $A_2$ and

$B_2$ concentrations increases the reaction rate by a factor of four. The rate of the reaction, then, is proportional to the product of the concentrations of $A_2$ and $B_2$.

$$\text{rate} = k[A_2][B_2]$$

The proportionality constant, $k$, is called the **rate constant;** its numerical value is dependent upon temperature, catalysis, and the terms in which the reaction rate is expressed.

The rate equation may also be expressed in terms of pressures,

$$\text{rate} = kp_{A_2}p_{B_2}$$

The value of $k$ in this equation is not the same as that of the previous rate equation.

The reaction

$$2N_2O(g) \rightarrow 2N_2(g) + O_2(g)$$

proceeds by means of collisions between two $N_2O$ molecules. For a given system of definite volume, the number of collisions per second for a *single* $N_2O$ molecule is proportional to the total number of molecules present less one $(n - 1)$. Inasmuch as this is true of all $n$ of the molecules present, the total number of collisions per second is proportional to $\frac{1}{2}n(n - 1)$. The factor $\frac{1}{2}$ must be included so that a given collision is not counted twice—once for each molecule in the collision. The number of molecules in a sample is so large that $(n - 1)$ is approximately equal to $n$; thus the total number of collisions per second is proportional to $\frac{1}{2}n^2$ to a good approximation. The constant $\frac{1}{2}$ may be incorporated into the rate constant and the number of molecules of $N_2O$ in the system is proportional to the concentration of $N_2O$. Thus

$$\text{rate} = k[N_2O]^2$$

In general, if the chemical equation for a single-step reaction or for one step of a multistep mechanism is indicated as

$$w\text{W} + x\text{X} + \cdots \rightarrow \text{products}$$

the rate equation is

$$\text{rate} = k[W]^w[X]^x \cdots$$

This is a statement of the **law of mass action** which was first expressed in general form by Cato Guldberg and Peter Waage in 1864. If the reactants are gases, similar expressions using partial pressures may be derived.

The **order** of a rate equation is given by the sum of the exponents of

Table 13.1 Rate equations for some gas phase reactions

FIRST ORDER

| | |
|---|---|
| $2N_2O_5 \longrightarrow 4NO_2 + O_2$ | rate $= k[N_2O_5]$ |
| $2N_2O \overset{Au}{\longrightarrow} 2N_2 + O_2$ | rate $= k[N_2O]$ |
| $SO_2Cl_2 \longrightarrow SO_2 + Cl_2$ | rate $= k[SO_2Cl_2]$ |
| $2NO_2Cl \longrightarrow 2NO_2 + Cl_2$ | rate $= k[NO_2Cl]$ |
| $N_2O_4 \longrightarrow 2NO_2$ | rate $= k[N_2O_4]$ |

SECOND ORDER

| | |
|---|---|
| $2N_2O \longrightarrow 2N_2 + O_2$ | rate $= k[N_2O]^2$ |
| $2NOCl \longrightarrow 2NO + Cl_2$ | rate $= k[NOCl]^2$ |
| $2NO_2 \longrightarrow 2NO + O_2$ | rate $= k[NO_2]^2$ |

THIRD ORDER

| | |
|---|---|
| $2NO + O_2 \longrightarrow 2NO_2$ | rate $= k[NO]^2[O_2]$ |
| $2NO + Cl_2 \longrightarrow 2NOCl$ | rate $= k[NO]^2[Cl_2]$ |
| $2NO + Br_2 \longrightarrow 2NOBr$ | rate $= k[NO]^2[Br_2]$ |

the concentrations appearing in the equation. Thus for the reaction between NO and $O_2$ listed in Table 13.1, the rate expression is said to be second order in NO, first order in $O_2$, and third order overall. Since experimentally determined rate expressions may represent a mathematical summary of several rate equations (one for each step of a multistep mechanism), overall rate equations of fractional order are sometimes obtained.

## Example 13.1

The data of the following table were obtained at 25°C for the reaction

$$S_2O_8^{2-}(aq) + 2I^-(aq) \rightarrow 2SO_4^{2-}(aq) + I_2(aq)$$

What is the rate equation for the reaction and the value of $k$?

| Experiment | Initial Conc. $S_2O_8^{2-}$ (M) | Initial Conc. $I^-$ (M) | Initial Rate — Increase in Conc. of $I_2$ (M/min) |
|---|---|---|---|
| A | $1.0 \times 10^{-4}$ | $1.0 \times 10^{-2}$ | $0.65 \times 10^{-6}$ |
| B | $2.0 \times 10^{-4}$ | $1.0 \times 10^{-2}$ | $1.30 \times 10^{-6}$ |
| C | $2.0 \times 10^{-4}$ | $0.5 \times 10^{-2}$ | $0.65 \times 10^{-6}$ |

## Solution

Comparison of the data from experiment A with that from experiment B shows that the rate is doubled when the concentration of $S_2O_8^{2-}$ is doubled. According to the data from experiments B and C, doubling the concentration of $I^-$ has the same effect on the reaction rate. The reaction, therefore, is first order in $S_2O_8^{2-}$, first order in $I^-$, and second order overall.

$$\text{rate} = k[S_2O_8^{2-}][I^-]$$

Note that the chemical equation for the reaction might lead one to expect a rate expression of third order.

The data from any of the experiments may be used to derive the value for the rate constant; $k = 0.65/M$ min.

## Example 13.2

The mechanism of the thermal decomposition of gaseous acetaldehyde, $CH_3CHO$, is complicated, and the order of the reaction changes with changes in the conditions under which the reaction is run. The following data pertain to the decomposition at 800°K. Find the order of the reaction under the conditions specified.

| Experiment | Initial Pressure of Acetaldehyde (atm) | Initial Rate of Increase in Total Pressure (atm/sec) |
|---|---|---|
| A | 0.450 | $1.04 \times 10^{-3}$ |
| B | 0.200 | $2.05 \times 10^{-4}$ |

### Solution

The rate of the thermal decomposition of acetaldehyde vapor

$$CH_3CHO(g) \rightarrow CH_4(g) + CO(g)$$

may be obtained by measuring the rate of increase in the total pressure of the system. At any time the total pressure, $P$, is equal to the sum of the partial pressures of all gases present.

$$P = p_{CH_3CHO} + p_{CH_4} + p_{CO}$$

If we let $a$ equal the original pressure of acetaldehyde and $x$ equal the decrease in the partial pressure of this gas, the partial pressure of acetaldehyde is $(a - x)$. At the same time, the partial pressures of $CH_4$ and $CO$ are each equal to $x$, since the decomposition of 1 mol of $CH_3CHO$ yields 1 mol of $CH_4$ and 1 mol of CO. Therefore,

$$P = (a - x) + x + x = a + x$$

Hence, the rate of *increase* in the total pressure is also the rate of *decrease* in the partial pressure of $CH_3CHO$ vapor and represents the rate of decomposition of $CH_3CHO$.

For a rate expression of the form,

$$\text{rate} = k[CH_3CHO]^n$$

$n$ represents the order of the reaction. If we take the logarithm of this equation, we derive

$$\log \text{rate} = \log k + n \log [CH_3CHO]$$

Both $k$ and $n$ constants, and for two different sets of data

$$\log \text{rate}_1 = \log k + n \log [CH_3CHO]_1$$

$$\log \text{rate}_2 = \log k + n \log [CH_3CHO]_2$$

Subtraction of the second equation from the first gives

$$\log \text{rate}_1 - \log \text{rate}_2 = n(\log [CH_3CHO]_1 - n\log [CH_3CHO]_2)$$

or

$$\log(\text{rate}_1/\text{rate}_2) = n \log([CH_3CHO]_1/[CH_3CHO]_2)$$

Therefore,

$$n = \frac{\log(\text{rate}_1/\text{rate}_2)}{\log([CH_3CHO]_1/[CH_3CHO]_2)}$$

Since the pressure of a gas is proportional to its concentration, we may substitute our values.

$$n = \frac{\log(1.04 \times 10^{-3}/2.05 \times 10^{-4})}{\log(0.450/0.200)} = \frac{\log 5.07}{\log 2.25} = \frac{0.705}{0.352} = 2.00$$

The reaction is second order under the conditions of the study and

$$\text{rate} = k[CH_3CHO]^2$$

**13.6 RATE EQUATIONS AND TEMPERATURE**

The rate constant, $k$, varies with temperature in a manner described by the following equation

$$k = Ae^{-E_a/RT}$$

where $A$ is a constant that is characteristic of the reaction being studied, $e$ is the base of natural logarithms (2.718), $E_a$ is the energy of activation for the reaction (in calories), $R$ is the molar gas constant (1.987 cal/mol °K), and $T$ is the absolute temperature. This equation was first proposed by Svanté Arrhenius in 1889 and is known as the **Arrhenius equation.**

● When $E_a$ values are expressed in J/mol, the value

$$R = 8.3143 \text{ J/K·mol}$$

is used in the Arrhenius equation.

The factor $e^{-E_a/RT}$ represents the fraction of molecules that have energies equal to or greater than the energy of activation, $E_a$, and may be derived from the Maxwell–Boltzmann distribution law (Sections 6.11 and 13.2). The constant $A$ incorporates terms for other factors that influence reaction rate, such as the frequency of molecular collisions and the geometric requirements for the alignment of colliding molecules that react. The equation is only approximate, but in most cases the approximation is a good one.

If we take the natural logarithm of the Arrhenius equation, we get

$$\ln k = \ln A - \frac{E_a}{RT}$$

which may be transformed into

$$2.303 \log k = 2.303 \log A - \frac{E_a}{RT}$$

or

$$\log k = \log A - \frac{E_a}{2.303RT}$$

---

*Example 13.3*

For the reaction

$$2N_2O_5(g) \rightarrow 4NO_2(g) + O_2(g)$$

$A$ is $4.3 \times 10^{13}$/sec, $E_a$ is 24,700 cal, and rate $= k[N_2O_5]$. What is the value of $k$ at 300°K?

*Solution*

$$\log k = \log A - \frac{E_a}{2.303RT}$$

$$= \log (4.3 \times 10^{13}) - \frac{24{,}700}{(2.30)(1.99)(300)}$$

$$= 13.63 - \frac{24{,}700}{1370}$$

$$= -4.34 = 0.64 - 5.00$$

$$k = 4.4 \times 10^{-5}/\text{sec}$$

**Table 13.2** Effect of temperature on some first-order rate constants

| REACTION | $A$ (1/sec) | $E_a$ (cal) | $k$ (1/sec) | $T$ (°K) |
|---|---|---|---|---|
| $2N_2O_5(g) \rightarrow 4NO_2(g) + O_2(g)$ | $4.3 \times 10^{13}$ | 24,700 | $4.4 \times 10^{-5}$ | 300 |
| | | | $1.7 \times 10^{-4}$ | 310 |
| $C_2H_5Cl(g) \rightarrow C_2H_4(g) + HCl(g)$ | $1.6 \times 10^{14}$ | 59,500 | $7.2 \times 10^{-30}$ | 300 |
| | | | $1.8 \times 10^{-28}$ | 310 |
| | | | $4.2 \times 10^{-5}$ | 700 |
| | | | $7.7 \times 10^{-5}$ | 710 |

Since the relation between $k$ and $T$ is exponential, a small change in $T$ causes a relatively large change in $k$. For example, as the data of Table 13.2 show, a 10° rise in temperature (from 300°K to 310°K) causes an approximately fourfold increase in the rate of decomposition of $N_2O_5$. The larger the activation energy, the greater the effect of a given temperature rise on $k$. For example, $E_a$ for the decomposition of $C_2H_5Cl$ is more than double $E_a$ for the decomposition of $N_2O_5$, and increasing the temperature from 300°K to 310°K causes the theoretical rate of decomposition of $C_2H_5Cl$ to increase by a factor of about 25 (Table 13.2).

The values of $k$ for the decomposition of $C_2H_5Cl$ at 300°K and 310°K

**Figure 13.9** Plot of log $k$ versus $1/T$ for the reaction $2NOCl(g) \rightarrow 2NO(g) + Cl_2(g)$.

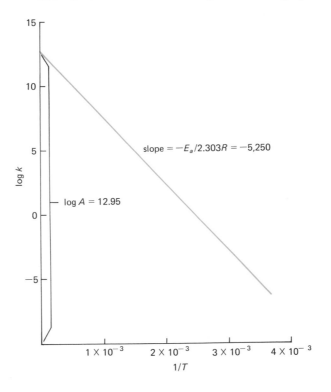

in Table 13.2 are of theoretical significance only; they are so small that the reaction would not occur at an appreciable rate at these temperatures. Nevertheless, the data of Table 13.2 illustrate not only the effect of the magnitude of $E_a$ on the variation of $k$ and $T$ but also the fact that the effect of a 10° temperature rise is greater at low temperatures than at high temperatures. Whereas an increase from 300°K to 310°K causes an approximate 25-fold increase in $k$ for the $C_2H_5Cl$ decomposition, an increase from 700°K to 710°K approximately doubles $k$.

Since the equation

$$\log k = -\frac{E_a}{2.303\,R}\left(\frac{1}{T}\right) + \log A$$

is in the form of the equation for a straight line ($y = mx + b$), a plot of $\log k$ against $1/T$ is a straight line with a slope of $-E_a/2.303R$ and a $y$-intercept of $\log A$ (Figure 13.9). Thus, if values of $k$ for a reaction are determined at several temperatures and the data plotted in this manner, $E_a$ for the reaction may be calculated from the slope of the curve and $A$ may be obtained by taking the antilogarithm of the intercept. For the curve shown in Figure 13.9, $E_a$ is 24,000 cal and $A$ is $9.0 \times 10^{12}$ liter/mol sec. The values of $E_a$ and $A$ are usually obtained from experimental data either graphically or by the method illustrated in the following example.

*Example 13.4*

For the reaction

$$2NOCl(g) \rightarrow 2NO(g) + Cl_2(g)$$

rate $= k[NOCl]^2$, and $k$ is $2.8 \times 10^{-5}$ liter/mol sec at 300°K and $7.0 \times 10^{-1}$ liter/mol sec at 400°K. What is the energy of activation for the reaction?

*Solution*

For two temperatures $T_1$ and $T_2$, where the corresponding rate constants are $k_1$ and $k_2$, we may write

$$\log k_2 = \log A - \frac{E_a}{2.303\,R T_2}$$

and

$$\log k_1 = \log A - \frac{E_a}{2.303\,R T_1}$$

Subtraction of the second equation from the first gives

$$\log k_2 - \log k_1 = \frac{E_a}{2.303\,R T_1} - \frac{E_a}{2.303\,R T_2}$$

or

$$\log\left(\frac{k_2}{k_1}\right) = \frac{E_a}{2.303\,R}\left(\frac{T_2 - T_1}{T_1 T_2}\right)$$

If this equation is solved for $E_a$, we get

$$E_a = 2.303\,R\left(\frac{T_1 T_2}{T_2 - T_1}\right)\log\left(\frac{k_2}{k_1}\right)$$

In this case,

$$E_a = 2.30(1.99)\left(\frac{(300)(400)}{(400-300)}\right)\log\left(\frac{7.0 \times 10^{-1}}{2.8 \times 10^{-5}}\right)$$

$$= 5490 \log (2.5 \times 10^4) = 5490(4.38)$$

$$= 24,000 \text{ cal} = 24 \text{ kcal}$$

**13.7 REACTION
MECHANISMS AND
RATE EQUATIONS**

Mechanisms of chemical reactions are proposed on the basis of experimentally derived rate equations and other experimental evidence (e.g., the detection of short-lived reaction intermediates). It is practically impossible to be completely certain of the mechanism of any reaction; mechanisms are only plausible hypotheses consistent with all known facts.

Table 13.1 lists the uncatalyzed decomposition of $N_2O$ as a second-order reaction, whereas the gold-catalyzed decomposition of the same compound is listed as a first-order reaction. The uncatalyzed reaction is thought to be a simple one-step process involving bimolecular collisions (Section 13.5).

$$2N_2O \rightarrow 2N_2 + O_2$$

Such a mechanism would call for a second-order rate expression.

$$rate = k[N_2O]^2$$

On the other hand, the gold-catalyzed decomposition (Section 13.3, Figure 13.7) is thought to occur by a three-step mechanism in which $N_2O$ molecules are chemisorbed on the surface of the gold in the first step.

(i) $\qquad\qquad N_2O(g) \rightarrow N_2O(\text{on Au})$

(ii) $\qquad\qquad N_2O(\text{on Au}) \rightarrow N_2(g) + O(\text{on Au})$ (slow)

(iii) $O(\text{on Au}) + O(\text{on Au}) \rightarrow O_2(g)$ $\qquad$ (rapid)

The rate expression for either step (i) or step (ii) is first order. The rate of the chemisorption is determined by the pressure of gaseous $N_2O$, and the rate of step (ii) is determined by the number of chemisorbed $N_2O$ molecules, which in turn is determined by the pressure of gaseous $N_2O$. There is evidence however, that step (ii) is the slower process and as such is the **rate-determining step.** Step (iii) occurs much more rapidly than step (ii). As fast as (ii) produces O atoms, step (iii) uses them up. The second step, therefore, is the bottleneck and determines the overall rate of the entire decomposition.

At relatively high pressures of $N_2O$ gas, the gold-catalyzed decomposition is observed to be **zero-order;** that is, the rate is presumably unaffected by the concentration of gaseous $N_2O$. When the surface of the gold holds all the $N_2O$ molecules that it can, the rate of decomposition is constant. Under these conditions, small changes in the pressure of gaseous $N_2O$ do not cause the chemisorbed $N_2O$ molecules to decompose any more slowly or rapidly, and there is always a supply of gaseous $N_2O$ molecules sufficient to keep the surface saturated.

The decomposition of $NO_2$ is second order.

$$2NO_2(g) \rightarrow 2NO(g) + O_2(g)$$

$$rate = k[NO_2]^2$$

Even though the chemical equation for the uncatalyzed decomposition of $N_2O_5$ is very similar,

$$2N_2O_5(g) \rightarrow 4NO_2(g) + O_2(g)$$

this reaction has been found to be first order.

$$\text{rate} = k[N_2O_5]$$

We could account for the observed rate expression for the decomposition of $N_2O_5$ by visualizing a simple one-step unimolecular decomposition or a rate-determining unimolecular decomposition followed by one or more rapid steps.

However, there is evidence that the following is the mechanism for the decomposition of $N_2O_5$, with the third step being the slowest one.

(i) $\qquad N_2O_5 \xrightarrow{k_1} NO_2 + NO_3$

(ii) $NO_2 + NO_3 \xrightarrow{k_2} N_2O_5$

(iii) $NO_2 + NO_3 \xrightarrow{k_3} NO + O_2 + NO_2$

(iv) $NO + NO_3 \xrightarrow{k_4} 2NO_2$

In the first step the $N_2O_5$ molecule splits into two fragments, $NO_2$ and $NO_3$. The second step is the reverse of the first; $NO_2$ and $NO_3$ recombine to give the starting molecule. In the third step, $NO_2$ and $NO_3$ react in a different way from step (ii), and in the collision of $NO_2$ and $NO_3$, each molecule gives up an O atom to form $O_2$, so that NO and $NO_2$ are left. Step (iv) is a bimolecular collision of NO and $NO_3$.

The substances NO and $NO_3$ are **reaction intermediates;** they do not appear as final products. The molecule NO is a well-known oxide of nitrogen, but $NO_3$ is an unstable substance that does not exist for any length of time under ordinary conditions. The detection of $NO_3$ in systems in which $N_2O_5$ is decomposing is one of the lines of evidence used to support the mechanism.

We can reconcile the mechanism with the experimentally determined, first-order rate expression by the use of a **steady-state approximation.** It is assumed that, after the reaction has proceeded for a while, each of the reaction intermediates, NO and $NO_3$, attains a steady state in which its concentration is approximately constant and, therefore, the *net* rate of change in its concentration is zero. The concentrations of the intermediates are low at all times, and the *net* rate of change in the concentration of an intermediate is negligible in comparison to the rate at which the intermediate is produced or consumed; therefore, the approximation is a good one.

Applying the law of mass action to the appropriate steps, we derive an expression for the rate of increase in the concentration of NO which, according to the steady-state approximation, is equal to zero. Nitrogen oxide is *produced* in the third step at a rate equal to $+k_3[NO_2][NO_3]$; it is *used* in the fourth step at a rate equal to $-k_4[NO][NO_3]$. Therefore the rate of *production* of NO is

$$+k_3[NO_2][NO_3] - k_4[NO][NO_3] = 0$$

From this we derive

$$[NO] = \frac{k_3}{k_4}[NO_2] \tag{1}$$

In like manner, we can derive an expression for the net rate of *production* of $NO_3$, which is produced in step (i) and used in steps (ii), (iii), and (iv).

$$+k_1[N_2O_5] - k_2[NO_2][NO_3] - k_3[NO_2][NO_3] - k_4[NO][NO_3] = 0$$

Solving for $[NO_3]$, we get

$$[NO_3] = \frac{k_1[N_2O_5]}{(k_2 + k_3)[NO_2] + k_4[NO]}$$

If we substitute equation (1) into this expression, we derive

$$[NO_3] = \frac{k_1[N_2O_5]}{(k_2 + 2k_3)[NO_2]} \tag{2}$$

We are now ready to find a single rate equation for the overall reaction in terms of the concentration of the starting material, $N_2O_5$. From the equations for the first two steps of the mechanism,

$$\text{rate of } decrease \text{ in } [N_2O_5] = +k_1[N_2O_5] - k_2[NO_2][NO_3]$$

Substitution of (2) yields

$$\text{rate of } decrease \text{ in } [N_2O_5] = k_1[N_2O_5] - k_2[NO_2]\left(\frac{k_1[N_2O_5]}{(k_2 + 2k_3)[NO_2]}\right)$$

$$= \frac{2k_1k_3}{k_2 + 2k_3}[N_2O_5]$$

If we combine all the rate constants into a single constant, we get

$$\text{rate of } decrease \text{ in } [N_2O_5] = k[N_2O_5]$$

which is the experimentally determined rate equation.

The mathematical treatment of mechanisms of other chemical reactions may be much more involved; in view of the complex nature of the mechanisms of some reactions, it is not surprising that data obtained from some experimental studies yield complicated rate equations, which may be of fractional order.

The study of reactions that are initiated by means of light is called **photochemistry.** The most famous example of a photochemical reaction is **photosynthesis,** which occurs in plants and converts carbon dioxide and water into plant carbohydrates. The green coloring matter in plants, chlorophyll, acts as a catalyst for photosynthesis.

Light also initiates the reaction between hydrogen and chlorine. The primary, light-induced step of the reaction is the dissociation of chlorine molecules into chlorine atoms.

$$Cl_2 \rightarrow 2Cl$$

These chlorine atoms initiate a **chain reaction** in which the cycle

$$Cl + H_2 \rightarrow HCl + H$$

$$H + Cl_2 \rightarrow HCl + Cl$$

is repeated again and again. A chain may be terminated by such reactions as

$$2H \rightarrow H_2$$

$$2Cl \rightarrow Cl_2$$

or

$$H + Cl \rightarrow HCl$$

Chain reactions usually proceed very rapidly. Many explosive reactions occur by a chain mechanism; atomic fission and atomic fusion are examples of this type. In the reaction of hydrogen and chlorine, the chain is sustained by the chlorine atoms and the hydrogen atoms; these extremely reactive intermediates are called chain propagators.

A branched-chain reaction, such as that between hydrogen and oxygen, is a reaction in which one or more of the steps produce more than one chain propagator. Following are some of the steps that have been postulated for the reaction between $H_2$ and $O_2$.

(i)   $H_2 + O_2 \rightarrow 2OH$

(ii)  $OH + H_2 \rightarrow H_2O + H$

(iii)  $H + O_2 \rightarrow OH + O$

(iv)   $O + H_2 \rightarrow OH + H$

Step (i) is a chain-initiating step, and step (ii) merely sustains the chain. The chain propagators are OH, H, and O. Steps (iii) and (iv) are chain-branching steps, and since each of these steps produces *two* chain propagators, *two* reactions follow from either reaction (iii) or reaction (iv). The result is a rapid increase in the overall rate of the reaction. Chains terminate upon collisions of the chain propagators—usually at the walls of the container. Generally, a third body is necessary in these collisions to absorb the energy evolved by the highly exothermic recombination.

**13.8 REVERSIBLE REACTIONS AND CHEMICAL EQUILIBRIUM**

In Sections 13.1 and 13.2 we discussed the reversible reaction

$$A_2(g) + B_2(g) \rightleftharpoons 2AB(g)$$

This equation may be read either forward or backward. If $A_2$ and $B_2$ are mixed, they react to produce AB. The rate of this forward reaction declines as the reaction proceeds, since $A_2$ and $B_2$ are used up and their concentrations decrease. If no AB is present at the start, the backward reaction is initially impossible. However, the forward reaction produces AB, and the backward reaction begins soon after the $A_2$ and $B_2$ are mixed. The backward reaction starts slowly (the concentration of AB is low) but picks up speed as the concentration of AB is built up by the more rapid forward reaction.

With time, the forward reaction rate decreases and the backward reaction rate increases until eventually a condition is reached in which the two rates are equal. At this point chemical equilibrium is established, and the concentrations of all chemical species are constant. The rate at which AB is produced by the forward reaction equals the rate at which AB is consumed by the backward reaction. In like manner, $A_2$ and $B_2$ are produced at rates exactly equal to the rates at which they are used.

It is important to note that equilibrium is a dynamic condition; equilibrium concentrations are constant because the rates of the opposing reactions are equal and not because all activity has ceased. Typical data for this reaction are plotted in Figure 13.10; equilibrium is attained at $t_e$.

The rate of the forward reaction is given by

$$rate_f = k[A_2][B_2]$$

and the rate of the backward reaction is

$$rate_b = k'[AB]^2$$

At equilibrium these rates are equal, and

$$k[A_2][B_2] = k'[AB]^2$$

or

$$\frac{k}{k'} = \frac{[AB]^2}{[A_2][B_2]}$$

Figure 13.10 Curves show-
ing changes in concentra-
tions of materials with time
for the reaction $A_2 + B_2 \rightleftharpoons$
2AB; equilibrium is attained
at time $t_e$.

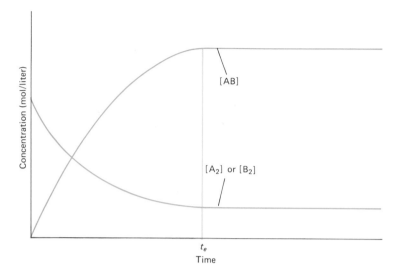

The rate constant of the forward reaction, $k$, divided by the rate constant of the backward reaction, $k'$, is equal to a third constant—the **equilibrium constant** $K$.

$$K = \frac{[AB]^2}{[A_2][B_2]}$$

The numerical value of $K$ varies with temperature. There are an infinite number of possible equilibrium systems for this reaction. However, the concentrations of $A_2$, $B_2$, and AB for any system in equilibrium will, when expressed in the preceding manner, equal $K$ for the particular temperature under consideration.

In general, for any reversible reaction

$$w\text{W} + x\text{X} + \cdots \rightleftharpoons y\text{Y} + z\text{Z} + \cdots$$

at equilibrium,

$$K = \frac{[\text{Y}]^y[\text{Z}]^z \cdots}{[\text{W}]^w[\text{X}]^x \cdots}$$

By convention, the concentration terms for the materials on the right of the chemical equation are written in the numerator of the expression for the equilibrium constant. If the equation is written in reverse form, the equilibrium constant becomes the reciprocal of that for the original equation. The preceding equation constitutes a statement of the **law of chemical equilibrium,** which Guldberg and Waage derived from their law of mass action (Section 13.5).

In the rate equation for the reaction of $A_2$ and $B_2$ and the rate equation for the decomposition of AB, the exponents of the concentrations correspond to the coefficients of the balanced chemical equation. In this instance both the forward and backward reactions occur through single-step bimolecular collisions, and the rate expressions may be derived from the chemical equation. It is not immediately apparent that the law of chemical equilibrium holds for a reaction that occurs by means of a mechanism of more than one step. However, an equilibrium constant may be derived for any reversible chemical change on the basis of the overall equation for the change.

For example, the forward reaction

$$2NO_2Cl \rightleftharpoons 2NO_2 + Cl_2$$

is first order; $rate_f = k[NO_2Cl]$. The reaction is assumed to follow the mechanism

$$NO_2Cl \underset{k_1'}{\overset{k_1}{\rightleftharpoons}} NO_2 + Cl$$

$$NO_2Cl + Cl \underset{k_2'}{\overset{k_2}{\rightleftharpoons}} NO_2 + Cl_2$$

Both steps of the mechanism must be reversible, and when equilibrium is established for the overall change, each step of the mechanism must exist in equilibrium. Thus

$$K_a = \frac{k_1}{k_1'} = \frac{[NO_2][Cl]}{[NO_2Cl]}$$

and

$$K_b = \frac{k_2}{k_2'} = \frac{[NO_2][Cl_2]}{[NO_2Cl][Cl]}$$

By combining these expressions, we get

$$K_a K_b = \frac{k_1 k_2}{k_1' k_2'} = \frac{[NO_2][Cl]}{[NO_2Cl]} \frac{[NO_2][Cl_2]}{[NO_2Cl][Cl]} = \frac{[NO_2]^2[Cl_2]}{[NO_2Cl]^2}$$

which is the same as the expression for the equilibrium constant that would have been derived directly from the equation for the overall change. In this case, the equilibrium constant for the overall change is the product of the equilibrium constants of each of the steps.

**13.9 EQUILIBRIUM CONSTANTS**

For the reaction

$$H_2(g) + I_2(g) \rightleftharpoons 2HI(g)$$

the equilibrium constant at 425°C is

$$K = \frac{[HI]^2}{[H_2][I_2]} = 54.8$$

The numerical value of $K$ must be determined experimentally. At 425°C the concentrations of the materials present in any equilibrium mixture when expressed in the manner prescribed by the equilibrium constant will equal 54.8; if this is not the case, the mixture is not in equilibrium.

The equilibrium condition may be approached from either direction; that is, an equilibrium mixture can be obtained by mixing hydrogen and iodine, by allowing pure hydrogen iodide to dissociate, or by mixing all three materials.

The magnitude of the value of the equilibrium constant gives an indication of the position of equilibrium. Recall that the concentration terms of materials on the right of the equation are written in the numerator of the expression for the equilibrium constant. For the reaction

$$CO(g) + Cl_2(g) \rightleftharpoons COCl_2(g)$$

at 100°C,

$$K = \frac{[COCl_2]}{[CO][Cl_2]} = 4.57 \times 10^9 \text{ liters/mol}$$

From this relatively large value of $K$, we conclude that equilibrium concentrations of CO and $Cl_2$ are small and that the synthesis of $COCl_2$ is virtually complete. In other words, the reaction to the right is fairly complete at equilibrium.

For the reaction

$$N_2(g) + O_2(g) \rightleftharpoons 2NO(g)$$

at 2000°C,

$$K = \frac{[NO]^2}{[N_2][O_2]} = 4.08 \times 10^{-4}$$

We conclude from this small value of $K$ that NO is largely dissociated into $N_2$ and $O_2$ at equilibrium; the reaction to the left is fairly complete.

Equilibria between substances in two or more phases are called **heterogeneous equilibria.** The concentration of a pure solid or a pure liquid is actually its density and, a density is constant under constant conditions. Hence, for any heterogeneous equilibrium, the values of the concentrations of solids or liquids involved are included in the value of $K$, and concentration terms for these substances do not appear in the expression for the equilibrium constant.

For example, for the reaction

$$CaCO_3(s) \rightleftharpoons CaO(s) + CO_2(g)$$

the values for the concentrations of CaO and $CaCO_3$ are included in the value of $K$, and the expression for the equilibrium constant is

$$K = [CO_2]$$

Hence, at any temperature the equilibrium concentration of $CO_2$ over a mixture of the solids is a definite value. The equilibrium constant for the reaction

$$3Fe(s) + 4H_2O(g) \rightleftharpoons Fe_3O_4(s) + 4H_2(g)$$

is expressed in the following terms

$$K = \frac{[H_2]^4}{[H_2O]^4}$$

Since the partial pressure of a gas is a measure of its concentration, equilibrium constants for reactions involving gases may be given in terms of the partial pressures of the reacting gases; such an equilibrium constant is given the designation $K_p$. For the calcium carbonate equilibrium, the constant in terms of partial pressures is

$$K_p = p_{CO_2}$$

For the equilibrium

$$N_2(g) + 3H_2(g) \rightleftharpoons 2NH_3(g)$$

the $K_p$ is

$$K_p = \frac{(p_{NH_3})^2}{(p_{N_2})(p_{H_2})^3}$$

There is a simple relation between the $K_p$ for a reaction and the equilibrium constant derived from concentrations. Consider the reaction

$$wW + xX \rightleftharpoons yY + zZ$$

If all these materials are gases,

$$K_p = \frac{(p_Y)^y(p_Z)^z}{(p_W)^w(p_X)^x}$$

If we assume that these gases follow the ideal gas law, the partial pressure of any gas is

$$p = \frac{n}{V} RT$$

The concentration of a gas in moles per liter is equal to $n/V$. Hence, for W

$$p_W = [W]RT$$

If we substitute terms such as this for the partial pressures in the expression for $K_p$, we get

$$K_p = \frac{[Y]^y(RT)^y[Z]^z(RT)^z}{[W]^w(RT)^w[X]^x(RT)^x} = K(RT)^{+y+z-w-x}$$

We shall designate the quantity $(+y + z - w - x)$, which is equal to the number of moles of products minus the number of moles of reactants, as $\Delta n$. Therefore,

$$K_p = K(RT)^{\Delta n}$$

or

$$K = K_p(RT)^{-\Delta n}$$

Generally, partial pressures are expressed in atmospheres; concentrations are expressed in moles per liter; $R$ is 0.08206 liter atm/°K mol; and $T$ is the absolute temperature.

It is obvious that $K = K_p$ when $\Delta n = 0$, as in the reaction

$$H_2(g) + I_2(g) \rightleftharpoons 2HI(g)$$

To be strictly accurate, equations for equilibrium constants should be given in terms of activities rather than concentrations or partial pressures. The activity of a substance may be regarded as its effective concentration and may be obtained by multiplying the actual concentration by an activity coefficient (Section 9.13). By this means, deviations, which are due principally to the existence of intermolecular forces of attraction, may be minimized. However, at low concentrations and pressures up to a few atmospheres, ideality may be assumed, and concentrations may be used with reasonable accuracy.

---

**Example 13.5**

For the reaction

$$N_2O_4(g) \rightleftharpoons 2NO_2(g)$$

the concentrations of an equilibrium mixture at 25°C are $[N_2O_4] = 4.27 \times 10^{-2}$ mol/liter and $[NO_2] = 1.41 \times 10^{-2}$ mol/liter. What is $K$?

**Solution**

$$K = \frac{[NO_2]^2}{[N_2O_4]}$$

$$= \frac{(1.41 \times 10^{-2} \text{ mol/liter})^2}{(4.27 \times 10^{-2} \text{ mol/liter})}$$

$$= 4.66 \times 10^{-3} \text{ mol/liter}$$

---

**Example 13.6**

What is $K$ for the reaction

$$N_2(g) + 3H_2(g) \rightleftharpoons 2NH_3(g)$$

at 500°C if $K_p = 1.50 \times 10^{-5}$/atm² at this temperature?

**Solution**

For the reaction, $\Delta n = -2$

$$K = K_p(RT)^{-\Delta n}$$

$$= (1.50 \times 10^{-5}/\text{atm}^2)[(0.08206 \text{ liter atm/°K mol})(773°K)]^2$$

$$= 6.04 \times 10^{-2} \text{ liter}^2/\text{mol}^2$$

*Example 13.7*

$K$ for the reaction

$$H_2(g) + CO_2(g) \rightleftharpoons H_2O(g) + CO(g)$$

at 750°C is 0.771. If 0.0100 mol of $H_2$ and 0.0100 mol of $CO_2$ are mixed in a 1 liter container at 750°C, what are the concentrations of all substances at equilibrium?

*Solution*

If $x$ mol of $H_2$ reacts with $x$ mol of $CO_2$, $x$ mol of $H_2O$ and $x$ mol of CO will be produced. Since the container has a volume of 1 liter, at equilibrium the concentrations are (in mol/liter):

$$\begin{array}{cccc} H_2(g) & + & CO_2(g) & \rightleftharpoons H_2O(g) + CO(g) \\ (0.0100 - x) & & (0.0100 - x) & \quad x \qquad\qquad x \end{array}$$

$$K = \frac{[H_2O][CO]}{[H_2][CO_2]} = 0.771$$

$$= \frac{x^2}{(0.0100 - x)(0.0100 - x)} = 0.771$$

If we extract the square root of both sides of this equation, we get

$$\frac{x}{(0.0100 - x)} = 0.878$$

$$x = 0.00468$$

Therefore, at equilibrium

$$[H_2] = [CO_2] = (0.0100 - x)$$
$$= 0.00532 \text{ mol/liter}$$
$$= 5.32 \times 10^{-3} \text{ mol/liter}$$

$$[H_2O] = [CO] = x = 0.00468 \text{ mol/liter}$$
$$= 4.68 \times 10^{-3} \text{ mol/liter}$$

---

**13.10 LE CHATELIER'S PRINCIPLE**

The effect of an alteration in a reaction condition on a system in equilibrium was summarized in 1884 by Henri Le Chatelier. **Le Chatelier's principle** may be stated as: an alteration in any condition that determines the state of a system in equilibrium will cause the position of equilibrium to shift in a manner that tends to counteract the alteration. This important and powerful generalization is very simple to apply.

Alterations in three conditions cause changes in the position of equilibrium: concentration, pressure, and temperature. Of these three, only a change in temperature will affect the value of the equilibrium constant. The presence or absence of a catalyst has no effect on the position of equilibrium, since a catalyst will affect the rate of the forward and backward reaction to an equal extent (Section 13.3). However, the introduction of a catalyst into a system that is not in equilibrium will cause the system to attain equilibrium more rapidly than it otherwise would.

If we have a system in equilibrium,

$$H_2(g) + I_2(g) \rightleftharpoons 2HI(g)$$

and we increase the concentration of $H_2$, the equilibrium will be upset and the system will react to establish a new equilibrium condition. When equilibrium is again attained; the concentration of HI will be increased over its initial value, the concentration of $I_2$ will be decreased in comparison to its initial value, and the concentration of $H_2$ will be lower than its value after the addition. In summary, the reaction will proceed to the right, thereby partially counteracting the alteration and decreasing the concentration of $H_2$.

If additional HI is added to a system in equilibrium, the reaction will proceed to the left, thus increasing the concentrations of $H_2$ and $I_2$ and using up HI. Removal of one of the substances from an equilibrium system also causes the position of equilibrium to shift. For example, removal of HI would cause the reaction to proceed to the right, thus producing more HI and decreasing the concentrations of $H_2$ and $I_2$.

*Example 13.8*

$K$ for the HI equilibrium

$$H_2(g) + I_2(g) \rightleftharpoons 2HI(g)$$

is 54.8 at 425°C, and for a particular equilibrium system the concentrations are: $[H_2] = [I_2] = 0.010$ mol/liter and $[HI] = 0.074$ mol/liter. If the concentration of HI is momentarily increased to 0.100 mol/liter, what will be the concentrations of all substances when a new equilibrium is established?

*Solution*

If we consider 1 liter of the reaction mixture and let $2x$ be the number of additional moles of HI dissociated at equilibrium, then $x$ additional moles of $H_2$ and $x$ additional moles of $I_2$ will be formed at equilibrium. The new equilibrium concentrations will be (in mol/liter):

$$H_2(g) \quad + \quad I_2(g) \quad \rightleftharpoons \quad 2HI(g)$$
$$(0.010+x) \qquad (0.010+x) \qquad (0.100-2x)$$

$$K = \frac{[HI]^2}{[H_2][I_2]} = 54.8$$

$$\frac{(0.100-2x)^2}{(0.010+x)^2} = 54.8$$

$$\frac{(0.100-2x)}{(0.010+x)} = 7.4$$

$$x = 0.0028$$

The new equilibrium concentrations are

$$[H_2] = [I_2] = 0.010 + x = 0.013 \text{ mol/liter}$$

$$[HI] = 0.100 - 2x = 0.094 \text{ mol/liter}$$

Le Chatelier's principle predicts that the position of equilibrium will shift to the left for a change of this type; this prediction is borne out by the calculation.

Le Chatelier's principle may also be used to make qualitative predictions of the effect of pressure changes on systems in equilibrium. Consider the effect of a pressure increase on an equilibrium mixture of $SO_2$, $O_2$, and $SO_3$.

$$2SO_2(g) + O_2(g) \rightleftharpoons 2SO_3(g)$$

In the forward reaction two gas molecules ($2SO_3$) are produced by the disappearance of three gas molecules ($2SO_2 + O_2$); two gas molecules do not exert as high a pressure as three gas molecules. Hence, when the pressure on an equilibrium mixture is increased (or the volume of the system decreased), the position of equilibrium shifts to the right and thus counteracts the change. Alternatively, decreasing the pressure (or increasing the volume) would cause the position of this equilibrium to shift to the left.

For reactions in which $\Delta n = 0$, pressure changes have no effect on the equilibria. For example, equilibria involving the systems

$$H_2(g) + I_2(g) \rightleftharpoons 2HI(g)$$

$$N_2(g) + O_2(g) \rightleftharpoons 2NO(g)$$

$$H_2(g) + CO_2(g) \rightleftharpoons H_2O(g) + CO(g)$$

are not influenced by changing the pressure, since there is no difference in the total volume in either the forward or backward direction for any one of these reactions.

For a system that involves only liquids and solids, the effect of pressure on the position of equilibrium is slight and may usually be ignored for ordinary changes in pressure. However, large pressure changes can significantly alter such equilibria, and, at times, even slight changes in such equilibria are of interest. For example, the equilibrium

$$H_2O(s) \rightleftharpoons H_2O(l)$$

is forced to the right by an increase in pressure because a given quantity of water occupies a smaller volume in the liquid state than in the solid state (its density is higher in the liquid state).

Pressure changes affect equilibria involving gases to a much greater degree. For example, a high pressure would favor the production of a high yield of ammonia from the equilibrium

$$N_2(g) + 3H_2(g) \rightleftharpoons 2NH_3(g)$$

Hence, Le Chatelier's principle is of practical importance as an aid in determining favorable reaction conditions for the production of a desired substance.

For heterogeneous equilibria the effect of pressure is predicted by counting the number of *gas* molecules on each side of the equation. For example, the equilibrium

$$3Fe(s) + 4H_2O(g) \rightleftharpoons Fe_3O_4(s) + 4H_2(g)$$

is virtually unaffected by pressure because there are four gas molecules on each side of the equation.

*Example 13.9*

For the equilibrium

$$PCl_5(g) \rightleftharpoons PCl_3(g) + Cl_2(g)$$

$K = 4.16 \times 10^{-2}$ mol/liter at 250°C. In a given system at 250°C, the equilibrium concentrations are $[PCl_5] = 1$ mol/liter and $[PCl_3] = [Cl_2] = 0.204$ mol/liter. If the pressure is reduced by half (volume doubled), what are the new equilibrium concentrations?

*Solution*

One would predict by Le Chatelier's principle that a decrease in pressure would favor the forward reaction, since there are two gas molecules on the right and only one gas molecule on the left. Immediately after the initial change the concentrations are each cut in half. For example, the concentration of $PCl_5$, instead of being 1 mol/liter, is now 1 mol/2 liters, or 0.5 mol/liter.

If $x$ additional mol/liter of $PCl_5$ has dissociated when equilibrium is reestablished, $x$ mol/liter of $PCl_3$ and $x$ mol/liter of $Cl_2$ have been formed. The new equilibrium concentrations are (in mol/liter):

$$\underset{(0.500 - x)}{PCl_5(g)} \rightleftharpoons \underset{(0.102 + x)}{PCl_3(g)} + \underset{(0.102 + x)}{Cl_2(g)}$$

$$K = \frac{[PCl_3][Cl_2]}{[PCl_5]} = 4.16 \times 10^{-2} \text{ mol/liter}$$

$$\frac{(0.102 + x)^2}{(0.500 - x)} = 4.16 \times 10^{-2} \text{ mol/liter}$$

$$x^2 + 0.246x - 0.0104 = 0$$

By use of the quadratic formula,[1]

$$x = 0.0365$$

Therefore the new equilibrium concentrations are

$$[PCl_5] = 0.464 \text{ mol/liter}$$

$$[PCl_3] = [Cl_2] = 0.138 \text{ mol/liter}$$

In order to predict the effect of a temperature change on a system in equilibrium, the thermochemical nature of the chemical reaction must be known. At 25°C the thermochemical equation for the synthesis of ammonia is

$$N_2(g) + 3H_2(g) \rightleftharpoons 2NH_3(g) \qquad \Delta H = -22.1 \text{ kcal}$$

The forward reaction is exothermic, and the backward reaction is endothermic. If the temperature of an equilibrium mixture is raised, the equilibrium will shift to the left — the direction that absorbs heat. Lowering the temperature favors the reaction to the right in which heat is produced. We conclude that the highest yields of ammonia are obtained at the lowest temperatures.

Thermochemical equations for reversible reactions may also be written so that the pertinent $\Delta H$ value has a positive sign. Increasing the termperature *always* favors the endothermic change, and decreasing the temperature *always* favors the exothermic change. The reaction ($\Delta H$ measured at 25°C)

---

[1] For an equation in the form $ax^2 + bx + c = 0$, $x = \dfrac{-b \pm \sqrt{b^2 - 4ac}}{2a}$.

$$CO_2(g) + H_2(g) \rightleftharpoons CO(g) + H_2O(g) \qquad \Delta H = +9.83 \text{ kcal}$$

is forced to the right by an increase in temperature.

The numerical value of the equilibrium constant changes with a change in temperature. For the preceding reaction the position of equilibrium is displaced to the right by an increase in temperature. Thus, the concentrations of the materials on the right are increased; concentration terms for these materials appear in the numerator of the expression for $K$. Hence, as the temperature is increased, the value of $K$ for this reaction increases; at 700°C, $K = 0.63$, and at 1000°C, $K = 1.66$.

An equation relating the equilibrium constant and temperature may be derived from the Arrhenius equation (Section 13.6). The Arrhenius equation for the rate constant for the forward reaction is

$$k = Ae^{-E_a/RT}$$

and the corresponding equation for the reverse equation is

$$k' = A'e^{-E'_a/RT}$$

Since the equilibrium constant, $K$, is equal to $k/k'$

$$K = \frac{Ae^{-E_a/RT}}{A'e^{-E'_a/RT}}$$

or

$$K = Ce^{(-E_a + E'_a)/RT}$$

Since $\Delta E$ for the forward reaction, or $\Delta H$ if the pressure is held constant, is equal to $E_a - E'_a$ (Section 13.2), we may write

$$-\Delta H = -E_a + E'_a$$

Therefore,

$$K = Ce^{-\Delta H/RT}$$

or

$$\ln K = \ln C - \frac{\Delta H}{RT}$$

$$\log K = \log C - \frac{\Delta H}{2.303RT}$$

For two temperatures $T_1$ and $T_2$ for which the equilibrium constants are $K_1$ and $K_2$,

$$\log K_2 = \log C - \frac{\Delta H}{2.303RT_2}$$

$$\log K_1 = \log C - \frac{\Delta H}{2.303RT_1}$$

Subtraction of the second equation from the first gives

$$\log K_2 - \log K_1 = -\frac{\Delta H}{2.303 R T_2} + \frac{\Delta H}{2.303 R T_1}$$

or

$$\log \left(\frac{K_2}{K_1}\right) = \frac{\Delta H}{2.303 R} \left(\frac{1}{T_1} - \frac{1}{T_2}\right)$$

which is

$$\log \left(\frac{K_2}{K_1}\right) = \frac{\Delta H}{2.303 R} \left(\frac{T_2 - T_1}{T_1 T_2}\right)$$

Since $\Delta H$ varies with temperature, the equation is only approximate. However, over a narrow temperature range the value of $\Delta H$ may be assumed to be constant and an average value for the temperature range used.

Notice the similarity between this equation and the Clausius–Clapeyron equation (Section 7.5). The vapor pressure of a substance is a type of $K_p$ that pertains to an equilibrium system composed of the substance and its vapor.

*Example 13.10*
For the reaction

$$CO_2(g) + H_2(g) \rightleftharpoons CO(g) + H_2O(g)$$

$K$ is 0.63 at 700°C and 1.66 at 1000°C. (a) What is the average $\Delta H$ for the temperature range considered? (b) What is the value of $K$ at 800°C?

*Solution*
(a) Let $T_1 = 973°K$, $K_1 = 0.63$, $T_2 = 1273°K$, and $K_2 = 1.66$.

$$\log \left(\frac{1.66}{0.63}\right) = \frac{\Delta H}{2.30(1.99)} \left(\frac{1273 - 973}{(973)(1273)}\right)$$

$$0.421 = \frac{\Delta H}{4.58} \left(\frac{3 \times 10^2}{1.24 \times 10^6}\right)$$

$$\Delta H = 8.0 \times 10^3 \text{ cal}$$

(b) Let the unknown equilibrium constant be $K_2$, $T_2 = 1073°K$, $K_1 = 0.63$, and $T_1 = 973°K$.

$$\log \left(\frac{K_2}{0.63}\right) = \frac{8.0 \times 10^3}{2.30(1.99)} \left(\frac{1073 - 973}{(973)(1073)}\right)$$

$$\log K_2 - (-0.201) = 0.167$$

$$\log K_2 = -0.033 = 0.967 - 1$$

$$K_2 = 0.93$$

## Problems

**13.1** List four factors that influence the rate of a chemical reaction and explain on the atomic-molecular level how each exerts its influence.

**13.2** Define the following: (a) activated complex, (b) energy of activation, (c) reaction order, (d) catalyst, (e) chemisorption, (f) rate-determining step, (g) reaction intermediate, (h) zero-order reaction, (i) steady state.

**13.3** Draw a potential energy diagram for the hypothetical reaction $A(g) + B(g) \rightleftharpoons 2C(g)$ for which $E_a$ is 37 kcal and $E_a'$ is 25 kcal. Indicate the magnitude of $\Delta H$ on the diagram. Is the reaction exothermic or endothermic? Draw a curve for a catalyzed reaction on the diagram.

**13.4** The systhesis of perbromates has only recently been accomplished. The best preparation involves the oxidation of bromates in alkaline solution by fluorine. For the half reaction

$$2e^- + 2H^+ + BrO_4^- \rightarrow$$
$$BrO_3^- + H_2O$$

$\mathscr{E}°$ is +1.763 V. Although this potential is fairly high, it alone cannot account for the difficulty encountered in the synthesis of perbromates. What explanation can you give for this difficulty, and how well would you expect perbromates to function as oxidizing agents?

**13.5** A common and serious mistake is to assume that the rate equation for reaction can be derived from the balanced chemical equation for the reaction by using the coefficients of the chemical equation as exponents in the rate expression. Why cannot rate equations be derived in this way? How do you reconcile your answer with the law of mass action?

**13.6** The rate equation for the reaction

$$2NO(g) + 2H_2(g) \rightarrow$$
$$N_2(g) + 2H_2O(g)$$

is second order in $NO(g)$ and first order in $H_2(g)$. (a) Write an equation for the rate of appearance of $N_2(g)$. (b) If concentrations are expressed in mol/liter, what units would the rate constant, $k$, have? (c) If concentrations are expressed in atmospheres, what units would $k$ have? (d) Give an equation for the rate of disappearance of $NO(g)$. Would $k$ in this equation have the same numerical value as $k$ in the equation of part (a)?

**13.7** For a reaction in which A and B react to form C, the following data were obtained from three experiments.

| [A] | [B] | Rate (Formation of C) |
|-----|-----|------------------------|
| 0.30 $M$ | 0.15 $M$ | $7.0 \times 10^{-3}$ $M$/min |
| 0.60 $M$ | 0.30 $M$ | $2.8 \times 10^{-2}$ $M$/min |
| 0.30 $M$ | 0.30 $M$ | $1.4 \times 10^{-2}$ $M$/min |

(a) What is the rate equation for the reaction? (b) What is the numerical value of the specific rate constant?

**13.8** For a reaction in which A and B react to form C, the following data were obtained from three experiments.

| $[A]$ | $[B]$ | Rate (Formation of C) |
|---|---|---|
| 0.03 M | 0.03 M | $0.3 \times 10^{-3}$ M/min |
| 0.06 M | 0.06 M | $1.2 \times 10^{-3}$ M/min |
| 0.06 M | 0.09 M | $2.7 \times 10^{-3}$ M/min |

(a) What is the rate equation for the reaction? (b) What is the numerical value of the specific rate constant?

**13.9** For the reaction

$$2NO(g) + Cl_2(g) \rightarrow 2NOCl(g)$$

the rate equation is second order in NO(g), first order in $Cl_2(g)$, and third order overall. Compare the initial rate of reaction of a mixture of 0.02 mol NO(g) and 0.02 mol $Cl_2(g)$ in a 1 liter container with (a) the rate of the reaction when half the NO(g) has been consumed, (b) the rate of the reaction when half the $Cl_2(g)$ has been consumed, (c) the rate of the reaction when two-thirds of NO(g) has been consumed, (d) the initial rate of a mixture of 0.04 mol NO(g) and 0.02 mol $Cl_2(g)$ in a 1 liter container, (e) the initial rate of a mixture of 0.02 mol NO(g) and 0.02 mol $Cl_2(g)$ in a 0.5 liter container.

**13.10** Calculate the value of the specific rate constant in (a) 1/atm sec and (b) liter/mol sec for the decomposition of acetaldehyde vapor at 800°K. Use the data of Example 13.2.

**13.11** Use the data of Table 13.2 to calculate the value of the specific rate constant for the decomposition of chloroethane, $C_2H_5Cl(g)$, at 600°K.

**13.12** The following data pertain to the decomposition of phosphine gas at 950°K:

$$4PH_3(g) \rightarrow P_4(g) + 6H_2(g)$$

| Experiment | Initial Pressure of $PH_3(g)$ | Initial Rate of Increase in Total Pressure |
|---|---|---|
| 1 | 0.120 atm | 0.00330 atm/sec |
| 2 | 0.160 atm | 0.00440 atm/sec |

(a) What is the order of the rate equation for the reaction? (b) What is the value of the specific rate constant?

**13.13** For the reaction

$$2NOCl \rightarrow 2NO + Cl_2$$

$A$ is $9.0 \times 10^{12}$/sec and $E_a$ is 24,000 cal. What is the value of the specific rate constant for the reaction conducted at 350°K?

**13.14** For the reaction

$$HI(g) + CH_3I(g) \rightarrow CH_4(g) + I_2(g)$$

the rate constant is $1.7 \times 10^{-2}$/sec at 430°K and $9.5 \times 10^{-2}$/sec at 450°K. What is the energy of activation for the reaction?

**13.15** For the reaction

$$C_2H_5Br(g) \rightarrow C_2H_4(g) + HBr(g)$$

the rate constant is $2.0 \times 10^{-5}$/sec at 650°K and the energy of activation is 53.9 kcal/mol. At what temperature is the rate constant $6.0 \times 10^{-5}$/sec?

**13.16** At 400°K a certain reaction is 50.0% complete in 1.50 minutes. At 430°K the same reaction is 50.0% complete in 0.50 minutes. Calculate the activation energy for the reaction.

**13.17** What is the activation energy

of a reaction that doubles in rate when the temperature is increased from 300°K to 310°K?

**13.18** The reaction

$$CH_4(g) + Cl_2(g) \rightarrow CH_3Cl(g) + HCl(g)$$

proceeds by a chain mechanism. The chain propagators are Cl atoms and $CH_3$ radicals, and it is believed that free H atoms are not involved. Write a series of equations showing the mechanism and identify the chain-initiating, chain-sustaining, and chain-terminating steps.

**13.19** The rate expression for the reaction

$$2NO(g) + O_2(g) \rightarrow 2NO_2(g)$$

is second order in NO(g) and first order in $O_2(g)$. The following mechanism has been suggested.

$$NO(g) + O_2(g) \rightleftharpoons NO_3(g)$$

$$NO_3(g) + NO(g) \rightarrow 2NO_2(g)$$

Show how this mechanism leads to the observed rate equation. The first step of the mechanism is a rapidly established equilibrium (for which an equilibrium constant should be derived), and the second step is slower and rate-determining. Notice that only bimolecular collisions are involved in the mechanism even though the rate equation is third order overall.

**13.20** (a) A proposed mechanism for the reaction

$$2NO_2Cl(g) \rightarrow 2NO_2(g) + Cl_2(g)$$

consists of the following steps:

(i)  $NO_2Cl(g) \xrightarrow{k_1} NO_2(g) + Cl(g)$

(ii) $NO_2Cl(g) + Cl(g) \xrightarrow{k_2} NO_2(g) + Cl_2(g)$

Assume that the concentration of Cl(g) reaches a steady state, and show that the rate equation for the overall reaction is first order. (b) What is the value of $k$ for the rate equation in terms of $k_1$ and $k_2$?

**13.21** The reaction A(g) + B(g) $\rightleftharpoons$ C(g) is exothermic as written. (a) What effect would an increase in temperature have on the numerical value of $K$? (b) How would the numerical values of $K$ and $K_p$ vary with an increase in total pressure?

**13.22** The reaction A(g) + B(g) $\rightleftharpoons$ C(g) is exothermic as written. Assume that an equilibrium system is established. How would the equilibrium concentration of C(g) change with (a) an increase in temperature, (b) an increase in pressure, (c) the addition of A(g), (d) the addition of a catalyst, (e) the removal of B(g)? How would the numerical value of $K$ change with (f) an increase in temperature, (g) an increase in pressure, (h) the addition of a catalyst, (i) the addition of A(g)?

**13.23** For the equilibrium

$$C(s) + CO_2(g) \rightleftharpoons 2CO(g)$$

$K_p$ is 167.5 atm at 1000°C. What is the partial pressure of CO(g) in equilibrium with $CO_2(g)$ at a partial pressure of 0.100 atm?

**13.24** $K_p$ for the equilibrium

$$FeO(s) + CO(g) \rightleftharpoons Fe(s) + CO_2(g)$$

at 1000°C is 0.403. If CO(g), at a pressure of 1.000 atm, and excess FeO(s) are confined in a container at 1000°C, what are the partial pressures of CO(g) and $CO_2$(g) when equilibrium is attained?

**13.25** At 800°K, $K_p$ is 0.220 atm for the equilibrium

$$CaCO_3(s) \rightleftharpoons CaO(s) + CO_2(g)$$

What is the concentration of $CO_2$(g), in mol/liter, that is in equilibrium with solid $CaCO_3$ and CaO at this temperature?

**13.26** At 425°C, $K$ is 54.8 for the equilibrium

$$H_2(g) + I_2(g) \rightleftharpoons 2HI(g)$$

If 1.000 mol of $H_2$(g), 1.000 mol of $I_2$(g), and 1.000 mol of HI(g) are placed in a liter container at 425°C, what are the concentrations of all gases present at equilibrium?

**13.27** At 425°C, $K$ is $1.82 \times 10^{-2}$ for the equilibrium

$$2HI(g) \rightleftharpoons H_2(g) + I_2(g)$$

Assume that an equilibrium is established at 425°C by adding only HI(g) to the reaction flask. (a) What are the concentrations of $H_2$(g) and $I_2$(g) in equilibrium with 0.0100 mol/liter of HI(g)? (b) What was the initial concentration of HI(g) before equilibrium was established? (c) What percent of the HI(g) added is dissociated at equilibrium?

**13.28** For the equilibrium

$$N_2(g) + O_2(g) \rightleftharpoons 2NO(g)$$

$K$ is $2.5 \times 10^{-3}$ at 2400°K. The partial pressure of $N_2$(g) is 0.50 atm and the partial pressure of $O_2$(g) is 0.50 atm in a mixture of these two gases. What is the partial pressure of NO(g) when equilibrium is established at 2400°K?

**13.29** For the equilibrium

$$2IBr(g) \rightleftharpoons I_2(g) + Br_2(g)$$

$K$ is $8.5 \times 10^{-3}$ at 150°C. If 0.0300 mol of IBr(g) is introduced into a one liter container, what is the concentration of this substance after equilibrium is established?

**13.30** At 500°K, NOCl(g) is 9.0% dissociated.

$$2NOCl(g) \rightleftharpoons 2NO(g) + Cl_2(g)$$

Calculate the value of $K$ for the equilibrium at 500°C. Assume that 1.00 mol of NOCl(g) in 1.00 liter was present before dissociation.

**13.31** A mixture consisting of 1.000 mol of CO(g) and 1.000 mol of $H_2O$(g) is placed in a 10.00 liter container at 800°K. At equilibrium, 0.665 mol of $CO_2$(g) and 0.665 mol of $H_2$(g) are present.

$$CO(g) + H_2O(g) \rightleftharpoons CO_2(g) + H_2(g)$$

(a) What are the equilibrium concentrations of all four gases? (b) What is the value of $K$ at 800°K? (c) What is the value of $K_p$ at 800°K?

**13.32** At 585°K and a total pressure of 1.00 atm, NOCl(g) is 56.4% dissociated.

$$2NOCl(g) \rightleftharpoons 2NO(g) + Cl_2(g)$$

Assume that 1.00 mol of NOCl(g) was present before dissociation. (a)

Nernst

How many moles of NOCl(g), NO(g) and $Cl_2$(g) are present at equilibrium? (b) What is the total number of moles of gas present at equilibrium? (c) What are the equilibrium partial pressures of the three gases? (d) What is the numerical value of $K_p$ at 585°K?

**13.33** At 100°C, $K$ is $4.57 \times 10^9$ liters/mol for the equilibrium

$$CO(g) + Cl_2(g) \rightleftharpoons COCl_2(g)$$

What is (a) $K$ and (b) $K_p$ for the equilibrium

$$COCl_2(g) \rightleftharpoons CO(g) + Cl_2(g)$$

at this temperature?

**13.34** At 1100°K, $K_p$ is 2.45 atm for the equilibrium

$$2SO_3(g) \rightleftharpoons 2SO_2(g) + O_2(g)$$

(a) What is $K$ for this equilibrium? (b) What is the value of $K$ and that of $K_p$ for the equilibrium

$$2SO_2(g) + O_2(g) \rightleftharpoons 2SO_3(g)$$

**13.35** For the equilibrium

$$4HCl(g) + O_2(g) \rightleftharpoons 2Cl_2(g) + 2H_2O(g)$$

$K$ is 889 liters/mol at 480°C. (a) What is the value of $K_p$ at this temperature? (b) Calculate the values of $K$ and $K_p$ at 480°C for the equilibrium

$$2Cl_2(g) + 2H_2O(g) \rightleftharpoons 4HCl(g) + O_2(g)$$

**13.36** For the equilibrium

$$Cl_2(g) \rightleftharpoons 2Cl(g)$$

$K_p = 4.80 \times 10^{-16}$ atm at 600°K and $1.04 \times 10^{-10}$ atm at 800°K. (a) What is the average value of $\Delta H$ for the temperature range considered? (b) What is the value of $K_p$ at 700°K?

**13.37** For the equilibrium

$$2NO(g) \rightleftharpoons N_2(g) + O_2(g)$$

$K$ is $2.45 \times 10^3$ at 2000°K and $1.46 \times 10^3$ at 2100°K. (a) What is the average value of $\Delta H$ for this temperature range? (b) What is the value of $K$ at 2200°K on the basis of the calculated value of $\Delta H$?

**13.38** At 25°C, $K_p$ is 0.114 atm for the equilibrium

$$N_2O_4(g) \rightleftharpoons 2NO_2(g)$$

If $\Delta H$ is +13.9 kcal, what is the value of $K_p$ at 75°C?

**13.39** At 25°C, $K_p$ is $5.6 \times 10^{-15}$/atm for the equilibrium

$$2F_2(g) + O_2(g) \rightleftharpoons 2F_2O(g)$$

Assume that $\Delta H$ is +11.0 kcal and calculate the value of $K_p$ at 100°C.

**13.40** At a given temperature and a total pressure of 1.00 atm, the partial pressures of an equilibrium mixture

$$N_2O_4(g) \rightleftharpoons 2NO_2(g)$$

are $p_{N_2O_4} = 0.50$ atm and $p_{NO_2} = 0.50$ atm. (a) What is $K_p$ at this temperature? (b) If the total pressure is increased to 2.00 atm and the temperature is constant, what are the partial pressures of the components of an equilibrium mixture?

**13.41** State the direction in which each of the following equilibrium systems would be shifted upon the application of the stress listed beside the equation.

(a) $2SO_2(g) + O_2(g) \rightleftharpoons 2SO_3(g)$            decrease temperature
  (exothermic)

(b) $C(s) + CO_2(g) \rightleftharpoons 2CO(g)$            increase temperature
  (endothermic)

(c) $N_2O_4(g) \rightleftharpoons 2NO_2(g)$            increase total pressure

(d) $CO(g) + H_2O(g) \rightleftharpoons CO_2(g) + H_2(g)$            decrease total pressure

(e) $2NOBr(g) \rightleftharpoons 2NO(g) + Br_2(g)$            decrease total pressure

(f) $3Fe(s) + 4H_2O(g) \rightleftharpoons Fe_3O_4(s) + 4H_2(g)$      add $Fe(s)$

(g) $2SO_2(g) + O_2(g) \rightleftharpoons 2SO_3(g)$            add a catalyst

(h) $CaCO_3(s) \rightleftharpoons CaO(s) + CO_2(g)$            remove $CO_2(g)$

(i) $N_2(g) + 3H_2(g) \rightleftharpoons 2NH_3(g)$            increase the concentration of $H_2(g)$

# 14

# Elements of Chemical Thermodynamics

We have repeatedly referred to the energy effects that accompany physical and chemical changes. **Thermodynamics** is the study of these energy effects; in particular, it summarizes the relations between heat, work, and other forms of energy that are involved in all types of change. The laws of thermodynamics can be used to predict whether a particular chemical or physical transformation is theoretically possible under a given set of conditions.

Thermodynamics, however, has nothing to say about the rate at which a predicted change will occur. This question is the concern of chemical kinetics (Chapter 13). The stable form of carbon under ordinary conditions is graphite, not diamond. According to thermodynamic principles, therefore, the transformation of a diamond crystal into a graphite crystal is spontaneous. This change, however, is so extremely slow that it is not observed at ordinary temperatures and pressures.

**14.1 FIRST LAW OF THERMODYNAMICS**

Many scientists of the late eighteenth and early nineteenth centuries studied the relation between mechanical work and heat; thermodynamics had its origin in these studies. By the 1840's it became clear that heat and work are two manifestations of a larger classification—energy, that different forms of energy are interconvertible, and that energy is conserved. The **first law of thermodynamics** is the law of conservation of energy: energy can be converted from one form into another but cannot be created or destroyed.

The careful and convincing experimental work of James Joule in the years 1840 to 1849 did much to establish this law. Joule studied the conversion of mechanical and electrical work into heat. He used the work done by a falling weight to turn a paddle wheel that was immersed in a container of water and determined the heat produced by measuring the increase in the temperature of the water. In a series of such experiments, he used different weights and different quantities of water, as well as mercury in place of the water. Joule also studied the heating effects of electric currents and the conversion of work into heat by the compression of gases. Joule found that a given amount of work always produces the same quantity of heat. In modern terms the relation is

4.1840 J = 1 cal

One joule (J) is the amount of work done when a force of one newton (1 N = 1 kg m/sec$^2$) acts through a distance of one meter. The calorie, which was originally defined as the amount of heat required to raise the temperature of 1 g of water from 14.5° to 15.5°C, is now defined by its joule equivalent. Since it is possible to make electrical measurements with greater precision than calorimetric measurements, the joule (which is a volt coulomb) is a better primary standard than the calorie.

In the International System of Units (SI) the joule is the unit for all energy measurements. However, thermodynamic values have been almost exclusively recorded in terms of calories and kilocalories in the past, and therefore one must be familiar with both units.

In applying thermodynamic concepts, we frequently confine our attention to the changes that occur within definite boundaries. That portion of the universe included within these boundaries is called a **system;** the remainder of the universe is called the **surroundings.** Work done on a system need not always result in an increase in the temperature of the system or in the conversion of work into heat. For example, the charging of a storage battery by an automobile engine results in an increase of the "chemical energy" of the battery; doing work on (or adding heat to) a sample of ice at 0°C could result in melting a part of the ice with no increase in the temperature of the system. A system is assumed to have an **internal energy,** $E$, which includes all possible forms of energy attributable to the system (some of which are the attractions and repulsions between the atoms, molecules, ions, and subatomic particles comprising the system and the kinetic energies of all of its parts).

According to the first law of thermodynamics, the internal energy of an *isolated* system is constant. The actual value of $E$ for any system is not known and cannot be determined. However, thermodynamics is concerned only with *changes* in internal energy, and such changes are measurable.

The internal energy of a system depends upon the state of the system and not upon how the system arrived at that state. Internal energy is therefore called a **state function.** Consider a sample of an ideal gas that occupies a volume of 1 liter at 100°K and 1 atm pressure (state A). At 200°K and 0.5 atm (state B), the sample occupies a volume of 4 liters. According to the first law, the internal energy of the system in state A, $E_A$, is a constant, as is the internal energy of the system in state B, $E_B$. It follows that the difference in the internal energies of the two states, $\Delta E$, is also a constant and is independent of the path taken between state A and state B. It makes no difference whether the gas is heated before the pressure change, whether the heating is done after the pressure change, or indeed whether the total change is brought about in several steps,

$$\Delta E = E_B - E_A$$

If $\Delta E$ were not independent of the manner in which the corresponding change was brought about, it would be possible to create or destroy energy, in violation of the first law, by taking a system from A to B by one route and then returning it to state A by a different route. Instead, in a cyclic process (in which a system is returned to its original state), the increase in $E$ in one direction exactly equals the decrease in $E$ in the opposite direction—there is no net gain or loss in energy for the cycle.

The change in internal energy of a system may be determined by measurement of the heat absorbed by the system from its surroundings, $q$, and the work done by the system on its surroundings, $w$. In these terms,

$$\Delta E = q - w$$

It is important to keep in mind the conventions regarding the signs of these quantities:

q, positive = heat *absorbed* by the system

q, negative = heat *evolved* by the system

w, positive = work done *by* the system

w, negative = work done *on* the system

The values of q and w involved in changing a system from state A to state B depend upon the way in which the change is carried out. However, the value of $(q - w)$ is a constant, equal to $\Delta E$, for the change no matter how it is brought about. If a system undergoes a change in which the internal energy of the system remains constant, the work done by the system equals the heat absorbed by the system. If a system undergoes a change that reduces its internal energy, the energy released will be the sum of the heat evolved and the work done.

**14.2 ENThALPY**

● The pascal (Pa) is the SI unit of pressure.

$$1 \text{ Pa} = 1 \text{ N/m}^2$$

Since

$$1 \text{ atm} = 1.01325 \times 10^5 \text{ N/m}^2$$

and 1 liter is $10^{-3}$ m$^3$, one liter atmosphere is

$$1.01325 \times 10^2 \text{ N·m (or J)}$$

In addition, since 1 cal = 4.184 J, one liter atmosphere equals

$$24.2173 \text{ cal}$$

For ordinary chemical reactions the work term generally arises as a consequence of pressure–volume changes—for example, the work done against the atmosphere if the system expands in the course of the reaction. The term $PV$ has the dimensions of work. Pressure, which is force per unit area, may be expressed in newtons per square meter; multiplying pressure by volume, which is expressed in cubic meters, gives a product in newton meters (or joules). A newton meter (or joule) is a dimension of work, since work is force times distance. In like manner, liter atmospheres are units of work. If the pressure is held constant, the work done in expansion from $V_A$ to $V_B$ is

$$w = P(V_B - V_A) = P\Delta V$$

No pressure–volume work can be done by a process carried out at constant volume, and $w = 0$. Thus, at constant volume the equation

$$\Delta E = q - w$$

becomes

$$\Delta E = q_V$$

where $q_V$ is the heat absorbed by the system at constant volume.

Processes carried out at constant pressure are far more common in chemistry than those conducted at constant volume. If we restrict our attention to pressure–volume work, the work done in constant pressure processes is $P\Delta V$. Thus, at constant pressure the equation

$$\Delta E = q - w$$

becomes

$$\Delta E = q_P - P\Delta V$$

or, by rearranging,

$$q_P = \Delta E + P\Delta V$$

where $q_P$ is the heat absorbed by the system at constant pressure. This relationship may be written

$$q_P = (E_B - E_A) + P(V_B - V_A)$$

or, by rearranging,

$$q_P = (E_B + PV_B) - (E_A + PV_A)$$

The thermodynamic function **enthalpy,** $H$, is defined by the equation

$$H = E + PV$$

Therefore,

$$q_P = H_B - H_A = \Delta H$$

Thus, the heat absorbed by a process conducted at constant pressure is equal to the change in enthalpy, and $\Delta H$ is related to the change in the internal energy of the system by the expression

$$\Delta H = \Delta E + P\Delta V$$

Enthalpy, like internal energy, is a function of the state of the system and is independent of the manner in which the state was achieved. The validity of the law of Hess rests on this fact (see Sections 5.7, 5.8, and 5.9).

The **standard enthalpy of formation** of a compound (given the symbol $\Delta H_f^\circ$) is the change in enthalpy for the reaction (at 25°C and 1 atm) in which one mole of the compound in its standard state is prepared from its constituent elements in their standard states. The standard state of a substance is the stable form of the substance at 25°C and 1 atm. The use of $\Delta H_f^\circ$ values is discussed in Section 5.8; a number of standard enthalpies of formation are listed in Table 14.1.

When a bomb calorimeter is used to make a calorimetric determination (Section 5.7), the heat effect is measured at constant volume; ordinarily, reactions are run at constant pressure. The relationship between change in enthalpy and change in internal energy is used to convert heats of reaction at constant volume ($q_V = \Delta E$) to heats of reaction at constant pressure ($q_P = \Delta H$). The appropriate conversion is made by considering the change in volume of the system—the total volume of the

Table 14.1 Enthalpy of formation at 25°C and 1 atm

| COMPOUND | $\Delta H_f^\circ$ (kcal/mol) | $\Delta H_f^\circ$ (kJ/mol)[a] | COMPOUND | $\Delta H_f^\circ$ (kcal/mol) | $\Delta H_f^\circ$ (kJ/mol)[a] |
|---|---|---|---|---|---|
| $H_2O(g)$ | $-57.80$ | $-241.8$ | $COCl_2(g)$ | $-53.3$ | $-223.$ |
| $H_2O(l)$ | $-68.32$ | $-285.9$ | $SO_2(g)$ | $-70.96$ | $-296.9$ |
| $HF(g)$ | $-64.2$ | $-269.$ | $CO(g)$ | $-26.42$ | $-110.5$ |
| $HCl(g)$ | $-22.06$ | $-92.30$ | $CO_2(g)$ | $-94.05$ | $-393.5$ |
| $HBr(g)$ | $-8.66$ | $-36.2$ | $NO(g)$ | $+21.60$ | $+90.37$ |
| $HI(g)$ | $+6.20$ | $+25.9$ | $NO_2(g)$ | $+8.09$ | $+33.8$ |
| $H_2S(g)$ | $-4.82$ | $-20.2$ | $HNO_3(l)$ | $-41.40$ | $-173.2$ |
| $HCN(g)$ | $+31.2$ | $+130.5$ | $NH_4NO_3(s)$ | $-87.27$ | $-365.1$ |
| $NH_3(g)$ | $-11.04$ | $-46.19$ | $NaCl(s)$ | $-98.23$ | $-411.0$ |
| $PH_3(g)$ | $+2.21$ | $+9.25$ | $MgO(s)$ | $-143.84$ | $-601.83$ |
| $CH_4(g)$ | $-17.89$ | $-74.85$ | $CaO(s)$ | $-151.9$ | $-635.5$ |
| $C_2H_6(g)$ | $-20.24$ | $-84.68$ | $Ca(OH)_2(s)$ | $-235.80$ | $-986.59$ |
| $C_2H_4(g)$ | $+12.50$ | $+52.30$ | $CaCO_3(s)$ | $-288.45$ | $-1206.9$ |
| $C_2H_2(g)$ | $+54.19$ | $+226.7$ | $Ca_3P_2(s)$ | $-120.50$ | $-504.17$ |
| $C_6H_6(l)$ | $+11.72$ | $+49.04$ | $BaO(s)$ | $-133.4$ | $-588.1$ |
| $CH_3OH(g)$ | $-48.08$ | $-201.2$ | $BaCO_3(s)$ | $-291.3$ | $-1218.$ |
| $CH_3OH(l)$ | $-57.02$ | $-238.6$ | $Al_2O_3(s)$ | $-399.09$ | $-1669.8$ |
| $CH_3NH_2(g)$ | $-6.7$ | $-28.$ | $Fe_2O_3(s)$ | $-196.5$ | $-822.2$ |
| $NF_3(g)$ | $-27.2$ | $-113.$ | $AgCl(s)$ | $-30.36$ | $-127.0$ |
| $CF_4(g)$ | $-218.3$ | $-913.4$ | $HgBr_2(s)$ | $-40.5$ | $-169.$ |
| $CHCl_3(l)$ | $-31.5$ | $-132.$ | $ZnO(s)$ | $-83.17$ | $-348.0$ |

[a] 1 kcal = 4.184 kJ (exactly).

reactants subtracted from the total volume of the products. The changes in the volumes of liquids and solids are so small that they are neglected; if only liquids and solids are involved in the reaction, the equation $q_V = q_P$ is sufficiently accurate for most purposes.

For reactions involving gases, however, volume changes may be significant. Let us say that $V_A$ is the total volume of gaseous reactants, $V_B$ is the total volume of gaseous products, $n_A$ is the number of moles of gaseous reactants, $n_B$ is the number of moles of gaseous products, and the pressure and temperature are constant.

$$PV_A = n_A RT \quad \text{and} \quad PV_B = n_B RT$$

● The values of $R$ in J/K·mol and cal/K·mol can be derived from

0.082056 l·atm/K·mol

by use of the values for the energy equivalents of the liter atmosphere that have been previously given.

Thus

$$
\begin{aligned}
P\Delta V &= PV_B - PV_A \\
&= n_B RT - n_A RT \\
&= (n_B - n_A)RT \\
&= (\Delta n)RT
\end{aligned}
$$

Therefore,

$$\Delta H = \Delta E + (\Delta n)RT$$

Table 14.2 Value of the gas constant, $R$, in various units

| $R$ | UNITS |
| --- | --- |
| 0.082056 | liter atm/°K mol |
| 8.3143 | J/°K mol |
| 1.9872 | cal/°K mol |

where $\Delta n$ is the number of moles of gaseous products minus the number of moles of gaseous reactants.

In order to solve problems using this equation, we must express the value of $R$ in appropriate units. As noted previously in this section, liter atmospheres are dimensions of energy; therefore, the value of $R$ in liter atm/°K mol may be converted to cal/°K mol or to J/°K mol (see Table 14.2).

---

*Example 14.1*

The heat of combustion at constant volume of $CH_4(g)$ is measured in a bomb calorimeter at 25°C and is found to be $-211,613$ cal/mol. What is the $\Delta H$?

*Solution*

For the reaction

$$CH_4(g) + 2O_2(g) \rightarrow CO_2(g) + 2H_2O(l) \qquad \Delta E = -211,613 \text{ cal}$$

$$\Delta n = 1 - (2 + 1) = -2$$

Therefore,

$$\Delta H = \Delta E + (\Delta n)RT$$

$$\Delta H = -211,613 \text{ cal} + (-2 \text{ mol})(1.9872 \text{ cal/°K mol})(298.2°K)$$
$$= -211,613 \text{ cal} - 1185 \text{ cal}$$
$$= -212,798 \text{ cal} = -212.798 \text{ kcal}$$

The change in internal energy could have been given in joules; $\Delta E = -885,389$ J

$$\Delta H = \Delta E + (\Delta n)RT$$
$$= -885,389 \text{ J} + (-2 \text{ mol})(8.3143 \text{ J/°K mol})(298.2°K)$$
$$= -885,389 \text{ J} - 4958 \text{ J}$$
$$= -890,347 \text{ J} = -890.347 \text{ kJ}$$

---

*Example 14.2*

Calculate $\Delta H$ and $\Delta E$ at 25°C for the reaction

$$OF_2(g) + H_2O(g) \rightarrow O_2(g) + 2HF(g)$$

The enthalpies of formation are: $OF_2(g)$, +5.5 kcal/mol; $H_2O(g)$, $-57.8$ kcal/mol; and HF(g), $-64.2$ kcal/mol.

*Solution*

The enthalpies of formation are used to calculate the $\Delta H$ for the reaction (Section 5.8).

$$\Delta H = 2\Delta H_f^\circ(HF) - [\Delta H_f^\circ(OF_2) + \Delta H_f^\circ(H_2O)]$$
$$= 2(-64.2) - [(+5.5) + (-57.8)]$$
$$= -128.4 + 52.3 = -76.1 \text{ kcal}$$

This value of $\Delta H$ is used to calculate $\Delta E$. For the reaction, $\Delta n = +1$ mol.

$$\Delta E = \Delta H - (\Delta n)RT$$
$$= -76.1 \text{ kcal} - (1 \text{ mol})(1.987 \text{ cal/°K mol})(298°K)$$
$$= -76.1 \text{ kcal} - 593 \text{ cal} = -76.7 \text{ kcal}$$

Therefore, the internal energy of the system decreases by 76.7 kcal ($\Delta E = -76.7$ kcal). At a constant pressure of 1 atm, only 76.1 kcal of this quantity are given off as heat ($\Delta H = -76.1$ kcal). The remainder is used for the work of expansion against the atmosphere.

*Example 14.3*

For the reaction

$$B_2H_6(g) + 3O_2(g) \rightarrow$$
$$B_2O_3(s) + 3H_2O(l)$$

$\Delta E = -2143.2$ kJ. (a) Calculate $\Delta H$ for the reaction in kJ. (b) Determine the value of the enthalpy of formation of $B_2H_6(g)$. For $B_2O_3(s)$, $\Delta H_f^\circ = -1264.0$ kJ/mol; and for $H_2O(l)$, $\Delta H_f^\circ = -285.9$ kJ/mol.

*Solution*

(a) $\Delta n = -4$. Therefore,

$$\Delta H = \Delta E + (\Delta n)RT$$
$$= -2143.2 \text{ kJ} + (-4 \text{ mol})(8.314 \times 10^{-3} \text{ kJ/}^\circ\text{K mol})(298^\circ\text{K})$$
$$= -2143.2 \text{ kJ} - 9.9 \text{ kJ}$$
$$= -2153.1 \text{ kJ}$$

(b)

$$\Delta H = \Delta H_f^\circ(B_2O_3) + 3\Delta H_f^\circ(H_2O) - \Delta H_f^\circ(B_2H_6)$$

$$-2153.1 \text{ kJ} = (-1264.0 \text{ kJ}) + 3(-285.9 \text{ kJ}) - \Delta H_f^\circ(B_2H_6)$$

$$\Delta H_f^\circ(B_2H_6) = +31.4 \text{ kJ}$$

## 14.3 BOND ENERGIES

The value of $\Delta H$ for the dissociation of a gaseous diatomic molecule into gaseous atoms is a measure of the strength of the covalent bond of the molecule and is called the **bond dissociation energy.** For example, 118.3 kcal are absorbed when 1 mol of $O_2(g)$ is dissociated into oxygen atoms.

$$O_2(g) \rightarrow 2O(g) \qquad \Delta H = +118.3 \text{ kcal}$$

This value is also called the **heat of atomization** of $O_2(g)$.

A bond dissociation energy of a polyatomic molecule refers to a process in which a specific bond of the molecule is broken. The dissociation of the molecule may be imagined to occur in steps; the $\Delta H$ values for the steps are not the same. For example,

$$HOH(g) \rightarrow H(g) + OH(g) \qquad \Delta H = +119.7 \text{ kcal}$$

$$OH(g) \rightarrow H(g) + O(g) \qquad \Delta H = +101.5 \text{ kcal}$$

The sum of these bond dissociation energies is the heat of atomization of gaseous $H_2O$.

$$H_2O(g) \rightarrow 2H(g) + O(g) \qquad \Delta H = +221.2 \text{ kcal}$$

In many instances the individual $\Delta H$ values pertaining to the successive dissociations of a polyatomic molecule are not known. However, the heat of atomization can be calculated by use of the heat of formation of the gaseous compound together with the heats of atomization of the elements that make up the compound (Example 14.4).

*Example 14.4*

The bond dissociation energies (or heats of atomization) of $H_2(g)$ and $O_2(g)$ are $+104.2$ kcal/mol and $+118.3$ kcal/mol, respectively, and the heat of formation of $H_2O(g)$ is $-57.8$ kcal/mol. What is the heat of atomization of $H_2O(g)$?

*Solution*

The following equations and corresponding $\Delta H$ values are derived from the thermochemical data given.

$$H_2O(g) \rightarrow H_2(g) + \tfrac{1}{2}O_2(g) \qquad \Delta H = +57.8 \text{ kcal}$$

$$H_2(g) \rightarrow 2H(g) \qquad\qquad\quad \Delta H = +104.2 \text{ kcal}$$

$$\tfrac{1}{2}O_2(g) \rightarrow O(g) \qquad\qquad\quad \Delta H = +59.2 \text{ kcal}$$

The sum of these enthalpy changes is the heat of atomization of $H_2O(g)$.

$$H_2O(g) \rightarrow 2H(g) + O(g) \qquad \Delta H = +221.2 \text{ kcal}$$

An **average bond energy** can be derived from the heat of atomization of a molecule in which all the bonds are alike. For the O—H bond of the water molecule, the average bond energy is one half the heat of atomization of the $H_2O$ molecule since there are two O—H bonds in the molecule ($221.2/2 = 110.6$ kcal/mol). The average bond energy of the N—H bond in the ammonia molecule is one third the heat of atomization of $NH_3$.

The bond energy of a given type of bond is not the same in all molecules containing that bond. However, the variation from compound to compound is small in most cases, and an average value is usually a satisfactory approximation. Hence, a heat of atomization for a molecule

Table 14.3 Average bond energies

| BOND | AVERAGE BOND ENERGY (kcal/mol) | AVERAGE BOND ENERGY (kJ/mol) | BOND | AVERAGE BOND ENERGY (kcal/mol) | AVERAGE BOND ENERGY (kJ/mol) |
|---|---|---|---|---|---|
| H—H | 104 | 435 | N—Cl | 48 | 201 |
| H—F | 135 | 565 | C—C | 83 | 347 |
| H—Cl | 103 | 431 | C=C | 148 | 619 |
| H—Br | 87 | 364 | C≡C | 194 | 812 |
| H—I | 71 | 297 | C—H | 99 | 414 |
| F—F | 37 | 155 | C—O | 80 | 335 |
| Cl—Cl | 58 | 243 | C=O | 169 | 707 |
| O—O | 33 | 138 | C—F | 116 | 485 |
| $O_2^a$ | 118 | 494 | C—Cl | 78 | 326 |
| O—H | 111 | 464 | C—N | 70 | 293 |
| O—F | 44 | 184 | C≡N | 210 | 879 |
| O—Cl | 49 | 205 | S—H | 81 | 339 |
| N—N | 38 | 159 | S—Cl | 66 | 276 |
| N≡N | 225 | 941 | P—H | 76 | 318 |
| N—H | 93 | 389 | P—Cl | 78 | 326 |

$^a$ Double bond of molecular oxygen.

together with established average bond energies can be used to derive a new average bond energy. For example, the hydrogen peroxide molecule (H—O—O—H) contains two H—O bonds and one O—O bond. The heat of atomization of hydrogen peroxide, which is the energy required to break all three bonds, is 254.3 kcal/mol. Subtraction of 221.2 kcal for the two H—O bonds from 254.3 kcal gives 33.1 kcal/mol as the bond energy of the O—O bond.

Average bond energies are given in Table 14.3. The values listed for diatomic molecules are, of course, also bond dissociation energies. Average bond energies may be used to obtain approximate heats of reaction.

*Example 14.5*

Use average bond energies to calculate the $\Delta H$ of the reaction

$$OF_2(g) + H_2O(g) \rightarrow O_2(g) + 2HF(g)$$

*Solution*

We can imagine the reaction to take place by a series of steps. Energy is absorbed ($\Delta H$, positive) when a bond is broken, and energy is evolved ($\Delta H$, negative) when a bond is formed.

$$F—O—F(g) \rightarrow 2F(g) + O(g)$$
$$\Delta H = 2(+44) = +88 \text{ kcal}$$

$$H—O—H(g) \rightarrow 2H(g) + O(g)$$
$$\Delta H = 2(+111) = +222 \text{ kcal}$$

$$2O(g) \rightarrow O_2(g)$$
$$\Delta H = -118 = -118 \text{ kcal}$$

$$2H(g) + 2F(g) \rightarrow 2H—F(g)$$
$$\Delta H = 2(-135) = -270 \text{ kcal}$$

The sum of these steps is

$$OF_2(g) + H_2O(g) \rightarrow O_2(g) + 2HF(g)$$
$$\Delta H = -78 \text{ kcal}$$

The result of Example 14.2 is a more reliable value for the $\Delta H$ of this reaction. Since most average bond energies are approximations, a $\Delta H$ obtained by use of these values must be regarded as an estimate.

The problem could be solved in the same way by using values expressed in kJ/mol. The $\Delta H$ value derived in this manner is $-328$ kJ.

**14.4 SECOND LAW OF THERMODYNAMICS**

A chemical reaction or a physical change is said to be **spontaneous** if it has the potential to proceed of its own accord under the conditions specified. The first law of thermodynamics puts only one restriction on chemical or physical changes—energy must be conserved. However, the first law provides no basis for determining whether a proposed change will occur spontaneously. The second law of thermodynamics establishes criteria for making this important prediction.

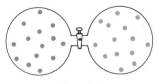

before mixing

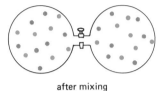

after mixing

**Figure 14.1** Spontaneous mixing of two gases.

The thermodynamic function **entropy**, $S$, is central to the second law. Entropy may be interpreted as a measure of the randomness, or disorder, of a system; a highly disordered system is said to have a high entropy. Since a disordered condition is more statistically probable than an ordered one, entropy may be regarded as a probability function (Section 9.4). One statement of the **second law of thermodynamics** is: every spontaneous change is accompanied by an increase in entropy.

As an example of a spontaneous change, consider the mixing of two ideal gases. The two gases, which are under the same pressure, are placed in bulbs that are joined by a stopcock (Figure 14.1). When the stopcock is opened, the gases spontaneously mix until each is evenly distributed throughout the entire apparatus. Why did this spontaneous change occur? The first law cannot help us answer this question. Throughout the mixing, the volume, total pressure, and temperature remain constant. Since the gases are ideal, no intermolecular forces exist, and neither the internal energy nor the enthalpy of the system is affected.

This change represents an increase in entropy. The final state is more random and hence more probable than the initial state. The random motion of the gas molecules has produced a more disordered condition. The fact that the gases mix spontaneously is not surprising; one would have predicted it from experience. Indeed, it would be surprising if the reverse were to be observed—a gaseous mixture spontaneously separating into two pure component gases, each occupying one of the bulbs.

For a given substance the solid, crystalline state is the state of lowest entropy (most ordered); the gaseous state is the state of highest entropy (most random); and the liquid state is intermediate between the other two. Hence, when a substance either melts or vaporizes, its entropy increases. The reverse changes, crystallization and condensation, are changes in which the entropy of the substance decreases. Why, then, should a substance spontaneously freeze at temperatures below its melting point, since this change represents a decrease in the entropy of the substance?

If entropy is to be used as a criterion of spontaneity, all the entropy effects that result from the proposed change must be considered. When two ideal gases mix by the process previously described, there is no exchange of matter or energy between the isolated system in which the change occurs and its surroundings; the only entropy effect is an increase in the entropy of the isolated system itself. Usually, however, a chemical reaction or a physical change is conducted in such a way that the system is not isolated from its surroundings, and the total change in entropy is equal to the sum of the change in the entropy of the system ($\Delta S_{\text{system}}$) and the change in entropy of the surroundings ($\Delta S_{\text{surroundings}}$).

$$\Delta S_{\text{total}} = \Delta S_{\text{system}} + \Delta S_{\text{surroundings}}$$

When a liquid freezes, the heat of fusion is evolved by the liquid and absorbed by the surroundings. This heat increases the random motion of the surrounding molecules and therefore increases the entropy of the surroundings. Hence, the spontaneous freezing of a liquid at a temperature below the melting point occurs because the decrease in entropy of the liquid ($\Delta S_{\text{system}}$, negative) is more than offset by the increase in entropy of the surroundings ($\Delta S_{\text{surroundings}}$, positive) so that there is a net increase in entropy.

The total change in entropy should always be considered to determine spontaneity. When a substance melts, the entropy increases, but this effect alone does not determine whether or not the transformation is spontaneous. The entropy of the surroundings must also be considered, and spontaneity is indicated only if the total entropy of system and surroundings taken together increases.

The data of Table 14.4 pertain to the freezing of water. Measurement of entropy changes and the meaning of the units of $\Delta S$ values will be discussed in later sections. For the moment, let us be concerned only with the numerical values listed in the table. At $-1°C$ the change is spontaneous; $\Delta S_{\text{total}}$ is positive. At $+1°C$, however, $\Delta S_{\text{total}}$ is negative, and freezing is a nonspontaneous change. On the other hand, the reverse change, melting, is spontaneous at $+1°C$ (the signs of all $\Delta S$ values would be reversed).

Table 14.4 Entropy changes for the transformation $H_2O(l) \rightarrow H_2O(s)$ at 1 atm

| TEMPERATURE (°C) | $\Delta S_{\text{system}}$ (cal/°K mol) | $\Delta S_{\text{surroundings}}$ (cal/°K mol) | $\Delta S_{\text{total}}$ (cal/°K mol) |
| --- | --- | --- | --- |
| +1 | −5.29 | +5.27 | −0.02 |
| 0 | −5.25 | +5.25 | 0 |
| −1 | −5.22 | +5.24 | +0.02 |

At $0°C$, the melting point, $\Delta S_{\text{total}}$ is zero, which means that neither freezing nor melting is spontaneous. At this temperature a water–ice system would be in equilibrium, and no *net* change would be observed. Note, however, that freezing or melting could be made to occur at $0°C$ by removing or adding heat, but neither change will occur spontaneously.

Thus, the $\Delta S_{\text{total}}$ of a postulated change may be used as a criterion for whether the change will occur spontaneously. The entropy of the universe is steadily increasing as spontaneous changes occur. Rudolf Clausius summarized the first and second laws of thermodynamics as: "The energy of the universe is constant; the entropy of the universe tends toward a maximum."

Entropy, like internal energy and enthalpy, is a state function. The entropy, or randomness, of a system in a given state is a definite value, and hence, $\Delta S$ for a change from one state to another is a definite value depending only on the initial and final states and not on the path between them.

It must be emphasized that, whereas thermodynamic concepts can be used to determine what changes are possible, thermodynamics has nothing to say about the rapidity of change. Some thermodynamically favored changes occur very slowly. Although reactions between carbon and oxygen, as well as between hydrogen and oxygen, at 25°C and 1 atm pressure are definitely predicted by theory, mixtures of carbon and oxygen and mixtures of hydrogen and oxygen can be kept for prolonged periods without significant reaction; such reactions are generally initiated by suitable means. Thermodynamics can authoritatively indicate postulated changes that will *not* occur and need not be attempted, and it can tell us how to alter the conditions of a presumably unfavored reaction in such a manner that the reaction will be thermodynamically possible.

## 14.5 GIBBS FREE ENERGY

The type of change of primary interest to the chemist is, of course, the chemical reaction. Determination of $\Delta S_{\text{system}}$ for a given reaction will be discussed in Sections 14.6 and 14.8. The $\Delta S_{\text{surroundings}}$ for a reaction conducted at constant temperature and pressure may be calculated by means of the equation

$$\Delta S_{\text{surroundings}} = -\frac{\Delta H}{T}$$

where $\Delta H$ is the enthalpy change of the reaction and $T$ is the absolute temperature. The change in the entropy of the surroundings is brought about by the heat transferred into or out of the surroundings because of the enthalpy change of the reaction. Since heat *evolved* by the reaction is *absorbed* by the surroundings (and *vice versa*), the sign of $\Delta H$ must be reversed. Hence, the larger the value of $-\Delta H$, the more disorder created in the surroundings and the larger the value of $\Delta S_{\text{surroundings}}$.

On the other hand, the change in the entropy of the surroundings is *inversely* proportional to the absolute temperature at which the change takes place. A given quantity of heat added to the surroundings at a low temperature (where the randomness is relatively low initially) will create a larger difference in the disorder of the surroundings than the same quantity of heat added at a high temperature (where the randomness is relatively high to begin with). Entropy is therefore measured in units of cal/°K or J/°K.

In the last section we noted that

$$\Delta S_{\text{total}} = \Delta S_{\text{system}} + \Delta S_{\text{surroundings}}$$

If $-\Delta H/T$ is substituted for $\Delta$ surroundings and if the symbol $\Delta S$ (without a subscript) is used to indicate the entropy change of the system, the following equation is obtained.

$$\Delta S_{\text{total}} = \Delta S - \frac{\Delta H}{T}$$

Multiplication by $T$ gives

$$T\Delta S_{\text{total}} = T\Delta S - \Delta H$$

Since $\Delta S$ is measured in units of cal/°K or J/°K, the term $T\Delta S$ is expressed in cal or J; therefore, all of the terms of this equation ($T\Delta S$ as well as $\Delta H$) are energy terms. By reversing the signs of the terms of this equation, we get

$$-T\Delta S_{\text{total}} = \Delta H - T\Delta S$$

The **Gibbs free energy** function, $G$, is defined by the equation

$$G = H - TS$$

For a change at constant temperature,

$$\Delta G = \Delta H - T\Delta S$$

Hence,

$$\Delta G = -T\Delta S_{\text{total}}$$

Since $\Delta S_{\text{total}}$ is greater than zero for a spontaneous change, $T\Delta S_{\text{total}}$ must also be greater than zero, and $-T\Delta S_{\text{total}}$ must be less than zero. Therefore, for a spontaneous change

$$\Delta G < 0$$

When a spontaneous change occurs at constant temperature and pressure, the free energy of the system decreases. Rather than use the positive value of $T\Delta S_{\text{total}}$ as a measure of spontaneity, the negative value of this quantity is employed so that values of $\Delta G$ conform to thermodynamic sign convention (negative values indicating energy liberated by the system). For a system in equilibrium,

$$\Delta G = 0$$

since $\Delta S_{\text{total}} = 0$ in such an instance. Notice that all of the terms used in defining $\Delta G$ pertain to changes in the properties of the *system;* therefore, the use of free energy values removes the necessity of considering changes in the surroundings.

What is the driving force of a chemical reaction? The earliest answer to the question was that reactions proceed because the reactants have a chemical affinity for each other; this "explanation" is still valid but unfortunately not very illuminating. In an effort to relate chemical affinity to some measurable quantity, Julius Thomsen and Marcellin Berthelot (1878) proposed that chemical reactions proceed spontaneously only if they evolve heat; presumably the heat evolved measures the chemical affinity of the reactants in going from a more energetic to a lesser energetic (more stable) state. Most chemical reactions are exothermic; however, the hypothesis of Thomsen and Berthelot is not tenable because spontaneous endothermic reactions are known and many spontaneous physical processes absorb heat (e.g., the melting of ice at room temperature).

The flaw of the hypothesis of Thomsen and Berthelot is that it ignores the role of entropy. The materials of a chemical reaction do indeed seek a minimum in energy, but they also seek a maximum in randomness (entropy). At times, these two factors work together; at other times, they oppose one another. The change in free energy takes into account both factors

$$\Delta G = \Delta H - T\Delta S$$

A system is more ordered at low temperatures than at high temperatures; as the temperature increases, the randomness increases. The factor $T\Delta S$ includes both temperature and change in entropy.

The most favorable circumstance for a negative value of $\Delta G$, which indicates a spontaneous reaction, is a negative value of $\Delta H$ together with a positive value of $\Delta S$ (reaction **a** in Table 14.5). However, a large negative value of $\Delta H$ can outweigh an unfavorable entropy change, resulting in a negative value of $\Delta G$ (reaction **b** in Table 14.5). In addition, a large positive value of $T\Delta S$ can overshadow an unfavorable enthalpy change, giving rise to a negative value of $\Delta G$ (reaction **c** in Table 14.5). For most chemical reactions at 25°C and 1 atm, the absolute value of $\Delta H$ is much larger than the value of $T\Delta S$; under these conditions exothermic reactions are usually spontaneous no matter how the entropy changes.

Table 14.5 Thermodynamic values for some chemical reactions at 25°C and 1 atm (kcal)

| REACTION | $\Delta H$ | $-$ | $(T\Delta S)$ | $=$ | $\Delta G$ |
|---|---|---|---|---|---|
| (a) $H_2(g) + Br_2(l) \rightarrow 2HBr(g)$ | $-17.32$ | $-$ | $(+8.13)$ | $=$ | $-25.45$ |
| (b) $2H_2(g) + O_2(g) \rightarrow 2H_2O(l)$ | $-136.64$ | $-$ | $(-23.25)$ | $=$ | $-113.39$ |
| (c) $Br_2(l) + Cl_2(g) \rightarrow 2BrCl(g)$ | $+7.02$ | $-$ | $(+7.45)$ | $=$ | $-0.43$ |
| (d) $2Ag_2O(s) \rightarrow 4Ag(s) + O_2(g)$ | $+14.62$ | $-$ | $(+9.44)$ | $=$ | $+5.18$ |

However, with increasing temperature the value of $T\Delta S$ increases and the influence of the entropy effect on $\Delta G$ increases. Both $\Delta H$ and $\Delta S$ usually do not change greatly with increasing temperature. However, the term $T\Delta S$ includes the temperature itself so that at high temperatures $T\Delta S$ can be sufficiently large to be the dominant influence on $\Delta G$. For example, consider reaction **d** in Table 14.5 (a typical decomposition reaction) for which both $\Delta H$ and $\Delta S$ are positive. At 25°C, $\Delta H$ is larger than $T\Delta S$ and therefore $\Delta G$ is positive. If we assume that $\Delta S$ does not change with increasing temperature (the actual change is small), at 300°C the value of $T\Delta S$ would be +18.94 kcal because of the increase in the value of $T$; therefore, $T\Delta S$ would be larger than $\Delta H$ (which also does not change greatly with increasing temperature), and $\Delta G$ would be negative.

The values in Table 14.5 pertain to the differences between the free energy of the products and the free energy of the reactants at 25°C and 1 atm. However, in some cases (notably reaction **c**) the reaction goes to an intermediate, equilibrium state rather than to completion because the free energy of the equilibrium state is lower than the free energy of the state at which the reaction is complete. Until equilibrium is discussed (Section 14.9), we should refrain from drawing conclusions as to the degree of completion of a reaction from such data.

**14.6 MEASUREMENT OF $\Delta G$ AND $\Delta S$**

Any spontaneous reaction has the capacity to do work. If work must be done on a system to bring about a chemical change, that change is not spontaneous. For a spontaneous reaction conducted at constant temperature and pressure, the decrease in Gibbs free energy is a measure of the maximum *useful* work that the system can do. The word *useful* is used to indicate that $\Delta G$ does not include pressure–volume work.

Some reactions are accompanied by an increase or decrease in volume (because gases are produced or used). If such reactions are run at constant pressure, work is done by the system to push back the atmosphere (in the case of an expanding volume) or work is done on the system by the surroundings (if the volume of the system contracts as the reaction proceeds). Such pressure–volume work must be done if the reaction is to take place at constant pressure, and therefore work of this type is not available for any other purpose. The change in free energy measures the maximum amount of work, other than pressure–volume work, that can be derived from a reaction under conditions of constant pressure. Since only a spontaneous reaction has the capacity to do work, $\Delta G$ is a measure of spontaneity.

Consider the following spontaneous reaction

$$2Ag(s) + Cl_2(g) \rightarrow 2AgCl(s) \qquad \Delta H = -60.72 \text{ kcal}$$

When this reaction occurs at 25°C and 1 atm, 60.72 kcal of heat is evolved, but no useful work is done. One way to obtain useful work from the reaction is to make it produce electrical work by having it serve as the cell reaction of a voltaic cell. However, in the normal operation of a spontaneously discharging cell, less than the maximum amount of useful work is obtained because of the internal resistance of the cell and the concentration changes within the cell that occur when it delivers current.

We can, however, use the value of the reversible emf of the cell (which can be derived from standard electrode potentials) to calculate the maximum amount of useful work that the cell can theoretically produce (Section 10.7). For the reaction under study,

anode: $\qquad 2Ag(s) + 2Cl^-(aq) \rightleftharpoons 2AgCl(s) + 2e^- \qquad \mathscr{E}_{ox} = -0.222$ V

cathode: $\qquad 2e^- + Cl_2(g) \rightleftharpoons 2Cl^-(aq) \qquad \mathscr{E} = +1.359$ V

cell reaction: $\quad 2Ag(s) + Cl_2(g) \rightarrow 2AgCl(s) \qquad \mathscr{E} = +1.137$ V

The reversible emf of the cell, $+1.137$ V, is the maximum voltage that the cell can develop.

The maximum quantity of electrical work that can be obtained from the reaction, which is the *decrease* in Gibbs free energy for the reaction, can be calculated from

$$\Delta G = -nF\mathscr{E}$$

where $n$ is the number of moles of electrons transferred in the complete reaction (or the number of faradays), $F$ is the value of the faraday, and $\mathscr{E}$ is the reversible emf of the cell in volts. If 96,487 coulombs is used as the value of the faraday, the answer is expressed in volt–coulombs, which are joules. On the other hand, the faraday also equals 23.061 kcal/V (since 1 J = 4.1840 cal), and if this value is used, the answer is expressed in kcal. For the complete reaction of silver and chlorine, 2 faradays of electricity is produced, and therefore

$$\Delta G = -2(23.06 \text{ kcal/V})(+1.137 \text{ V})$$
$$= -52.44 \text{ kcal}$$

Gibbs free energy, like the other thermodynamic functions we have discussed, is a state function; the value of $\Delta G$ depends only upon the final and initial states of the system and not upon the path taken between those states. When a spontaneous reaction occurs, the free energy of the system declines. Even though the reaction is conducted in such a way that no useful work is obtained from it, the possibility of getting the work defined by $\Delta G$ is lost. The value of $\Delta G$ for the reaction at 25°C and 1 atm is $-52.44$ kcal no matter how the reaction is conducted.

Therefore, for the reaction

$$2Ag(s) + Cl_2(g) \rightarrow 2AgCl(s)$$

at 25°C and 1 atm, $\Delta H$ is $-60.72$ kcal, and $\Delta G$ is $-52.44$ kcal. From these two values we can derive a value for the entropy change of the reaction.

Since

$$\Delta G = \Delta H - T\Delta S$$

$$-52.44 \text{ kcal} = -60.72 \text{ kcal} - T\Delta S$$

$$T\Delta S = -8.28 \text{ kcal}$$

But $T = 298°\text{K}$; therefore,

$$\Delta S = \frac{T\Delta S}{T}$$

$$= \frac{-8.28 \text{ kcal}}{298°\text{K}}$$

$$= -0.0278 \text{ kcal/}°\text{K} = -27.8 \text{ cal/}°\text{K}$$

The fact that $\Delta S$ is negative indicates that the system becomes more ordered (less random) as the reaction proceeds. Notice that a mole of gas is consumed during the course of the reaction; the decrease in the entropy of the system is therefore not surprising.

The relations between these thermodynamic functions are summarized in Figure 14.2. If the reaction is conducted outside the cell (a), heat equivalent to $-\Delta H$ is evolved. In the case of the ideal, reversible cell (b), the maximum useful work is obtained ($-\Delta G$) and heat equivalent to $-T\Delta S$ is evolved (which arises from the entropy change of the reaction). Since $\Delta G = \Delta H - T\Delta S$,

**Figure 14.2** Relations between thermodynamic functions for the reaction 2 Ag(s) + Cl$_2$(g) → 2 AgCl(s) at 25°C and 1 atm.

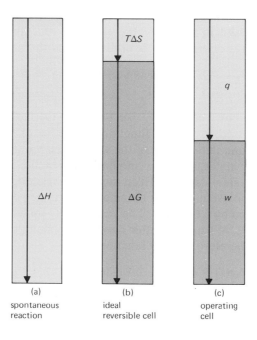

$$\Delta H = \Delta G + T\Delta S$$

as the figure illustrates. The ideal, reversible cell is an abstraction, not an operating device. When the reversible emf of a cell is measured, the emf of the cell is balanced against an external emf in such a way that no current flows; in this way the maximum voltage that the cell is capable of producing is measured. In an operating cell (c), less than the maximum amount of work is done $(w)$ and an amount of heat greater than $-T\Delta S$ is evolved $(q)$.

---

*Example 14.6*

For the standard cell

$$\text{Ag(s)}|\text{AgCl(s)}|\text{Cl}^-\text{(aq)}|\text{Hg}_2\text{Cl}_2\text{(s)}|\text{Hg(l)}|\text{Pt}$$

$\mathscr{E}$ is 0.0454 V and the cell reaction is

$$2\text{Ag(s)} + \text{Hg}_2\text{Cl}_2\text{(s)} \rightarrow 2\text{Hg(l)} + 2\text{AgCl(s)}$$

The heat of formation of AgCl(s) is $-30.36$ kcal/mol and that of $\text{Hg}_2\text{Cl}_2$ is $-63.32$ kcal/mol. Calculate (a) $\Delta H$, (b) $\Delta G$, and (c) $\Delta S$ for the cell reaction at 25°C.

*Solution*

(a) The enthalpy change for the reaction may be obtained from the $\Delta H_f$ values.

$$\Delta H = 2\Delta H_f(\text{AgCl}) - \Delta H_f(\text{Hg}_2\text{Cl}_2)$$
$$= 2(-30.36) - (-63.32)$$
$$= +2.60 \text{ kcal}$$

The reaction is endothermic.

(b) The free energy change may be calculated from the emf of the cell.

$$\Delta G = -nF\mathscr{E}$$
$$= -2(23.1 \text{ kcal/V})(0.0454 \text{ V})$$
$$= -2.10 \text{ kcal}$$

The cell reaction is spontaneous.

(c) The value of $\Delta S$ may be obtained from the relation

$$\Delta G = \Delta H - T\Delta S$$
$$-2.10 \text{ kcal} = +2.60 \text{ kcal} - T\Delta S$$
$$T\Delta S = +4.70 \text{ kcal}$$
$$\Delta S = \frac{+4,700 \text{ cal}}{298°\text{K}} = +15.8 \text{ cal/}°\text{K}$$

The $\Delta G$ value for the reaction is negative because this favorable entropy change outweighs an unfavorable change in enthalpy.

---

**14.7 STANDARD FREE ENERGIES**  A standard free energy change, which is given the symbol $\Delta G°$, is the free energy change for a reaction at 25°C and 1 atm in which the reactants in their standard states are converted to the products in their standard states. The value of $\Delta G°$ for a reaction can be derived from standard free energies of formation in the same way that $\Delta H°$ values can be calculated from standard enthalpies of formation.

The **standard free energy of formation** of a compound, $\Delta G_f°$, is defined as the change in standard free energies when one mole of the compound is prepared from its constituent elements. According to this definition, the standard free energy of formation of any element is zero. The value of $\Delta G°$ for a reaction is equal to the sum of the standard free energies of formation of the products minus the sum of the standard free energies of formation of the reactants. Some standard free energies of formation are given in Table 14.6. A negative value means that the compound can be prepared from its constituent elements under standard conditions.

Table 14.6 Gibbs free energy of formation at 25°C and 1 atm

| COMPOUND | $\Delta G_f^\circ$ (kcal/mol) | (kJ/mol)[a] | COMPOUND | $\Delta G_f^\circ$ (kcal/mol) | (kJ/mol)[a] |
|---|---|---|---|---|---|
| $H_2O(g)$ | −54.64 | −228.61 | $CO(g)$ | −32.81 | −137.28 |
| $H_2O(l)$ | −56.69 | −237.19 | $CO_2(g)$ | −94.26 | −394.38 |
| $HF(g)$ | −64.7 | −270.7 | $NO(g)$ | +20.72 | +86.69 |
| $HCl(g)$ | −22.77 | −95.27 | $NO_2(g)$ | +12.39 | +51.84 |
| $HBr(g)$ | −12.72 | −53.22 | $NaCl(s)$ | −97.79 | −384.05 |
| $HI(g)$ | +0.31 | +1.30 | $CaO(s)$ | −144.4 | −604.2 |
| $H_2S(g)$ | −7.89 | −33.0 | $Ca(OH)_2(s)$ | −214.33 | −896.76 |
| $NH_3(g)$ | −3.98 | −16.7 | $CaCO_3(s)$ | −269.78 | −1128.76 |
| $CH_4(g)$ | −12.14 | −50.79 | $BaO(s)$ | −126.3 | −528.4 |
| $C_2H_6(g)$ | −7.86 | −32.89 | $BaCO_3(s)$ | −272.2 | −1138.9 |
| $C_2H_4(g)$ | +16.28 | +68.12 | $Al_2O_3(s)$ | −376.77 | −1576.41 |
| $C_2H_2(g)$ | +50.00 | +209.20 | $Fe_2O_3(s)$ | −177.1 | −741.0 |
| $C_6H_6(l)$ | +30.99 | +129.66 | $AgCl(s)$ | −26.22 | −109.70 |
| $SO_2(g)$ | −71.79 | −300.37 | $ZnO(s)$ | −76.05 | −318.19 |

[a] 1 kcal = 4.184 kJ (exactly).

*Example 14.7*
Use $\Delta G_f^\circ$ values in Table 14.6 to calculate $\Delta G^\circ$ for the reaction

$$2NO(g) + O_2(g) \rightarrow 2NO_2(g)$$

*Solution*

$$\begin{aligned}\Delta G^\circ &= 2\Delta G_f^\circ(NO_2) - 2\Delta G_f^\circ(NO) \\ &= 2(+12.39 \text{ kcal}) - 2(+20.72 \text{ kcal}) \\ &= -16.66 \text{ kcal}\end{aligned}$$

The following illustrates how this calculation relates to the reactions involved.

| | |
|---|---|
| $N_2(g) + 2O_2(g) \rightarrow 2NO_2(g)$ | $\Delta G^\circ = 2(12.39 \text{ kcal}) = 24.78 \text{ kcal}$ |
| $2NO(g) \rightarrow N_2(g) + O_2(g)$ | $\Delta G^\circ = -2(20.72 \text{ kcal}) = -41.44 \text{ kcal}$ |
| $2NO(g) + O_2(g) \rightarrow 2NO_2(g)$ | $\Delta G^\circ = -16.66 \text{ kcal}$ |

Only free energy *changes* can be determined; the absolute values of free energies are not known and cannot be measured. The $\Delta G$ values used refer to the differences between the free energies of the products and the free energies of the reactants; these $\Delta G$ values do *not* imply that the *absolute* free energies of the elements are zero but rather that the standard free energies of formation of the elements are zero. When the two equations given in this problem are added, cancellation of terms gives the desired final equation so that the fact that absolute values of free energies (of compounds as well as of elements) are not known is of no importance.

---

**Example 14.8**

(a) What is $\Delta G°$ (in kJ) for the reaction

$$Cl_2(g) + 2I^-(aq) \rightarrow 2Cl^-(aq) + I_2(s)$$

For $I^-(aq)$, $\Delta G_f° = -51.7$ kJ/mol; for $Cl^-(aq)$, $\Delta G_f° = -131.2$ kJ/mol. (b) Use the value of $\Delta G°$ to calculate $\mathscr{E}°$ for a cell utilizing this reaction as the cell reaction.

**Solution**

(a)

$$\Delta G° = 2\Delta G_f°(Cl^-) - 2\Delta G_f°(I^-)$$
$$= 2(-131.2 \text{ kJ}) = 2(51.7 \text{ kJ})$$
$$= -159.0 \text{ kJ}$$

(b)

$$\Delta G° = -nF\mathscr{E}°$$

$$\mathscr{E}° = \frac{-159.0 \text{ kJ}}{-(2)(96.5 \text{ kJ/V})}$$

$$\mathscr{E}° = +0.824 \text{ V}$$

The value of $\mathscr{E}°$ derived from standard electrode potentials is +0.82 V.

---

**14.8 ABSOLUTE ENTROPIES**

The addition of heat to a substance results in an increase in molecular randomness. Hence, the entropy of a substance increases as the temperature increases. Conversely, cooling a substance makes it more ordered and decreases its entropy. At absolute zero the entropy of a perfect crystalline substance may be taken as zero. This statement is sometimes called the **third law of thermodynamics** and was first formulated by Walther Nernst in 1906. The entropy of an imperfect crystal, a glass, or a solid solution is not zero at 0°K.

On the basis of the third law, absolute entropies can be calculated from heat capacity data by extrapolating to absolute zero. The **standard absolute entropy** of a substance, $S°$, is the entropy of the substance in its standard state at 25°C and 1 atm; some $S°$ values are given in Table 14.7. The $\Delta S°$ value for a reaction is equal to the sum of the absolute entropies of the products minus the sum of the absolute entropies of the reactants. Note that the absolute entropy of an element is *not* equal to zero and that the absolute entropy of a compound is *not* the entropy change when the compound is formed from its constituent elements.

Table 14.7 Absolute entropy at 25°C and 1 atm

| SUBSTANCE | $S°$ (cal/°K mol) | $S°$ (J/°K mol)[a] | SUBSTANCE | $S°$ (cal/°K mol) | $S°$ (J/°K mol)[a] |
|---|---|---|---|---|---|
| $H_2(g)$ | 31.21 | 130.6 | $HBr(g)$ | 47.44 | 198.5 |
| $F_2(g)$ | 48.6 | 203.3 | $HI(g)$ | 49.31 | 206.3 |
| $Cl_2(g)$ | 53.29 | 223.0 | $H_2S(g)$ | 49.15 | 205.6 |
| $Br_2(l)$ | 36.4 | 152.3 | $NH_3(g)$ | 46.01 | 192.5 |
| $I_2(s)$ | 27.9 | 116.7 | $CH_4(g)$ | 44.50 | 186.2 |
| $O_2(g)$ | 49.003 | 205.03 | $C_2H_6(g)$ | 54.85 | 229.5 |
| S(rhombic) | 7.62 | 31.9 | $C_2H_4(g)$ | 52.45 | 219.5 |
| $N_2(g)$ | 45.77 | 191.5 | $C_2H_2(g)$ | 48.00 | 200.8 |
| C(graphite) | 1.36 | 5.69 | $SO_2(g)$ | 59.40 | 248.5 |
| Li(s) | 6.7 | 28.0 | $CO(g)$ | 47.30 | 197.9 |
| Na(s) | 12.2 | 51.0 | $CO_2(g)$ | 51.06 | 213.6 |
| Ca(s) | 9.95 | 41.6 | $NO(g)$ | 50.34 | 210.6 |
| Al(s) | 6.77 | 28.3 | $NO_2(g)$ | 57.47 | 240.5 |
| Ag(s) | 10.21 | 42.72 | NaCl(s) | 17.30 | 72.38 |
| Fe(s) | 6.49 | 27.2 | CaO(s) | 9.5 | 39.8 |
| Zn(s) | 9.95 | 41.6 | $Ca(OH)_2(s)$ | 18.2 | 76.1 |
| Hg(l) | 18.5 | 77.4 | $CaCO_3(s)$ | 22.2 | 92.9 |
| La(s) | 13.7 | 57.3 | $Al_2O_3(s)$ | 12.19 | 51.00 |
| $H_2O(g)$ | 45.11 | 188.7 | $Fe_2O_3(s)$ | 21.5 | 90.0 |
| $H_2O(l)$ | 16.72 | 69.96 | HgO(s) | 17.2 | 72.0 |
| HF(g) | 41.47 | 173.5 | AgCl(s) | 22.97 | 96.11 |
| HCl(g) | 44.62 | 186.7 | ZnO(s) | 10.5 | 43.9 |

[a] 1 cal = 4.184 J (exactly).

*Example 14.9*
(a) The standard absolute entropies of the substances in the reaction

$$2Hg(l) + O_2(g) \rightarrow 2HgO(s)$$

are given in Table 14.7. Calculate $\Delta S°$ for the reaction. (b) The heat of formation, $\Delta H_f°$, of HgO is $-21.68$ kcal/mol. What is the standard free energy of formation of HgO(s)?

*Solution*
(a)

$$\Delta S° = 2S°(HgO) - [2S°(Hg) + S°(O_2)]$$
$$= 2(17.2) - [2(18.5) + 49.0]$$
$$= -51.6 \text{ cal/°K}$$

(b) For the reaction as written

$$\Delta H° = 2\Delta H_f° = -43,400 \text{ cal}$$

$$\Delta G° = \Delta H° - T\Delta S°$$
$$= -43,400 - (298)(-51.6)$$
$$= -28,000 \text{ cal}$$

The equation shows the formation of two moles of HgO; for one mole, $\Delta G_f° = -14.0$ kcal.

**14.9 FREE ENERGY AND CHEMICAL EQUILIBRIUM**

The standard free energy change, $\Delta G°$, for the reaction

$$N_2O_4(g) \rightleftharpoons 2NO_2(g)$$

is $+1.29$ kcal. Since $\Delta G°$ is positive, one might predict that $N_2O_4$ in its standard state at 25°C and 1 atm would not dissociate into $NO_2$ at all and that the reverse reaction, the formation of $N_2O_4$ from $NO_2$, would go to completion. Both these predictions are incorrect. The reaction is reversible, and as we have seen, reversible reactions tend to go to equilibrium, not to completion. Furthermore, equilibrium can be approached from either direction; both the forward reaction (to the equilibrium state) and the reverse reaction (to the equilibrium state) should be spontaneous.

The free energy of a system in which this reaction occurs at 25°C and 1 atm is plotted against the fraction of $N_2O_4$ dissociated in Figure 14.3. Point A represents the standard free energy of one mole of $N_2O_4$, point B represents the standard free energy of two moles of $NO_2$, and the intervening points on this curve represent the free energies of mixtures of $N_2O_4$ and $NO_2$ such that a material balance exists. Absolute values of free energies are not known, and no scale is indicated on the vertical axis of the diagram; however, differences in free energies can be calculated so that the shape of the curve is accurately represented.

The free energy curve exhibits a minimum at the equilibrium point, E, where 16.6% of the $N_2O_4$ is dissociated. The difference between the standard free energy of two moles of $NO_2$ (point B) and the standard free

**Figure 14.3** Free energy of a system that contains the equivalent of one mole of $N_2O_4$ as the reaction $N_2O_4(g) \rightleftharpoons 2\ NO_2(g)$ occurs (25°C and 1 atm).

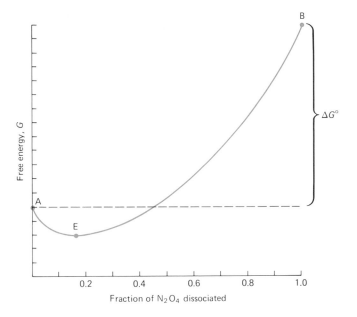

energy of one mole of $N_2O_4$ (point A) is $\Delta G°$ for the reaction (+1.29 kcal) and is indicated on the diagram. However, $\Delta G$ for the preparation of the equilibrium mixture (point E) from one mole of $N_2O_4$ (point A) is −0.20 kcal, which indicates that $N_2O_4$ will spontaneously dissociate until equilibrium is reached.

The graph shows that equilibrium can be approached from either direction. Thus, $\Delta G = -1.49$ kcal for the preparation of the equilibrium mixture (point E) from two moles of pure $NO_2$ (point B). The negative values of $\Delta G$ for both changes (from A to E and from B to E) indicate that both changes are spontaneous.

Why does the free energy curve exhibit a minimum? The value of $\Delta G$ for the change from pure $N_2O_4$ to the equilibrium state, in which 16.6% of the $N_2O_4$ has dissociated, might be expected to be 0.166 times the $\Delta G°$ value (+1.29 kcal), or +0.22 kcal; instead, $\Delta G$ for this change is −0.20 kcal. The value of $\Delta H$ for this change ($N_2O_4$ to equilibrium) is indeed 0.166 times $\Delta H°$, but $\Delta S$ for this change is larger than 0.166 times $\Delta S°$. The value 0.166 $\Delta S°$ is equal to the difference between (a) the weighted average of the entropies of pure $N_2O_4$ in its standard state and pure $NO_2$ in its standard state (weighted according to the relative amounts of each gas present in the equilibrium mixture) and (b) the entropy of one mole of $N_2O_4$ in its standard state.

The entropy of the equilibrium mixture is larger than this weighted average because of the entropy of mixing—an "extra" increment of randomness brought about by the mixing of unlike gases. Notice that neither gas is in its standard state at equilibrium; the partial pressure of each gas is less than 1 atm (the total pressure is 1 atm). The entropy (randomness) of a gas increases when its volume expands and pressure decreases. Since $\Delta G = \Delta H - T\Delta S$, the term $T\Delta S$ is *subtracted* from $\Delta H$ in the calculation of $\Delta G$, and the increase in $\Delta S$ over the "expected" produces a minimum in the free energy curve.

One criterion of chemical equilibrium is that $\Delta G = 0$ for a reaction in equilibrium. If we start with one mole of $N_2O_4$ and allow it to dissociate, the free energy of the system declines until equilibrium is attained. At that point the free energy of the system is a minimum, no further change in free energy is observed, and $\Delta G = 0$. At equilibrium there is no *net* change in the system that requires work or that can be harnessed to do work. The macroscopic properties of the system do not change with time (the concentrations of all substances present are constant) even though the reaction is proceeding in both directions.

If we assign the Gibbs free energy of a mole of $NO_2$ in its standard state (unit activity) the symbol $G°_{NO_2}$, the free energy of a mole of $NO_2$ at another activity, $G_{NO_2}$, is given by the equation

$$G_{NO_2} = G°_{NO_2} + RT \ln a_{NO_2}$$

where $R$ is the gas constant, $T$ is the absolute temperature, and $\ln a_{NO_2}$ is the natural logarithm of the activity of $NO_2$. Notice that since the natural logarithm of 1 is zero, $G = G°$ when the activity of $NO_2$ is 1, which means that the standard state is characterized by unit activity. For two moles of $NO_2$,

$$2G_{NO_2} = 2G°_{NO_2} + 2RT \ln a_{NO_2}$$

or

$$2G_{NO_2} = 2G°_{NO_2} + RT \ln(a_{NO_2})^2 \tag{1}$$

Similarly, for one mole of $N_2O_4$,

$$G_{N_2O_4} = G°_{N_2O_4} + RT \ln a_{N_2O_4} \tag{2}$$

The free energy change for the reaction

$$N_2O_4(g) \rightleftharpoons 2NO_2(g)$$

is given by the equation

$$\Delta G = 2G_{NO_2} - G_{N_2O_4} \tag{3}$$

and the standard free energy change is

$$\Delta G° = 2G°_{NO_2} - G°_{N_2O_4} \tag{4}$$

By substituting equations (1) and (2) into equation (3), we get

$$\Delta G = 2G°_{NO_2} + RT \ln (a_{NO_2})^2 - G°_{N_2O_4} - RT \ln a_{N_2O_4}$$

Combining terms gives

$$\Delta G = 2G°_{NO_2} - G°_{N_2O_4} + RT \ln \left( \frac{(a_{NO_2})^2}{a_{N_2O_4}} \right)$$

From equation (4),

$$\Delta G = \Delta G° + RT \ln \left( \frac{(a_{NO_2})^2}{a_{N_2O_4}} \right)$$

In this equation the logarithmic term corrects the standard free energy change, $\Delta G°$, to the free energy change for a more general condition in which the activities of the substances involved are not unity. We shall use the approximation that the activities of gases are given by their partial pressures.

At equilibrium $\Delta G = 0$, and therefore

$$0 = \Delta G° + RT \ln \left( \frac{(p_{NO_2})^2}{p_{N_2O_4}} \right)$$

Since the system is at equilibrium, the fraction in this equation is the equilibrium constant $K_p$. Therefore,

$$\Delta G^\circ = -RT \ln K_p$$

or

$$\Delta G^\circ = -2.303 \ RT \log K_p$$

where $R$ is the gas constant (1.987 cal/$^\circ$K mol or 8.3143 J/$^\circ$K mol) and $T$ is the absolute temperature.

For the dissociation of $N_2O_4$

$$\Delta G^\circ = -2.303 \ RT \log K_p$$

$$+1290 \text{ cal/mol} = -(2.303)(1.987 \text{ cal/}^\circ\text{K mol})(298^\circ\text{K}) \log K_p$$

$$\log K_p = -0.946 = 0.054 - 1$$

$$K_p = 1.13 \times 10^{-1} = 0.113$$

The numerical value of $K$ obtained by use of this equation depends upon the definition of the standard states used in determining $\Delta G^\circ$. Since the standard state of a gas is defined in terms of its pressure (in atm), $K_p$ is obtained in this case.

The equation relating $\Delta G^\circ$ with the equilibrium constant is an important one. It may be used to derive values of equilibrium constants from thermodynamic data. In addition, since $\Delta G^\circ = -nF\mathcal{E}^\circ$, electrochemical data can be used to calculate equilibrium constants for reactions that can be studied in voltaic cells. This equation also gives us the ability to interpret more fully the meaning of $\Delta G^\circ$ in relation to reaction spontaneity.

Values of $K$ corresponding to various $\Delta G^\circ$ values are listed in Table 14.8. A large *negative* value of $\Delta G^\circ$ means that $K$ for the reaction is a large *positive* value, and therefore the reaction from left to right will go virtually to completion. On the other hand, if $\Delta G^\circ$ is a large positive value,

Table 14.8 Values of $K$ corresponding to $\Delta G^\circ$ values according to the equation $\Delta G^\circ = -RT \ln K$.

| $\Delta G^\circ$ | | |
|---|---|---|
| kcal | kJ | $K$ |
| −50 | −209.2 | $4.5 \times 10^{36}$ |
| −25 | −104.6 | $2.1 \times 10^{18}$ |
| −10 | −41.84 | $2.1 \times 10^{7}$ |
| −5 | −20.92 | 4600 |
| −1 | −4.184 | 5.4 |
| 0 | 0 | 1.0 |
| +1 | +4.184 | 0.18 |
| +5 | +20.92 | 0.00022 |
| +10 | +41.84 | $4.7 \times 10^{-8}$ |
| +25 | +104.6 | $4.7 \times 10^{-19}$ |
| +50 | +209.2 | $2.2 \times 10^{-37}$ |

$K$ will be extremely small, thus indicating that the reverse reaction, from right to left, will go virtually to completion. Only if the value of $\Delta G^\circ$ is neither very large nor very small (see Table 14.8) will the value of $K$ indicate a situation in which the reaction will not go essentially to completion in one direction or the other.

---

**Example 14.10**

Calculate $K_p$ for the following reaction at 25°C.

$$2SO_2(g) + O_2(g) \rightleftharpoons 2SO_3(g)$$

$\Delta G_f^\circ(SO_2) = -71.79$ kcal/mol;

$\Delta G_f^\circ(SO_3) = -88.52$ kcal/mol.

**Solution**

The standard free energy change for the reaction is

$\Delta G^\circ = 2$ mol $(-88.52$ kcal/mol$) - 2$ mol $(-71.79$ kcal/mol$)$

$\Delta G^\circ = -33.46$ kcal

If substitutions are made for the constants in the equation $\Delta G^\circ = -2.303\ RT \log K_p$, in terms of a temperature of 25°C and energy units of kcal, the equation becomes

$$\Delta G^\circ = -(1.364\ \text{kcal/mol}) \log K_p$$

If energy units of kJ are used, at 25°C

$$\Delta G^\circ = -(5.709\ \text{kJ/mol}) \log K_p$$

Thus,

$$\log K_p = \frac{\Delta G^\circ}{-1.364\ \text{kcal/mol}} = \frac{-33.46\ \text{kcal/mol}}{-1.364\ \text{kcal/mol}} = 22.09$$

$$K_p = 1.2 \times 10^{22}$$

---

**Example 14.11**

Using electrochemical data, calculate $K$ for the reaction

$$Fe^{2+}(aq) + Ag^+(aq) \rightleftharpoons Fe^{3+}(aq) + Ag(s)$$

**Solution**

The half reactions are

$$Fe^{2+} \rightarrow Fe^{3+} + e^- \qquad \mathscr{E}_{ox}^\circ = -0.771\ \text{V}$$

$$e^- + Ag^+ \rightarrow Ag(s) \qquad \mathscr{E}^\circ = +0.799\ \text{V}$$

Therefore, $\mathscr{E}^\circ = +0.028$ V and 1 F of electricity is involved. The standard change in free energy is

$$\Delta G^\circ = -nF\mathscr{E}^\circ$$

$$= -1(23.1\ \text{kcal/V})(0.028\ \text{V})$$

$$= -0.65\ \text{kcal}$$

The equilibrium constant may be calculated from the equation

$$\Delta G^\circ = -(1.36\ \text{kcal/mol}) \log K$$

$$\log K = \frac{-0.65\ \text{kcal/mol}}{-1.36\ \text{kcal/mol}} = 0.478$$

$$K = 3.0$$

Since the standard states and activities of solutes in aqueous solutions are defined in terms of ideal concentrations, $K$ obtained in this calculation pertains to activities so defined. If energy units of kJ are used, $\Delta G^\circ = -2.70$ kJ/mol, and

$$\Delta G^\circ = -(5.71\ \text{kJ/mol}) \log K$$

$$\log K = \frac{-2.70\ \text{kJ/mol}}{-5.71\ \text{kJ/mol}} = 0.473$$

$$K = 3.0$$

## Problems

Thermochemical problems may be solved using either calories or joules as energy units. Thermochemical values expressed in joules appear in brackets.

**14.1** The combustion of 1.000 g of ethyl alcohol, $C_2H_5OH(l)$, in a bomb calorimeter (at constant volume) evolves 7.08 kcal [or 29.62 kJ] of heat at 25°C. The products of the combustion are $CO_2(g)$ and $H_2O(l)$. The molecular weight of ethyl alcohol is 46.06. (a) What is $\Delta E°$ for the combustion of one mole of ethyl alcohol? (b) Write the equation for the reaction. What is the value of $\Delta H°$ for the combustion? (c) Use values from Table 14.1 and your answer to part (b) to calculate $\Delta H_f°$ for ethyl alcohol.

**14.2** The combustion of 1.000 g of cyclohexane, $C_6H_{12}(l)$, in a bomb calorimeter (at constant volume) evolves 11.11 kcal [or 46.48 kJ] of heat at 25°C. The products of the reaction are $CO_2(g)$ and $H_2O(l)$. The molecular weight of cyclohexane is 84.16. (a) What is $\Delta E°$ for the combustion of one mole of cyclohexane? (b) Write the equation for the reaction. What is the value of $\Delta H°$ for the combustion? (c) Use values from Table 14.1 and your answer to part (b) to calculate the value of $\Delta H_f°$ for cyclohexane.

**14.3** Calculate $\Delta E°$ for the combustion of octane, $C_8H_{18}(l)$, to $CO_2(g)$ and $H_2O(l)$ at 25°C. The value of $\Delta H°$ for this reaction at 25°C is $-1307.53$ kcal per mole of octane [or $-5470.71$ kJ per mole of octane].

**14.4** Calculate $\Delta H°$ and $\Delta E°$ for the reaction

$$3NO_2(g) + H_2O(l) \rightarrow 2HNO_3(l) + NO(g)$$

Use the values given in Table 14.1.

**14.5** Calculate $\Delta H°$ and $\Delta E°$ for the reaction of $Ca_3P_2(s)$ and $H_2O(l)$. The reaction yields $Ca(OH)_2(s)$ and $PH_3(g)$. The enthalpy of formation values needed may be found in Table 14.1.

**14.6** For the reaction

$$CaCN_2(s) + 3H_2O(l) \rightarrow CaCO_3(s) + 2NH_3(g)$$

$\Delta E°$ is $-62.56$ kcal [or $-261.75$ kJ]. Use this fact and values from Table 14.1 to calculate $\Delta H_f°$ for $CaCN_2(s)$.

**14.7** For the reaction

$$NH_4NO_3(s) \rightarrow N_2O(g) + 2H_2O(l)$$

$\Delta E°$ is $-30.47$ kcal [or $-127.49$ kJ]. Use values from Table 14.1 to calculate $\Delta H_f°$ for $N_2O(g)$.

**14.8** The enthalpy of formation of $P(g)$ from solid phosphorus is $+75$ kcal/mol [or $+314$ kJ/mol]. Use this fact plus pertinent bond energies from Table 14.3 to calculate the enthalpy of formation of $PCl_3(g)$.

**14.9** For $CF_4(g)$, $\Delta H_f°$ is $-218.3$ kcal/mol [or $-913.4$ kJ/mol]. Use bond energies from Table 14.3 to calculate the heat of sublimation of graphite.

**14.10** (a) Use bond energies (Table 14.3) to calculate $\Delta H$ for the reaction

$$H—C\equiv N(g) + 2H_2(g) \rightarrow$$

$$\begin{array}{ccc} & H & H \\ & | & | \\ H—&C—N& (g) \\ & | & | \\ & H & H \end{array}$$

(b) Calculate $\Delta H°$ for the reaction using enthalpies of formation from Table 14.1. How well do the two values agree?

**14.11** (a) Use bond energies (Table 14.3) to calculate $\Delta H$ for the reaction

$$CH_4(g) + 4F_2(g) \rightarrow$$
$$CF_4(g) + 4HF(g)$$

(b) Calculate $\Delta H°$ for the reaction using enthalpies of formation (Table 14.1)

**14.12** (a) Use enthalpies of formation (Table 14.1) to calculate $\Delta H°$ for the reaction

$$CO(g) + 2H_2(g) \rightarrow$$

$$\begin{array}{c} H \\ | \\ H—C—O—H \ (g) \\ | \\ H \end{array}$$

(b) Use bond energies (Table 14.3) to calculate the heat of atomization of $CH_3OH(g)$. (c) Use your answers to parts (a) and (b) and the bond energy of the H—H bond to calculate a value for the bond energy of the C to O bond found in carbon monoxide.

**14.13** (a) Use enthalpies of formation (Table 14.1) to calculate $\Delta H°$ for the reaction

$$NH_3(g) + 3F_2(g) \rightarrow$$
$$NF_3(g) + 3HF(g)$$

(b) Use your answer to part (a) and bond energy values from Table 14.3 to calculate the bond energy of the N—F bond.

**14.14** Consider the standard cell

$$Cu(s) \mid CuCl(s) \mid Cl^-(aq) \mid$$
$$Cl_2(g) \mid Pt(s)$$

(a) Write the equation for the cell reaction. (b) Calculate $\mathscr{E}°$ for the half reaction

$$e^- + CuCl(s) \rightarrow Cu(s) + Cl^-(aq)$$

The free energy of formation of $CuCl(s)$ is $-28.20$ kcal/mol [or $-117.99$ kJ/mol], and $\mathscr{E}°$ for the $Cl^-/Cl_2$ electrode is $+1.360$ V.

**14.15** The following half reactions take place in a standard cell

$$Te(s) + 2H_2O \rightarrow$$
$$TeO_2(s) + 4H^+(aq) + 4e^-$$
$$\mathscr{E}°_{ox} = ?$$

$$4e^- + 4H^+(aq) + O_2(g) \rightarrow 2H_2O$$
$$\mathscr{E}° = +1.229 \text{ V}$$

(a) Write the equation for the cell reaction. (b) The free energy of formation of $TeO_2(s)$ is $-64.60$ kcal/mol [or $-270.29$ kJ/mol]. What is $\mathscr{E}°$ for the $TeO_2/Te$ electrode?

**14.16** For the half reaction

$$XeF_2(aq) + 2H^+(aq) + 2e^- \rightarrow$$
$$Xe(g) + 2HF(aq)$$

$\mathscr{E}°$ has been reported to equal $+2.64$ V. (a) Calculate $\Delta G°$ for the reaction

$$XeF_2(aq) + H_2(g) \rightarrow$$
$$Xe(g) + 2HF(aq)$$

(b) The free energy of formation, $\Delta G_f^\circ$, of HF(aq) is $-66.08$ kcal/mol [or $-276.48$ kJ/mol]. What is $\Delta G_f^\circ$ of $XeF_2$(aq)?

**14.17** Given the following electrode potentials

$$H_3BO_3(s) + 3H^+(aq) + 3e^- \rightarrow$$
$$B(s) + 3H_2O$$
$$\mathscr{E}^\circ = -0.869 \text{ V}$$

$$4H^+(aq) + O_2(g) + 4e^- \rightarrow 2H_2O$$
$$\mathscr{E}^\circ = +1.229 \text{ V}$$

and the value of $\Delta G_f^\circ$ for $H_2O(l)$ in Table 14.6, determine $\Delta G_f^\circ$ of $H_3BO_3$?

**14.18** (a) Use electrode potentials (Appendix D) to calculate the emf of a standard cell utilizing the reaction

$$Cl_2(g) + 2Br^-(aq) \rightarrow$$
$$2Cl^-(aq) + Br_2(l)$$

(b) What is $\Delta G^\circ$ for the reaction? (c) If $\Delta H^\circ$ for the reaction is $-22.25$ kcal [or $-93.09$ kJ], what is $\Delta S^\circ$? (d) What is $\Delta E^\circ$ for the reaction?

**14.19** (a) Use electrode potentials (Appendix D) to calculate the emf of a standard cell utilizing the reaction

$$2H_2O_2(aq) \rightarrow O_2(g) + 2H_2O$$

(b) What is $\Delta G^\circ$ for the reaction? (c) The $\Delta H_f^\circ$ of $H_2O(l)$ is $-68.32$ kcal/mol [or $-285.85$ kJ/mol], and the $\Delta H_f^\circ$ of $H_2O_2$(aq) is $-45.68$ kcal/mol [or $-191.13$ kJ/mol]. What is $\Delta H^\circ$ for the reaction? (d) What is $\Delta S^\circ$ for the reaction? (e) What is $\Delta E^\circ$ for the reaction?

**14.20** (a) Use electrode potentials (Appendix D) to calculate the emf of a standard cell utilizing the reaction

$$I_2(s) + H_2S(aq) \rightarrow$$
$$S(rhombic) + 2H^+(aq) + 2I^-(aq)$$

(b) What is $\Delta G^\circ$ for the reaction? (c) The absolute entropy of $H_2S$(aq) is 29.2 cal/$^\circ$K mol [or 122.2 J/$^\circ$K mol], that of $I^-$(aq) is 26.14 cal/$^\circ$K mol [or 109.37 J/$^\circ$K mol], and that of $H^+$(aq) is 0.0 cal/$^\circ$K mol; other values can be found in Table 14.7. What is $\Delta S^\circ$ for the reaction? (d) What is $\Delta H^\circ$ for the reaction?

**14.21** For the reaction

$$2Li(s) + 2H^+(aq) \rightarrow$$
$$2Li^+(aq) + H_2(g)$$

$\Delta H^\circ$ is $-133.1$ kcal [or $-556.9$ kJ]. The value of $S^\circ$ for $Li^+$(aq) is 3.4 cal/$^\circ$K mol [or 14.2 J/$^\circ$K mol], and that of $H^+$(aq) is 0.0; other values can be found in Table 14.7. Calculate (a) $\Delta S^\circ$ and (b) $\Delta G$ for the reaction. (c) Determine $\mathscr{E}^\circ$ for the half reaction

$$Li^+(aq) + e^- \rightarrow Li(s)$$

**14.22** For the reaction

$$2La(s) + 6H^+(aq) \rightarrow$$
$$2La^{3+}(aq) + 3H_2(g)$$

$\Delta H^\circ = -352.4$ kcal [or $-1474.4$ kJ]. The value of $S^\circ$ for $La^{3+}$(aq) is $-39.0$ cal/$^\circ$K mol [or $-163.2$ J/$^\circ$K mol] and that for $H^+$(aq) is 0.0; other values can be found in Table 14.7. Calculate (a) $\Delta S^\circ$ and (b) $\Delta G^\circ$ for the reaction. (c) Determine $\mathscr{E}^\circ$ for the half reaction

$$La^{3+}(aq) + 3e^- \rightarrow La(s)$$

**14.23** Perbromates have only recently been prepared. For the half reaction

$$BrO_4^-(aq) + 8H^+(aq) + 7e^- \rightarrow$$
$$\tfrac{1}{2}Br_2(l) + 4H_2O(l)$$

$\mathscr{E}°$ is reported to be $+1.59$ V. (a) For $H_2O(l)$, $\Delta G_f°$ is $-56.69$ kcal/mol [or $-237.19$ kJ/mol]; for $H^+(aq)$, $\Delta G_f° = 0.0$. Calculate $\Delta G_f°$ for $BrO_4^-(aq)$. (b) Use values from (a) and the fact that $\Delta G_f°$ for $BrO_3^-(aq)$ is $+5.19$ kcal/mol [or $+21.71$ kJ/mol] to calculate $\mathscr{E}°$ for the half reaction

$$BrO_4^-(aq) + 2H^+(aq) + 2e^- \rightarrow$$
$$BrO_3^-(aq) + H_2O$$

(c) What oxidizing agents listed in Appendix D *should* be capable of oxidizing $BrO_3^-(aq)$ to $BrO_4^-(aq)$?

**14.24** The heat of vaporization of benzene, $C_6H_6$, is 94.14 cal/g [or 393.88 J/g] at 80.1°C, which is the normal boiling point of benzene. For the vaporization of one mole of liquid benzene calculate (a) $\Delta H$, (b) $\Delta E$, (c) $\Delta S$, (d) $\Delta G$.

**14.25** The signs of the standard free energies of formation of all oxides of nitrogen are positive. What implications does this fact have regarding (a) preparation of these oxides from the elements and (b) the products of the combustion of a nitrogen-containing compound in oxygen?

**14.26** For oxygen difluoride, $OF_2(g)$, $\Delta G_f° = +9.7$ kcal/mol [or $+40.6$ kJ/mol]. (a) Is the preparation of $OF_2(g)$ from the constituent elements a spontaneous reaction at 25°C? (b) For ozone, $O_3(g)$, $\Delta G_f° = +39.06$ kcal/mol [or $+163.43$ kJ/mol]. Is it theoretically possible to prepare $OF_2(g)$ from $F_2(g)$ and $O_3(g)$ at 25°C?

**14.27** (a) For $PH_3(g)$, $\Delta H_f° = +2.21$ kcal/mol [or $+9.25$ kJ/mol] and $S° = 50.2$ cal/°K mol [or 210.0 J/°K mol]. For the standard state of $P(s)$, $S° = 10.6$ cal/°K mol [or 44.4 J/°K mol], and for $H_2(g)$, $S° = 31.2$ cal/°K mol [or 130.5 J/°K mol]. What is the standard free energy of formation of $PH_3(g)$? (b) Is the formation of $PH_3(g)$ from the elements a spontaneous reaction at 25°C?

**14.28** Will both $BF_3(g)$ and $BCl_3(l)$ hydrolyze at 25°C according to the following equation (where $X = F$ or Cl)?

$$BX_3 + 3H_2O(l) \rightarrow$$
$$H_3BO_3(aq) + 3HX(aq)$$

The pertinent $\Delta G_f°$ values are: $H_3BO_3(aq)$, $-230.24$ kcal/mol [or $-963.32$ kJ/mol]; HF(aq), $-66.08$ kcal/mol [or $-276.48$ kJ/mol]; HCl(aq), $-31.35$ kcal/mol [or $-131.17$ kJ/mol]; $H_2O(l)$, $-56.69$ kcal/mol [or $-237.19$ kJ/mol]; $BF_3(g)$, $-261.30$ kcal/mol [or $-1093.28$ kJ/mol]; $BCl_3(l)$, $-90.60$ kcal/mol [or $-379.07$ kJ/mol].

**14.29** For the reaction

$$HCO_2H(l) \rightarrow CO_2(g) + H_2(g)$$

$\Delta H° = +3.75$ kcal [or $+15.69$ kJ] and $\Delta S° = +51.45$ cal/°K [or $+215.27$ J/°K]. Is the decomposition of formic acid as shown in the equation a spontaneous reaction at 25°C?

**14.30** Is the formation of the poisonous gas phosgene, $COCl_2$, from chloroform, $CHCl_3$, and oxygen a spontaneous reaction at 25°C?

$$CHCl_3(l) + \tfrac{1}{2}O_2(g) \rightarrow$$
$$COCl_2(g) + HCl(g)$$

Enthalpies of formation needed to find $\Delta H°$ for the reaction are found in Table 14.1. The absolute entropy of $CHCl_3(l)$ is 48.5 cal/°K mol [or 202.9 J/°K mol], and that of $COCl_2(g)$ is 69.1 cal/°K mol [or 289.1 J/°K mol]; other values are found in Table 14.7.

**14.31** In a table of thermodynamic values, $\Delta H_f° = -93.0$ kcal/mol [or $-389.1$ kJ/mol] and $\Delta G_f° = -75.0$ kcal/mol [or $-313.8$ kJ/mol] are listed for $SO_2Cl_2(l)$, but no value is given for $S°$ of the compound. The value of $\Delta H_f°$, $\Delta G_f°$, and $S°$ for $SO_2(g)$ can be found in tables in this chapter. The value of $S°$ for $Cl_2$ can be found in Table 14.7. Find a value for the absolute entropy, $S°$, of $SO_2Cl_2(l)$.

**14.32** For HgBr(s), $\Delta H_f° = -40.5$ kcal/mol [or $-169.5$ kJ/mol], and $\Delta G_f° = -38.8$ kcal/mol [or $-162.3$ kJ/mol]. The absolute entropies of Hg(l) and $Br_2(l)$ can be found in Table 14.7. What is $S°$ for $HgBr_2(l)$?

**14.33** Graphite is the standard state of carbon, and $S°$ for graphite is 1.361 cal/°K mol [or 5.694 J/°K mol]. For diamond, $\Delta H_f° = +0.453$ kcal/mol [or $+1.895$ kJ/mol], and $\Delta G_f° = +0.685$ kcal/mol [or+ 2.866 kJ/mol]. What is the absolute entropy of diamond at 25°C? Which form of carbon is the more ordered?

**14.34** The combustion of cyanogen, $C_2N_2(g)$, in oxygen (to form carbon dioxide and nitrogen) produces an extremely hot flame. For $C_2N_2(g)$,

$\Delta G_f° = +70.81$ kcal/mol [or $+296.27$ kJ/mol], and $S° = 57.86$ cal/°K mol [or 242.09 J/°K mol]. Use this data together with values from Tables 14.6 and 14.7 to calculate $\Delta H°$ for the reaction.

**14.35** When plaster of Paris, $(CaSO_4)_2 \cdot H_2O$, is mixed with water, it rapidly hardens to form a more insoluble hydrate, gypsum, $CaSO_4 \cdot 2H_2O$.

$$(CaSO_4)_2 \cdot H_2O(s) + 3H_2O(l) \rightarrow$$
$$2(CaSO_4 \cdot 2H_2O)(s)$$

Calculate (a) $\Delta G°$, (b) $\Delta S°$, and (c) $\Delta H°$ for the reaction. For $(CaSO_4)_2 \cdot H_2O(s)$, $\Delta G_f° = -686.0$ kcal/mol [or $-2870.2$ kJ/mol], and $S° = 62.4$ cal/°K mol [or 261.1 J/°K mol]. For $CaSO_4 \cdot 2H_2O(s)$, $\Delta G_f° = -429.2$ kcal/mol [or $-1795.8$ kJ/mol], and $S° = 46.4$ cal/°K mol [or 194.1 J/°K mol]. For $H_2O(l)$, $\Delta G_f° = -56.7$ kcal/mol [or $-237.2$ kJ/mol], and $S° = 16.7$ cal/°K mol [or 69.9 J/°K mol].

**14.36** Draw a rough diagram similar to that in Figure 14.2 to show the relation between the thermodynamic functions for reaction (a) in Table 14.5.

**14.37** The value of $\Delta G_f°$ for $ZnCO_3(s)$ is $-174.80$ kcal/mol [or $-731.36$ kJ/mol]. Use this fact plus values from Table 14.6 to determine $K_p$ for the reaction

$$ZnCO_3(s) \rightleftharpoons ZnO(s) + CO_2(g)$$

**14.38** For urea, $CO(NH_2)_2(s)$, $\Delta G_f°$ is $-47.12$ kcal/mol [or $-197.15$ kJ/mol]. Use this value together with values from Table 14.6 to determine $K_p$ for the reaction

$$CO_2(g) + 2NH_3(g) \rightleftharpoons$$
$$H_2O(g) + CO(NH_2)_2(s)$$

**14.39** For $SO_3(g)$, $\Delta G_f^\circ = -88.52$ kcal/mol [or $-370.37$ kJ/mol]. Use this value together with values from Table 14.6 to determine $K_p$ for the reaction

$$SO_2(g) + NO_2(g) \rightleftharpoons$$
$$SO_3(g) + NO(g)$$

**14.40** For the reaction

$$Br_2(l) + Cl_2(g) \rightleftharpoons 2BrCl(g)$$

$\Delta G^\circ = -0.43$ kcal/mol [or $-1.80$ kJ/mol]. What is $K_p$ for the reaction?

**14.41** Given that the value of $\Delta \dot{G}_f^\circ$ for $Br_2(g)$ is $+0.75$ kcal/mol [or $+3.14$ kJ/mol], what is the vapor pressure of $Br_2(l)$ at 25°C? For $Br_2(l)$, $\Delta G_f^\circ = 0.0$.

**14.42** At 25°C, $K_p = 0.108$ for the reaction

$$NH_4HS(s) \rightleftharpoons NH_3(g) + H_2S(g)$$

Values of $\Delta G_f^\circ$ for $NH_3(g)$ and $H_2S(g)$ are given in Table 14.6. What is the value of $\Delta G_f^\circ$ for $NH_4HS(s)$?

**14.43** For the equilibrium

$$PCl_5(g) \rightleftharpoons PCl_3(g) + Cl_2(g)$$

at 25°C, $K_p = 1.8 \times 10^{-7}$. What is $\Delta G^\circ$ for the reaction?

**14.44** When the substances of the equation

$$I_2(s) + Br_2(l) \rightleftharpoons 2IBr(g)$$

are at equilibrium at 25°C, the partial pressure of $IBr(g)$ is 0.214 atm. The $S^\circ$ of $IBr(g)$ is 61.8 cal/°K mol [or 258.6 J/°K mol]. Other absolute entropies may be found in Table 14.7. Calculate a value for the enthalpy change, $\Delta H$, of the reaction.

**14.45** Use electrode potentials (given in Appendix D) to calculate the equilibrium constant at 25°C for the reaction

$$3Au^+(aq) \rightleftharpoons 2Au(s) + Au^{3+}(aq)$$

**14.46** Use standard electrode potentials (given in Appendix D) to determine the equilibrium constant at 25°C for the reaction

$$Cl_2(g) + H_2O \rightleftharpoons$$
$$H^+(aq) + Cl^-(aq) + HOCl(aq)$$

**14.47** Use standard electrode potentials (given in Appendix D) to calculate the equilibrium constant at 25°C for the reaction

$$4H^+(aq) + 4Br^-(aq) + O_2(g) \rightleftharpoons$$
$$2Br_2(l) + 2H_2O$$

**14.48** Calculate $\mathscr{E}^\circ$ for the half reaction

$$PbBr_2(s) + 2e^- \rightarrow$$
$$Pb(s) + 2Br^-(aq)$$

from the following data:

$$PbBr_2(s) \rightleftharpoons Pb^{2+}(aq) + 2Br^-(aq)$$
$$K = 4.60 \times 10^{-6}$$

$$Pb^{2+}(aq) + 2e^- \rightarrow Pb(s)$$
$$\mathscr{E}^\circ = -0.126$$

# 15

# Acids and Bases

Throughout the history of chemistry various acid–base concepts have been proposed and used. In this chapter four concepts in current use are reviewed. Each of the definitions can be applied with advantage in appropriate circumstances. In a given situation the chemist uses the concept that best suits his purpose.

The earliest criteria for the characterization of acids and bases were the experimentally observed properties of aqueous solutions. An acid was defined as a substance that in water solution tastes sour, turns litmus red, neutralizes bases, and so on. A substance was a base if its aqueous solution tastes bitter, turns litmus blue, neutralizes acids, and so on. Concurrent with the development of generalizations concerning the structure of matter, scientists searched for a correlation between acidic and basic properties and the structure of compounds that exhibit these properties.

**15.1 THE ARRHENIUS CONCEPT**

Acids, bases, and salts were classified as electrolytes by Faraday in 1834. Justus von Liebig proposed (1838) that acids are compounds containing hydrogen that can be replaced by metals. The role of water as an ionizing solvent was emphasized by Svanté Arrhenius in his theory of electrolytic dissociation (1884), which led to the definition of acids and bases in terms of the ions of water; Arrhenius' views were vigorously championed and developed by Wilhelm Ostwald.

At the time of Arrhenius the distinction between ionic and covalent compounds was not clear, and the nature of $H^+(aq)$ was not understood. Consequently, the **water system** (or **Arrhenius concept**) has been modified throughout its history. In present-day terms an **acid** is a substance that ionizes in water solution to produce $H^+(aq)$, which is called the hydronium ion; a **base** is a substance that dissolves in water to produce hydroxide ion, $OH^-(aq)$. The strength of an acid is defined in terms of the concentration of $H^+(aq)$ that is present in a water solution of given concentration of acid, and the strength of a base depends upon the relative concentration of $OH^-(aq)$ in an aqueous solution of base.

**Neutralization,** then, may be represented by the equation

● The enthalpy of neutralization for the reaction of a strong acid with a strong base in a dilute solution is −57.3 kJ/mol.

$$H^+(aq) + OH^-(aq) \rightarrow H_2O \qquad \Delta H = -13.7 \text{ kcal}$$

As indicated in the preceding equation, the enthalpy for the neutralization of a dilute solution of any strong acid by a dilute solution of any strong base is a constant value. If a weak acid or a weak base is involved in the neutralization, or if a slightly soluble salt is formed from the cation derived from the base and the anion derived from the acid, the enthalpy of the reaction is different; in such cases the total enthalpy includes heat effects due to the ionization of the weak acid or base or due to the precipitation of a salt, as well as the enthalpy of neutralization.

The properties that all Arrhenius acids have in common are those of $H^+(aq)$; chemical properties include such reactions as those of acids with reactive metals and with the bicarbonate ion.

$$2H^+(aq) + Zn(s) \rightarrow H_2(g) + Zn^{2+}(aq)$$

$$H^+(aq) + HCO_3^-(aq) \rightarrow H_2O + CO_2(g)$$

The properties that are characteristic of Arrhenius bases are due to the $OH^-(aq)$ ion; among the reactions of strong bases are

$$2OH^-(aq) + H_2O + Si(s) \rightarrow SiO_3^{2-}(aq) + 2H_2(g)$$

$$OH^-(aq) + NH_4^+(aq) \rightarrow H_2O + NH_3(g)$$

The oxides of many nonmetals react with water to form acids and are called **acidic oxides** or **acid anhydrides.**

$$N_2O_5(s) + H_2O \rightarrow 2H^+(aq) + 2NO_3^-(aq)$$

Many oxides of metals dissolve in water to form hydroxides; such compounds are called **basic oxides.**

$$Na_2O(s) + H_2O \rightarrow 2Na^+(aq) + 2OH^-(aq)$$

Acidic oxides and basic oxides react to produce salts in the absence of water. It must be noted, however, that not all acids and bases may be derived from oxides (e.g., HCl and $NH_3$).

The Arrhenius concept is severely limited by its emphasis on water and reactions in aqueous solution. Later definitions are more general, serve to correlate more reactions, and are applicable to reactions in nonaqueous media.

**15.2 THE SOLVENT SYSTEM CONCEPT**

The principles of the Arrhenius-water concept can be used to devise acid–base schemes for many solvents. In a **solvent system** an **acid** is a substance that gives the cation characteristic of the solvent, and a **base** is a substance that yields the anion characteristic of the solvent. Thus, the reaction of an acid and a base, a **neutralization,** yields the solvent as one of its products. Many solvent systems of acids and bases have been developed (Table 15.1); the water concept is but a single example of a solvent system.

The ammonia system has been investigated more extensively than any other with the exception of the water system. The properties of liquid ammonia (boiling point, $-33.4°C$) are strikingly similar to those of

Table 15.1  Some solvent systems

| SOLVENT | ACID ION | BASE ION | TYPICAL ACID | TYPICAL BASE |
|---------|----------|----------|--------------|--------------|
| $H_2O$ | $H_3O^+$ $(H^+ \cdot H_2O)$ | $OH^-$ | $HCl$ | $NaOH$ |
| $NH_3$ | $NH_4^+$ $(H^+ \cdot NH_3)$ | $NH_2^-$ | $NH_4Cl$ | $NaNH_2$ |
| $NH_2OH$ | $NH_3OH^+$ $(H^+ \cdot NH_2OH)$ | $NHOH^-$ | $NH_2OH \cdot HCl$ $(NH_3OH^+, Cl^-)$ | $K(NHOH)$ |
| $HC_2H_3O_2$ | $H_2C_2H_3O_2^+$ $(H^+ \cdot HC_2H_3O_2)$ | $C_2H_3O_2^-$ | $HCl$ | $NaC_2H_3O_2$ |
| $SO_2$ | $SO^{2+}$ | $SO_3^{2-}$ | $SOCl_2$ | $Cs_2SO_3$ |
| $N_2O_4$ | $NO^+$ | $NO_3^-$ | $NOCl$ | $AgNO_3$ |
| $COCl_2$ | $COCl^+$ | $Cl^-$ | $(COCl)AlCl_4$ | $CaCl_2$ |

water. Liquid ammonia is associated through hydrogen bonding (Section 8.9), and the $NH_3$ molecule is polar. Hence, liquid ammonia is an excellent solvent for ionic and polar compounds, and it functions as an ionizing solvent for electrolytes. Many compounds form ammoniates, which are analogous to hydrates (e.g., $BaBr_2 \cdot 8NH_3$ and $CaCl_2 \cdot 6NH_3$), and ions are solvated in liquid ammonia solutions [e.g., $Ag(NH_3)_2^+$ and $Cr(NH_3)_6^{3+}$]. Whereas solutions of electrolytes in ammonia are good conductors of electricity, pure liquid ammonia, like water, has a relatively low conductance.

The autoionization of ammonia

$$2NH_3 \rightleftharpoons NH_4^+ + NH_2^-$$

which occurs only to a low degree, is responsible for the electrical conductivity of pure solvent, just as the autoionization of water

$$2H_2O \rightleftharpoons H_3O^+ + OH^-$$

is responsible for the electrical properties of this compound.

Any compound that produces ammonium ion, $NH_4^+$, in liquid ammonia solution is an acid, and any compound that yields amide ion, $NH_2^-$, is a base. Thus, the neutralization reaction is the reverse of the autoionization reaction.

$$NH_4^+ + NH_2^- \rightarrow 2NH_3$$

Indicators may be used to follow an acid–base reaction in liquid ammonia. For example, phenolphthalein is red in a liquid ammonia solution of potassium amide, $KNH_2$, and becomes colorless after a stoichiometrically equivalent amount of ammonium chloride has been added.

In addition to the neutralization reaction, the ammonium ion in liquid ammonia undergoes other reactions analogous to the reactions of the hydronium ion in water. For example, metals such as sodium react with the ammonium ion to liberate hydrogen.

$$2Na(s) + 2NH_4^+ \rightarrow 2Na^+ + H_2(g) + 2NH_3(l)$$

The reactions of the amide ion are analogous to those of the hydroxide ion.

$$Zn(OH)_2(s) + 2OH^- \rightarrow Zn(OH)_4^{2-}$$

$$Zn(NH_2)_2(s) + 2NH_2^- \rightarrow Zn(NH_2)_4^{2-}$$

$$Hg^{2+} + 2OH^- \rightarrow HgO(s) + H_2O$$

$$3Hg^{2+} + 6NH_2^- \rightarrow Hg_3N_2(s) + 4NH_3$$

Many properties and reactions of compounds belonging to the ammonia system have been predicted and correlated by comparison to the better known chemistry of the compounds of the water system. In fact, the study of various solvent systems has been responsible for greatly increasing our knowledge of the reactions that occur in solvents other than water. The solvent system approach, however, is compartmentalized and does not offer extensive correlation or generalization.

**15.3 THE BRØNSTED-LOWRY CONCEPT**

In 1923 Johannes Brønsted and Thomas Lowry independently proposed a broader concept of acids and bases. According to the Brønsted–Lowry definitions, an **acid** is a substance that can donate protons, and a **base** is a substance that can accept protons. In these terms, the reaction of an acid with a base constitutes a transfer of a proton from the acid to the base; this is the only characteristic reaction of these substances treated by the Brønsted–Lowry concept. We shall see, however, that many reactions are included in this classification.

The dissolution of ammonia in water may be represented by the equation

$$H_2O + NH_3 \rightleftharpoons NH_4^+ + OH^-$$

In the reaction as written, $H_2O$ is serving as an acid and is releasing a proton to the base, $NH_3$. Although the solution resulting from the addition of $NH_3$ to $H_2O$ is alkaline, conductivity and $i$ factor measurements show that the ionization is incomplete. Furthermore, if a solution of an ammonium salt is made strongly alkaline, ammonia gas is released. The reaction, therefore, is reversible, and the system exists in equilibrium.

According to the Brønsted–Lowry concept, acids and bases may be molecules or ions. In the reverse reaction, $NH_4^+$ is serving as an acid and $OH^-$ is a base since the equation, read from right to left, shows $NH_4^+$ releasing a proton to $OH^-$. It follows, then, that in this Brønsted acid-base reaction, two acids ($H_2O$ and $NH_4^+$) and two bases ($OH^-$ and $NH_3$) are involved; the reaction is actually a competition between the two bases for a proton.

The base $NH_3$ gains a proton and thereby forms the acid $NH_4^+$, and the acid $NH_4^+$ upon the loss of a proton forms the base $NH_3$. Such an acid–base pair, which is related through the loss or gain of a proton, is called a **conjugate pair**; $NH_4^+$ is the conjugate acid of the base $NH_3$, and $NH_3$ is the conjugate base of the acid $NH_4^+$. In like manner, the acid $H_2O$ and the base $OH^-$ constitute a second conjugate pair in the preceding reaction. We may indicate conjugate relationships by the use of subscripts in the following manner:

$$\overset{\text{Acid}_1}{H_2O} + \overset{\text{Base}_2}{NH_3} \rightleftharpoons \overset{\text{Acid}_2}{NH_4^+} + \overset{\text{Base}_1}{OH^-}$$

There are many molecules and ions that can function as acids in certain reactions and as bases in other reactions; such species are called

**amphiprotic.** For example, in the reaction with ammonia, water acts as an acid (conjugate base, $OH^-$); in the reaction with acetic acid, water is a base (conjugate acid, $H_3O^+$).

$$\underset{\text{Acid}_1}{HC_2H_3O_2} + \underset{\text{Base}_2}{H_2O} \rightleftharpoons \underset{\text{Acid}_2}{H_3O^+} + \underset{\text{Base}_1}{C_2H_3O_2^-}$$

In like manner, $NH_3$ acts as a base (conjugate acid, $NH_4^+$) in its reaction with water. In the reaction of the hydride ion, $H^-$, with liquid ammonia, $NH_3$ acts as an acid (conjugate base, $NH_2^-$).

$$\underset{\text{Acid}_1}{NH_3} + \underset{\text{Base}_2}{H^-} \rightarrow \underset{\text{Acid}_2}{H_2} + \underset{\text{Base}_1}{NH_2^-}$$

Several amphiprotic substances are listed in Table 15.2.

Table 15.2 Some amphiprotic substances

| AMPHIPROTIC SUBSTANCE | TYPICAL REACTION | | | |
| --- | --- | --- | --- | --- |
| | ACID$_1$ | BASE$_2$ | ACID$_2$ | BASE$_1$ |
| $H_2O$ | $H_2O$ | $+ H^-$ | $\rightleftharpoons H_2$ | $+ OH^-$ |
| | $HCN$ | $+ H_2O$ | $\rightleftharpoons H_3O^+$ | $+ CN^-$ |
| $NH_3$ | $H_2O$ | $+ NH_2^-$ | $\rightleftharpoons NH_3$ | $+ OH^-$ |
| | $HCl$ | $+ NH_3$ | $\rightleftharpoons NH_4^+$ | $+ Cl^-$ |
| $H_2SO_4$ | $H_2SO_4$ | $+ H_2O$ | $\rightleftharpoons H_3O^+$ | $+ HSO_4^-$ |
| | $HClO_4$ | $+ H_2SO_4$ | $\rightleftharpoons H_3SO_4^+$ | $+ ClO_4^-$ |
| $HSO_4^-$ | $H_2SO_4$ | $+ H_2O$ | $\rightleftharpoons H_3O^+$ | $+ HSO_4^-$ |
| | $HSO_4^-$ | $+ H_2O$ | $\rightleftharpoons H_3O^+$ | $+ SO_4^{2-}$ |

The neutralization reactions of the Arrhenius system and certain solvent systems may therefore be interpreted in terms of the Brønsted definitions. Such neutralizations are merely acid–base reactions between the conjugate acid and the conjugate base of an amphiprotic solvent. Thus,

$$\underset{\text{Acid}_1}{H_3O^+} + \underset{\text{Base}_2}{OH^-} \rightleftharpoons \underset{\text{Acid}_2}{H_2O} + \underset{\text{Base}_1}{H_2O}$$

$$NH_4^+ + NH_2^- \rightleftharpoons NH_3 + NH_3$$

**15.4 STRENGTHS OF BRØNSTED ACIDS AND BASES**

In Brønsted terms the strength of an acid is determined by its tendency to donate protons, and the strength of a base is dependent upon its tendency to receive protons. The reaction

$$\underset{\text{Acid}_1}{HCl} + \underset{\text{Base}_2}{H_2O} \rightleftharpoons \underset{\text{Acid}_2}{H_3O^+} + \underset{\text{Base}_1}{Cl^-}$$

proceeds virtually to completion (from left to right). We must conclude, therefore, that HCl is a stronger acid than $H_3O^+$, since it has the stronger tendency to lose protons and the equilibrium is displaced far to the right. In addition, it is apparent that $H_2O$ is a stronger base than $Cl^-$, since, in the competition for protons, water molecules succeed in holding practically all of them. The strong acid, HCl, has a weak conjugate base, $Cl^-$.

A strong acid, which has a great tendency to lose protons, is necessarily conjugate to a weak base, which has a small tendency to gain and hold protons. Hence, the stronger the acid, the weaker its conjugate base. In like manner, a strong base attracts protons strongly and is necessarily conjugate to a weak acid, one that does not readily lose protons. The stronger the base, the weaker its conjugate acid.

Acetic acid in $1.0M$ solution is 0.42% ionized at 25°C (Section 16.1). The equilibrium

$$\overset{\text{Acid}_1}{HC_2H_3O_2} + \overset{\text{Base}_2}{H_2O} \rightleftharpoons \overset{\text{Acid}_2}{H_3O^+} + \overset{\text{Base}_1}{C_2H_3O_2^-}$$

is displaced to the left. This equation may be said to represent a competition between bases, acetate ions and water molecules, for protons. The position of the equilibrium shows that the $C_2H_3O^-$ ion is a stronger base than $H_2O$; at equilibrium more protons form $HC_2H_3O_2$ molecules than form $H_3O^+$ ions. We may also conclude that $H_3O^+$ is a stronger acid than $HC_2H_3O_2$; at equilibrium more $H_3O^+$ ions than $HC_2H_3O_2$ molecules have lost protons. In the preceding example we note again that the stronger acid, $H_3O^+$, is conjugate to the weaker base, $H_2O$, and the stronger base, $C_2H_3O_2^-$, is conjugate to the weaker acid, $HC_2H_3O_2$.

One further conclusion should be stated. In a given reaction, the position of equilibrium favors the formation of the weaker acid and the weaker base. Thus, in the reaction of HCl and $H_2O$, the equilibrium concentrations of $H_3O^+$ and $Cl^-$ (the *weaker* acid and base, respectively) are *high,* whereas in the solution of acetic acid, the equilibrium concentrations of $H_3O^+$ and $C_2H_3O_2^-$ (the *stronger* acid and base, respectively) are *low.*

Notice that the Brønsted concept is an extension of the Arrhenius concept. According to the Arrhenius scheme, strong acids, such as HCl, are strong electrolytes—they are virtually 100% ionized in water solution and produce solutions with relatively high concentrations of $H_3O^+$. According to the Arrhenius concept, weak acids, such as acetic acid, are weak electrolytes—they are incompletely ionized in water solution and produce solutions with relatively low concentrations of $H_3O^+$. Although the Brønsted system classifies many more substances as acids, compounds that qualify as acids according to the Arrhenius definition are also acids in the Brønsted concept; that is, they are proton donors.

In the Brønsted system, acids are classified according to their ability to donate protons to the specific base under consideration. If water is used as a reference base, the acid strengths of the Arrhenius-water concept may be explained in terms of the Brønsted concept. Thus, strong

Arrhenius acids are those compounds that are stronger acids than $H_3O^+$, and weak Arrhenius acids are those compounds that are weaker acids than $H_3O^+$.

It is apparent that acid strengths are influenced by the solvent water. Acids that are stronger than $H_3O^+$ are essentially completely ionized in water solution.

$$HClO_4 + H_2O \rightleftharpoons H_3O^+ + ClO_4^-$$

$$HCl + H_2O \rightleftharpoons H_3O^+ + Cl^-$$

$$HNO_3 + H_2O \rightleftharpoons H_3O^+ + NO_3^-$$

Aqueous solutions of $HClO_4$, $HCl$, and $HNO_3$ of the same concentration appear to be of the same acid strength. The acid properties of the solutions are due to the $H_3O^+$ ion, which the compounds produce to an equivalent extent in their reactions with water. Water is said to have a **leveling effect** on acids stronger than $H_3O^+$. The strongest acid that can exist in water solution is the conjugate acid of water, $H_3O^+$. Acids that are weaker than $H_3O^+$ are not leveled by water. Thus, $HC_2H_3O_2$, $H_3PO_4$, $HNO_2$, $H_2S$, and other weak acids show a wide variation in their degree of ionization—the extent to which they form $H_3O^+$ in their reactions with water (Section 16.1).

The leveling effect is also observed for solvents other than water. The strongest acid in liquid ammonia solutions is the conjugate acid of ammonia, $NH_4^+$. Acetic acid in liquid ammonia solution is essentially completely ionized, since $HC_2H_3O_2$ is a stronger acid than $NH_4^+$.

$$HC_2H_3O_2 + NH_3 \rightleftharpoons NH_4^+ + C_2H_3O_2^-$$

Nitric acid in methanol ($CH_3OH$) solution is incompletely dissociated, since $HNO_3$ is a weaker acid than the conjugate acid of methanol, $CH_3OH_2^+$.

$$HNO_3 + CH_3OH \rightleftharpoons CH_3OH_2^+ + NO_3^-$$

Solvents that can function as acids exert a leveling effect on bases. The strongest base capable of existing in water solution is the conjugate base of water, $OH^-$. Many substances, such as $NH_2^-$ and $H^-$, are stronger bases than $OH^-$. However, in water solution these strongly basic substances accept protons from water to form $OH^-$ ions; these reactions are essentially complete. The apparent basicity of strongly basic materials in water solution is reduced to the level of the $OH^-$ ion.

$$H_2O + NH_2^- \rightleftharpoons NH_3 + OH^-$$

$$H_2O + H^- \rightleftharpoons H_2 + OH^-$$

Materials, such as ammonia, that are less basic than $OH^-$ are not leveled by water and show varying degrees of ionization in aqueous solution (Section 16.1).

Other solvents, in addition to water, level bases. Since $H^-$ and $NH_2^-$ appear equally strong in water solution, we must examine their reactions in other solvents to compare their relative basicities. It would be surprising if both ions had exactly the same basic character. In liquid ammonia the hydride ion reacts rapidly and essentially completely to form hydrogen and the amide ion.

$$NH_3 + H^- \rightleftharpoons H_2 + NH_2^-$$

We conclude that the hydride ion is a stronger base than the amide ion; liquid ammonia reduces $H^-$ to the level of the conjugate base of ammonia, $NH_2^-$ — the strongest base possible in liquid ammonia solution.

The Brønsted definition enlarges the Arrhenius concept even more for bases than it does for acids. In essence, the Arrhenius concept treats only one Brønsted base — the hydroxide ion. According to the Arrhenius concept, strong bases are completely ionic in water solution and form solutions with high concentrations of $OH^-$ ion. In fact, a compound such as NaOH is ionic when pure; molecules of NaOH do not exist. Solutions of weak bases, according to the Arrhenius concept, are prepared from compounds that produce low concentrations of $OH^-$ ion in water (such as $NH_3$).

The Brønsted system classifies many more substances as bases, and in this system it is the $OH^-$ ion of NaOH that is the base, not the compound itself. From the Arrhenius view, solutions resulting from the addition of such compounds as NaH or $Na_2O$ to water owe their basic character to the fact that they are merely solutions of NaOH after the reaction of $H^-$ or $O^{2-}$ with water.

$$H_2O + H^- \rightarrow H_2 + OH^-$$

$$H_2O + O^{2-} \rightarrow OH^- + OH^-$$

**15.5 HYDROLYSIS**  The reaction of an ion with water in which either $H_3O^+$ or $OH^-$ is produced is called **hydrolysis.** Such reactions are Brønsted acid–base reactions. Anions that function as bases in water solution ($B^-$ in the equation that follows) are hydrolyzed.

$$\underset{\text{Acid}_1}{H_2O} + \underset{\text{Base}_2}{B^-} \rightleftharpoons \underset{\text{Acid}_2}{HB} + \underset{\text{Base}_1}{OH^-}$$

In the reaction water acts as an acid, and a proton is transferred to the anion. The extent of hydrolysis of a given anion depends upon the base strength of the anion; an indication of the degree of hydrolysis is given by the concentration of $OH^-$ relative to the concentration of anion present in the solution. Anions that are extensively hydrolyzed are strong bases with weak conjugate acids.

Hydrolysis is complete in the reaction

$$H_2O + H^- \rightarrow H_2 + OH^-$$

The anion undergoing hydrolysis, $H^-$, is a stronger base than $OH^-$, and the conjugate acid of the anion, $H_2$, is a much weaker acid than $H_2O$; complete hydrolysis is characteristic of such anions. This equation, in fact, illustrates the leveling effect of water on bases stronger than $OH^-$.

At the other extreme, some anions, such as $Cl^-$, do not hydrolyze at all.

$$H_2O + Cl^- \leftarrow HCl + OH^-$$

The $Cl^-$ ion is a very weak base, and its conjugate acid, HCl, is not only stronger than $H_2O$ but is also stronger than $H_3O^+$. In fact, HCl is leveled in water solution to $H_3O^+$; therefore, HCl molecules do not exist in measurable quantities in aqueous solutions of moderate concentrations. The chloride ion is simply too weak a base to accept a proton from water to a significant extent. Anions that have conjugate acids stronger than $H_3O^+$ do not hydrolyze.

The behavior of many anions in water solution is intermediate between the two extremes represented by $H^-$ and $Cl^-$. For example,

$$H_2O + C_2H_3O_2^- \rightleftharpoons HC_2H_3O_2 + OH^-$$

The $C_2H_3O_2^-$ ion is a weaker base than $OH^-$ but not so weak as $Cl^-$. Even though $HC_2H_3O_2$, the conjugate acid of $C_2H_3O_2^-$, is a stronger acid than water, it is a weaker acid than HCl or $H_3O^+$. Acetic acid is not leveled by water; molecules of the weak acid $HC_2H_3O_2$ can exist in aqueous solution. The hydrolysis represented by the equation occurs only to a slight extent; the equilibrium, as written, is displaced to the left because the substances written on the left are the weaker acid and the weaker base of the reaction. Nevertheless, in a solution of sodium acetate there is a measurable concentration of $OH^-$ from the hydrolysis of the acetate ion—enough to cause red litmus to turn blue. Anions, such as $C_2H_3O_2^-$, $CN^-$, $NO_2^-$, and $S^{2-}$, that have conjugate acids stronger than water but weaker than $H_3O^+$ are measurably, but not completely, hydrolyzed in water solution.

There are some hydrogen-containing cations, of which the ammonium ion is the principal example, that hydrolyze to produce $H_3O^+$.

$$NH_4^+ + H_2O \rightleftharpoons H_3O^+ + NH_3$$

Ammonia is a stronger base than water, and therefore the position of equilibrium is to the left. However, water is a sufficiently strong base to gain some of the protons in the competition; although the hydrolysis is far from complete, a solution of $NH_4Cl$ gives an acid reaction toward litmus because of the hydrolysis of the $NH_4^+$ ion.

There are a few relatively unimportant cations that hydrolyze completely. For example, the phosphonium ion, $PH_4^+$, is a stronger acid than the hydronium ion; in water solution the phosphonium ion is leveled.

$$PH_4^+ + H_2O \rightarrow H_3O^+ + PH_3$$

Metal cations are hydrated in water solution, and some of these hydrated cations hydrolyze. For example, solutions of aluminum chloride, zinc chloride, and iron(III) chloride show acidic characteristics because of the hydrolysis of the cations. The hydrolysis of the zinc ion [which we shall indicate as $Zn(H_2O)_4^{2+}$] may be represented as

$$Zn(H_2O)_4^{2+} + H_2O \rightleftharpoons H_3O^+ + Zn(H_2O)_3(OH)^+$$

In the hydrolysis, $Zn(H_2O)_4^{2+}$ functions as an acid and donates a proton to $H_2O$, which acts as a base. A water molecule coordinated to the $Zn^{2+}$ ion has an enhanced acidic character because the O—H bonds of

are weakened by the displacement of electrons toward the positively charged central ion. Thus, the coordinated water molecules release protons and act as Brønsted acids.

In general, small ions with high positive charges form hydrated species that act as Brønsted acids in hydrolysis reactions. The cations of the group I A and group II A metals have comparatively low charges for their relatively large sizes; the hydrated species of these ions have no significant acid character and consequently do not hydrolyze appreciably. The hydrolysis of ions is discussed in more detail in Section 17.6.

The amphoteric nature of some hydroxides [e.g., $Zn(OH)_2$ and $Al(OH)_3$] may be explained by means of the Brønsted theory. These substances, which are normally insoluble in water, dissolve in both acidic solution and alkaline solution. In acid a typical reaction is

$$H_3O^+ + Zn(OH)_2(H_2O)_2(s) \rightleftharpoons Zn(OH)(H_2O)_3^+ + H_2O$$

A proton is transferred from $H_3O^+$ to the hydrated hydroxide, which acts as a base. A coordinated OH of the zinc complex accepts the proton and is transformed into a coordinated $H_2O$. The other coordinated OH groups can also accept protons, and the reaction continues until $Zn(H_2O)_4^{2+}$ is formed.

In alkaline solution the reaction is

$$Zn(OH)_2(H_2O)_2(s) + OH^- \rightleftharpoons H_2O + Zn(OH)_3(H_2O)^-$$

The hydrated zinc hydroxide acts as an acid, and a coordinated $H_2O$ of the zinc compound loses a proton to the $OH^-$ ion. The loss of a proton transforms this $H_2O$ into an OH group, and the $Zn(OH)_3(H_2O)^-$ ion results. Whereas all hydroxides react with $H_3O^+$, only those hydroxides that are classed as amphoteric react with $OH^-$. Amphoterism is discussed further in Section 17.5.

## 15.6 ACID AND BASE STRENGTHS AND STRUCTURE

The correlation between molecular structure and acid or base strength is complex and involves many factors. Polar molecules with a hydrogen atom situated at the positive end of the dipole are acids; in such molecules, electrons are withdrawn from the hydrogen atom, thereby facilitating its release as a proton. For the hydrogen compounds of the elements of a given period, increasing acid strength parallels increasing electronegativity of the atom combined with hydrogen. For the elements of the second period, the order of increasing acid strength of the hydrogen compounds of the last three elements is

$$NH_3 < H_2O < HF$$

However, increasing acid strength does not always parallel increasing electronegativity of the atom bonded to hydrogen; evidently other factors are also involved. There is an increase in acid strength of the hydro acids of the elements in any group of the periodic classification with increasing atomic size of the electronegative element. For example, the hydrogen compounds of the group VI A and group VII A elements arranged according to increasing acid strength are

$$H_2O < H_2S < H_2Se < H_2Te$$

$$HF < HCl < HBr < HI$$

In each series the last compound is the strongest acid and is formed by the element of lowest electronegativity.

When the hydrogen compounds of a period are compared, the small differences in the atomic radius of the electronegative elements are unimportant. However, in the hydrogen halide series, the bond distance increases from 1.0 Å for HF to 1.7 Å for HI, and with it the bond energy decreases from 135 kcal/mol for HF to 71 kcal/mol for HI. The bond energy pertains to a process in which the molecule is broken into atoms, not ions. However, the energy required for the removal of a proton from each hydrogen halide may be calculated by adding the bond energy

$$HX(g) \rightarrow H(g) + X(g)$$

the ionization energy of hydrogen

$$H(g) \rightarrow H^+(g) + e^-$$

and the electron affinity of the halogen

$$e^- + X(g) \rightarrow X^-(g)$$

The values for the hydrogen halides fall in the same order as do the bond energies themselves. If the reaction of HX with water is of interest, the heats of hydration of the ions may be added without altering the observed order. The actual process

$$HX + H_2O \rightleftharpoons H_3O^+ + X^-$$

does not, of course, occur by the mechanism used for the analysis of the energy effect.

A base must have an unshared electron pair in order to attract and hold a proton; a polar molecule that has an electron pair situated at the negative end of the dipole is a base. Trends in the base strength of anions are readily derived from conjugate relationships. Thus, since $H_2S$ is a stronger acid than $H_2O$, $S^{2-}$ is a weaker base than $O^{2-}$; for monatomic anions of similar charge, base strength decreases with increasing size.

The base strength of the uninegative ions of elements of the second period

$$NH_2^- > OH^- > F^-$$

parallels a decrease in the electronegativity for the element of the second period; a similar trend is observed for the hydrogen-containing molecules of these elements

$$NH_3 > H_2O > HF$$

In these series, size differences are small and unimportant; the trend is set by the decreasing electronegativity of the central element. Both small size and low electronegativity are found in the hydride ion, $H^-$, which is a powerful base.

A third factor that influences the base strength of anions is the charge on the ion. Thus, the base strength of the monatomic anions of the elements of the second period

$$N^{3-} > O^{2-} > F^-$$

decreases with increasing electronegativity and with decreasing negative charge on the ion.

The oxy acids have been studied more extensively than any other type of acids. For acids with the structure

$$H\!-\!O\!-\!Z$$

the acid strength increases with increasing electronegativity of Z. The higher the electronegativity of Z, the more the electrons of the molecule are displaced toward Z, and the more readily the proton is removed. For example,

HOCl > HOBr > HOI

is the order of decreasing acidity of the hypohalous acids.

In compounds in which additional oxygen atoms are bonded to Z, the electron-withdrawing power of the group bonded to hydrogen is increased; thus the proton is more readily removed. This effect is illustrated by the series

$$H\!-\!\ddot{\underset{..}{O}}\!-\!\ddot{\underset{..}{Cl}}\!: \ <\ H\!-\!\ddot{\underset{..}{O}}\!-\!\overset{\oplus}{\underset{..}{Cl}}\!-\!\ddot{\underset{..}{O}}\!:^{\ominus}\ <\ H\!-\!\ddot{\underset{..}{O}}\!-\!\overset{:\ddot{O}:^{\ominus}}{\underset{:\underset{..}{O}:^{\ominus}}{\overset{|}{\underset{|}{Cl}}}}\!-\!\ddot{\underset{..}{O}}\!:^{\ominus}\ <\ H\!-\!\ddot{\underset{..}{O}}\!-\!\overset{:\ddot{O}:^{\ominus}}{\underset{:\underset{..}{O}:^{\ominus}}{\overset{|}{\underset{|}{Cl}}}}\!-\!\ddot{\underset{..}{O}}\!:^{\ominus}$$

in which acidity increases with increasing oxidation state of chlorine (from 1+ in HOCl to 7+ in $HClO_4$). However, the formal charge of the central atom of an oxy acid (determined from a structure written in conformity with the octet principle) is a better indicator of acid strength than oxidation number. For the oxy acids of chlorine, increasing formal charge on the chlorine atom (indicated in the preceding structures) parallels an increase in the oxidation number of the chlorine atom, and either serves to indicate increasing acid strength. However, for the oxy acids of phosphorus, oxidation number fails to give a true indication of acid strength. The acids

$$H\!-\!\ddot{\underset{..}{O}}\!-\!\overset{:\ddot{O}:^{\ominus}}{\underset{\underset{H}{|}}{\overset{|}{\underset{\oplus}{P}}}}\!-\!H \qquad H\!-\!\ddot{\underset{..}{O}}\!-\!\overset{:\ddot{O}:^{\ominus}}{\underset{\underset{H}{|}}{\overset{|}{\underset{\oplus}{P}}}}\!-\!\ddot{\underset{..}{O}}\!-\!H \qquad H\!-\!\ddot{\underset{..}{O}}\!-\!\overset{:\ddot{O}:^{\ominus}}{\underset{\underset{H}{\overset{|}{\underset{|}{O}}}}{\overset{|}{\underset{\oplus}{P}}}}\!-\!\ddot{\underset{..}{O}}\!-\!H$$

are approximately of equal strength, and in each case the formal charge on the phosphorus atom is 1+. Note, however, that the oxidation number of P in $H_3PO_2$ is 1+, in $H_3PO_3$ it is 3+, and in $H_3PO_4$ it is 5+.

The number of O atoms bonded to the central atom but *not* bonded to H atoms influences the formal charge of the central atom, and therefore a qualitative indication of the strength of acids of general formula $(HO)_mZO_n$ is provided by the value of $n$ in the formula. In general, acids are very strong when $n = 3$ ($HClO_3$, $HIO_3$), strong when $n = 2$ [$HClO_2$, $(HO)_2SO_2$, $HONO_2$], weak when $n = 1$ [$HOClO$, $(HO)_3PO$, HONO], and very weak when $n = 0$ [HOCl, $(HO)_3B$]. There are, of course, variations within any of the groups; the electronegativity of the central atom provides the key to these variations. Thus, for the halic acids the oxidation state of each halogen atom is 5+, and $n = 2$; the acidity of the compounds in the series increases with increasing electronegativity of the halogen atom.

$$HOIO_2 < HOBrO_2 < HOClO_2$$

Increasing acid strength parallels decreasing base strength of the conjugate base. Thus, for the series

$$HOCl < HOClO < HOClO_2 < H\dot{O}ClO_3$$

the order of base strengths of the anions is

$$OCl^- > ClO_2^- > ClO_3^- > ClO_4^-$$

The addition of oxygen atoms provides a larger volume to accommodate the charge of the ion. A large uninegative ion attracts and holds a proton less strongly than a small uninegative ion.

**15.7 THE LEWIS CONCEPT**

In reality, the Brønsted concept enlarges the definition of a base much more than it does that of an acid. In the Brønsted system a base is a molecule or ion that has an unshared electron pair with which it can attract and hold a proton, and an acid is a substance that can supply a proton to a base. If a molecule or ion can share an electron pair with a proton, it can do the same thing with other substances as well.

Gilbert N. Lewis proposed a broader concept of acids and bases that liberated acid–base phenomena from the proton; although Lewis first proposed his system in 1923, he did little to develop it until 1938. Lewis defined a **base** as a substance that has an unshared electron pair with which it can form a covalent bond with an atom, molecule, or ion. An **acid** is a substance that can form a covalent bond by accepting an electron pair from a base. The emphasis has been shifted by the Lewis concept from the proton to the electron pair and covalent-bond formation.

An example of an acid–base reaction that is not treated as such by any other acid–base concept is

Many Lewis acids and bases of this type can be titrated against one another by the use of suitable indicators in the same way that traditional acids and bases can be titrated.

Substances that are bases in the Brønsted system are also bases according to the Lewis concept. However, the Lewis definition of an acid considerably expands the number of substances that are classified as acids. A Lewis acid must have an empty orbital capable of receiving the electron pair of the base; the proton is but a single example of a Lewis acid.

Lewis acids include molecules or atoms that have incomplete octets.

$$
\begin{array}{c}
:\ddot{F}: \\
| \\
:\ddot{F}-B \\
| \\
:\ddot{F}:
\end{array}
+ :\ddot{F}:^- \rightarrow
\left[
\begin{array}{c}
:\ddot{F}: \\
| \\
:\ddot{F}-B-\ddot{F}: \\
| \\
:\ddot{F}:
\end{array}
\right]^-
$$

$$
:\ddot{S} +
\left[
\begin{array}{c}
:\ddot{O}: \\
| \\
:\ddot{S}-\ddot{O}: \\
| \\
:\ddot{O}:
\end{array}
\right]^{2-}
\rightarrow
\left[
\begin{array}{c}
:\ddot{O}: \\
| \\
:\ddot{S}-\ddot{S}-\ddot{O}: \\
| \\
:\ddot{O}:
\end{array}
\right]^{2-}
$$

$$
\begin{array}{c}
:\ddot{Cl}: \\
| \\
:\ddot{Cl}-Al \\
| \\
:\ddot{Cl}:
\end{array}
+ :\ddot{Cl}:^- \rightarrow
\left[
\begin{array}{c}
:\ddot{Cl}: \\
| \\
:\ddot{Cl}-Al-\ddot{Cl}: \\
| \\
:\ddot{Cl}:
\end{array}
\right]^-
$$

Aluminum chloride, although it reacts as $AlCl_3$, is actually a dimer—$Al_2Cl_6$. The formation of the dimer from the monomer may be regarded as a Lewis acid–base reaction in itself, since a chlorine atom in each $AlCl_3$ unit supplies an electron pair to the aluminum atom of the other $AlCl_3$ unit to complete the octet of the aluminum atom; these bonds are indicated by $\cdots$ in the diagram

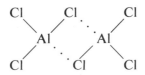

Many simple cations can function as Lewis acids; for example,

$$Cu^{2+} + 4:NH_3 \rightarrow Cu(:NH_3)_4^{2+}$$

$$Fe^{3+} + 6:C\equiv N:^- \rightarrow Fe(:C\equiv N:)_6^{3-}$$

Some metals atoms can function as acids in the formation of compounds such as the carbonyls, which are produced by the reaction of the metal with carbon monoxide.

$$Ni + 4:C\!\!\equiv\!\!O: \rightarrow Ni(:C\!\!\equiv\!\!O:)_4$$

Compounds that have central atoms capable of expanding their valence shells are Lewis acids in reactions in which this expansion occurs. Examples are

$$SnCl_4 + 2Cl^- \rightarrow SnCl_6^{2-}$$

$$SiF_4 + 2F^- \rightarrow SiF_6^{2-}$$

$$PF_5 + F^- \rightarrow PF_6^-$$

In each of the first two reactions, the valence shell of the central atom (Sn and Si) is expanded from eight to twelve electrons, and in the third reaction the valence shell of P goes from ten to twelve electrons.

In addition, some compounds have an acidic site because of one or more multiple bonds in the molecule. Examples are

The reactions of silica, $SiO_2$, with metal oxides are analogous to the reaction of carbon dioxide with the oxide ion, although both silica and the silicate products (compounds of $SiO_3^{2-}$) are polymeric. This reaction is important in high-temperature metallurgical processes in which a basic oxide is added to an ore to remove silica in the form of silicates (slag). Many of the processes used in the manufacture of glass, cement, and ceramics involve the reaction of the base $O^{2-}$ (from metal oxides, carbonates, and so forth) with acid oxides (such as $SiO_2$, $Al_2O_3$, and $B_2O_3$).

Arrhenius and Brønsted acid–base reactions may be interpreted in Lewis terms by focusing attention on the proton, as a Lewis acid

$$H^+(aq) + OH^-(aq) \rightarrow H_2O$$

in which case the Brønsted acid is termed a **secondary** Lewis acid since it serves to provide the **primary** Lewis acid, the proton. Probably a better interpretation is that which classifies Brønsted acid–base reactions as Lewis base displacements. The Brønsted acid is interpreted as a complex in which the Lewis acid (proton) is already combined with a base; the reaction is viewed as a displacement of this base by another, stronger base.

$$H_3O^+ + OH^- \rightarrow H_2O + H_2O$$

In this reaction the base $OH^-$ displaces the weaker base $H_2O$ from its combination with the acid, the proton.

All Brønsted acid–base reactions are Lewis base displacements. In the reaction

$$HCl + H_2O \rightarrow H_3O^+ + Cl^-$$

the base $H_2O$ displaces the weaker base, $Cl^-$. A base supplies an electron pair to a nucleus and is therefore called **nucleophilic** (from Greek, meaning "nucleus loving"). Base displacements are nucleophilic displacements.

Nucleophilic displacements may be identified among reactions that are not Brønsted acid–base reactions. The formation of $Cu(NH_3)_4^{2+}$ has been previously used as an illustration of a Lewis acid–base reaction. Since the reaction occurs in water, the formation of this complex is more accurately interpreted as the displacement of the base $H_2O$ from the complex $Cu(H_2O)_4^{2+}$ by the stronger base $NH_3$.

$$Cu(H_2O)_4^{2+} + 4NH_3 \rightarrow Cu(NH_3)_4^{2+} + 4H_2O$$

Lewis acids accept an electron pair in a reaction with a base; they are **electrophilic** (from Greek, meaning "electron loving"). Acid displacements, or electrophilic displacements, are not so common as base displacements, but this type of reaction is known. For example, if $COCl_2$ is viewed as a combination of $COCl^+$ (an acid) with $Cl^-$ (a base), the reaction

$$COCl_2 + AlCl_3 \rightarrow COCl^+ + AlCl_4^-$$

is an electrophilic displacement in which the acid $AlCl_3$ displaces the weaker acid $COCl^+$ from its complex with the base $Cl^-$ (Table 15.1). The reaction

$$SeOCl_2 + BCl_3 \rightarrow SeOCl^+ + BCl_4^-$$

may be similarly interpreted.

The Lewis theory is frequently used to interpret reaction mechanisms. Examples of such interpretations are found in Sections 20.5 and 20.6.

## Problems

**15.1** Give one-sentence descriptions of an *acid*, a *base*, and a *neutralization reaction* based on (a) the Arrhenius concept, (b) the solvent system concept, (c) the Brønsted–Lowry concept, and (d) the Lewis concept.

**15.2** What is the conjugate base of (a) $H_3AsO_3$, (b) $H_2AsO_4^-$, (c) $HNO_2$, (d) $HS^-$, (e) $Cr(H_2O)_6^{3+}$?

**15.3** What is the conjugate base of (a) $NH_3$, (b) $HC_2O_4^-$, (c) $HOBr$, (d) $H_3PO_4$, (e) $Zn(OH)(H_2O)_5^+$?

**15.4** What is the conjugate acid of (a) $H_2AsO_4^-$, (b) $PH_3$, (c) $C_2H_3O_2^-$, (d) $S^{2-}$, (e) $PO_4^{3-}$?

**15.5** What is the conjugate acid of (a) $NH_3$, (b) $HC_2O_4^-$, (c) $OCl^-$, (d) $NH_2^-$, (e) $Zn(OH)(H_2O)_5^+$?

**15.6** Identify all the Brønsted acids and bases in the following equations.

(a) $HF + HF \rightleftharpoons H_2F^+ + F^-$
$\qquad\qquad\qquad$ (in liquid HF)

(b) $HNO_3 + HF \rightleftharpoons H_2NO_3^+ + F^-$
$\qquad\qquad\qquad$ (in liquid HF)

(c) $HF + CN^- \rightleftharpoons HCN + F^-$
$\qquad\qquad\qquad$ (in liquid HF)

(d) $H_2PO_4^- + CO_3^{2-} \rightleftharpoons$
$\quad HPO_4^{2-} + HCO_3^-$ $\quad$ (in water)

(e) $N_2H_4 + HSO_4^- \rightleftharpoons$
$\quad N_2H_5^+ + SO_4^{2-}$ $\quad$ (in water)

(f) $HC_2O_4^- + HS^- \rightleftharpoons$
$\quad C_2O_4^{2-} + H_2S$ $\quad$ (in water)

**15.7** Write chemical equations to illustrate the behavior of the following as Brønsted acids: (a) $HOCl$, (b) $H_2O$, (c) $HCO_3^-$, (d) $Fe(H_2O)_6^{3+}$.

**15.8** Write chemical equations to illustrate the behavior of the fol-

lowing as Brønsted bases: (a) $H_2NOH$, (b) $N^{3-}$, (c) $H^-$, (d) $HSO_3^-$.

**15.9** (a) What is an amphiprotic substance? (b) Give four examples of such substances. Select both molecules and ions as your examples. (c) Write chemical equations to illustrate the characteristic behavior of the substances that you listed in part (b).

**15.10** What is the leveling effect of a solvent? Use chemical equations in your answer, and describe the leveling of both acids and bases.

**15.11** What is the difference in meaning between the terms amphiprotic and amphoteric?

**15.12** Each of the following reactions is displaced to the right. (a) Arrange all the Brønsted acids that appear in these equations according to decreasing acid strength. (b) Make a similar list for the Brønsted bases.

(a) $H_3O^+ + H_2PO_4^- \rightleftharpoons$
$\quad H_3PO_4 + H_2O$

(b) $HCN + OH^- \rightleftharpoons$
$\quad H_2O + CN^-$

(c) $H_3PO_4 + CN^- \rightleftharpoons$
$\quad HCN + H_2PO_4^-$

(d) $H_2O + NH_2^- \rightleftharpoons$
$\quad NH_3 + OH^-$

**15.13** On the basis of your lists from Problem 15.12, would you expect an appreciable reaction (over 50%) between the species listed in each of the following?

(a) $H_3O^+ + CN^- \rightarrow$

(b) $NH_3 + CN^- \rightarrow$

(c) $HCN + H_2PO_4^- \rightarrow$

(d) $H_3PO_4 + NH_2^- \rightarrow$

**15.14** Each of the following reactions is displaced to the right. (a) Arrange all the Brønsted acids that appear in these equations according to decreasing acid strength. (b) Make a similar list for the Brønsted bases.

(a) $HCO_3^- + OH^- \rightleftharpoons$
$\phantom{xxx} H_2O + CO_3^{2-}$

(b) $HC_2H_3O_2 + HS^- \rightleftharpoons$
$\phantom{xxx} H_2S + C_2H_3O_2^-$

(c) $H_2S + CO_3^{2-} \rightleftharpoons$
$\phantom{xxx} HCO_3^- + HS^-$

(d) $HSO_4^- + C_2H_3O_2^- \rightleftharpoons$
$\phantom{xxx} HC_2H_3O_2 + SO_4^{2-}$

**15.15** On the basis of your lists from Problem 15.14, would you expect an appreciable reaction (over 50%) between the species listed in each of the following?

(a) $HCO_3^- + C_2H_3O_2^- \rightarrow$

(b) $HSO_4^- + HS^- \rightarrow$

(c) $HC_2H_3O_2 + CO_3^{2-} \rightarrow$

(d) $H_2S + C_2H_3O_2^- \rightarrow$

**15.16** Briefly discuss, using chemical equations, how the compound $H_2O$ is classified according to the (a) Arrhenius concept, (b) the Brønsted–Lowry concept, and (c) the Lewis concept.

**15.17** Do acid–base reactions ever involve oxidation and reduction? Consider all of the acid–base theories presented in this chapter.

**15.18** Write equations for the hydrolysis of (a) $C_2H_3O_2^-$, (b) $CN^-$, (c) $NO_2^-$, (d) $NH_4^+$, (e) $PH_4^+$, (f) $Cu(H_2O)_4^{2+}$.

**15.19** Ammonium chloride ($NH_4Cl$) reacts with sodium amide ($NaNH_2$) in liquid ammonia to produce so-dium chloride and ammonia. Interpret this reaction in terms of the solvent system, Brønsted, and Lewis theories of acids and bases. State clearly what acid(s) and base(s) are involved in each case.

**15.20** Which compound of each of the following pairs is the stronger acid?

(a) $H_3PO_4$ or $H_3AsO_4$

(b) $H_3AsO_3$ or $H_3AsO_4$

(c) $H_2SO_4$ or $H_2SO_3$

(d) $H_3BO_3$ or $H_2CO_3$

(e) $H_2Se$ or $HBr$

**15.21** Which compound of each of the following pairs is the stronger base?

(a) $P^{3-}$ or $S^{2-}$

(b) $PH_3$ or $NH_3$

(c) $SiO_3^{2-}$ or $SO_3^{2-}$

(d) $NO_2^-$ or $NO_3^-$

(e) $Br^-$ or $F^-$

**15.22** Perchloric acid, $HClO_4$, is a stronger acid than chloric acid, $HClO_3$. On the other hand, periodic acid, $H_5IO_6$, is a weaker acid than iodic acid, $HIO_3$. (a) Draw Lewis structures for the four compounds, and assign formal charges to the atoms of the molecules. (b) Explain the differences in acid strength on the basis of your structures of part (a).

**15.23** Draw Lewis structures (include formal charges) for selenic acid, $H_2SeO_4$, and telluric acid, $H_6TeO_6$. (b) Which is the stronger acid?

**15.24** Determine a value for the $\Delta H$ pertaining to the process:

$$HX(g) \rightarrow H^+(aq) + X^-(aq)$$

for each of the hydrogen halides

by using the Born–Haber approach outlined in Section 15.6. Bond energies are given in Table 3.6; electron affinities appear in Table 3.1 (1 eV = 23.06 kcal/mol); heats of hydration of $X^-(aq)$ are listed in Table 11.5 (for 2 mol); the ionization energy of $H(g)$ is $+314$ kcal/mol; the heat of hydration of $H^+(g)$ is $-261$ kcal/mol; 1 kcal = 4.184 kJ.

**15.25** Interpret the following reactions in terms of the Lewis theory:

(a) $S=C=S + SH^- \rightarrow S_2CSH^-$
(b) $Fe(CO)_5 + H^+ \rightarrow FeH(CO)_5^+$
(c) $Ag^+ + 2NH_3 \rightarrow Ag(NH_3)_2^+$
(d) $Mn(CO)_5^- + H^+ \rightarrow HMn(CO)_5$
(e) $HF + F^- \rightarrow HF_2^-$

**15.26** Interpret the following reactions in terms of the Lewis theory.

(a) $BeF_2 + 2F^- \rightarrow BeF_4^{2-}$
(b) $H_2O + BF_3 \rightarrow H_2OBF_3$
(c) $S + S^{2-} \rightarrow S_2^{2-}$
(d) $H^- + H_2C=O \rightarrow H_3CO^-$
(e) $AuCN + CN^- \rightarrow Au(CN)_2^-$

**15.27** Interpret the following as Lewis displacement reactions. For each, state the type of displacement, what is displaced, and the agent for the displacement.

(a) $NH_3 + H^- \rightarrow H_2 + NH_2^-$
(b) $HONO_2 + H_2SO_4 \rightarrow$
$\qquad NO_2^+ + H_2O + HSO_4^-$
(c) $[Co(NH_3)_5(H_2O)]^{3+} + Cl^- \rightarrow$
$\qquad [Co(NH_3)_5Cl]^{2+} + H_2O$
(d) $Ge + GeS_2 \rightarrow 2GeS$
(e) $O^{2-} + H_2O \rightarrow 2OH^-$
(f) $Br_2 + FeBr_3 \rightarrow Br^+ + FeBr_4^-$
(g) $CH_3Cl + AlCl_3 \rightarrow$
$\qquad CH_3^+ + AlCl_4^-$
(h) $CH_3I + OH^- \rightarrow CH_3OH + I^-$

**15.28** Sulfur dioxide is a significant contributor to atmospheric pollution. The sulfurous acid produced by the reaction of $SO_2$ and atmospheric moisture adversely effects the eyes, skin, and respiratory systems of humans and causes the corrosion of metals and the deterioration of other materials. The reaction of $SO_2$ and water can be roughly indicated as

$$H_2O + SO_2 \rightarrow H_2OSO_2 \rightarrow (HO)_2SO$$

(a) Draw Lewis structures (complete with formal charges) for the compounds given. (b) Interpret the two steps of the sequence in terms of the Lewis acid–base theory. (c) Why does the proton migration of the second step occur?

# 16

# Ionic Equilibria, Part I

The principles of chemical equilibrium can be applied to equilibrium systems involving molecules and ions in aqueous solution. In pure water hydronium and hydroxide ions, which are derived from $H_2O$, exist in equilibrium with water molecules. Other molecular substances (weak electrolytes) are partially ionized in aqueous solution and exist in equilibrium with their component ions. Equilibria of this type are important in preparative chemistry (many ionic reactions in water are reversible and tend toward a condition of equilibrium), in analytical chemistry (both qualitative and quantitative determinations), in biological systems (for example, the chemistry of the blood), and in industrial processes (examples are electroplating and dyeing). In this chapter systems involving weak acids or weak bases are discussed. Other aspects of ionic equilibria are described in Chapter 17.

**16.1 WEAK ELECTROLYTES**

Strong electrolytes are completely ionic in water solution. For example, a $0.01M$ solution of $CaCl_2$ is $0.01M$ in calcium ions and $0.02M$ in chloride ions. Interionic attractions (the Debye–Hückel effect, Section 9.13) account for the observed deviations in the properties of solutions of strong electrolytes from the properties calculated on the basis of complete dissociation; such deviations are not large.

Weak electrolytes, however, are incompletely ionized in water solution; dissolved molecules exist in equilibrium with ions in such solutions. For example, the following equation represents the dissociation of acetic acid in water

$$HC_2H_3O_2 + H_2O \rightleftharpoons H_3O^+ + C_2H_3O_2^-$$

and the equilibrium constant for this reaction as written is

$$K' = \frac{[H_3O^+][C_2H_3O_2^-]}{[HC_2H_3O_2][H_2O]}$$

In dilute solutions the molar concentration of water is virtually a constant (approximately 1000/18, or 55.5, $M$); the number of moles of water consumed in the formation of hydronium ion (approximately $10^{-3}$ mole per liter in a $0.1M$ acetic acid solution) is negligible in comparison with the large number of moles of water present. Thus,

$$K = K'[H_2O] = \frac{[H_3O^+][C_2H_3O_2^-]}{[HC_2H_3O_2]}$$

We shall follow the practice of representing the concentration of hydronium ion by the symbol $[H^+]$. Thus, the expression for the equilibrium constant reduces to a form consistent with the simplified equation

$$HC_2H_3O_2 \rightleftharpoons H^+ + C_2H_3O_2^-$$

$$K = \frac{[H^+][C_2H_3O_2^-]}{[HC_2H_3O_2]}$$

For complete accuracy, equilibrium constants should be expressed in terms of activities instead of concentrations. However, the concentrations of ions present in dilute solutions of weak electrolytes are so small that interionic attractions are negligible. Thus, under these conditions molar concentrations may be used rather than activities, and reasonably accurate results can be obtained. By convention the ions are written on the right of an equation for the reversible dissociation of a weak electrolyte; hence, the concentration terms for ions appear in the numerator of the expression for the equilibrium constant.

The degree of dissociation, $\alpha$, of a weak electrolyte in aqueous solution may be determined by conductance measurements (Section 10.5) or less accurately by measurements of colligative properties (Section 9.12). Potentiometric methods for such determinations are also used (Sections 16.4 and 17.8). From the degree of dissociation the value of the equilibrium constant (or ionization constant) may be calculated.

---

### Example 16.1

At 25°C a 0.100$M$ solution of acetic acid is 1.34% ionized. What is the ionization constant for acetic acid?

*Solution*

Since $\alpha = 0.0134$, in 1 liter of solution

$$(0.0134)(0.100) = 0.00134 \text{ mol}$$

of acetic acid would be in ionic form, and

$$0.100 - 0.00134 = 0.09866 \text{ mol}$$

of acetic acid would remain in molecular form. According to the chemical equation for the ionization, one mole of $H^+$ and one mole of $C_2H_3O_2^-$ are produced for every mole of acetic acid that ionizes. Therefore, the equilibrium concentrations are

$$\underset{0.099M}{HC_2H_3O_2} \rightleftharpoons \underset{0.00134\,M}{H^+} + \underset{0.00134M}{C_2H_3O_2^-}$$

These concentrations may be used to find the numerical value of the equilibrium constant.

$$K = \frac{[H^+][C_2H_3O_2^-]}{[HC_2H_3O_2]}$$

$$= \frac{(0.00134)(0.00134)}{(0.099)}$$

$$= 1.81 \times 10^{-5}$$

In future problem work we shall express equilibrium constants to two significant figures; higher accuracy is generally not warranted when molar concentrations are used instead of activities.

---

### Example 16.2

What are the concentrations of all species present in 1.0$M$ acetic acid at 25°C? What is the degree of ionization?

*Solution*

If we let $x$ equal the number of moles of acetic acid in ionic form in 1 liter of solution, the equilibrium concentrations are

$$\underset{(1.0-x)M}{HC_2H_3O_2} \rightleftharpoons \underset{(x)M}{H^+} + \underset{(x)M}{C_2H_3O_2^-}$$

Hence, we find

$$1.8 \times 10^{-5} = \frac{[H^+][C_2H_3O_2^-]}{[HC_2H_3O_2]}$$

$$= \frac{x^2}{(1.0-x)}$$

which can be rearranged to

$$x^2 + 1.8 \times 10^{-5}x - 1.8 \times 10^{-5} = 0$$

This equation may be solved by means of the quadratic formula (Appendix B.3); we find

$$x = [H^+] = [C_2H_3O_2^-] = 4.2 \times 10^{-3}M$$
$$(1.0-x) = [HC_2H_3O_2] = 9.958 \times 10^{-1}M$$

(Note that this last concentration is 1.0$M$ to two significant figures.) The degree of ionization can be calculated by dividing the number of ionized moles of acetic acid in 1 liter of solution by the total number of moles of acetic acid present in 1 liter.

$$\alpha = \frac{4.2 \times 10^{-3}}{1.0} = 4.2 \times 10^{-3}$$

The use of the quadratic formula in problem solving may be avoided by making an approximation that is frequently employed in calculations involving aqueous equilibria. The subtraction of a very small number from a large number does not significantly alter the value of the large number and may be neglected. Thus, in the preceding example such a small amount of acetic acid is ionized ($x$) that the quantity ($1.0 - x$) used to represent the concentration of undissociated acetic acid molecules is for all practical purposes equal to 1.0 (as previously noted). By using 1.0 instead of ($1.0 - x$) for the concentration of $HC_2H_3O_2$, we find

$$1.8 \times 10^{-5} = \frac{[H^+][C_2H_3O_2^-]}{[HC_2H_3O_2]}$$

$$= \frac{x^2}{1.0}$$

$$x = 4.2 \times 10^{-3}M$$

which is the same as the result obtained through the use of the quadratic formula.

The subtraction of a small number from another small number may not be neglected; therefore, the quadratic formula must be employed to solve some problems. As a rule of thumb, if the number to be subtracted is more than 5% of the number from which it is to be subtracted, the simplified procedure should not be used because the final result of the calculation would not have two significant figures. Thus, the result of a calculation obtained by the use of the simplified procedure may be used to check whether the use of the procedure was justified.

In Table 16.1 the molar concentrations of ions are listed for solutions of acetic acid of various concentrations. The percent ionized increases with dilution; at infinite dilution all electrolytes are completely dissociated. This is consistent with Le Chatelier's principle. Addition of water to an equilibrium system of a weak electrolyte,

$$HC_2H_3O_2 + H_2O \rightleftharpoons H_3O^+ + C_2H_3O_2^-$$

shifts the equilibrium to the right so that proportionately more of the electrolyte is found in ionic form.

The values of ionization constants change with temperature; most values are reported at 25°C. At 25°C, $K$ for acetic acid is $1.8 \times 10^{-5}$; at

Table 16.1 Ion concentrations and percent ionization of solutions of acetic acid at 25°C

| CONCENTRATION OF SOLUTION ($M$) | [H$^+$] OR [C$_2$H$_3$O$_2^-$] ($M$) | PERCENT IONIZATION |
|---|---|---|
| 1.00 | 0.00420 | 0.420 |
| 0.100 | 0.00134 | 1.34 |
| 0.0100 | 0.000420 | 4.20 |
| 0.00100 | 0.000125 | 12.5 |

100°C the numerical value of the constant for this equilibrium is $1.1 \times 10^{-5}$.

An equilibrium system of a weak electrolyte may not only be prepared from the pure compound but also from compounds that supply the component ions of the electrolyte; this is illustrated in the following example.

*Example 16.3*
What are the concentrations of all species present in a solution made by diluting 0.10 mol of HCl and 0.50 mol of $NaC_2H_3O_2$ to 1.0 liter?

*Solution*
Hydrochloric acid and sodium acetate are strong electrolytes. We may assume, therefore, that before equilibrium is attained, the concentration of $H^+$ is 0.10$M$, and the concentration of $C_2H_3O_2^-$ is 0.50$M$. If we let $x$ equal the number of moles per liter of acetic acid at equilibrium, the equilibrium concentrations are

$$HC_2H_3O_2 \rightleftharpoons H^+ + C_2H_3O_2^-$$
$$(x)M \quad (0.10-x)M \quad (0.50-x)M$$

If we solve the problem using these quantities, the quadratic formula must be employed since $x$ is not negligible in comparison to 0.10 or 0.50.

A simpler way to solve the problem is to assume that the reaction goes as far to the left as is possible and that then a portion of the acetic acid thus formed dissociates into ions. In this instance let $y$ equal the number of moles of acetic acid that is dissociated at equilibrium.

$$HC_2H_3O_2 \rightleftharpoons H^+ + C_2H_3O_2^-$$

| | | | |
|---|---|---|---|
| after mixing: | | 0.10$M$ | 0.50$M$ |
| reaction to left: | 0.10$M$ | | 0.40$M$ |
| equilibrium: | $(0.10-y)M$ | $(y)M$ | $(0.40+y)M$ |

Now, $y$ is indeed negligible in comparison to both 0.10 and 0.40, and we may simplify the problem by neglecting $y$ in the terms for the concentrations of acetic acid and acetate ion. Thus,

$$1.8 \times 10^{-5} = \frac{[H^+][C_2H_3O_2^-]}{[HC_2H_3O_2]}$$

$$= \frac{y(0.40)}{(0.10)}$$

$$y = 4.5 \times 10^{-6}M$$

We can see from the value obtained for $y$ that the approximations applied to $[C_2H_3O_2^-]$ and $[HC_2H_3O_2]$ are warranted. To two significant figures, the equilibrium concentrations are

$$[H^+] = 4.5 \times 10^{-6}M$$

$$[C_2H_3O_2^-] = 0.40M$$

$$[HC_2H_3O_2] = 0.10M$$

The hydroxides of most metals are either strong electrolytes or slightly soluble compounds. There are some water-soluble compounds, however, that produce alkaline solutions in which equilibria exist between molecules and ions. The most important such compound is ammonia, and the chemical equation for the reversible reaction is

$$H_2O + NH_3 \rightleftharpoons NH_4^+ + OH^-$$

The expression for the equilibrium constant for this reaction

$$K' = \frac{[NH_4^+][OH^-]}{[NH_3][H_2O]}$$

simplifies to the following if the concentration of water is assumed to be a constant,

$$K = K'[H_2O] = \frac{[NH_4^+][OH^-]}{[NH_3]}$$

Formerly it was assumed that ammonia gas dissolves in water to produce ammonium hydroxide molecules, $NH_4OH$, and that these molecules dissociate into $NH_4^+$ and $OH^-$ ions. Currently, however, the existence of such ammonium hydroxide molecules is questioned. Considerable experimental evidence suggests that in aqueous solutions of ammonia each $NH_3$ molecule is hydrogen bonded to several $H_2O$ molecules; if $NH_4OH$ molecules exist at all in such solutions, they constitute a minor component. The ammonia equilibrium, therefore, is probably best represented in the manner employed in the preceding equation.

Table 16.2 lists the ionization constants of some weak acids and some weak bases. Since the terms for the ion concentrations appear in the numerator of the ionization-constant expression, the magnitude of the value of $K$ gives a qualitative idea of the strength of the electrolyte. Thus, HOCN ($K = 1.2 \times 10^{-4}$) is a stronger acid than HCN ($K = 4.0 \times 10^{-10}$). The relative acidity of solutions of the two acids, however, must be found by comparing the concentrations of hydrogen ion in the solutions and not by comparing the equilibrium constants.

**Table 16.2** Ionization constants at 25°C

| WEAK ACIDS | | |
|---|---|---|
| acetic | $HC_2H_3O_2 \rightleftharpoons H^+ + C_2H_3O_2^-$ | $1.8 \times 10^{-5}$ |
| benzoic | $HC_7H_5O_2 \rightleftharpoons H^+ + C_7H_5O_2^-$ | $6.0 \times 10^{-5}$ |
| chlorous | $HClO_2 \rightleftharpoons H^+ + ClO_2^-$ | $1.1 \times 10^{-2}$ |
| cyanic | $HOCN \rightleftharpoons H^+ + OCN^-$ | $1.2 \times 10^{-4}$ |
| formic | $HCHO_2 \rightleftharpoons H^+ + CHO_2^-$ | $1.8 \times 10^{-4}$ |
| hydrazoic | $HN_3 \rightleftharpoons H^+ + N_3^-$ | $1.9 \times 10^{-5}$ |
| hydrocyanic | $HCN \rightleftharpoons H^+ + CN^-$ | $4.0 \times 10^{-10}$ |
| hydrofluoric | $HF \rightleftharpoons H^+ + F^-$ | $6.7 \times 10^{-4}$ |
| hypobromous | $HOBr \rightleftharpoons H^+ + BrO^-$ | $2.1 \times 10^{-9}$ |
| hypochlorous | $HOCl \rightleftharpoons H^+ + ClO^-$ | $3.2 \times 10^{-8}$ |
| nitrous | $HNO_2 \rightleftharpoons H^+ + NO_2^-$ | $4.5 \times 10^{-4}$ |
| WEAK BASES | | |
| ammonia | $NH_3 + H_2O \rightleftharpoons NH_4^+ + OH^-$ | $1.8 \times 10^{-5}$ |
| aniline | $C_6H_5NH_2 + H_2O \rightleftharpoons C_6H_5NH_3^+ + OH^-$ | $4.6 \times 10^{-10}$ |
| dimethylamine | $(CH_3)_2NH + H_2O \rightleftharpoons (CH_3)_2NH_2^+ + OH^-$ | $7.4 \times 10^{-4}$ |
| hydrazine | $N_2H_4 + H_2O \rightleftharpoons N_2H_5^+ + OH^-$ | $9.8 \times 10^{-7}$ |
| methylamine | $CH_3NH_2 + H_2O \rightleftharpoons CH_3NH_3^+ + OH^-$ | $5.0 \times 10^{-4}$ |
| pyridine | $C_5H_5N + H_2O \rightleftharpoons C_5H_5NH^+ + OH^-$ | $1.5 \times 10^{-9}$ |
| trimethylamine | $(CH_3)_3N + H_2O \rightleftharpoons (CH_3)_3NH^+ + OH^-$ | $7.4 \times 10^{-5}$ |

*Example 16.4*
Compare the acidity of 0.10*M* HOCN with the acidity of 0.10*M* HCN.

*Solution*
Let $x$ equal the concentration of hydrogen ion in 0.10*M* HOCN and $y$ equal the concentration of hydrogen ion in 0.10*M* HCN.

$$HOCN \rightleftharpoons H^+ + OCN^-$$
$$\quad 0.10M \quad (x)M \quad (x)M$$

$$HCN \rightleftharpoons H^+ + CN^-$$
$$\quad 0.10M \quad (y)M \quad (y)M$$

$$1.2 \times 10^{-4} = \frac{[H^+][OCN^-]}{[HOCN]}$$

$$4.0 \times 10^{-10} = \frac{[H^+][CN^-]}{[HCN]}$$

$$= \frac{x^2}{0.10}$$

$$= \frac{y^2}{0.10}$$

$$x = \sqrt{1.2 \times 10^{-4}(0.10)}$$

$$y = \sqrt{4.0 \times 10^{-10}(0.10)}$$

$$\frac{\text{acidity } 0.10M \text{ HOCN}}{\text{acidity } 0.10M \text{ HCN}} = \frac{x}{y} = \frac{\sqrt{1.2 \times 10^{-4}(0.10)}}{\sqrt{4.0 \times 10^{-10}(0.10)}}$$

$$= \sqrt{0.30 \times 10^6}$$

$$= 5.5 \times 10^2$$

The solution of HOCN is 550 times more acidic than the solution of HCN.

**16.2 THE IONIZATION OF WATER**

Pure water is itself a very weak electrolyte and ionizes according to the equation

$$H_2O + H_2O \rightleftharpoons H_3O^+ + OH^-$$

In simplified form this is

$$H_2O \rightleftharpoons H^+ + OH^-$$

The expression for the ionization constant derived from this equation is

$$K = \frac{[H^+][OH^-]}{[H_2O]}$$

In dilute solutions the concentration of water is virtually a constant, and we may combine $[H_2O]$ with the constant $K$. Thus,

$$K[H_2O] = [H^+][OH^-]$$

This constant, $K[H_2O]$, is called the ion product of water, or the water constant, and is given the symbol $K_w$. At 25°C,

$$K_w = 1.0 \times 10^{-14} = [H^+][OH^-]$$

In pure water,

$$[H^+] = [OH^-] = x$$

$$[H^+][OH^-] = 1.0 \times 10^{-14}$$

$$x^2 = 1.0 \times 10^{-14}$$

$$x = 1.0 \times 10^{-7}M$$

Thus, the concentrations of both of the ions of water are equal to $1.0 \times 10^{-7}M$ in pure water or in any neutral solution at 25°C. Hence, in 1 liter only $10^{-7}$ mol of water is in ionic form out of a total of approximately 55.5 mol.

In any aqueous solution both hydronium and hydroxide ions exist. In an acid solution the concentration of hydronium ions is larger than $1.0 \times 10^{-7}M$ and larger than the hydroxide ion concentration. In an alkaline solution the hydroxide ion concentration is larger than $1.0 \times 10^{-7}M$ and larger than the hydronium ion concentration.

---

*Example 16.5*

What are $[H^+]$ and $[OH^-]$ in a $0.020M$ solution of HCl?

*Solution*

The quantity of hydronium ion obtained from the ionization of water is negligible compared to that derived from the hydrochloric acid. Furthermore, since HCl is a strong electrolyte, $[H^+] = 0.020M$.

$$[H^+][OH^-] = 1.0 \times 10^{-14}$$

$$(2.0 \times 10^{-2})[OH^-] = 1.0 \times 10^{-14}$$

$$[OH^-] = \frac{1.0 \times 10^{-14}}{2.0 \times 10^{-2}}$$

$$[OH^-] = 5.0 \times 10^{-13}M$$

Thus, there are hydroxide ions present in this acidic solution. Notice, however, that $[OH^-]$ is extremely small; in this solution there would be one hydroxide ion for every 40 billion hydronium ions.

---

*Example 16.6*

What are $[H^+]$ and $[OH^-]$ in a $0.0050M$ solution of NaOH?

*Solution*

Sodium hydroxide is a strong electrolyte, and therefore $[OH^-] = 5.0 \times 10^{-3}M$.

$$[H^+][OH^-] = 1.0 \times 10^{-14}$$

$$[H^+](5.0 \times 10^{-3}) = 1.0 \times 10^{-14}$$

$$[H^+] = \frac{1.0 \times 10^{-14}}{5.0 \times 10^{-3}}$$

$$[H^+] = 2.0 \times 10^{-12}M$$

**16.3 pH**     A convenient method for expressing the hydronium ion concentration of a solution is in terms of the pH scale. The **pH** of a solution may be defined as the logarithm of the reciprocal of the hydronium ion concentration. Since the logarithm of 1 is 0, pH may also be defined as the negative logarithm of the hydronium ion concentration. Thus,

$$pH = \log\left(\frac{1}{[H^+]}\right) = -\log\ [H^+]$$

or

$$[H^+] = 10^{-pH} = antilog\ (-pH)$$

For a neutral solution, therefore,

$$[H^+] = 1.0 \times 10^{-7}$$
$$pH = -\log\ (10^{-7}) = -(-7)$$
$$pH = 7$$

For a $0.001M$ solution of HCl,

$$[H^+] = 1 \times 10^{-3}$$
$$pH = -\log\ (10^{-3}) = 3$$

We can define **pOH** in the same terms—as the negative logarithm of the hydroxide ion concentration. Such values generally are not quoted; the pH value of a solution is used to define the acidity or alkalinity of the solution. However, it is frequently convenient to use pOH's in calculations involving alkaline solutions or solutions for which the hydroxide ion concentration is known. The water constant is

$$[H^+][OH^-] = 10^{-14}$$

By taking the logarithm of each term and multiplying through by $-1$, we find

$$-\log[H^+] - \log[OH^-] = -\log(10^{-14})$$

or

$$pH + pOH = 14$$

Thus, for a $0.01M$ solution of NaOH the pOH is 2. Since the sum of the pH and the pOH equals 14, the pH of this solution is 12.

*Example 16.7*
What is the pH of a solution 0.050*M* in H$^+$?

*Solution*

$$[H^+] = 5.0 \times 10^{-2}$$

$$\log[H^+] = \log 5.0 + \log 10^{-2}$$

$$= 0.70 - 2.00 = -1.30$$

$$pH = 1.30$$

*Example 16.8*
What is the pH of a solution for which [OH$^-$] = 0.030*M?*

*Solution*

$$[OH^-] = 3.0 \times 10^{-2}$$

$$\log[OH^-] = \log 3.0 + \log 10^{-2}$$

$$= 0.48 - 2.00 = -1.52$$

$$pOH = 1.52$$

$$pH = 12.48$$

An alternative solution is

$$[H^+][OH^-] = 1.0 \times 10^{-14}$$

$$[H^+] = \frac{1.0 \times 10^{-14}}{3.0 \times 10^{-2}} = 3.3 \times 10^{-13}$$

$$\log[H^+] = \log 3.3 + \log 10^{-13}$$

$$= 0.52 - 13.00 = -12.48$$

$$pH = 12.48$$

*Example 16.9*
What is the [H$^+$] of a solution with a pH of 10.60?

*Solution*

$$\log[H^+] = -10.60 = 0.40 - 11.00$$

$$[H^+] = \text{antilog } 0.40 \times \text{antilog}(-11)$$

$$[H^+] = 2.5 \times 10^{-11} \, M$$

It should be kept in mind that pH relates to a power of 10. Hence, a solution of pH = 1 has a hydronium ion concentration 100 times that of a solution of pH = 3 (not three times). Furthermore, since the pH is related to a *negative* exponent, the lower the pH value, the larger the concentration of hydronium ion. At pH = 7, a solution is neutral. Solutions with pH's below 7 are acidic; those with pH's above 7 are alkaline. These relationships are summarized in Table 16.3.

**16.4 DETERMINATION OF pH**

Potentiometric methods are commonly employed for pH determination. Since the emf of a cell is dependent upon activities and not upon concentrations, we redefine pH as the negative logarithm of the activity of the hydronium ion concentration. However, this redefinition introduces a problem since it is experimentally impossible to determine activities of

Table 16.3 The pH scale

| pH | $[H^+]$ | $[OH^-]$ | |
|----|---------|----------|---|
| 14 | $10^{-14}$ | $10^0$ | |
| 13 | $10^{-13}$ | $10^{-1}$ | |
| 12 | $10^{-12}$ | $10^{-2}$ | |
| 11 | $10^{-11}$ | $10^{-3}$ | increasing |
| 10 | $10^{-10}$ | $10^{-4}$ | alkalinity |
| 9 | $10^{-9}$ | $10^{-5}$ | |
| 8 | $10^{-8}$ | $10^{-6}$ | |
| 7 | $10^{-7}$ | $10^{-7}$ | neutrality |
| 6 | $10^{-6}$ | $10^{-8}$ | |
| 5 | $10^{-5}$ | $10^{-9}$ | |
| 4 | $10^{-4}$ | $10^{-10}$ | |
| 3 | $10^{-3}$ | $10^{-11}$ | increasing |
| 2 | $10^{-2}$ | $10^{-12}$ | acidity |
| 1 | $10^{-1}$ | $10^{-13}$ | |
| 0 | $10^0$ | $10^{-14}$ | |

single ions, and as we shall see, it is necessary to introduce an assumption into the treatment.

The following cell might be used for the determination of pH's

$$Pt \mid H_2(1\ atm) \mid H^+(a = ?) \parallel Cl^-(1N) \mid Hg_2Cl_2(s) \mid Hg$$

This cell is diagramed in Figure 16.1. The left-hand electrode is a hydrogen electrode (Section 10.8) in which the electrolyte is the solution being studied. This electrode is connected by means of a salt bridge to a normal calomel electrode. The calomel electrode utilizes as the electrolyte a

**Figure 16.1** Hydrogen electrode and normal calomel electrode.

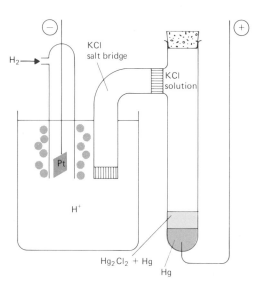

normal solution of potassium chloride that is saturated with mercurous chloride ($Hg_2Cl_2$); the electrode proper consists of mercury metal, which is at the bottom of the tube, overlayed with a paste of mercurous chloride (calomel) and mercury.

The half reactions for the cell are:

$$H_2(g) \rightarrow 2H^+(aq) + 2e^-$$

$$2e^- + Hg_2Cl_2(s) \rightarrow 2Hg(l) + 2Cl^-(aq)$$

and the corresponding net chemical change is

$$H_2(g) + Hg_2Cl_2(s) \rightarrow 2H^+(aq) + 2Cl^-(aq) + 2Hg(l)$$

The cell is not thermodynamically reversible (Section 10.7). When the cell is discharging. $H^+(aq)$ ions leave the test solution and enter the salt bridge, and $Cl^-$ ions leave the salt bridge and enter the test solution. When the operation of the cell is reversed, $K^+$ ions enter the test solution from the salt bridge, and ions from the test solution (not necessarily $Cl^-$ ions) enter the salt bridge. Hence, there is an indeterminable liquid-junction potential at the salt bridge. The dependence of the emf of the cell upon the activities of the species present may be expressed by adapting the Nernst equation (Section 10.9),

$$\mathscr{E} = \mathscr{E}^\circ - \frac{0.05916}{2} \log \left( \frac{(a_{H^+})^2 (a_{Cl^-})^2}{(a_{H_2})} \right) + \mathscr{E}_{\text{liquid junction}}$$

Upon expansion this becomes

$$\mathscr{E} = \mathscr{E}^\circ + \frac{0.05916}{2} \log(a_{H_2}) - 0.05916 \log(a_{Cl^-}) + \mathscr{E}_{\text{liquid junction}} - 0.05916 \log(a_{H^+})$$

We shall assume that the pressure of $H_2$ gas is maintained at 1 atm, that the salt bridge is effective in maintaining a constant $Cl^-$ ion concentration in the calomel electrode, and that the liquid-junction potential is a constant value that is negligibly small. Hence, the first four terms of the expression for the emf of the cell may be combined into an experimentally determined constant, $\mathscr{E}'$.

$$\mathscr{E} = \mathscr{E}' - 0.05916 \log(a_{H^+})$$

For the normal calomel electrode in air at 25°C, $\mathscr{E}'$ is 0.2825 V.[1] Therefore,

---

[1] In the absence of air the normal potassium chloride – calomel electrode has an $\mathscr{E}'$ of 0.2812 V. Frequently a saturated solution of KCl is employed; the generally accepted value of $\mathscr{E}'$ for such an electrode is 0.2415 V.

$$\mathscr{E} = 0.2825 + 0.05916(\text{pH})$$

$$\text{pH} = \frac{\mathscr{E} - 0.2825}{0.05916}$$

This expression may be regarded as an operational definition of pH. For problem work we shall use pH's based on concentrations and activities interchangeably, since the activities of $H^+(aq)$ approximate the concentrations of $H^+(aq)$ in dilute solutions.

In place of the hydrogen electrode, a glass electrode is usually employed for the determination of pH's. The hydrogen electrode is easily poisoned and requires the use of gaseous hydrogen. A glass electrode consists of a reference electrode that contains a solution of known pH sealed in a membrane of special glass. In the determination of the pH of a solution, the glass electrode and another reference electrode are immersed in the solution being studied.

The emf of the complete cell depends upon the difference in potential across the glass membrane that separates solutions of different pH's and is found to respond to changes in pH in the same way that an assembly using a hydrogen electrode does. The value of $\mathscr{E}'$ for a cell using a glass electrode depends upon the electrode used in the construction of the glass electrode as well as the other reference electrode employed in the complete cell. Instruments known as pH meters employ glass electrodes and are usually calibrated in pH units rather than volts; such instruments are standardized with solutions of known pH.

---

*Example 16.10*

For a cell containing a solution of unknown pH and utilizing a hydrogen electrode and a normal calomel electrode, the potential is 0.608 V at 25°C. Calculate the pH of the solution and the approximate concentration of $H^+(aq)$.

*Solution*

$$\text{pH} = \frac{\mathscr{E} - 0.283}{0.0592}$$

$$= \frac{0.608 - 0.283}{0.0592}$$

$$\text{pH} = 5.49$$

$$\log[H^+] = -5.49 = 0.51 - 6.00$$

$$[H^+] = 3.2 \times 10^{-6} M$$

---

An approximate value of the ionization constant of a weak acid or a weak base may be determined by measuring the pH of a solution of known concentration of the weak electrolyte. A more exact method for the determination of ionization constants based on pH determinations is described in Section 17.7.

*Example 16.11*

The pH of a $0.10\,M$ solution of a weak base, B, is 10.60. What is the ionization constant of B?

*Solution*

$$\log[H^+] = -10.60 = 0.40 - 11.00$$

$$[H^+] = 2.5 \times 10^{-11}$$

$$[OH^-] = \frac{1.0 \times 10^{-14}}{2.5 \times 10^{-11}} = 4.0 \times 10^{-4}M$$

Since the solution was prepared from the base alone,

$$[BH^+] = [OH^-] = 4.0 \times 10^{-4}M$$

The concentration of B in the solution is for all practical purposes equal to $0.10\,M$; the small quantity of B dissociated need not be considered. Thus, the equilibrium concentrations are

$$\begin{array}{ccccc} B & + H_2O \rightleftharpoons & BH^+ & + & OH^- \\ 0.10M & & 4.0 \times 10^{-4}M & & 4.0 \times 10^{-4}M \end{array}$$

The value of the ionization constant is

$$K = \frac{[BH^+][OH^-]}{[B]}$$

$$= \frac{(4.0 \times 10^{-4})^2}{1.0 \times 10^{-1}} = \frac{16 \times 10^{-8}}{1.0 \times 10^{-1}}$$

$$K = 1.6 \times 10^{-6}$$

**16.5 INDICATORS**  Indicators are colored, organic compounds of complex structure that change color in solution as the pH of the solution changes. For example, methyl orange is red in solutions of pH below 3.1 and yellow in solutions of pH above 4.5; the color of this indicator is a varying mixture of yellow and red in the pH range between 3.1 and 4.5. Many indicators have been described and used; a few are listed in Table 16.4.

Table 16.4 Some indicators

| INDICATOR | ACID COLOR | pH RANGE OF COLOR CHANGE | ALKALINE COLOR |
|---|---|---|---|
| thymol blue | red | 1.2–2.8 | yellow |
| methyl orange | red | 3.1–4.5 | yellow |
| bromcresol green | yellow | 3.8–5.5 | blue |
| methyl red | red | 4.2–6.3 | yellow |
| litmus | red | 5.0–8.0 | blue |
| bromthymol blue | yellow | 6.0–7.6 | blue |
| thymol blue | yellow | 8.0–9.6 | blue |
| phenolphthalein | colorless | 8.3–10.0 | red |
| alizarin yellow | yellow | 10.0–12.1 | lavender |

By the use of indicators one may determine the pH of a solution. If thymol blue is yellow in a test solution and methyl orange is red in another sample of the same solution, the pH of the solution is between 2.8 and 3.1. Reference to Table 16.4 shows that thymol blue is yellow only in solutions of pH greater than 2.8, and methyl orange is red only in solutions of pH less than 3.1. If enough indicators are employed, it is possible to determine pH values that are accurate to the first decimal place; potentiometric measurements of pH's are, however, more accurate.

Indicators are weak acids or weak bases. Since they are intensely colored, only a few drops of a dilute solution of an indicator need be employed in any determination; hence, the acidity of the solution in question is not significantly altered by the addition of the indicator.

If we let the symbol HIn stand for the litmus molecule (which is red) and the symbol In⁻ stand for the anion (which is blue) derived from the weak acid, the equation for the litmus equilibrium may be written

$$\underset{\text{red}}{HIn} \rightleftharpoons H^+ + \underset{\text{blue}}{In^-}$$

According to the principle of Le Chatelier, increasing the concentration of $H^+$ shifts the equilibrium to the left, and the red (or acid) color of HIn is observed. On the other hand, addition of $OH^-$ decreases the concentration of $H^+$; the equilibrium shifts to the right, and the blue (or alkaline) color of In⁻ is observed.

The ionization constant for litmus is approximately equal to $10^{-7}$.

$$10^{-7} = \frac{[H^+][In^-]}{[HIn]}$$

We may rearrange this expression in the following manner

$$\frac{10^{-7}}{[H^+]} = \frac{[In^-]}{[HIn]}$$

At a pH of 5 or below, the red color of litmus is observed. If we substitute $[H^+] = 10^{-5}$, which corresponds to pH = 5, into the preceding expression, we get

$$\frac{10^{-7}}{10^{-5}} = \frac{1}{100} = \frac{[In^-] \leftarrow \text{blue}}{[HIn] \leftarrow \text{red}}$$

Thus, the mixture appears red to the eye when the concentration of the red HIn is 100 times (or more) that of the blue In⁻.

The blue color of litmus is observed in solutions of pH = 8 or higher. If $[H^+] = 10^{-8}$,

$$\frac{10^{-7}}{10^{-8}} = \frac{10}{1} = \frac{[In^-] \leftarrow \text{blue}}{[HIn] \leftarrow \text{red}}$$

When the concentration of the blue In⁻ ion is 10 times that of the red HIn, or approximately 91% of the indicator is in ionic form, the blue color of the In⁻ ions completely masks the red color of the HIn molecules, and the mixture appears blue.

Thus, the blue color predominates when the concentration of the blue species is 10 times that of the red, whereas the red color predominates only when the concentration of the red form is 100 times that of the blue.

This is not surprising since the blue color of litmus is a much stronger color than the red. Only when $[H^+] = K = 10^{-7}$ will the factor $K/[H^+] = 1$ and $[In^-] = [HIn]$. Hence, at pH = 7, litmus exhibits its "neutral" color (purple).

Therefore, the pH range over which a given indicator changes color depends upon the ionization constant of the indicator. For indicators that are weak acids, the smaller the value of $K$, the higher is the pH range of the color change.

**16.6** THE COMMON-ION EFFECT

In a $0.1 M$ solution of acetic acid, methyl orange assumes its acid color, red. If sodium acetate is added to this solution, the color changes to yellow, showing that the addition causes the acidity of the solution to decrease. This experimental observation is readily explained on the basis of Le Chatelier's principle. The equilibrium

$$HC_2H_3O_2 \rightleftharpoons H^+ + C_2H_3O_2^-$$

is shifted to the left by the addition of the acetate ion from the sodium acetate, and the concentration of hydronium ion correspondingly decreases. Since acetic acid and sodium acetate have the acetate ion in common, this phenomenon is called the **common-ion effect**.

The concentration of hydronium ion in a solution of acetic acid to which acetate ion has been added can be calculated by means of the ionization constant of acetic acid.

---

*Example 16.12*

What is the concentration of hydronium ion in a $0.10 M$ solution of acetic acid that has been made $0.15 M$ in sodium acetate?

*Solution*

Sodium acetate is a strong electrolyte; the concentration of acetate ion derived from this source is therefore $0.15 M$. In a $0.10 M$ solution of pure acetic acid $[H^+] = [C_2H_3O_2^-] = 0.00134 M$ (Table 16.1). In a $0.10 M$ solution of acetic acid to which sodium acetate has been added, the concentration of acetate ion derived from the ionization of acetic acid is much less than $0.00134 M$ because of the common-ion effect. Therefore, the concentration of acetate ion may be regarded as $0.15 M$ and the contribution from acetic acid ignored. Likewise, the concentration of acetic acid may be taken as equal to the total concentration of acetic acid present, $0.10 M$, since such a small amount of the acetic acid is ionized. Thus,

$$HC_2H_3O_2 \rightleftharpoons H^+ + C_2H_3O_2^-$$
$$0.10 M \qquad (x) M \qquad 0.15 M$$

$$1.8 \times 10^{-5} = \frac{[H^+][C_2H_3O_2^-]}{[HC_2H_3O_2]}$$

$$= \frac{[H^+](0.15)}{(0.10)}$$

$$[H^+] = 1.2 \times 10^{-5} M$$

Therefore, when a $0.10 M$ solution of acetic acid is made $0.15 M$ in sodium acetate, the concentration of hydronium ion is reduced from $1.3 \times 10^{-3} M$ to $1.2 \times 10^{-5} M$.

*Example 16.13*
What is the concentration of hydroxide ions in a solution made by dissolving 0.020 mol of ammonium chloride in 100 ml of $0.15\,M$ ammonia? Assume that the addition of the solid does not cause the volume of the solution to change.

*Solution*
Ammonium chloride is a strong electrolyte. The concentration of $NH_4^+$ from the $NH_4Cl$ added is

$$? \text{ mol } NH_4^+ = 1000 \text{ ml soln.} \left( \frac{0.020 \text{ mol } NH_4^+}{100 \text{ ml soln.}} \right)$$

$$= 0.20 \text{ mol } NH_4^+$$

Therefore,

$$NH_3 + H_2O \rightleftharpoons NH_4^+ + OH^-$$
$$0.15\,M \qquad\qquad 0.20\,M \quad (x)\,M$$

$$1.8 \times 10^{-5} = \frac{[NH_4^+][OH^-]}{[NH_3]}$$

$$= \frac{(0.20)[OH^-]}{(0.15)}$$

$$[OH^-] = 1.4 \times 10^{-5}\,M$$

**16.7 BUFFERS**  It is sometimes necessary that a solution of definite pH be prepared and stored. The preservation of such a solution is even more difficult than its preparation. If the solution comes in contact with the air, it will absorb carbon dioxide (an acid anhydride) and become more acidic. If the solution is stored in a glass bottle, alkaline impurities leached from the glass may alter the pH. **Buffer solutions** are capable of maintaining their pH at some fairly constant value even when small amounts of acid or base are added.

A buffer may be prepared from a weak acid or a weak base and a salt of the weak electrolyte. For example, a buffer can be prepared from acetic acid and sodium acetate. If both these materials are present in the same concentration, for example, $1.00\,M$,

$$1.8 \times 10^{-5} = \frac{[H^+][C_2H_3O_2^-]}{[HC_2H_3O_2]}$$

$$[H^+] = 1.8 \times 10^{-5} \left( \frac{[HC_2H_3O_2]}{[C_2H_3O_2^-]} \right) = 1.8 \times 10^{-5}\,M$$

$$pH = -\log(1.8 \times 10^{-5}) = 4.74$$

The $pK$ of a weak electrolyte may be defined in a manner analogous to pH or pOH,

$$pK = -\log K$$

A solution of a weak acid in which the concentration of the anion is the same as the concentration of the undissociated acid has a pH equal to the $pK$ of the acid.

In a sample of the buffer previously described, the quantities of acetic acid molecules and acetate ion are much larger than the quantity of hydronium ion present—approximately 50,000 times larger.

$$HC_2H_3O_2 \rightleftharpoons H^+ + C_2H_3O_2^-$$

$$1.0M \qquad 1.8 \times 10^{-5}M \qquad 1.0M$$

If a small quantity of hydronium ion is added, the large reservoir of acetate ion will quickly convert it to acetic acid. If a small amount of hydroxide ion is added to the buffer, it will neutralize hydronium ion, but the large quantity of acetic acid present will by dissociation replace any hydronium ion removed and maintain the pH at a fairly constant value.

The action of buffers may be interpreted in terms of the Brønsted concept. Thus, an acetic acid–acetate buffer

$$\overset{\text{Acid}_1}{HC_2H_3O_2} + \overset{\text{Base}_2}{H_2O} \rightleftharpoons \overset{\text{Acid}_2}{H_3O^+} + \overset{\text{Base}_1}{C_2H_3O_2^-}$$

effectively neutralizes small additions of hydronium ion because of the presence of a relatively large concentration of the base $C_2H_3O_2^-$. Similarly, small additions of hydroxide ion are effectively neutralized by the acid $HC_2H_3O_2$, which is present in relatively large concentration. In either instance the change in the pH of the solution brought about by the addition is slight.

Buffers cannot withstand the addition of large amounts of acids or alkalies. The addition of 0.01 mole per liter of $H^+$ or $OH^-$ is about the maximum that any buffer can be expected to withstand. How successfully a buffer withstands such an addition is illustrated in the following example.

---

*Example 16.14*

The ionization constant of acetic acid to three significant figures is $1.81 \times 10^{-5}$. A buffer containing $1.00M$ concentrations of acetic acid and sodium acetate has a pH of 4.742. (a) What is the pH of the solution after 0.01 mol of HCl has been added to 1 liter of the buffer? (b) What is the pH of the solution after the addition of 0.01 mol of NaOH?

*Solution*

(a) The 0.01 mol of $H^+$ from HCl converts an equivalent amount of acetate ion into acetic acid. Thus, the concentrations are

$$HC_2H_3O_2 \rightleftharpoons H^+ + C_2H_3O_2^-$$

| | | | |
|---|---|---|---|
| buffer: | $1.00M$ | $1.81 \times 10^{-5}$ | $1.00M$ |
| after $H^+$ addition: | $1.01M$ | ? | $0.99M$ |

$$\frac{[H^+][C_2H_3O_2^-]}{[HC_2H_3O_2]} = 1.81 \times 10^{-5}$$

$$\frac{[H^+](0.99)}{(1.01)} = 1.81 \times 10^{-5}$$

$$[H^+] = 1.81 \times 10^{-5} \frac{(1.01)}{(0.99)}$$

$$= 1.85 \times 10^{-5}M$$

$$pH = 4.733$$

Thus, the addition causes the pH to change 0.009 pH units. A similar addition to pure water would change the pH from 7.0 to 2.0 – a change of 5.0 pH units.

(b) The addition of 0.0100 mol of $OH^-$ would change the concentration of acetate ion to $1.01 M$ and the concentration of acetic acid to $0.99 M$. Thus,

$$\frac{[H^+][C_2H_3O_2^-]}{[HC_2H_3O_2]} = 1.81 \times 10^{-5}$$

$$\frac{[H^+](1.01)}{(0.99)} = 1.81 \times 10^{-5}$$

$$[H^+] = 1.81 \times 10^{-5}\frac{(0.99)}{(1.01)}$$

$$= 1.77 \times 10^{-5} M$$

$$pH = 4.752$$

A similar addition to water would have caused the pH to rise from 7.0 to 12.0.

Alkaline buffers may also be prepared. If the base and its derived ion are present in equal concentrations,

$$pOH = pK$$

$$pH = 14.0 - pK$$

A solution with $[NH_3] = 0.10 M$ and $[NH_4^+] = 0.10 M$ is an example of this type of buffer.

$$\underset{0.10M}{NH_3} + H_2O \rightleftharpoons \underset{0.10M}{NH_4^+} + \underset{?}{OH^-}$$

$$\frac{[NH_4^+][OH^-]}{[NH_3]} = 1.8 \times 10^{-5}$$

$$[OH^-] = 1.8 \times 10^{-5} M$$

$$pOH = 4.74$$

$$pH = 9.26$$

Buffers may also be prepared in which the ratio of the concentration of weak electrolyte to the concentration of common ion is not 1 : 1; this technique may be used to obtain a buffer that has a pH (or pOH) different from the pK of the weak acid (or base). Let us assume that a buffer is to be prepared from a hypothetical weak acid, HA,

$$HA \rightleftharpoons H^+ + A^-$$

for which

$$\frac{[H^+][A^-]}{[HA]} = K$$

The concentration of hydronium ion is

$$[H^+] = \frac{K[HA]}{[A^-]}$$

Taking the negative logarithm of each term, we get

$$-\log[H] = -\log K - \log\left(\frac{[HA]}{[A^-]}\right)$$

or

$$pH = pK - \log\left(\frac{[HA]}{[A^-]}\right)$$

In general, for an effective buffer the ratio of the concentration of the molecular species to the concentration of the ionic species should be between 1/10 and 10/1. This concentration range is equivalent to the following pH range.

$$pH = pK - \log\left(\frac{1}{10}\right) = pK - \log 10^{-1}$$

$$= pK + 1$$

$$pH = pK - \log\left(\frac{10}{1}\right)$$

$$= pK - 1$$

Therefore, an efficient buffer that has any desired pH between 3.7 and 5.7 can be prepared from acetic acid ($pK = 4.7$).

The use of buffers is an important part of many industrial processes; examples are electroplating and the manufacture of leather, photographic materials, and dyes. In bacteriological research, culture media are generally buffered to maintain the pH required for the growth of the bacteria being studied. Buffers are used extensively in analytical chemistry and are used to calibrate pH meters. Human blood is buffered to a pH of 7.4 by means of bicarbonate, phosphate, and complex protein systems.

## Example 16.15

What concentrations should be used to prepare a cyanic acid–cyanate buffer of pH = 3.50?

*Solution*

$$pH = 3.50$$

$$\log[H^+] = -3.50 = 0.50 - 4.00$$

$$[H^+] = 3.1 \times 10^{-4} M$$

$$HOCN \rightleftharpoons H^+ + OCN^-$$

$$1.2 \times 10^{-4} = \frac{[H^+][OCN^-]}{[HOCN]}$$

$$= \frac{3.1 \times 10^{-4}[OCN^-]}{[HOCN]}$$

$$\frac{[HOCN]}{[OCN^-]} = \frac{3.1 \times 10^{-4}}{1.2 \times 10^{-4}}$$

$$\frac{[HOCN]}{[OCN^-]} = 2.6$$

Any solution in which $[HOCN]/[OCN^-]$ is 2.6 will have a pH of 3.5. For example, if $[HOCN] = 0.26M$, $[OCN^-]$ must be $0.10M$; or if $[HOCN] = 0.52M$, $[OCN^-] = 0.20M$.

An alternative solution of the problem utilizes the relationship

$$pH = pK - \log \frac{[HOCN]}{[OCN^-]}$$

$$3.50 = 3.92 - \log \frac{[HOCN]}{[OCN^-]}$$

$$\log \frac{[HOCN]}{[OCN^-]} = 0.42$$

$$\frac{[HOCN]}{[OCN^-]} = 2.6$$

## Example 16.16

What is the pH of a solution made by mixing 100 ml of 0.15$M$ HCl and 200 ml of 0.20$M$ aniline ($C_6H_5NH_2$)? Assume that the volume of the final solution is 300 ml.

*Solution*

The number of moles of HCl used and the number of moles of aniline used are

$$? \text{ mol HCl} = 100 \text{ ml soln.} \left(\frac{0.15 \text{ mol HCl}}{1000 \text{ ml soln.}}\right)$$

$$= 0.015 \text{ mol HCl}$$

$$? \text{ mol aniline} = 200 \text{ ml soln.} \left(\frac{0.20 \text{ mol aniline}}{1000 \text{ ml soln.}}\right)$$

$$= 0.040 \text{ mol aniline}$$

One mole of aniline reacts with one mole of $H^+$. Thus,

$$C_6H_5NH_2 + H^+ \rightarrow C_6H_5NH_3^+$$

| | | | |
|---|---|---|---|
| before reaction: | 0.040 mol | 0.015 mol | — |
| after reaction: | 0.025 mol | — | 0.015 mol |

The molar concentrations in the final solution are

$$? \text{ mol } C_6H_5NH_3^+ = 1000 \text{ ml soln.} \left(\frac{0.015 \text{ mol } C_6H_5NH_3^+}{300 \text{ ml soln.}}\right)$$

$$= 0.050 \text{ mol } C_6H_5NH_3^+$$

$$? \text{ mol } C_6H_5NH_2 = 1000 \text{ ml soln.} \left(\frac{0.025 \text{ mol } C_6H_5NH_2}{300 \text{ ml soln.}}\right)$$

$$= 0.083 \text{ mol } C_6H_5NH_2$$

Therefore, the concentrations in the solution are

$$C_6H_5NH_2 + H_2O \rightleftharpoons C_6H_5NH_3^+ + OH^-$$
$$0.083\,M \qquad\qquad 0.050\,M \qquad ?$$

$$\frac{[C_6H_5NH_3^+][OH^-]}{[C_6H_5NH_2]} = 4.6 \times 10^{-10}$$

$$\frac{(0.050)[OH^-]}{(0.083)} = 4.6 \times 10^{-10}$$

$$[OH^-] = 7.6 \times 10^{-10}\,M$$

$$[H^+] = 1.3 \times 10^{-5}\,M$$

$$pH = 4.89$$

## 16.8 POLYPROTIC ACIDS

**Polyprotic acids** are acids that contain more than one acid hydrogen per molecule; examples include sulfuric acid ($H_2SO_4$), oxalic acid ($H_2C_2O_4$), phosphoric acid ($H_3PO_4$), and arsenic acid ($H_3AsO_4$). Polyprotic acids ionize in a stepwise manner, and there is an ionization constant for each step. Subscripts are added to the symbol $K$ in order to specify the step to which the constant applies.

Phosphoric acid is triprotic and ionizes in three steps.

$$H_3PO_4 \rightleftharpoons H^+ + H_2PO_4^- \qquad \frac{[H^+][H_2PO_4^-]}{[H_3PO_4]} = K_1 = 7.5 \times 10^{-3}$$

$$H_2PO_4^- \rightleftharpoons H^+ + HPO_4^{2-} \qquad \frac{[H^+][HPO_4^{2-}]}{[H_2PO_4^-]} = K_2 = 6.2 \times 10^{-8}$$

$$HPO_4^{2-} \rightleftharpoons H^+ + PO_4^{3-} \qquad \frac{[H^+][PO_4^{3-}]}{[HPO_4^{2-}]} = K_3 = 1 \times 10^{-12}$$

Thus, in a solution of phosphoric acid three equilibria occur together with the water equilibrium; $H_3PO_4$, $H_2PO_4^-$, $HPO_4^{2-}$, $PO_4^{3-}$, $H^+$, $OH^-$, and $H_2O$ are present.

The ionization of phosphoric acid is typical of all polyprotic acids in that the primary ionization is stronger than the secondary, and the secondary ionization is stronger than the tertiary. This trend in the value of the ionization constant is consistent with the nature of the particle that ionizes in each step. One would predict that a proton would be released more readily by an uncharged molecule than by a uninegative ion and more readily by a uninegative ion than by a binegative ion.

No polyprotic acid is known for which all ionizations are strong. The primary ionization of sulfuric acid is essentially complete,

$$H_2SO_4 \rightarrow H^+ + HSO_4^-$$

but the secondary ionization is weak,

$$HSO_4^- \rightleftharpoons H^+ + SO_4^{2-} \qquad \frac{[H^+][SO_4^{2-}]}{[HSO_4^-]} = K_2 = 1.3 \times 10^{-2}$$

Solutions of carbon dioxide are acidic. Carbon dioxide reacts with water to form carbonic acid, $H_2CO_3$; however, the reaction is not complete — most of the carbon dioxide exists in solution as $CO_2$ molecules. Therefore, we shall indicate the primary ionization as

$$CO_2 + H_2O \rightleftharpoons H^+ + HCO_3^- \qquad \frac{[H^+][HCO_3^-]}{[CO_2]} = K_1 = 4.2 \times 10^{-7}$$

where the symbol $[CO_2]$ is used to represent the total concentration of $CO_2(aq)$ and $H_2CO_3$. The second ionization step is

$$HCO_3^- \rightleftharpoons H^+ + CO_3^{2-} \qquad \frac{[H^+][CO_3^{2-}]}{[HCO_3^-]} = K_2 = 4.8 \times 10^{-11}$$

An analogous situation exists for solutions of sulfur dioxide in water. The acidity of aqueous $SO_2$ has been attributed to the ionization of sulfurous acid, $H_2SO_3$. However, $H_2SO_3$ has never been isolated in pure form; in solution it apparently exists in equilibrium with $SO_2(aq)$

$$SO_2(aq) + H_2O \rightleftharpoons H_2SO_3(aq)$$

Table 16.5 Ionization constants of some polyprotic acids

| acid | reaction | constant |
|---|---|---|
| arsenic | $H_3AsO_4 \rightleftharpoons H^+ + H_2AsO_4^-$ | $K_1 = 2.5 \times 10^{-4}$ |
| | $H_2AsO_4^- \rightleftharpoons H^+ + HAsO_4^{2-}$ | $K_2 = 5.6 \times 10^{-8}$ |
| | $HAsO_4^{2-} \rightleftharpoons H^+ + AsO_4^{3-}$ | $K_3 = 3 \times 10^{-13}$ |
| carbonic | $CO_2 + H_2O \rightleftharpoons H^+ + HCO_3^-$ | $K_1 = 4.2 \times 10^{-7}$ |
| | $HCO_3^- \rightleftharpoons H^+ + CO_3^{2-}$ | $K_2 = 4.8 \times 10^{-11}$ |
| hydrosulfuric | $H_2S \rightleftharpoons H^+ + HS^-$ | $K_1 = 1.1 \times 10^{-7}$ |
| | $HS^- \rightleftharpoons H^+ + S^{2-}$ | $K_2 = 1.0 \times 10^{-14}$ |
| oxalic | $H_2C_2O_4 \rightleftharpoons H^+ + HC_2O_4^-$ | $K_1 = 5.9 \times 10^{-2}$ |
| | $HC_2O_4^- \rightleftharpoons H^+ + C_2O_4^{2-}$ | $K_2 = 6.4 \times 10^{-5}$ |
| phosphoric | $H_3PO_4 \rightleftharpoons H^+ + H_2PO_4^-$ | $K_1 = 7.5 \times 10^{-3}$ |
| | $H_2PO_4^- \rightleftharpoons H^+ + HPO_4^{2-}$ | $K_2 = 6.2 \times 10^{-8}$ |
| | $HPO_4^{2-} \rightleftharpoons H^+ + PO_4^{3-}$ | $K_3 = 1 \times 10^{-12}$ |
| phosphorous (diprotic) | $H_3PO_3 \rightleftharpoons H^+ + H_2PO_3^-$ | $K_1 = 1.6 \times 10^{-2}$ |
| | $H_2PO_3^- \rightleftharpoons H^+ + HPO_3^{2-}$ | $K_2 = 7 \times 10^{-7}$ |
| sulfuric | $H_2SO_4 \rightarrow H^+ + HSO_4^-$ | strong |
| | $HSO_4^- \rightleftharpoons H^+ + SO_4^{2-}$ | $K_2 = 1.3 \times 10^{-2}$ |
| sulfurous | $SO_2 + H_2O \rightleftharpoons H^+ + HSO_3^-$ | $K_1 = 1.3 \times 10^{-2}$ |
| | $HSO_3^- \rightleftharpoons H^+ + SO_3^{2-}$ | $K_2 = 5.6 \times 10^{-8}$ |

We shall represent the primary ionization of sulfurous acid as

$$SO_2 + H_2O \rightleftharpoons H^+ + HSO_3^-$$

The ionization constants of some polyprotic acids are listed in Table 16.5.

Polyprotic acids form more than one salt. Depending upon the stoichiometric ratio of reactants, the reaction of NaOH and $H_2SO_4$ yields either the normal salt, $Na_2SO_4$ (sodium sulfate) or the acid salt $NaHSO_4$ (sodium bisulfate or sodium hydrogen sulfate). Three salts may be derived from phosphoric acid: $NaH_2PO_4$ (sodium dihydrogen phosphate), $Na_2HPO_4$ (sodium hydrogen phosphate), and $Na_3PO_4$ (sodium phosphate).

*Example 16.17*

Calculate $[H^+]$, $[H_2PO_4^-]$, $[HPO_4^{2-}]$, $[PO_4^{3-}]$, and $[H_3PO_4]$ in a $0.10M$ solution of phosphoric acid.

*Solution*

The principal source of $H^+$ is the primary ionization; the $H^+$ produced by the other ionizations, as well as that from the ionization of water, is negligible in comparison. Furthermore, the concentration of $H_2PO_4^-$ derived from the primary ionization is not significantly diminished by the secondary ionization. Thus, we write

$$\underset{(0.10-x)M}{H_3PO_4} \rightleftharpoons \underset{(x)M}{H^+} + \underset{(x)M}{H_2PO_4^-}$$

The problem must be solved by means of the quadratic formula.

$$\frac{[H^+][H_2PO_4^-]}{[H_3PO_4]} = 7.5 \times 10^{-3}$$

$$\frac{x^2}{(0.10-x)} = 7.5 \times 10^{-3}$$

$$x = [H^+] = [H_2PO_4^-] = 2.4 \times 10^{-2}M$$

$$(0.10-x) = [H_3PO_4] = 7.6 \times 10^{-2}M$$

The $[H^+]$ and $[H_2PO_4^-]$ apply to the secondary ionization. Therefore,

$$\underset{2.4 \times 10^{-2}M}{H_2PO_4^-} \rightleftharpoons \underset{2.4 \times 10^{-2}M}{H^+} + \underset{?}{HPO_4^{2-}}$$

$$\frac{[H^+][HPO_4^{2-}]}{[H_2PO_4^-]} = 6.2 \times 10^{-8}$$

$$\frac{(2.4 \times 10^{-2})[HPO_4^{2-}]}{(2.4 \times 10^{-2})} = 6.2 \times 10^{-8}$$

$$[HPO_4^{2-}] = 6.2 \times 10^{-8}M$$

In any solution of $H_3PO_4$ that does not contain ions derived from another electrolyte, the concentration of the secondary ion is equal to $K_2$.

For the tertiary ionization,

$$\underset{6.2 \times 10^{-8}M}{HPO_4^{2-}} \rightleftharpoons \underset{2.4 \times 10^{-2}M}{H^+} + \underset{?}{PO_4^{3-}}$$

$$\frac{[H^+][PO_4^{3-}]}{[HPO_4^{2-}]} = 1 \times 10^{-12}$$

$$\frac{(2.4 \times 10^{-2})[PO_4^{3-}]}{(6.2 \times 10^{-8})} = 1 \times 10^{-12}$$

$$[PO_4^{3-}] = 3 \times 10^{-18}M$$

*Example 16.18*

What are $[H^+]$, $[HS^-]$, $[S^{2-}]$, and $[H_2S]$ in a $0.10M$ solution of $H_2S$?

*Solution*

$K_1$ for $H_2S$ is $1.1 \times 10^{-7}$; therefore, the small amount of $H_2S$ that ionizes is negligible in comparison to the original concentration of $H_2S$. In addition, the concentrations of $H^+$ and $HS^-$ are not significantly altered by the secondary ionization ($K_2 = 1.0 \times 10^{-14}$). Thus,

$$H_2S \rightleftharpoons H^+ + HS^-$$
$$0.10M \quad (x)M \quad (x)M$$

$$\frac{[H^+][HS^-]}{[H_2S]} = 1.1 \times 10^{-7}$$

$$\frac{x^2}{0.10} = 1.1 \times 10^{-7}$$

$$x = [H^+] = [HS^-] = 1.0 \times 10^{-4}M$$

These concentrations also apply to the secondary ionization.

$$HS^- \rightleftharpoons H^+ + S^{2-}$$
$$1.0 \times 10^{-4}M \quad 1.0 \times 10^{-4}M \quad ?$$

$$\frac{[H^+][S^{2-}]}{[HS^-]} = 1.0 \times 10^{-14}$$

$$\frac{(1.0 \times 10^{-4})[S^{2-}]}{(1.0 \times 10^{-4})} = 1.0 \times 10^{-14}$$

$$[S^{2-}] = 1.0 \times 10^{-14}M$$

The concentration of the secondary ion is equal to $K_2$ in any solution of $H_2S$ that does not contain ions derived from another electrolyte.

The product of the expressions for the two ionizations of $H_2S$ is

$$\left(\frac{[H^+][HS^-]}{[H_2S]}\right)\left(\frac{[H^+][S^{2-}]}{[HS^-]}\right) = K_1K_2$$

$$\frac{[H^+]^2\,[S^{2-}]}{[H_2S]} = (1.1 \times 10^{-7})(1.0 \times 10^{-14}) = 1.1 \times 10^{-21}$$

This very convenient relationship can be misleading. Superficially it looks as though it applies to a process in which one sulfide ion is produced for every two hydronium ions. However, the ionization of $H_2S$ does not proceed in this manner; in any solution of $H_2S$ the concentration of hydronium ion is much larger than the concentration of sulfide ion (Example 16.18). The majority of the $H_2S$ molecules that ionize do so only to the $HS^-$ stage, and $S^{2-}$ ions result only from the small ionization of the secondary ion.

At 25°C a saturated solution of $H_2S$ is $0.10M$. For a *saturated solution*, therefore,

$$\frac{[H^+]^2[S^{2-}]}{(0.10)} = 1.1 \times 10^{-21}$$

$$[H^+]^2\,[S^{2-}] = 1.1 \times 10^{-22}$$

This relation can be used to calculate the sulfide ion concentration of a solution of known pH that has been saturated with $H_2S$. We shall see later that acidic solutions of $H_2S$ are frequently employed laboratory reagents.

A similar relationship may be derived for carbonic acid.

$$\left(\frac{[H^+][HCO_3^-]}{[CO_2]}\right)\left(\frac{[H^+][CO_3^{2-}]}{[HCO_3^-]}\right) = K_1K_2$$

$$\frac{[H^+]^2\ [CO_3^{2-}]}{[CO_2]} = (4.2 \times 10^{-7})(4.8 \times 10^{-11})$$

$$= 2.0 \times 10^{-17}$$

Carbon dioxide is less soluble than $H_2S$; a saturated solution is $0.034M$ in $CO_2$.

$$\frac{[H^+]^2\ [CO_3^{2-}]}{(3.4 \times 10^{-2})} = 2.0 \times 10^{-17}$$

$$[H^+]^2\ [CO_3^{2-}] = 6.8 \times 10^{-19}$$

---

*Example 16.19*

What is the sulfide ion concentration of a dilute HCl solution that has been saturated with $H_2S$ if the pH of the solution is 3.00?

*Solution*

Since the pH = 3.00,

$$[H^+] = 1.0 \times 10^{-3}M$$

Therefore,

$$[H^+]^2\ [S^{2-}] = 1.1 \times 10^{-22}$$

$$(1.0 \times 10^{-3})^2\ [S^{2-}] = 1.1 \times 10^{-22}$$

$$[S^{2-}] = 1.1 \times 10^{-16}M$$

In a saturated solution of pure $H_2S$ (Example 16.18), $[S^{2-}] = 1.0 \times 10^{-14}M$. In the $H_2S$ solution described in the problem, the common ion, $H^+$, has repressed the ionization of $H_2S$. In addition, since the solution contains $H^+$ ions from a source other than $H_2S$, $[H^+]$ does not equal $[HS^-]$, and consequently $[S^{2-}]$ does not equal $K_2$.

---

### Problems

**16.1** A $0.200M$ solution of dichloroacetic acid, $H(O_2CCHCl_2)$, a weak monobasic acid, is 33% ionized. What is the ionization constant of this acid?

**16.2** A solution is prepared by dissolving 0.300 mol of cyanoacetic acid, $H(O_2CCH_2CN)$, in sufficient water to make exactly one liter. The concentration of $H^+(aq)$ in the resulting solution is $0.032M$. What is the ionization constant of cyanoacetic acid?

**16.3** In a $0.25M$ solution of benzyl amine, $C_7H_7NH_2$, the concentration of $OH^-(aq)$ is $2.4 \times 10^{-3}M$.

What is the ionization constant for this weak base? The reaction is

$$C_7H_7NH_2 + H_2O \rightleftharpoons$$
$$C_7H_7NH_3^+ + OH^-$$

**16.4** Morphine is a weak base, and the aqueous ionization can be indicated as

$$M(aq) + H_2O \rightleftharpoons$$
$$MH^+(aq) + OH^-(aq)$$

In a $0.0050M$ solution of morphine the ratio of $OH^-(aq)$ ions to morphine molecules is $1.0/83$. What is the value of $K$ for the aqueous ionization of morphine?

**16.5** (a) What concentration of $HClO_2$ molecules is in equilibrium with $0.030M$ $H^+(aq)$ in a solution prepared from pure chlorous acid? (b) How many moles of $HClO_2$ should be used to prepare one liter of this solution?

**16.6** The ionization constant for cacodylic acid, a monobasic, arsenic-containing, organic acid, is $6.4 \times 10^{-7}$. What is the concentration of $H^+(aq)$ in a $0.30M$ solution of cacodylic acid?

**16.7** (a) What are the concentrations of $H^+(aq)$, $OCl^-(aq)$ and $HOCl(aq)$ in a $0.20M$ solution of $HOCl$? (b) What is the degree of ionization of $HOCl$ in this solution?

**16.8** What are the concentrations of $C_6H_5NH_3^+(aq)$, $OH^-(aq)$, and $C_6H_5NH_2(aq)$ in a $0.30M$ solution of aniline?

**16.9** What are the concentrations of $H^+(aq)$, $NO_2^-(aq)$, and $HNO_2(aq)$ in a $0.20M$ solution of nitrous acid? Note that the quadratic formula must be used to solve this problem.

**16.10** What are the concentrations of $(CH_3)_2NH_2^+(aq)$, $OH^-(aq)$, and $(CH_3)_2NH(aq)$ in a $0.15M$ solution of dimethylamine? Note that the quadratic formula must be used to solve this problem.

**16.11** A weak acid, $HX$, is $0.50\%$ ionized in $0.030M$ solution. What percent of $HX$ is ionized in a $0.30M$ solution?

**16.12** A weak acid, $HX$, is $0.20\%$ ionized in $0.25M$ solution. At what concentration is the acid $2.0\%$ ionized?

**16.13** What are the concentrations of $H^+(aq)$, $N_3^-(aq)$, and $HN_3(aq)$ in a solution prepared from $0.23$ mol of $NaN_3$ and $0.10$ mol of $HCl$ in a total volume of $1.0$ liter?

**16.14** What are the concentrations of $NH_3(aq)$, $NH_4^+(aq)$ and $OH^-(aq)$ in a solution prepared from $0.15$ mol of $NH_4Cl$ and $0.25$ mol of $NaOH$ in a total volume of $1.0$ liter?

**16.15** What are the concentrations of $H^+(aq)$, $C_7H_5O_2^-(aq)$, and $HC_7H_5O(aq)$ in a solution prepared by adding $10$ ml of $0.30M$ sodium benzoate, $NaC_7H_5O_2$, to $10$ ml of $0.20M$ $HCl$? Assume that the total volume of the solution is $20$ ml.

**16.16** What are the concentrations of $OH^-(aq)$, $NH_4^+(aq)$, and $NH_3(aq)$ in a solution prepared by adding $150$ ml of $0.45M$ $NH_4Cl$ to $300$ ml of $0.30M$ $NaOH$? Assume that the total volume of the solution is $450$ ml.

**16.17** What are the concentrations of $H^+(aq)$ and $OH^-(aq)$ in (a) a $0.020M$ solution of $HNO_3$, (b) a $0.020M$ solution of $Ba(OH)_2$?

**16.18** What pH corresponds to

each of the following? (a) $[H^+] = 0.35M$, (b) $[OH^-] = 0.15M$, (c) $[OH^-] = 6.0 \times 10^{-6}M$, (d) $[H^+] = 2.5 \times 10^{-8}M$.

**16.19** What pH corresponds to each of the following? (a) $[H^+] = 7.3 \times 10^{-5}M$, (b) $[H^+] = 0.084M$, (c) $[OH^-] = 3.3 \times 10^{-4}M$, (d) $[OH^-] = 0.042M$.

**16.20** What concentration of $H^+(aq)$ corresponds to each of the following? (a) pH = 3.33, (b) pOH = 3.33, (c) pH = 6.78, (d) pOH = 11.11.

**16.21** What concentration of $OH^-(aq)$ corresponds to each of the following? (a) pH = 0.55, (b) pOH = 4.32, (c) pH = 12.13, (d) pOH = 12.34.

**16.22** For a cell containing a solution of unknown pH and utilizing a hydrogen electrode and a normal calomel electrode, the potential is +0.533 V. Calculate the pH of the solution and the approximate concentration of $H^+(aq)$.

**16.23** The cell described in Problem 16.22 is filled with a solution in which the concentration of $H^+(aq)$ is $3.3 \times 10^{-3}M$. What is the emf of the cell?

**16.24** Given the following standard electrode potentials:

$$2H_2O + 2e^- \rightleftharpoons H_2 + 2OH^-$$
$$\mathscr{E}° = -0.828 \text{ V}$$

$$O_2 + 4H^+ + 4e^- \rightleftharpoons 2H_2O$$
$$\mathscr{E}° = +1.229 \text{ V}$$

calculate the potentials corresponding to half reactions that occur in a solution with a pH of 7.00. Assume that the gas pressures are maintained at exactly 1 atm.

**16.25** $\mathscr{E}°$ is +0.536 V for the following cell:

$$Pt|H_2(g)|H^+(aq), I^-(aq)|I_2(s)|Pt$$

(a) Write the equation for the cell reaction. (b) The emf of the cell is +0.818 V when the pressure of $H_2(g)$ is maintained at 1 atm and a HI solution of unknown concentration is employed. What is the pH of the HI solution?

**16.26** How many moles of benzoic acid, $HC_7H_5O_2$, must be used to prepare 500 ml of solution that has a pH of 2.44?

**16.27** A $0.23M$ solution of a weak acid, HX, has a pH of 2.89. What is the ionization constant of the acid?

**16.28** What is the pH of a $0.30M$ $NH_3$ solution?

**16.29** Cocaine is a weak base; the aqueous ionization can be indicated as

$$C(aq) + H_2O \rightleftharpoons$$
$$CH^+(aq) + OH^-(aq)$$

A $5.0 \times 10^{-3}M$ solution of cocaine has a pH of 10.04. What is the ionization constant for this substance?

**16.30** An indicator, HIn, has an ionization constant of $9.0 \times 10^{-9}$. The acid color of the indicator is yellow, and the alkaline color is red. The yellow color is visible when the ratio of yellow form to red form is 30 to 1, and the red color is predominant when the ratio of red form to yellow form is 2 to 1. What is the pH range of color change for the indicator?

**16.31** In each of the following parts, a given solution affects the indicators listed in the way shown. Refer to Table 16.4, and describe the pH of each solution as completely as possible.

(a) Thymol blue turns blue and alizarin yellow turns yellow.

(b) Bromcresol green turns blue and bromthymol blue turns yellow.

(c) Methy orange turns yellow and litmus turns red.

**16.32** A solution is prepared by adding 0.010 mol of sodium nitrite, $NaNO_2$, to 100 ml of $0.035M$ nitrous acid, $HNO_2$. Assume that the volume of the final solution is 100 ml. Calculate (a) the pH of the solution and (b) the percent ionization of $HNO_2$.

**16.33** A solution is prepared from $3.0 \times 10^{-3}$ mol of a weak acid, HX, and $6.0 \times 10^{-4}$ mol of NaX diluted to 200 ml. The solution has a pH of 4.80. What is the ionization constant of HX?

**16.34** A $0.050M$ solution of formic acid, $HCHO_2$, containing an unknown concentration of formate ion, $CHO_2^-$, derived from sodium formate has a pH of 3.70. What is the concentration of formate ion in the solution?

**16.35** A $0.24M$ solution of dimethylamine, $(CH_3)_2NH$, containing an unknown concentration of dimethylamine hydrochloride, $(CH_3)_2NH_2^+ Cl^-$, has a pH of 10.40. What is the concentration of dimethylamine hydrochloride?

**16.36** A 1.00 liter solution prepared from 0.049 mol of a weak acid, HX, has a pH of 3.55. What is the pH of the solution after 0.020 mol of solid NaX is dissolved in it? Assume that the volume of the solution does not change when the NaX is dissolved in it.

**16.37** What concentrations should be used to prepare a benzoic acid–benzoate ion buffer with a pH of 5.00?

**16.38** What concentrations should be used to prepare an ammonia–ammonium ion buffer with a pH of 10.20?

**16.39** How many moles of sodium hypochlorite, NaOCl, should be added to 200 ml of $0.22M$ hypochlorous acid, HOCl, to produce a buffer with a pH of 6.75? Assume that no volume change occurs when the NaOCl is added to the solution.

**16.40** What is the pH of a buffer made by adding 50 ml of $0.20M$ sodium hydroxide to 100 ml of $0.50M$ acetic acid? Assume that the final volume of the solution is 150 ml.

**16.41** What is the pH of a buffer made by adding 100 ml of $0.44M$ HCl to 100 ml of $0.64M$ methylamine? Assume that the final volume of the solution is 200 ml.

**16.42** Calculate the concentrations of $H^+(aq)$, $H_2AsO_4^-(aq)$, $HAsO_4^{2-}(aq)$, $AsO_4^{3-}(aq)$, and $H_3AsO_4(aq)$ in a $0.30M$ solution of arsenic acid.

**16.43** (a) What are the concentrations of $H^+(aq)$, $HCO_3^-(aq)$, $CO_3^{2-}(aq)$, and $CO_2(aq)$ in a saturated solution of carbonic acid (0.034$M$ in $CO_2$)? (b) What is the pH of the solution?

**16.44** What are the concentrations of $H^+$(aq), $HSO_4^-$(aq), $SO_4^{2-}$(aq), and $H_2SO_4$(aq) in a 0.30$M$ solution of sulfuric acid? What is the pH of the solution?

**16.45** What are the concentrations of $H^+$(aq), $HSO_3^-$(aq), $SO_3^{2-}$(aq), and $SO_2$(aq) in a 0.050$M$ solution of $SO_2$?

**16.46** A 0.15$M$ solution of HCl is saturated with $H_2S$. (a) What is the concentration of $S^{2-}$(aq)? (b) What is the concentration of $HS^-$(aq)?

**16.47** A 0.050$M$ solution of HCl is saturated with $CO_2$. (a) What is the concentration of $CO_3^{2-}$(aq)? (b) What is the concentration of $HCO_3^-$(aq)?

**16.48** A solution with a pH of 3.30 is saturated with $H_2S$. What is the concentration of $S^{2-}$(aq)?

**16.49** What should the concentration of $H^+$(aq) be in order to have a sulfide ion concentration of $3.0 \times 10^{-17}M$ when the solution is saturated with $H_2S$?

# 17

# Ionic Equilibria, Part II

Homogeneous, aqueous equilibrium systems involving weak acids and bases were discussed in Chapter 16. This chapter begins with a study of heterogeneous systems in which slightly-soluble solids are in equilibrium with their constituent ions in solution. Equilibrium principles apply to the precipitation of such solids in the qualitative or quantitative determinations of dissolved ions. Another aspect of analytical chemistry — the acid–base titrations that are a part of volumetric analysis — is also discussed. Several additional types of homogeneous systems are considered, including equilibria between oxidizing and reducing agents in solution, the equilibria that occur in the formation of complex ions in solution, and equilibria involving the hydrolysis of ions.

**17.1 THE SOLUBILITY PRODUCT**

Most substances are soluble in water to at least some slight extent. If an "insoluble" or "slightly soluble" material is placed in water, an equilibrium is established when the rate of dissolution of ions from the solid equals the rate of precipitation of ions from the *saturated solution*. Thus, an equilibrium exists between solid silver chloride and a saturated solution of silver chloride.

$$AgCl(s) \rightleftharpoons Ag^+(aq) + Cl^-(aq)$$

The equilibrium constant is

$$K' = \frac{[Ag^+][Cl^-]}{[AgCl]}$$

Since the concentration of a pure solid is a constant, $[AgCl]$ may be combined with $K'$ to give

$$K_{SP} = K'[AgCl] = [Ag^+][Cl^-]$$

The constant $K_{SP}$ is called a **solubility product;** the ionic concentrations of the expression are those for a saturated solution at the reference temperature.

Since the solubility of a salt usually varies widely with temperature, the numerical value of $K_{SP}$ for a salt changes with temperature; values are usually recorded at 25°C. The ionic concentrations of a saturated solution of a slightly soluble material are low; therefore, concentrations may be satisfactorily employed in equilibrium expressions. For the more soluble salts interionic attractions cause the activity coefficients to deviate significantly from unity, and activities rather than concentrations must be employed for the mathematical analysis of these systems. Fortunately, it is the slightly soluble substances that are of primary interest in analytical chemistry. A table of solubility products at 25°C is given in the appendix.

The numerical value of $K_{SP}$ for a salt may be found from the molar solubility of the salt.

---

*Example 17.1*

At 25°C, 0.00188 g of AgCl dissolves in 1 liter of water. What is the $K_{SP}$ of AgCl?

*Solution*

The molar solubility of AgCl (molecular weight, 143) is

$$? \text{ mol AgCl} = 0.00188 \text{ g AgCl}\left(\frac{1 \text{ mol AgCl}}{143 \text{ g AgCl}}\right)$$

$$= 1.31 \times 10^{-5} \text{ mol AgCl}$$

For each mole of AgCl dissolving, one mole of $Ag^+$ and one mole of $Cl^-$ are formed.

$$AgCl(s) \rightleftharpoons \underset{1.31 \times 10^{-5}M}{Ag^+} + \underset{1.31 \times 10^{-5}M}{Cl^-}$$

$$K_{SP} = [Ag^+][Cl^-]$$
$$= (1.31 \times 10^{-5})^2$$
$$= 1.7 \times 10^{-10}$$

---

The $K_{SP}$ of a substance may also be found by conductance measurements (if the equivalent conductances of the ions are known) or by potentiometric measurements.

---

**Example 17.2**

Calculate the $K_{SP}$ of AgCl from the electrode potentials

$$e^- + AgCl(s) \rightleftharpoons Ag(s) + Cl^-(aq)$$
$$\mathscr{E}° = +0.222 \text{ V}$$

$$e^- + Ag^+(aq) \rightleftharpoons Ag(s)$$
$$\mathscr{E}° = +0.799 \text{ V}$$

**Solution**

The theoretical cell

$$Ag|Ag^+, Cl^-|AgCl|Ag$$

would have the following electrode reactions

$$Ag \rightarrow Ag^+ + e^- \qquad \mathscr{E}°_{ox} = -0.799 \text{ V}$$

$$e^- + AgCl \rightarrow Ag + Cl^- \qquad \mathscr{E}° = +0.222 \text{ V}$$

The cell reaction is

$$AgCl \rightarrow Ag^+ + Cl^- \qquad \mathscr{E}° = -0.577 \text{ V}$$

Notice that the reaction for the standard cell is not spontaneous as written. For this reaction, $n = 1$.

$$\Delta G° = -nF\mathscr{E}°$$

$$= -1(23.1 \text{ kcal/V})(-0.577 \text{ V})$$

$$= +13.3 \text{ kcal}$$

$$\Delta G° = -RT \ln K$$

$$+13.3 \text{ kcal} = -1.36 \text{ kcal (log } K)$$

$$\log K = -9.78$$

$$K = 1.7 \times 10^{-10}$$

---

For salts that have more than two ions per formula unit, the ion concentrations must be raised to the powers indicated by the coefficients of the balanced chemical equation.

$$Mg(OH)_2(s) \rightleftharpoons Mg^{2+} + 2OH^- \qquad K_{SP} = [Mg^{2+}][OH^-]^2$$

$$Bi_2S_3(s) \rightleftharpoons 2Bi^{3+} + 3S^{2-} \qquad K_{SP} = [Bi^{3+}]^2 [S^{2-}]^3$$

$$Hg_2Cl_2(s) \rightleftharpoons Hg_2^{2+} + 2Cl^- \qquad K_{SP} = [Hg_2^{2+}][Cl^-]^2$$

For a salt of this type the calculation of the $K_{SP}$ from the molar solubility requires that the chemical equation representing the dissociation process be carefully interpreted.

---

**Example 17.3**

At 25°C, $7.8 \times 10^{-5}$ mol of silver chromate dissolves in 1 liter of water. What is the $K_{SP}$ of $Ag_2CrO_4$?

**Solution**

For each mole of $Ag_2CrO_4$ that dissolves, two moles of $Ag^+$ and one mole of $CrO_4^{2-}$ are formed.

Therefore,

$$Ag_2CrO_4(s) \rightleftharpoons \underset{2(7.8 \times 10^{-5})M}{2Ag^+} + \underset{7.8 \times 10^{-5}M}{CrO_4^{2-}}$$

$$K_{SP} = [Ag^+]^2 [CrO_4^{2-}]$$
$$= (1.56 \times 10^{-4})^2 (7.8 \times 10^{-5})$$
$$= 1.9 \times 10^{-12}$$

*Example 17.4*

The $K_{SP}$ of $CaF_2$ is $3.9 \times 10^{-11}$ at 25°C. What is the concentration of $Ca^{2+}$ and of $F^-$ in the saturated solution? How many grams of calcium fluoride will dissolve in 100 ml of water at 25°C?

*Solution*

Let $x$ equal the molar solubility of $CaF_2$.

$$CaF_2(s) \rightleftharpoons \underset{x}{Ca^{2+}} + \underset{2x}{2F^-}$$

$$K_{SP} = [Ca^{2+}][F^-]^2 = 3.9 \times 10^{-11}$$

$$x(2x)^2 = 3.9 \times 10^{-11}$$

$$4x^3 = 3.9 \times 10^{-11}$$

$$x = 2.1 \times 10^{-4}M$$

Therefore,

$$[Ca^{2+}] = x = 2.1 \times 10^{-4}M$$

$$[F^-] = 2x = 4.2 \times 10^{-4}M$$

$$? \text{ g } CaF_2 = 100 \text{ ml } H_2O \left(\frac{2.1 \times 10^{-4} \text{ mol } CaF_2}{1000 \text{ ml } H_2O}\right)\left(\frac{78 \text{ g } CaF_2}{1 \text{ mol } CaF_2}\right)$$

$$= 1.6 \times 10^{-3} \text{ g } CaF_2$$

There is evidence that some salts ionize in a stepwise manner. For example, lead chloride

$$PbCl_2(s) \rightleftharpoons Pb^{2+}(aq) + 2Cl^-(aq) \qquad K_{SP} = [Pb^{2+}][Cl^-]^2$$

is thought to ionize according to the equations

$$PbCl_2(s) \rightleftharpoons PbCl_2(aq) \qquad\qquad K_1 = [PbCl_2(aq)]$$

$$PbCl_2(aq) \rightleftharpoons PbCl^+(aq) + Cl^-(aq) \qquad K_2 = \frac{[PbCl^+][Cl^-]}{[PbCl_2(aq)]}$$

$$PbCl^+(aq) \rightleftharpoons Pb^{2+}(aq) + Cl^-(aq) \qquad K_3 = \frac{[Pb^{2+}][Cl^-]}{[PbCl^+]}$$

The product of the stepwise constants is

$$K_1 K_2 K_3 = [PbCl_2(aq)] \left(\frac{[PbCl^+][Cl^-]}{[PbCl_2(aq)]}\right)\left(\frac{[Pb^{2+}][Cl^-]}{[PbCl^+]}\right)$$

$$= [Pb^{2+}][Cl^-]^2 = K_{SP}$$

In these expressions $PbCl_2(aq)$ is a neutral molecule in solution.

Thus, the solubility product principle applies to all solutions of slightly soluble materials whether they dissociate in a stepwise manner or not. However, the $K_{SP}$'s of salts that dissociate in steps must be carefully interpreted. For example, in a saturated solution of $PbCl_2$ the concentration of $Cl^-$ is not twice the concentration of $Pb^{2+}$ as the expression for the $K_{SP}$ might lead one to expect; such an erroneous deduction ignores the existence of $PbCl_2$(aq) molecules and $PbCl^+$ ions. The value of the $K_{SP}$ applies only if the *actual* concentrations of $Pb^{2+}$ and $Cl^-$ are employed; the stepwise mechanism must be considered in order to deduce the correct concentration terms.

Other factors introduce errors into solubility calculations for certain salts. The solubility of lead chloride is enhanced in moderately concentrated solutions of chloride ion because of the formation of the *complex ion,* $PbCl_3^-$.

$$PbCl_2(s) + Cl^-(aq) \rightleftharpoons PbCl_3^-(aq)$$

Also, the hydrolysis of the $Pb^{2+}$ ion

$$Pb^{2+}(aq) + H_2O \rightleftharpoons Pb(OH)^+(aq) + H^+(aq)$$

reduces the concentration of $Pb^{2+}$ so that the solubility of $PbCl_2$ is actually higher than the value obtained from a calculation that ignores hydrolysis. Complex-ion and hydrolysis equilibria are discussed in Sections 17.4 and 17.6.

The solubility of any solid is increased by the presence of the ions of another salt in the solution; thus, $PbCl_2$ is more soluble in moderately concentrated solutions of $NaNO_3$ than it is in pure water. This **salt effect** also increases the degree of ionization of soluble weak electrolytes. The effect is due to interionic attractions, which do not depend upon the nature of the dissolved ions but upon their concentrations and charges. Therefore, foreign ions reduce the activities of the ions of the substance under study. Since the equilibrium constant is properly a function of activities (rather than concentrations), more of the substance ionizes to realize the activity required to satisfy the mathematical relationship.

**17.2 PRECIPITATION AND THE SOLUBILITY PRODUCT**

The numerical value of the solubility product of a salt is a quantitative statement of the limit of solubility of the salt. For a specific solution of a salt, the product of the concentrations of the ions, each raised to the proper power, is called the **ion product.** Thus, for a saturated solution in equilibrium with excess solid, the ion product equals the $K_{SP}$. If the ion product of a solution is less than the $K_{SP}$, the solution is unsaturated; additional solid can dissolve in this solution. On the other hand, if the ion product is greater than the $K_{SP}$, the solution is momentarily supersaturated; precipitation will occur until the ion product equals the $K_{SP}$.

*Example 17.5*

Will a precipitate form if 10 ml of $0.010M$ $AgNO_3$ and 10 ml of $0.00010M$ NaCl are mixed? Assume that the final volume of the solution is 20 ml. For AgCl, $K_{SP} = 1.7 \times 10^{-10}$.

*Solution*

Diluting a solution to twice its original volume reduces the concentrations of ions in the solution to half their original value. Therefore, if there were no reaction, the ion concentrations would be

$$[Ag^+] = 5.0 \times 10^{-3} M$$

$$[Cl^-] = 5.0 \times 10^{-5} M$$

The ion product is

$$[Ag^+][Cl^-] = ?$$

$$(5.0 \times 10^{-3})(5.0 \times 10^{-5}) = 2.5 \times 10^{-7}$$

Therefore, the ion product is larger than the $K_{SP}$ ($1.7 \times 10^{-10}$), and precipitation of AgCl will occur.

*Example 17.6*

Will a precipitate of $Mg(OH)_2$ form in a $0.0010 M$ solution of $Mg(NO_3)_2$ if the pH of the solution is adjusted to 9.0? The $K_{SP}$ of $Mg(OH)_2$ is $8.9 \times 10^{-12}$.

*Solution*

If the pH = 9.0,

$$[OH^-] = 1.0 \times 10^{-5} M$$

Since $[Mg^{2+}] = 1.0 \times 10^{-3}$, the ion product is

$$[Mg^{2+}][OH^-]^2 = ?$$

$$(1 \times 10^{-3})(1 \times 10^{-5})^2 = 1 \times 10^{-13}$$

Since the ion product is less than $8.9 \times 10^{-12}$, no precipitate will form.

The common-ion effect pertains to solubility equilibria. As an example, consider the system

$$BaSO_4(s) \rightleftharpoons Ba^{2+}(aq) + SO_4^{2-}(aq)$$

The addition of sulfate ion, from sodium sulfate, to a saturated solution of barium sulfate will cause the equilibrium to shift to the left; the concentration of $Ba^{2+}$ will decrease, and $BaSO_4$ will precipitate. Since the product $[Ba^{2+}][SO_4^{2-}]$ is a constant, increasing $[SO_4^{2-}]$ will cause $[Ba^{2+}]$ to decrease.

The amount of barium ion in a solution may be determined by precipitating the $Ba^{2+}$ as $BaSO_4$. The precipitate is then removed by filtration and is dried and weighed. The concentration of $Ba^{2+}$ left in solution after the precipitation may be reduced to a very low value if excess sulfate ion is employed in the precipitation. As a general rule, however, too large an excess of the common ion should be avoided. At high ionic concentrations the salt effect increases the solubility of a salt, and for certain precipitates the formation of a complex ion may lead to enhanced solubility.

*Example 17.7*

At 25°C a saturated solution of $BaSO_4$ is $3.9 \times 10^{-5} M$; the $K_{SP}$ of $BaSO_4$ is $1.5 \times 10^{-9}$.

What is the solubility of $BaSO_4$ in $0.050 M$ $Na_2SO_4$?

*Solution*
The sulfate ion derived from $BaSO_4$ may be ignored. Thus,

$$BaSO_4(s) \rightleftharpoons Ba^{2+} + SO_4^{2-}$$
$$\phantom{BaSO_4(s) \rightleftharpoons} ? \qquad 5.0 \times 10^{-2}$$

$$[Ba^{2+}][SO_4^{2-}] = 1.5 \times 10^{-9}$$
$$[Ba^{2+}](5.0 \times 10^{-2}) = 1.5 \times 10^{-9}$$
$$[Ba^{2+}] = 3.0 \times 10^{-8}M$$

The solubility of $BaSO_4$ has been reduced from $3.9 \times 10^{-5}M$ to $3.0 \times 10^{-8}M$ by the common ion effect.

Frequently, a solution contains more than one ion capable of forming a precipitate with another ion which is to be added to the solution. For example, a solution might contain both $Cl^-$ and $CrO_4^{2-}$ ions, both of which form insoluble salts with $Ag^+$. When $Ag^+$ is added to the solution, the less soluble silver salt will precipitate first. If the addition is continued, eventually a point will be reached where the more soluble salt will begin to precipitate along with the less soluble.

**Example 17.8**
A solution is $0.10M$ in $Cl^-$ and $0.10M$ in $CrO_4^{2-}$. If solid $AgNO_3$ is gradually added to this solution, which will precipitate first, $AgCl$ or $Ag_2CrO_4$? Assume that the addition causes no change in volume. For $AgCl$, $K_{SP} = 1.7 \times 10^{-10}$; for $Ag_2CrO_4$, $K_{SP} = 1.9 \times 10^{-12}$.

$$AgCl(s) \rightleftharpoons Ag^+ + Cl^-$$
$$\phantom{AgCl(s) \rightleftharpoons} ? \qquad 0.10M$$

$$[Ag^+][Cl^-] = 1.7 \times 10^{-10}$$
$$[Ag^+](0.10) = 1.7 \times 10^{-10}$$
$$[Ag^+] = 1.7 \times 10^{-9}M$$

*Solution*
When a precipitate *begins* to form, the pertinent ion product *just* exceeds the $K_{SP}$ of the solid. Therefore, we calculate the concentrations of $Ag^+$ needed to precipitate $AgCl$ and $Ag_2CrO_4$.

$$Ag_2CrO_4(s) \rightleftharpoons 2Ag^+ + CrO_4^{2-}$$
$$\phantom{Ag_2CrO_4(s) \rightleftharpoons 2Ag^+} 0.10M$$

$$[Ag^+]^2[CrO_4^{2-}] = 1.9 \times 10^{-12}$$
$$[Ag^+]^2(0.10) = 1.9 \times 10^{-12}$$
$$[Ag^+]^2 = 1.9 \times 10^{-11}$$
$$[Ag^+] = 4.4 \times 10^{-6}M$$

Therefore, $AgCl$ will precipitate first.

**Example 17.9**
(a) In the experiment described in Example 17.8, what will be the concentration of the $Cl^-$ ion when $Ag_2CrO_4$ begins to precipitate? (b) At this point, what percent of the chloride ion originally present remains in solution?

*Solution*
(a) From the preceding example, we see that $[Ag^+] = 4.4 \times 10^{-6}M$ when $Ag_2CrO_4$ starts to precipitate. At this point, the concentration of chloride ion will be

$$[Ag^+][Cl^-] = 1.7 \times 10^{-10}$$

$$(4.4 \times 10^{-6})[Cl^-] = 1.7 \times 10^{-10}$$

$$[Cl^-] = \frac{1.7 \times 10^{-10}}{4.4 \times 10^{-6}} = 3.9 \times 10^{-5} M$$

Thus, until the $[Cl^-]$ is reduced to $3.9 \times 10^{-5} M$, no $Ag_2CrO_4$ will form.

(b) Since the original concentration of chloride was $0.10 M$, the percent of $Cl^-$ remaining in solution when $Ag_2CrO_4$ starts to precipitate is

$$\frac{3.9 \times 10^{-5}}{1.0 \times 10^{-1}} \times 100 = 0.039\%$$

---

Chromate ion is used as a **precipitation indicator.** The concentration of chloride ion in a solution can be determined by titrating a sample of the solution against a standard $AgNO_3$ solution using a few drops of $K_2CrO_4$ as an indicator. In the course of the titration white $AgCl$ precipitates. The appearance of red $Ag_2CrO_4$ indicates that the precipitation of chloride ion is essentially complete. In a quantitative procedure of this type, the concentration of $CrO_4^{2-}$ used is much less than that employed in the preceding problem. Hence, a higher concentration of $Ag^+$ is required to start the precipitation of $Ag_2CrO_4$, and a lower concentration of $Cl^-$ will be present in the solution when the $Ag_2CrO_4$ begins to precipitate.

---

*Example 17.10*

What concentration of $NH_4^+$, derived from $NH_4Cl$, is necessary to prevent the formation of a $Mg(OH)_2$ precipitate in a solution that is $0.050 M$ in $Mg^{2+}$ and $0.050 M$ in $NH_3$? The $K_{SP}$ of $Mg(OH)_2$ is $8.9 \times 10^{-12}$.

*Solution*

We first calculate the maximum concentration of hydroxide ion that can be present in the solution without causing $Mg(OH)_2$ to precipitate.

$$[Mg^{2+}][OH^-]^2 = 8.9 \times 10^{-12}$$

$$(5.0 \times 10^{-2})[OH^-]^2 = 8.9 \times 10^{-12}$$

$$[OH^-]^2 = 1.8 \times 10^{-10}$$

$$[OH^-] = 1.3 \times 10^{-5} M$$

From the expression for the ionization constant for $NH_3$, we can derive the concentration of $NH_4^+$ that will maintain the concentration of $OH^-$ at this level.

$$\frac{[NH_4^+][OH^-]}{[NH_3]} = 1.8 \times 10^{-5}$$

$$\frac{[NH_4^+](1.3 \times 10^{-5})}{(5.0 \times 10^{-2})} = 1.8 \times 10^{-5}$$

$$[NH_4^+] = 6.9 \times 10^{-2} M$$

Thus, the minimum concentration of $NH_4^+$ that must be present is $0.069 M$.

---

**17.3** PRECIPITATION OF SULFIDES

The sulfide ion concentration in an acid solution that has been saturated with $H_2S$ is extremely low. In 10 ml of a saturated $H_2S$ solution that has been made $0.3 M$ in $H^+$, there are approximately seven $S^{2-}$ ions. Nevertheless, when $Pb^{2+}$ ions are added to such a solution, PbS precipitates immediately. It seems unlikely that the precipitate forms as a result of the reaction of $Pb^{2+}$ ions with $S^{2-}$ ions.

There is evidence that in this and in similar sulfide precipitations, a hydrosulfide forms initially and then decomposes to give the normal sulfide.

$$Pb(HS)_2(s) \rightleftharpoons PbS(s) + H_2S(aq)$$

The hydroxides of many metals are known to produce oxides in a parallel manner. In fact, what is known as lead hydroxide, $Pb(OH)_2$, is in reality hydrous lead oxide, $PbO \cdot xH_2O$. The hydrosulfide of lead could form as a result of the reaction of $Pb^{2+}$ ions with either $H_2S$ molecules or $HS^-$ ions; the concentrations of $H_2S$ and $HS^-$ are much higher than the concentration of $S^{2-}$.

However, an equilibrium constant does not depend upon the reaction mechanism by which the equilibrium is attained (Section 13.8). Provided the system is in equilibrium, the relationship expressed by the solubility product principle is valid no matter what series of reactions produces the precipitate. Thus, we may use $K_{SP}$'s to calculate favorable reaction conditions for the formation of a desired precipitate or reaction conditions that will prevent the formation of a precipitate.

---

*Example 17.11*

A solution that is $0.30M$ in $H^+$, $0.050M$ in $Pb^{2+}$, and $0.050M$ in $Fe^{2+}$ is saturated with $H_2S$; should PbS and/or FeS precipitate? The $K_{SP}$ of PbS is $7 \times 10^{-29}$ and the $K_{SP}$ of FeS is $4 \times 10^{-19}$.

*Solution*

For any saturated solution of $H_2S$,

$$[H^+]^2[S^{2-}] = 1.1 \times 10^{-22}$$

Since this solution is $0.30M$ in $H^+$,

$$(3.0 \times 10^{-1})^2[S^{2-}] = 1.1 \times 10^{-22}$$
$$[S^{2-}] = 1.2 \times 10^{-21}M$$

Both $Pb^{2+}$ and $Fe^{2+}$ are $2+$ ions, and the form of the ion product is

$$[M^{2+}][S^{2-}]$$

where $M^{2+}$ stands for either metal ion. Since both are present in concentrations of $0.050M$,

$$[M^{2+}][S^{2-}]$$
$$(5.0 \times 10^{-2})(1.2 \times 10^{-21}) = 6.0 \times 10^{-23}$$

This ion product is greater than the $K_{SP}$ of PbS; therefore, PbS will precipitate. However, the ion product is less than the $K_{SP}$ of FeS; the solubility of FeS has not been exceeded; no FeS will form.

---

*Example 17.12*

What must be the hydronium ion concentration of a solution that is $0.050M$ in $Ni^{2+}$ to prevent the precipitation of NiS when the solution is saturated with $H_2S$? The $K_{SP}$ of NiS is $3 \times 10^{-21}$.

*Solution*

$$[Ni^{2+}][S^{2-}] = 3 \times 10^{-21}$$
$$(0.050)[S^{2-}] = 3 \times 10^{-21}$$
$$[S^{2-}] = 6 \times 10^{-20}M$$

Therefore, the $[S^{2-}]$ must be less than $6 \times 10^{-20}M$ if NiS is not to precipitate.

For a solution saturated with $H_2S$,

$$[H^+]^2[S^{2-}] = 1.1 \times 10^{-22}$$

$$[H^+]^2(6 \times 10^{-20}) = 1.1 \times 10^{-22}$$

$$[H^+] = 0.04M$$

The $[H^+]$ must be greater than $0.04M$ to prevent the precipitation of NiS.

---

The preceding examples illustrate an important analytical technique. In the usual qualitative analysis scheme, certain cations are separated into groups on the basis of whether their sulfides form in acidic solution. Thus, in a solution that has a $H^+$ concentration of $0.3M$, the sulfides of $Hg^{2+}$, $Pb^{2+}$, $Cu^{2+}$, $Bi^{3+}$, $Cd^{2+}$, and $Sn^{2+}$ are insoluble, whereas the sulfides of $Fe^{2+}$, $Co^{2+}$, $Ni^{2+}$, $Mn^{2+}$, and $Zn^{2+}$ are soluble.

In considering systems that involve sulfide precipitation, one must note that the $H^+$ concentration of a solution increases when a sulfide precipitates from the solution.

---

### Example 17.13

A solution that is $0.050M$ in $Cd^{2+}$ and $0.10M$ in $H^+$ is saturated with $H_2S$. What concentration of $Cd^{2+}$ remains in solution after CdS has precipitated? The $K_{SP}$ of CdS is $1.0 \times 10^{-28}$.

#### Solution

For each $Cd^{2+}$ ion precipitated, two $H^+$ ions are added to the solution.

$$Cd^{2+}(aq) + H_2S(aq) \rightleftharpoons CdS(s) + 2H^+(aq)$$

We shall assume that virtually all the $Cd^{2+}$ precipitates as CdS. Hence, the precipitation introduces $0.10$ mol of $H^+$ per liter of solution, and the final $[H^+]$ is $0.20M$. Therefore,

$$[H^+]^2[S^{2-}] = 1.1 \times 10^{-22}$$

$$(0.20)^2[S^{2-}] = 1.1 \times 10^{-22}$$

$$[S^{2-}] = 2.8 \times 10^{-21}$$

The concentration of $Cd^{2+}$ after the CdS has precipitated may be derived from the $K_{SP}$ of CdS.

$$[Cd^{2+}][S^{2-}] = 1.0 \times 10^{-28}$$

$$[Cd^{2+}](2.8 \times 10^{-21}) = 1.0 \times 10^{-28}$$

$$[Cd^{2+}] = 3.6 \times 10^{-8}M$$

From our answer we can see that our assumption was justified. The value of $[H^+]$ that we used was well within our limits of accuracy.

---

It must be emphasized that calculations such as the preceding, although illustrative of principles, are only approximate. Exact mathematical treatment of sulfide systems is complicated and must take into account complex-ion formation (Section 17.4) and hydrolysis (Section 17.6). In addition, the solubility-product constants for the sulfides are generally less reliable than those for other slightly soluble solids. The attainment of equilibrium in sulfide systems is slow, and measurement of equilibrium constants for these systems is difficult. The precipitated sulfides of some cations (notably $Ni^{2+}$, $Co^{2+}$, and $Zn^{2+}$) become less soluble upon standing. The decrease in solubility may be caused by the transition from hydrosulfide to sulfide, but more probably it is the result of the rearrangement of the ions into a more stable crystalline form. Thus, for all practical purposes the $K_{SP}$'s for some sulfides change with time. The values of the solubility products of the sulfides listed in the appendix are for the freshly precipitated, or more soluble, modifications.

**17.4 EQUILIBRIA INVOLVING COMPLEX IONS**

Complex ions are discussed in Chapter 19. However, these ions take part in some aqueous equilibria that properly constitute a part of the topic of this chapter. A **complex ion** is an aggregate consisting of a central metal cation (usually a transition-metal ion) surrounded by a number of ligands. The **ligands** of a complex may be anions, molecules, or a combination of the two. The charge of a complex ion is obtained by adding the charges of its constituent particles. Examples of complex ions are: $Co(NH_3)_6^{3+}$, $Cu(H_2O)_4^{2+}$, $CdCl_4^{2-}$, $Fe(CN)_6^{4-}$, $Ag(S_2O_3)_2^{3-}$, and $Co(NH_3)_5Cl^{2+}$.

A ligand must have an unshared pair of electrons with which it can bond to the central ion. Thus, the ammonia molecule, $NH_3$, functions as a ligand in the formation of complex ions—the ammonium ion, $NH_4^+$, does not.

In addition to ammonia, the requirement is met by many molecules and anions.

The stabilities of complex ions vary widely. The group I A and group II A cations only form aggregates that are loosely held together by weak ion–dipole or ion–ion electrostatic attractions; these are probably better described as either ion pairs or ion clusters rather than complex ions. At the other extreme are the very stable complex ions, such as $Co(NH_3)_6^{3+}$ and $Fe(CN)_6^{4-}$, in which the strength of the attractions between the central ions and the ligands is of the same order as the covalent bond. We shall restrict our attention to those complex ions that are covalently bonded in definite stoichiometric ratios.

All ions are hydrated in water solution, and most hydrated metal ions should be regarded as complex ions since each metal ion has a definite number of water molecules tightly bonded to it. This number is difficult to determine and is not known with certainty for some species. Nevertheless, in water solution, complexes probably form by the replacement of water ligands by other ligands. For example,

$$Cu(H_2O)_6^{2+} + NH_3 \rightleftharpoons Cu(H_2O)_5(NH_3)^{2+} + H_2O$$

For simplicity, however, the coordinated water molecules are usually not shown.

$$Cu^{2+} + NH_3 \rightleftharpoons Cu(NH_3)^{2+}$$

The dissociation, as well as the formation, of a complex ion occurs in steps. Thus, the $Ag(NH_3)_2^+$ ion dissociates as follows:

$$Ag(NH_3)_2^+ \rightleftharpoons Ag(NH_3)^+ + NH_3 \qquad K_1 = \frac{[Ag(NH_3)^+][NH_3]}{[Ag(NH_3)_2^+]}$$

$$= 1.4 \times 10^{-4}$$

$$Ag(NH_3)^+ \rightleftharpoons Ag^+ + NH_3 \qquad K_2 = \frac{[Ag^+][NH_3]}{[Ag(NH_3)^+]}$$

$$= 4.3 \times 10^{-4}$$

The product of the two dissociation constants is called the **instability constant** of the $Ag(NH_3)_2^+$ ion. This instability corresponds to the overall dissociation of the ion.

$$Ag(NH_3)_2^+ \rightleftharpoons Ag^+ + 2NH_3$$

$$\left(\frac{[Ag(NH_3)^+][NH_3]}{[Ag(NH_3)_2^+]}\right)\left(\frac{[Ag^+][NH_3]}{[Ag(NH_3)^+]}\right) = \frac{[Ag^+][NH_3]^2}{[Ag(NH_3)_2^+]}$$

$$K_{inst} = K_1 K_2 = (1.4 \times 10^{-4})(4.3 \times 10^{-4}) = 6.0 \times 10^{-8}$$

The reciprocals of dissociation constants, which pertain to equilibrium reactions written in reverse form, are called **formation constants** or **stability constants.**

For many complexes the values of the equilibrium constants for the individual steps of the dissociation are not known, although the value for the overall dissociation may have been determined. However, an overall instability constant for a complex ion must be cautiously interpreted since the chemical equation to which it applies does not take into consideration any intermediate species. Some instability constants are given in Appendix E.

*Example 17.14*

A $1.0 \times 10^{-3} M$ solution of $AgNO_3$ is made $0.50 M$ in $NH_3$. What are the concentrations of $Ag^+$, $Ag(NH_3)^+$, and $Ag(NH_3)_2^+$ in the resulting solution?

*Solution*

We shall assume that the $Ag^+$ is practically completely converted into the higher complex, $Ag(NH_3)_2^+$, by this excess of $NH_3$. Therefore,

$$[Ag(NH_3)_2^+] = 1.0 \times 10^{-3} M$$

The formation of this concentration of $Ag(NH_3)_2^+$ would reduce the concentration of $NH_3$ by $2.0 \times 10^{-3} M$, but this, as well as the amount of $NH_3$ that reacts with $H_2O$ to form $NH_4^+$, is negligible in comparison with the original concentration of $NH_3$. From the instability constant we see

$$\frac{[Ag^+][NH_3]^2}{[Ag(NH_3)_2^+]} = 6.0 \times 10^{-8}$$

$$\frac{[Ag^+](5.0 \times 10^{-1})^2}{(1.0 \times 10^{-3})} = 6.0 \times 10^{-8}$$

$$[Ag^+] = 2.4 \times 10^{-10} M$$

The concentration of $Ag(NH_3)^+$ may be obtained from either of the stepwise dissociation constants. Thus,

$$\frac{[Ag(NH_3)^+][NH_3]}{[Ag(NH_3)_2^+]} = 1.4 \times 10^{-4}$$

$$\frac{[Ag(NH_3)^+](5.0 \times 10^{-1})}{(1.0 \times 10^{-3})} = 1.4 \times 10^{-4}$$

$$[Ag(NH_3)^+] = 2.8 \times 10^{-7} M$$

By comparing the concentrations of all of the cations in the solution, we can see that the assumption that $Ag(NH_3)_2^+$ is the principal silver-containing ion is a valid one.

*Example 17.15*

What is the solubility of AgCl in $0.10 M$ $NH_3$?

*Solution*

If we let $x$ equal the solubility of AgCl in moles per liter,

$$[Cl^-] = x.$$

From the $K_{SP}$ for AgCl we derive

$$[Ag^+][Cl^-] = 1.7 \times 10^{-10}$$

$$[Ag^+] = \frac{1.7 \times 10^{-10}}{x}$$

If we assume that the majority of the dissolved $Ag^+$ goes into solution as $Ag(NH_3)_2^+$,

$$[Ag(NH_3)_2^+] = x$$

Since two molecules of $NH_3$ are required for every $Ag(NH_3)_2^+$ ion formed,

$$[NH_3] = 0.10 - 2x$$

Therefore,

$$\frac{[Ag^+][NH_3]^2}{[Ag(NH_3)_2^+]} = 6.0 \times 10^{-8}$$

$$\frac{\left(\dfrac{1.7 \times 10^{-10}}{x}\right)(0.10 - 2x)^2}{x} = 6.0 \times 10^{-8}$$

from which we derive

$$\frac{(0.10 - 2x)^2}{x^2} = \frac{6.0 \times 10^{-8}}{1.7 \times 10^{-10}} = 3.5 \times 10^2$$

By extracting the square root of both sides of this equation, we get the relation

$$\frac{(0.10 - 2x)}{x} = 19$$

$$x = [Ag(NH_3)_2^+] = 4.8 \times 10^{-3} M.$$

which is the solubility of AgCl in $0.10 M$ $NH_3$.

The concentration of $NH_3$ is

$$[NH_3] = 0.10 - 2x = 9.0 \times 10^{-2} M$$

The concentration of $Ag^+$ is

$$[Ag^+] = \frac{1.7 \times 10^{-10}}{x}$$

$$= 3.5 \times 10^{-8} M$$

The concentration of $Ag(NH_3)^+$ can be derived from

$$\frac{[Ag(NH_3)^+][NH_3]}{[Ag(NH_3)_2^+]} = 1.4 \times 10^{-4}$$

$$\frac{[Ag(NH_3)^+](9.0 \times 10^{-2})}{4.8 \times 10^{-3}} = 1.4 \times 10^{-4}$$

$$[Ag(NH_3)^+] = 7.5 \times 10^{-6} M$$

Comparison of the concentrations of all the species in the solution will show that our assumptions were justified.

As an alternative solution to this problem, we write the equation for the complete transformation and let $x$ equal the molar solubility of AgCl.

$$\underset{0.10 - 2x}{AgCl(s) +} \ \underset{}{2NH_3} \ \rightleftharpoons \underset{x}{Ag(NH_3)_2^+} + \underset{x}{Cl^-}$$

An equilibrium constant corresponding to this equation may be derived by dividing the $K_{SP}$ of AgCl by the $K_{inst}$ of $Ag(NH_3)_2^+$.

$$K_{SP}\left(\frac{1}{K_{inst}}\right) = [Ag^+][Cl^-]\left(\frac{[Ag(NH_3)_2^+]}{[Ag^+][NH_3]^2}\right)$$

$$(1.7 \times 10^{-10})\left(\frac{1}{6.0 \times 10^{-8}}\right) = 2.8 \times 10^{-3}$$

Therefore,

$$\frac{[Ag(NH_3)_2^+][Cl^-]}{[NH_3]^2} = 2.8 \times 10^{-3}$$

$$\frac{x^2}{(0.10 - 2x)^2} = 2.8 \times 10^{-3}$$

$$\frac{x}{(0.10 - 2x)} = 5.3 \times 10^{-2}$$

$$x = 4.8 \times 10^{-3} M$$

If a large excess of ligand is employed and if each of the stepwise dissociation constants is small, the assumption may be made that the principal complex in solution is the one with the highest number of ligands. Such an assumption greatly simplifies problem solving (Examples 17.14 and 17.15). Unfortunately, there are situations in which such an assumption may not be made because either or both of the criteria are not satisfied. For example, consider the complexes formed by the $Ag^+$ ion and the thiosulfate ion.

$$Ag(S_2O_3)_3^{5-} \rightleftharpoons Ag(S_2O_3)_2^{3-} + S_2O_3^{2-} \qquad K_1 = 2.0 \times 10^{-1}$$

$$Ag(S_2O_3)_2^{3-} \rightleftharpoons Ag(S_2O_3)^- + S_2O_3^{2-} \qquad K_2 = 3.3 \times 10^{-5}$$

$$Ag(S_2O_3)^- \rightleftharpoons Ag^+ + S_2O_3^{2-} \qquad K_3 = 1.5 \times 10^{-9}$$

It is incorrect to assume that of all the silver-containing ions only the concentration of $Ag(S_2O_3)_3^{5-}$ is significant at high concentrations of thiosulfate ion. The constant for the first step of the dissociation is comparatively large. Therefore, at high concentrations of $S_2O_3^{2-}$ significant concentrations of both $Ag(S_2O_3)_2^{3-}$ and $Ag(S_2O_3)_3^{5-}$ exist. In solutions that are $1 M$ in $S_2O_3^{2-}$ approximately 16.7% of the silver is in the form of the $Ag(S_2O_3)_2^{3-}$ ion and 83.3% is in the form of the $Ag(S_2O_3)_3^{5-}$ ion. Notice that these conclusions can not be reached by use of the instability constant ($9.9 \times 10^{-15}$) alone.

Slightly soluble substances can often be dissolved through the formation of complex ions. The dissolution of AgCl in $NH_3$ (Example 17.15) illustrates this technique, which is an important one in analytical chemistry. The equilibrium between solid AgCl and its ions

$$AgCl(s) \rightleftharpoons Ag^+(aq) + Cl^-(aq)$$

is forced to the right by the removal of $Ag^+$ ions through the formation of the silver–ammonia complex ion, and therefore AgCl dissolves.

A few metals form thio complex ions (e.g., $HgS_2^{2-}$, $AsS_3^{3-}$, $SbS_2^{3-}$, $SnS_3^{2-}$); the true structure of these anions may involve the $HS^-$ ion. The formation of a thio complex ion may be used to separate sulfide precipitates. For example, a mixture of CuS (which does not form a thio complex) and $As_2S_3$ may be separated by dissolving the $As_2S_3$ in an alkaline solution containing $S^{2-}$ ion. The solution must have a high pH to maintain a relatively high concentration of $S^{2-}$.

$$CuS(s) + S^{2-}(aq) \rightarrow \text{no reaction}$$

$$As_2S_3(s) + 3S^{2-}(aq) \rightleftharpoons 2AsS_3^{3-}(aq)$$

The reaction of $As_2S_3$ may involve $HS^-$ ions or $H_2S$ molecules. The actual structure of the complex ion may be $As(SH)_4^-$.

If a solution containing $Al^{3+}$ and $Zn^{2+}$ is treated with a buffer of $NH_3$ at a controlled alkaline pH, $Al(OH)_3$ will precipitate, but $Zn^{2+}$ will stay in solution as $Zn(NH_3)_4^{2+}$. The precipitation of $Zn(OH)_2$ is prevented by the formation of the complex ion; $Al^{3+}$ does not form an ammonia complex.

$$Al^{3+}(aq) + 3OH^-(aq) \rightleftharpoons Al(OH)_3(s)$$

$$Zn^{2+}(aq) + 4NH_3(aq) \rightleftharpoons Zn(NH_3)_4^{2+}(aq)$$

Complex ions are frequently highly colored; $Fe(SCN)^{2+}$ is deep red. Thus, the production of a deep red color when $SCN^-$ is added to a solution serves as a test for the $Fe^{3+}$ ion.

**17.5 AMPHOTERISM**    The hydroxides of certain metals, called **amphoteric hydroxides,** can function as both acids and bases. These water-insoluble compounds dissolve in solutions of high pH or low pH. For example, zinc hydroxide dissolves in excess hydrochloric acid to produce solutions of zinc chloride, $ZnCl_2$.

$$Zn(OH)_2(s) + 2H^+(aq) \rightarrow Zn^{2+}(aq) + 2H_2O$$

Zinc hydroxide will also dissolve in sodium hydroxide solutions; the reaction yields sodium zincate, $Na_2Zn(OH)_4$.

$$Zn(OH)_2(s) + 2OH^-(aq) \rightarrow Zn(OH)_4^{2-}(aq)$$

Other examples of amphoteric hydroxides are: $Al(OH)_3$, $Sn(OH)_2$, $Cr(OH)_3$, $Be(OH)_2$, $Sb(OH)_3$ (or $Sb_2O_3$), and $As(OH)_3$ (or $As_2O_3$); many other compounds including $Cu(OH)_2$ and $AgOH$ (or $Ag_2O$) exhibit this property to a lesser degree. The oxides that correspond to amphoteric hydroxides react in the same way.

The cation of an amphoteric hydroxide has the ability to form complex ions with $H_2O$ and $OH^-$. For example, $Zn(H_2O)_6^{2+}$ has been identified as one of the complex cations present in solutions of zinc chloride. It is assumed that the aquo complex ion forms other complexes, depending upon the pH, in which one or more of the $H_2O$ molecules are replaced by $OH^-$ ions. Thus,

$$Zn(H_2O)_6^{2+}(aq) + OH^-(aq) \rightleftharpoons Zn(OH)(H_2O)_5^+(aq) + H_2O$$

The interpretation of reactions such as this in terms of the Brønsted theory is considered in Section 15.5.

Usually, however, such equations are written without showing the coordinated water (and in most cases the number of coordinated $H_2O$ molecules is in doubt). Hence, the following series of equations is used to describe the reactions that occur when the $OH^-$ concentration of a $Zn^{2+}(aq)$ solution is gradually increased.

$$Zn^{2+}(aq) + OH^-(aq) \rightleftharpoons Zn(OH)^+(aq)$$

$$Zn(OH)^+(aq) + OH^-(aq) \rightleftharpoons Zn(OH)_2(s)$$

$$Zn(OH)_2(s) + OH^-(aq) \rightleftharpoons Zn(OH)_3^-(aq)$$

$$Zn(OH)_3^-(aq) + OH^-(aq) \rightleftharpoons Zn(OH)_4^{2-}(aq)$$

The precipitation of $Zn(OH)_2$ is observed when the proper pH is reached, and upon further increase in pH the precipitate dissolves. The formulas of the amphoterate anions used here are based upon the fact that $NaZn(OH)_3$ and $NaZn(OH)_4$ have been isolated from such solutions.

The process can be reversed. Acidification of solutions of the zincate ion results in the stepwise removal of $OH^-$. For example,

$$Zn(OH)_4^{2-}(aq) + H^+(aq) \rightleftharpoons Zn(OH)_3^+(aq) + H_2O$$

In general, amphoteric systems are very complicated and much remains to be learned about them. The equilibrium constants are of doubtful validity, and the stepwise constants for many hydroxide systems are not known. Indeed, there is evidence that not all the ions and equilibria in systems of this type have been identified. Probably the reactions of most systems are much more complicated than we have depicted them.

We have considered only mononuclear complexes (complexes with a single coordinated metal atom); however, polynuclear complexes appear to be common — $Sn_2(OH)_2^{2+}$ and $Sn_3(OH)_4^{2+}$, as well as $Sn(OH)^+$, have been postulated as occuring in acidic solutions of the $Sn^{2+}$ ion. The formula of the aluminate ion is customarily written $Al(OH)_4^-$, and such a tetrahedral ion has been identified in aluminate solutions with an $OH^-$ concentration of from $0.1 M$ to $1.5 M$. However, at alkaline hydroxide concentrations below and above this range, the principle complex ions are thought to be polynuclear.

Advantage is taken of the amphoteric nature of some hydroxides in analytical chemistry and in some commercial processes. For example, $Mg^{2+}$ and $Zn^{2+}$ may be separated from a solution containing the two ions by making the solution alkaline.

$$Mg^{2+} + 2OH^- \rightleftharpoons Mg(OH)_2(s)$$

$$Zn^{2+} + 4OH^- \rightleftharpoons Zn(OH)_4^{2-}$$

The insoluble magnesium hydroxide may be removed by filtration from the solution containing the zincate ion.

In the production of aluminum metal from bauxite (impure hydrated $Al_2O_3$), the ore is purified prior to its reduction to aluminum metal. This purification is accomplished by dissolving the aluminum oxide in a solution of sodium hydroxide and removing the insoluble impurities by filtration.

$$Al_2O_3(s) + 2OH^- + 3H_2O \rightarrow 2Al(OH)_4^-$$

When the filtered solution is acidified by $CO_2$, aluminum hydroxide precipitates and is recovered.

**17.6 HYDROLYSIS**  Certain ions react with water to produce acidic or alkaline solutions (Section 15.5). For example, the acetate ion can accept a proton from a water molecule to form an acetic acid molecule (a weak electrolyte) and a hydroxide ion.

$$C_2H_3O_2^- + H_2O \rightleftharpoons HC_2H_3O_2 + OH^-$$

This reaction of the acetate ion is similar to that of any other weak base, such as $NH_3$, with water. The fact that the acetate ion has a charge and the ammonia molecule does not is unimportant. Nevertheless, the reaction of an ion with water is customarily called a hydrolysis reaction and interpreted by reference to the weak electrolyte from which the ion is derived.

Ions derived from acids or alkalies that are strong electrolytes (such as $Cl^-$ and $Na^+$) do not hydrolyze. The weaker the electrolyte from which an ion is derived, the more extensive is its hydrolysis. A $0.1 M$ solution of sodium acetate has a pH of 8.9 (Example 17.16), and the pH of a $0.1M$ solution of sodium cyanide is 11.2. In each case it is the hydrolysis of the anion that causes the solution to be alkaline; the sodium ion does not undergo hydrolysis.

$$C_2H_3O_2^- + H_2O \rightleftharpoons HC_2H_3O_2 + OH^-$$

$$CN^- + H_2O \rightleftharpoons HCN + OH^-$$

Since HCN (ionization constant, $4.0 \times 10^{-10}$) is a weaker electrolyte than $HC_2H_3O_2$ (ionization constant, $1.8 \times 10^{-5}$), HCN does a better job of tying up protons than $HC_2H_3O_2$ does. Therefore, the hydrolysis of $CN^-$ is more complete than that of $C_2H_3O_2^-$, and the concentration of $OH^-$ is higher in the NaCN solution than in the $NaC_2H_3O_2$ solution. Note, however, that in both of these systems the position of equilibrium is such that the reverse reaction, which may be regarded as a neutralization, proceeds to a greater extent than the forward reaction since $H_2O$ is a weaker electrolyte than either HCN or $HC_2H_3O_2$.

The equilibrium constant for a hydrolysis equilibrium such as

$$C_2H_3O_2^- + H_2O \rightleftharpoons HC_2H_3O_2 + OH^-$$

may be obtained by dividing the water constant by the ionization constant for the weak acid.

$$[H^+][OH^-] \left( \frac{[HC_2H_3O_2]}{[H^+][C_2H_3O_2^-]} \right) = \frac{[HC_2H_3O_2][OH^-]}{[C_2H_3O_2^-]}$$

$$K_w \left( \frac{1}{K_{HC_2H_3O_2}} \right) = K_H$$

$$(1.0 \times 10^{-14}) \left( \frac{1}{1.8 \times 10^{-5}} \right) = 5.6 \times 10^{-10}$$

---

**Example 17.16**

What is the pH of a $0.10M$ solution of $NaC_2H_3O_2$?

*Solution*

If we let $x$ equal the equilibrium concentration of $HC_2H_3O_2$, we derive the following concentration terms.

$$\begin{array}{ccc} C_2H_3O_2^- + H_2O \rightleftharpoons & HC_2H_3O_2 & + OH^- \\ 0.10M & (x)M & (x)M \end{array}$$

In arriving at these concentration expressions, we assume that the decrease in the concentration of $C_2H_3O_2^-$ because of hydrolysis is negligible in comparison with the original concentration of $C_2H_3O_2^-$; also, we assume that the contribution to the concentration of $OH^-$ from the ionization of water is negligible in comparison to that derived from the hydrolysis.

$$\frac{[HC_2H_3O_2][OH^-]}{[C_2H_3O_2^-]} = 5.6 \times 10^{-10}$$

$$\frac{x^2}{0.10} = 5.6 \times 10^{-10}$$

$$x^2 = 5.6 \times 10^{-11}$$

$$x = [HC_2H_3O_2] = [OH^-] = 7.5 \times 10^{-6}M$$

We can see that our assumptions were justified; $7.5 \times 10^{-6}$ is indeed negligible in comparison to 0.10, and for all practical purposes $[C_2H_3O_2^-] = 0.10M$. Also, the concentration of $OH^-$ in pure water is only approximately 1.3% of the concentration of $OH^-$ determined here; in a solution of sodium acetate the $OH^-$ produced by the hydrolysis would repress the ionization of water so that any additional $OH^-$ from this source would be even more insignificant.

Since

$$[OH^-] = 7.5 \times 10^{-6}M$$

$$[H^+] = 1.3 \times 10^{-9}M$$

$$pH = 8.89$$

---

The hydrolysis of cations produces acidic solutions.

$$NH_4^+ + H_2O \rightleftharpoons NH_3 + H_3O^+$$

This reaction of $NH_4^+$ is similar to the ionization of any other weak acid, such as $HC_2H_3O_2$, and rests on the ability of $H_2O$ to accept a proton from a donor. The reverse reaction, a neutralization, proceeds to a greater extent than the forward reaction. Ammonia is a stronger base than water, and in the competition for protons $NH_3$ is more effective than $H_2O$. Therefore, the concentration of $NH_4^+$ is higher than the concentration of $H_3O^+$.

The equilibrium constant for this hydrolysis equilibrium is obtained by dividing the water constant by the ionization constant for ammonia.

$$K_H = \frac{[NH_3][H_3O^+]}{[NH_4^+]}$$

$$K_H = \frac{K_w}{K_{NH_3}} = \frac{1.0 \times 10^{-14}}{1.8 \times 10^{-5}} = 5.6 \times 10^{-10}$$

---

*Example 17.17*

What is the pH of a $0.30M$ solution of $NH_4Cl$?

*Solution*

$$H_2O + NH_4^+ \rightleftharpoons NH_3 + H_3O^+$$
$$\quad\quad 0.30M \quad\ (x)M \quad (x)M$$

$$\frac{[NH_3][H_3O^+]}{[NH_4^+]} = 5.6 \times 10^{-10}$$

$$\frac{x^2}{0.30} = 5.6 \times 10^{-10}$$

$$x^2 = 1.7 \times 10^{-10}$$

$$x = [NH_3] = [H_3O^+] = 1.3 \times 10^{-5}M$$

$$pH = 4.89$$

---

In the hydrolysis of a metal cation a coordinated water molecule of the hydrated cation donates a proton to a free water molecule.

$$Fe(H_2O)_6^{3+}(aq) + H_2O \rightleftharpoons Fe(OH)(H_2O)_5^{2+}(aq) + H_3O^+(aq)$$

Such equations are usually written without indicating the coordinated water.

$$Fe^{3+}(aq) + H_2O \rightleftharpoons Fe(OH)^{2+}(aq) + H^+(aq)$$

Additional steps in the hydrolysis of the iron(III) ion produce $Fe(OH)_2^+$, $[Fe_2(OH)_2(H_2O)_8]^{4+}$, polynuclear ions of a higher order, and ultimately a precipitate of hydrous $Fe_2O_3$.

Mathematical analysis of the hydrolysis of metal cations is complicated by several factors. As in the hydrolysis of the iron(III) ion, more than one hydrolytic product usually exists, and many of these are polynuclear. For many systems reliable values for the equilibrium constants are not available, and in many instances not all of the hydrolysis equilibria have been identified. Occasionally, hydrolysis proceeds to the point where the metal hydroxide or the hydrous metal oxide precipitates.

*Example 17.18*

What is the pH of a $0.10M$ solution of $Fe(NO_3)_2$?

*Solution*

The stepwise constants for the ionization of $Fe(OH)_2$ are

$$Fe(OH)_2(s) \rightleftharpoons Fe(OH)_2(aq)$$
$$K_1 = 7.2 \times 10^{-6}$$

$$Fe(OH)_2(aq) \rightleftharpoons Fe(OH)^+ + OH^-$$
$$K_2 = 1.0 \times 10^{-4}$$

$$Fe(OH)^+ \rightleftharpoons Fe^{2+} + OH^-$$
$$K_3 = 2.5 \times 10^{-6}$$

The species $Fe(OH)_2(aq)$ is a neutral molecule in solution. From $K_2$ and $K_3$ we derive the constants for the steps of the hydrolysis.

$$Fe^{2+} + H_2O \rightleftharpoons Fe(OH)^+ + H^+$$

$$K_{H_1} = \frac{K_w}{K_3} = 4.0 \times 10^{-9}$$

$$Fe(OH)^+ + H_2O \rightleftharpoons Fe(OH)_2(aq) + H^+$$

$$K_{H_2} = \frac{K_w}{K_2} = 1.0 \times 10^{-10}$$

From the values of the constants we see that the first step of the hydrolysis proceeds to a limited extent only and that the concentration of $Fe(OH)^+$ from this step is low. The hydrolysis of this low concentration of $Fe(OH)^+$, described in the second hydrolysis step (which also occurs to a very limited extent), is therefore negligible, and we shall ignore the second step in our calculations.

$$\underset{(0.10-x)M}{Fe^{2+}} + H_2O \rightleftharpoons \underset{(x)M}{Fe(OH)^+} + \underset{(x)M}{H^+}$$

The $x$ is negligible in comparison to 0.10 in the term for the concentration of $Fe^{2+}$.

$$\frac{[Fe(OH)^+][H^+]}{[Fe^{2+}]} = 4.0 \times 10^{-9}$$

$$\frac{x^2}{0.10} = 4.0 \times 10^{-9}$$

$$x = [H^+] = 2.0 \times 10^{-5}M$$

$$pH = 4.70$$

The concentrations of the species of interest present in this solution are

$$[H^+] = 2.0 \times 10^{-5}M$$

$$[Fe^{2+}] = 1.0 \times 10^{-1}M$$

$$[Fe(OH)^+] = 2.0 \times 10^{-5}M$$

$$[Fe(OH)_2(aq)] = 1.0 \times 10^{-10}M$$

The concentration of $Fe(OH)_2(aq)$ is readily calculated by means of $K_{H_2}$. From these concentration figures we see that we were justified in neglecting the second step of the hydrolysis.

Notice that the equilibrium

$$Fe(OH)_2(s) \rightleftharpoons Fe(OH)_2(aq) \quad K_1 = 7.2 \times 10^{-6}$$

defines the concentration of $Fe(OH)_2(aq)$ that exists in equilibrium with solid $Fe(OH)_2$ in a saturated solution. Since the concentration of $Fe(OH)_2(aq)$ in this $0.10M$ solution of $Fe(NO_3)_2$ is less than $7.2 \times 10^{-6}M$, the solution is not saturated with respect to $Fe(OH)_2$ and $Fe(OH)_2$ does not precipitate. The same conclusion could be reached by use of the $K_{SP}$ of $Fe(OH)_2$.

The hydrolysis of an anion derived from a weak polyprotic acid proceeds in several steps. The ionization constants for $H_2S$ are

$$H_2S \rightleftharpoons H^+ + HS^- \qquad K_1 = 1.1 \times 10^{-7}$$

$$HS^- \rightleftharpoons H^+ + S^{2-} \qquad K_2 = 1.0 \times 10^{-14}$$

Therefore, the hydrolysis constants for the sulfide ion are

$$S^{2-} + H_2O \rightleftharpoons HS^- + OH^- \qquad K_{H_1} = \frac{K_w}{K_2} = 1.0$$

$$HS^- + H_2O \rightleftharpoons H_2S + OH^- \qquad K_{H_2} = \frac{K_w}{K_1} = 9.1 \times 10^{-8}$$

Notice that the *first* hydrolysis constant is obtained by dividing the water constant by the *second* ionization constant of $H_2S$.

In solutions of a soluble sulfide the first step of the hydrolysis of the sulfide ion is so nearly complete that it far overshadows the second, and the acidity of the solution may be calculated by neglecting the hydrolysis of the $HS^-$ ion.

---

**Example 17.19**

What is the pH of a $0.10M$ solution of $Na_2S$?

**Solution**

$$S^{2-} \;+\; H_2O \rightleftharpoons HS^- + OH^-$$
$$(0.10 - x)M \qquad\quad (x)M \quad\; (x)M$$

$$\frac{[HS^-][OH^-]}{[S^{2-}]} = 1.0$$

$$\frac{x^2}{(0.10 - x)} = 1.0$$

$$x = 9.16 \times 10^{-2}$$

$$(0.10 - x) = [S^{2-}] = 8.4 \times 10^{-3}M$$

$$[HS^-] = [OH^-] = 9.2 \times 10^{-2}M$$

$$[H^+] = 1.1 \times 10^{-13}M$$

$$pH = 12.96$$

The pH of a solution of an acid salt (such as NaHS, $NaH_2PO_4$, $Na_2HPO_4$, and $NaHCO_3$) is affected not only by the hydrolysis of the anion but also by the dissociation of the anion. In a solution of NaHA, a salt of a hypothetical weak acid $H_2A$, the following equilibria are of importance.

$$HA^- \rightleftharpoons H^+ + A^{2-} \tag{1}$$

$$HA^- + H_2O \rightleftharpoons H_2A + OH^- \tag{2}$$

Since the concentrations of $H_2A$ and $A^{2-}$ are low, the ionization of $H_2A$ and the hydrolysis of $A^{2-}$ may be neglected.

The $[H^+]$ produced by the dissociation of $HA^-$ (equation 1) is equal to $[A^{2-}]$. However, the $OH^-$ produced by the hydrolysis of $HA^-$ (equation 2) reduces the concentration of $H^+$ by an amount equal to $[H_2A]$. If we neglect the $[H^+]$ from the ionization of $H_2O$,

$$[H^+] = [A^{2-}] - [H_2A] \tag{3}$$

Expressions for $[A^{2-}]$ and $[H_2A]$ may be derived from the appropriate ionization constants.

$$K_1 = \frac{[H^+][HA^-]}{[H_2A]}$$

therefore,

$$[H_2A] = \frac{[H^+][HA^-]}{K_1} \tag{4}$$

$$K_2 = \frac{[H^+][A^{2-}]}{[HA^-]}$$

and

$$[A^{2-}] = \frac{K_2[HA^-]}{[H^+]} \tag{5}$$

Substitution of equations (4) and (5) in (3) gives

$$[H^+] = \frac{K_2[HA^-]}{[H^+]} - \frac{[H^+][HA^-]}{K_1}$$

By clearing fractions we derive

$$[H^+]^2 K_1 = K_1 K_2 [HA^-] - [H^+]^2 [HA^-]$$

Rearranging terms gives

$$[H^+]^2 K_1 + [H^+]^2 [HA^-] = K_1 K_2 [HA^-]$$

$$[H^+]^2 (K_1 + [HA^-]) = K_1 K_2 [HA^-]$$

If the concentration of $HA^-$ is moderately large, $K_1$ will be negligible in comparison to $[HA^-]$ and $(K_1 + [HA^-]) \cong [HA^-]$. Therefore,

$$[H^+]^2 [HA^-] = K_1 K_2 [HA^-]$$

$$[H^+]^2 = K_1 K_2$$

$$[H^+] = \sqrt{K_1 K_2}$$

Thus, for any solution of NaHS of moderate concentration, $[H^+]$ is equal to the square root of the product of the ionization constants of $H_2S$. For a solution of $NaH_2PO_4$, $[H^+] = \sqrt{K_1 K_2}$, and for a solution of $Na_2HPO_4$, $[H^+] = \sqrt{K_2 K_3}$, where $K_1$, $K_2$, and $K_3$ are the successive ionization constants of $H_3PO_4$.

The hydrolysis of ions influences the solubility of slightly soluble substances. For example, in a solution of lead sulfide,

$$PbS(s) \rightleftharpoons Pb^{2+} + S^{2-}$$

the concentrations of $Pb^{2+}$ and $S^{2-}$ are reduced by hydrolysis (in which $Pb(OH)^+$, $HS^-$, and $H_2S$ are formed), and therefore more PbS dissolves than would be the case if hydrolysis did not occur.

---

*Example 17.20*

What is the molar solubility of PbS?

*Solution*

If the hydrolysis of ions is ignored,

$$K_{SP} = [Pb^{2+}][S^{2-}] = 7.0 \times 10^{-29}$$

$$x^2 = 7.0 \times 10^{-29}$$

$$x = 8.4 \times 10^{-15} M$$

As a first approximation, therefore, $[Pb^{2+}] = [S^{2-}] = 8.4 \times 10^{-15} M$, and any $H^+$ or $OH^-$ obtained from the hydrolysis of such low concentrations of ions may be ignored in comparison to that derived from the ionization of water. Therefore,

$$[H^+] = [OH^-] = 1.0 \times 10^{-7} M$$

The hydrolysis equilibria in this solution are:

$$Pb^{2+} + H_2O \rightleftharpoons Pb(OH)^+ + H^+$$

$$K_H = \frac{1.0 \times 10^{-14}}{1.5 \times 10^{-8}} = 6.7 \times 10^{-7}$$

$$S^{2-} + H_2O \rightleftharpoons HS^- + OH^-$$

$$K_{H_1} = \frac{1.0 \times 10^{-14}}{1.0 \times 10^{-14}} = 1.0$$

$$HS^- + H_2O \rightleftharpoons H_2S + OH^-$$

$$K_{H_2} = \frac{1.0 \times 10^{-14}}{1.1 \times 10^{-7}} = 9.1 \times 10^{-8}$$

The sum of the concentrations of all sulfur-containing species must equal the sum of the concentrations of all lead-containing species since both types were derived from PbS. Therefore,

$$[Pb^{2+}] + [Pb(OH)^+] = [S^{2-}] + [HS^-] + [H_2S] \tag{6}$$

By use of the hydrolysis constants we can derive expressions in terms of $[Pb^{2+}]$ or $[S^{2-}]$ for the concentrations of all species. Thus,

$$\frac{[Pb(OH)^+][H^+]}{[Pb^{2+}]} = 6.7 \times 10^{-7}$$

$$\frac{[Pb(OH)^+](1.0 \times 10^{-7})}{[Pb^{2+}]} = 6.7 \times 10^{-7}$$

Therefore,

$$[Pb(OH)^+] = 6.7[Pb^{2+}] \qquad (7)$$

$$\frac{[HS^-][OH^-]}{[S^{2-}]} = 1.0$$

$$\frac{[HS^-](1.0 \times 10^{-7})}{[S^{2-}]} = 1.0$$

and

$$[HS^-] = 1.0 \times 10^7[S^{2-}] \qquad (8)$$

$$\frac{[H_2S][OH^-]}{[HS^-]} = 9.1 \times 10^{-8}$$

$$\frac{[H_2S](1.0 \times 10^{-7})}{1.0 \times 10^7[S^{2-}]} = 9.1 \times 10^{-8}$$

Thus,

$$[H_2S] = 9.1 \times 10^6[S^{2-}] \qquad (9)$$

Substitution of equations (7), (8), and (9) into (6) gives

$$[Pb^{2+}] + 6.7[Pb^{2+}] = [S^{2-}] + 1.0 \times 10^7[S^{2-}] \\ + 9.1 \times 10^6[S^{2-}]$$

$$7.7[Pb^{2+}] = 1.9 \times 10^7[S^{2-}] \qquad (10)$$

From the $K_{SP}$ of PbS,

$$[S^{2-}] = \frac{7.0 \times 10^{-29}}{[Pb^{2+}]} \qquad (11)$$

Substitution of (11) into (10) gives

$$7.7[Pb^{2+}] = 1.9 \times 10^7 \frac{7.0 \times 10^{-29}}{[Pb^{2+}]}$$

$$[Pb^{2+}]^2 = 1.7 \times 10^{-22}$$

$$[Pb^{2+}] = 1.3 \times 10^{-11}M$$

From (7),

$$[Pb(OH)^+] = 6.7(1.3 \times 10^{-11}) = 8.7 \times 10^{-11}M$$

The molar solubility of PbS is

$$[Pb^{2+}] + [Pb(OH)^+] = 1.0 \times 10^{-10}M$$

This value is 12,000 times that obtained by ignoring hydrolysis.

---

**17.7 TITRATION OF ACIDS AND ALKALIES**

We are now in a position to study acid–alkali titrations in some detail. Let us consider the titration of a 50.0 ml sample of 0.100N HCl with a solution of 0.100N NaOH. Since HCl and NaOH are both strong electrolytes, the only equilibrium to be considered is the water equilibrium. Furthermore, for either solution the molarity is equal to the normality. We shall assume in the calculations that follow that when a sample of NaOH solution is added to 50.0 ml of the HCl solution, the resulting volume is equal to the sum of the volumes of the two solutions.

The concentration of $H^+$ in the original 50.0 ml sample of acid in the titration flask is $0.100M$, and therefore the pH is 1.00.

After 10.0 ml of 0.100N NaOH has been added from the buret, the total volume of the solution is 60.0 ml, and the equivalent of 40.0 ml of 0.100N HCl remains unneutralized.

$$VN = VN$$

$$40.0(0.100) = 60.0x$$

$$x = [H^+] = 0.0667N$$

$$pH = 1.18$$

When 50.0 ml of 0.100$N$ NaOH has been added, the "end point" of the titration is reached; all the acid is neutralized and the pH = 7.00.

As NaOH solution is added beyond the equivalent point, the solution in the titration flask becomes increasingly alkaline. For example, when 60.0 ml of 0.100$N$ NaOH has been added (total volume, 110 ml), the solution contains the equivalent of 10.0 ml of 0.100$N$ NaOH.

$$VN = VN$$

$$10.0(0.100) = 110.0x$$

$$x = [OH^-] = 0.00909N$$

$$pOH = 2.04$$

$$pH = 11.96$$

The values in Table 17.1 were obtained from calculations such as these. The data of Table 17.1 are plotted in Figure 17.1. Notice that the curve rises sharply in the section around the equivalence point. Whereas the first 49.9 ml of NaOH solution added causes the pH to change by three units, the next 0.2 ml added causes a change of *six* units in the pH.

Table 17.1 Titration of 50.0 ml of 0.100$N$ HCl with 0.100$N$ NaOH

| VOLUME OF 0.100$N$ NaOH ADDED (ml) | pH |
|---|---|
| 0.0 | 1.00 |
| 10.0 | 1.18 |
| 20.0 | 1.37 |
| 30.0 | 1.60 |
| 40.0 | 1.96 |
| 49.0 | 3.00 |
| 49.9 | 4.00 |
| 50.0 | 7.00 |
| 50.1 | 10.00 |
| 51.0 | 11.00 |
| 60.0 | 11.96 |
| 70.0 | 12.22 |
| 80.0 | 12.36 |
| 90.0 | 12.46 |
| 100.0 | 12.52 |

**Figure 17.1** Titration of 50.0 ml of 0.100 *N* HCl with 0.100 *N* NaOH.

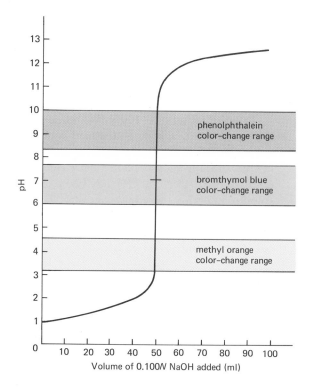

If an indicator is to be used in this titration to locate the equivalence point, any indicator may be employed that changes color in the pH range of the straight portion of the curve. Around this point, the addition of one drop of NaOH solution causes a sharp increase in the pH. Any one of the three indicators indicated in Figure 17.1 would be satisfactory.

Let us now consider the titration of 50.0 ml of $0.100N$ acetic acid — a weak acid — with $0.100N$ sodium hydroxide. The concentration of $H^+$ in the original 50.0 ml sample of acid may be calculated by using the equilibrium constant (Example 16.2). From Table 16.1 we see that $[H^+] = 1.34 \times 10^{-3}M$. The pH of the solution is therefore 2.87.

The solution resulting from the addition of 10.0 ml of $0.100\,N$ NaOH to the 50.0 ml sample of $0.100N$ $HC_2H_3O_2$ is, in effect, a buffer since it contains a mixture of $C_2H_3O_2^-$ ions, produced by the neutralization, together with unneutralized $HC_2H_3O_2$. The pH of a buffer is conveniently calculated by the use of the relation

$$pH = pK - \log\left(\frac{[HA]}{[A^-]}\right)$$

which was derived in Section 16.7. For acetic acid the p$K$ is 4.74 (the negative logarithm of $1.8 \times 10^{-5}$). The ratio $[HC_2H_3O_2]/[C_2H_3O_2^-]$ is easily calculated. After 10.0 ml of NaOH is added, 40/50 of the acid of the original 50.0 ml sample remains unneutralized and 10/50 has, in effect, been converted into the salt sodium acetate. The ratio is therefore 4 to 1. Thus,

$$pH = pK - \log\left(\frac{[HC_2H_3O_2]}{[C_2H_3O_2^-]}\right)$$
$$= 4.74 - \log(4/1)$$
$$= 4.14$$

At the equivalence point all of the acid has been neutralized, and the solution (100 ml) is effectively $0.0500M$ in sodium acetate. The calculation of the pH of this solution must take into account the hydrolysis of the $C_2H_3O_2^-$ ion (Example 17.16).

$$C_2H_3O_2^- + H_2O \rightleftharpoons HC_2H_3O_2 + OH^-$$

0.0500M $\qquad\qquad$ $(x)M$ $\qquad$ $(x)M$

$$\frac{[HC_2H_3O_2][OH^-]}{[C_2H_3O_2^-]} = 5.56 \times 10^{-10}$$

$$\frac{x^2}{5.00 \times 10^{-2}} = 5.56 \times 10^{-10}$$

$$x = [OH^-] = 5.27 \times 10^{-6}$$

$$pOH = 5.28$$

$$pH = 8.72$$

Notice that the equivalence point in this titration does not occur at the "neutral" pH, 7.

After the equivalence point the addition of NaOH causes the solution to become increasingly alkaline. The added $OH^-$ represses the hydrolysis of the acetate ion, and the contribution of the hydrolysis to the concentration of $OH^-$ is negligible in comparison to the added $OH^-$. Thus, the calculations from this point on are identical to those for the HCl–NaOH titration.

Data for a $HC_2H_3O_2$–NaOH titration are summarized in Table 17.2 and plotted in Figure 17.2. The equivalence point of this titration occurs at a higher pH than that of the preceding titration, and all pH values on the acid side of the equivalence point are higher. Therefore, the rapidly ascending portion of the curve around the equivalence point is reduced in length. From the curve of Figure 17.2 we can see that methyl orange is not a suitable indicator for this titration. Neither is bromthymol blue suitable, since its color change would start after 47.34 ml of NaOH has been added and continue until 49.97 ml has been added; this would hardly

**Table 17.2** Titration of 50.0 ml of 0.100$N$ HC$_2$H$_3$O$_2$ with 0.100$N$ NaOH

| VOLUME OF 0.100$N$ NaOH ADDED (ml) | pH |
|---|---|
| 0.0 | 2.87 |
| 10.0 | 4.14 |
| 20.0 | 4.56 |
| 30.0 | 4.92 |
| 40.0 | 5.34 |
| 49.0 | 6.44 |
| 49.9 | 7.45 |
| 50.0 | 8.72 |
| 50.1 | 10.00 |
| 51.0 | 11.00 |
| 60.0 | 11.96 |
| 70.0 | 12.22 |
| 80.0 | 12.36 |
| 90.0 | 12.46 |
| 100.0 | 12.52 |

constitute a sharp end point for a titration. Phenolphthalein, however, would be a satisfactory indicator to employ.

Other titration curves may be drawn following the general line of approach outlined here. Figure 17.3 represents the titration curve for the

**Figure 17.2** Titration of 50.0 ml of 0.100 $N$ HC$_2$H$_3$O$_2$ with 0.100 $N$ NaOH.

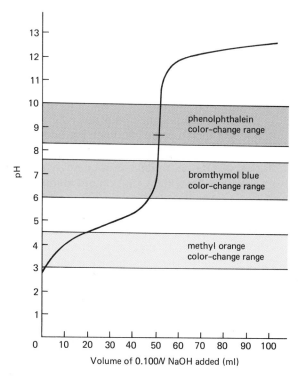

**Figure 17.3** Titration of 50.0 ml of 0.100 *N* NH$_3$ with 0.100 *N* HCl.

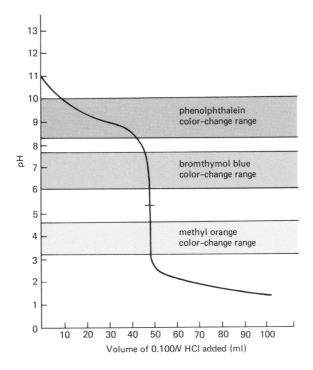

titration of 50.0 ml of 0.100$N$ NH$_3$ with 0.100$N$ HCl. In this instance methyl orange could be used as the indicator. Figure 17.4 is the titration curve for the titration of 50.0 ml of 0.100$N$ HC$_2$H$_3$O$_2$ with 0.100$N$ NH$_3$; both solutes are weak electrolytes. No indicator can be found that would function satisfactorily for this titration, and such titrations — between weak electrolytes — are not usually run.

Titrations may be conducted potentiometrically. For example, a titration may be performed with the electrodes of a pH meter immersed in the solution being analyzed. The pH of the solution is determined after successive additions of reagent. The equivalence point of the titration is indicated by an abrupt change in the pH, but additions and readings are continued beyond this point. The equivalence point may be determined by graphing the data and estimating the volume corresponding to the midpoint of the steeply rising portion of the titration curve. A plot of cell potential has the same shape as a titration curve based on pH, and titrations may be conducted using such measurements instead of pH determinations. The same considerations apply for the determination of the equivalence point for curves based on either type of measurement.

The potentiometric titration of a weak electrolyte serves as an important method of determining the dissociation constant of the electrolyte. Since

$$pH = pK - \log([HA]/[A^-])$$

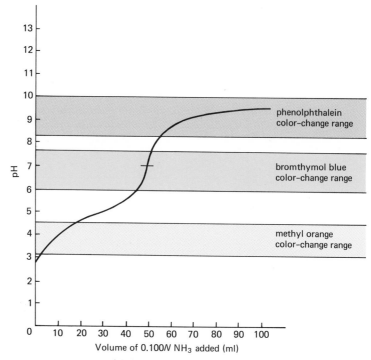

**Figure 17.4** Titration of 50.0 ml of 0.100 $N$ HC$_2$H$_3$O$_2$ with 0.100 $N$ NH$_8$.

the p$K$ of the electrolyte is equal to the pH of the solution at half neutralization (where $[\text{HA}] = [\text{A}^-]$), but the p$K$, and hence the $K$ itself, may be determined from any point on the titration curve.

*Example 17.21*

The equivalence point in the titration of 40.00 ml of a solution of a weak monoprotic acid occurs when 35.00 ml of a 0.100$N$ NaOH solution has been added. The pH of the solution is 5.75 after 20.00 ml of the NaOH solution has been added. What is the dissociation constant of the acid?

*Solution*

Since 35.00 ml of NaOH is required for complete neutralization, the acid has been 20/35 neutralized after 20.00 ml of NaOH has been added. In other words, 15/35 of the acid is in the form HA, and 20/35 is in the form A$^-$. The ratio $[\text{HA}]/[\text{A}^-]$ is therefore 15 to 20, or 0.75.

$$\text{pH} = \text{p}K - \log\left(\frac{[\text{HA}]}{[\text{A}^-]}\right)$$

$$5.75 = \text{p}K - \log(0.75)$$

$$\text{p}K = 5.63$$

$$K = 2.4 \times 10^{-6}$$

**17.8 OXIDATION–REDUCTION EQUILIBRIA**

The ionic equilibria we have thus far considered in this chapter have not involved oxidation and reduction. Electrochemical measurements are the principal source of data for the study of oxidation–reduction equilibria.

As we have seen, electrochemical measurements are used in the study of certain equilibria that do not involve oxidation–reduction — such as the determination of pH (Section 16.4), $K_{SP}$'s (Section 17.1), and titration curves (Section 17.7).

The equilibrium constant for an oxidation–reduction equilibrium is readily calculated from the standard cell potential of the reaction which in turn may be calculated from the standard electrode potentials for the half reactions. The change in standard Gibbs free energy for the reaction under consideration is given by the equation

$$\Delta G^\circ = -nF\mathscr{E}^\circ \tag{12}$$

and $\Delta G^\circ$ is related to the equilibrium constant by the expression

$$\Delta G^\circ = -RT \ln K \tag{13}$$

An expression that relates the equilibrium constant directly to the cell potential is obtained by combining equations (12) and (13).

$$\mathscr{E}^\circ = \left(\frac{RT}{nF}\right) \ln K$$

At 25°C,

$$\mathscr{E}^\circ = \frac{0.05916}{n} \log K$$

This expression may also be derived from the Nernst equation (Section 10.9). A convenient and equivalent expression is

$$\log K = 16.90 \, n\mathscr{E}^\circ$$

As a rule, an ionic reaction that does not involve oxidation–reduction reaches equilibrium very rapidly — usually in the time that it takes to prepare the solution. Although many oxidation–reduction reactions occur rapidly in solution, a large number do not; in this latter instance, equilibrium is attained very slowly. Therefore, even a large value for an equilibrium constant does not necessarily mean that the reaction to which it pertains will occur at an appreciable rate.

Oxidation–reduction reactions can be followed potentiometrically in much the same way as acid–alkali reactions. Redox titrations are extensively used in analysis. For example, the amount of iron in a sample of an ore can be determined by dissolving the ore, converting all of the iron in the resulting solution into the ferrous state, and titrating this solution against a standard solution of an oxidizing agent (such as ceric sulfate, potassium permanganate, or potassium dichromate). Let us consider the titration of 50.0 ml of $0.10M$ $Fe^{2+}$ solution with a $0.10M$ solution of $Ce^{4+}$.

*Example 17.22*

A 50.0 ml sample of 0.10$M$ ferrous sulfate is titrated with 0.10$M$ ceric sulfate. (a) What is the equilibrium constant for the reaction? (b) What are the concentrations of ions in equilibrium at the equivalence point?

*Solution*

(a) The electrode potentials for the half reactions are

$$e^- + Fe^{3+} \rightleftharpoons Fe^{2+} \qquad \mathscr{E}^\circ = +0.77 \text{ V}$$

$$e^- + Ce^{4+} \rightleftharpoons Ce^{3+} \qquad \mathscr{E}^\circ = +1.61 \text{ V}$$

and the potential for the reaction is

$$Fe^{2+} + Ce^{4+} \rightleftharpoons Fe^{3+} + Ce^{3+} \qquad \mathscr{E}^\circ = +0.84 \text{ V}$$

$$\log K = 16.9 \; n\mathscr{E}^\circ$$

$$= 16.9 \; (1) \; (0.84)$$

$$K = 1.6 \times 10^{14}$$

(b) To reach the equivalence point, as much $Ce^{4+}$ will have to be added as there is $Fe^{2+}$ in the original solution. At the equivalence point, therefore,

$$[Fe^{3+}] = [Ce^{3+}]$$

and

$$[Fe^{2+}] = [Ce^{4+}]$$

From the equilibrium constant,

$$\frac{[Fe^{3+}][Ce^{3+}]}{[Fe^{2+}][Ce^{4+}]} = 1.6 \times 10^{14}$$

$$\frac{[Fe^{3+}]^2}{[Fe^{2+}]^2} = 1.6 \times 10^{14}$$

$$\frac{[Fe^{3+}]}{[Fe^{2+}]} = 1.3 \times 10^{7}$$

We can see that the reaction proceeds essentially to completion, and on this basis we can calculate the concentrations of ions. Since the volume of the solution equals 100 ml at the equivalence point, $[Fe^{3+}] = [Ce^{3+}] = 0.050M$. If we let $x = [Fe^{2+}] = [Ce^{4+}]$,

$$\frac{(0.050)}{x} = 1.3 \times 10^{7}$$

$$x = [Fe^{2+}] = [Ce^{4+}] = 3.8 \times 10^{-9}M$$

and

$$[Fe^{3+}] = [Ce^{3+}] = 5.0 \times 10^{-2}M$$

At every point in the titration of $Fe^{2+}$ with $Ce^{4+}$, the $\mathscr{E}$ of a theoretical cell

$$Pt \mid Ce^{4+}, Ce^{3+} \mid Fe^{3+}, Fe^{3+} \mid Pt$$

would be zero since equilibrium is established at each point. However, if the titration is run in a cell containing a standard hydrogen electrode (the reference electrode) and a platinum electrode (the indicator electrode), a difference in potential may be read as the titration proceeds.

The $\mathscr{E}$ of the cell

$$Pt \mid H_2 \mid H^+ \mid Fe^{3+}, Fe^{2+} \mid Pt$$

where the left-hand electrode is a standard hydrogen electrode, varies with the concentrations of the iron ions

$$\mathscr{E} = \mathscr{E}^\circ - \left(\frac{0.059}{1}\right) \log \frac{[Fe^{2+}]}{[Fe^{3+}]}$$

Initially, before any $Ce^{4+}$ is added to the 50.0 ml sample of $0.10M$ $Fe^{2+}$, $\mathscr{E}$ is theoretically equal to $-\infty$, although traces of $Fe^{3+}$ due to air oxidation will give a small positive reading.

After 10.0 ml of $0.10M$ $Ce^{4+}$ is added, 40/50 of the iron is in the original form, $Fe^{2+}$, and 10/50 of the iron is in the oxidized form, $Fe^{3+}$. The ratio of $[Fe^{2+}]$ to $[Fe^{3+}]$ is therefore 4 to 1, and

$$\mathscr{E} = +0.77 - (0.059) \log 4$$
$$= +0.77 - (0.059)(0.60)$$
$$= +0.73 \text{ V}$$

At the half-titration point, when 25.0 ml of $Ce^{4+}$ has been added, $[Fe^{2+}] = [Fe^{3+}]$, and $[Fe^{2+}]/[Fe^{3+}] = 1$. Since the logarithm of 1 is 0,

$$\mathscr{E} = +0.77 \text{ V}$$

At the equivalence point, $[Fe^{3+}]/[Fe^{2+}] = 1.3 \times 10^7$ (Example 17.22). Therefore,

$$\frac{[Fe^{2+}]}{[Fe^{3+}]} = \frac{1}{1.3 \times 10^7} = 7.7 \times 10^{-8}$$

and

$$\mathscr{E} = +0.77 - 0.059 \log(7.7 \times 10^{-8})$$
$$= +0.77 - 0.059(-7.11)$$
$$= +1.19 \text{ V}$$

After the equivalence point the concentration of $Fe^{2+}$ becomes vanishingly small. We may calculate points for the titration curve by remembering that at each point of the curve the $\mathscr{E}$ for the reduction of $Ce^{4+}$ equals the $\mathscr{E}$ for the reduction of $Fe^{3+}$, since $\mathscr{E}$ for the couple is zero whenever equilibrium is established. Prior to the equivalence point the $Ce^{4+}$ is consumed in the oxidation of $Fe^{2+}$ as fast as it is added; after the equivalence point $Ce^{4+}$ is present in excess. The cell, in effect, is

$$Pt \mid H_2 \mid H^+ \mid Ce^{4+}, Ce^{3+} \mid Pt$$

After 60.0 ml of $0.100M$ $Ce^{4+}$ has been added, 10.0 ml is present in excess. The concentration of $Ce^{4+}$ is proportional to the 10.0 ml added above the amount needed for equivalence, and the concentration of $Ce^{3+}$ is proportional to 50.0 ml (the volume needed to react with the 50.0 ml of $Fe^{2+}$). Therefore, the ratio of $[Ce^{3+}]$ to $[Ce^{4+}]$ is equal to 50/10, or 5.

$$\mathscr{E} = +1.61 - 0.059 \log \left( \frac{[Ce^{3+}]}{[Ce^{4+}]} \right)$$
$$= +1.61 - 0.059 \log(5)$$
$$= +1.56 \text{ V}$$

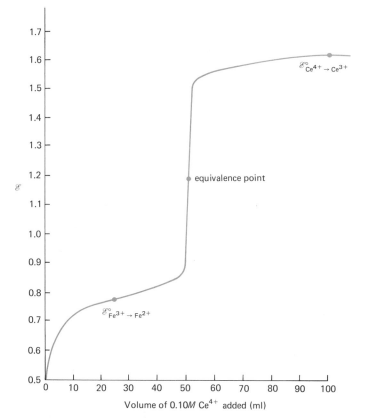

Volume of 0.10$M$ $Ce^{4+}$ added (ml)

Notice that $\mathscr{E}^\circ$ for the $Ce^{4+}$–$Ce^{3+}$ couple (1.61 V) is reached when 100.0 ml of $Ce^{4+}$ solution has been added.

The titration curve of Figure 17.5 was plotted from data obtained from calculations such as the foregoing. The equivalence point of an actual titration may be obtained by taking readings of the potential after successive additions of the oxidizing agent, plotting the data, and finding the inflection point of the titration curve in the same manner as that employed for acid–alkali titrations.

In addition to potentiometric determinations, oxidation–reduction indicators may be used to locate the end point of an oxidation–reduction titration. For the titration of $Fe^{2+}$ by $Ce^{4+}$, ferroin is a suitable indicator. Ferroin undergoes a conversion from a pale blue form to a red form at approximately 1.11 V, a value suitably close to the equivalence point voltage of 1.19 V.

The data plotted in Figure 17.5 were calculated for a cell utilizing a standard hydrogen electrode ($\mathscr{E}^\circ = 0.00$ V) as a reference electrode. In actual practice calomel electrodes (Section 16.4) or silver–silver chloride electrodes are generally employed in place of hydrogen electrodes since the latter are inconvenient to use. If a normal calomel electrode is employed as the reference electrode ($\mathscr{E} = +0.283$ V), every potential measured will be displaced by 0.283 V. For example, the equivalence point of the titration would occur at 0.91 V instead of 1.19 V. The actual $\mathscr{E}$ measurements are of no concern, since the equivalence point is derived from the titration curve and the curve is merely displaced to lower $\mathscr{E}$ values by the use of a calomel electrode in place of a hydrogen electrode.

As a matter of fact, for a given oxidation–reduction reaction, a titration curve based on actual $\mathscr{E}$ measurements would be slightly different from a titration curve derived from $\mathscr{E}°$ values. $\mathscr{E}°$ values are based on activities rather than concentrations; the difference between the activity and the concentration of a multivalent ion can be appreciable. If so-called "formal potentials" are used instead of standard potentials, a titration curve is obtained that is more in harmony with that plotted from the data obtained from an actual titration. Formal potentials are based on concentrations rather than activities. In any event, the equivalence point is located from the titration curve rather than from a calculated equivalence-point potential, and the principles of an oxidation–reduction titration are adequately illustrated by the use of $\mathscr{E}°$'s.

## Problems

**17.1** At 25°C, $1.7 \times 10^{-5}$ mol of $Cd(OH)_2$ dissolves in one liter of water. Calculate the $K_{SP}$ of $Cd(OH)_2$.

**17.2** The $K_{SP}$ of $NiCO_3$ is $1.4 \times 10^{-7}$. What is the molar solubility of $NiCO_3$? Neglect the hydrolysis of ions.

**17.3** At 25°C, $5.2 \times 10^{-6}$ mol of $Ce(OH)_3$ dissolves in one liter of water. Calculate the $K_{SP}$ of $Ce(OH)_3$.

**17.4** The $K_{SP}$ of $CaF_2$ is $3.9 \times 10^{-11}$. What is the molar solubility of $CaF_2$? Neglect the hydrolysis of ions.

**17.5** Which carbonate has the lower molar solubility, $Ag_2CO_3$ ($K_{SP}$, $8.2 \times 10^{-12}$) or $CuCO_3$ ($K_{SP}$, $2.5 \times 10^{-10}$)? Neglect the hydrolysis of ions.

**17.6** Which sulfide has the lower molar solubility, $Ag_2S$ ($K_{SP}$, $5.5 \times 10^{-51}$) or $CuS$ ($K_{SP}$, $8.0 \times 10^{-37}$)? Neglect the hydrolysis of ions.

**17.7** A saturated solution of a slightly-soluble hydroxide, $M(OH)_2$, has a pH of 9.53. What is the $K_{SP}$ of $M(OH)_2$?

**17.8** Calculate the $K_{SP}$ of thallium(I) iodide, $TlI$, from the following standard electrode potentials.

$$e^- + TlI(s) \rightleftharpoons Tl(s) + I^-(aq)$$
$$\mathscr{E}° = -0.752 \text{ V}$$

$$e^- + Tl^+(aq) \rightleftharpoons Tl(s)$$
$$\mathscr{E}° = +0.336 \text{ V}$$

**17.9** Calculate the $K_{SP}$ of mercury(I) sulfate, $Hg_2SO_4$, from the following standard electrode potentials.

$$2e^- + Hg_2SO_4 \text{ (s)} \rightleftharpoons$$
$$2Hg(l) + SO_4^{2-}(aq)$$
$$\mathscr{E}° = +0.615 \text{ V}$$

$$2e^- + Hg_2^{2+}(aq) \rightleftharpoons 2Hg(l)$$
$$\mathscr{E}° = +0.788 \text{ V}$$

**17.10** Calculate the $K_{SP}$ of yttrium(III) hydroxide, $Y(OH)_3$, from the following standard electrode potentials.

$$3e^- + Y(OH)_3(s) \rightleftharpoons$$
$$Y(s) + 3OH^-(aq)$$
$$\mathscr{E}° = -2.810 \text{ V}$$

$$3e^- + Y^{3+}(aq) \rightleftharpoons Y(s)$$
$$\mathscr{E}° = -2.372 \text{ V}$$

**17.11** The $K_{SP}$ of $Ca_3(PO_4)_2$ is $1.3 \times 10^{-32}$ and the $Ca^{2+}/Ca$ electrode potential is

$$2e^- + Ca^{2+}(aq) \rightleftharpoons Ca(s)$$
$$\mathscr{E}° = -2.866 \text{ V}$$

Calculate $\mathscr{E}°$ for the half reaction

$$6e^- + Ca_3(PO_4)_2(s) \rightleftharpoons$$
$$3Ca(s) + 2PO_4^{3-}(aq)$$

**17.12** How many moles of $Ni(OH)_2$ will dissolve in one liter of a solution of NaOH with a pH of 12.34? Neglect hydrolysis of ions.

**17.13** How many moles of $Ag_2CO_3$ will dissolve in 150 ml of $0.15M$ $Na_2CO_3$ solution? Neglect hydrolysis of ions.

**17.14** Calculate the final concentrations of $Na^+(aq)$, $C_2O_4^{2-}(aq)$, $Ba^{2+}(aq)$ and $Cl^-(aq)$ in a solution prepared by adding 100 ml of $0.20M$ $Na_2C_2O_4$ to 150 ml of $0.25M$ $BaCl_2$. Neglect the hydrolysis of ions.

**17.15** Calculate the final concentrations of $Sr^{2+}(aq)$, $NO_3^-(aq)$, $Na^+(aq)$, and $F^-(aq)$ in a solution prepared by adding 50 ml of $0.30M$ $Sr(NO_3)_2$ to 150 ml of $0.12M$ NaF. Neglect the hydrolysis of ions.

**17.16** The formula of the mercurous ion is $Hg_2^{2+}$, and the $K_{SP}$ of mercurous carbonate, $Hg_2CO_3$, is $9.0 \times 10^{-17}$. (a) Determine the molar solubility of $Hg_2CO_3$. (b) What would the $K_{SP}$ of this compound be if the mercurous ion were $Hg^+$?

**17.17** (a) A solution is $0.30M$ in $Ag^+(aq)$ and $0.050M$ in $Ba^{2+}(aq)$. If solid $Na_2SO_4$ is very slowly added to this solution, which will precipitate first, $Ag_2SO_4$ or $BaSO_4$? Assume that the change in volume brought about by the addition of $Na_2SO_4$ is negligible. (b) The addition of $Na_2SO_4$ is continued until the second cation just starts to precipitate as the sulfate. What is the concentration of the first cation at this point?

**17.18** (a) A solution is $0.050M$ in $I^-$ and $0.24M$ in $Cl^-$. Solid $AgNO_3$ is gradually added to the solution. Neglect volume changes and calculate the concentration of $Ag^+(aq)$ required to start the precipitation of AgI and AgCl. Which compound precipitates first? (b) What is the concentration of $I^-(aq)$ when AgCl just begins to precipitate? (c) What is the concentration of $I^-(aq)$ when half the $Cl^-(aq)$ has been precipitated as AgCl?

**17.19** What concentration of

$SO_4^{2-}$(aq) is necessary to start the precipitation of $BaSO_4$ from a saturated solution of $BaF_2$?

**17.20** What minimum concentration of $NH_4^+$ is necessary to prevent formation of $Mn(OH)_2$(s) from a solution that is $0.030M$ in $Mn^{2+}$(aq) and $0.030M$ in $NH_3$?

**17.21** A solution is $0.090M$ in $Mg^{2+}$(aq) and $0.33M$ in $NH_4^+$(aq). What minimum concentration of $NH_3$ will cause $Mg(OH)_2$ to start to precipitate?

**17.22** A solution is prepared by mixing 10 ml of $0.50M$ $CaCl_2$ with 10 ml of a solution that is $0.50M$ in $NH_3$ and $0.050M$ in $NH_4^+$(aq). Will $Ca(OH)_2$ precipitate?

**17.23** Calculate the number of $S^{2-}$ ions in 10 ml of a solution that is $0.30M$ in $H^+$ and that has been saturated with $H_2S$.

**17.24** A solution that is $0.25M$ in $H^+$(aq) and $0.10M$ in $Co^{2+}$(aq) is saturated with $H_2S$. Will CoS precipitate?

**17.25** A solution that is $0.50M$ in $H^+$(aq) and $0.030M$ in $Cd^{2+}$(aq) is saturated with $H_2S$. Will CdS precipitate?

**17.26** What is the lowest concentration of $H^+$(aq) that must be present in a $0.50M$ solution of $Zn^{2+}$(aq) to prevent the precipitation of ZnS when the solution is saturated with $H_2S$?

**17.27** A solution that is $0.20M$ in $H^+$(aq) and $0.20M$ in $Pb^{2+}$(aq) is saturated with $H_2S$. What is the concentration of $Pb^{2+}$(aq) after PbS has precipitated? Note that it is necessary to take into account the increase in acidity caused by the precipitation.

**17.28** A solution that is $0.10M$ in $H^+$(aq), $0.30M$ in $Cu^{2+}$(aq), and $0.30M$ in $Fe^{2+}$(aq) is saturated with $H_2S$. Calculate the concentrations of $H^+$(aq), $S^{2-}$(aq), $Cu^{2+}$(aq), and $Fe^{2+}$(aq). Note that it is necessary to take into account any increase in acidity caused by precipitation.

**17.29** What concentration of $H^+$(aq) should be present in a solution that is $0.20M$ in $Ni^{2+}$(aq) and $0.20M$ in $Cd^{2+}$ so that when the solution is saturated with $H_2S$ the maximum amount of CdS will precipitate but no NiS will precipitate at any stage?

**17.30** (a) Will a precipitate of MnS form when a solution that is $0.10M$ in acetic acid, $HC_2H_3O_2$, and $0.10M$ in $Mn^{2+}$ is saturated with $H_2S$? (b) If the solution described in part (a) also is $0.10M$ in sodium acetate, $NaC_2H_3O_2$, will MnS precipitate?

**17.31** A $0.010M$ solution of $Zn(NO_3)_2$ is made $0.50M$ in $NH_3$. The principal zinc-containing ion in the resulting solution is $Zn(NH_3)_4^{2+}$. What is the concentration of $Zn^{2+}$?

**17.32** A $0.0010M$ solution of $Hg^{2+}$(aq) is made $0.40M$ in $I^-$(aq). The principal mercury-containing species in the resulting solution is $HgI_4^{2-}$. What is the concentration of $Hg^{2+}$(aq)?

**17.33** Compare the molar solubilities of (a) AgCl, (b) AgBr, and (c) AgI in $0.50M$ $NH_3$ solution.

**17.34** A $0.010M$ solution of $AgNO_3$ is made $0.50M$ in $NH_3$, thus forming the $Ag(NH_3)_2^+$ complex ion. (a) Will AgCl precipitate if sufficient NaCl is added to make the solution $0.010M$ in $Cl^-$? (b) Will AgI precipitate if the solution is made $0.010M$ in $I^-$?

**17.35** Use the stepwise dissociation constants for the silver-thiosulfate complex ions to derive expressions for the concentration ratios $[Ag(S_2O_3)_2^{3-}]/[Ag(S_2O_3)_3^{5-}]$, $[Ag(S_2O_3)^-]/[Ag(S_2O_3)_2^{3-}]$, and $[Ag^+]/[Ag(S_2O_3)^-]$ in terms of the concentration of thiosulfate ion. Evaluate the ratios for thiosulfate concentrations of $1M$, $10^{-3}M$, $10^{-7}M$, and $10^{-11}M$. Identify the principal silver-containing ions at each of the thiosulfate concentrations listed.

**17.36** Given the following electrode potentials:

$$3e^- + RhCl_6^{3-} \rightleftharpoons Rh + 6Cl^-$$
$$\mathscr{E}° = +0.431 \text{ V}$$

$$3e^- + Rh^{3+} \rightleftharpoons Rh$$
$$\mathscr{E}° = +0.800 \text{ V}$$

calculate the value of the instability constant of the $RhCl_6^{3-}$ ion.

**17.37** Use the constants for the stepwise dissociation of the $Ag(S_2O_3)_3^{5-}$ ion given in Section 17.4 to determine the concentrations of $Ag^+$, $Ag(S_2O_3)^-$, $Ag(S_2O_3)_2^{3-}$, $Ag(S_2O_3)_3^{5-}$, and $S_2O_3^{2-}$ in a solution prepared by dissolving 1.0 mol of $Ag(S_2O_3)_3^{5-}$ in 1.0 liter of water.

**17.38** What is the pH of a $0.10M$ solution of sodium nitrite ($NaNO_2$)?

**17.39** What is the pH of a $0.10M$ solution of aniline hydrochloride ($C_6H_5NH_3^+Cl^-$)?

**17.40** The pH of a $0.15M$ solution of NaX is 9.77. What is the ionization constant of the weak acid HX?

**17.41** The pH of a $0.30M$ solution of NaY is 9.50. What is the ionization constant of the weak acid HY?

**17.42** A solution of sodium benzoate ($NaC_7H_5O_2$) has a pH of 9.00. What is the concentration of sodium benzoate?

**17.43** What concentration of ammonium chloride, $NH_4Cl$, will produce a solution with a pH of 5.20?

**17.44** The $Zn(OH)^+$ ion is believed to be the only zinc-containing ion resulting from the hydrolysis of $Zn^{2+}(aq)$ in dilute solutions. What is the pH of a $0.010M$ solution of $Zn(NO_3)_2$? The $K_{inst}$ of $Zn(OH)^+$ is $4.1 \times 10^{-5}$.

**17.45** The $Cd(OH)^+$ ion is believed to be the only cadmium-containing ion resulting from the hydrolysis of $Cd^{2+}(aq)$ in dilute solutions. What is the concentration of $Cd^{2+}(aq)$ in a solution of $Cd(ClO_4)_2$ that has a pH of 5.48? The $K_{inst}$ of $Cd(OH)^+$ is $6.9 \times 10^{-5}$.

**17.46** Determine the pH of (a) a $0.10M$ solution of $Na_2CO_3$, (b) a $0.10M$ solution of $NaHCO_3$.

**17.47** Determine the pH of a $0.10M$ solution of (a) $Na_3AsO_4$, (b) $Na_2HAsO_4$, (c) $NaH_2AsO_4$.

**17.48** Determine the molar solubility of $ZnS(s)$: (a) neglect hydrolysis, (b) take hydrolysis into account. Assume that $Zn^{2+}(aq)$ hydrolyzes to $Zn(OH)^+(aq)$ only.

**17.49** Determine pH values for the titration curve pertaining to the titration of 30.00 ml of $0.100N$ benzoic acid, $HC_7H_5O_2$, with $0.100N$ NaOH (a) after 10.00 ml of NaOH solution has been added, (b) after 30.00 ml of NaOH solution has been added, (c) after 40.00 ml of NaOH solution has been added.

**17.50** Determine pH values for the titration curve pertaining to the titration of 25.00 ml of $0.100N$ $NH_3$ solution with a $0.100N$ solution of HCl (a) after 10.00 ml of the HCl solution has been added, (b) after 25.00 ml of HCl solution has been added, (c) after 35.00 ml of HCl solution has been added.

**17.51** In the titration of 25.00 ml of a solution of a weak acid HX with $0.250N$ NaOH, the pH of the solution is 4.50 after 5.00 ml of the NaOH solution have been added. The equivalence point in the titration occurs when 30.40 ml of the NaOH solution have been added. What is the equilibrium constant for the ionization of HX?

**17.52** The equivalence point in the titration of 50.00 ml of a solution of acetic acid occurs when 33.00 ml of $0.200N$ NaOH have been added. How many ml of the NaOH solution should be added to 50.00 ml of the acetic acid solution to produce a solution with a pH of 5.04?

**17.53** (a) Calculate the equilibrium constant for the reaction

$$Cu(s) + 2Ag^+ \rightleftharpoons Cu^{2+} + 2Ag(s)$$

(b) If metallic Cu is added to a $0.10M$ $Ag^+(aq)$ solution, what are the concentrations of $Cu^{2+}(aq)$ and $Ag^+$ (aq) at equilibrium?

**17.54** (a) Calculate the equilibrium constant for the reaction

$$Ni(s) + Sn^{2+} \rightleftharpoons Ni^{2+} + Sn(s)$$

(b) Solid Ni is added to a $0.10M$ solution of $Sn^{2+}(aq)$. What are the equilibrium concentrations of $Ni^{2+}(aq)$ and $Sn^{2+}(aq)$?

**17.55** (a) Calculate the equilibrium constant for the following reaction (*as written*).

$$Ce^{4+} + Co^{2+} \rightleftharpoons Co^{3+} + Ce^{3+}$$

(b) Calculate the equilibrium concentrations of ions in a solution prepared by mixing 10.0 ml of a $0.20M$ $Co^{2+}(aq)$ solution with 10.0 ml of a $0.20M$ $Ce^{4+}(aq)$ solution.

**17.56** (a) Calculate the equilibrium constant for the reaction

$$Cr^{2+} + Fe^{3+} \rightleftharpoons Cr^{3+} + Fe^{2+}$$

(b) Determine the concentrations of ions present at equilibrium in a solution prepared by the addition of 50.0 ml of a $0.050M$ $Cr^{2+}$ solution to 50.0 ml of a $0.050M$ $Fe^{3+}$ solution.

# 18

# Metals

Several physical and chemical properties that are typical of metals are used to define this classification of elements—although the extent of each of these properties varies widely from metal to metal. Metals have superior electrical and thermal conductivities, characteristic luster, and the ability to be deformed under stress without cleaving. The tendency of these elements toward the formation of cations through electron loss and their formation of basic oxides are among the chemical characteristics of metals.

More than three-quarters of all known elements are metals. The stepped line that appears in the periodic table marks an approximate division between metals and nonmetals—the nonmetals appear in the upper right corner of the table. The division, however, is slightly arbitrary; elements that appear close to the line have intermediate properties.

Since most compounds of metals contain nonmetals as well, much of the chemistry of the metals is included in the discussions of the nonmetals in Chapters 8, 11, and 12.

**18.1 PHYSICAL PROPERTIES OF METALS**

The densities of metals show a wide variation. Of all the *solid* elements, lithium has the lowest density and osmium has the highest. The majority of metals, however, have relatively high densities in comparison to the densities of the solid nonmetals. The metals of groups I A and II A are referred to as the light metals because they are exceptions to the preceding generalization. The close-packed arrangement of the atoms of most metallic crystals helps to explain the relatively high densities of metals.

The trend in the magnitude of the density across a period (Figure 18.1) reflects the variation in atomic radius of the elements of the period. Thus, the atomic radii of the elements of the fourth, fifth, and sixth periods reach a minimum approximately at group VIII, and the densities are at a maximum at this point. The I A metals have the largest atomic radii and the smallest atomic masses of their periods; these factors coupled with their crystal structure give them comparatively low densities.

The densities of the elements of a group increase with increasing atomic number; this trend corresponds to an increase in atomic mass

**Figure 18.1** Densities of the metals of the fourth, fifth, and sixth periods.

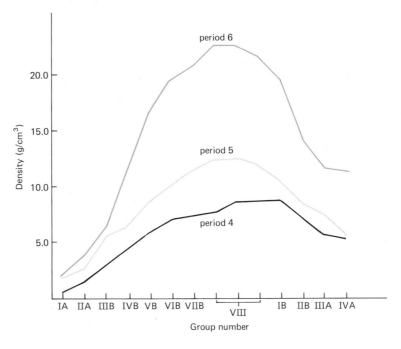

which is not offset by a proportionate increase in atomic radius. Data for the lanthanides of period 6 are not plotted in Figure 18.1, but the lanthanide contraction causes the pronounced difference between the curves for period 5 and period 6.

The melting points and boiling points of the nonmetals show extreme variations—from the very low values of the gaseous elements to the extremely high values of the elements that crystallize in covalent network lattices, such as the diamond. There is considerable variation in the melting and boiling points of the metals; however, low values, such as those of the gaseous nonmetals, are not observed. In general, most metals have comparatively high melting and boiling points.

The striking feature of the curves of Figure 18.2 is the maximum that occurs at approximately the center of each curve. Curves for the boiling point, heat of fusion, heat of vaporization, and hardness of metals have approximately the same appearance. For a given period, therefore, the

**Figure 18.2** Melting points of the metals of the fourth, fifth, and sixth periods.

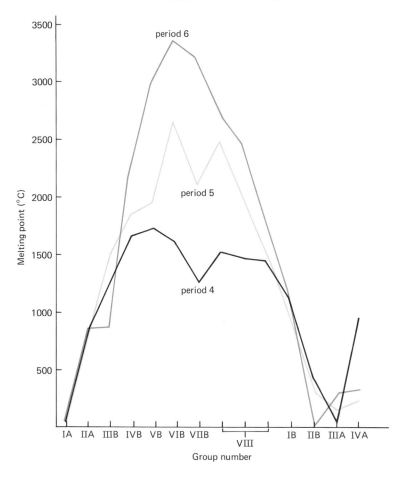

strength of the metallic bonding must reach a maximum around the center of the transition series.

The strength of the metallic bonding is, of course, related to the number of delocalized electrons per atom used in the bonding. If we count only the $ns$, $np$, and unpaired $(n-1)$ $d$ electrons of a metal as bonding electrons, we get an order that approximates that of the properties under discussion. This analysis is a decided oversimplification; other factors, such as atomic radius, nuclear charge, number of bonding orbitals, overlap of orbital energies, and crystal form, are involved.

The deformability, luster, thermal conductivity, and electrical conductivity of metals are the properties most characteristic of metallic bonding and the metallic state (Section 4.7). Electrical conductivity is measured in units of 1/ohm cm; a conductivity of 1.0/ohm cm means that a current of 1.0 amp flows when a potential of 1.0 V is applied to opposite faces of a 1.0 cm$^3$ cube of the material being studied (Section 10.1). The conductivities of metals, or conductors, lie in the general range of 10$^4$ to 10$^6$/ohm cm at room temperature and decrease with increasing temperature. These conductivities are approximately 10$^4$ to 10$^{11}$ times the conductivities of the semiconductors (Section 7.14), which range from 10$^{-5}$ to 10/ohm cm at room temperature and increase with rising temperature. The nonmetals, or insulators, have negligible conductivities— from 10$^{-22}$ to 10$^{-10}$/ohm cm at room temperature; for these materials conductivity increases with increasing temperature.

The conductivities of some metals are listed in Table 18.1. The metals of group I B (copper, silver, and gold) are outstanding electrical conductors. Except for similarities in the relative standing of elements of the same group within a period, periodic trends in the conductivities are difficult to discern. Nor is it easy to explain the lack of a close correlation

Table 18.1 Electrical conductivities of the metals at 0°C in units of 10$^4$/ohm cm

| Li 11.8 | Be 18 | | | | | | | | | | | | | | |
|---|---|---|---|---|---|---|---|---|---|---|---|---|---|---|---|
| Na 23 | Mg 25 | | | | | | | | | | | | Al 40 | | |
| K 15.9 | Ca 23 | Sc | Ti 1.2 | V 0.6 | Cr 6.5 | Mn 20 | Fe 11.2 | Co 16 | Ni 16 | Cu 65 | Zn 18 | Ga 2.2 | | | |
| Rb 8.6 | Sr 3.3 | Y | Zr 2.4 | Nb 4.4 | Mo 23 | Tc | Ru 8.5 | Rh 22 | Pd 10 | Ag 66 | Cd 15 | In 12 | Sn 10 | Sb 2.8 | |
| Cs 5.6 | Ba 1.7 | La 1.7 | Hf 3.4 | Ta 7.2 | W 20 | Re 5.3 | Os 11 | Ir 20 | Pt 10 | Au 49 | Hg 4.4 | Tl 7.1 | Pb 5.2 | Bi 1 | |

between the general trend in melting points and the pattern of the conductivities.

The thermal conductivity, luster, and deformability of metals are discussed in Section 4.7. In general, thermal conductivities parallel electrical conductivities. The nondirectional character of metallic bonding accounts for the ease with which planes of atoms slide across one another under stress and thus explains why a crystal may be deformed without shattering. When the planes of an ionic crystal are displaced (Figure 4.32), the new alignment brings ions of the same charge into proximity and results in the cleavage of the crystal. Covalent network crystals are deformed only by breaking the covalent bonds of the crystal, a process which also results in the fragmentation of the crystal.

**18.2 NATURAL OCCURRENCE OF METALS**

An **ore** is a naturally occurring material from which one or more metals can be profitably extracted. The principal types of ores and examples of each are listed in Table 18.2. A few of the less reactive metals occur in nature in elementary form; for many of these metals, native ores constitute the most important source. The greatest tonnage of metals is derived from oxides—either oxide ores or metal oxides that are produced by the roasting of carbonate or sulfide ores.

Table 18.2 Occurrence of metals

| TYPE OF ORE | EXAMPLES |
| --- | --- |
| native metals | Cu, Ag, Au, As, Sb, Bi, Pd, Pt |
| oxides | $Al_2O_3$, $Fe_2O_3$, $Fe_3O_4$, $SnO_2$, $MnO_2$, $TiO_2$, $FeO \cdot Cr_2O_3$, $FeO \cdot WO_3$, $Cu_2O$, ZnO |
| carbonates | $CaCO_3$, $CaCO_3 \cdot MgCO_3$, $MgCO_3$, $FeCO_3$, $PbCO_3$, $BaCO_3$, $SrCO_3$, $ZnCO_3$, $MnCO_3$, $CuCO_3 \cdot Cu(OH)_2$, $2CuCO_3 \cdot Cu(OH)_2$ |
| sulfides | $Ag_2S$, $Cu_2S$, CuS, PbS, ZnS, HgS, $FeS \cdot CuS$, $FeS_2$, $Sb_2S_3$, $Bi_2S_3$, $MoS_2$, NiS, CdS |
| halides | NaCl, KCl, AgCl, $KCl \cdot MgCl_2 \cdot 6H_2O$, NaCl and $MgCl_2$ in sea water |
| sulfates | $BaSO_4$, $SrSO_4$, $PbSO_4$, $CaSO_4 \cdot 2H_2O$, $CuSO_4 \cdot 2Cu(OH)_2$ |
| silicates | $Be_3AlSi_6O_{18}$, $ZrSiO_4$, $Sc_2Si_2O_7$, $(NiSiO_3, MgSiO_3)^a$ |
| phosphates | $[CePO_4, LaPO_4, NdPO_4, PrPO_4, Th_3(PO_4)_4]^a$, $LiF \cdot AlPO_4$ |

$^a$ Occur in a single mineral but not in fixed proportions.

Silicate minerals are abundant in nature. However, the extraction of metals from silicates is difficult, and the cost of such processes may be prohibitive. Consequently, only the less common metals are commercially derived from silicate ores. Phosphate minerals are, in general, rare and occur in low concentrations.

A number of metals occur as impurities in the ores of other metals so that both metals are derived from the same commercial operation. For example, cadmium metal is obtained as a by-product in the production of zinc.

Ores, as mined, generally contain variable amounts of unwanted materials (such as silica, clay, and granite), which are called **gangue**. The concentration of the desired metal must be sufficiently high to make its extraction chemically feasible and economically competitive. Ores of low concentration are worked only if they can be processed comparatively easily and inexpensively or if the metal product is scarce and valuable. The required concentration varies greatly from metal to metal. For aluminum or iron it should be 30% or more; for copper it may be 1% or less.

**18.3 METALLURGY: PRELIMINARY TREATMENT OF ORES**

**Metallury** is the science of extracting metals from their ores and preparing them for use. Metallurgical processes may be conveniently divided into three principal operations: (1) **preliminary treatment,** in which the desired component of the ore is concentrated, specific impurities removed, and/or the mineral is put into a suitable form for subsequent treatment, (2) **reduction,** in which the metal compound is reduced to the free metal, and (3) **refining,** in which the metal is purified and, in some cases, substances added to give desired properties to the final product. Since the problems encountered in each step vary from metal to metal, many different metallurgical procedures exist.

The processing of many ores requires, as a first step, that most of the gangue be removed. Such **concentration** procedures, which are usually carried out on ores that have been crushed and ground, may be based on physical or chemical properties. **Physical separations** are based on differences between the physical properties of the mineral and the gangue. Thus, through washing with water the particles of rocky impurities may often be separated from the heavier mineral particles. This separation may be accomplished by shaking the crushed ore in a stream of water on an inclined table; the heavier mineral particles settle to the bottom and are collected. Adaptations of this process exist.

**Flotation** is a method of concentration applied to many ores, especially those of copper, lead, and zinc. The finely crushed ore is mixed with a suitable oil and water in large tanks. The mineral particles are wetted by the oil, whereas the gangue particles are wetted by the water. Agitation of the mixture with air produces a froth which contains the oil and mineral particles. This froth floats on the top of the water and is skimmed off.

**Magnetic separation** is employed to separate $Fe_3O_4$ (the mineral magnetite) from gangue. The ore is crushed, and electromagnets are used to attract the particles of $Fe_3O_4$, which is ferromagnetic.

The removal of free metals from native ores may be considered to be a type of concentration. Certain ores, such as native bismuth and native copper, are heated to a temperature just above the melting point of the metal, and the liquid metal is poured away from the gangue. The process is called **liquation.**

Mercury will dissolve silver and gold to form what are called **amalgams.** Hence, the native ores of silver and gold are treated with mercury, the resulting liquid amalgam collected, and the free silver or gold recovered from the amalgam by distilling the mercury away.

The chemical properties of minerals are the basis of **chemical separations.** An important example is the **Bayer method** of obtaining pure aluminum oxide from the ore bauxite. Since aluminum hydroxide is amphoteric, the alumina is dissolved away from the ore impurities (principally ferric oxide and silicates) by treatment of the crushed ore with hot sodium hydroxide solution.

$$Al_2O_3(s) + 2OH^-(aq) + 3H_2O \rightarrow 2Al(OH)_4^-(aq)$$

The solution of sodium aluminate is filtered and cooled; its pH is adjusted downward by dilution and/or neutralization with carbon dioxide so that aluminum hydroxide precipitates (Section 17.5). Pure aluminum oxide, ready for reduction, is obtained by heating the aluminum hydroxide.

$$2Al(OH)_3(s) \rightarrow Al_2O_3(s) + 3H_2O(g)$$

The first step in the extraction of magnesium from sea water is a chemical concentration. The $Mg^{2+}$ in sea water (approximately 0.13%) is precipitated as magnesium hydroxide by treatment of the sea water with a slurry of calcium hydroxide. The calcium hydroxide is obtained from CaO that is produced by roasting oyster shells ($CaCO_3$).

$$CaCO_3(s) \xrightarrow{\text{heat}} CaO(s) + CO_2(g)$$

$$CaO(s) + H_2O \rightarrow Ca(OH)_2(s)$$

$$Mg^{2+}(aq) + Ca(OH)_2(s) \rightarrow Mg(OH)_2(s) + Ca^{2+}(aq)$$

The magnesium hydroxide is converted to magnesium chloride by reaction with hydrochloric acid, and the magnesium chloride solution is evaporated to produce the dry $MgCl_2$ necessary for electrolytic reduction.

The metal component of some ores may be obtained by **leaching.** Thus, low-grade carbonate and oxide ores of copper may be leached with dilute sulfuric acid

$$CuO(s) + 2H^+(aq) \rightarrow Cu^{2+}(aq) + H_2O$$

$$CuCO_3(s) + 2H^+(aq) \rightarrow Cu^{2+}(aq) + CO_2(g) + H_2O$$

and the resulting copper sulfate solutions may be directly subjected to electrolysis.

Silver and gold ores are leached with solutions of sodium cyanide in the presence of air, since these metals form very stable complex ions with the cyanide ion. For native silver, argenitite ($Ag_2S$), and cerargyrite ($AgCl$), the reactions are

$$4Ag(s) + 8CN^-(aq) + O_2(g) + 2H_2O \rightarrow 4Ag(CN)_2^-(aq) + 4OH^-(aq)$$

$$Ag_2S(s) + 4CN^-(aq) \rightarrow 2Ag(CN)_2^-(aq) + S^{2-}(aq)$$

$$AgCl(s) + 2CN^-(aq) \rightarrow Ag(CN)_2^-(aq) + Cl^-(aq)$$

After concentration many ores are **roasted** in air. The sulfide ores of the less reactive metals are directly reduced to the free metal by heating (see Section 18.4). However, the majority of the sulfide ores, as well as the carbonate ores, are converted into oxides by roasting.

$$2ZnS(s) + 3O_2(g) \rightarrow 2ZnO(s) + 2SO_2(g)$$

$$PbCO_3(s) \rightarrow PbO(s) + CO_2(g)$$

The free metals are more readily obtained from oxides than from sulfides or carbonates.

**18.4 METALLURGY: REDUCTION**

By far the largest quantity of metals, as well as the largest number of metals, are produced by **smelting** operations—high-temperature reduction processes in which the metal is usually secured in a molten state. In most of these processes a **flux** (such as limestone, $CaCO_3$) is used to remove the gangue that remains after ore concentration. The flux forms a **slag** with silica and silicate impurities; for limestone and silica the simplified equations are

$$CaCO_3(s) \rightarrow CaO(s) + CO_2(g)$$

$$CaO(s) + SiO_2(s) \rightarrow CaSiO_3(l)$$

The slag, which is a liquid at the smelting temperatures, generally floats on top of the molten metal and is readily separated from the metal.

The reducing agent employed for a given smelting operation is the least expensive material that is capable of yielding a product of the required purity. For the ores of the metals of low reactivity—for example, the sulfide ores of mercury, copper, and lead—no chemical agent is required. Mercury is produced by roasting cinnabar, $HgS$, in air.

$$HgS(s) + O_2(g) \rightarrow Hg(g) + SO_2(g)$$

The mercury vapor is condensed in a receiver and requires no further purification.

Copper sulfide ores, after concentration by flotation, are smelted to a material called **matte,** which is essentially cuprous sulfide, $Cu_2S$. The matte is then reduced by blowing air through the molten material.

$$Cu_2S(l) + O_2(g) \rightarrow 2Cu(l) + SO_2(g)$$

Any cuprous oxide that forms in this process is reduced by stirring the molten metal by poles of green wood. The copper produced by this method (blister copper) is about 99% pure and is refined by electrolysis.

Lead sulfide ores (galena) are subjected to a roasting operation in which a portion of the PbS is converted into lead oxide and lead sulfate.

$$2PbS(s) + 3O_2(g) \rightarrow 2PbO(s) + 2SO_2(g)$$

$$PbS(s) + 2O_2(g) \rightarrow PbSO_4(s)$$

The resulting PbS, PbO, and $PbSO_4$ mixture, together with fluxes, is smelted in the absence of air. The reactions, in which sulfur is oxidized and lead are reduced, are

$$PbS(s) + 2PbO(s) \rightarrow 3Pb(l) + SO_2(g)$$

$$PbS(s) + PbSO_4(s) \rightarrow 2Pb(l) + 2SO_2(g)$$

Some lead is also produced by smelting PbO with coke. The lead oxide is obtained by roasting the sulfide ore in an excess of air so that virtually complete conversion to the oxide results.

Iron, zinc, tin, cadmium, antimony, nickel, cobalt, molybdenum, lead, and other metals are produced by **carbon reduction** of oxides, which are obtained as ores or as the products of roasting operations.

The reactions that occur in a high-temperature carbon reduction are complex. In most cases the reduction is effected principally by carbon monoxide, not carbon. Since both the mineral and coke are solids that are not readily fusible, contact between them is poor and direct reaction slow.

$$MO(s) + C(s) \rightarrow M(l) + CO(g)$$

(M stands for a metal.) However, a gas and a solid make better contact and react more readily.

$$2C(s) + O_2(g) \rightarrow 2CO(g)$$

$$MO(s) + CO(g) \rightarrow M(l) + CO_2(g)$$

The carbon dioxide that is produced is converted into carbon monoxide by reaction with coke.

$$C(s) + CO_2(g) \rightarrow 2CO(g)$$

The most important commercial metal, iron, is produced by carbon reduction in a **blast furnace** designed to operate continuously (Figure 18.3). The ore, coke, and a limestone flux are charged into the top of the

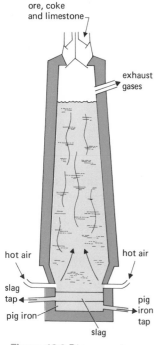

ore, coke and limestone

exhaust gases

hot air    hot air

slag tap

pig iron tap

pig iron

slag

**Figure 18.3** Diagram of a blast furnace (schematic).

furnace and heated air, which is sometimes enriched with oxygen, is blown in at the bottom. The incoming air reacts with carbon to form carbon monoxide and to liberate considerable amounts of heat; at this point the temperature of the furnace is the highest (ca. 1500°C).

The iron ore (usually $Fe_2O_3$) is reduced in stages depending upon the temperature. Near the top of the furnace, where the temperature is lowest, $Fe_3O_4$ is the reduction product.

$$3Fe_2O_3(s) + CO(g) \rightarrow 2Fe_3O_4(s) + CO_2(g)$$

The descending $Fe_3O_4$ is reduced to FeO in a lower, hotter zone.

$$Fe_3O_4(s) + CO(g) \rightarrow 3FeO(s) + CO_2(g)$$

In the hottest zone, reduction to metallic iron occurs.

$$FeO(s) + CO(g) \rightarrow Fe(l) + CO_2(g)$$

The molten iron collects in the bottom of the furnace. Molten slag, principally calcium silicate which is produced by the ultimate action of the flux on the gangue, also collects on the bottom. The slag floats on top of the molten iron, thus protecting the metal from oxidation by the incoming air. The slag and iron are periodically drawn off. The hot exhaust gases are used to heat incoming air.

The impure iron produced by the blast furnace (**pig iron**) contains up to 4% carbon, up to 2% silicon, some phosphorus, and a trace of sulfur. In the manufacture of steel (Section 18.5), these impurities are removed or their concentrations are adjusted; in addition, certain metallic ingredients are added. The presence of some carbon in steel is desirable, and the amount of carbon in the final product is readily controlled in the refining process.

The carbon reduction of some oxides (e.g., those of chromium, manganese, and tungsten) yields products that contain an appreciable quantity of carbon. If these metals are to be used in the manufacture of steel, the carbon content does not matter. However, since carbon impurities are difficult to remove from many metals, carbon reduction is not a satisfactory method for preparing these metals in a pure form. In such instances, as well as those in which carbon is incapable of effecting a reduction, reactive metals (such as Na, Mg, and Al) may be used as reducing agents. Processes that use metals as reducing agents are more expensive than those that employ carbon because the cost of the production of the metal used as a reducing agent must be taken into account.

The reduction of an oxide by aluminum is called a **Goldschmidt process** or a **thermite process**.

$$Cr_2O_3(s) + 2Al(s) \rightarrow 2Cr(l) + Al_2O_3(l)$$

$$3MnO_2(s) + 4Al(s) \rightarrow 3Mn(l) + 2Al_2O_3(l)$$

$$3BaO(s) + 2Al(s) \rightarrow 3Ba(l) + Al_2O_3(l)$$

The reactions are highly exothermic, and molten metals are produced. The reaction of $Fe_2O_3$ and aluminum is used at times for the production of molten iron in welding operations and as the basis of incendiary (thermite) bombs. Other oxides commercially reduced by metals include: $UO_3$ (by Al or Ca), $V_2O_5$ (by Al), $Ta_2O_5$ (by Na), $MoO_3$ (by Al), $ThO_2$ (by Ca), and $WO_3$ (by Al).

The **Kroll process** involves the reduction of metal halides (such as $TiCl_4$, $ZrCl_4$, $UF_4$, and $LaCl_3$) by magnesium, sodium, or calcium. In the production of titanium, titanium tetrachloride is prepared by the reaction of titanium dioxide, carbon, and chlorine. The chloride is a liquid and is readily purified by distillation. Purified $TiCl_4$ is passed into molten magnesium or sodium at approximately 700°C under an atmosphere of argon or helium (which prevents oxidation of the product).

$$TiCl_4(g) + 2Mg(l) \rightarrow Ti(s) + 2MgCl_2(l)$$

A **metal displacement** reaction employing zinc is used to obtain silver and gold from the solutions obtained from cyanide leaching operations.

$$2Ag(CN)_2^-(aq) + Zn(s) \rightarrow 2Ag(s) + Zn(CN)_4^{2-}(aq)$$

**Hydrogen reduction** is used for the preparation of some metals (at high temperatures) when carbon reduction yields an unsatisfactory product. Examples are

$$GeO_2(s) + 2H_2(g) \rightarrow Ge(s) + 2H_2O(g)$$

$$WO_3(s) + 3H_2(g) \rightarrow W(s) + 3H_2O(g)$$

$$MoO_3(s) + 3H_2(g) \rightarrow Mo(s) + 3H_2O(g)$$

The metals are produced as powders by this method. Germanium is melted and cast into ingots, but tungsten (wolfram) and molybdenum have high melting points. They are heated and pounded into compact form. Hydrogen cannot be used to reduce the oxides of some metals because of the formation of undesirable hydrides.

The most reactive metals are prepared by **electrolytic reduction** of molten compounds. Sodium and magnesium (as well as the other I A and II A metals) are prepared by electrolysis of the fused chlorides. Sodium chloride is electrolyzed in a **Downs cell** (Figure 18.4), which is designed to keep the products separate from one another so that they do not react.

anode: $\quad\quad 2Cl^- \rightarrow Cl_2(g) + 2e^-$

cathode: $\quad e^- + Na^+ \rightarrow Na(l)$

Purified $Al_2O_3$ is electrolyzed in the **Hall process** for the production of aluminum (Figure 18.5). Alumina dissolved in cryolite, $Na_3AlF_6$, pro-

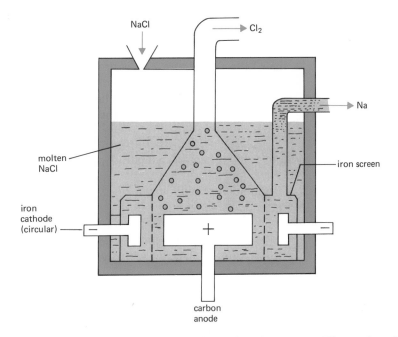

**Figure 18.4** Schematic diagram of the Downs cell for the production of sodium and chlorine.

duces a solution that will conduct electric current. The carbon lining of the cell serves as the cathode, and carbon anodes are used. The primary reaction at the anode may be considered to be the discharge of oxygen from oxide ions. However, the oxygen reacts with the carbon anodes so that the cell reactions are

anode:   $C(s) + 2O^{2-} \rightarrow CO_2(g) + 4e^-$

cathode:   $3e^- + Al^{3+} \rightarrow Al(l)$

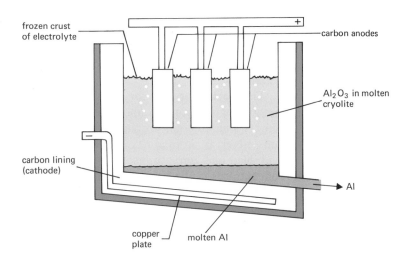

**Figure 18.5** Schematic diagram of an electrolytic cell for the production of aluminum.

The molten aluminum collects at the bottom of the cell and is periodically withdrawn.

Some metals are prepared by the electrolysis of aqueous solutions of their salts. Very pure zinc is produced by the electrolysis of zinc sulfate solutions. The zinc sulfate is obtained by treating zinc oxide (obtained from zinc sulfide ores by roasting) with sulfuric acid. The sulfuric acid for the process is made from the sulfur dioxide derived from the roasting of the sulfide ore.

anode: $$2H_2O \rightarrow O_2(g) + 4H^+(aq) + 4e^-$$

cathode: $$2e^- + Zn^{2+}(aq) \rightarrow Zn(s)$$

Aqueous solutions of salts of copper, cadmium, chromium, cobalt, gallium, indium, manganese, and thorium are electrolyzed to produce the corresponding metals. Electrolytically prepared metals generally require no further refining.

**18.5 METALLURGY: REFINING**

Most of the metals obtained from reduction operations require refining to rid them of objectionable impurities. Refining processes vary widely from metal to metal, and for a given metal the method employed may vary with the proposed use of the final product. Along with the removal of material that impart undesirable properties to the metal, the refining step may include the addition of substances to give the product desired characteristics. Some refining processes are designed to recover valuable metal impurities, such as gold, silver, and platinum.

Crude tin, lead, and bismuth are purified by **liquation.** Ingots of the impure metals are placed at the top of a sloping hearth that maintained at a temperature slightly above the melting point of the metal. The metal melts and flows down the inclined hearth into a well, leaving behind the solid impurities.

Some low-boiling metals, such as zinc and mercury, are purified by **distillation.**

The **Parkes process** for refining lead, which also is a concentration method for silver, relies upon the selective dissolution of silver in molten zinc. A small amount of zinc (1 to 2%) is added to molten lead that contains silver as an impurity. Silver is much more soluble in zinc than in lead; lead and zinc are insoluble in each other. Hence, most of the silver concentrates in the zinc, which comes to the top of the molten lead. The zinc layer solidifies first upon cooling and is removed. The silver is obtained by remelting the zinc layer and distilling away the zinc, which is collected and used over again.

In the **Mond process,** impure nickel is treated with carbon monoxide at 60°C to 80°C. The nickel reacts to form gaseous nickel carbonyl, leaving solid impurities, including cobalt, behind.

$$Ni(s) + 4CO(g) \rightarrow Ni(CO)_4(g)$$

Pure nickel is recovered from the carbonyl by heating the gas to temperatures of from 180°C to 200°C.

$$Ni(CO)_4(g) \rightarrow Ni(s) + 4CO(g)$$

The carbon monoxide is recirculated.

The **Van Arkel process** is also based upon the thermal decomposition of a metal compound. The method, which is used for the purification of titanium, hafnium, and zirconium, involves the decomposition of a metal iodide on a hot metal filament. For example, impure zirconium is heated with a limited amount of iodine in an evacuated glass apparatus.

$$Zr(s) + 2I_2(g) \rightarrow ZrI_4(g)$$

The gaseous zirconium tetraiodide is decomposed upon contact with a hot filament, and pure zirconium metal is deposited upon the filament.

$$ZrI_4(g) \rightarrow Zr(s) + 2I_2(g)$$

The regenerated iodine reacts with more zirconium. The process is very expensive and is employed for the preparation of limited amounts of very pure metals for special uses.

Another process capable of producing metals of very high purity is **zone refining.** A circular heater is fitted around a rod of an impure metal, such as germanium (Figure 18.6). The heater, which is slowly moved down the rod, melts a band of the metal. As the heater moves along, pure metal recrystallizes out of the melt, and impurities are swept along in the molten zone to one end of the rod, which is subsequently discarded. More than one pass of the heater may be made on the same rod.

Electrolytic refining is an important and widely used method of purification. Many metals, including copper, tin, lead, gold, zinc, chromium, and nickel, are refined electrolytically. Plates of the impure metal are used as anodes, and the electrolyte is a solution of a salt of the metal.

**Figure 18.6** Schematic diagram of zone refining.

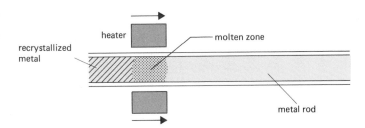

The pure metal plates out on the cathode. For copper, copper sulfate is employed as the electrolyte, and the electrode reactions are

anode: $\qquad$ $Cu(s) \rightarrow Cu^{2+}(aq) + 2e^-$

cathode: $\quad 2e^- + Cu^{2+}(aq) \rightarrow Cu(s)$

The more reactive metals in the crude copper anode, such as iron, are oxidized and pass into solution where they remain; they are not reduced at the cathode. The less reactive metals, such as silver, gold, and platinum, are not oxidized. As the copper anode dissolves away, they fall to the bottom of the cell from where they are recovered as a valuable "anode sludge."

Pig iron is refined into steel by the **Bessemer process** or by the **open hearth process.** In both processes the impurities (carbon from the coke used in the reduction, and such substances as silicon, phosphorus, and sulfur from the ore) are oxidized and volatilize out of the iron or go into the slag.

In the Bessemer process, molten pig iron from the blast furnace is poured into egg-shaped converters, and a blast of air is blown through the melt to oxidize the impurities. The heat required to keep the charge molten is supplied by these oxidations. The process is rapid (10 to 15 min); however, it is difficult to control, and the quality of the product is variable.

By far the largest percentage of steel is manufactured by the open hearth process. A charge of pig iron, scrap iron, and iron oxide is heated in a shallow hearth lined with calcium oxide or magnesium oxide. A blast of burning fuel and hot air is directed on the molten iron, and the oxidation of the impurities is accomplished by reaction with excess air or with iron oxide. The carbon monoxide escapes and burns, and the oxides of silicon, phosphorus, and sulfur combine with the oxides of the hearth lining to form a slag.

Since it takes about 8 to 10 hours to produce a batch of steel by the open hearth process, the quality of the final product is readily controlled and is uniform. Alloying metals, such as manganese, chromium, nickel, tungsten, molybdenum, and vanadium, may be added before the charge is poured.

**18.6 THE GROUP I A METALS**

The I A metals, also called the **alkali-metals,** constitute the most reactive group of metals. None of these elements is found free in nature, and all are prepared by the electrolysis of dry, molten salts. The element francium, $Z = 87$, is formed in certain natural radioactive processes; all of its isotopes are radioactive with short half-lives, and the element is extremely rare.

The group I A elements are silvery metals; cesium has a slight golden yellow cast. The elements are comparatively soft (they can be cut by a knife) and have low melting points and boiling points (Table 18.3). Melting point, boiling point, and hardness decrease with increasing atomic number. The metals are good conductors of heat and electricity. They have very low densities. Compare the densities given in Table 18.3 with the densities of the metals of the first transition series, which range from 2.5 g/cm$^3$ for $_{21}$Sc to 8.9 g/cm$^3$ for $_{30}$Cu.

Table 18.3 Some properties of the I A metals

|  | LITHIUM | SODIUM | POTASSIUM | RUBIDIUM | CESIUM |
|---|---|---|---|---|---|
| outer electronic configuration | $2s^1$ | $3s^1$ | $4s^1$ | $5s^1$ | $6s^1$ |
| melting point (°C) | 179 | 97.5 | 63.7 | 39.0 | 28.5 |
| boiling point (°C) | 1336 | 880 | 760 | 700 | 670 |
| density (g/cm$^3$) | 0.53 | 0.97 | 0.86 | 1.53 | 1.90 |
| atomic radius (Å) | 1.23 | 1.57 | 2.03 | 2.16 | 2.35 |
| ionic radius, $M^+$ (Å) | 0.60 | 0.95 | 1.33 | 1.48 | 1.69 |
| ionization potential (eV) |  |  |  |  |  |
|   first | 5.4 | 5.1 | 4.3 | 4.2 | 3.9 |
|   second | 75.6 | 47.3 | 31.8 | 27.4 | 23.4 |
| electrode potential, $\mathscr{E}°$, |  |  |  |  |  |
|   ($M^+/M$) (V) | −3.05 | −2.71 | −2.93 | −2.93 | −2.92 |

The alkali metals emit electrons when irradiated (the photoelectric effect). Cesium, which ejects electrons most readily, is used in the manufacture of photocells (employed in light meters and electric eyes), which convert light signals into electric signals.

The electronic configuration of each of the alkali metals is that of the preceding noble gas (a noble gas core) plus a single $s$ valence electron in the outer shell. These valence electrons are easily lost to give 1+ ions that are isoelectronic with noble gases. Thus, an element of group I A has the lowest first ionization potential of any element of its period. The ease with which these elements lose electrons makes them extremely strong reducing agents.

The second ionization potentials of the I A metals are so much higher than the first ionization potentials that the 1+ oxidation state is the only one observed for these metals. With few exceptions, alkali metal compounds are ionic.

In general, the reactivity of the elements increases with increasing atomic number—paralleling a decrease in first ionization potential. Thus, in most cases cesium is the most reactive element of the group and lithium is the least reactive. This order of reactivity is expected since a large atom holds its valence electron less tightly than a small atom; in a large atom the electron is farther from the nucleus and is screened from the positive nuclear charge by a large number of underlying electron shells.

Each I A element has the largest atomic radius and largest ionic radius of any element of its period. This comparatively large size combined with the low charge of the I A ions leads to species that have little polarizing ability (and hence form strongly ionic compounds) and that do not readily form complex ions.

Since the alkali metal ions have comparatively low charge/size ratios, the hydration energies of the ions are low. The lithium ion, which is the smallest ion and has the least effectively screened nuclear charge of the group, has the largest hydration energy. The hydration energies decrease with increasing ion size (Table 18.4).

Table 18.4 Enthalpy changes of some I A metal transformations (kcal)

|  | Li | Na | K | Rb | Cs |
|---|---|---|---|---|---|
| $M(s) \rightarrow M(g)$ | +37 | +26 | +22 | +20 | +19 |
| $M(g) \rightarrow M^+(g) + e^-$ | +124 | +118 | +100 | +96 | +89 |
| $M^+(g) \rightarrow M^+(aq)$ | −121 | −95 | −76 | −69 | −62 |
| $M(s) \rightarrow M^+(aq) + e^-$ | +40 | +49 | +46 | +47 | +46 |

Although all the I A ions are loosely hydrated in water solution, only salts of lithium and sodium (the two smallest ions) form hydrates in which the water may be considered to be coordinated around the cation, and in these substances the metal ion–water attractions are electrostatic. Hydrates of salts of the other I A metals are known, but the water is not present in these crystals as coordination water.

The trend in the magnitudes of the hydration energies of the ions helps to explain why the order of the electrode potentials is different from that of the ionization potentials (Table 18.3). Of all the alkali metals lithium is the most difficult to oxidize to the 1+ state, and lithium has the highest ionization potential of any member of the group. On the other hand, lithium is the I A metal most readily converted to *hydrated* 1+ ions, and lithium has the lowest standard electrode potential.

The relationship between ionization potential and electrode potential may be clarified by means of a Born–Haber treatment of the half reaction .

$$M(s) \rightarrow M^+(aq) + e^-$$

For the purpose of analysis, the enthalpy change for this half reaction may be considered to be derived from the enthalpy of sublimation (energy absorbed),

$$M(s) \rightarrow M(g)$$

the ionization energy (energy absorbed),

$$M(g) \rightarrow M^+(g) + e^-$$

and the enthalpy of hydration (energy evolved),

$$M^+(g) \rightarrow M^+(aq)$$

The sums of these energy effects actually apply to transformations in which the electrons are left in the gaseous state; however, the values derived, which are listed in Table 18.4, permit some valid comparisons to be made. Each of the factors listed in Table 18.4 varies in a regular and predictable manner. The sums of the energy effects, however, indicate that the oxidation of lithium to produce hydrated lithium ions requires less energy than the oxidation of any other I A metal and that the corresponding oxidation of sodium requires more energy than the oxidation of any other I A metal. This conclusion is supported by the electrode potentials of the alkali metals.

Electrode potentials can only be derived from changes in Gibbs free energy, which include not only changes in enthalpy but also changes in entropy. These entropy effects are not large; however, since we have ignored them and also used $\Delta H$ values correct only to 1 kcal, not much importance should be attached to the order of the values for K, Rb, and Cs recorded in Table 18.4, other than to note that they lie between the values for Li and Na.

All the hydroxides of the alkali metals are strong electrolytes. Alkali metal compounds are generally very soluble in water. However, the hydroxide, carbonate, phosphate, and fluoride of lithium are much less soluble than the corresponding salts of the other I A metals. Sodium ion may be precipitated from solution as sodium zinc uranyl acetate, $NaZn(UO_2)_3(C_2H_3O_2)_9 \cdot 6H_2O$. The perchlorate ion, $ClO_4^-$, the hexachloroplatinate ion, $[PtCl_6]^{2-}$, and the cobaltinitrite ion, $[Co(NO_2)_6]^{3-}$, may be used to precipitate $K^+$, $Rb^+$, and $Cs^+$.

The unipositive ammonium ion has a radius that lies between those of $K^+$ and $Rb^+$, and ammonium salts have solubilities that resemble those of the alkali metal salts. The ionic radius of the thallous ion, $Tl^+$, is similar to that of the rubidium ion. Not only do thallous compounds resemble rubidium compounds in solubility, but TlOH is also a strong electrolyte.

Some of the reactions of the I A metals are summarized in Table 18.5. Whereas the properties of lithium are similar in most respects to those of the other members of group I A, there are some dissimilarities which may be traced to the comparatively small size of Li and $Li^+$. The first members of the groups of the periodic table often deviate slightly in character from the remaining members of their groups. In addition, similarities may be observed between these first members and elements that adjoin them diagonally on the periodic table (diagonal relationships).

Li   Be   B   C

Na  Mg  Al  Si

Thus, lithium resembles magnesium in some of its properties. The magnesium ion has a slightly larger ionic radius (0.65 Å) than the lithium ion (0.60 Å), but the charge on the magnesium ion is 2+.

**Table 18.5** Reaction of the I A metals[a]

| REACTION | REMARKS |
|---|---|
| $2M + X_2 \rightarrow 2MX$ | $X_2$ = all halogens |
| $4Li + O_2 \rightarrow 2Li_2O$ | excess oxygen |
| $2Na + O_2 \rightarrow Na_2O_2$ | |
| $M + O_2 \rightarrow MO_2$ | M = K, Rb, Cs |
| $2M + S \rightarrow M_2S$ | also with Se and Te |
| $6Li + N_2 \rightarrow 2Li_3N$ | Li only |
| $12M + P_4 \rightarrow 4M_3P$ | also with As, Sb |
| $2M + 2C \rightarrow M_2C_2$ | M = Li and Na; other I A metals give non-stoichiometric interstitial compounds |
| $2M + H_2 \rightarrow 2MH$ | |
| $2M + 2H_2O \rightarrow 2MOH + H_2$ | room temperature |
| $2M + 2H^+ \rightarrow 2M^+ + H_2$ | violent reaction |
| $2M + 2NH_3 \rightarrow 2MNH_2 + H_2$ | liquid $NH_3$ presence of catalysts such as Fe; gaseous $NH_3$, heated |

[a] M = any I A metal, except where noted.

Lithium resembles magnesium and differs from its congeners in the following ways. The carbonate, phosphate, and fluoride of lithium are only slightly soluble in water. Lithium forms a normal oxide rather than a peroxide or a superoxide upon burning in oxygen. Lithium ions are more strongly hydrated than those of any other I A element. Lithium reacts directly with nitrogen to give a nitride. Lithium hydroxide may be decomposed to $Li_2O$ and water upon heating, and lithium carbonate decomposes to $Li_2O$ and $CO_2$ upon ignition. Lithium nitrate decomposes upon heating.

$$4LiNO_3(s) \rightarrow 2Li_2O(s) + 4NO_2(g) + O_2(g)$$

The nitrates of the other alkali metals form nitrites upon heating,

$$2MNO_3(s) \rightarrow 2MNO_2(s) + O_2(g)$$

and the hydroxides and carbonates of Na, K, Rb, and Cs are thermally stable.

**18.7 THE GROUP II A METALS**

The group II A metals, called the **alkaline earth metals,** are highly electropositive and constitute the second most reactive group of metals. They are not found free in nature and are commonly produced by the electrolysis of molten chlorides. Radium is a comparatively scarce element and all its isotopes are radioactive.

Because of its larger nuclear charge, each II A metal has a smaller atomic radius than that of the I A metal of its period. Since atoms of the II A metals are smaller and have two valence electrons instead of one, the II A metals have higher melting and boiling points and greater densities than the I A metals (Table 18.6). In addition, the alkaline earth metals are harder than the alkali metals. Beryllium is hard enough to scratch glass and has a tendency toward brittleness; the degree of hardness declines with increasing atomic number. The alkaline earth metals are white metals with a silvery luster. They are good conductors of electricity.

Table 18.6 Some properties of the II A metals

| | BERYLLIUM | MAGNESIUM | CALCIUM | STRONTIUM | BARIUM |
|---|---|---|---|---|---|
| outer electronic configuration | $2s^2$ | $3s^2$ | $4s^2$ | $5s^2$ | $6s^2$ |
| melting point (°C) | 1280 | 651 | 851 | 800 | 850 |
| boiling point (°C) | 1500 | 1107 | 1440 | 1366 | 1537 |
| density (g/cm³) | 1.86 | 1.75 | 1.55 | 2.6 | 3.59 |
| atomic radius (Å) | 0.89 | 1.36 | 1.74 | 1.91 | 1.98 |
| ionic radius, $M^{2+}$ (Å) | 0.31 | 0.65 | 0.99 | 1.13 | 1.35 |
| ionization potential (eV) | | | | | |
| first | 9.3 | 7.6 | 6.1 | 5.7 | 5.2 |
| second | 18.2 | 15.0 | 11.9 | 11.0 | 10.0 |
| third | 153.9 | 80.1 | 51.2 | — | — |
| electrode potential, $\mathscr{E}°$, ($M^{2+}$/M) (V) | −1.85 | −2.36 | −2.87 | −2.89 | −2.91 |

Each of the II A metals has an electronic configuration consisting of a noble gas core plus two $s$ electrons in the outer, valence level. The loss of the two valence electrons produces an ion that is isoelectronic not only with a noble gas but also with the I A metal ion of the same period. However, the alkaline earth metal ions have a larger nuclear charge than the alkali metal ions and are therefore considerably smaller. The beryllium ion is an exceptionally small cation.

In comparison to the I A metal ions, the II A metal ions have considerably higher ratios of ionic charge to ionic radius. There are several important consequences of this fact. The hydration energy of a given alkaline earth ion is about five times that of the alkali metal ion of the same period. Furthermore, the compounds of the smaller members of the group, magnesium and beryllium, have appreciable covalent character because the cations of these metals have a strong polarizing effect on anions. This tendency for covalent bond formation is particularly pronounced for beryllium. All compounds of this metal, even those with the most electronegative elements such as oxygen and fluorine, have significant covalent character. Beryllium has the greatest tendency of the group toward the formation of complex ions; examples are $[Be(H_2O)_4]^{2+}$, $[Be(NH_3)_4]^{2+}$, $BeF_4^{2-}$, and $[Be(OH)_4]^{2-}$. Beryllium hydroxide is the only amphoteric hydroxide formed by a group II A metal.

The covalent nature of beryllium halides is indicated by the poor electrolytic conduction of the anhydrous molten compounds; NaCl is generally added to anhydrous molten $BeCl_2$ in the electrolytic preparation of beryllium. Gaseous $BeCl_2$ molecules are linear and presumably of the *sp* hybrid type. In most solid compounds beryllium is tetrahedrally bonded through $sp^3$ hybrid orbitals. This configuration is observed in the complex ions of beryllium. Beryllium chloride is polymeric; each beryllium atom is joined to four chlorine atoms.

The BeO crystal may be considered to be composed of $BeO_4$ tetrahedra.

The ionization potentials of the II A metals are higher than those of the I A metals because of differences in atomic size and nuclear charge. The hydration energies of the II A cations, however, are also larger than those of the I A cations because of the smaller size and higher charge of the II A cations. Consequently, the electrode potentials of the II A metals are, in general, similar in magnitude to those of the I A metals.

The hydration energy is largest for $Be^{2+}$ and smallest for $Ba^{2+}$, but the electrode potentials fall in the same order as the ionization potentials; Ba is the strongest reducing agent of the group and Be is the weakest. This order of reactivity is illustrated by the reactions of the alkaline earth metals with water. Beryllium fails to react at red heat. Magnesium will react with boiling water or steam. Calcium, strontium, and barium react vigorously with cold water. Some chemical reactions of the group II A metals are summarized in Table 18.7.

Table 18.7 Reactions of the II A metals[a]

| REACTION | REMARKS |
|---|---|
| $M + X_2 \rightarrow MX_2$ | $X_2 =$ all halogens |
| $2M + O_2 \rightarrow 2MO$ | Ba also gives $BaO_2$ |
| $M + S \rightarrow MS$ | also with Se and Te |
| $3M + N_2 \rightarrow M_3N_2$ | at high temperatures |
| $6M + P_4 \rightarrow 2M_3P_2$ | at high temperatures |
| $M + 2C \rightarrow MC_2$ | all except Be which forms $Be_2C$; high temperatures |
| $M + H_2 \rightarrow MH_2$ | M = Ca, Sr, Ba; high temperatures; Mg with $H_2$ under pressure |
| $M + 2H_2O \rightarrow M(OH)_2 + H_2$ | M = Ca, Sr, Ba; room temperature |
| $Mg + H_2O \rightarrow MgO + H_2$ | steam; Be does not react at red heat |
| $M + 2H^+ \rightarrow M^{2+} + H_2$ | |
| $Be + 2OH^- + 2H_2O \rightarrow Be(OH)_4^{2-} + H_2$ | Be only |
| $M + 2NH_3 \rightarrow M(NH_2)_2 + H_2$ | M = Ca, Sr, Ba; liquid ammonia in presence of catalysts |
| $3M + 2NH_3 \rightarrow M_3N_2 + 3H_2$ | gaseous $NH_3$; high temperatures |

[a] M = any II A metal, except where noted.

Almost all the salts of the I A metals are very soluble in water. In contrast, a number of II A metal compounds are not appreciably water soluble; the solubility products of some of these insoluble compounds are listed in Table 18.8.

Table 18.8 Solubility product constants of some II A metal salts

| | $OH^-$ | $SO_4^{2-}$ | $CO_3^{2-}$ | $C_2O_4^{2-}$ | $F^-$ | $CrO_4^{2-}$ |
|---|---|---|---|---|---|---|
| $Be^{2+}$ | $1.6 \times 10^{-26}$ | — | — | — | — | — |
| $Mg^{2+}$ | $8.9 \times 10^{-12}$ | — | $10^{-5}$ | $8.6 \times 10^{-5}$ | $8 \times 10^{-8}$ | — |
| $Ca^{2+}$ | $1.3 \times 10^{-6}$ | $2.4 \times 10^{-5}$ | $4.7 \times 10^{-9}$ | $1.3 \times 10^{-9}$ | $1.7 \times 10^{-10}$ | $7.1 \times 10^{-4}$ |
| $Sr^{2+}$ | $3.2 \times 10^{-4}$ | $7.6 \times 10^{-7}$ | $7 \times 10^{-10}$ | $5.6 \times 10^{-8}$ | $7.9 \times 10^{-10}$ | $3.6 \times 10^{-5}$ |
| $Ba^{2+}$ | $5.0 \times 10^{-3}$ | $1.5 \times 10^{-9}$ | $1.6 \times 10^{-9}$ | $1.5 \times 10^{-8}$ | $2.4 \times 10^{-5}$ | $8.5 \times 10^{-11}$ |

The solubility of a salt depends upon the lattice energy of the salt (energy absorbed in the solution process)

$$MA(s) \rightarrow M^{2+}(g) + A^{2-}(g)$$

and the hydration energies of the ions (energy evolved)

$$M^{2+}(g) \rightarrow M^{2+}(aq)$$

$$A^{2-}(g) \rightarrow A^{2-}(aq)$$

When the solubilities of salts containing the same anion are compared, the hydration energy of the anion may be neglected and solubility differences may be attributed to the combination of the other two factors.

The solubility of the alkaline earth sulfates decreases with increasing cation size; $BeSO_4$ is very soluble, and $BaSO_4$ is very insoluble. The lattice energies of the sulfates do not change greatly in the sequence from $BeSO_4$ to $BaSO_4$, presumably because the anion is so much larger than any II A cation. The trend in the solubility of the sulfates therefore parallels the trend in hydration energies of the ions. The hydration of the small $Be^{2+}$ ion is by far the most exothermic of any ion of the group, and $BeSO_4$ is by far the most soluble sulfate formed by any ion of the group.

The trend in the solubility of hydroxides is the reverse of that of the sulfates; $Be(OH)_2$ is the least soluble alkaline earth hydroxide, and the solubility increases down the group. For hydroxides the lattice energy is dependent upon cation size. The strength of the crystal forces decreases with increasing cation size; $Be(OH)_2$ has the largest lattice energy of any alkaline earth hydroxide. Apparently the trend in lattice energy (energy required) overshadows the trend in hydration energy (energy released).

Solubility data cannot always be so simply interpreted. In the preceding solubility considerations we have ignored entropy effects. In addition, even though the lattice energies and hydration energies of a group of salts may vary regularly, the sum of the two energy effects and hence the solubilities of the salts may vary in an irregular manner in the same way that the standard electrode potentials of the I A metals vary.

The lattice energies of the oxides of the II A metals decrease regularly from BeO to BaO, and the effect of this trend is seen in the series of reactions of the alkaline earth oxides with water. Beryllium oxide is insoluble in and unreactive toward water. Magnesium oxide reacts very slowly with water; MgO that has been ignited at high temperatures, however, is practically inert. The oxides of calcium, strontium, and barium readily react with water to form hydroxides

$$MO + H_2O \rightarrow M(OH)_2$$

The carbonates of the alkaline earth metals decompose upon heating

$$MCO_3(s) \rightarrow MO(s) + CO_2(g)$$

The thermal stability of the carbonates varies directly with the size of the cation. Beryllium carbonate is very unstable and can be prepared only in an atmosphere of carbon dioxide presumably because of the enhanced stability of BeO over $BeCO_3$.

The tendency of the II A metal ions toward hydration causes a number of the compounds of these ions to hydrate with ease; $Mg(ClO_4)_2$, $CaCl_2$, $CaSO_4$, and $Ba(ClO_4)_2$ are used as desiccants.

The beryllium ion hydrolyzes in solution

$$[Be(H_2O)_4]^{2+} + H_2O \rightleftharpoons [Be(H_2O)_3(OH)]^+ + H_3O^+$$

but the other cations do not. The hydroxides of beryllium and magnesium may be regarded as insoluble (Table 18.8). Even though the hydroxides of calcium, strontium, and barium are limitedly soluble in water, these compounds are completely dissociated in aqueous solution.

The sulfides of the group are water soluble, and solutions of these compounds are alkaline due to the hydrolysis of the sulfide ion. In quantitative determinations barium ion is usually precipitated as barium sulfate; strontium ion, as strontium sulfate or strontium oxalate; and calcium ion, as calcium oxalate. Magnesium ion is precipitated as $Mg(NH_4)PO_4 \cdot 6H_2O$ by the $HPO_4^{2-}$ ion in the presence of ammonia. Beryllium may be determined as the hydroxide.

Because of the extremely small size of Be and $Be^{2+}$, beryllium is even more exceptional with regard to the other members of group II A than lithium is with regard to the other members of group I A. The diagonal relationship between beryllium and aluminum is particularly striking. Although $Al^{3+}$ is larger than $Be^{2+}$, the two ions have similar electric fields because of the higher charge of $Al^{3+}$.

Both elements have a strong tendency toward the formation of covalent compounds. The halides are largely covalent in nature, are soluble in organic solvents, and are strong Lewis acids. Both aluminum hydroxide and beryllium hydroxide are amphoteric, and both metals dissolve in solutions of hydroxides to give hydrogen. The standard electrode potential of beryllium is similar in magnitude to that of aluminum but much smaller than those of the other II A metals.

The carbides of beryllium and aluminum yield methane, $CH_4$, on hydrolysis.

$$Be_2C(s) + 4H_2O \rightarrow 2Be(OH)_2(s) + CH_4(g)$$

$$Al_4C_3(s) + 12H_2O \rightarrow 4Al(OH)_3(s) + 3CH_4(g)$$

The hydrolysis of the other II A metal carbides yields acetylene, $C_2H_2$.

Aluminum and beryllium (as well as magnesium) have thin protective oxide coatings, which make them resistant to attack by dilute nitric acid. Beryllium oxide and aluminum oxide are hard, extremely high melting, insoluble solids. Beryllium and aluminum form the stable fluoro complex anions $BeF_4^{2-}$ and $AlF_6^{3-}$. The other II A metals do not form fluoro complexes that are stable in solution. Normal beryllium carbonate is unstable, and aluminum carbonate cannot be made at all.

## 18.8 THE TRANSITION METALS

The transition metals are found in groups III B to II B in the periodic table. Some chemists do not include group II B (the zinc group) in this classification, and some exclude group I B (the copper group) as well.

However, there are advantages in classifying the copper- and zinc-group elements as transition elements, and we shall follow this practice.

In general, the transition metals have high melting and boiling points, high heats of fusion, and high heats of vaporization. The group II B elements (zinc, cadmium, and mercury) are exceptions to this generalization. Mercury is a liquid under ordinary conditions, and all of the II B elements are comparatively low melting and relatively easy to volatilize (distillation is a method of refining the metals). Most transition metals are good conductors of heat and electricity; the I B elements are outstanding in these respects (Table 18.1).

The significant feature of the electronic configurations of the transition elements (Table 18.9) is the gradual build up of the $d$ subshell of the shell adjacent to the outer shell. Included in the sixth and seventh periods are the lanthanides and actinides, respectively. These are the inner-transition elements in which inner $f$ orbitals are being filled (Section 18.9).

**Table 18.9** Postulated electronic configurations of the valence subshells of the transition metals

| | | | | | |
|----|----------------|----|----------------|----|----------------|
| Sc | $3d^1\,4s^2$   | Y  | $4d^1\,5s^2$   | La | $5d^1\,6s^2$   |
| Ti | $3d^2\,4s^2$   | Zr | $4d^2\,5s^2$   | Hf | $5d^2\,6s^2$   |
| V  | $3d^3\,4s^2$   | Nb | $4d^4\,5s^1$   | Ta | $5d^3\,6s^2$   |
| Cr | $3d^5\,4s^1$   | Mo | $4d^5\,5s^1$   | W  | $5d^4\,6s^2$   |
| Mn | $3d^5\,4s^2$   | Tc | $4d^6\,5s^1$   | Re | $5d^5\,6s^2$   |
| Fe | $3d^6\,4s^2$   | Ru | $4d^7\,5s^1$   | Os | $5d^6\,6s^2$   |
| Co | $3d^7\,4s^2$   | Rh | $4d^8\,5s^1$   | Ir | $5d^7\,6s^2$   |
| Ni | $3d^8\,4s^2$   | Pd | $4d^{10}$      | Pt | $5d^9\,6s^1$   |
| Cu | $3d^{10}4s^1$  | Ag | $4d^{10}5s^1$  | Au | $5d^{10}6s^1$  |
| Zn | $3d^{10}4s^2$  | Cd | $4d^{10}5s^2$  | Hg | $5d^{10}6s^2$  |

The transition elements exhibit a wide variation in chemical properties. Many compounds of the transition elements are colored and paramagnetic because of unpaired electrons. Since the inner $d$ orbitals are energetically close to the $s$ orbital of the outer level, both $ns$ and $(n-1)d$ electrons are involved in compound formation. With the exceptions of zinc, cadmium, and the elements of group III B, all transition elements exhibit more than one oxidation state in compound formation. The largest number of oxidation states, as well as the maximum oxidation number, is observed for the fifth transition element of period 4 (Mn) and the sixth elements of periods 5 and 6 (Ru and Os, respectively). For these elements and the ones preceding them in each transition series, the maximum oxidation number corresponds to the total number of $ns$ and $(n-1)d$ electrons (Table 18.10). The higher oxidation numbers of a given element are most frequently seen in compounds containing the more electronegative elements: fluorine, oxygen, and chlorine.

Table 18.10 Oxidation states of the transition elements (less common or unstable states in parentheses)

| Sc | Ti | V | Cr | Mn | Fe | Co | Ni | Cu | Zn |
|----|----|---|----|----|----|----|----|----|----|
|    |    | (1+) | (1+) | (1+) |    | (1+) | (1+) | 1+ | (1+) |
|    | (2+) | (2+) | 2+ | 2+ | 2+ | 2+ | 2+ | 2+ | 2+ |
| 3+ | 3+ | 3+ | 3+ | (3+) | 3+ | 3+ | (3+) | (3+) |  |
|    | 4+ | 4+ | (4+) | 4+ | (4+) | (4+) | (4+) |  |  |
|    |    | 5+ | (5+) | (5+) | (5+) |    |    |  |  |
|    |    |    | 6+ | (6+) | (6+) |    |    |  |  |
|    |    |    |    | 7+ |    |    |    |  |  |

| Y | Zr | Nb | Mo | Tc | Ru | Rh | Pd | Ag | Cd |
|---|----|----|----|----|----|----|----|----|----|
|    | (1+) | (1+) | (1+) |    | (1+) | (1+) |    | 1+ | (1+) |
|    | (2+) | (2+) | (2+) | (2+) | 2+ | (2+) | 2+ | (2+) | 2+ |
| 3+ | (3+) | (3+) | 3+ | (3+) | 3+ | 3+ | (3+) | (3+) |  |
|    | 4+ | (4+) | 4+ | 4+ | 4+ | 4+ | 4+ |  |  |
|    |    | 5+ | 5+ | 5+ | (5+) | (5+) |    |  |  |
|    |    |    | 6+ | (6+) | (6+) | (6+) |    |  |  |
|    |    |    |    | 7+ | (7+) |    |    |  |  |
|    |    |    |    |    | (8+) |    |    |  |  |

| La | Hf | Ta | W | Re | Os | Ir | Pt | Au | Hg |
|----|----|----|---|----|----|----|----|----|----|
|    | (1+) |    |    | (1+) | (1+) | (1+) |    | 1+ | 1+ |
|    | (2+) | (2+) | (2+) | (2+) | (2+) | (2+) | 2+ |    | 2+ |
| 3+ | (3+) | (3+) | (3+) | 3+ | (3+) | 3+ |    | 3+ |  |
|    | 4+ | (4+) | 4+ | 4+ | 4+ | 4+ | 4+ |  |  |
|    |    | 5+ | 5+ | 5+ | (5+) | (5+) | (5+) |  |  |
|    |    |    | 6+ | (6+) | 6+ | (6+) | (6+) |  |  |
|    |    |    |    | 7+ |    |    |    |  |  |
|    |    |    |    |    | 8+ |    |    |  |  |

After the maximum oxidation state of a period is reached in each transition series, there is a decrease in the highest oxidation number observed for each subsequent element; these oxidation states are, in general, difficult to obtain and unstable. Less common and unstable oxidation states are enclosed in parentheses in Table 18.10.

Compounds of $Cu^+$, $Ag^+$, $Au^+$, and $Hg_2^{2+}$ are the only compounds in which the 1+ oxidation state is important. The mercury(I) ion is unique; evidence for its dimeric structure includes the following. Mercurous salts are diamagnetic in agreement with the formula $Hg_2^{2+}$; the ion $Hg^+$ would have one unpaired electron. X-ray studies of mercurous salts indicate discrete $Hg_2^{2+}$ ions. The interpretations of certain equilibria involving the mercurous ion are in agreement with experimentally derived values for the equilibrium constants only if the formula $Hg_2^{2+}$ is used.

Within each group the higher oxidation states become more stable and more important with increasing atomic number; a corresponding decline in the stability and importance of the lower oxidation states is also observed. The gain in stability of the higher states with increasing atomic number may be attributed to the increasing atomic size, which makes the $d$ electrons more available for compound formation.

Thus, the 2+ ions of many of the fourth period transition elements are stable and characteristic of the elements, whereas the 2+ states of the heavier elements are not particularly common or stable. The most important oxidation states of iron are 2+ and 3+; for osmium, the heaviest member of the same group, the most important states are 4+, 6+, and 8+.

The maximum oxidation number of manganese (which is the first member of its group) is attained in the permanganate ion, $MnO_4^-$. This ion is a strong oxidizing agent in acidic solution.

$$5e^- + 8H^+ + MnO_4^- \rightarrow Mn^{2+} + 4H_2O \qquad \mathscr{E}° = +1.51 \text{ V}$$

In contrast, the analogous perrhenate ion, $ReO_4^-$, formed by the heaviest member of the group, does not react in this way under comparable conditions.

The transition elements of group VI B form series of oxy anions in which the central metal ion is in an oxidation state of 6+; the simplest members of this series are the chromate ion, $CrO_4^{2-}$, the molybdate ion, $MoO_4^{2-}$, and the tungstate ion, $WO_4^{2-}$. Only a few polynuclear complexes of chromium have been reported; the most important of these is the dichromate ion, $Cr_2O_7^{2-}$, which results when chromate solutions are acidified.

$$2H^+ + CrO_4^{2-} \rightarrow Cr_2O_7^{2-} + H_2O$$

The dichromate ion is easily reduced in acid and consequently is a strong oxidizing agent.

$$6e^- + 14H^+ + Cr_2O_7^{2-} \rightarrow 2Cr^{3+} + 7H_2O \qquad \mathscr{E}° = +1.33 \text{ V}$$

In contrast, molybdenum and tungsten form very extensive series of polynuclear molybdates and tungstates, and these compounds, as well as the corresponding acids, are very stable and are without significant oxidizing power. The formation of such polynuclear oxy anions is not a common phenomenon in transition metal chemistry.

With increasing oxidation number the oxides of a given element become less basic and more acidic. Thus, $CrO$ is an exclusively basic oxide and dissolves in acids to form $Cr^{2+}$ salts; $Cr_2O_3$ is amphoteric and forms $Cr^{3+}$ salts with acids and the chromite ion, $Cr(OH)_4^-$, in alkaline solution; $CrO_3$ is entirely acidic in character and gives rise to chromates, $CrO_4^{2-}$, and dichromates, $Cr_2O_7^{2-}$.

The atomic and ionic radii of the transition elements are, in general, smaller than those of the representative elements of the same period (Figure 3.1) since, for the transition elements, electrons are being added to an inner $d$ sublevel where the effect of the nuclear charge is greater than it is in the outer shell. This relatively small size causes the transition element ions to have comparatively high densities; this, plus the availability of $d$ orbitals for bonding, accounts for the pronounced tendency of most transition elements to form numerous stable complex ions (Chapter 19).

There are horizontal (or period) similarities as well as vertical (or group) similarities between the transition metals. Since the elements Cr through Cu, which are a part of the first transition series, have very similar atomic radii and ionic radii for ions of the same charge (Table 18.11), horizontal similarities occur (such as the formation of compounds of formula MO, MS, $MCl_2$, and so on). Among the heavier elements, horizontal similarities are also evident (e.g., the formation of $MO_2$ compounds). Similarities of this type are particularly striking among the transition triads of group VIII (Fe, Co, and Ni; Ru, Rh, and Pd; and Os, Ir, and Pt). The elements of the two heavier triads are similar both physically and chemically.

**Table 18.11** Atomic and ionic radii of the transition elements of the fourth period

|  | Sc | Ti | V | Cr | Mn | Fe | Co | Ni | Cu | Zn |
|---|---|---|---|---|---|---|---|---|---|---|
| atomic radii (Å) | 1.44 | 1.32 | 1.22 | 1.17 | 1.17 | 1.17 | 1.16 | 1.15 | 1.17 | 1.25 |
| ionic radii (Å) | | | | | | | | | | |
| $M^{2+}$ | — | 0.90 | 0.88 | 0.84 | 0.80 | 0.76 | 0.74 | 0.72 | 0.72 | 0.74 |
| $M^{3+}$ | 0.81 | 0.76 | 0.74 | 0.69 | 0.66 | 0.64 | 0.63 | — | — | — |

The lanthanides occur at the beginning of the transition series of the sixth period. For the lanthanides, electrons are being added to the $4f$ sublevel without significant change in the $5s$, $5p$, $5d$, and $6s$ sublevels. The nuclear charge increases as this $4f$ sublevel (which is deep within the atom) is being filled, and there is a consequent decrease in the atomic and ionic radii of the lanthanides, which is known as the **lanthanide contraction.**

The lanthanide contraction has an important effect on the properties of the transition elements that follow the lanthanides in the sixth period. Because of the lanthanide contraction, the atomic radii of these elements are not much different from those of the corresponding elements of the fifth period (Figure 18.7). Consequently, the second and third members of each group resemble each other much more than they resemble the first member of the group; there is no regular increase in size with increasing atomic number such as is observed for the A family elements.

Even though 32 elements intervene, hafnium ($Z = 72$) is notably similar to zirconium ($Z = 40$). Both elements have approximately the same atomic and ionic radii, and the chemical properties of the two elements

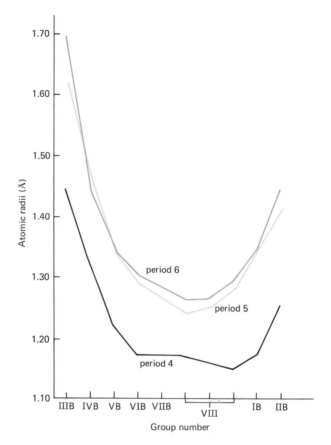

**Figure 18.7** Atomic radii of the transition elements.

are very similar; the separation of the two elements, which occur together in nature, is extremely difficult.

Some of the standard electrode potentials of the transition elements are listed in Table 18.12. For the most part, the potentials listed for the heavier elements, since they are for lower oxidation states, are not particularly important; however, they do serve to illustrate several trends.

Many of the metals react with dilute acids, as well as with water or steam, to liberate hydrogen. Some of the metals, however, are poor reducing agents—notably, mercury, the group I B elements (copper, silver and gold), and the transition triads of the fifth and sixth periods (ruthenium, rhodium, palladium; osmium, iridium, and platinum). Frequently these six elements of the transition triads, silver, and gold are referred to as noble metals.

Thus, the elements of low reactivity appear to be concentrated toward the end of the transition series, particularly those of the fifth and sixth periods. In general, there is a decrease in electropositive character (or strength of the metals as reducing agents) across a period and down a

Table 18.12 Standard electrode potentials of the transition elements (volts)

|  | Sc | Ti | V | Cr | Mn | Fe | Co | Ni | Cu | Zn |
|---|---|---|---|---|---|---|---|---|---|---|
| $M^+/M$ | — | — | — | — | — | — | — | — | +0.52 | — |
| $M^{2+}/M$ | — | -1.63 | -1.19 | -0.91 | -1.18 | -0.44 | -0.28 | -0.25 | +0.34 | -0.76 |
| $M^{3+}/M$ | -2.08 | -1.21 | -0.88 | -0.74 | -0.28 | -0.04 | +0.42 | — | — | — |

|  | Y | Zr | Nb | Mo | Tc | Ru | Rh | Pd | Ag | Cd |
|---|---|---|---|---|---|---|---|---|---|---|
| $M^+/M$ | — | — | — | — | — | — | +0.6 | — | +0.80 | — |
| $M^{2+}/M$ | — | — | — | — | +0.4 | +0.45 | +0.6 | +0.99 | — | -0.40 |
| $M^{3+}/M$ | -2.37 | — | -1.1 | -0.2 | — | — | +0.8 | — | — | — |

|  | La | Hf | Ta | W | Re | Os | Ir | Pt | Au | Hg |
|---|---|---|---|---|---|---|---|---|---|---|
| $M^+/M$ | — | — | — | — | — | — | — | — | +1.69 | +0.79 |
| $M^{2+}/M$ | — | — | — | — | — | +0.85 | — | +1.2 | — | +0.85 |
| $M^{3+}/M$ | -2.52 | — | — | -0.11 | +0.3 | — | +1.15 | — | +1.50 | — |

group. The decreasing size of the atoms across a period causes the electrons to be held more tightly. For *transition* elements the increase in nuclear charge on the atoms down a group is not accompanied by a large increase in atomic size, and the electrons are held more tightly. In the case of the transition elements of the sixth period that follow the lanthanides, the unusually small sizes of the atoms cause the electrons to be very tightly held. With the exceptions of La and Hf, the least electropositive member of each group is the sixth period element.

Table 18.13 Products of some reactions of the transition elements of the fourth period

|  | Sc | Ti | V | Cr | Mn |
|---|---|---|---|---|---|
| $O_2$ | $Sc_2O_3$ | $TiO_2$ | $V_2O_5$, $VO_2$ | $Cr_2O_3$ | $Mn_3O_4$ |
| $X_2$ | $ScX_3$ | $TiX_4$ | $VF_5$, $VCl_4$, $VBr_3$, $VI_3$ | $CrX_3$ (except $CrI_2$) | $MnX_2$ |
| S | $Sc_2S_3$ | $TiS_2$ | $V_2S_5$, $VS_2$ | CrS | MnS |
| $N_2$ | ScN | TiN | VN | CrN | $Mn_3N_2$ |
| HCl | $Sc^{3+}$ + $H_2$ | $Ti^{3+}$ + $H_2$ | — | $Cr^{2+}$ + $H_2$ | $Mn^{2+}$ + $H_2$ |
| $H_2O$ | $Sc(OH)_3$ + $H_2$ | $TiO_2$ + $H_2^a$ | — | $Cr_2O_3$ + $H_2^a$ | $Mn(OH)_2$ + $H_2$ |
| NaOH | — | — | — | $Cr(OH)_6^{3-}$ + $H_2$ | — |

|  | Fe | Co | Ni | Cu | Zn |
|---|---|---|---|---|---|
| $O_2$ | $Fe_3O_4$, $Fe_2O_3$ | $Co_3O_4$ | NiO | $Cu_2O$, CuO | ZnO |
| $X_2$ | $FeX_3$ (except $FeI_2$) | $CoX_2$ | $NiX_2$ | $CuX_2$ (except CuI) | $ZnX_2$ |
| S | FeS | CoS | NiS | $Cu_2S$ | ZnS |
| $N_2$ | — | — | — | — | — |
| HCl | $Fe^{2+}$ + $H_2$ | $Co^{2+}$ + $H_2$ | $Ni^{2+}$ + $H_2$ | — | $Zn^{2+}$ + $H_2$ |
| $H_2O$ | $Fe_3O_4$ + $H_2^a$ | CoO + $H_2^a$ | NiO + $H_2^a$ | — | ZnO + $H_2^a$ |
| NaOH | — | — | — | — | $[Zn(OH)_4]^{2-}$ + $H_2$ |

[a] Steam.

Although a given electrode potential may indicate a thermodynamic tendency for a reaction to occur, at times the rates at which some transition elements react are extremely slow. Thus, chromium is a moderately strong reducing agent and reacts with nonoxidizing acids, such as HCl, to liberate hydrogen. Chromium does not, however, react with nitric acid, a strong oxidizing agent. It is said that the metal is made passive by the nitric acid. The phenomenon of passivity, which is displayed by many transition elements, is not well understood. In some instances it may be that the metal is protected by a thin, transparent, impervious oxide coating.

The products of some of the reactions of the transition metals of the fourth period, which are the most common of the transition elements, are given in Table 18.13.

**18.9 THE LANTHANIDES**

Postulated electronic configurations for the lanthanides are given in Table 18.14. The differentiating electrons of the lanthanides, or inner-transition elements, are lodged in the third shell from the outside, in $4f$ orbitals. This factor accounts for some of the unusual aspects of the chemistry of these elements. The atomic and ionic radii of the elements show a regular decrease with increasing atomic number (the lanthanide contraction).

The outstanding feature of the chemistry of the lanthanides is their similarity. The differentiating $4f$ electrons lie deep within the lanthanide atoms; thus, these $f$ electrons are shielded from the surroundings of the atom and do not cause strong variations in properties. In contrast, the

Table 18.14 Some properties of the lanthanides

| | $Z$ | POSTULATED ELECTRONIC CONFIGURATIONS OF VALENCE SUBSHELLS | OXIDATION STATES | ATOMIC RADIUS, Å | IONIC RADIUS, $M^{3+}$, Å | $\mathscr{E}°$ $3e^- + M^{3+} \rightarrow M$ (volts) |
|---|---|---|---|---|---|---|
| La | 57 | $5d^1\ 6s^2$ | 3+ | 1.69 | 1.06 | −2.52 |
| Ce | 58 | $4f^2\ 6s^2$ | 3+, 4+ | 1.65 | 1.03 | −2.48 |
| Pr | 59 | $4f^3\ 6s^2$ | 3+, 4+ | 1.65 | 1.01 | −2.46 |
| Nd | 60 | $4f^4\ 6s^2$ | 2+, 3+, 4+ | 1.64 | 1.00 | −2.43 |
| Pm | 61 | $4f^5\ 6s^2$ | 3+ | − | 0.98 | −2.42 |
| Sm | 62 | $4f^6\ 6s^2$ | 2+, 3+ | 1.66 | 0.96 | −2.41 |
| Eu | 63 | $4f^7\ 6s^2$ | 2+, 3+ | 1.85 | 0.95 | −2.41 |
| Gd | 64 | $4f^7\ 5d^1 6s^2$ | 3+ | 1.61 | 0.94 | −2.40 |
| Tb | 65 | $4f^9\ 6s^2$ | 3+, 4+ | 1.59 | 0.92 | −2.39 |
| Dy | 66 | $4f^{10}\ 6s^2$ | 3+, 4+ | 1.59 | 0.91 | −2.35 |
| Ho | 67 | $4f^{11}\ 6s^2$ | 3+ | 1.58 | 0.89 | −2.32 |
| Er | 68 | $4f^{12}\ 6s^2$ | 3+ | 1.57 | 0.88 | −2.30 |
| Tm | 69 | $4f^{13}\ 6s^2$ | 2+, 3+ | 1.56 | 0.87 | −2.28 |
| Yb | 70 | $4f^{14}\ 6s^2$ | 2+, 3+ | 1.70 | 0.86 | −2.27 |
| Lu | 71 | $4f^{14}\ 5d^1 6s^2$ | 3+ | 1.56 | 0.85 | −2.26 |

properties of the transition elements vary widely. The $d$ electrons project from transition element atoms and ions and interact with the surroundings (Chapter 19). Hence, the number and arrangement of the $d$ electrons of the transition elements is significant, whereas the number and arrangement of the $f$ electrons of the inner-transition elements cause only minor variations in properties.

All lanthanides form ions in the characteristic group III B oxidation state, 3+ (Table 18.14). For some of the elements other oxidation states are known, but these are less stable. The 3+ ions are formed through the loss of the two $6s$ electrons and one $4f$ electron (or one $5d$ electron, if available).

There is some evidence that $f^0$, $f^7$ (half-filled), and $f^{14}$ (filled) ions have stable configurations. Thus, $La^{3+}$, ($f^0$), $Gd^{3+}$ ($f^7$), and $Lu^{3+}$ ($f^{14}$) are the only ions that these elements form. The most stable 2+ and 4+ ions are $Eu^{2+}$ ($f^7$), $Yb^{2+}$ ($f^{14}$), $Ce^{4+}$ ($f^0$), and $Tb^{4+}$ ($f^7$). The ceric ion, $Ce^{4+}$, is a good oxidizing agent.

The elements occur together in nature because of their great chemical similarity. Promethium, which is radioactive and probably occurs only in trace amounts, is an exception. The elements are extremely difficult to separate; repeated fractional crystallization and ion-exchange techniques have been employed to effect separations.

All the lanthanides are silvery white, very reactive metals. The electrode potentials for the reduction of the 3+ ions are strikingly similar (Table 18.14). Some reactions of the lanthanides are given in Table 18.15; a number of the metals are also known to react with carbon to produce saltlike carbides and with hydrogen to give saltlike hydrides. The $M_2O_3$ oxides react with water to form insoluble hydroxides, $M(OH)_3$, that are not amphoteric. Insoluble carbonates, $M_2(CO_3)_3$, may be produced by the reactions of carbon dioxide with the oxides or hydroxides. The halides (with the exception of fluorides), nitrates, chlorates, acetates, and sulfates of the 3+ ions are water soluble; the phosphates, fluorides, oxalates, hydroxides, and carbonates are insoluble. The precipitation of the elements as oxalates serves as the basis of an analytical determination. In general, the compounds are paramagnetic and highly colored.

**Table 18.15** Reactions of the lanthanides[a]

| REACTION | REMARKS |
|---|---|
| $2M + 3X_2 \rightarrow 2MX_3$ | $X_2$ = all halogens; Ce gives $CeF_4$ with $F_2$ |
| $4M + 3O_2 \rightarrow 2M_2O_3$ | Ce gives $CeO_2$ |
| $2M + 3S \rightarrow M_2S_3$ | not Eu; reactions with Se similar |
| $2M + N_2 \rightarrow 2MN$ | high temperatures; reactions with $P_4$, As, Sb, and Bi similar |
| $2M + 6H^+ \rightarrow 2M^{3+} + 3H_2$ | |
| $2M + 6H_2O \rightarrow 2M(OH)_3 + 3H_2$ | slow with cold water |

[a] M = any lanthanide, except where noted.

**18.10** THE METALS OF GROUP III A

The characteristics of the group III A elements have been discussed in Section 12.15, and some of the properties of the metals are listed in Table 12.9.

Aluminum, the most abundant metal of the earth's crust (approximately 8%), is obtained by the electrolysis of molten $Al_2O_3$ (the Hall process, Section 18.4). Gallium, indium, and thallium are widely distributed in nature but occur only in trace amounts; they may be prepared by the electrolysis of aqueous solutions of salts of the metals. They are soft, white metals with relatively low melting points (Figure 18.2). Gallium has an unusually low melting point (30°C). Since its boiling point is not abnormally low (2070°C), gallium has an exceptional liquid range and has found use as a thermometer fluid.

The metals are fairly reactive (Table 18.16). Aluminum, gallium, and indium (but not thallium) have protective oxide coatings and are passive toward nitric acid. However, all the metals react with nonoxidizing acids to liberate hydrogen. As expected from their $ns^2 np^1$ electronic configurations, the most important oxidation state is 3+. Most of the compounds of the metals in the 3+ oxidation state are covalent. In water, however, the $M^{3+}$ ions are stabilized through hydration (Section 12.15), and the heats of hydration are high.

Table 18.16 Reactions of aluminum, gallium, indium, and thallium[a]

| REACTION | REMARKS |
|----------|---------|
| $2M + 3X_2 \rightarrow 2MX_3$ | $X_2$ = all halogens; Tl also gives TlX; no iodide of $Tl^{3+}$ |
| $4M + 3O_2 \rightarrow 2M_2O_3$ | high temperatures; Tl also gives $Tl_2O$ |
| $2M + 3S \rightarrow M_2S_3$ | high temperatures; Tl also gives $Tl_2S$; also with Se and Te |
| $2Al + N_2 \rightarrow 2AlN$ | Al only; GaN and InN may be prepared indirectly |
| $2M + 6H^+ \rightarrow 2M^{3+} + 3H_2$ | M = Al, Ga, and In; Tl gives $Tl^+$ |
| $2M + 2OH^- + 6H_2O \rightarrow 2M(OH)_4^- + 3H_2$ | M = Al and Ga |

[a] M = Al, Ga, In, and Tl, except where noted.

The sulfates, nitrates, and halides are water soluble, but the $M^{3+}$ ions hydrolyze readily.

$$M(H_2O)_6^{3+} + H_2O \rightarrow M(H_2O)_5(OH)^{2+} + H_3O^+$$

For this reason, salts of weak acids (such as acetates, carbonates, sulfides, and cyanides) do not exist in water solution. Such salts are completely hydrolyzed.

$$M(H_2O)_6^{3+} + 3C_2H_3O_2^- \rightarrow M(OH)_3(s) + 3H_2O + 3HC_2H_3O_2$$

In addition, complexes with ammonia do not exist in water.

$$M(H_2O)_6^{3+} + NH_3 \rightarrow M(H_2O)_5(OH)^{2+} + NH_4^+$$

The hydroxides, $M(OH)_3$, are insoluble in water. Aluminum hydroxide and gallium hydroxide are amphoteric.

$$Al(OH)_3 + OH^- \rightarrow Al(OH)_4^-$$

$$Ga(OH)_3 + OH^- \rightarrow Ga(OH)_4^-$$

Aluminum sulfate forms an important series of double salts called alums, $MAl(SO_4)_2 \cdot 12H_2O$ (or $M_2SO_4 \cdot Al_2(SO_4)_3 \cdot 24H_2O$), where M may be almost any univalent cation ($Na^+$, $K^+$, $Rb^+$, $Cs^+$, $NH_4^+$, $Ag^+$, and $Tl^+$) except $Li^+$ (which is too small). In addition to $Al^{3+}$, other $M^{3+}$ species form series of such double sulfates: $Fe^{3+}$, $Cr^{3+}$, $Mn^{3+}$, $Ti^{3+}$, $Co^{3+}$, $Ga^{3+}$, $In^{3+}$, $Rh^{3+}$, and $Ir^{3+}$.

Hydrides analogous to those of boron are not formed by aluminum, gallium, indium, and thallium. However, $MH_4^-$ ions, which are analogous to the borohydride ion, are formed.

Among the heavier members of the group, compounds with the metal in a 1+ oxidation state are known. The $np^1$ electron is more readily removed than the $ns^2$ electrons, which is sometimes called an inert pair. This effect is also seen for the $M^{2+}$ ions of group IV A. The low reactivity of mercury has been ascribed to the $6s^2$ inert pair of its valence level.

Unlike the transition elements, however, the 1+ state of the III A elements (their lowest state) is most stable among the heavier members of the group. No 1+ compounds of aluminum are known. Gallium and indium form a few compounds (such as $Ga_2S$, $Ga_2O$ at high temperatures, InCl, InBr, and $In_2O$). The $Ga^+$ and $In^+$ ions are not stable in water solution.

For thallium, however, the 1+ state is important and stable. In fact, $Tl^+$ is more stable in water than $Tl^{3+}$ (which is a strong oxidizing agent).

$$2e^- + Tl^{3+} \rightleftharpoons Tl^+ \qquad \mathscr{E}° = +1.25 \text{ V}$$

The oxide $Tl_2O$ dissolves in water to form the soluble hydroxide TlOH. Whereas the $Tl^{3+}$ ion is extensively hydrolyzed in aqueous solution, the $Tl^+$ ion is not. A wide variety of $Tl^+$ compounds are known. The sulfate, nitrate, acetate, and fluoride are water soluble; the chloride, bromide, iodide, sulfide, and chromate are insoluble.

**18.11 THE METALS OF GROUP IV A**

Of the elements of group IV A, germanium, tin, and lead are classified as metals. The characteristics of the group as a whole are discussed in Section 12.9, and some of the properties of the metals are listed in Table 12.6. The metals are not abundant in nature; germanium is a rare element.

Germanium is a semiconductor and is used in the manufacture of transistors. It is a hard, brittle, white metal and has the highest melting point of the metals of the group. Tin and lead have relatively low melting points and low tensile strengths. Lead is especially soft and malleable.

The metals are fairly reactive (Table 18.17). Lead frequently appears less reactive than the electrode potential

$$2e^- + Pb^{2+} \rightleftharpoons Pb \qquad \mathscr{E}° = -0.13 \text{ V}$$

would indicate because of the formation of surface coatings. Thus, the reaction of lead with sulfuric acid or hydrochloric acid is impeded by the formation of insoluble lead sulfate or lead chloride on the surface of the metal.

**Table 18.17** Reactions of germanium, tin, and lead[a]

| REACTION | REMARKS |
|---|---|
| $M + 2X_2 \rightarrow MX_4$ | $X_2$ = any halogen; M = Ge and Sn; Pb yields $PbX_2$ |
| $M + O_2 \rightarrow MO_2$ | M = Ge and Sn; high temperatures; Pb yields PbO or $Pb_3O_4$ |
| $M + 2S \rightarrow MS_2$ | M = Ge and Sn; high temperatures; Pb yields PbS |
| $M + 2H^+ \rightarrow M^{2+} + H_2$ | M = Sn and Pb |
| $3M + 4H^+ + 4NO_3^- \rightarrow 3MO_2 + 4NO + 2H_2O$ | M = Ge and Sn |
| $3Pb + 8H^+ + 2NO_3^- \rightarrow 3Pb^{2+} + 2NO + 4H_2O$ | |
| $M + OH^- + 2H_2O \rightarrow M(OH)_3^- + H_2$ | M = Sn and Pb; slow |
| $Ge + 2OH^- + 4H_2O \rightarrow Ge(OH)_6^{2-} + 2H_2$ | |

[a] M = Ge, Sn, and Pb, except where noted.

Since the elements have $ns^2 \, np^2$ valence shell configurations, two oxidation states are observed: 4+ and 2+ (inert pair species). The 4+ state declines in importance and the 2+ state becomes increasingly important down the series: Ge, Sn, Pb. Thus, only a few 2+ germanium compounds are important (GeO, GeS, $GeCl_2$, $GeBr_2$, and $GeI_2$), and only a few 4+ lead compounds are important [$PbO_2$, $Pb(C_2H_3O_2)_4$, $PbF_4$, and $PbCl_4$]. The $Sn^{2+}$ [stannous, or tin(II) ion] and $Pb^{2+}$ [plumbous, or lead(II) ion] are the only cationic species of the group that exist in water; 4+ ions probably do not exist. Most of the pure 2+ and 4+ compounds are covalent, although $PbF_2$ is known to be ionic.

All the metals form dioxides; $GeO_2$ and $SnO_2$ are the products of the reactions of the metals with oxygen. Lead dioxide may be prepared by the oxidation of PbO or $Pb^{2+}$ salts in alkaline solution.

$$Pb(OH)_3^-(aq) + OCl^-(aq) \rightarrow PbO_2(s) + Cl^-(aq) + OH^-(aq) + H_2O$$

Lead dioxide is a strong oxidizing agent.

$$2e^- + 4H^+ + PbO_2 \rightarrow Pb^{2+} + 2H_2O \qquad \mathscr{E}° = +1.46 \text{ V}$$

In acid solution $PbO_2$ oxidizes $Mn^{2+}$ to $MnO_4^-$ and $Cl^-$ to $Cl_2$. In contrast to $GeO_2$ and $SnO_2$, $PbO_2$ is thermally unstable. Upon gentle heating $PbO_2$ decomposes to $Pb_3O_4$ (red lead); stronger heating gives $PbO$ (litharge). The compound $Pb_3O_4$ contains lead in two oxidation states and may be represented as $Pb_2^{II}Pb^{IV}O_4$ like the compounds $Fe_3O_4$ and $Co_3O_4$, both of which conform to the formula $M^{II}(M^{III}O_2)_2$.

Germanates, stannates, and plumbates may be derived from the dioxides by reactions with aqueous alkali or in the case of $PbO_2$ by fusion with alkali metal oxides or alkaline earth oxides. The stannates and plumbates form trihydrates in which the anion is an octahedral hydroxy complex: $[M(OH)_6]^{2-}$; a few germanates of similar structure are known (e.g., $Fe[Ge(OH)_6]$). There are no hydroxides of formula $M(OH)_4$, but hydrous $MO_2$ oxides may be prepared.

Treatment of the $Sn^{2+}$ and $Pb^{2+}$ ions in water solution with $OH^-$ ion gives hydrous-oxide precipitates that are commonly assigned the formulas $Sn(OH)_2$ and $Pb(OH)_2$. By heating these products, $SnO$ and $PbO$ can be prepared; $PbO$ is also the product of the reaction of lead and oxygen as well as of the decompositions of $Pb_3O_4$ and $PbO_2$. The compound $GeO$ may be prepared by the hydrolysis of $GeCl_2$.

All the monoxides, as well as the hydrous oxides (or hydroxides), are amphoteric and dissolve in either acid or alkaline solution. In excess alkali, germanites, stannites, and plumbites are produced. For example,

$$PbO(s) + OH^-(aq) + H_2O \rightarrow Pb(OH)_3^-(aq)$$

The stannite ion is a strong reducing agent.

$$2e^- + Sn(OH)_6^{2-} \rightarrow Sn(OH)_3^- + 3OH^- \qquad \mathscr{E}° = -0.93 \text{ V}$$

All the metals form monosulfides, $MS$, but only germanium and tin form disulfides, $MS_2$. The disulfides of germanium and tin may be prepared by the reactions of the metals and sulfur and may be precipitated from solutions of $Ge^{IV}$ or $Sn^{IV}$ compounds by $H_2S$. The disulfides, like the dioxides, are amphoteric. In solutions of alkali metal sulfides or ammonium sulfide, $GeS_2$ and $SnS_2$ dissolve to form thioanions; compounds of the $SnS_3^{2-}$ and $SnS_4^{4-}$ ions have been isolated from such solutions, but thiogermanates have not been obtained in pure form.

$$SnS_2(s) + S^{2-}(aq) \rightarrow SnS_3^{2-}(aq)$$

Insoluble $PbS$ and $SnS$ may be precipitated from solutions by means of $H_2S$. Lead(II) sulfide is also the product of the direct union of the elements; $SnS$ may be prepared by the thermal decomposition of $SnS_2$.

Germanium(II) sulfide may be derived from the disulfide of germanium by the reaction

$$GeS_2 + Ge \rightarrow 2GeS$$

None of the monosulfides are soluble in sulfide solutions.

Complete series of the tetrahalides of germanium, tin, and lead are known except for $PbBr_4$ and $PbI_4$. Lead in a 4+ state has strong oxidizing power and cannot exist in a compound with bromine and iodine in 1− states (which have reducing abilities). Lead(IV) chloride readily decomposes at 100°C.

$$PbCl_4(l) \rightarrow PbCl_2(s) + Cl_2(g)$$

The tetrahalides, in general, are volatile covalent substances. They hydrolyze readily to produce hydrous dioxides.

All the dihalides of germanium, tin, and lead are known. Germanium(II) halides, as well as tin(II) halides, may be prepared by the reaction of the appropriate tetrahalide and metal.

$$GeCl_4 + Ge \rightarrow 2GeCl_2$$

The reactions of tin and the hydrohalic acids yield tin(II) halides. Lead(II) halides may be produced by the direct reaction of the elements. Since all $PbX_2$ compounds are insoluble in water, they may be precipitated from solutions containing $Pb^{2+}$ by the addition of halide ions.

The dihalides are much less volatile than the tetrahalides, a fact which indicates an increased degree of ionic character; $PbF_2$, $PbCl_2$, and $PbBr_2$ are ionic in the solid state. The dihalides of germanium are not particularly stable. Complex anions, such as $GeCl_3^-$, $SnCl_4^{2-}$, $PbCl_3^-$, $PbCl_4^{2-}$, and $PbCl_6^{4-}$, are known. Both $Sn^{2+}$ and $Pb^{2+}$ form such ions in aqueous solution in the presence of halide ions; in general, the ions $MX^+$ and $MX_3^-$ are most important. Insoluble lead(II) halides dissolve in solutions containing excess halide ions because of the formation of such complexes.

All three metals form covalent, volatile hydrides: $GeH_4$, germane; $SnH_4$, stannane; and $PbH_4$, plumbane. The hydride of lead is thermally unstable and decomposes to the elements at 0°C; $SnH_4$ decomposes at approximately 150°C. Continuing the trend established by carbon and silicon, germanium forms catenated hydrides of formula $Ge_nH_{2n+2}$, where $n$ is any number from 2 to 8. Tin or lead form only the simple hydrides.

The $Sn^{2+}$ and $Pb^{2+}$ ions are extensively hydrolyzed in water.

$$[Sn(H_2O)_6]^{2+} + H_2O \rightleftharpoons [Sn(H_2O)_5(OH)]^+ + H_3O^+ \qquad K_H \cong 10^{-2}$$

$$[Pb(H_2O)_6]^{2+} + H_2O \rightleftharpoons [Pb(H_2O)_5(OH)]^+ + H_3O^+ \qquad K_H \cong 10^{-8}$$

The hydrolysis is particularly pronounced in the case of tin(II) compounds; excess acid is generally added to aqueous solutions of such compounds to inhibit hydrolysis and prevent the precipitation of basic salts, such as Sn(OH)Cl.

Lead(II) nitrate and lead(II) acetate are soluble in water; the acetate is only slightly dissociated. The sulfate, chromate, carbonate, sulfide, and all of the halides of $Pb^{2+}$ are only slightly soluble in water.

## Problems

**18.1** What properties distinguish metals from nonmetals?

**18.2** What evidence supports the belief that the strength of the metallic bonding of the metals of the fourth and subsequent periods reaches a maximum around the center of the transition series? How can the trend be explained?

**18.3** What diameter of iron wire must be used so that a given length will have an electrical conductivity equal to that of the same length of copper wire that is 1.0 cm in diameter? Electrical conductivities are listed in Table 18.1.

**18.4** What are the principal types of ores? Give an example of each type.

**18.5** Explain why it is not surprising that the following ores are found in nature: (a) native ores of platinum, gold, and silver, (b) sulfate ores of barium, strontium, and lead, (c) $Na^+$ and $Mg^{2+}$ in sea water, (d) sulfide ores of lead, bismuth, and nickel.

**18.6** What types of ores are roasted? Write a chemical equation for the roasting of an ore of each type. Why is this procedure used?

**18.7** Give equations for reactions that occur in the blast furnace for the production of pig iron.

**18.8** Describe the following metallurgical processes. Use equations when possible. (a) Froth flotation, (b) Parkes process, (c) Mond process, (d) Van Arkel process, (e) Kroll process, (f) liquation, (g) zone refining, (h) thermite process.

**18.9** Identify the following: (a) ore, (b) gangue, (c) alum, (d) amalgam, (e) flux, (f) slag, (g) smelting.

**18.10** Describe, with equations, the steps in the production of aluminum metal from bauxite.

**18.11** Use equations to show how magnesium is obtained from sea water.

**18.12** Compare the metallurgical processes used to obtain copper from (a) low-grade $CuCO_3$ ores, (b) CuS ores.

**18.13** Explain, using chemical equations, how silver is obtained from low-grade ores of native silver.

**18.14** Use equations to compare the processes used to obtain lead from (a) $PbCO_3$ by roasting and carbon reduction, (b) PbS by incomplete roasting and smelting in the absence of air.

**18.15** Write an equation for the reaction that occurs between Na and (a) $H_2$, (b) $N_2$, (c) $O_2$, (d) $Cl_2$, (e) S, (f) $P_4$, (g) C, (h) $H_2O$, (i) $NH_3$.

**18.16** According to first ionization potentials, the most reactive I A metal is cesium. According to electrode potentials, the most reactive I A metal is lithium. Reconcile these observations.

**18.17** Discuss the ways in which the chemistry of lithium and its compounds differs from the chemistry of sodium and its compounds.

**18.18** Write an equation for the reaction that occurs between Ca and (a) $H_2$, (b) $N_2$, (c) $O_2$, (d) $Cl_2$, (e) S, (f) $P_4$, (g) C, (h) $H_2O$, (i) $NH_3$.

**18.19** The second ionization potentials of the II A metals are much larger than the first ionization potentials. Why do these elements not form 1+ ions instead of 2+ ions in their reactions?

**18.20** Explain why the compounds of beryllium are far more covalent than those of any other II A metal.

**18.21** Write equations to compare the reactions of Fe, Cr, and Zn with (a) $Cl_2$, (b) $O_2$, (c) S, (d) $N_2$, (e) $H_2O$, (f) $H^+(aq)$, (g) $OH^-(aq)$.

**18.22** In what ways do the properties of the transition metals differ from those of the A family metals?

**18.23** Group IV B consists of Ti, Zr, and Hf. Why do Zr and Hf resemble each other chemically much more than either of them resemble Ti?

**18.24** Write equations for the reactions of CrO, $Cr_2O_3$, and $CrO_3$ with $H^+(aq)$ and with $OH^-(aq)$.

**18.25** What reasons can you give to explain the low order of reactivity of the noble metals?

**18.26** The standard electrode potential, $\mathscr{E}°$, for the $Hg_2^{2+}/Hg$ half reaction is +0.788 V. For the cell

$$Pt \mid Hg \mid Hg_2^{2+}(10^{-3}M) \parallel Hg_2^{2+}(10^{-2}M) \mid Hg \mid Pt$$

$\mathscr{E}$ equals +0.0296 V. (a) What would be the emf of this concentration cell if the mercurous ion had the formula $Hg^+$? Notice that the ion concentrations given apply to the formula $Hg_2^{2+}$. (b) What would be the standard electrode potential of the mercurous ion/mercury couple if the formula of the mercurous ion were $Hg^+$? Notice that a liter of a $1M$ $Hg^+$ solution would contain one-half the weight of "mercurous" ion that a liter of $1M$ $Hg_2^{2+}$ solution would contain. (c) Explain why measurement of the emf of the concentration cell proves that the formula of the mercurous ion is $Hg_2^{2+}$, whereas measurement of the $\mathscr{E}°$ for the mercurous ion/mercury couple does *not* establish the formula of the ion.

**18.27** The crystalline structures of three oxides of iron are related. (a) In the first, oxide ions form a face-centered cubic lattice and iron ions fill all of the octahedral holes. What type of iron ion, $Fe^{2+}$ or $Fe^{3+}$, satisfies this requirement? What is the formula of this oxide? (b) The crystalline structure of the second type of oxide can be derived from the first structure by removing three-quarters of the iron ions and replacing them with the electrically equivalent number of ions of the other type. This second group of ions, however, is distributed equally between octahedral holes and tetrahedral holes. What is the formula of this oxide? What fraction of the total number of octahedral holes is filled? What fraction of the total number of tetrahedral holes is filled? (c) The crystalline structure of the third oxide can be derived by removing all the iron ions from the first crystal and replacing them with the electrically equivalent number of ions of the other type. This second group of ions is distributed randomly between octahedral and tetrahedral holes. What is the formula of this oxide? What fraction of the total number of octahedral and tetrahedral holes taken together is filled?

**18.28** Write an equation for each reaction that occurs between La and (a) $Cl_2$, (b) $O_2$, (c) S, (d) $N_2$, (e) $H_2O$, (f) $H^+(aq)$.

**18.29** The chemical properties of the transition metals vary widely from element to element. In contrast, the inner-transition metals markedly resemble each other chemically. Explain why this difference exists.

**18.30** Write an equation for the reaction that occurs between Al and (a) $Cl_2$, (b) $O_2$, (c) S, (d) $N_2$, (e) $H_2O$, (f) $H^+(aq)$.

**18.31** Why do Al and Pb appear at times to be less reactive than electrode potentials would indicate?

**18.32** A "diagonal relationship" exists between Al and Be. Explain the atomic-ionic properties that cause such a relationship and cite evidence for its existence.

**18.33** Why is it impossible to precipitate $Al_2S_3$ from aqueous solutions containing $Al^{3+}(aq)$ ions?

**18.34** Write an equation for the reaction that occurs between Sn and (a) $Cl_2$, (b) $O_2$, (c) S, (d) $H^+(aq)$, (e) $OH^-(aq)$, (f) $HNO_3$.

**18.35** Complete the following equations:

(a) $PbO_2(s) + H_2O + OH^-(aq) \rightarrow$

(b) $PbO_2(s) \xrightarrow[\text{heating}]{\text{mild}}$

(c) $PbO_2(s) \xrightarrow[\text{heating}]{\text{strong}}$

(d) $PbO(s) + H_2O + OH^-(aq) \rightarrow$

(e) $SnS(s) + S^{2-}(aq) \rightarrow$

(f) $SnS_2(s) + S^{2-}(aq) \rightarrow$

(g) $PbCl_2(s) + Cl^-(aq) \rightarrow$

(h) $SnF_4 + F^-(aq) \rightarrow$

# 19

# Complex Compounds

Chemists of the late nineteenth century had difficulty in understanding how "molecular compounds" or "compounds of higher order" are bonded. The formation of a compound such as $CoCl_3 \cdot 6NH_3$ was baffling — particularly in this case since simple $CoCl_3$ does not exist. In 1893 Alfred Werner proposed a theory to account for compounds of this type. Werner wrote the formula of the cobalt compound as $[Co(NH_3)_6]Cl_3$. He assumed that the six ammonia molecules are symmetrically "coordinated" to the central cobalt atom by "subsidiary valencies" of cobalt while the "principal valencies" of cobalt are satisfied by the chloride ions. Werner spent over twenty years preparing and studying coordination compounds and perfecting and proving his theory.

Although modern work has amplified his theory, it has required little modification.

Many practical applications have been derived from the study of complex compounds; advances have resulted in such fields as metallurgy, analytical chemistry, biochemistry, water purification, textile dyeing, electrochemistry, and bacteriology. In addition, the study of these compounds has enlarged our understanding of chemical bonding, certain physical properties (e.g., spectral and magnetic properties), minerals (many minerals are complex compounds), and metabolic processes (both heme of blood and chlorophyll of plants are complex compounds).

**19.1 STRUCTURE**    A complex ion or complex compound consists of a central metal cation to which several anions and/or molecules (called **ligands**) are bonded. With few exceptions, *free* ligands have at least one electron pair that is not engaged in bonding.

$$:\overset{..}{\underset{..}{Cl}}:^- \qquad :C\equiv N:^- \qquad H-\overset{..}{\underset{..}{O}}: \qquad H-\overset{\displaystyle H}{\underset{\displaystyle H}{\overset{|}{\underset{|}{N}}}}-H$$

These electron pairs may be considered to be donated to the electron-deficient metal ions in the formation of complexes; ligands, therefore, are substances that are capable of acting as Lewis bases. The bonding of complexes, however, shows a wide variation in character—from strongly covalent to predominantly ionic (Section 19.5).

The ligands are said to be coordinated around the central cation in a **first coordination sphere.** In the formulas of complex compounds, such as $K_3[Fe(CN)_6]$ and $[Cu(NH_3)_4]Cl_2$, the first coordination sphere is indicated by square brackets. The ligands are disposed about the central ion in a regular geometric manner (Figure 19.1). The number of atoms *directly* bonded to the central metal ion, or the number of coordination positions, is called the **coordination number** of the central ion.

The charge of a complex is the sum of the charges of the constituent parts; complexes may be cations, anions, or neutral molecules. Thus, in each of the complexes of platinum(IV)—$[Pt(NH_3)_5Cl]^{3+}$, $[Pt(NH_3)_2Cl_4]^0$, and $[PtCl_6]^{2-}$—the platinum contributes 4+, each chlorine contributes 1−, and the coordinated ammonia molecules do not contribute to the charge of the complex.

An interesting series of platinum(IV) complexes appears in Table 19.1. The list is headed by the chloride of the $[Pt(NH_3)_6]^{4+}$ ion, and in each subsequent entry an ammonia molecule of the coordination sphere is replaced by a chloride ion. The coordinated ammonia molecules and chloride ions are tightly held and do not dissociate in water solution; however, those chloride ions of the compound that are not coordinated

Figure 19.1 Common configurations of complex ions: (a) linear, (b) square planar, (c) tetrahedral, and (d) octahedral.

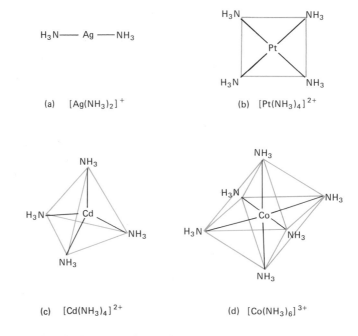

(a) $[Ag(NH_3)_2]^+$

(b) $[Pt(NH_3)_4]^{2+}$

(c) $[Cd(NH_3)_4]^{2+}$

(d) $[Co(NH_3)_6]^{3+}$

to the platinum are ionizable. Hence, aqueous solutions of the last three compounds of the table do not precipitate silver chloride upon the addition of silver nitrate, whereas solutions of the first four precipitate AgCl in amounts proportional to 4/4, 3/4, 2/4, and 1/4, respectively, of their total chlorine content. In each case the total number of ions per formula unit derived from conductance data agrees with the formula listed in the table.

In general, the most stable complexes are formed by metal ions that have a high positive charge and a small ionic radius. The transition elements and the metals immediately following (notably the III A and IV A metals) have a marked tendency to form complexes. Few complexes are

Table 19.1 Some platinum(IV) complex compounds

| | MOLAR CONDUCTANCE 0.001$N$ SOLUTION$^a$ (25°C) | NUMBER OF IONS PER FORMULA UNIT | NUMBER OF Cl$^-$ IONS PER FORMULA UNIT |
|---|---|---|---|
| $[Pt(NH_3)_6]Cl_4$ | 523 | 5 | 4 |
| $[Pt(NH_3)_5Cl]Cl_3$ | 404 | 4 | 3 |
| $[Pt(NH_3)_4Cl_2]Cl_2$ | 228 | 3 | 2 |
| $[Pt(NH_3)_3Cl_3]Cl$ | 97 | 2 | 1 |
| $[Pt(NH_3)_2Cl_4]$ | 0 | 0 | 0 |
| $K[Pt(NH_3)Cl_5]$ | 108 | 2 | 0 |
| $K_2[PtCl_6]$ | 256 | 3 | 0 |

$^a$ cm$^2$/ohm mol.

known for the lanthanides and the I A and II A metals (with the exception of beryllium). The bonding of transition-metal complexes involves the *d* orbitals of the central metal atom.

Complexes containing metal ions with coordination numbers ranging from two to nine are known. However, the majority of complexes are two-, four-, and sixfold coordinate (Figure 19.1), and the coordination number six is by far the most common.

Six-coordinate complexes are octahedral. The regular octahedral arrangement of atoms is frequently represented

This representation is a convenient way to create a three-dimensional illusion. However, no difference between the bonds of the vertical axis and the other bonds should be inferred from this drawing; *all* the bonds are equivalent. Tetragonal geometry is a distorted form of the octahedral in which the bond distances along one of the axes are longer or shorter than the remaining bonds.

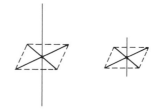

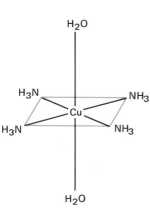

**Figure 19.2** Configuration of $[Cu(NH_3)_4(H_2O)_2]^{2+}$.

Four-coordinate complexes are known in tetrahedral and square-planar configurations. The square-planar configuration is the usual one for $Pt^{II}$, $Pd^{II}$, and $Au^{III}$ and is assumed by many complexes of $Ni^{II}$ and $Cu^{II}$. Examples include $[Pt(NH_3)_4]^{2+}$, $[PdCl_4]^{2-}$, $[AuCl_4]^-$, $[Ni(CN)_4]^{2-}$, and $[Cu(NH_3)_4]^{2+}$. In some instances there is evidence that square complexes may be tetragonal forms with two groups, along a vertical axis, located at greater distances from the central ion than the ligands of the plane. Thus, $[Cu(NH_3)_4]^{2+}$ in water solution may have two water molecules coordinated in the manner shown in Figure 19.2. Hence, square-planar and octahedral geometries may be considered to merge.

For four-coordinate complexes the tetrahedral configuration is encountered more frequently than the square-planar and is particularly common for complexes of the nontransition elements. Certain complexes of $Cu^I$, $Ag^I$, $Au^I$, $Be^{II}$, $Zn^{II}$, $Cd^{II}$, $Hg^{II}$, $Al^{III}$, $Ga^{III}$, $In^{III}$, $Fe^{III}$, $Co^{II}$, and

$Ni^0$ are tetrahedral; examples include $[Cu(CN)_4]^{3-}$, $[BeF_4]^{2-}$, $[AlF_4]^-$, $[FeCl_4]^-$, $[Cd(CN)_4]^{2-}$, $[ZnCl_4]^{2-}$, and $[Ni(CO)_4]^0$. The oxyanions of certain transition metals (such as $VO_4^{3-}$, $CrO_4^{2-}$, $FeO_4^{2-}$, and $MnO_4^-$) are tetrahedral, resembling the tetrahedral oxyanions of nonmetals (such as $SiO_4^{4-}$, $PO_4^{3-}$, $AsO_4^{3-}$, $SO_4^{2-}$ and $ClO_4^-$). Although a number of tetrahedral complexes of transition elements are known, the majority of complexes of these elements are octahedral.

Linear, two-coordinate complexes are not so common as the other forms previously mentioned. However, well-characterized complexes of this type are known for $Cu^I$, $Ag^I$, and $Hg^{II}$; examples are $[CuCl_2]^-$, $[Ag(NH_3)_2]^+$, $[Au(CN)_2]^-$, $[Hg(NH_3)_2]^{2+}$, and $[Hg(CN)_2]^0$.

In general, each metal exhibits more than one coordination number and geometry in its complexes. Although all known complexes of $Co^{III}$ are octahedral, most cations form more than one type of complex. For example, $Al^{III}$ forms tetrahedral and octahedral complexes, $Cu^I$ forms linear and tetrahedral complexes, and $Ni^{II}$ forms square-planar, tetrahedral, and octahedral complexes.

Complexes of the transition elements are frequently highly colored. Examples of complexes that exhibit a wide range of colors are $[Co(NH_3)_6]^{3+}$ (yellow), $[Co(NH_3)_5(H_2O)]^{3+}$ (pink), $[Co(NH_3)_5Cl]^{2+}$ (violet), $[Co(H_2O)_6]^{3+}$ (purple), and $[Co(NH_3)_4Cl_2]^+$ (a violet form and a green form—see Section 19.4).

The ligands we have discussed thus far have been capable of forming only one bond with the central ion; they are referred to as **unidentate** (from Latin, meaning "one toothed") ligands. Certain ligands are capable of occupying more than one coordination position of a metal ion. Ligands that coordinate through two bonds from different parts of the molecule or anion are called **bidentate**. Examples are

carbonate ion            oxalate ion            ethylenediamine

The carbonate and oxalate ions each coordinate through two oxygen atoms; the ethylenediamine molecule (abbreviated *en*) coordinates through both nitrogen atoms. These positions are marked with arrows. Bidentate ligands form rings with the central metal ions and the resulting metal complexes are called **chelates** (from Greek, meaning "claw"). The formation of five- or six-membered rings is generally favored.

Multidentate ligands have been prepared that can coordinate at 2, 3, 4, 5, or 6 positions. In general, chelates are more stable than complexes containing only unidentate ligands. Thus, the sexadentate complexing agent ethylenediaminetetraacetate ion (EDTA)

$$^-O_2C-CH_2 \qquad\qquad\qquad CH_2-CO_2^-$$
$$N-CH_2-CH_2-N$$
$$^-O_2C-CH_2 \qquad\qquad\qquad CH_2-CO_2^-$$

is capable of forming a very stable complex with the calcium ion—an ion with one of the least tendencies toward the formation of complexes (Figure 19.3).

Heme of hemoglobin is a chelate of $Fe^{2+}$, and chlorophyll is a chelate of $Mg^{2+}$. In both these substances the metal atom is coordinated to a quadridentate ligand, which may be considered to be derived from a porphin structure (Figure 19.4) by the substitution of various groups for the H atoms of the porphin. The substituted porphins are called **porphyrins.**

Coordination around the iron atom of heme is octahedral. Four of the coordination positions are utilized in the formation of the heme (which is essentially planar), the fifth is used to bond the heme to a protein molecule (globin), and the sixth is used to coordinate either

**Figure 19.3** Ethylenediaminetetraacetate complex of $Ca^{2+}$ ($[Ca(EDTA)]^{2-}$).

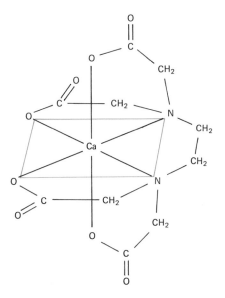

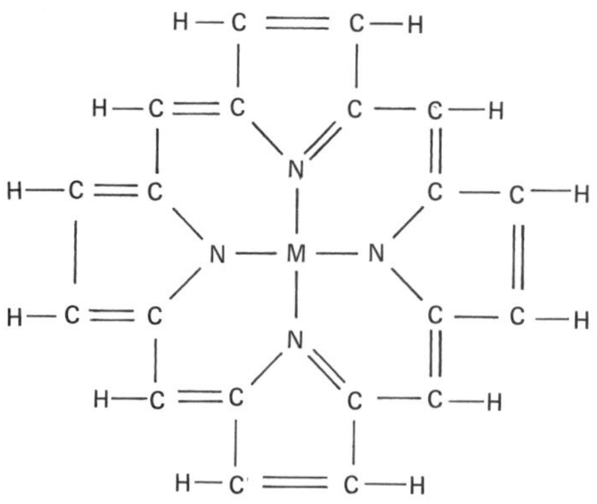

Figure 19.4 **Metal porphin.**

$H_2O$ (hemoglobin) or $O_2$ (oxyhemoglobin). Coordination about this sixth position is reversible

$$\text{hemoglobin} + O_2 \rightleftharpoons \text{oxyhemoglobin} + H_2O$$

and dependent upon the pressure of $O_2$. Hence, hemoglobin picks up $O_2$ in the lungs and releases it in the body tissues, where it is used for the oxidation of food. Hemoglobin reacts with carbon monoxide to form a complex that is more stable than oxyhemoglobin, and therefore CO is toxic.

Chlorophyll, the green pigment of plants, serves as a catalyst for the process of photosynthesis in which $CO_2$ and $H_2O$ are converted into a carbohydrate (glucose) and $O_2$. The energy for photosynthesis comes from sunlight, and the chlorophyll molecule initiates this process by absorbing a quantum of light.

**19.2 LABILE AND INERT COMPLEXES**

Certain complexes (which are called **labile**) rapidly undergo reactions in which ligands are replaced; other complexes (**nonlabile,** or **inert, complexes**) do not undergo these substitution reactions or do so slowly. This distinction applies to the *rate* of attainment of equilibrium and has no bearing on the position of equilibrium. With the exception of the complexes of $Cr^{III}$ and $Co^{III}$, most octahedral complexes of the fourth period transition elements are very labile; exchange reactions come to equilibrium almost as fast as the reagents are mixed. The reason that the complexes of $Co^{III}$ have been studied more than any other group of complexes is that these substances undergo ligand exchange reactions at a slower, more convenient rate.

The inertness of a complex should not be confused with its thermodynamic stability. Although $[Co(NH_3)_6]^{3+}$ is stable in aqueous solution

$$[Co(NH_3)_6]^{3+} + 6H_2O \rightleftharpoons [Co(H_2O)_6]^{3+} + 6NH_3 \qquad K \cong 10^{-34}$$

the complex is unstable in aqueous acid

$$[Co(NH_3)_6]^{3+} + 6H_3O^+ \rightleftharpoons [Co(H_2O)_6]^{3+} + 6NH_4^+ \qquad K \cong 10^{22}$$

Nevertheless, the $[Co(NH_3)_6]^{3+}$ ion can exist in dilute acid for weeks; the latter reaction must have a high activation energy. Thus, the complex is thermodynamically unstable and at the same time inert.

Examples of the reverse situation are also known. The stable complex $[FeCl_6]^{3-}$ is very labile and undergoes rapid exchange with radioactive chloride in aqueous solution.

In many cases complex formation results in the stabilization of metal ions toward oxidation or reduction. Electrode potentials for $Zn^{2+}$ and some complexes of this ion are

$$2e^- + Zn^{2+}(aq) \rightleftharpoons Zn(s) \qquad \mathscr{E}° = -0.763 \text{ V}$$

$$2e^- + [Zn(NH_3)_4]^{2+} \rightleftharpoons Zn(s) + 4NH_3 \qquad \mathscr{E}° = -1.04 \text{ V}$$

$$2e^- + [Zn(CN)_4]^{2-} \rightleftharpoons Zn(s) + 4CN^- \qquad \mathscr{E}° = -1.26 \text{ V}$$

The increasing stability toward reduction observed in this series parallels increasing stability of the complexes toward dissociation in aqueous solution. The instability constant of $[Zn(NH_3)_4]^{2+}$ is approximately $10^{-10}$ and that of the $[Zn(CN)_4]^{2-}$ ion is about $10^{-18}$.

It is tempting to ascribe the lower electrode potentials of the complexes to the decreased concentrations of $Zn^{2+}(aq)$ in the solutions of these complexes. Such an interpretation is in agreement with the concentration effects predicted by the Nernst equation (Section 10.9). However, this explanation implies a mechanism for the reductions that has not been confirmed. The "free" zinc ion, $Zn^{2+}(aq)$, is an aquo complex, and it is difficult to see why the reduction of every other complex of $Zn^{2+}$ should be required to proceed through and be dependent upon the aquo complex. Mechanism studies of the oxidation–reduction reactions of complexes indicate that the actual situation is not this simple. There is, however, a relationship between the stability of a complex toward dissociation and its tendency to undergo oxidation or reduction.

In many instances complex formation results in the stabilization of a metal in a rare or otherwise unknown oxidation state. A classic example is afforded by the complexes of $Co^{III}$. Simple compounds containing cobalt in an oxidation state of $3+$ are rare. The electrode potential

$$e^- + [Co(H_2O)_6]^{3+} \rightleftharpoons [Co(H_2O)_6]^{2+} \qquad \mathscr{E}° = +1.81 \text{ V}$$

indicates that the hydrated $Co^{3+}$ ion is a strong oxidizing agent which is capable of oxidizing water to oxygen and hence incapable of prolonged existence in water.

In the presence of many complexing agents, such as $NH_3$, the $3+$ oxidation state of cobalt is much more stable toward reduction.

$$e^- + [Co(NH_3)_6]^{3+} \rightleftharpoons [Co(NH_3)_6]^{2+} \qquad \mathscr{E}^\circ = +0.11 \text{ V}$$

The $[Co(NH_3)_6]^{3+}$ ion can exist in aqueous solution. It does not oxidize water to oxygen, and, in fact, the reverse reaction—the oxidation of $Co^{II}$ complexes by a stream of air—is used to prepare $Co^{III}$ complexes. The ammonia complex of $Co^{III}$, which has an instability constant of about $10^{-34}$, is much more stable toward dissociation than the ammonia complex of $Co^{II}$, which has an instability constant of approximately $10^{-5}$.

The electrode potentials for copper and its ions may be diagramed

$$Cu^{2+} \xrightarrow{\text{+0.153 V}} Cu^+ \xrightarrow{\text{+0.521 V}} Cu$$

From this diagram we can see that the $Cu^+$ ion is unstable in water solution and disproportionates (Section 10.8).

$$2Cu^+(aq) \rightleftharpoons Cu^{2+}(aq) + Cu(s) \qquad \mathscr{E}^\circ = +0.368 \text{ V}$$

The equilibrium constant for this disproportionation may be derived from the cell potential.

$$K = \frac{[Cu^{2+}]}{[Cu^+]^2} \cong 10^6$$

With certain complexing agents the stability of the 1+ state is enhanced more than that of the 2+ state. For example, with ammonia $K \cong 10^{-2}$, and the $[Cu(NH_3)_2]^+$ ion is stable toward disproportionation. Other complexing agents, such as ethylenediamine, have a greater affinity for $Cu^{II}$ than for $Cu^I$. For ethylenediamine, $K \cong 10^5$, and an ethylenediamine complex of $Cu^I$ does not exist in water solution.

A similar situation is observed for the ions of gold; $Au^+$ is theoretically unstable toward disproportionation into $Au^{3+}$ and Au metal. In addition, both gold ions are strong oxidizing agents and are capable of oxidizing water. However, stable complexes of both $Au^I$ and $Au^{III}$ are known.

The stability of complex ions toward dissociation in aqueous solution is discussed in Section 17.4.

**19.3 NOMENCLATURE** Since thousands of complexes are known and the number is constantly expanding, a system of nomenclature has been adopted for these compounds. The following paragraphs summarize the important rules of the system; they are adequate for naming the simple and frequently encountered complexes.

1. If the complex compound is a salt, the cation is named first whether or not it is the complex ion.

2. The constituents of the complex are named in the following order: anions, neutral molecules, central metal ion.

3. Anionic ligands are given -*o* endings; examples are: $OH^-$, hydroxo; $O^{2-}$, oxo; $S^{2-}$, thio; $Cl^-$, chloro; $F^-$, fluoro; $CO_3^{2-}$, carbonato; $CN^-$, cyano; $CNO^-$, cyanato; $C_2O_4^{2-}$, oxalato; $NO_3^-$, nitrato; $NO_2^-$, nitro; $SO_4^{2-}$, sulfato; and $S_2O_3^{2-}$, thiosulfato.

4. The names of neutral ligands are not changed. Exceptions to this rule are: $H_2O$, aquo; $NH_3$, ammine; CO, carbonyl; and NO, nitrosyl.

5. The number of ligands of a particular type is indicated by a prefix: di-, tri-, tetra-, penta-, and hexa- (for two to six). For complicated ligands (such as ethylenediamine), the prefixes bis-, tris, and tetrakis- (two to four) are employed.

6. The oxidation number of the central ion is indicated by a Roman numeral, which is set off by parentheses and placed after the name of the complex.

7. If the complex is an anion, the ending -*ate* is employed. If the complex is a cation or a neutral molecule, the name is not changed. Examples follow:

| | |
|---|---|
| $[Ag(NH_3)_2]Cl$ | Diamminesilver(I) chloride |
| $[Co(NH_3)_3Cl_3]$ | Trichlorotriamminecobalt(III) |
| $K_4[Fe(CN)_6]$ | Potassium hexacyanoferrate(II) |
| $[Ni(CO)_4]$ | Tetracarbonylnickel(0) |
| $[Cu(en)_2]SO_4$ | Bis(ethylenediamine)copper(II) sulfate |
| $[Pt(NH_3)_4][PtCl_6]$ | Tetraammineplatinum(II) hexachloroplatinate(IV) |
| $[Co(NH_3)_4(H_2O)Cl]Cl_2$ | Chloroaquotetraamminecobalt(III) chloride |

Common names are frequently employed when they are clearly more convenient than the systematic name (e.g., ferrocyanide rather than hexacyanoferrate(II) for $[Fe(CN)_6]^{4-}$) or when the structure of the complex is not certain (e.g., the aluminate ion).

**19.4 ISOMERISM**

Two compounds with the same molecular formula but different arrangements of atoms are called **isomers;** such compounds differ in their chemical and physical properties. **Structural isomerism** is displayed by compounds that have different ligands within their coordination spheres; several types of structural isomers may be identified.

The following pair of compounds of $Co^{III}$ serve as an example of **ionization isomers.**

(a) $[Co(NH_3)_5(SO_4)]Br$     (b) $[Co(NH_3)_5Br]SO_4$
     red                             violet

Conductance data show that both compounds dissociate into two ions in aqueous solution. In the first compound, the $SO_4^{2-}$ ion is a part of the coordination sphere, and the $Br^-$ ion is ionizable. Hence, an aqueous solution of compound (a) gives an immediate precipitate of AgBr upon the addition of $AgNO_3$, but since the $SO_4^{2-}$ ion is not free, no precipitate forms upon the addition of $BaCl_2$. For compound (b), the reverse is true. An aqueous solution of this compound gives a precipitate of $BaSO_4$ but not AgBr, since the $SO_4^{2-}$ is ionizable and the $Br^-$ is coordinated. Note that $SO_4^{2-}$ functions as a unidentate ligand and that the charge on the complex ion of compound (a) is $1 +$ and that of compound (b) is $2 +$.

There are numerous additional examples of ionization isomers, for example,

$$[Pt(NH_3)_4Cl_2]Br_2 \qquad [Pt(NH_3)_4Br_2]Cl_2$$

**Hydrate isomerism** is analogous to ionization isomerism and is probably best illustrated by the following series of compounds that have the formula $CrCl_3 \cdot 6H_2O$.

   (a) $[Cr(H_2O)_6]Cl_3$
          violet

   (b) $[Cr(H_2O)_5Cl]Cl_2 \cdot H_2O$
           green

   (c) $[Cr(H_2O)_4Cl_2]Cl \cdot 2H_2O$
          green

In a mole of each of these compounds there are six moles of water. However, in compound (a), six water molecules are coordinated; in compound (b), five; and in compound (c), four. The uncoordinated water, which probably occurs in the crystals as lattice water, is readily lost when compounds (b) and (c) are exposed to desiccants; however, the coordinated water is not so easily removed. Further evidence for the structures of the compounds is afforded by conductance data (the compounds are composed of four, three, and two ions, respectively) and the quantity of AgCl precipitated (the compounds have three, two, and one ionizable chloride ions, respectively).

Another example of hydrate isomerism is given by the following pair of compounds.

$$[Co(NH_3)_4(H_2O)Cl]Cl_2 \qquad [Co(NH_3)_4Cl_2]Cl \cdot H_2O$$

**Coordination isomers** may exist in compounds that have two or more centers of coordination. Isomers arise through the exchange of ligands between these coordination centers. In simple examples which involve only two complex ions per compound, the coordinated metal ions may be the same

$$[Cr(NH_3)_6][Cr(NCS)_6] \qquad [Cr(NH_3)_4(NCS)_2][Cr(NH_3)_2(NCS)_4]$$

Figure 19.5 (a) *Cis* and (b) *trans* isomers of dichlorodiammineplatinum(II).

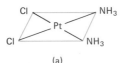

or different

$$[Cu(NH_3)_4][PtCl_4] \qquad [Pt(NH_3)_4][CuCl_4]$$

and the oxidation state of the metals may vary

$$[Pt(NH_3)_4][PtCl_6] \qquad [Pt(NH_3)_4Cl_2][PtCl_4]$$

tetraammineplatinum(II)-hexachloroplatinate(IV)     dichlorotetraammineplatinum(IV)-tetrachloroplatinate(II)

**Linkage isomerism** is a rare but interesting type; it arises when ligands are capable of coordinating in two ways. For example, the nitrite ion, $NO_2^-$, can coordinate through an oxygen atom (—ONO, nitrito compounds) or through the nitrogen atom (—$NO_2$, nitro compounds).

$$[Co(NH_3)_5(NO_2)]Cl_2 \qquad [Co(NH_3)_5(ONO)]Cl_2$$

(yellow)     (red)
nitropentaamminecobalt(III)-chloride     nitritopentaamminecobalt(III)-chloride

Theoretically, many ligands might be capable of forming linkage isomers: $CN^-$ might coordinate either through the C atom or the N atom; $SCN^-$, through N or S; CO, through C or O. However, few authentic cases of linkage isomerism are known.

**Stereoisomerism** is a second general classification of isomers. Compounds are stereoisomers when they both contain the same ligands in their coordination spheres but differ in the way that these ligands are arranged in space. One type of stereoisomerism is **geometric,** or ***cis-trans*, isomerism.**

An example of geometric isomerism is afforded by the *cis* and *trans* isomers of the square planar dichlorodiammineplatinum(II); see Figure 19.5. In the *cis* isomer the chlorine atoms are situated on adjacent corners of the square (along an edge), whereas in the *trans* isomer they occupy opposite corners (along the diagonal).

Since all the ligands of a tetrahedral complex have the same relationship to one another, *cis–trans* isomerism does not exist for this geometry.

Figure 19.6 (a) *Cis* and (b) *trans* isomers of the dichlorotetraamminecobalt(III) ion.

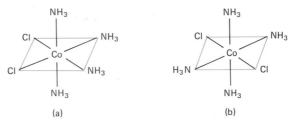

Figure 19.7 Dextro and levo forms of the tris(ethylene-diamine)cobalt(III) ion.

Many geometric isomers are known for octahedral complexes, however. There are two isomers of the dichlorotetraamminecobalt(III) ion: a violet *cis* form and a green *trans* form (Figure 19.6).

A second type of stereoisomerism is **optical isomerism.** Some molecules and ions can exist in two forms that are not superimposable and that bear the same relationship that a right hand bears to a left hand. That the hands are not superimposable is readily demonstrated by attempting to put a left-handed glove on a right hand. Such molecules and ions are spoken of as being **dissymmetric,** and the forms are called **enantiomorphs** (from Greek, meaning opposite forms) or **mirror images** (since the one may be considered to be a mirror reflection of the other).

Enantiomorphs have identical physical properties except for their effects on plane-polarized light. Light that has been passed through a polarizer consists of waves that vibrate in a single plane. One enantiomorph (the **dextro** form), whether pure or in solution, will rotate the plane of the plane-polarized light to the right; the other (the **levo** form) will rotate the plane an equal extent to the left.

For this reason enantiomorphs are called optical isomers or **optical antipodes.** An equimolar mixture of enantiomorphs, called a **racemic modification,** has no effect on plane-polarized light since it contains an equal number of *dextro*rotary and *levo*rotary forms.

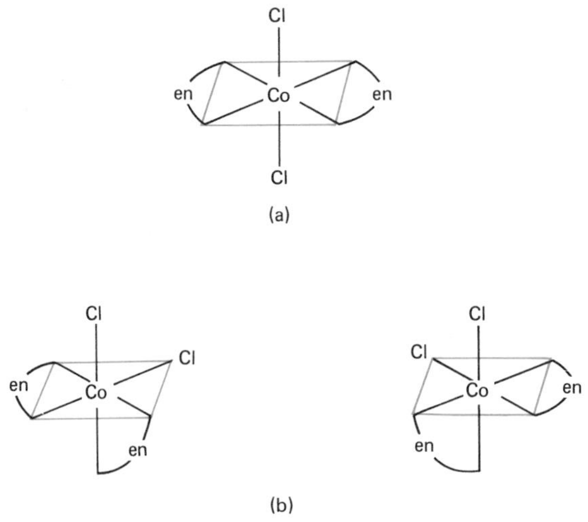

Figure 19.8 Isomers of the dichlorobis(ethylene-diamine)cobalt(III)ion: (a) *trans* isomer, (b) optical isomers of the *cis* form.

Cl
|
en — Co — en
|
Cl

(a)

Cl
|
en — Co — Cl
|
en

Cl
|
Cl — Co — en
|
en

(b)

The tris(ethylenediamine)cobalt(III) ion exists in enantiomorphic forms. Examination of the diagrams of Figure 19.7 confirms that the enantiomorphs are not superimposable and that the ions are dissymmetric. Note that bidentate chelating agents can span only *cis* positions.

Both types of stereoisomerism—geometric and optical—are illustrated by the isomers of the dichlorobis(ethylenediamine)cobalt(III) ion (Figure 19.8). The *trans* configuration of this ion is optically inactive, and a mirror image would be identical to the original. The *cis* arrangement, however, is dissymmetric and exists in *dextro* and *levo* forms. The *trans* modification is said to be a *diastereoisomer* of either the *dextro* or *levo* *cis* modification.

**19.5 THE BONDING IN COMPLEXES**

There are several approaches to the theoretical treatment of the bonding in complexes. The **valence bond theory** describes the bonding in terms of hybridized orbitals of the central metal ion. The metal ion is assumed to have a number of unoccupied orbitals available for complex formation equal to its coordination number. Each ligand donates a pair of electrons toward the formation of a coordinate covalent bond, and a bond arises from the overlap of a vacant hybrid orbital of the metal atom and a filled orbital of a ligand. The bonds usually have an appreciable degree of polarity. The common types of hybridization (Section 4.2) are

| | |
|---|---|
| $sp$ | linear |
| $sp^3$ | tetrahedral |
| $dsp^2$ | square planar |
| $d^2sp^3$ | octahedral |

For octahedral coordination, $d_{z^2}$ and $d_{x^2-y^2}$ orbitals, as well as $p_x$, $p_y$, and $p_z$ orbitals, are directed toward ligands; these orbitals are combined with an $s$ orbital into a $d^2sp^3$ set (Figure 4.7). Square-planar ($dsp^2$) hybridization utilizes the $d_{x^2-y^2}$, $p_x$, $p_y$, and $s$ orbitals (Figure 4.4).

The valence bond diagrams of some simple complexes together with the electronic configurations of the central metal ions are shown in Figure 19.9. Hund's rule applies to the electrons occupying the nonbonding $d$ orbitals, and, accordingly, some of these electrons are unpaired. In each of the complexes shown, the number of unpaired electrons determined by magnetic susceptibility measurement agrees with the structure diagramed. Thus, $[Co(en)_3]^{3+}$ and $[Ni(CN)_4]^{2-}$ are diamagnetic, and the others are paramagnetic. In addition, the experimentally determined geometry of each complex agrees with that predicted on the basis of the assumed type of hybridization.

There are some defects in this attractively simple picture. From the diagrams of Figure 19.9 one might suppose that all of the octahedral

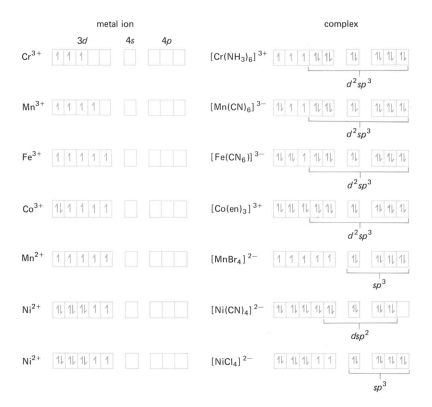

**Figure 19.9** Valence-bond diagrams for some simple complexes.

complexes of $Fe^{III}$ would have one unpaired electron (as in $[Fe(CN)_6]^{3-}$) and that all of the octahedral complexes of $Co^{III}$ would be diamagnetic (like $[Co(en)_3]^{3+}$). Such is not the case. The magnetic moments of $[FeF_6]^{3-}$ and $[CoF_6]^{3-}$ indicate the existence of five and four unpaired electrons, respectively, in these complexes; these numbers of unpaired electrons are the same as those found in the free $Fe^{3+}$ and $Co^{3+}$ ions. At one time, the bonding in this type of complex was thought to be ionic with the $d$ orbitals of the central ion undisturbed by complex formation. According to a later explanation, the $4d$, rather than the $3d$, orbitals are assumed to be used for the octahedral hybridization. Hence, **inner** ($d^2sp^3$) and **outer** ($sp^3d^2$) **complexes** are postulated (Figure 19.10). Since it is impossible to clear two inner $d$ orbitals of the $Ni^{2+}$ ion for inner $d^2sp^3$ hybridization, all octahedral $Ni^{II}$ complexes must be assumed to be of the outer type.

A further complication arises in the case of $[Co(NO_2)_6]^{4-}$ — an octahedral complex with one unpaired electron. In order to fit this complex into the $d^2sp^3$ category, it is necessary to postulate the promotion of an electron into the $4d$ sublevel.

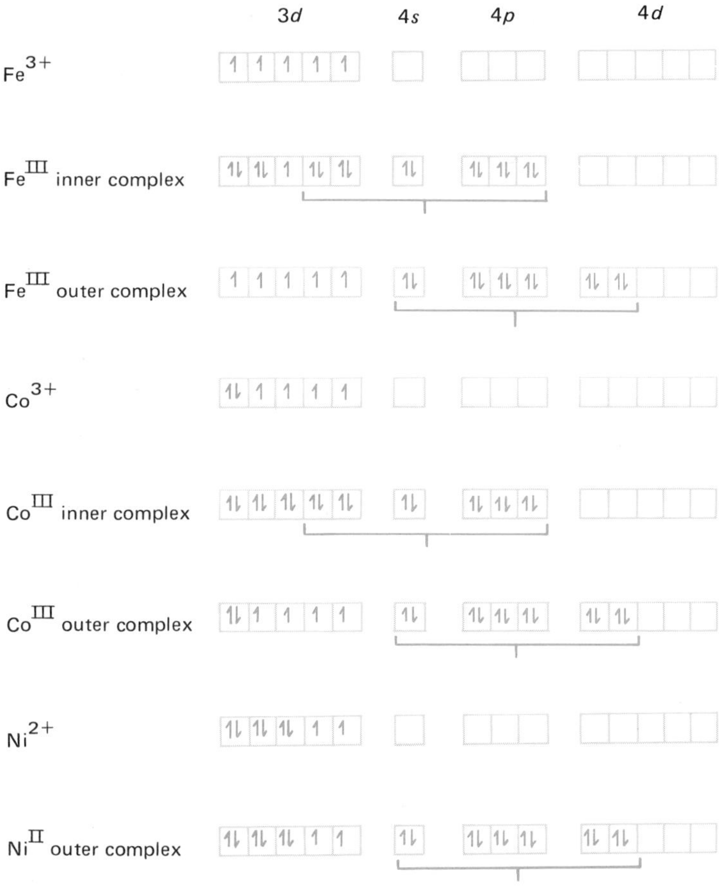


**Figure 19.10** Inner and outer complexes of $Fe^{III}$, $Co^{III}$, and $Ni^{II}$.

---

| | 3d | 4s | 4p | 4d |
|---|---|---|---|---|
| $Co^{2+}$ | ⥮ ⥮ ↿ ↿ ↿ | _ | _ _ _ | _ _ _ _ _ |

| | 3d | $d^2sp^3$ | 4d |
|---|---|---|---|
| $[Co(NO_2)_6]^{4-}$ | ⥮ ⥮ ⥮ | ⥮ ⥮ ⥮ ⥮ ⥮ ⥮ | ↿ _ _ _ _ |

It is also necessary to assume such an unlikely promotion for all of the square complexes of $Cu^{II}$.

| | 3d | 4s | 4p |
|---|---|---|---|
| $Cu^{2+}$ | ⥮ ⥮ ⥮ ⥮ ↿ | _ | _ _ _ |

| | 3d | $dsp^2$ | 4p |
|---|---|---|---|
| $[CuCl_4]^{2-}$ | ⥮ ⥮ ⥮ ⥮ | ⥮ ⥮ ⥮ ⥮ | ↿ |

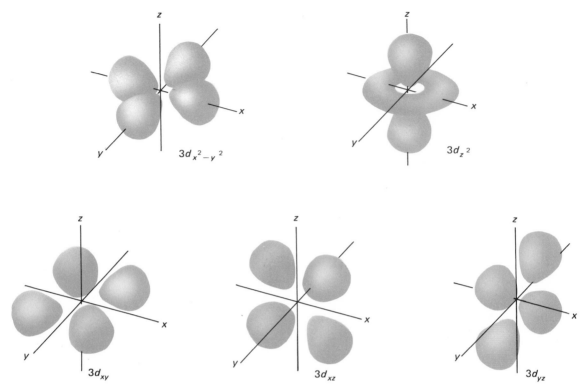

**Figure 19.11** Boundary surface diagrams for the *d* orbitals.

The fundamental defect of the valence bond theory is that it fails to take into account the existence of the antibonding molecular orbitals that are produced, along with bonding molecular orbitals, in the formation of the complexes. Many of the experimental observations that are difficult to explain on the basis of the valence bond approach are readily understood in terms of electrons occupying antibonding orbitals.

The **crystal field theory** and its outgrowth, the **ligand field theory,** center on the *d* orbitals of the metal ion. In a free transition-metal ion all five of the *d* orbitals have equal energies—they are **degenerate.** All the *d* orbitals are not equivalent, however, when the metal ion engages in complex formation; the degeneracy is split.

Consider the relation of the *d* orbitals (Figure 19.11) to the arrangement of the ligands of an octahedral complex (Figure 19.12). The $d_{z^2}$ and $d_{x^2-y^2}$ orbitals have lobes that point toward ligands, whereas the lobes of the $d_{xy}$, $d_{xz}$, and $d_{yz}$ orbitals lie between ligands. Thus, in the complex two sets of *d* orbitals exist; the $d_{xy}$, $d_{xz}$, and $d_{yz}$ orbitals (or $t_{2g}$ orbitals) are equivalent to each other, and the $d_{z^2}$ and $d_{x^2-y^2}$ orbitals (or $e_g$ orbitals) are equivalent to each other and different from the first three. The symbols

**Figure 19.12** Arrangement of ligands of (a) octahedral and (b) square planar complexes in relation to sets of Cartesian coordinates.

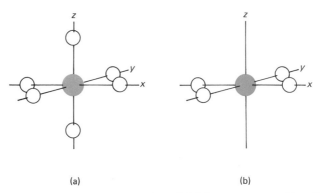

(a)                    (b)

$t_{2g}$ and $e_g$ are applied to threefold degenerate and twofold degenerate sets of orbitals, respectively.

It is not immediately obvious that the $d_{z^2}$ orbital is perfectly equivalent to the $d_{x^2-y^2}$ orbital. The $d_{z^2}$ orbital may be regarded as a combination, in equal parts, of two hypothetical orbitals, $d_{z^2-y^2}$ and $d_{z^2-x^2}$, which have shapes exactly like that of the $d_{x^2-y^2}$ orbital (see Figure 19.13). Since the number of $d$ orbitals is limited to five, the $d_{z^2-y^2}$ and $d_{z^2-x^2}$ orbitals have no independent existence.

In the crystal field theory the assumption is made that an electrostatic field surrounding the central metal ion is produced by the negative ends of the dipolar molecules or by the anions that function as ligands. A metal-ion electron in a $d$ orbital that has lobes directed toward ligands has a higher energy (owing to electrostatic repulsion) than an electron in an orbital with lobes that point between ligands. In octahedral complexes the orbitals of the $e_g$ group, therefore, have higher energies than those of the $t_{2g}$ group. The difference between the energies of the $t_{2g}$ and $e_g$ orbitals in an octahedral complex is given the symbol $\Delta_o$.

In a square-planar complex (see Figure 19.12), the $d$ orbitals exhibit four different relationships. The lobes of the $d_{x^2-y^2}$ orbital point toward

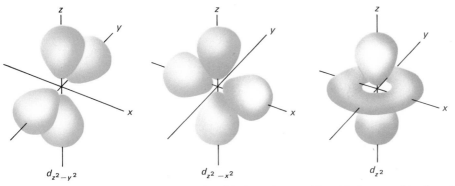

$d_{z^2-y^2}$              $d_{z^2-x^2}$              $d_{z^2}$

**Figure 19.13** Diagram showing that the $d_{z^2}$ orbital may be considered to be a combination of the $d_{z^2-y^2}$ and $d_{z^2-x^2}$ orbitals.

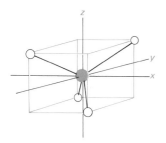

**Figure 19.14** The relation of tetrahedrally arranged ligands to a set of Cartesian coordinates.

ligands, and this orbital has the highest energy. The lobes of the $d_{xy}$ orbital lie between orbitals but are coplanar with them, and hence this orbital is next highest in energy. The lobes of the $d_{z^2}$ orbital point out of the plane of the complex, but the belt around the center of the orbital (which contains about a third of the electron density) lies in the plane; therefore, the $d_{z^2}$ orbital is next highest in energy. The $d_{xz}$ and $d_{yz}$ orbitals, which are degenerate, are least affected by the electrostatic field of the ligands since their lobes point out of the plane of the complex; these orbitals are lowest in energy.

The order of splitting of the $d$ orbitals in a tetrahedral complex may be derived from an examination of Figure 19.14, which shows the relation of tetrahedrally arranged ligands to a system of Cartesian coordinates and a hypothetical cube. The lobes of the $d_{xy}$, $d_{xz}$, and $d_{yz}$ orbitals point toward cube edges, and the lobes of the $d_{z^2}$ and $d_{x^2-y^2}$ orbitals point toward the centers of cube faces. Notice that the distance from the center of a cube face to a ligand is farther than the distance from the center of a cube edge to a ligand. Therefore, the order of orbital energies is the reverse of that for octahedral coordination, and the threefold degenerate set, $t_{2g}$, is of higher energy than the twofold degenerate set, $e_g$. The difference in energy is given the symbol $\Delta_t$.

The splitting of $d$-orbital energies in tetrahedral, octahedral, and square-planar complexes is summarized in Figure 19.15. The crystal field theory fails to take into account the mixing of metal and ligand orbitals and the covalent character of the bonding in complexes, and hence it is ultimately unsatisfactory. The ligand field theory offers a more realistic explanation, which is derived from molecular orbital theory. In both theories the orders of splitting are those shown in Figure 19.15.

**Figure 19.15** The splitting of $d$-orbital energies by ligand fields of three different geometries.

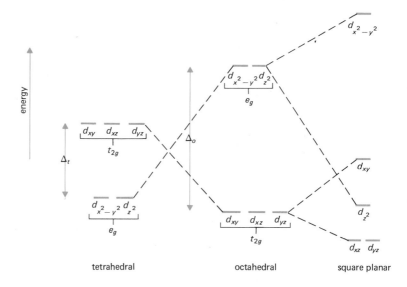

**Figure 19.16** Diagram indicating the formation of molecular orbitals for an octahedral complex with no $\pi$ binding.

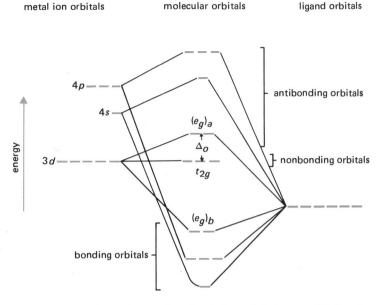

metal ion orbitals     molecular orbitals     ligand orbitals

antibonding orbitals

nonbonding orbitals

bonding orbitals

In an octahedral complex the $3d_{z^2}$ and $3d_{x^2-y^2}$ orbitals along with the $4s$ and three $4p$ orbitals are assumed to overlap the six ligand $\sigma$ orbitals with the attendant formation of six bonding molecular orbitals and six antibonding molecular orbitals (Section 4.4). The $d_{xy}$, $d_{xz}$, and $d_{yz}$ orbitals (the $t_{2g}$ set), which do not overlap the $\sigma$ orbitals of the ligands, are essentially nonbonding. (The $t_{2g}$ set can be used in $\pi$ bonding, however.)

A bonding molecular orbital concentrates electron density between the atoms and is of relatively low energy in comparison to an antibonding molecular orbital, which has a low electron density between the atoms and acts as a disruptive force. A molecular orbital energy level diagram for an octahedral complex with no $\pi$ bonding is given in Figure 19.16.

Whenever two atomic orbitals of different energies combine, the character of the resulting bonding molecular orbital is predominantly that of the atomic orbital of lower energy, and the antibonding molecular orbital has mainly the character of the higher energy atomic orbital. In an octahedral complex the bonding orbitals have predominantly the character of ligand orbitals. The antibonding orbitals resemble metal orbitals more than ligand orbitals. The $t_{2g}$ set, which are nonbonding, may be considered as purely metal orbitals.

In an octahedral complex the six electron pairs from the ligands completely occupy the bonding molecular orbitals. The $d$ electrons of the central metal ion are accommodated in the $t_{2g}$ nonbonding orbitals and the $(e_g)_a$ antibonding orbitals; the difference between the energies of these sets is $\Delta_o$. The four remaining antibonding orbitals are never occupied in the ground states of any known complex.

Hence, the conclusion reached by this treatment is much the same as that postulated by the crystal field theory: in an octahedral complex the degeneracy of the metal $d$ orbitals may be considered to be split into a threefold degenerate set, $t_{2g}$, and a higher energy, metal-like, twofold degenerate set, which may be labeled $(e_g)_a$ or simply $e_g$.

The molecular orbital treatment may be applied to tetrahedral and square-planar complexes, but the applications are more complicated. The conclusions reached are in essential agreement with the splittings diagramed in Figure 19.15.

For the octahedral complexes of a given metal ion, the magnitude of $\Delta_o$ is different for each set of ligands, and the electronic configurations of many complexes depend upon the size of $\Delta_o$.

For complexes of transition-element ions with one, two, or three $d$ electrons (which are referred to as $d^1$, $d^2$, or $d^3$ ions), the orbital occupancy is certain and is independent of the magnitude of $\Delta_o$. The electrons enter the lower energy $t_{2g}$ orbitals singly with their spins parallel. For $d^4$, $d^5$, $d^6$, and $d^7$ ions, a choice of two configurations is possible (see Figure 19.17).

In the case of an octahedral complex of a $d^4$ ion, the fourth electron can singly occupy a higher energy $e_g$ orbital or it can enter a $t_{2g}$ orbital and pair with an electron already present. The former configuration has four unpaired electrons and is called the **high-spin state;** the latter configuration, which has two unpaired electrons, is called the **low-spin state.** Which configuration is assumed depends upon which is energetically more favorable.

If $\Delta_o$ is small, the electron may be promoted to the $e_g$ level where it occupies an orbital singly. However, if $\Delta_o$ is large, the electron may be forced to pair with an electron in a $t_{2g}$ orbital even though this pairing requires the expenditure of a pairing energy, $P$, to overcome the interelectronic repulsion. Thus, the high-spin configuration results when

$$\Delta_o < P$$

and the low-spin configuration results when

$$\Delta_o > P$$

The value of $P$ depends upon the metal ion; $\Delta_o$ is different for each complex.

This conclusion is also valid for complexes of $d^5$, $d^6$, and $d^7$ ions. For a complex of a $d^8$ ion there is only one possible configuration: six electrons paired in the $t_{2g}$ orbitals and two unpaired electrons in the $e_g$ orbitals. Likewise, complexes of $d^9$ and $d^{10}$ ions exist in only one configuration.

The complexes that were fitted into the valence-bond scheme only with difficulty are readily classified by the ligand field theory. Hence, the inner complex $[Fe(CN)_6]^{3-}$ with one unpaired electron is a low-spin state of this $d^5$ ion, and the outer complex $[FeF_6]^{3-}$ with five unpaired

**Figure 19.17** Arrangements of electrons in high-spin and low-spin octahedral complexes of $d^4$, $d^5$, $d^6$, and $d^7$ ions.

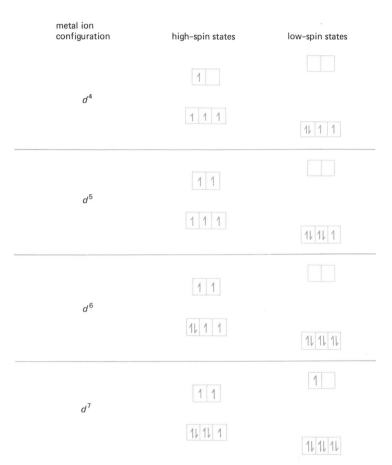

electrons has the high-spin configuration of a $d^5$ ion. The octahedral complexes of Ni$^{II}$, all of which must be assumed to be of the outer valence-bond type, have the only configuration possible for a complex of a $d^8$ ion: six electrons in the $t_{2g}$ orbitals and two unpaired electrons in the $e_g$ orbitals. The complex $[Co(NO_2)_6]^{4-}$, for which it was necessary to postulate the promotion of an electron to a $4d$ atomic orbital in the valence-bond treatment, is a low-spin $d^7$ complex according to the ligand field theory.

Values of $\Delta_o$ can be obtained from studies of the absorption spectra of complexes. In the complex $[Ti(H_2O)_6]^{3+}$, there is only one electron to be accommodated in either the $t_{2g}$ or $e_g$ orbitals. In the ground state this electron occupies a $t_{2g}$ orbital. However, excitation of the electron to an $e_g$ orbital is possible when the energy required for this transition, $\Delta_o$, is supplied; the absorption of light by the complex can bring about such excitations. The wavelength of light absorbed most strongly by the $[Ti(H_2O)_6]^{3+}$ ion is approximately 4900 Å, which corresponds to a $\Delta_o$ of about 58 kcal/mol. The single absorption band of this complex spreads out over a considerable portion of the visible spectrum; however, most of the red and violet light is not absorbed, and this causes the red-violet color of the complex.

The interpretation of the absorption spectra of complexes with more than one $d$ electron is considerably more complicated, since more than two arrangements of $d$ electrons are possible. In general, for a given metal ion the replacement of one set of ligands by another causes a change in the energy difference between the $t_{2g}$ and $e_g$ orbitals, $\Delta_o$, which gives rise to different light-absorption properties. Hence, in many instances a striking color change is observed when the ligands of a complex are replaced by other ligands.

Ligands may be arranged in a **spectrochemical series** according to the magnitude of $\Delta_o$ they bring about. From the experimental study of the spectra of many complexes, it has been found that the order is generally the same for the complexes of all of the transition elements in their common oxidation states with only occasional inversions of order between ligands that stand near to one another on the list. The order of some common ligands is

$$I^- < Br^- < Cl^- < F^- < OH^- < C_2O_4^{2-} < H_2O < NH_3 < en < NO_2^- < CN^-$$

The values of $\Delta_o$ induced by the halide ions are generally low, and complexes of these ligands usually have high-spin configurations. The cyanide ion, which stands at the opposite end of the series from the halide ions, induces the largest $d$-orbital splittings of any ligand listed; cyano complexes generally have low-spin configurations.

A given ligand, however, does not always produce complexes of the same spin type. Thus, the hexaammine complex of $Fe^{II}$ has a high-spin configuration, whereas the hexaammine complex of $Co^{III}$ which is isoelectronic with $Fe^{II}$) has a low-spin configuration.

For each metal ion there is a point in the series that corresponds to the change from ligands that produce high-spin complexes to ligands that form low-spin complexes. For example, $Co^{II}$ forms high-spin complexes with $NH_3$ and ethylenediamine, but the $NO_2^-$ and $CN^-$ complexes of $Co^{II}$ have low-spin configurations. The actual position in the series at which this change from high- to low-spin complex formation occurs depends upon the electron-pairing energy, $P$, for the metal ion, as well as the values of $\Delta_o$ for the complexes under consideration.

## Problems

Problems may be solved using either kilocalories or kilojoules as energy units. Values expressed in kilojoules appear in brackets.

**19.1** Define the following terms: (a) coordination number, (b) ligand, (c) chelate, (d) enantiomorph, (e) inert complex, (f) outer complex, (g) low-spin state.

**19.2** State the oxidation number of the central atom in each of the following complexes:
(a) $[Co(NO_2)_6]^{3-}$
(b) $[Au(CN_4)]^-$
(c) $[V(CO)_6]$
(d) $[Co(NH_3)_4Br_2]^+$
(e) $[Co(en)Cl_4]^{2-}$

**19.3** Write formulas for (a) potassium pentachloroaquorhodate(III), (b) sulfatotetraamminecobalt(III) nitrate, (c) sodium dioxotetracyanorhenate(V), (d) diamminebis(ethylenediamine)cobalt(II) chloride, (e) potassium tetracyanonickelate(0), (f) potassium hexacyanonickelate(II), (g) tetraamminecopper(II) hexachlorochromate(III).

**19.4** Name the following compounds:
(a) $[Co(NO)(CO)_3]$
(b) $[Pt(NH_3)_2Cl_4]$
(c) $K_4[Pt(CN)_4]$
(d) $[Co(NH_3)_6]_4[Co(NO_2)_6]_3$
(e) $Na[Au(CN)_2]$
(f) $[Co(en)_2(SCN)Cl]Cl$

**19.5** Addition of a solution of potassium hexacyanoferrate(II) to $Fe^{3+}$(aq) yields a precipitate called Prussian blue. Addition of a solution of potassium hexacyanoferrate(III) to $Fe^{2+}$(aq) also yields a blue precipitate (Turnbull's blue). The blue precipitates are now known to be identical and are called potassium iron(III) hexacyanoferrate(II). Write the formula for (a) Prussian blue, (b) iron(III) hexacyanoferrate(III), a brown precipitate, (c) copper(II) hexacyanoferrate(II), a purple precipitate, (d) potassium iron(II) hexacyanoferrate(II), a white precipitate.

**19.6** Two compounds have the same empirical formula: $Co(NH_3)_3(H_2O)_2ClBr_2$. One mole of compound A readily loses one mole of water in a desiccator, whereas compound B does not lose any water under the same conditions. An aqueous solution of A has a conductivity equivalent to that of a compound with two ions per formula unit; the conductivity of an aqueous solution of B corresponds to that of a compound with three ions per formula unit. When $AgNO_3$ is added to a solution of compound A, one mole of AgBr is precipitated per mole of A; a solution of compound B yields two moles of AgBr per mole of B. (a) Write the formulas of A and B. (b) What type of isomers are A and B?

**19.7** Write a formula for an example of (a) an ionization isomer of $[Co(NH_3)_5(NO_3)]SO_4$, (b) a linkage isomer of $[Co(en)_2(NO_2)_2]$, (c) a coordination isomer of $[Pt(NH_3)_4][PtCl_4]$, (d) a hydrate isomer of $[Co(en)_2(H_2O)_2]Br_3$.

**19.8** Write a formula for an example of (a) an ionization isomer of $[Pt(NH_3)_4(OH)_2]SO_4$, and for (b) a linkage isomer of $[Pd(dipy)(SCN)_2]$ (dipy is a bidentate ligand — 2,2'-dipyridine), and for (c) a hydrate isomer of $[Co(NH_3)_4(H_2O)Cl]Cl_2$, and for (d) a coordination isomer of $[Cr(NH_3)_4(C_2O_4)][Cr(NH_3)_2(C_2O_4)]$.

**19.9** Identify all the possible

isomers (stereoisomers as well as coordination isomers) of the compound $[Pt(NH_3)_4][PtCl_6]$.

**19.10** Diagram the structures of all of the possible stereoisomers of each of the following octahedral complexes. Classify the structures as geometric or optical isomers.
(a) $[Cr(NH_3)_2(NCS)_4]^-$
(b) $[Co(NH_3)_3(NO_2)_3]$
(c) $[Co(en)(NH_3)_2Cl_2]^+$
(d) $[Co(en)Cl_4]^-$
(e) $[Co(en)_2ClBr]^+$
(f) $[Cr(NH_3)_2(C_2O_4)_2]^-$
(g) $[Cr(C_2O_4)_3]^{3-}$

**19.11** Diagram all the stereoisomers of each of the following molecules. In the formulas, py stands for pyridine, $C_5H_5N$, a unidentate ligand.
(a) $[Pt(NH_3)(py)ClBr]$ (square planar)
(b) $[Pt(NH_3)(py)Cl_2]$ (square planar)
(c) $[Pt(py)_2Cl_2]$ (square planar)
(d) $[Co(py)_2Cl_2]$ (tetrahedral)

**19.12** How can dipole-moment measurement distinguish between the *cis*- and *trans*-isomers of the square planar $[Pt(NH_3)_2Cl_2]$?

**19.13** Diagram the four stereoisomers of $[Pt(en)(NO_2)_2Cl_2]$.

**19.14** Draw valence bond diagrams for the bonding in the following, diamagnetic ions. What geometric configuration does each have? (a) $AuCl_2^-$, (b) $PdCl_4^{2-}$, (c) $ZnCl_4^{2-}$, (d) $BeCl_4^{2-}$, (e) $RhCl_6^{3-}$.

**19.15** The complex $[Ni(CN)_4]^{2-}$ is square planar, and the complex $[NiCl_4]^{2-}$ is tetrahedral. Draw valence bond diagrams for these complexes and predict the number

of unpaired electrons in each.

**19.16** The octahedral complexes $[Fe(CN)_6]^{3-}$ and $[FeF_6]^{3-}$ have one and five unpaired electrons, respectively. Offer an explanation for these observations based on (a) the valence bond theory, (b) the ligand field theory.

**19.17** (a) Contrast the valence bond and ligand field explanations for the electronic structures of the two types of octahedral $Co^{2+}$ complexes (one with three unpaired electrons and another with only one unpaired electron). (b) Why is the valence bond description of the complex with one unpaired electron unsatisfactory?

**19.18** For $Fe^{2+}$ the electron-pairing energy, $P$, is approximately 50 kcal/mol [or 210 kJ/mol]. Approximate values of $\Delta_o$ for the complexes $[Fe(H_2O)_6]^{2+}$ and $[Fe(CN)_4]^{4-}$ are 30 kcal/mol [or 120 kJ/mol] and 94 kcal/mol [or 390 kJ/mol], respectively. (a) Do these complexes have high-spin or low-spin configurations? (b) Draw a *d*-orbital splitting diagram for each.

**19.19** For $[Mn(H_2O)_6]^{3+}$, which is a high-spin complex, $\Delta_o$ is approximately 60 kcal/mol [or 250 kJ/mol]. For $[Mn(CN)_6]^{3-}$, which is a low-spin complex, $\Delta_o$ is approximately 110 kcal/mol [or 460 kJ/mol]. (a) What conclusion can you reach in regard to the magnitude of the electron-pairing energy, $P$, for $Mn^{3+}$? (b) Would you expect the complex $[Mn(C_2O_4)_3]^{3-}$ to be a high-spin or a low-spin complex?

**19.20** Draw *d*-orbital splitting diagrams for $[Co(NH_3)_6]^{2+}$ and $[Co(NH_3)_6]^{3+}$; the $\Delta_o$ values for these two complexes are approximately 29 kcal/mol [or 120 kJ/mol] and 66 kcal/mol [or 270 kJ/mol], respectively. The pairing energy of $Co^{2+}$ is approximately 64 kcal/mol [or 270 kJ/mol] and of $Co^{3+}$ is approximately 50 kcal/mol [or 210 kJ/mol].

**19.21** Draw *d*-orbital splitting diagrams for the high-spin and low-spin octahedral complexes of (a) $Zn^{2+}$, (b) $Cr^{2+}$, (c) $Ni^{2+}$, (d) $Ni^{3+}$, (e) $Mn^{2+}$, (f) $Fe^{3+}$, (g) $V^{2+}$, (h) $V^{3+}$, (i) $V^{4+}$, (j) $V^{5+}$.

**19.22** Draw *d*-orbital splitting diagrams for (a) the octahedral complexes of $Cr^{3+}$, (b) the octahedral complexes of $Ni^{2+}$, (c) the square planar complexes of $Ni^{2+}$ (all of which are diamagnetic), (d) the square planar complexes of $Co^{2+}$ (all of which have one unpaired electron), (e) the tetrahedral complexes of $Co^{2+}$, (f) the octa-hedral complexes of $Ru^{2+}$ (all of which are diamagnetic), (g) the octahedral complexes of $Ir^{4+}$ (all of which have one unpaired electron).

**19.23** The $[Ti(H_2O)_6]^{3+}$ complex is red-violet. What change in color would be expected if the ligands of this complex were replaced by ligands that induce a larger $\Delta_o$? Note that the color of the complex corresponds to light transmitted, not absorbed.

**19.24** A complex of $Fe^{3+}$ has a magnetic moment of 5.9 Bohr magnetons (see Section 2.16). (a) How many unpaired electrons are present in the complex? (b) Is this a high- or low-spin complex?

**19.25** Given the following data, calculate the instability constant, $K_{inst}$, of the $Al(OH)_4^-$ ion.

$$3e^- + Al(OH)_4^- \rightleftharpoons Al + 4OH^-$$
$$\mathscr{E}^\circ = -2.330 \text{ V}$$

$$3e^- + Al^{3+} \rightleftharpoons Al$$
$$\mathscr{E}^\circ = -1.662 \text{ V}$$

# 20
# Organic Chemistry

The name organic chemistry derives from the early concept that substances of plant or animal origin (organic substances) were different from those of mineral origin (inorganic substances). In the middle of the nineteenth century, however, the idea that organic substances could only be synthesized by living organisms was gradually discounted. Not only has a large number of natural products been synthesized in the laboratory, but also countless related materials have been made that do not occur in nature. All these compounds contain carbon. Over a million carbon compounds are known.

Because it represents a convenient division of chemistry, the term organic chemistry is retained, but it is now commonly defined as the chemistry of carbon and its compounds. However, since some carbon compounds (such as carbonates, carbides, and cyanides) are traditionally classed as inorganic compounds, organic chemistry is probably better defined as the chemistry of the hydrocarbons (compounds containing only carbon and hydrogen) and their derivatives.

Carbon forms an unusually large number of compounds because of its exceptional ability to catenate (Section 12.9). In addition, the carbon atom can form four, very stable, single-covalent bonds; it also has the ability to form multiple bonds with other carbon atoms or with atoms of other elements.

**20.1 THE ALKANES**    In the first four sections of this chapter various types of hydrocarbons are characterized. The reactions of the hydrocarbons is the topic of Section 20.5. The simplest hydrocarbon is methane, $CH_4$. This molecule contains four equivalent carbon–hydrogen bonds arranged tetrahedrally (Figure 20.1). The bonding may be considered to arise through the use of $sp^3$ hybrid orbitals of the carbon atom. Each of the bonds of methane is of the same length (1.095 Å), and each of the H—C—H bond angles is 109° 28′ (the so-called tetrahedral angle). The commonly employed structural formula of methane,

$$\begin{array}{c} H \\ | \\ H-C-H \\ | \\ H \end{array}$$

does not accurately represent the tetrahedral arrangement of the molecule.

The **alkanes** are hydrocarbons in which all of the carbon–carbon bonds are single bonds; they may be considered to be derived from methane by

**Figure 20.1** Representations of the structure of methane, $CH_4$. The bonds in diagram (a) are of exaggerated length in comparison to the atomic sizes.

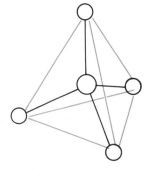

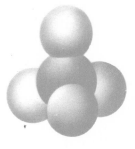

(a)                    (b)

the successive addition of —$CH_2$— units. Such a series of compounds is said to be **homologous.** Thus, the formula of the second member of the family, ethane

$$\begin{array}{c}\text{H} \quad \text{H} \\ | \quad\quad | \\ \text{H—C—C—H} \\ | \quad\quad | \\ \text{H} \quad \text{H}\end{array}$$

may be formally derived by the introduction of a —$CH_2$— unit between the carbon and a hydrogen of methane. This molecule is also represented by the formula $CH_3$—$CH_3$.

The alkanes conform to the general formula $C_nH_{2n+2}$, where $n$ is the number of carbon atoms in the compound. A few subsequent members of the family, and their names, are

$$\begin{array}{c}\text{H} \quad \text{H} \quad \text{H} \\ | \quad\quad | \quad\quad | \\ \text{H—C—C—C—H} \\ | \quad\quad | \quad\quad | \\ \text{H} \quad \text{H} \quad \text{H}\end{array} \qquad \begin{array}{c}\text{H} \quad \text{H} \quad \text{H} \quad \text{H} \\ | \quad\quad | \quad\quad | \quad\quad | \\ \text{H—C—C—C—C—H} \\ | \quad\quad | \quad\quad | \quad\quad | \\ \text{H} \quad \text{H} \quad \text{H} \quad \text{H}\end{array}$$

$$CH_3CH_2CH_3 \qquad\qquad CH_3CH_2CH_2CH_3$$
$$\text{propane} \qquad\qquad\qquad \text{butane}$$

$$\begin{array}{c}\text{H} \quad \text{H} \quad \text{H} \quad \text{H} \quad \text{H} \\ | \quad\quad | \quad\quad | \quad\quad | \quad\quad | \\ \text{H—C—C—C—C—C—H} \\ | \quad\quad | \quad\quad | \quad\quad | \quad\quad | \\ \text{H} \quad \text{H} \quad \text{H} \quad \text{H} \quad \text{H}\end{array}$$

$$CH_3CH_2CH_2CH_2CH_3$$
$$\text{pentane}$$

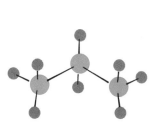

**Figure 20.2** Representation of the structure of propane, $CH_3CH_2CH_3$.

These compounds are spoken of as **straight-chain** compounds even though the carbon chains are far from linear (Figure 20.2); all the bond angles are approximately tetrahedral, and axial rotation of any carbon atom around a single bond of the chain is possible.

There is only one compound for each of the formulas $CH_4$, $C_2H_6$, and $C_3H_8$. However, there are two compounds with the formula $C_4H_{10}$: the compound with a straight chain and one with a **branched chain.**

$$\begin{array}{c}CH_3CHCH_3 \\ | \\ CH_3\end{array}$$
$$\text{methylpropane}$$

The two compounds of formula $C_4H_{10}$ are **structural isomers** (compounds with the same molecular formula but different structural formulas); they have different properties.

There are three structural isomers of $C_5H_{12}$: pentane (the straight-chain compound),

$$CH_3CH_2CHCH_3 \quad and \quad CH_3\overset{\overset{\displaystyle CH_3}{|}}{\underset{\underset{\displaystyle CH_3}{|}}{C}}CH_3$$

$$\underset{\displaystyle |}{CH_3}$$

methylbutane       dimethylpropane

In the alkane series the number of possible structural isomers increases rapidly: there are five structural isomers for $C_6H_{14}$, nine for $C_7H_{16}$, and 75 for $C_{10}H_{22}$. It has been calculated that more than $4 \times 10^9$ isomers are possible for $C_{30}H_{62}$.

A branched-chain compound may be named in terms of the longest straight chain in the molecule. The side chains are named as alkyl radicals, and their positions on the straight chain are usually indicated by a numbering system. Alkyl radicals are fragments of alkane molecules from which a hydrogen atom has been removed; their names are derived from the name of the parent alkane with the ending changed to -*yl*. A list of common alkyl radicals appears in Table 20.1.

Table 20.1 Simple alkyl radicals

| FORMULA | NAME |
|---|---|
| $CH_3—$ | methyl |
| $CH_3CH_2—$ | ethyl |
| $CH_3CH_2CH_2—$ | *normal*-propyl or *n*-propyl |
| $CH_3\underset{\underset{\displaystyle CH_3}{|}}{CH}—$ | isopropyl |
| $CH_3CH_2CH_2CH_2—$ | *normal*-butyl or *n*-butyl |
| $CH_3\underset{\underset{\displaystyle CH_3}{|}}{CH}CH_2—$ | isobutyl |
| $CH_3CH_2\underset{\underset{\displaystyle CH_3}{|}}{CH}—$ | *secondary*-butyl or *sec*-butyl |
| $CH_3\overset{\overset{\displaystyle CH_3}{|}}{\underset{\underset{\displaystyle CH_3}{|}}{C}}—$ | *tertiary*-butyl or *tert*-butyl |

For example, the longest straight chain in the molecule

$$CH_3\overset{\overset{\displaystyle CH_3}{|}}{\underset{\underset{\displaystyle CH_3}{|}}{C}}\overset{}{\underset{\underset{\underset{\underset{\displaystyle CH_3}{|}}{CH_2}}{|}}{CH}}CH_2CH_3$$

consists of five carbon atoms, and the compound is named as a derivative of pentane. The pentane chain is numbered starting at the end that will give the side chains the lowest numbers.

$$
\begin{array}{c}
CH_3 \\
| \\
CH_3C\!-\!\!-\!\!-\!CHCH_2CH_3 \\
| \qquad\quad | \\
CH_3 \quad CH_2 \\
| \\
CH_3 \\
1 \quad 2 \qquad 3 \quad 4 \quad 5
\end{array}
$$

*how about 3 tert-butylpentane?*

In the name a number is used to indicate the position of each substituent radical. Thus, the name of the compound is

2,2-dimethyl-3-ethylpentane

No numbers appear in the names of our earlier examples of structural isomers. The name methylpropane requires no number to indicate the position of the methyl radical with regard to the propane chain since there is only one possible position for the methyl group (number 2). If the methyl group were placed on either terminal carbon atom (number 1 or number 3), the compound would be the straight-chain isomer, butane. In like manner, the names methylbutane and dimethylpropane do not require the use of numbers; there is only one possible structure corresponding to each name.

Names for some of the higher straight-chain homologs may be obtained from Table 20.2, which lists the melting points and boiling points of some of the straight-chain alkanes. The melting point and boiling point increase as the length of the chain (and hence the molecular weight) of the hydrocarbon increases. The first four compounds are gases under ordinary conditions; higher homologs are liquids ($C_5H_{12}$ to $C_{15}H_{32}$) and solids (from $C_{16}H_{34}$ on).

Table 20.2 Physical properties of some straight-chain alkanes

| COMPOUND | FORMULA | MELTING POINT (°C) | BOILING POINT (°C) |
|---|---|---|---|
| methane | $CH_4$ | −183 | −162 |
| ethane | $C_2H_6$ | −172 | −88 |
| propane | $C_3H_8$ | −187 | −42 |
| butane | $C_4H_{10}$ | −135 | 1 |
| pentane | $C_5H_{12}$ | −131 | 36 |
| hexane | $C_6H_{14}$ | −94 | 69 |
| heptane | $C_7H_{16}$ | −91 | 99 |
| octane | $C_8H_{18}$ | −57 | 126 |
| nonane | $C_9H_{20}$ | −54 | 151 |
| decane | $C_{10}H_{22}$ | −30 | 174 |
| hexadecane | $C_{16}H_{34}$ | 20 | 288 |
| heptadecane | $C_{17}H_{36}$ | 23 | 303 |

**Aliphatic hydrocarbons** are open-chain structures; in addition, **cyclic hydrocarbons** are known. The **cycloalkanes** are ring structures that contain only single carbon–carbon bonds; they have the general formula $C_nH_{2n}$. Examples are

cyclopropane

cyclohexane

The cyclohexane ring is not planar but is puckered in such a way that the C–C–C bond angles are 109°28′, the tetrahedral angle. Three- and four-membered rings are strained because the C–C–C bond angles deviate significantly from the tetrahedral angle. The strain is greatest in cyclopropane, in which the C atoms form an equilateral triangle and the bond angles are 60°. Thus, the cyclopropane ring is readily broken by $H_2$ (Ni catalyst) to give propane. Cyclobutane reacts with $H_2$ in an analogous manner (to give butane), but a higher temperature is required.

Petroleum is the principle source of alkanes and cycloalkanes. In the refining of petroleum the crude oil is first separated into fractions by distillation (Table 20.3). Some higher-boiling fractions are subjected to **cracking processes** in which large molecules are broken down into smaller molecules, thus increasing the yield of the most valuable fraction, gasoline. In addition, small molecules are converted into larger ones in what are called **alkylation processes.** Gases that are by-products of these processes, together with the petroleum fractions and substances obtained from the fractions, comprise important starting materials for the manufacture of many chemical products. More than 2000 such products (such as rubber, detergents, plastics, and textiles) are derived from the substances obtained from petroleum and natural gas.

**Table 20.3** Petroleum fractions

| FRACTION | BOILING RANGE (°C) | CARBON CONTENT OF COMPOUNDS | USE |
|---|---|---|---|
| gas | below 20 | $C_1$–$C_4$ | fuel |
| petroleum ether | 20–90 | $C_5$–$C_7$ | solvent |
| gasoline | 35–220 | $C_5$–$C_{12}$ | motor fuel |
| kerosene | 200–315 | $C_{12}$–$C_{16}$ | jet fuel |
| fuel oil | 250–375 | $C_{15}$–$C_{18}$ | diesel fuel |
| lubricating oils, greases | 350 up | $C_{16}$–$C_{20}$ | lubrication |
| paraffin wax | 50–60 (m.p.) | $C_{20}$–$C_{30}$ | candles |
| asphalt | viscous liquid | | paving |
| residue | solid | | fuel |

**20.2 THE ALKENES**   The **alkenes,** which are also called **olefins,** are aliphatic hydrocarbons that have a carbon–carbon double bond somewhere in their molecular structure. The first member of the series is ethene (also called ethylene),

$$
\begin{array}{ccc}
\text{H} & & \text{H} \\
\diagdown & & \diagup \\
& \text{C}=\text{C} & \\
\diagup & & \diagdown \\
\text{H} & & \text{H}
\end{array}
$$

Each carbon atom of ethene uses $sp^2$ hybrid orbitals to form three sigma bonds: one with the other C atom and two with H atoms. Thus, all six atoms lie in the same plane, and all bond angles are approximately 120°. In addition, each carbon atom has an electron in a $p$ orbital that is not engaged in the formation of $sp^2$ $\sigma$ bonds. These two $p$ orbitals are coplanar and perpendicular to the plane of the molecule; they overlap to form a $\pi$ bonding orbital with regions of charge density above and below the plane of the molecule (Section 4.5, Figure 4.22).

The $p$ orbitals can overlap and form a $\pi$ bonding orbital only when they lie in the same plane. Consequently, free rotation around the carbon–carbon double bond is not possible without breaking the $\pi$ bond. The carbon–carbon bond distance in ethene (1.33 Å) is shorter than that in ethane (1.54 Å) because of the double bond. The $\pi$ bond is not so strong as the $\sigma$ bond; the carbon–carbon bond energy in ethane is approximately 81 kcal/mol, and the bond energy of the carbon–carbon double bond in ethene is about 145 kcal/mol.

Alkanes are called **saturated** compounds because all the valence electrons of the carbon atoms are engaged in single bond formation, and no more hydrogen atoms or atoms of other elements can be accommodated by the carbon atoms of the chain. **Unsaturated** compounds, such as the alkenes, can undergo **addition reactions** (Section 20.5) because of the availability of the $\pi$ electrons of the double bond. In one such reaction hydrogen adds to ethene to form ethane.

$$CH_2{=}CH_2 + H_2 \rightarrow CH_3{-}CH_3$$

Alkenes conform to the general formula $C_nH_{2n}$. The name of an alkene is derived from the name of the corresponding alkane by changing the ending from *-ane* to *-ene*; a number is used, when necessary, to indicate the position of the double bond. The second member of the series is propene.

$$CH_2{=}CH{-}CH_3$$

New types of isomerism arise in the alkene series. The type of structural isomerism displayed by the alkanes (e.g., butane and methylpropane) is known as **chain isomerism.** In addition to chain isomerism, **position isomerism** occurs in the olefin series. There is only one straight-chain butane. There are, however, two straight-chain butenes, and they differ in the **position** of the double bond.

$$CH_2{=}CH{-}CH_2{-}CH_3 \qquad CH_3{-}CH{=}CH{-}CH_3$$

1-butene                2-butene

In the first compound the double bond is located between carbon atoms number 1 and number 2, and the lower number (1) is used to indicate its position. In like manner, the lower number is used to indicate the position of the double bond (which occurs between carbon atoms number 2 and number 3) in the name of the second compound. In each case the chain is numbered from the end that gives the lowest possible number for the name. The name of a branched-chain alkene is derived from that of the longest straight chain that contains the double bond.

The alkenes also exhibit **geometric** (or **cis-trans**) **isomerism,** which is a type of **stereoisomerism.** Stereoisomers have the same structural formula but differ in the arrangement of the atoms in space. An example is provided by the isomers of 2-butene. Because of the restricted rotation around the double bond, one isomer exists with both methyl groups on the same side of the double bond (the *cis* isomer), and another isomer exists with the methyl groups on opposite sides of the double bond (the *trans* isomer). All carbon atoms of the molecule lie in the same plane.

*cis*-2-butene
boiling point, 1°C

*trans*-2-butene
boiling point, 2.5°C

The properties of the 2-butenes do not differ greatly. Other *cis-trans* isomers exhibit wide variation in their physical properties.

*cis*-1,2-dichloroethene
boiling point, 60.1°C

*trans*-1,2-dichloroethene
boiling point, 48.4°C

The physical properties of the alkenes are very similar to the alkanes. The $C_2H_4$, $C_3H_6$, and $C_4H_8$ compounds are gases under ordinary conditions; the $C_5H_{10}$ to $C_{18}H_{36}$ compounds are liquids; and the higher alkenes are solids. All the compounds are only slightly soluble in water.

Molecules can contain more than one double bond. For example,

$$CH_2{=}CH{-}CH{=}CH_2$$

1,3-butadiene

which is used to produce a type of synthetic rubber (Section 20.12). Cycloalkenes are known, for example,

$$H_2C—CH_2$$
$$H_2C \qquad CH_2$$
$$HC=CH$$

cycloxhexene

**20.3 THE ALKYNES**  Molecules that contain carbon–carbon triple bonds are known as alkynes. These unsaturated hydrocarbons have the general formula $C_nH_{2n-2}$ and constitute the **alkyne, or acetylene, series.** The first member of the series is ethyne, or acetylene.

$$H—C≡C—H$$

In acetylene each carbon atom uses $sp$ hybrid orbitals to form a $\sigma$ bond with a hydrogen atom and a $\sigma$ bond with the other carbon atom. Consequently, the four atoms of the molecule lie in a straight line. In addition to the carbon–carbon $\sigma$ bond, two $\pi$ bonds are formed between the carbon atoms by the overlap of $p$ orbitals (Section 4.5, Figure 4.23). The resultant carbon–carbon bond distance (1.21 Å) is shorter than the carbon–carbon double-bond distance (1.33 Å). The $\pi$ bonds are weaker than the C–C $\sigma$ bond; the bond energy of the triple bond is approximately 198 kcal/mol (compared to the single bond energy of 81 kcal/mol). Alkynes readily undergo addition reactions across the triple bond because of the availability of the four electrons of the $\pi$ bonds.

Alkynes are named in a manner analogous to that employed in naming alkenes; the ending *-yne* is used in place of *-ene*. Their physical properties are similar to those of the other aliphatic hydrocarbons. Both chain and position isomerism occur in the alkyne series, but *cis-trans* isomerism is not possible because of the linear geometry of the group X—C≡C—X, where X is a carbon atom or an atom of another element.

Notice that a branched-chain isomer of $C_4H_6$ is impossible; only two isomers of $C_4H_6$ exist

$$HC≡CCH_2CH_3 \qquad CH_3C≡CCH_3$$

1-butyne          2-butyne

Both branched-chain and straight-chain isomers of $C_5H_8$ are known.

$$HC≡CCH_2CH_2CH_3 \qquad CH_3C≡CCH_2CH_3 \qquad HC≡CCHCH_3$$
$$\qquad\qquad\qquad\qquad\qquad\qquad\qquad\qquad\qquad\qquad\qquad | $$
$$\qquad\qquad\qquad\qquad\qquad\qquad\qquad\qquad\qquad\qquad\quad CH_3$$

1-pentyne                2-pentyne              methyl-1-butyne

**20.4 AROMATIC HYDROCARBONS**

**Aromatic hydrocarbons** are compounds that have molecular structures based on that of benzene, $C_6H_6$. The six carbon atoms of benzene are arranged in a ring from which the hydrogen atoms are radially bonded (Figure 20.3). The entire structure is planar, and all of the bond angles are 120°. The carbon–carbon bond distance is 1.39 Å, which is between the single-bond distance of 1.54 Å and the double-bond distance of 1.33 Å.

The electronic structure of benzene may be represented as a resonance hybrid

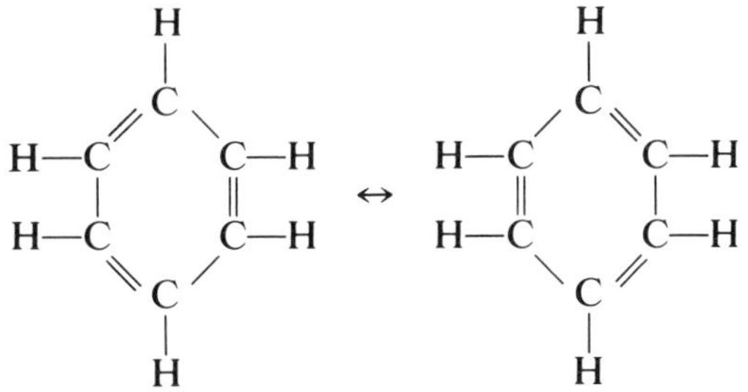

This representation correctly shows that all of the carbon–carbon bonds are equivalent and that each one is intermediate between a single bond and a double bond.

Each carbon atom of the ring uses $sp^2$ hybrid orbitals to form $\sigma$ bonds with two adjacent carbon atoms and with a hydrogen atom. Therefore, the resulting framework of the molecule is planar with bond angles of 120° (Figure 20.3). A $2p$ electron of each carbon atom is not employed in the formation of the $sp^2$-hybrid, $\sigma$ bonds. The axes of these $p$ orbitals are perpendicular to the plane of the molecule, and each $p$ orbital can

**Figure 20.3** Geometry of the benzene molecule.

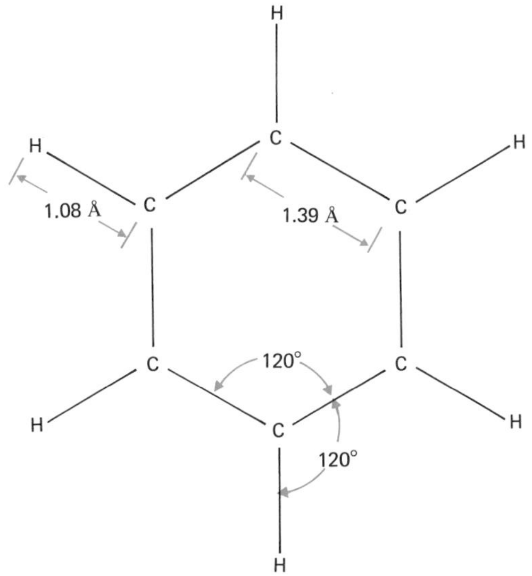

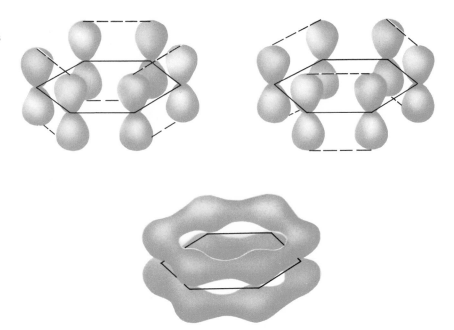

overlap two adjacent $p$ orbitals. The result is the formation of a system of $\pi$ orbitals with regions of charge density above and below the plane of the ring (Figure 20.4).

The electron pairs of the $\sigma$ bonds of the benzene molecule (as well as the electron pair of an ordinary $\pi$ bond) are localized with respect to the two nuclei that they serve to bond. However, the six electrons that occupy the $\pi$ bonding system (one contributed by each carbon atom) are delocalized. On the average, the $\pi$ electrons tend to be distributed evenly about the benzene ring, which explains the bond length and equivalence of all the C to C bonds. Since delocalization minimizes electron–electron repulsion, the structure, which is reminiscent of that of graphite (Section 12.10), is very stable and is responsible for the properties that are typical of aromatic compounds. The benzene ring does not readily undergo addition reactions in the way that alkenes and alkynes do.

Notice that the resonance and molecular-orbital pictures of benzene are qualitatively equivalent. If the overlap of the $p$ orbitals is imagined to occur only between orbitals of adjacent carbon atoms in such a way that traditional electron-pair bonds between only two atoms result, the two resonance forms are derived (Figure 20.4). According to the resonance theory, neither of these forms is correct; the true structure is a hybrid of both of them.

The benzene ring is usually represented as

The carbon and hydrogen atoms of the ring are not shown. In formulas for benzene derivatives, such as

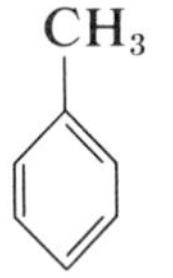

CH$_3$ and CH$_2$CH$_3$

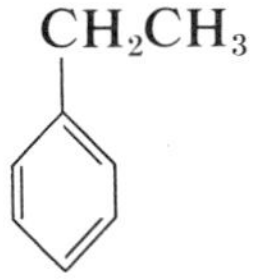

toluene
methylbenzene

ethylbenzene

it is understood that the substituents replace hydrogen atoms of the ring. The radical formed from benzene by the removal of a hydrogen atom

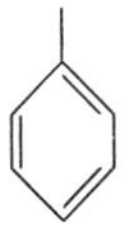

is called a **phenyl radical.**

There are three position isomers of any disubstituted benzene derivative. For example,

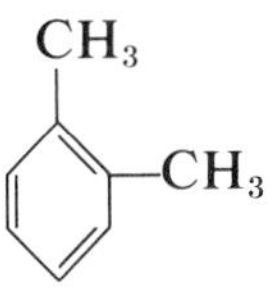

CH$_3$ —CH$_3$

CH$_3$ CH$_3$

CH$_3$ CH$_3$

o-xylene

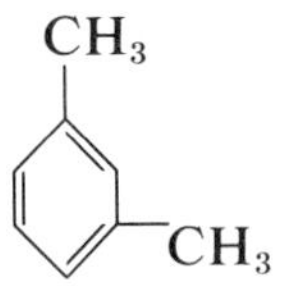

m-xylene

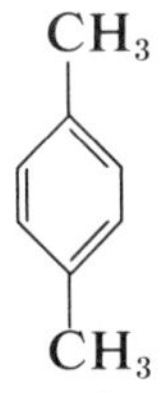

p-xylene

The prefixes *ortho-* (*o-*), *meta-* (*m-*), and *para-* (*p-*) are used to designate the positions of the two substituent groups. Numbers are used when more than two substituents are present on the ring.

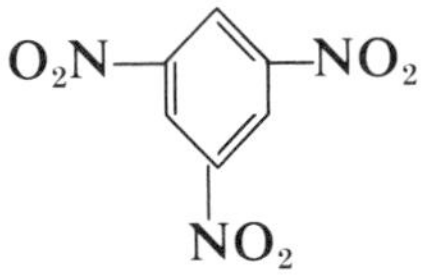

O$_2$N— —NO$_2$

NO$_2$

1,3,5-trinitrobenzene

Compounds are known in which several rings are fused together. For example, naphthalene, C$_{10}$H$_8$, is a planar molecule that is a resonance hybrid,

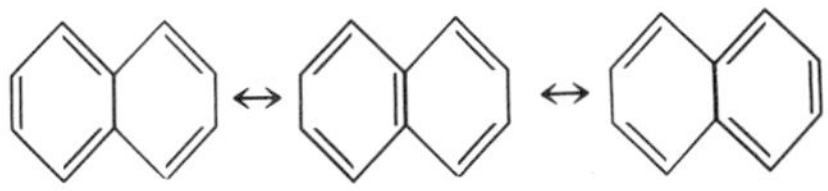

There are no hydrogen atoms bonded to the carbon atoms through which the rings are fused.

Coal is a major source of aromatic compounds. In the preparation of coke for industrial use (principally for the reduction of iron ore), bituminous coal is heated out of contact with air in retorts. Coal tar, a byproduct of this process, is a black, viscous material from which various aromatic compounds may be separated. Aromatic compounds are also obtained from processes that utilize the alkanes and cycloalkanes derived from petroleum.

## 20.5 REACTIONS OF THE HYDROCARBONS

Combustion of any hydrocarbon in excess oxygen yields carbon dioxide and water.

$$CH_4(g) + 2O_2(g) \rightarrow CO_2(g) + 2H_2O(g)$$

The reactions are highly exothermic, a fact which accounts for the use of hydrocarbons as fuels. Carbon compounds exist in oxidation states intermediate between the hydrocarbons and carbon dioxide. These compounds are discussed in sections that follow; they are not generally prepared by direct reaction with elementary oxygen.

The replacement of a hydrogen atom of a hydrocarbon molecule by another atom or group of atoms is called a **substitution reaction.** The alkanes react with chlorine or bromine in the presence of sunlight or ultraviolet light by means of a free-radical chain mechanism (Section 13.7). The chain is initiated by the light-induced dissociation of a chlorine molecule into atoms.

$$Cl_2 \rightarrow 2Cl \cdot$$

Propagation of the chain occurs by the reactions

$$Cl \cdot + CH_4 \rightarrow CH_3 \cdot + HCl$$

$$CH_3 \cdot + Cl_2 \rightarrow CH_3Cl + Cl \cdot$$

The chain is terminated by the reactions

$$CH_3 \cdot + Cl \cdot \rightarrow CH_3Cl$$

$$Cl \cdot + Cl \cdot \rightarrow Cl_2$$

$$CH_3 \cdot + CH_3 \cdot \rightarrow CH_3CH_3$$

The reactions that terminate the chain occur at the walls of the container. The energy liberated by the formation of the new bond is transferred to the walls of the container. In a collision between radicals in the gas phase, the united fragments fly apart almost immediately because a third body is not present to carry away the energy released by bond formation. Hence, chain-terminating reactions do not occur with great frequency, and the initial reactants are rapidly consumed.

The overall reaction for the preceding mechanism is

$$CH_4 + Cl_2 \rightarrow CH_3Cl + HCl$$

In addition to chloromethane, $CH_3Cl$, the reaction produces dichloromethane, $CH_2Cl_2$, trichloromethane (chloroform), $CHCl_3$, and tetrachloromethane (carbon tetrachloride), $CCl_4$. In terms of the chain mechanism, these compounds arise through collisions of chlorine atoms with already chlorinated methane molecules. For example,

$$Cl\cdot + CH_3Cl \rightarrow CH_2Cl\cdot + HCl$$

$$CH_2Cl\cdot + Cl_2 \rightarrow CH_2Cl_2 + Cl\cdot$$

Complex mixtures of mono- and polysubstituted isomers are obtained as products of the chlorination of higher alkanes.

**Addition reactions** are characteristic of unsaturated hydrocarbons; examples follow.

$$\underset{}{\overset{H_2C-CH_2}{\underset{HC=CH}{H_2C \qquad CH_2}}} + H_2 \xrightarrow{Pt} \underset{\text{cyclohexane}}{\overset{H_2C-CH_2}{\underset{H_2C-CH_2}{H_2C \qquad CH_2}}}$$

$$CH_2{=}CH_2 + Br_2 \longrightarrow \underset{\text{1,2-dibromoethane}}{CH_2Br-CH_2Br}$$

$$HC{\equiv}CH + HCl \longrightarrow \underset{\text{1-chloroethene}}{CH_2{=}CHCl}$$

$$CH_2{=}CHCl + HCl \longrightarrow \underset{\text{1,1-dichloroethane}}{CH_3-CHCl_2}$$

$$\underset{}{\overset{CH_3}{\underset{}{CH_3CH_2C{=}CH_2}}} + HOH \xrightarrow{H_2SO_4} \underset{\text{2-methyl-2-butanol}}{\overset{CH_3}{\underset{OH}{CH_3CH_2CCH_3}}}$$

$$CH_3CH{=}CH_2 + HBr \longrightarrow \underset{\text{2-bromopropane}}{\overset{}{\underset{Br}{CH_3CHCH_3}}}$$

When unsymmetrical molecules (such as HBr) add to the double bond of an olefin, two isomeric compounds are sometimes produced. Thus, in the last reaction of the previous series, two compounds are obtained.

$$\underset{\text{2-bromopropane}}{\overset{}{\underset{Br}{CH_3CHCH_3}}} \qquad \underset{\text{1-bromopropane}}{CH_3CH_2CH_2Br}$$

Over 90% of the product, however, is the 2-bromo-isomer. The principal product of such an addition is that in which the hydrogen atom of the addend is bonded to the carbon that originally had the larger number of hydrogen atoms (**Markovnikov's rule**).

The initial step of the mechanism of the addition of HBr is thought to be an electrophilic attack by a proton (a Lewis acid) on the $\pi$ electrons of the double bond (a Lewis base). The product of this step is a positive ion called a **carbonium ion;** for propene two such ions are possible.

$$CH_3-\underset{H}{\overset{H}{C}}=\underset{H}{\overset{H}{C}}-H + H^+$$

$$CH_3-\underset{\underset{+}{H}}{\overset{H}{C}}-\underset{H}{\overset{H}{C}}-H$$

$$CH_3-\underset{H}{\overset{H}{C}}-\underset{+}{\overset{H}{C}}-H$$

The $\pi$ electrons are used to bond the incoming proton to one of the carbon atoms of the double bond which leaves the other carbon atom with a positive charge. The stability of carbonium ions is known to decrease in the order

$$R-\underset{+}{\overset{R}{C}}-R \; > \; R-\underset{+}{\overset{H}{C}}-R \; > \; R-\underset{+}{\overset{H}{C}}-H$$

where R is an alkyl radical. Consequently, in the reaction of propene with HBr, the carbonium ion produced in larger quantity is the one shown on top.

Carbonium ions are highly reactive Lewis acids; they exist only transiently and rapidly combine with bromide ion (a Lewis base).

$$CH_3\underset{+}{C}HCH_3 + Br^- \rightarrow CH_3\underset{\underset{Br}{|}}{C}HCH_3$$

$$CH_3CH_2CH_2^+ + Br^- \rightarrow CH_3CH_2CH_2Br$$

Thus, the relative stabilities of the carbonium ions are responsible for the fact that 2-bromopropane is the principal product of the reaction.

In contrast to unsaturated aliphatic hydrocarbons, the principal reactions of benzene are substitutions, not additions.

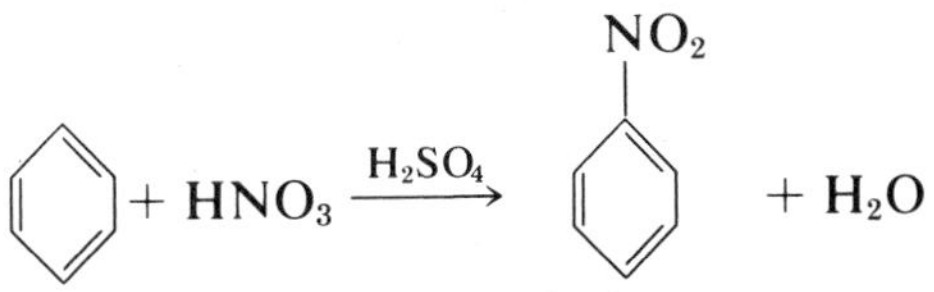

$$\text{C}_6\text{H}_6 + \text{HNO}_3 \xrightarrow{\text{H}_2\text{SO}_4} \text{C}_6\text{H}_5\text{NO}_2 + \text{H}_2\text{O}$$

nitrobenzene

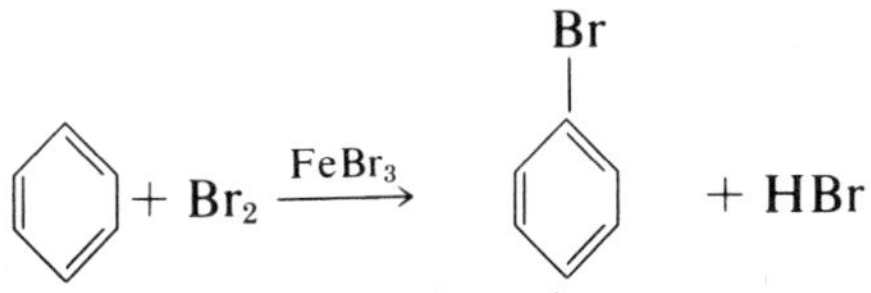

$$\text{C}_6\text{H}_6 + \text{Br}_2 \xrightarrow{\text{FeBr}_3} \text{C}_6\text{H}_5\text{Br} + \text{HBr}$$

bromobenzene

$$\text{C}_6\text{H}_6 + \text{CH}_3\text{Cl} \xrightarrow{\text{AlCl}_3} \text{C}_6\text{H}_5\text{CH}_3 + \text{HCl}$$

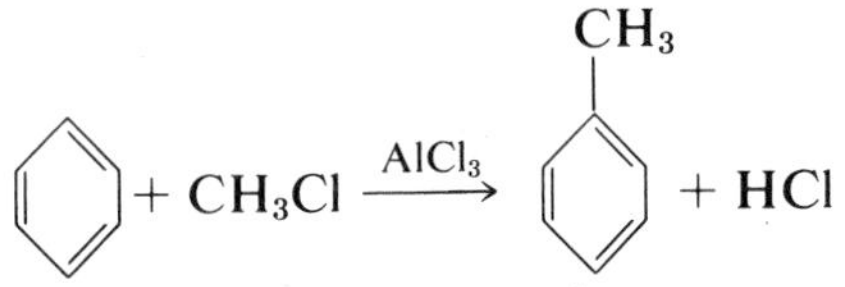

toluene

The last reaction is the Friedel–Crafts synthesis for the preparation of aromatic hydrocarbons.

The catalysts, which are indicated over the arrows in the preceding equations, produce powerful Lewis acids ($NO_2^+$, $Br^+$, and $CH_3^+$) as follows:

$$HNO_3 + 2H_2SO_4 \rightarrow NO_2^+ + H_3O^+ + 2HSO_4^-$$

$$Br_2 + FeBr_3 \rightarrow Br^+ + FeBr_4^-$$

$$CH_3Cl + AlCl_3 \rightarrow CH_3^+ + AlCl_4^-$$

These cationic Lewis acids, which are short-lived reaction intermediates, are electrophilic, and we shall indicate them as $E^+$ in the mechanism that follows.

The benzene ring is attacked by the electrophilic group, $E^+$, which bonds to a carbon atom by means of a pair of $\pi$ electrons and creates an activated complex with a positive charge.

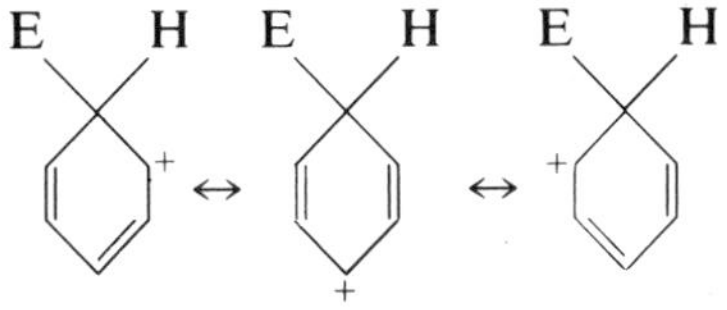

Notice that the carbon atom to which E is bonded holds two groups in the complex; the remaining five carbon atoms hold only their customary hydrogen atoms. The complex is a resonance hybrid, and the charge is delocalized, thus providing a degree of stability. The $\pi$ bonding system of benzene is more stable, however, and is restored by the loss of a proton.

$$E \quad H \qquad E$$

The proton is lost to the anion produced in the reaction that generated the cationic electrophile.

$$H^+ + HSO_4^- \rightarrow H_2SO_4$$

$$H^+ + FeBr_4^- \rightarrow HBr + FeBr_3$$

$$H^+ + AlCl_4^- \rightarrow HCl + AlCl_3$$

**20.6 ALCOHOLS AND ETHERS** The alcohols may be considered as derivatives of hydrocarbons in which a hydroxyl group, OH, replaces a hydrogen. The hydroxyl group is one of a number of **functional groups,** which are groups of atoms that give organic compounds bearing them characteristic chemical and physical properties.

The alcohols are named as derivatives of a hydrocarbon consisting of the longest straight chain that contains the OH group. The ending of the parent hydrocarbon is changed from *-ane* to *-anol,* and numbers are used to indicate the positions of substituents, the lowest possible number being assigned to the OH group. Alcohols may be classified as *primary, secondary,* or *tertiary,* according to the number of alkyl radicals on the carbon atom holding the OH. Thus,

$$CH_3CH_2CH_2CH_2OH$$

1-butanol
a primary alcohol

$$CH_3{-}\overset{\displaystyle CH_3}{\underset{\displaystyle OH}{\overset{|}{\underset{|}{CH}}}}{-}CHCH_3$$

3-methyl-2-butanol
a secondary alcohol

$$CH_3\overset{\displaystyle CH_3}{\underset{\displaystyle OH}{\overset{|}{\underset{|}{C}}}}CH_2CH_3$$

2-methyl-2-butanol
a tertiary alcohol

Alcohols associate through hydrogen bonding,

$$\overset{\displaystyle R}{\underset{\displaystyle O-H}{\overset{|}{\phantom{O}}}}\cdots\overset{\displaystyle R}{\underset{\displaystyle O-H}{\overset{|}{\phantom{O}}}}\cdots\overset{\displaystyle R}{\underset{\displaystyle O-H}{\overset{|}{\phantom{O}}}}$$

For this reason, the melting points and boiling points of the alcohols are higher than those of alkanes of corresponding molecular weight (Table 20.4). The lower alcohols are miscible with water in all proportions because of intermolecular hydrogen bonding.

$$\overset{\displaystyle H}{\underset{\displaystyle O-H}{\overset{|}{\phantom{O}}}}\cdots\overset{\displaystyle R}{\underset{\displaystyle O-H}{\overset{|}{\phantom{O}}}}\cdots\overset{\displaystyle H}{\underset{\displaystyle O-H}{\overset{|}{\phantom{O}}}}$$

However, as the size of the alkyl radical increases, the alcohols become more like alkanes in their physical properties, and the higher alcohols are only slightly soluble in water.

Table 20.4 Physical properties of some alcohols

| NAME | FORMULA | MELTING POINT (°C) | BOILING POINT (°C) | SOLUBILITY IN WATER (g/100 g $H_2O$, 20°C) |
|---|---|---|---|---|
| methanol | $CH_3OH$ | $-98$ | 65 | miscible |
| ethanol | $CH_3CH_2OH$ | $-115$ | 78 | miscible |
| 1-propanol | $CH_3(CH_2)_2OH$ | $-127$ | 97 | miscible |
| 1-butanol | $CH_3(CH_2)_3OH$ | $-90$ | 117 | 7.9 |
| 1-pentanol | $CH_3(CH_2)_4OH$ | $-79$ | 138 | 2.4 |
| 1-hexanol | $CH_3(CH_2)_5OH$ | $-52$ | 157 | 0.6 |

Alcohols, like water, are amphiprotic substances. They can act as Brønsted bases (proton acceptors) with very strong acids.

$$H_2SO_4 + ROH \rightleftharpoons ROH_2^+ + HSO_4^-$$

With very strong bases they are Brønsted acids (proton donors).

$$ROH + H^- \rightarrow OR^- + H_2$$

Alcohols, however, are more weakly acidic and basic than water; consequently, the conjugate acids ($ROH_2^+$) and conjugate bases ($OR^-$) are stronger than $H_3O^+$ and $OH^-$. The anion $OR^-$, known as an alkoxide ion, results from the reaction of an alcohol with a reactive metal (compare the reaction of water with sodium).

$$2CH_3CH_2OH + 2Na \rightarrow 2CH_3CH_2ONa + H_2$$

The first member of the series, methanol (or methyl alcohol), is known as wood alcohol because it can be obtained by the destructive distillation of wood. The principal commercial source of this alcohol is the catalytic hydrogenation of carbon monoxide.

$$CO + 2H_2 \rightarrow CH_3OH$$

The alcohol of alcoholic beverages is ethanol, or ethyl alcohol. It is prepared by the fermentation of starches or sugars. Ethanol is also commercially prepared from ethene. The indirect addition of water to olefins by means of sulfuric acid is a general method of preparing alcohols. The synthesis of ethanol proceeds by the following steps.

$$CH_2{=}CH_2 + H_2SO_4 \rightarrow CH_3CH_2OSO_2OH$$

$$CH_3CH_2OSO_2OH + H_2O \rightarrow CH_3CH_2OH + H_2SO_4$$

Alcohols can be prepared from alkyl halides by displacement, or nucleophilic substitution, reactions.

$$RX + OH^- \rightarrow ROH + X^-$$

This is an important classification of organic reactions; it includes the reactions of the alkyl halides, as well as the alcohols themselves, with a wide variety of nucleophilic substances. The reactions are usually reversible and are run under conditions that favor the formation of the desired compound; an excess of the nucleophilic reagent may be employed or the product may be removed from the reaction mixture as it forms (e.g., by distillation).

Primary and secondary halides or alcohols generally react by what is called an $S_N2$ mechanism, which can be illustrated by the reaction of hydroxide ion with methyl bromide. The hydroxide ion, a Lewis base, attacks the carbon atom and displaces the bromide ion (also a Lewis base), which takes along the electron pair with which it had been bonded.

$$OH^- + H{-}\overset{\displaystyle H}{\underset{\displaystyle H}{C}}{-}Br \rightarrow \left[ HO{-}{-}{-}\overset{\displaystyle H}{\underset{\displaystyle H}{C}}{-}{-}{-}Br \right]^- \rightarrow HO{-}\overset{\displaystyle H}{\underset{\displaystyle H}{C}}{-}H + Br^-$$

activated complex

The attack of the $OH^-$ ion takes place on the side of the carbon atom that is opposite the bromine atom. As the $OH^-$ ion approaches, it begins to form a covalent bond, and the bromide ion begins to break away. The activated complex of the reaction is the form in which both nucleophilic groups are partially bonded to the carbon atom. The reaction causes inversion of the geometric arrangement of the groups attached to the carbon atom, a process that is customarily compared to the inversion of an umbrella in a high wind. The mechanism is given the designation $S_N2$ because it is the substitution of one nucleophilic group for another, and the rate-determining step (the formation of the activated complex) is bimolecular. The nucleophilic substitution reaction of $HBr$ with $CH_3OH$ proceeds through the formation of $CH_3OH_2^+$ and the displacement of $H_2O$ from this ion by $Br^-$.

In the case of tertiary alkyl halides, the three alkyl groups bonded to the carbon atom bearing the halogen inhibit the rearward approach of the $OH^-$ ion, and it is thought that tertiary alkyl halides undergo displacement reactions by a different mechanism. The rate-determining step is thought to be the formation of a carbonium ion,

$$CH_3{-}\overset{\displaystyle CH_3}{\underset{\displaystyle CH_3}{C}}{-}Br \rightarrow CH_3\overset{\displaystyle CH_3}{\underset{\displaystyle CH_3}{C^+}} + Br^-$$

The central carbon atom of the carbonium ion has only six electrons in its valence level and displays $sp^2$ hybridization; the carbonium ion is therefore planar. It is a powerful Lewis acid and rapidly adds hydroxide ion (a Lewis base).

$$CH_3-\underset{\underset{CH_3}{|}}{\overset{\overset{CH_3}{|}}{C}}{}^+ + OH^- \rightarrow CH_3-\underset{\underset{CH_3}{|}}{\overset{\overset{CH_3}{|}}{C}}-OH$$

This mechanism is given the designation $S_N1$ because it is a nucleophilic substitution in which the rate-determining step is unimolecular. The $S_N1$ mechanism is favored when the reaction is run in polar solvents (such as water), which aid the first-step ionization of the halide.

Tertiary alcohols also undergo $S_N1$ reactions. The steps of a typical substitution involving HBr are

$$(CH_3)_3COH + H^+ \rightleftharpoons (CH_3)_3COH_2^+ \qquad \text{(rapid, equilibrium)}$$

$$(CH_3)_3COH_2^+ \rightarrow (CH_3)_3C^+ + H_2O \quad \text{(rate determining)}$$

$$(CH_3)_3C^+ + Br^- \rightarrow (CH_3)_3CBr$$

In all these nucleophilic substitution reactions, appreciable quantities of olefins are formed along with the substitution products; this is particularly true of the reactions of the tertiary compounds. Olefin formation is thought to proceed by the elimination of a proton by a carbonium ion.

$$H-\underset{\underset{H}{|}}{\overset{\overset{H}{|}}{C}}-\underset{\underset{CH_3}{|}}{\overset{\overset{CH_3}{|}}{C}}{}^+ \rightarrow H^+ + \underset{\underset{H}{|}}{\overset{\overset{H}{|}}{C}}=\underset{\underset{CH_3}{|}}{\overset{\overset{CH_3}{|}}{C}}$$

Tertiary carbonium ions are formed more readily than any other type (Section 20.5); elimination reactions of primary and secondary alcohols are thought to occur by a mechanism that does not involve the formation of carbonium ions.

The elimination of one water molecule from one molecule of an alcohol produces an olefin.

$$CH_3CH_2OH \xrightarrow[\text{above 150°C}]{H_2SO_4} CH_2{=}CH_2 + H_2O$$

Under milder conditions only one molecule of water is removed from two molecules of alcohol, and an ether is produced.

$$CH_3CH_2OH + HOCH_2CH_3 \xrightarrow[\text{130-140°C}]{H_2SO_4} CH_3CH_2OCH_2CH_3 + H_2O$$
$$\text{diethyl ether}$$

Ethers, ROR, may be regarded as derivatives of water, HOH, in which both hydrogen atoms are replaced by alkyl groups, R. The alkyl groups may be alike or different. Ethers are also produced by nucleophilic substitution reactions involving alkoxide ions and alkyl halides.

$$CH_3CH_2O^- + CH_3CH_2CH_2Br \rightarrow CH_3CH_2OCH_2CH_2CH_3 + Br^-$$
ethyl propyl ether

Unlike alcohols, ethers do not associate by hydrogen bonding. Therefore, the boiling points of ethers are much lower than those of alcohols of corresponding molecular weight, and the ethers are much less soluble in water.

$CH_3CH_2OH$　　　　$CH_3OCH_3$

ethanol　　　　　dimethyl ether
boiling point, 78°C　　boiling point, −24°C

Notice that ethanol and dimethyl ether are isomeric; the type of isomerism displayed by this pair of compounds is known as **functional group isomerism.** The ethers are less reactive than the alcohols.

Polyhydroxy alcohols contain more than one OH group. Examples include 1,2-ethanediol (ethylene glycol),

$$\begin{array}{cc} CH_2 & \!\!\!-CH_2 \\ | & | \\ OH & OH \end{array}$$

and 1,2,3-propanetriol (glycerol), which is derived from fats,

$$\begin{array}{ccc} CH_2 & \!\!\!-CH & \!\!\!-CH_2 \\ | & | & | \\ OH & OH & OH \end{array}$$

Unlike aliphatic alcohols, which do not dissociate in water solution, the aromatic alcohols are weak acids in water.

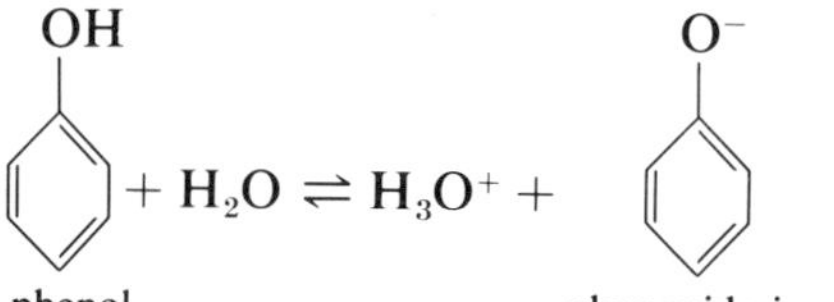

phenol　　　　　　　　phenoxide ion

The acidity of phenol (carbolic acid) is attributed to the stability of the phenoxide ion in which the charge is delocalized by the $\pi$ bonding system of the aromatic ring.

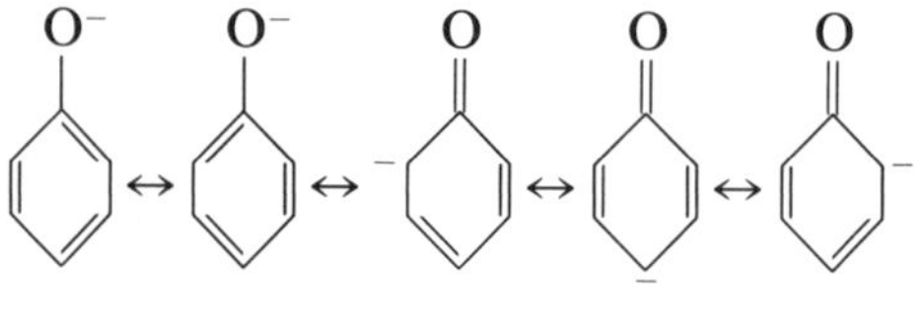

Aromatic halides do not readily undergo displacement reactions. Sodium phenoxide is commercially obtained from chlorobenzene by reaction with NaOH at elevated temperatures and under high pressure.

Phenol is derived by acidification of the phenoxide.

**20.7 CARBONYL COMPOUNDS**

A carbon atom can form a double bond with an oxygen atom to produce what is called a **carbonyl group**

$$\overset{O}{\underset{\phantom{C}}{\overset{\|}{-C-}}}$$

The double bond, like the carbon–carbon double bond, consists of a $\sigma$ bond and a $\pi$ bond between the bonded atoms. The carbon atom of the carbonyl group and the atoms bonded to it are coplanar and form bond angles of 120°, a geometry that is typical of $sp^2$ hybridized species. The carbonyl double bond differs from the olefinic double bond in that it is markedly polar (oxygen is more electronegative than carbon).

If the carbonyl group is bonded to one alkyl group and one hydrogen or to two hydrogens the compound is an **aldehyde.**

$$\overset{O}{\underset{\substack{\text{formaldehyde}\\\text{methanal}}}{\overset{\|}{H-C-H}}} \qquad \overset{O}{\underset{\substack{\text{acetaldehyde}\\\text{ethanal}}}{\overset{\|}{CH_3-C-H}}} \qquad \overset{CH_3 \quad\; O}{\underset{\text{3-methylbutanal}}{\overset{|\qquad\;\|}{CH_3-C-CH_2-C-H}}}$$

Systematic nomenclature of aldehydes employs the ending *-al;* substituent groups are given numbers based on the assignment of the number 1 (understood) to the carbonyl carbon.

In a **ketone** the carbonyl group is bonded to two alkyl groups, which may be alike or different.

$$\underset{\substack{\text{acetone}\\\text{propanone}}}{\overset{O}{\overset{\|}{CH_3-C-CH_3}}} \quad \underset{\substack{\text{methyl ethyl ketone}\\\text{butanone}}}{\overset{O}{\overset{\|}{CH_3-C-CH_2CH_3}}} \quad \underset{\text{5-methyl-3-hexanone}}{\overset{O \qquad\quad CH_3}{\overset{\|\qquad\qquad\;|}{CH_3CH_2-C-CH_2CHCH_3}}}$$

The ending *-one* is used to designate a ketone. Numbers are employed to indicate the positions of substituents on the longest continuous chain containing the carbonyl group, which is assigned the lowest possible number.

Aldehydes may be prepared by the mild oxidation of primary alcohols. An oxidizing agent such as potassium dichromate in dilute sulfuric acid may be used.

$$3RCH_2OH + Cr_2O_7^{2-} + 8H^+ \rightarrow 3R\overset{\displaystyle O}{\overset{\displaystyle \|}{C}}-H + 2Cr^{3+} + 7H_2O$$

Equations for a reaction such as this are usually written with the oxidizing agent indicated over the arrow and only the organic reactant and product shown. Thus,

$$CH_3CH_2OH \xrightarrow{Cr_2O_7^{2-}/H^+} CH_3\overset{\displaystyle O}{\overset{\displaystyle \|}{C}}-H$$

When an aldehyde is prepared by the oxidation of a primary alcohol, provision must be made to prevent the aldehyde from being destroyed by further oxidation, since aldehydes are easily oxidized to carboxylic acids (Section 20.8). Many aldehydes may be distilled out of the reaction mixture as they are formed.

Some aldehydes are made commercially by reacting alcohol vapors with air over a copper catalyst at elevated temperatures.

$$2CH_3OH + O_2 \xrightarrow{Cu} 2H-\overset{\displaystyle O}{\overset{\displaystyle \|}{C}}-H + 2H_2O$$

The oxidation of alcohols may also be conducted by passing the hot alcohol vapor over a heated copper catalyst in the absence of oxygen. This process results in the removal of a molecule of hydrogen from an alcohol molecule and is called a **dehydrogenation**; it avoids the danger of secondary oxidation of the aldehyde product.

$$CH_3CH_2CH_2OH \xrightarrow{Cu} CH_3CH_2\overset{\displaystyle O}{\overset{\displaystyle \|}{C}}-H + H_2$$

The oxidation of secondary alcohols produces ketones.

$$CH_3CH_2\overset{\displaystyle OH}{\overset{\displaystyle |}{C}}HCH_3 \xrightarrow{Cr_2O_7^{2-}/H^+} CH_3CH_2\overset{\displaystyle O}{\overset{\displaystyle \|}{C}}CH_3$$

Acetone (propanone) may be prepared by the oxidation of 2-propanol with oxygen or by the dehydrogenation of the alcohol.

$$CH_3\overset{\displaystyle OH}{\overset{\displaystyle |}{C}}HCH_3 \xrightarrow{Cu} CH_3\overset{\displaystyle O}{\overset{\displaystyle \|}{C}}CH_3 + H_2$$

The oxidation of tertiary alcohols results in the destruction of the carbon skeleton of the molecule.

The double bond of the carbonyl group readily undergoes many addition reactions. Reduction of aldehydes or ketones with hydrogen in the presence of a nickel or platinum catalyst yields primary or secondary alcohols.

$$\underset{\substack{\| \\ O}}{R-C-H} + H_2 \rightarrow \underset{\substack{| \\ OH}}{R-CH_2}$$

$$\underset{\substack{\| \\ O}}{R-C-R} + H_2 \rightarrow \underset{\substack{| \\ OH}}{R-CH-R}$$

When an unsymmetrical reagent (such as HCN) adds to the double bond, the positive part of the addend (the hydrogen) adds to the oxygen, and the negative part of the addend (the cyanide group) bonds to the carbon. This mode of addition reflects the polarity of a carbonyl double bond; the oxygen is negative with respect to the carbon.

$$R-\overset{O}{\underset{\|}{C}}-R + H-C\equiv N \rightarrow R-\overset{OH}{\underset{\underset{C\equiv N}{|}}{C}}-R$$

a cyanohydrin

The proposed mechanism for additions of this type consists of the electrophilic attack of the anionic Lewis base

$$R-\overset{O}{\underset{\|}{C}}-R + CN^- \rightarrow R-\overset{O^-}{\underset{\underset{CN}{|}}{C}}-R$$

followed by combination with a proton

$$R-\overset{O^-}{\underset{\underset{CN}{|}}{C}}-R + H^+ \rightarrow R-\overset{OH}{\underset{\underset{CN}{|}}{C}}-R$$

An important addition reaction involves organomagnesium compounds known as **Grignard reagents.** Alkyl halides react with magnesium metal in dry diethyl ether

$$R-X + Mg \rightarrow R-Mg-X$$

These reagents are usually given the formula RMgX; however, the materials possess a high degree of ionic character and may be mixtures of magnesium dialkyls ($MgR_2$) and magnesium halides ($MgX_2$). Water must be excluded from a Grignard reaction because these reagents are easily hydrolyzed.

$$R—Mg—X + HOH \rightarrow R—H + Mg(OH)X$$

The formula $Mg(OH)X$ stands for an equimolar mixture of magnesium hydroxide and magnesium halide.

A Grignard reagent adds to an aldehyde as follows

$$
\begin{array}{ccc}
\overset{\displaystyle O}{\underset{\phantom{}}{\|}} & & \overset{\displaystyle OMgX}{\underset{\phantom{}}{|}} \\
R'—C—H + R—Mg—X \rightarrow & & R'—\overset{|}{\underset{|}{C}}—H \\
& & R
\end{array}
$$

where the alkyl groups R and R' may be alike or different. The addition compound is readily decomposed by water or by dilute acid.

$$
\begin{array}{ccc}
\overset{\displaystyle OMgX}{\underset{\phantom{}}{|}} & & \overset{\displaystyle OH}{\underset{\phantom{}}{|}} \\
R'—\overset{|}{\underset{|}{C}}—H + HX \rightarrow & & R'—\overset{|}{\underset{|}{C}}—H + MgX_2 \\
R & & R
\end{array}
$$

Thus, the ultimate product of the reaction of a Grignard reagent and an aldehyde is a secondary alcohol.

Hydrolysis of an addition compound formed by a Grignard reagent and a ketone yields a tertiary alcohol.

$$
\begin{array}{ccc}
\overset{\displaystyle O}{\underset{\phantom{}}{\|}} & & \overset{\displaystyle OMgX}{\underset{\phantom{}}{|}} \\
R'—C—R'' + RMgX \rightarrow & & R'—\overset{|}{\underset{|}{C}}—R'' \\
& & R
\end{array}
$$

$$
\begin{array}{ccc}
\overset{\displaystyle OMgX}{\underset{\phantom{}}{|}} & & \overset{\displaystyle OH}{\underset{\phantom{}}{|}} \\
R'—\overset{|}{\underset{|}{C}}—R'' + HX \rightarrow & & R'—\overset{|}{\underset{|}{C}}—R'' + MgX_2 \\
R & & R
\end{array}
$$

The alkyl groups may be alike or different.

In these reactions of Grignard reagents, new carbon–carbon bonds are formed, and the reactions are frequently useful as steps in organic syntheses. The alcohols produced may be oxidized to carbonyl compounds, subjected to displacement reactions, or dehydrated to olefins. Thus, a series of reactions may be employed to synthesize a desired compound from compounds of lower molecular weight.

**20.8 CARBOXYLIC ACIDS AND ESTERS**

Oxidation of the aldehyde group yields a **carboxyl group**

$$-\overset{\overset{\displaystyle O}{\|}}{C}-H \rightarrow -\overset{\overset{\displaystyle O}{\|}}{C}-OH$$

Compounds containing the —COOH group are weak acids (**carboxylic acids**); ionization constants for some of these acids are listed in Table 20.5.

$$R-\overset{\overset{\displaystyle O}{\|}}{C}-OH + H_2O \rightleftharpoons R-\overset{\overset{\displaystyle O}{\|}}{C}-O^- + H_3O^+$$

The charge of the carboxylate anion is delocalized.

$$R-\overset{\overset{\displaystyle O}{\|}}{C}-O^- \leftrightarrow R-\overset{\overset{\displaystyle O^-}{|}}{C}=O$$

The acids are associated through hydrogen bonding. Lower members of the series form dimers in the vapor state.

$$R-C\overset{O---H-O}{\underset{O-H---O}{}}C-R$$

Table 20.5 Properties of some carboxylic acids

| ACID | FORMULA | MELTING POINT (°C) | BOILING POINT (°C) | IONIZATION CONSTANT AT 25°C |
|---|---|---|---|---|
| methanoic (formic) | HCOOH | 8 | 101 | $1.8 \times 10^{-4}$ |
| ethanoic (acetic) | $CH_3COOH$ | 17 | 118 | $1.8 \times 10^{-5}$ |
| propanoic (propionic) | $CH_3CH_2COOH$ | −22 | 141 | $1.4 \times 10^{-5}$ |
| butanoic (butyric) | $CH_3(CH_2)_2COOH$ | −8 | 164 | $1.5 \times 10^{-5}$ |
| pentanoic (valeric) | $CH_3(CH_2)_3COOH$ | −35 | 187 | $1.6 \times 10^{-5}$ |
| benzoic | $C_6H_5COOH$ | 122 | 249 | $6.0 \times 10^{-5}$ |

According to systematic nomenclature, the name of a carboxylic acid is derived from the parent hydrocarbon by elision of the final -*e*, addition of the ending -*oic*, and addition of the separate word *acid*. The numbers used to designate the positions of substituents on the carbon chain are derived by numbering the chain starting with the carbon of the carboxyl group.

$$\underset{\text{propanoic acid}}{CH_3CH_2\overset{\overset{\displaystyle O}{\|}}{C}-OH} \qquad \underset{\text{3-methylbutanoic acid}}{CH_3\overset{\overset{\displaystyle CH_3}{|}}{CH}CH_2\overset{\overset{\displaystyle O}{\|}}{C}-OH}$$

Carboxylic acids may be prepared by the oxidation of primary alcohols or aldehydes. Potassium dichromate or potassium permanganate are frequently employed as the oxidizing agent.

$$CH_3CH_2CH_2CH_2OH \xrightarrow{Cr_2O_7^{2-}/H^+} CH_3CH_2CH_2\overset{\overset{\displaystyle O}{\|}}{C}-OH$$

Acetic acid is commercially prepared by the air oxidation of acetaldehyde.

$$2CH_3\overset{\overset{\displaystyle O}{\|}}{C}-H + O_2 \xrightarrow{Mn(C_2H_3O_2)_2} 2CH_3\overset{\overset{\displaystyle O}{\|}}{C}-OH$$

Vigorous oxidation of an alkyl-substituted aromatic compound converts the side chain to a carbonyl group and thus yields benzoic acid.

toluene            benzoic acid

These oxidations do not touch the aromatic ring, an illustration of the stability of this structure.

The oxidation of an unsaturated hydrocarbon yields a variety of products, including carboxylic acids, depending upon conditions. Dilute aqueous permanganate at room temperature oxidizes olefins to glycols.

$$R-CH=CH-R \xrightarrow{MnO_4^-} \underset{\qquad \overset{\displaystyle |}{OH} \overset{\displaystyle |}{OH}}{R-CH-CH-R}$$

Under more vigorous conditions (more concentrated solutions and heating), the carbon chain is cleaved at the double bond and carboxylic acids are produced.

$$R-CH=CH-R \xrightarrow{MnO_4^-} R-\overset{\overset{\displaystyle O}{\|}}{C}-OH + HO-\overset{\overset{\displaystyle O}{\|}}{C}-R$$

Alkynes also may be cleaved to yield acids.

$$R-C\equiv C-R \xrightarrow{MnO_4^-} R-\overset{\overset{\displaystyle O}{\|}}{C}-OH + HO-\overset{\overset{\displaystyle O}{\|}}{C}-R$$

Oxidation of some olefins yields ketones.

$$R-\underset{\underset{R}{|}}{C}=\underset{\underset{R}{|}}{C}-R \xrightarrow{MnO_4^-} R-\underset{\underset{R}{|}}{C}=O + O=\underset{\underset{R}{|}}{C}-R$$

Salts of carboxylic acids may be produced by the alkaline hydrolysis of alkyl cyanides; the cyanides are the product of nucleophilic substitution reactions of the $CN^-$ ion with alkyl halides.

$$CH_3CH_2Br + CN^- \rightarrow CH_3CH_2CN + Br^-$$

$$CH_3CH_2CN + OH^- + H_2O \rightarrow CH_3CH_2\overset{\overset{O}{\|}}{C}-O^- + NH_3$$

Some compounds contain more than one carboxyl group. Examples include

$$\underset{\text{oxalic acid}}{\underset{\underset{COOH}{|}}{COOH}} \quad \text{and} \quad \underset{\text{citric acid}}{HO-\underset{\underset{CH_2COOH}{|}}{\overset{\overset{CH_2COOH}{|}}{C}}-COOH}$$

Carboxylic acids react with alcohols to produce compounds known as esters.

$$CH_3\overset{\overset{O}{\|}}{C}-OH + CH_3CH_2OH \rightarrow \left[CH_3\underset{\underset{OCH_2CH_3}{|}}{\overset{\overset{OH}{|}}{C}}-OH\right] \rightarrow$$

$$\underset{\text{ethyl acetate}}{CH_3\overset{\overset{O}{\|}}{C}-OCH_2CH_3 + H_2O}$$

Esterification reactions are reversible and proceed to equilibrium; they may be forced in either direction by the appropriate choice of conditions. The name of an ester reflects the alcohol and acid from which it is derived, the ending -*ate* being employed with the base of the name of the acid to give the second portion of the name of the ester.

The lower molecular weight esters have pleasant, fruity odors. Esters of high-molecular-weight carboxylic acids and glycerol are animal and vegetable **fats:** glyceryl esters that are liquids at room temperature are commonly called **oils.**

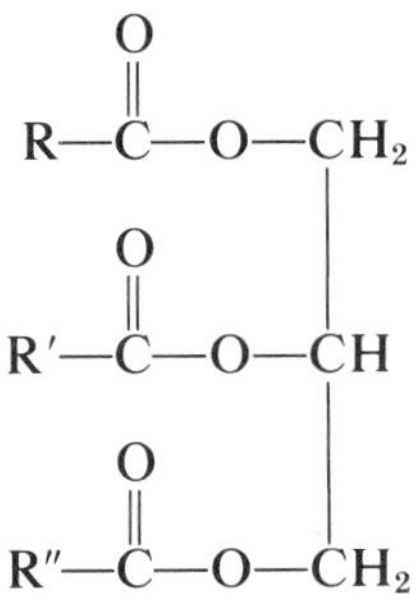

$$
\begin{array}{c}
\text{O} \\
\parallel \\
\text{R—C—O—CH}_2 \\
\\
\text{O} \\
\parallel \\
\text{R'—C—O—CH} \\
\\
\text{O} \\
\parallel \\
\text{R''—C—O—CH}_2
\end{array}
$$

The alkyl groups indicated in the preceding general formula are part of the carboxylic acids from which the compound may be considered to have been derived; they may be alike or different. The carboxylic acids are long-chain structures and may be saturated

$$CH_3(CH_2)_{14}COOH$$
palmitic acid

or unsaturated

$$CH_3(CH_2)_7CH{=}CH(CH_2)_7COOH$$
oleic acid

Liquid fats, or oils, contain a high percentage of glyceryl esters of unsaturated acids. Hydrogenation of vegetable oils produces solid fats. In commercial practice the process is run only until a solid of the desired consistency is obtained; not all of the olefinic linkages are saturated.

The alkaline hydrolysis of an ester produces an alcohol and a salt of a carboxylic acid.

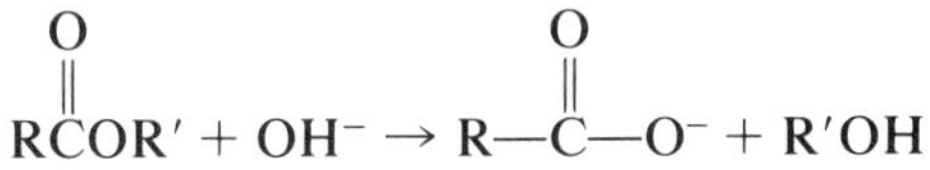

$$
\text{RCOR}' + \text{OH}^- \rightarrow \text{R—C—O}^- + \text{R'OH}
$$

The salts of long-chain carboxylic acids are soaps. These compounds, together with glycerol, are produced by the alkaline hydrolysis of fats; this process is commonly called **saponification** (soap making).

**20.9 AMINES AND AMIDES**

The **amines** may be considered as derivatives of ammonia with one, two, or three hydrogen atoms replaced by alkyl groups.

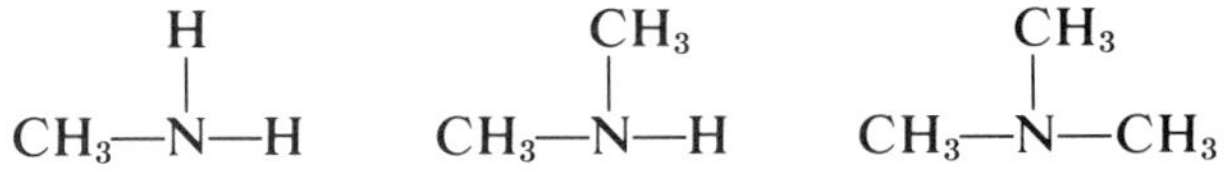

$$
\begin{array}{ccc}
\text{H} & \text{CH}_3 & \text{CH}_3 \\
| & | & | \\
\text{CH}_3\text{—N—H} & \text{CH}_3\text{—N—H} & \text{CH}_3\text{—N—CH}_3 \\
\text{methylamine} & \text{dimethylamine} & \text{trimethylamine} \\
\text{a primary amine} & \text{a secondary amine} & \text{a tertiary amine}
\end{array}
$$

The amines resemble ammonia in that they are weak bases.

$$CH_3NH_2 + H_2O \rightleftharpoons CH_3NH_3^+ + OH^-$$

$$(CH_3)_2NH + H_2O \rightleftharpoons (CH_3)_2NH_2^+ + OH^-$$

$$(CH_3)_3N + H_2O \rightleftharpoons (CH_3)_3NH^+ + OH^-$$

Ionization constants for some amines are listed in Table 20.6. The amines form ionic salts with acids

$$CH_3NH_2 + HCl \rightarrow CH_3NH_3^+ + Cl^-$$

which may be decomposed by hydroxides

$$CH_3NH_3^+ + OH^- \rightarrow CH_3NH_2 + H_2O$$

The last reaction is the reverse of the ionization of methylamine in water; it is forced to yield the free amine by employment of an excess of $OH^-$ or by warming.

Table 20.6 Properties of some amines

| AMINE | FORMULA | MELTING POINT (°C) | BOILING POINT (°C) | IONIZATION CONSTANT AT 25°C |
|---|---|---|---|---|
| methylamine | $CH_3NH_2$ | −93 | −7 | $5 \times 10^{-4}$ |
| dimethylamine | $(CH_3)_2NH$ | −96 | 7 | $7.4 \times 10^{-4}$ |
| trimethylamine | $(CH_3)_3N$ | −124 | 4 | $7.4 \times 10^{-5}$ |
| ethylamine | $CH_3CH_2NH_2$ | −81 | 17 | $5.6 \times 10^{-4}$ |
| propylamine | $CH_3CH_2CH_2NH_2$ | −83 | 49 | $4.7 \times 10^{-4}$ |
| aniline | $C_6H_5NH_2$ | −6 | 184 | $4.6 \times 10^{-10}$ |

Amines may be prepared by reacting alkyl halides with ammonia. These reactions are nucleophilic substitutions.

$$NH_3 + CH_3CH_2Br \rightarrow CH_3CH_2NH_3^+ + Br^-$$

Treatment of the amine salt with hydroxide ion yields the free amine. Mixtures of primary, secondary, and tertiary amines are produced by this type of reaction.

Primary amines may be produced by the catalytic hydrogenation of alkyl cyanides.

$$CH_3CH_2C{\equiv}N + 2H_2 \rightarrow CH_3CH_2CH_2{-}NH_2$$

Aniline is produced by the reduction of nitrobenzene. Iron and steam, with a trace of hydrochloric acid, serve as the reducing agent.

nitrobenzene          aniline

Some compounds that contain more than one amino group are important. Ethylenediamine (1,2-diaminoethane, $H_2N-CH_2-CH_2-NH_2$) is a useful chelating agent (Section 19.1), and hexamethylenediamine (1,6-diaminohexane, $H_2N(CH_2)_6NH_2$) is used in the manufacture of nylon.

**Amides** may be produced from ammonia and carboxylic acids. The preparation proceeds through the formation of ammonium salts which eliminate water upon heating.

$$CH_3\overset{\overset{\displaystyle O}{\|}}{C}-OH + NH_3 \rightarrow CH_3\overset{\overset{\displaystyle O}{\|}}{C}-O^-NH_4^+$$

$$\underset{\text{ammonium acetate}}{CH_3\overset{\overset{\displaystyle O}{\|}}{C}-O^-NH_4^+} \rightarrow \underset{\text{acetamide}}{CH_3\overset{\overset{\displaystyle O}{\|}}{C}-NH_2} + H_2O$$

Amides may also be prepared by the nucleophilic substitution reaction of ammonia on an ester.

$$NH_3 + CH_3\overset{\overset{\displaystyle O}{\|}}{C}-OCH_2CH_3 \rightarrow CH_3\overset{\overset{\displaystyle O}{\|}}{C}-NH_2 + CH_3CH_2OH$$

Primary and secondary amines react with acids in a similar manner to produce substituted amides, compounds in which one or two alkyl radicals replace hydrogen atoms of the $-NH_2$ group.

**20.10 AMINO ACIDS AND PROTEINS**    Carboxylic acid molecules that contain an amino group, $-NH_2$, on the carbon atom adjacent to the carboxyl carbon (the so-called $\alpha$-carbon atom) are called **amino acids.** The simplest such compound is aminoacetic acid, or glycine.

$$\underset{\underset{\displaystyle NH_2}{|}}{CH_2}-\overset{\overset{\displaystyle O}{\|}}{C}-OH$$

Amino acids are both acidic and basic

$$H_2N-CH_2\overset{\overset{\displaystyle O}{\|}}{C}-OH + OH^- \rightleftharpoons H_2N-CH_2\overset{\overset{\displaystyle O}{\|}}{C}-O^- + H_2O$$

$$H_2N-CH_2\overset{\overset{\displaystyle O}{\|}}{C}-OH + H_3O^+ \rightleftharpoons {}^+H_3N-CH_2\overset{\overset{\displaystyle O}{\|}}{C}-OH + H_2O$$

Thus, internal neutralization, which produces a **zwitter ion,** can occur.

$$^+H_3N—CH_2—\overset{\overset{\textstyle O}{\|}}{C}—O^-$$

For a given compound the acid strength and base strength, which depend upon molecular structure, are generally not exactly equal. The **isoelectric point** is the pH at which the internal neutralization of a given amino acid is at a maximum.

The acid hydrolysis of proteins yields amino acids. Many different amino acids have been synthesized, but only about two dozen have been obtained from proteins. The general formula of these compounds is

$$H_2N—\underset{\underset{\textstyle R}{|}}{CH}—\overset{\overset{\textstyle O}{\|}}{C}—OH$$

The radicals, R in the preceding formula, vary considerably in structure. Some of the radicals contain additional amino groups or carboxyl groups, a number include rings (some of them aromatic rings), some incorporate sulfur atoms or hydroxyl groups in their structures, and one contains iodine.

In protein molecules many amino acid units are linked together in long chains. The molecular weights of proteins range from approximately $10^4$ to about $10^7$. When two amino acids combine, the amino group of one molecule joins, with the elimination of water, to the carboxyl group of the second molecule, and a **peptide linkage**, $—\overset{\overset{\textstyle O}{\|}}{C}—\overset{\overset{\textstyle H}{|}}{N}—$, is formed.

$$H_2N—\underset{\underset{\textstyle R}{|}}{CH}—\overset{\overset{\textstyle O}{\|}}{C}—OH + H_2N—\underset{\underset{\textstyle R}{|}}{CH}—\overset{\overset{\textstyle O}{\|}}{C}—OH \rightarrow$$

$$H_2N—\underset{\underset{\textstyle R}{|}}{CH}—\overset{\overset{\textstyle O}{\|}}{C}—\overset{\overset{\textstyle H}{|}}{N}—\underset{\underset{\textstyle R}{|}}{CH}—\overset{\overset{\textstyle O}{\|}}{C}—OH + H_2O$$

This product has a free amino group and a free carboxyl group and can condense with other amino acid molecules to form long chains called polypeptides. The hydrolysis of a polypeptide chain is the reverse of the condensation process and produces amino acids.

The amino acids provide examples of a type of stereoisomerism called **optical isomerism** (Section 19.4). The simplest optical isomers are compounds that contain asymmetric carbon atoms — carbon atoms to which

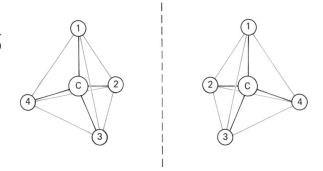

four different types of groups are bonded (Figure 20.5). Such molecules are dissymmetric — that is, the mirror image (enantiomorph) of the molecule is not superimposable on the original. Thus, the structures shown in Figure 20.5 are mirror reflections of one another. A study of the figure reveals that the two configurations cannot be superimposed. If groups 1 and 3 of the structures are superimposed, group 4 of one structure coincides with group 2 of the other. No matter how the two structures are oriented, it is impossible to make all four groups of one structure coincide with the same four groups of the other structure; two different isomers exist.

The physical properties of optical isomers are identical except for their effects on plane-polarized light. The isomers are said to be optically active because they rotate the plane of plane-polarized light in either a clockwise or a counterclockwise direction. The isomer that rotates the plane to the right (clockwise) is said to be **dextrorotatory** (+); the other isomer rotates the plane an equal amount to the left and is said to be **levorotatory** (−).

With the exception of glycine (aminoacetic acid), the $\alpha$-carbon atoms of all $\alpha$-amino acids are asymmetric.

$$\text{H}_2\text{N}-\overset{\overset{\text{H}}{|}}{\underset{\underset{\text{R}}{|}}{\text{C}}}-\overset{\overset{\text{O}}{\|}}{\text{C}}-\text{OH}$$

The two-dimensional representation of optical isomers is difficult; structures of the optical isomers of amino acids are usually diagramed

$$\underset{\text{L-amino acid}}{\text{H}_2\text{N}-\overset{\overset{\overset{\text{O}}{\|}}{\text{C}-\text{OH}}}{\underset{\underset{\text{R}}{|}}{\text{C}}}-\text{H}} \qquad \underset{\text{D-amino acid}}{\text{H}-\overset{\overset{\overset{\text{O}}{\|}}{\text{C}-\text{OH}}}{\underset{\underset{\text{R}}{|}}{\text{C}}}-\text{NH}_2}$$

L-amino acid          D-amino acid

The diagrams represent structures in which the asymmetric carbon atoms are in the plane of the paper, the —COOH and —R groups are behind the plane of the paper, and the —H and —NH₂ groups are in front of the plane of the paper. With such an arrangement of groups, the L-form is the one in which the amino group is to the left of the chain, and the D-form is the one in which the amino group is to the right. The structures can be related to the diagrams of Figure 20.5; the L-form is the one on the left in the figure if group 1 = —COOH, 2 = —R, 3 = —H, and 4 = —NH₂.

All the amino acids derived from proteins are L-amino acids. Only the L-forms can be utilized in the metabolic processes of the body. The sign of rotation should not be confused with the symbols D- or L- that are used to designate the configuration of the asymmetric carbon atom. Thus, the L-isomer of alanine (2-aminopropanoic acid) is dextrorotatory.

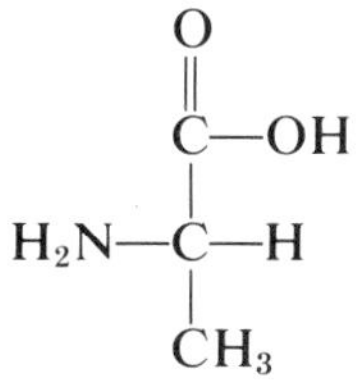

$$
\begin{array}{c}
\quad\quad O \\
\quad\quad \| \\
\quad\quad C\!-\!OH \\
\quad\quad | \\
H_2N\!-\!C\!-\!H \\
\quad\quad | \\
\quad\quad CH_3
\end{array}
$$

L-(+)-alanine

**20.11 CARBO-HYDRATES**

Sugars, starches, and cellulose belong to a group of organic compounds called **carbohydrates;** the name derives from the fact that most (but not all) compounds of this type have the general formula $C_x(H_2O)_y$. Carbohydrates are hydroxy aldehydes, hydroxy ketones, or substances that yield hydroxy carbonyl compounds upon hydrolysis.

The optical isomers of glucose, an important simple sugar, have the formulas

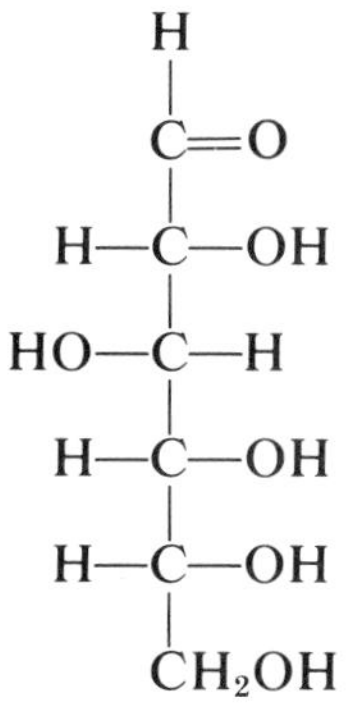

$$
\begin{array}{cc}
\begin{array}{c}
H \\
| \\
C\!=\!O \\
| \\
H\!-\!C\!-\!OH \\
| \\
HO\!-\!C\!-\!H \\
| \\
H\!-\!C\!-\!OH \\
| \\
H\!-\!C\!-\!OH \\
| \\
CH_2OH
\end{array}
&
\begin{array}{c}
H \\
| \\
C\!=\!O \\
| \\
HO\!-\!C\!-\!H \\
| \\
H\!-\!C\!-\!OH \\
| \\
HO\!-\!C\!-\!H \\
| \\
HO\!-\!C\!-\!H \\
| \\
CH_2OH
\end{array}
\\
\text{D-(+)-glucose} & \text{L-(−)-glucose}
\end{array}
$$

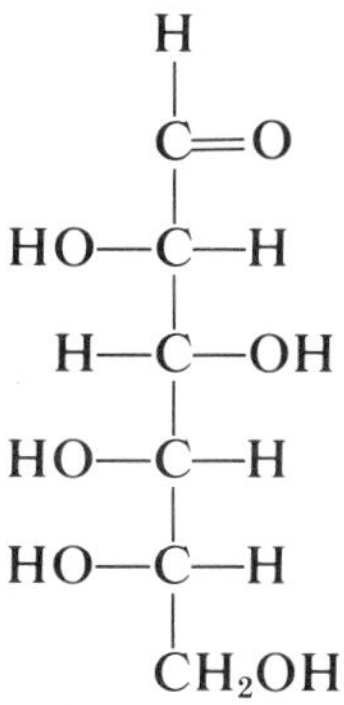

There are four asymmetric carbon atoms in the molecule (the terminal carbon atoms do not hold four different groups and are not asymmetric).

Sixteen optical isomers (eight D,L pairs) of this formula exist; only the preceding pair are called glucose.

D-glucose forms a cyclic molecule by an addition reaction involving the carbonyl group and a hydroxy group.

Ring formation generates a new asymmetric center, and two ring isomers of D-glucose exist which differ in the orientation of the new OH group.

α-D-glucose            β-D-glucose

In these diagrams the carbon atoms of the ring are not shown. The rings are actually puckered, not planar. Notice that in the α form the OH group of the extreme right-hand carbon atom is on the same side of the ring as the OH group of the adjacent carbon atom; these OH groups may be said to be *cis* to each other. In aqueous solution, α and β forms of D-glucose exist in equilibrium, together with a low concentration of the open-chain form.

Fructose is a hydroxy ketone.

D-(−)-fructose            L-(+)-fructose

There are three asymmetric carbon atoms in this molecule and eight stereoisomers (or four D,L pairs) of this formula are known. Cyclic forms of D-fructose are known with five- or six-membered rings.

Glucose and fructose are known as monosaccharides. Sucrose, which is cane sugar, is a disaccharide. Acid hydrolysis of sucrose yields the two simple sugars D-glucose and D-fructose. In the sucrose molecule an $\alpha$ D-glucose unit and the $\beta$ five-membered ring form of a D-fructose unit are joined through two OH groups by the elimination of a molecule of water.

D-glucose unit          D-fructose unit

Cellulose and starch are polysaccharides that yield only D-glucose upon hydrolysis. It is estimated that the number of D-glucose units in the molecular structures of these substances may be as high as several thousand. The D-glucose units of cellulose are linked in long chains in $\beta$ combination.

The D-glucose units of starch are linked in a different manner; there are occasional cross links between long chains of D-glucose units arranged in $\alpha$ combination.

Starch is an important food material, but cellulose cannot be digested by man — an important distinction brought about by difference in the way that the D-glucose units are linked together.

**20.12 POLYMERS**  Starch, cellulose, and proteins are examples of natural **polymers**—molecules of high molecular weight that are formed from simpler molecules, called **monomers.** Important polymers, or **macromolecules,** that do not occur in nature have been synthesized; most of these are linear, or chain-type, structures although some are cross-linked.

Many polymers are formed from compounds that contain carbon–carbon double bonds by a process that is called **addition polymerization.** For example, ethene (ethylene) polymerizes upon heating (100°C–400°C) under high pressure (1000 atm); the product is called polyethylene.

$$
\begin{array}{ccc}
\overset{\displaystyle H}{\underset{\displaystyle H}{|}}\overset{\displaystyle H}{\underset{\displaystyle H}{|}} & \overset{\displaystyle H}{\underset{\displaystyle H}{|}}\overset{\displaystyle H}{\underset{\displaystyle H}{|}} & \overset{\displaystyle H\ H\ H\ H}{\underset{\displaystyle H\ H\ H\ H}{|\ |\ |\ |}} \\
C\!=\!C & + \quad C\!=\!C & \rightarrow \quad -C\!-\!C\!-\!C\!-\!C\!-
\end{array}
$$

The preceding equation shows the combination of only two molecules of ethene; the actual polymerization process continues by successive additions of $CH_2\!=\!CH_2$ molecules until chains of hundreds or thousands of $-CH_2-CH_2-$ units are produced. The product, polyethylene, is a tough, waxy solid.

Important polymers have been made from ethene derivatives. Orlon and acrilan are polymers of acrylonitrile, $CH_2\!=\!CH-C\!\equiv\!N$; the polymers have the structure

$$
-CH_2-\underset{\underset{CN}{|}}{CH}\!\!\left(\!CH_2-\underset{\underset{CN}{|}}{CH}\!\right)_{\!n}\!\!CH_2-\underset{\underset{CN}{|}}{CH}-
$$

Teflon is a polymer of tetrafluoroethene, $CF_2\!=\!CF_2$. Vinyl chloride, $CH_2\!=\!CHCl$, and vinyl acetate, $CH_2\!=\!CH-O-\overset{\overset{\displaystyle O}{\displaystyle\|}}{C}-CH_3$, are important monomers in the preparation of vinyl plastics. Lucite, or Plexiglas, is a polymer of methyl methacrylate, $CH_2\!=\!C(CH_3)COOCH_3$.

Natural rubber is a polymer of the diolefin 2-methyl-1,3-butadiene, or isoprene.

$$
CH_2\!=\!\underset{\underset{CH_3}{|}}{C}\!-\!CH\!=\!CH_2
$$

Such a system of alternating double and single bonds is called a **conjugated system.** Rubber consists of thousands of these units joined in a chain.

$$
-CH_2-\underset{\underset{CH_3}{|}}{C}\!=\!CH-CH_2\!\!\left(\!CH_2-\underset{\underset{CH_3}{|}}{C}\!=\!CH-CH_2\!\right)_{\!n}\!\!CH_2-\underset{\underset{CH_3}{|}}{C}\!=\!CH-CH_2-
$$

Notice that the polymer contains double bonds. Crude rubber is vulcanized by heating with sulfur. It is thought that the sulfur atoms add to some of the double bonds, thereby linking adjacent chains into a complex network. This process adds strength to the final product.

A number of types of synthetic rubber have been made utilizing such monomers as 1,3-butadiene ($CH_2$=CH—CH=$CH_2$), 2-chloro-1,3-butadiene (chloroprene, $CH_2$=C(Cl)—CH=$CH_2$), and 2,3-dimethyl-1,3-butadiene ($CH_2$=C($CH_3$)—C($CH_3$)=$CH_2$) alone or in combination. The product formed by the polymerization of two different monomers is called a **copolymer;** the molecular structures and properties of such materials depend upon the proportions of the two monomers employed. Buna S rubber is a copolymer of 1,3-butadiene and styrene ($C_6H_5$CH=$CH_2$). A section of the chain of this material has the structure

—$CH_2$—CH=CH—$CH_2$—CH—$CH_2$—

An addition polymerization is thought to proceed by either a free-radical or a carbonium-ion chain mechanism. The free-radical type may be initiated by small amounts of hydrogen peroxide or organic peroxides, which are readily split into free radicals (indicated by $Z\cdot$ in the equations that follow).

$$Z_2 \rightarrow 2Z\cdot$$

A free radical from an initiator combines with one of the $\pi$ electrons of a molecule of the monomer to form a new free radical.

$$Z\cdot + CH_2{=}CH_2 \rightarrow Z{-}CH_2{-}CH_2\cdot$$

Chain propagation occurs by the combination of this free radical with another molecule of the monomer

$$Z{-}CH_2{-}CH_2\cdot + CH_2{=}CH_2 \rightarrow Z{-}CH_2{-}CH_2{-}CH_2{-}CH_2\cdot$$

and the molecule grows by repeated additions. Chain termination is caused by the combination of two free radicals or by other means, such as the elimination of a hydrogen atom.

A carbonium-ion chain polymerization is initiated by a Lewis acid, such as $AlCl_3$ or $BF_3$. In the equations that follow, the proton is used to indicate the Lewis acid, or electrophilic, initiator, although stronger Lewis acids are more frequently employed. Combination of the initiator with a molecule of the monomer produces a carbonium ion.

$$H^+ + CH_2{=}CH_2 \rightarrow CH_3{-}CH_2^+$$

Chain propagation occurs through the combination of the carbonium ion with both $\pi$ electrons of a molecule of the monomer.

$$CH_3—CH_2^+ + CH_2{=}CH_2 \rightarrow CH_3—CH_2—CH_2—CH_2^+$$

Chain termination may occur by the elimination of a proton from a long-chain carbonium ion.

$$CH_3—CH_2(CH_2—CH_2)_nCH_2—CH_2^+ \rightarrow$$

$$CH_3—CH_2(CH_2—CH_2)_nCH{=}CH_2 + H^+$$

**Condensation polymerization** occurs between molecules of monomers by the elimination of a small molecule, usually water. Proteins and polysaccharides are condensation polymers. Nylon, like the proteins, is a polyamide. It is formed by the condensation of a diamine [such as hexamethylenediamine, $H_2N(CH_2)_6NH_2$] and a dicarboxylic acid [such as adipic acid, $HOOC(CH_2)_4COOH$]. The polymerization is accompanied by the loss of water, and the chain has the structure

$$—\underset{\underset{H}{|}}{N}—\underset{\overset{O}{||}}{C}—(CH_2)_4—\underset{\overset{O}{||}}{C}—\underset{\underset{H}{|}}{N}—(CH_2)_6—\underset{\underset{H}{|}}{N}—\underset{\overset{O}{||}}{C}—$$

Dacron is a polyester formed by the elimination of water between ethylene glycol ($HOCH_2CH_2OH$) and a dicarboxylic acid (such as terephthalic acid, $p\text{-}HOOCC_6H_4COOH$).

$$—\underset{\overset{O}{||}}{C}—O—CH_2—CH_2—O—\underset{\overset{O}{||}}{C}{-}\!\!\!\left\langle\bigcirc\right\rangle\!\!\!{-}\underset{\overset{O}{||}}{C}—O—$$

Bakelite is a cross-linked polymer formed from phenol and formaldehyde by the elimination of water. Notice that the formaldehyde units condense in the *ortho* and *para* positions of phenol.

## Problems

**20.1** Write structural formulas for the following compounds. (a) 2.4-dimethyl-3-ethylpentane; (b) 2,4-dimethyl-1,4-pentadiene; (c) 3-isopropyl-1-hexyne; (d) 2,2,3-trichlorobutane; (e) 2-methyl-2-butanol; (f) methylcyclopropane.

**20.2** Write structural formulas for (a) butanal, (b) 2-butanone, (c) methyl ethyl ketone, (d) methyl ethyl amine, (e) methyl ethyl ether, (f) methyl ethanoate.

**20.3** Write structural formulas for (a) *p*-xylene; (b) 2,3,6-trinitrotoluene; (c) naphthalene; (d) sodium benzoate; (e) *o*-bromophenol; (f) diphenyl amine.

**20.4** Name each of the following compounds.

(a)
$$CH_3—CH—CH_2—CH_2—CH_2OH$$
$$\phantom{CH_3—}\underset{CH_3}{|}$$

(b) $CH_3—CH—CH_2—CH—CH_3$
$$\phantom{(b) CH_3—}\underset{CH_3}{|}\phantom{—CH_2—}\underset{OH}{|}$$

(c) $CH_3—\overset{OH}{\underset{CH_3}{\overset{|}{\underset{|}{C}}}}—CH_2—CH_2—CH_3$

(d) $CH_3—CH—\overset{O}{\overset{||}{C}}—CH_2CH_3$
$$\phantom{(d) CH_3—}\underset{CH_3}{|}$$

(e) $CH_3—CH—O—CH_2CH_3$
$$\phantom{(e) CH_3—}\underset{CH_3}{|}$$

(f)
$$CH_3—CH—CH_2—CH_2—\overset{O}{\overset{||}{C}}—H$$
$$\phantom{CH_3—}\underset{CH_3}{|}$$

**20.5** Name the following compounds.

(a) 
$$\overset{O}{\overset{||}{C}}—NH_2$$
(benzene ring)

(b) 
$$\overset{O}{\overset{||}{C}}—CH_3$$
(benzene ring)

(c) 
$$\overset{H}{\underset{|}{N}}—CH_3$$
(benzene ring)

(d) 
$$\overset{O}{\overset{||}{C}}$$
$$H_2C\phantom{xxx}CH_2$$
$$H_2C\phantom{xxx}CH_2$$
$$CH_2$$

(e) 
$$\overset{O}{\overset{||}{C}}—O—CH_3$$
(benzene ring)

(f)

$$
\begin{array}{c}
O \\
\parallel \\
C-OH
\end{array}
$$

$-NO_2$

**20.6** What is incorrect about each of the following names? (a) 3-pentene; (b) 1,2-dimethylpropane; (c) 2-methyl-2-butyne; (d) 2-methyl-3-butyne; (e) 3,3-dimethyl-2-butene; (f) 2-ethylpentane.

**20.7** The general formula for the alkanes is $C_nH_{2n+2}$. What are the general formulas for the (a) alkenes, (b) alkynes, (c) alkadienes, (d) cycloalkanes, (e) alcohols?

**20.8** Write structural formulas for all the mono-, di-, and tri-substituted chloro derivatives of propane.

**20.9** Write structural formulas for all the isomers that have the formula $C_4H_8$.

**20.10** Write structural formulas for all the isomers that have the formula $C_4H_{10}O$.

**20.11** Write structural formulas for all the isomers that have the formula $C_4H_8Cl_2$.

**20.12** Give an example of an isomer of (a) butane, (b) methyl methanoate (methyl formate), (c) dimethyl ether, (d) dimethyl amine, (e) acetone, (f) aminoacetic acid.

**20.13** When the terms primary, secondary, and tertiary are applied to amines, their meanings are not the same as when they are applied to alcohols. Explain the distinction.

**20.14** Give examples of each of the following types of isomers: (a) chain, (b) position, (c) functional group, (d) geometric, (e) optical.

**20.15** Write structural formulas for all the isomers of (a) bromophenol, (b) dibromophenol, (c) tribromobenzene.

**20.16** *Ortho-, meta-,* and *para-* isomers can be identified by determining the number of compounds that can be derived from each isomer when a third substituent is introduced into the ring (Körner's method). If one nitro group is introduced into the benzene ring of each of the following compounds, how many isomeric products would be secured from each? (a) *o*-dibromobenzene, (b) *m*-dibromobenzene, (c) *p*-dibromobenzene.

**20.17** Write structural formulas for the ten dichloro substitution isomers of naphthalene.

**20.18** *Cis-* and *trans*-isomers are known for certain alkenes. Why do not the alkynes display this type of isomerism?

**20.19** For which of the following compounds are *cis-* and *trans*-isomers possible? (a) 2,3-dimethyl-2-butene, (b) 3-methyl-2-pentene, (c) 2-methyl-2-pentene, (d) CH(COOH)=CH(COOH).

**20.20** For which of the following are optical isomers possible? (a) 2-hydroxypropanoic acid, (b) hydroxyacetic acid, (c) 2-bromo-2-chloropropane, (d) 1-bromo-1-chloroethane, (e) the cyanohydrin prepared from butanone.

**20.21** Define the following: (a) olefin, (b) homologous series, (c) cracking, (d) conjugated system, (e) saponification, (f) carbohydrate, (g) cyanohydrin, (h) zwitter ion.

**20.22** Why do alcohols have higher boiling points than hydrocarbons of corresponding molecular weight?

**20.23** What is the difference between (a) an amine and an amide, (b) an ester and an ether, (c) a fat and an oil, (d) an aliphatic hydrocarbon and an aromatic hydrocarbon, (e) a saturated hydrocarbon and an unsaturated hydrocarbon.

**20.24** What is the product of the reaction of 1-bromobutane with each of the following? (a) Mg, (b) NaCN, (c) $NH_3$, (d) $CH_3NH_2$, (e) $OH^-$(aq), (f) $NaOCH_2CH_3$.

**20.25** What is the product of the reaction of 2-propanol with each of the following? (a) Na, (b) $H_2SO_4$ (high temperature), (c) $H_2SO_4$ (moderate temperature), (d) propanoic acid, (e) $K_2Cr_2O_7$ in acid solution.

**20.26** What are the products when the following compounds are oxidized under vigorous conditions? (a) 2-hexene, (b) 2-methyl-2-butene, (c) 3-hexyne, (d) 1-propanol, (e) 2-propanol, (f) propanal, (g) *p*-nitrotoluene. What are the products when the following compounds are oxidized using mild conditions? (a) 2-hexene, (b) 1-propanol.

**20.27** What are the products when the following compounds are reacted with hydrogen in the presence of a suitable catalyst? (a) pentanal, (b) 2-pentanone, (c) 2-pentene, (d) 2-pentyne, (e) ethyl cyanide, (f) *p*-nitrotoluene.

**20.28** What is the product(s) of the reaction of bromine with each of the following? (a) methane, (b) 2-butene, (c) 2-butyne, (d) benzene (in the presence of $FeBr_3$).

**20.29** State the formulas of all products formed by the reaction of HBr with each of the following. If more than one compound is formed, tell which is produced in the greatest quantity.
(a) $CH_3CH_2OH$
(b) $CH_3CH_2MgBr$
(c) $(CH_3CH_2)_2C{=}CH_2$
(d) $CH_3C{\equiv}CH$
(e) $CH_3CH_2COO^-Na^+$
(f) $(CH_3CH_2)_2NH$

**20.30** State the compound formed when each of the following is treated with aqueous NaOH.
(a) $(CH_3)_2CHCOOH$
(b) $(CH_3)_2CHBr$
(c) $CH_3CH_2CH_2CN$
(d) $CH_3CH_2CH_2COOCH_2CH_3$
(e) $CH_3CH_2NH_3^+Cl^-$

**20.31** Write equations to show the reactions of $NH_3$ with (a) 2-bromopropane, (b) propanoic acid, (c) methyl propanoate.

**20.32** What is the final product (obtained by hydrolysis of the initial product) of the reaction of ethyl magnesium bromide with (a) methanal, (b) butanal, (c) butanone.

**20.33** What Grignard reagent and what carbonyl compound should be used to prepare the following alcohols? (a) 3-methyl-2-butanol, (b) 2-methyl-2-butanol, (c) 2-methyl-1-butanol.

**20.34** Write equations to show how the following can be made by means of addition reactions: (a) 1,2-dibromoethane from ethene; (b) 1,1-dibromoethane from acetylene, (c) cyclohexanol from cyclohexene.

**20.35** Write equations to show how ethanol may be converted into the following substances. More than one step may be required for a given preparation. (a) ethene, (b) ethanal, (c) diethyl ether, (d) ethane, (e) acetic acid, (f) bromoethane, (g) ethyl cyanide, (h) 2-hydroxypropanoic acid.

**20.36** Using 1-propanol and 2-propanol as the only organic starting materials, write a series of equations to show how to prepare: (a) 3-hexanol, (b) 2,3-dimethyl-2-butanol, (c) 2-methyl-2-pentanol, (d) 2-methyl-3-pentanol. Use a Grignard reaction as one of the steps in each preparation.

**20.37** Show by a series of reactions how to convert 1-propanol into 2-propanol.

**20.38** Write an equation to show how a soap may be made from a fat.

**20.39** Why are vegetable oils hydrogenated? What occurs chemically in the process?

**20.40** Alanine is 2-aminopropanoic acid. Write the structural formula of a polypeptide containing three alanine units.

**20.41** What is the difference in meaning between the designations D and (+) as applied to amino acids?

**20.42** Describe with examples: (a) addition polymers, (b) copolymers, (c) condensation polymers.

**20.43** (a) What is Markovnikov's rule? (b) How does the mechanism of the addition of HBr to an alkene explain why such an addition follows Markovnikov's rule?

**20.44** Compare and contrast the mechanism of a substitution reaction of an alkane with the mechanism of a substitution reaction of an aromatic hydrocarbon.

**20.45** Interpret the following in terms of the Lewis theory: (a) an $S_N1$ reaction of a alkyl halide, (b) an $S_N2$ reaction of an alcohol, (c) the addition of HBr to an olefin, (d) the addition of HCN to a ketone.

**20.46** Describe the function of $AlCl_3$ in (a) a Friedel–Crafts reaction, (b) a polymerization.

**20.47** Arrange the following compounds in decreasing order of strength as Brønsted acids: $CH_3CH_2OH$, $CH_3CH_2OH_2^+$, $H_2O$, $H_3O^+$, $CH_3COOH$, $CH_3CH_2NH_2$.

**20.48** Arrange the following compounds in decreasing order of strength as Brønsted bases: $CH_3CH_2OH$, $CH_3CH_2O^-$, $OH^-$, $CH_3COO^-$, $CH_3CH_2NH_2$.

**20.49** Draw the resonance structures of (a) benzene, (b) naphthalene, (c) the acetate ion.

**20.50** (a) Why is phenol a stronger acid then methanol? (b) Why is aniline a weaker base than methylamine?

# 21

# Nuclear Chemistry

Ordinary chemical reactions involve only the extranuclear electrons. In such reactions the nucleus is important only insofar as it influences the electrons. However, matter does undergo important transformations that involve the nucleus directly. The study of nuclear transformations has enlarged our understanding of the nature of matter and the courses of many chemical and biological processes; many technological applications have resulted from these investigations.

When Wilhelm Röntgen discovered X rays in 1895, he noticed that these invisible rays expose photographic plates and cause fluorescent salts to glow. Henri Becquerel in 1896 investigated the hypothesis that salts of this type emit invisible radiation independent of any external stimulus. Becquerel found that a double sulfate of potassium and uranium (that he happened to have on hand) emits radiation capable of exposing a photographic plate that is well protected from light. He subsequently identified uranium as the source of the "radioactive" rays. Following Becquerel's discovery, other radioactive elements were identified and isolated (notably by Marie and Pierre Curie), and the nature of the rays was elucidated (principally by Ernest Rutherford). Radioactive rays originate from transformations that take place within the nucleus.

**21.1 THE NUCLEUS**

The nuclei of atoms are thought to contain protons and neutrons, particles that are collectively called **nucleons.** The number of protons in a specific atomic nucleus corresponds to the atomic number (or nuclear charge), $Z$, and the total number of nucleons is given by the mass number, $A$. Thus, the number of neutrons equals $(A-Z)$. This information is indicated on the chemical symbol of a given nuclide by appending the atomic number as a subscript and the mass number as a superscript. The symbols for the two naturally occurring isotopes of lithium are

$$^6_3\text{Li} \quad \text{and} \quad ^7_3\text{Li}$$

● A metric ton (t) is $10^3$ kg. Although this unit is not a part of SI, it may be used with SI units. The pound (lb) and short ton (2000 lb) are units of the English system. One pound equals 0.453 592 37 kg; therefore, one short ton equals 9.071 847 4 × $10^2$ kg or 0.907 184 74 t. One cm³ of nuclear matter would weigh 2.44 × $10^8$ metric tons or 2.69 × $10^8$ short tons.

Much is known about the structure of the nucleus, but much more remains to be learned. Determination of the radii of a large number of atomic nuclei shows that the radius of a given nucleus, $r$, is directly related to the cube root of its mass number.

$$r = (1.3 \times 10^{-13} \text{ cm})A^{1/3}$$

The volume of a sphere is $\frac{4}{3}\pi r^3$, and if we assume a spherical nucleus, it follows that nuclear volume varies directly with mass number. In other words, the mass of a nucleus determines its volume; nuclear density is therefore approximately constant for all atomic nuclei. This density is about $2.44 \times 10^{14}$ g/cm³, an amazingly high value; 1 cm³ of this nuclear matter would weigh more than 250 million tons.

The density of a liquid is constant and independent of the size of any drop considered. Since all nuclei have approximately the same density, a fluid-droplet model of the nucleus has been proposed. The nature of the cohesive forces holding the nucleons into a nuclear fluid is far from being completely understood.

It is clear, however, that powerful cohesive forces between nucleons exist and that they effectively overcome the electrostatic forces of repulsion between the protons of the nucleus. The attractive forces between nucleons are charge independent, that is, proton–proton, proton–neutron, and neutron–neutron attractions are identical. The range of the force is believed to be on the order of $2 \times 10^{-13}$ cm, the diameter of a nucleon.

In 1935 Hidekei Yukawa postulated that neutrons and protons are bound together by the very rapid exchange of a nuclear particle, which was identified as a $\pi$ meson, or pion. Three types of pions are known to exist: positive, $\pi^+$, negative, $\pi^-$, and neutral, $\pi^0$. According to the exchange theory, a neutron, $n_A^0$, is converted into a proton, $p_A^+$, by the emission of a negative pion. The emitted $\pi^-$ is accepted by a proton, $p_B^+$, which is converted into a neutron, $n_B^0$.

$$n_A^0 \rightarrow p_A^+ + \pi^-$$

$$\pi^- + p_B^+ \rightarrow n_B^0$$

If we consider the deuterium nucleus, which consists of one proton and one neutron, this exchange then reverses, the neutron $n_B^0$ becoming a proton, $p_B^+$, and the proton $p_A^+$ becoming a neutron, $n_A^0$. The exchange is extremely rapid; approximately $10^{24}$ transfers occur in one second.

The same type of interaction can be postulated for the positive pion,

$$p_A^+ \rightarrow n_A^0 + \pi^+$$

$$\pi^+ + n_B^0 \rightarrow p_B^+$$

or the neutral pion,

$$p_A^+ \rightarrow p_A^+ + \pi^0$$

$$\pi^0 + n_B^0 \rightarrow n_B^0$$

The exchange of a neutral pion can also be used to account for the existence of neutron–neutron and proton–proton forces.

In Figure 21.1 the number of neutrons is plotted against the number of protons for the naturally occurring, nonradioactive nuclei. The points, which represent stable combinations of protons and neutrons, lie in what may be called a **zone of stability.** Nuclei that have compositions represented by points that lie outside of this zone spontaneously undergo radioactive transformations that tend to bring their compositions into or closer to this zone (Section 21.2).

The stable nuclei of the lighter elements contain approximately equal numbers of neutrons and protons, a neutron/proton ratio of 1. The heavier nuclei contain more neutrons than protons. With increasing atomic number more and more protons are packed into a tiny nucleus, and the electrostatic forces of repulsion increase sharply. A larger and larger excess of neutrons is required to diminish the effect of these repulsion forces, and the neutron/proton ratio increases with increasing atomic number until the ratio is approximately 1.5 at the end of the curve of Figure 21.1. There appears to be an upper limit to the number of protons that can be packed into a nucleus, no matter how many neutrons are present. The largest stable nucleus is $^{209}_{83}\text{Bi}$; nuclei that are larger than this exist, but all of them are radioactive.

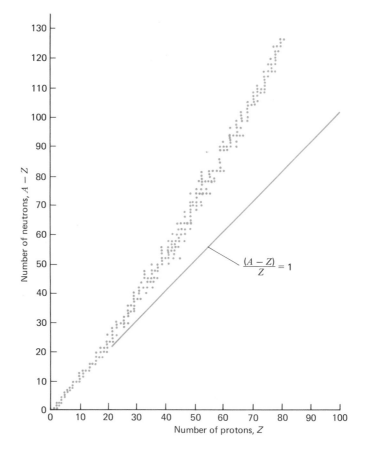

**Figure 21.1** Neutron/proton ratio.

Most naturally occurring stable nuclides have an even number of protons and an even number of neutrons; only five ($_1^2$H, $_3^6$Li, $_5^{10}$B, $_7^{14}$N, and $_{73}^{180}$Ta) have an odd number of protons and an odd number of neutrons (Table 21.1). For each odd atomic number there are never more than two stable nuclides, whereas for an even atomic number as many as ten stable nuclides may occur. The two elements of atomic number less than 83 that have never been proven to be naturally existing ($_{43}$Tc and $_{61}$Pm) have odd atomic numbers. Empirical observations such as these suggest that there is a periodicity in nuclear structure similar to the periodicity of atomic structure. A nuclear shell model, as yet incompletely developed, has been suggested.

Periodic variations are observed in many nuclear properties. Certain nuclides have relatively high binding energies (which are indicative of comparatively great stability; Sections 2.9 and 21.6) compared to nuclides of close atomic number or mass number. Exceptional nuclear stability is also shown by certain nuclides that have a poor ability to capture neutrons

Table 21.1 Distribution of naturally occurring stable nuclides

| PROTONS | NEUTRONS | NUMBER OF NUCLIDES |
|---------|----------|--------------------|
| even | even | 157 } 209 |
| even | odd | 52 |
| odd | even | 50 } 55 |
| odd | odd | 5 |

(Section 21·5). Comparison of data from these and other studies of nuclear properties indicates that unusual nuclear stability is associated with nuclides having either a number of protons or a number of neutrons equal to a magic number: 2, 8, 20, 28, 50, 82, and 126. It is thought that the magic numbers indicate closed nuclear shells in the same way that the atomic numbers of the noble gases, 2, 10, 18, 36, 54, and 86, indicate stable electronic configurations. In general, elements that have an atomic number equal to a magic number have a larger number of stable isotopes than neighboring elements. The nuclides with a magic number of protons as well as a magic number of neutrons, $^4_2$He, $^{16}_8$O, $^{40}_{20}$Ca, and $^{208}_{82}$Pb, have notably high stabilities in terms of neutron capture (Section 21.5), binding energy (Section 21.6), and relative abundance.

**21.2 RADIOACTIVITY**   Unstable nuclei spontaneously undergo certain changes that result in the attainment of more stable nuclear compositions. Some unstable nuclides are naturally occurring, others are man-made. Certain synthetic radioactive nuclides undergo some types of radioactive decay that have not been observed for any naturally occurring unstable nuclide.

**Alpha emission** consists of the ejection of $\alpha$ particles, which have an atomic number of 2 and a mass number of 4 and may be considered to be $^4_2$He nuclei. Both synthetic and natural nuclides undergo $\alpha$ decay, which is common only for nuclides of mass number greater than 209 and atomic number greater than 82. Nuclides such as these have too many protons for stability. Points indicating the composition of these nuclides fall outside the plot of Figure 21.1 to the upper right, beyond the zone of stability. The emission of $\alpha$ particles reduces the number of protons by two and the number of neutrons by two and adjusts the composition of the nucleus downward, closer to the stability zone.

An example of $\alpha$ decay is

$$^{210}_{84}\text{Po} \rightarrow {}^{206}_{82}\text{Pb} + {}^4_2\text{He}$$

Notice that the equation indicates the conservation of mass number (superscripts) and atomic number (subscripts). Equations such as these are written to indicate nuclear changes only; the extranuclear electrons are customarily ignored. An $\alpha$ particle is emitted as a $^4_2$He nucleus without electrons and with a 2+ charge. Subsequent to its emission, however, the particle attracts electrons from other atoms (which become cations), and the $\alpha$ particle becomes a neutral atom of $^4_2$He. The $^{206}_{82}$Pb is left therefore with a surplus of two electrons and a 2− charge; these excess electrons are rapidly lost to surrounding cations.

The energy released in this process is readily calculated in the same way that binding energy is calculated (Section 2.9); it is the energy equivalent of the difference in mass between the reactant (the parent nucleus) and products (the daughter nucleus and the $\alpha$ particle). The masses of neutral *atoms,* rather than the masses of *nuclei,* are generally recorded, but this causes no trouble in the calculation. If the mass of the $^{210}_{84}$Po *atom* is used, the mass of 84 electrons is included; for the products the mass of the $^{206}_{82}$Pb *atom* includes the mass of 82 electrons, and the mass of the $^4_2$He *atom* includes the mass of two electrons. Thus, the masses of the extranuclear electrons cancel when atomic masses are employed for the calculation.

$$(\text{mass}\,^{210}_{84}\text{Po}) - (\text{mass}\,^{206}_{82}\text{Pb} + \text{mass}\,^4_2\text{He})$$

$$209.9829\ \text{u} - (205.9745 + 4.0026\ \text{u}) = 0.0058\ \text{u}$$

● The symbol M- stands for the prefix *mega-* (which means $10^6 \times$). Hence,

$$1\ \text{MeV} = 10^6\ \text{eV}$$

The electron volt is not an SI unit, but it may be used with SI units.

$$1\ \text{eV} = 1.60219 \times 10^{-19}\ \text{J}$$
$$1\ \text{MeV} = 1.60219 \times 10^{-13}\ \text{J}$$

● The unified atomic mass unit (u), which is defined as 1/12 the mass of a $^{12}_6$C atom, may be used with SI units.

$$1\ \text{u} = 1.6605 \times 10^{-27}\ \text{kg}$$

The speed of light ($c$) is $2.9979 \times 10^8$ m/s. Substitution of these values into $E = mc^2$, gives the energy equivalent of 1 u in kg·m²/s² (which are joules).

$$1\ \text{u} = 1.4924 \times 10^{-10}\ \text{J}$$

Since 1 MeV is $1.60219 \times 10^{-13}$ J,

$$1\ \text{u} = 9.3148 \times 10^2\ \text{MeV}$$

The energy equivalent of this mass difference may be calculated by means of Einstein's equation, $E = mc^2$. The energy unit customarily employed is the MeV (which is a million electron volts). An electron volt (eV) is the energy acquired by an electron when it is accelerated through a potential difference of 1 V; it is the unit we employed in discussing ionization potentials (Section 3.2). The energy equivalent of 1 u is 931.478 MeV, and therefore the energy released by the $\alpha$ decay of $^{210}_{84}$Po is

$$0.0058\ \text{u} \times 931\ \text{MeV/u} = 5.4\ \text{MeV}$$

If the daughter nucleus in an $\alpha$-decay process is left in the ground state, the kinetic energy of the emitted $\alpha$ particle accounts for most of the energy released, and the kinetic energy of the recoiling daughter nucleus accounts for the remainder. If all the daughter nuclei are left in the ground state, all the $\alpha$ particles emitted have the same energy. In some $\alpha$-decay processes, however, several energy groups of $\alpha$ particles are emitted. In these cases the $\alpha$ particles of highest energy correspond to daughter nuclei in the ground state, and the $\alpha$ particles of lower energies originate from nuclei left in excited states. A nucleus in an excited state subsequently emits energy in the form of $\gamma$ radiation to reach the ground state. The sum of the kinetic energy of the $\alpha$ particle, the recoil energy of the daughter nucleus, and the energy of the $\gamma$ radiation equals the decay energy.

**Gamma radiation** is electromagnetic radiation of very short wavelength; its emission is caused by energy changes within the nucleus. Its emission alone does not cause changes in the mass number or in the atomic number of the nucleus. At times, nuclides are produced in excited states by nuclear reactions (Section 21.5), and such nuclides revert to their ground states by the emission of the excess energy in the form of $\gamma$ radiation.

$$[^{125}_{52}\text{Te}]^* \rightarrow \quad ^{125}_{52}\text{Te} \quad + \gamma$$
$$\text{excited state} \qquad \text{ground state}$$

The $\gamma$ rays emitted by a specific nucleus have a definite energy value or set of energy values because they correspond to transitions between discrete energy levels of the nucleus. Thus, an emission spectrum of $\gamma$ radiation is analogous to the line spectrum that results from transitions of electrons between energy levels in an excited atom.

Gamma radiation frequently accompanies all other types of radioactive decay. The following $\alpha$-decay process is an example.

$$^{240}_{94}\text{Pu} \rightarrow [^{236}_{92}\text{U}]^* + ^{4}_{2}\text{He}$$

$$[^{236}_{92}\text{U}]^* \rightarrow ^{236}_{92}\text{U} + \gamma$$

In cases such as this, deductions can be made concerning the energy levels of the daughter nucleus. The emission of a 5.16 MeV $\alpha$ particle results directly in the production of $^{236}_{92}\text{U}$ in the ground state; no $\gamma$ radiation accompanies such $\alpha$ particles. However, for the process outlined in the set of equations, a 5.12 MeV $\alpha$ particle is emitted along with a 0.04 MeV $\gamma$ ray. It is assumed that the energy of the $\gamma$ ray corresponds to the energy difference between the ground state and the first excited state of $^{236}_{92}\text{U}$. Alpha particles of other energies are also emitted, and the complete analysis of these energies results in a more detailed picture of the energy levels of the $^{236}_{92}\text{U}$ nucleus.

**Beta emission** is observed for nuclides that have too high a neutron/proton ratio for stability; points representing such nuclides lie to the left of the zone of stability of Figure 21.1. The $\beta$ particle is an electron, indicated by the symbol $_{-1}^{0}e$, which may be considered to result from the transformation of a nuclear neutron into a nuclear proton; electrons, as such, do not exist in the nucleus. The net effect of $\beta$ emission is that the number of neutrons is decreased by 1 and the number of protons is increased by 1. Thus, the neutron/proton ratio is decreased; the mass number does not change.

Beta decay is a very common mode of radioactive disintegration and is observed for both natural and synthetic nuclides. Examples include

$$^{186}_{73}\text{Ta} \rightarrow ^{186}_{74}\text{W} + _{-1}^{0}e$$

$$^{82}_{35}\text{Br} \rightarrow ^{82}_{36}\text{Kr} + _{-1}^{0}e$$

$$^{27}_{12}\text{Mg} \rightarrow ^{27}_{13}\text{Al} + _{-1}^{0}e$$

$$^{14}_{6}\text{C} \rightarrow ^{14}_{7}\text{N} + _{-1}^{0}e$$

Notice that the sums of the subscripts and superscripts on the right side of the equation equal the subscript and superscript of the parent nucleus on the left.

The last equation represents the mode of decay of the radioactive carbon isotope that is present in small amount in the atmosphere. The energy released by this process may be calculated by using atomic masses instead of nuclear masses. If we add six orbital electrons to both sides of the equation, we have the equivalent of the *atomic* mass of $^{14}_{6}$C (which includes the mass of six electrons) on the left and, taking into account the electron ejected as a $\beta$ particle, the equivalent of the *atomic* mass $^{14}_{7}$N (which includes the mass of seven electrons) on the right. Therefore,

$$\text{(mass } ^{14}_{6}\text{C)} - \text{(mass } ^{14}_{7}\text{N)}$$

$$14.00324 \text{ u} - 14.00307 \text{ u} = 0.00017 \text{ u} = 0.16 \text{ MeV}$$

In $\beta$ decay the recoil energy of the daughter nucleus is negligible since the ejected electron has a small mass. One would expect that the decay energy would be taken up by the kinetic energy of the $\beta$ particle and that all the $\beta$ particles emitted would have energies corresponding to this value. Instead, however, a continuous spectrum of $\beta$-particle energies is observed with the highest energy almost equal to the decay energy. It is postulated that when a $\beta$ particle of less than maximum energy is ejected, a **neutrino,** $\nu$, that carries off the excess energy is emitted at the same time.

$$^{14}_{6}\text{C} \rightarrow \, ^{14}_{7}\text{N} + \, ^{0}_{-1}e + \nu$$

The neutrino is assumed to be a chargeless particle of a vanishingly small mass.

Unstable nuclides that have neutron/proton ratios below those required for stability (points below the zone of stability of Figure 21.1) do not occur in nature. Many such artificial nuclides are known, however; two types of radioactive processes are observed that increase the neutron/proton ratio of this type of nuclide: positron emission and electron capture.

**Positron emission,** or $\beta^+$ emission, consists of the ejection of a positive electron, which is called a positron and indicated by the symbol $^{0}_{1}e$, from the nucleus. A positron has the same mass as an electron but an opposite charge; it arises from the conversion of a nuclear proton into a neutron. Positron emission results in a decrease of *one* in the number of protons and an increase of *one* in the number of neutrons; no change in mass number occurs. Hence, positron emission raises the numerical value of the neutron/proton ratio.

Examples of this mode of radioactive decay include

$$^{122}_{53}\text{I} \rightarrow \,^{122}_{52}\text{Te} + \,^{0}_{1}e$$

$$^{38}_{19}\text{K} \rightarrow \,^{38}_{18}\text{Ar} + \,^{0}_{1}e$$

$$^{23}_{12}\text{Mg} \rightarrow \,^{23}_{11}\text{Na} + \,^{0}_{1}e$$

$$^{15}_{8}\text{O} \rightarrow \,^{15}_{7}\text{N} + \,^{0}_{1}e$$

We can use atomic masses to calculate the energy released by the process described by the last equation. If we add eight orbital electrons to both sides of the equation, the equivalent of the atomic mass of $^{15}_{8}\text{O}$ (eight electrons) would be indicated on the left. The equivalent of the atomic mass of $^{15}_{7}\text{N}$ (seven electrons) plus the mass of one orbital electron (making up the eight added) plus the mass of the ejected positron would be indicated on the right. Thus, the mass difference is

$$(\text{mass } ^{15}_{8}\text{O}) - (\text{mass } ^{15}_{7}\text{N} + \text{mass } _{-1}^{0}e + \text{mass } ^{0}_{1}e)$$

Since the mass of a positron is identical to that of an electron,

$$(\text{mass } ^{15}_{8}\text{O}) - [\text{mass } ^{15}_{7}\text{N} + 2(\text{mass } _{-1}^{0}e)]$$

$$15.00308 \text{ u} - [15.00011 \text{ u} + 2(0.00055 \text{ u})]$$

$$15.00308 \text{ u} - 15.00121 \text{ u} = 0.00187 \text{ u} = 1.74 \text{ MeV}$$

For spontaneous positron emission the atomic mass of the parent must exceed the atomic mass of the daughter by at least 0.00110 u, the mass of two electrons.

In positron emission a spectrum of $\beta^+$ energies is observed similar to that observed for $\beta$ emission, and the simultaneous ejection of neutrinos is postulated.

$$^{15}_{8}\text{O} \rightarrow \,^{15}_{7}\text{N} + \,^{0}_{1}e + \nu$$

The positron was the first antiparticle to be observed. It is similar to the electron in all respects except charge. When a positron and an electron collide, they annihilate each other, and $\gamma$ radiation equivalent to the masses of the two particles is produced. It is believed that antiparticles exist for all particles except the neutral pion, $\pi^0$, and the photon. The negative proton (antiproton) has been detected; the annihilation of proton with antiproton usually produces several high-energy pions. Since the antineutron, like the neutron, has no charge, its detection rests on annihilation phenomena.

**Electron capture** (ec), sometimes called $K$ capture, is another process through which the neutron/proton ratio of an unstable, proton-rich nuclide may be increased. Unlike positron emission, electron capture can occur when the mass difference between parent and daughter does not exceed 0.00110 u. In this process the nucleus captures an orbital electron from the $K$ or $L$ shell, and the captured electron converts a nuclear proton into a neutron. The transformation results in a daughter nuclide with *one* less proton and *one* more neutron than the parent; consequently, the atomic number of the daughter is *one* less than that of the parent, and the mass number does not change. Examples are

$$_{-1}^{0}e + {}_{80}^{197}\text{Hg} \xrightarrow{\text{ec}} {}_{79}^{197}\text{Au}$$

$$_{-1}^{0}e + {}_{47}^{106}\text{Ag} \xrightarrow{\text{ec}} {}_{46}^{106}\text{Pd}$$

$$_{-1}^{0}e + {}_{18}^{37}\text{Ar} \xrightarrow{\text{ec}} {}_{17}^{37}\text{Cl}$$

$$_{-1}^{0}e + {}_{4}^{7}\text{Be} \xrightarrow{\text{ec}} {}_{3}^{7}\text{Li}$$

$$_{-1}^{0}e + {}_{26}^{55}\text{Fe} \xrightarrow{\text{ec}} {}_{25}^{55}\text{Mn}$$

The energy released can be calculated directly from the atomic masses of the parent and daughter. Thus, for the last example,

(mass ${}_{26}^{55}$Fe) − (mass ${}_{25}^{55}$Mn)
54.93830 u − 54.93805 u = 0.00025 u = 0.23 MeV

The recoil energy of the daughter nucleus is negligible, and if the daughter nucleus is left in the ground state, all the available energy is carried away by a neutrino that is ejected in the process. In addition, electron capture is accompanied by the production of X rays. The capture of an orbital electron leaves a vacancy in the $K$ or $L$ shell, and when an outer electron falls into this vacancy, X-ray emission follows (Section 2.12).

**21.3 RATE OF RADIOACTIVE DECAY**

Many techniques are employed to study the emissions of radioactive substances. Radiations from these materials affect photographic film in the same way that ordinary light does. Photographic techniques for the qualitative and quantitative detection of radiation are employed, but they are not very accurate nor are they suitable for rapid analysis.

The energy of emissions from radioactive sources is absorbed by some materials (e.g., zinc sulfide) and transformed into radiant energy of visible wavelength. The zinc sulfide is said to **fluoresce,** and a little flash of light may be observed from the impact of each particle from the radioactive source. This property has been put to use in an instrument known as a **scintillation counter.** The window of a sensitive photoelectric tube is coated with ZnS, and the flash of light emitted by the ZnS when it is struck by a particle causes a pulse of electric current to pass through the photoelectric tube. These signals are amplified and made to operate various kinds of counting devices.

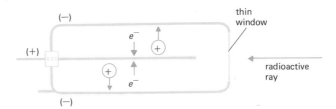

**Figure 21.2** Essential features of a Geiger–Müller counter tube.

Figure 21.2 Essential features of a Geiger–Müller counter tube.

The **Wilson cloud chamber** enables the path of ionizing radiation to be seen. The chamber contains air saturated with water vapor. By the movement of a piston the air in the chamber is suddenly expanded and cooled; this causes droplets of water to condense on the ions that are formed by the particles as they move through the vapor, and makes the paths of the particles visible. Photographs of these cloud tracks may be made and studied. Such photographs provide information on the length of the paths, collisions undergone by the particles, the speed of the particles, and the effects of external forces on the behavior of the particles.

The essential features of a **Geiger–Müller counter** are diagramed in Figure 21.2. The radiation enters the tube through a thin window. As a particle or a $\gamma$ ray traverses the tube, which contains argon gas, it knocks electrons off the argon atoms in its path and forms $Ar^+$ ions. A potential of about 1000 to 1200 V is applied between the electrodes of the tube, and the electrons and $Ar^+$ ions cause a pulse of electric current to flow through the circuit. This pulse is amplified to cause a clicker to sound or an automatic counting device to operate.

The rates of decay of all radioactive substances have been found to be first order (Section 13.5) and to be independent of temperature. This lack of temperature dependence implies that the activation energy of any radioactive-decay process is zero. The rate of decay, therefore, depends upon the amount of radioactive material present – a relationship similar to the rate at which money accumulates at compound interest, with the distinction that radioactive decay represents a loss, not an accumulation. We may express this relationship as

$$-\frac{\Delta N}{\Delta t} = kN$$

where $N$ is the number of atoms of radioactive material, $t$ is time, and $k$ is the rate constant. The rate expression is negative because it represents the disappearance of the radioactive substance.

Rearrangement of the rate expression gives

$$-\frac{\Delta N}{N} = k\Delta t$$

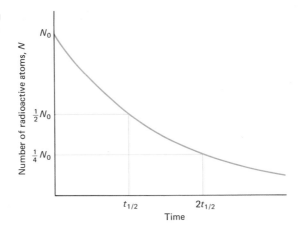

which states that the fraction lost ($-\Delta N/N$) in a given interval ($\Delta t$) is directly proportional to the length of the time interval. Therefore, the time required for half of the sample to decay (the **half-life**, $t_{1/2}$) is a constant.

The curve of Figure 21.3, which shows number of radioactive atoms versus time, is typical of first-order processes. After a single half-life period has elapsed, one-half of the original number of atoms remain ($\frac{1}{2}N_0$). This number is reduced by half (to $\frac{1}{4}N_0$) by the time that another half-life has passed. Each radioactive isotope has a characteristic half-life, and these vary widely; for example, $^5_3\text{Li}$ has a half-life estimated to be $10^{-21}$ sec, and $^{238}_{92}\text{U}$ has a half-life of $4.51 \times 10^9$ years.

The rate equation may be written in its differential form,

$$-\frac{dN}{dt} = kN$$

and by means of the calculus this equation may be integrated,

$$-2.303 \log \left(\frac{N}{N_0}\right) = kt$$

where $N_0$ is the amount present at time zero, and $N$ is the amount present at time $t$. The last equation may be rearranged to give

$$\log \left(\frac{N_0}{N}\right) = \frac{kt}{2.303}$$

An expression for the half-life may be derived by allowing

$$N = \tfrac{1}{2}N_0$$

$$\frac{N_0}{N} = 2$$

Therefore,

$$\log 2 = \frac{k(t_{1/2})}{2.303}$$

$$t_{1/2} = \frac{2.303 \log 2}{k} = \frac{0.693}{k}$$

According to the rate law, a radioactive substance never completely disappears. However, the rate law describes the behavior of a sample containing a large number of atoms. When the number of atoms in the sample declines to a very small value, the rate law is no longer followed exactly. How can a fraction of an atom disintegrate?

---

**Example 21.1**

The isotope $^{60}_{27}Co$ has a half-life of 5.27 years. What amount of a 0.0100 g sample of $^{60}_{27}Co$ re-remains after 1.00 year?

**Solution**

The rate constant for this disintegration is

$$k = \frac{0.693}{t_{1/2}}$$

$$= \frac{0.693}{5.27 \text{ years}} = 0.132/\text{year}$$

The fraction remaining undecomposed at the end of 1.00 year may be found in the following way.

$$\log\left(\frac{N_0}{N}\right) = \frac{kt}{2.30}$$

$$= \frac{(0.132/\text{year})(1.00 \text{ year})}{2.30} = 0.0573$$

$$\log\left(\frac{N}{N_0}\right) = -0.0573$$

$$\frac{N}{N_0} = 0.876$$

Therefore, the amount remaining after 1.00 year is

$$0.876 \times 0.0100 \text{ g} = 0.0088 \text{ g}$$

---

The radioactive isotope $^{14}_{6}C$ is produced in the atmosphere by the action of cosmic-ray neutrons on $^{14}_{7}N$.

$$^{14}_{7}N + ^{1}_{0}n \rightarrow ^{14}_{6}C + ^{1}_{1}H$$

The $^{14}_{6}C$ is oxidized to $CO_2$, and this radioactive $CO_2$ mixes with non-radioactive $CO_2$. The radiocarbon disappears through radioactive decay, but it is also constantly being made. The steady state is reached when the proportion is one atom of $^{14}_{6}C$ to $10^{12}$ ordinary carbon atoms, which represents $15.3 \pm 0.1$ disintegrations per minute per gram of carbon.

The $CO_2$ of the atmosphere is absorbed by plants through the process of photosynthesis, and the ratio of $^{14}_{6}C$ to ordinary carbon in plant materials that are alive and growing is the same ratio as that in the atmosphere. When the plant dies, however, the amount of $^{14}_{6}C$ diminishes through radioactive decay and is not replenished by the assimilation of atmospheric $CO_2$ by the plant.

The half-life of $^{14}_{6}C$, a $\beta$ emitter, is 5770 years. The age of a wooden object can be determined by comparing the radiocarbon activity of the object with that of growing trees. This method of **radiocarbon dating** has been applied to many archeological finds and objects of historical interest. By this means the Dead Sea scrolls were determined to be $1917 \pm 200$ years old.

---

*Example 21.2*

A sample of carbon from a wooden artifact is found to give 7.00 $^{14}_{6}C$ counts per minute per gram of carbon. What is the approximate age of the artifact?

*Solution*

The half-life of $^{14}_{6}C$ is 5770 years. Therefore,

$$k = \frac{0.693}{t_{1/2}} = \frac{0.693}{5770 \text{ years}} = 1.20 \times 10^{-4}/\text{year}$$

The $^{14}_{6}C$ from wood recently cut down decays at the rate of 15.3 disintegrations per minute per gram of carbon. Therefore,

$$\log \left(\frac{N_0}{N}\right) = \frac{kt}{2.30}$$

$$\log \left(\frac{15.3 \text{ disintegrations/min}}{7.00 \text{ disintegrations/min}}\right) = \frac{(1.20 \times 10^{-4}/\text{year})t}{2.30}$$

$$t = \frac{2.30 \log 2.19}{1.20 \times 10^{-4}/\text{year}}$$

$$= 6520 \text{ years}$$

---

● The SI derived unit of radioactive activity is 1 per second (symbol, $s^{-1}$). Use of the curie (official symbol, Ci) is permitted for a limited time.

$$1 \text{ Ci} = 3.7 \times 10^{10} \text{ s}^{-1}$$

The amount of radiation emanating from a source per unit time is termed the **activity** of the source.

$$\text{activity} = -\frac{dN}{dt} = kN$$

Activities are generally expressed in curies; 1 curie (c) is defined as $3.70 \times 10^{10}$ disintegrations per second, and 1 microcurie ($\mu c$) is $3.70 \times 10^{4}$ disintegrations per second.

---

*Example 21.3*

The half-life of $^{100}_{43}Tc$, a $\beta$ emitter, is 16 sec. How many atoms of $^{100}_{43}Tc$ are present in a sample with an activity of 0.200 $\mu c$? What is the weight of the sample?

*Solution*

For this radioactive decay the rate constant, $k$, is

$$k = \frac{0.693}{t_{1/2}}$$

$$= \frac{0.693}{16 \text{ sec}} = 0.0433/\text{sec}$$

The activity of the sample in terms of disintegrations per second is

$$0.200(3.70 \times 10^4 \text{ disintegrations/sec}) = 7.40 \times 10^3 \text{ disintegrations/sec}$$

This represents the decay of $7.40 \times 10^3$ atoms/sec and

$$\text{activity} = kN$$

$$(7.40 \times 10^3 \text{ atoms/sec}) = (4.33 \times 10^{-2}/\text{sec})N$$

$$N = 1.71 \times 10^5 \text{ atoms}$$

The weight of the sample can be derived from the fact that the atomic weight of $^{100}Tc$ to three significant figures is 100; thus,

$$? \text{ g Tc} = 1.71 \times 10^5 \text{ atoms Tc} \left(\frac{100 \text{ g Tc}}{6.02 \times 10^{23} \text{ atoms Tc}}\right)$$

$$= 2.84 \times 10^{-19} \text{ g Tc}$$

**21.4 RADIOACTIVE DISINTEGRATION SERIES**

The examples of $\alpha$, $\beta^-$, and $\beta^+$ emission and of electron capture given in Section 21.2 are one-step processes that lead to stable nuclides. Frequently, however, the daughter nucleus produced by a radioactive process is itself radioactive. The repetition of this situation creates a chain, or series, of disintegration processes involving many radioactive nuclides and leading ultimately to the production of a stable nuclide.

Three such **disintegration series,** which involve only $\alpha$ and $\beta$ emission, occur in nature: the $^{232}_{90}$Th, $^{238}_{92}$U, and $^{235}_{92}$U series. The $^{238}_{92}$U series, which leads finally to the stable nuclide $^{206}_{82}$Pb, is diagramed in Figure 21.4. In several places in the series branching occurs, and the series proceeds by two different routes. The branches, however, always rejoin at a later point, and commonly one branch is preferred over the other (note the percentage figures in Figure 21.4). By any given route the $^{238}_{92}$U series consists of 14 steps—8 involving $\alpha$ decay and 6 involving $\beta$ decay.

The radioactive nuclides that occur in nature are those that have very long half-lives or those that are constantly being produced by the disintegration of other nuclides. This last type of naturally occurring radioactive nuclide eventually exists in a steady state at which time the amount of the nuclide remains essentially constant because the material is being produced at the same rate that it is decomposing.

The natural disintegration series serve as the basis of a method of geological dating. A sample of rock may be analyzed for its $^{206}_{82}$Pb and $^{238}_{92}$U content, and the length of time to produce this ratio of lead to uranium may be calculated from the decay constants of the series. Results from

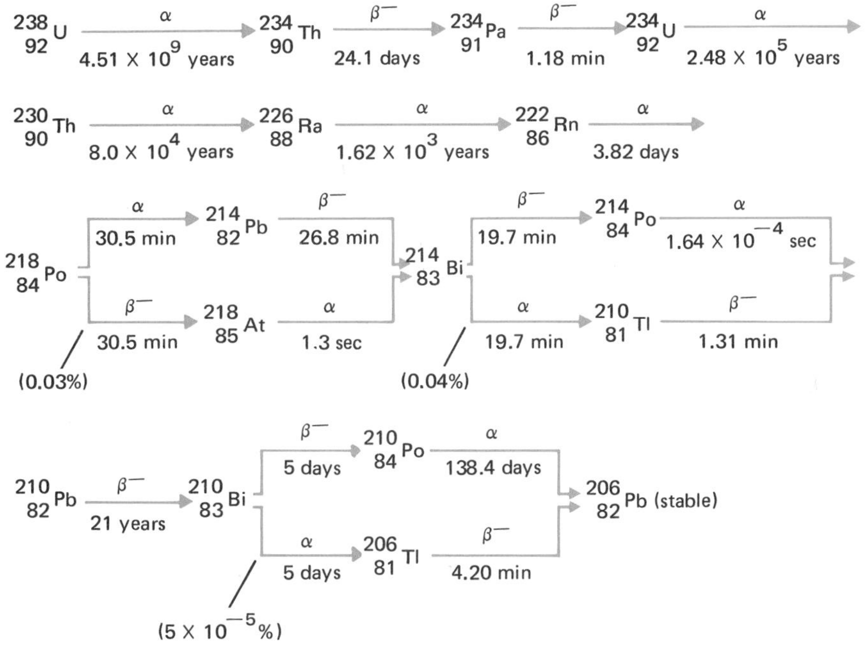

**Figure 21.4** Disintegration series of $^{238}_{92}$U. Half-lives of isotopes are indicated.

$^{238}_{92}$U $\xrightarrow[4.51 \times 10^9 \text{ years}]{\alpha}$ $^{234}_{90}$Th $\xrightarrow[24.1 \text{ days}]{\beta^-}$ $^{234}_{91}$Pa $\xrightarrow[1.18 \text{ min}]{\beta^-}$ $^{234}_{92}$U $\xrightarrow[2.48 \times 10^5 \text{ years}]{\alpha}$

$^{230}_{90}$Th $\xrightarrow[8.0 \times 10^4 \text{ years}]{\alpha}$ $^{226}_{88}$Ra $\xrightarrow[1.62 \times 10^3 \text{ years}]{\alpha}$ $^{222}_{86}$Rn $\xrightarrow[3.82 \text{ days}]{\alpha}$

$^{218}_{84}$Po
$\xrightarrow[30.5 \text{ min}]{\alpha}$ $^{214}_{82}$Pb $\xrightarrow[26.8 \text{ min}]{\beta^-}$ $^{214}_{83}$Bi
$\xrightarrow[30.5 \text{ min}]{\beta^-}$ $^{218}_{85}$At $\xrightarrow[1.3 \text{ sec}]{\alpha}$

(0.03%)

$\xrightarrow[19.7 \text{ min}]{\beta^-}$ $^{214}_{84}$Po $\xrightarrow[1.64 \times 10^{-4} \text{ sec}]{\alpha}$
$\xrightarrow[19.7 \text{ min}]{\alpha}$ $^{210}_{81}$Tl $\xrightarrow[1.31 \text{ min}]{\beta^-}$

(0.04%)

$^{210}_{82}$Pb $\xrightarrow[21 \text{ years}]{\beta^-}$ $^{210}_{83}$Bi
$\xrightarrow[5 \text{ days}]{\beta^-}$ $^{210}_{84}$Po $\xrightarrow[138.4 \text{ days}]{\alpha}$ $^{206}_{82}$Pb (stable)
$\xrightarrow[5 \text{ days}]{\alpha}$ $^{206}_{81}$Tl $\xrightarrow[4.20 \text{ min}]{\beta^-}$

$(5 \times 10^{-5}\%)$

studies such as these place the age of some rocks at from 3 to 3.5 billion years; the age of the earth is estimated to be 4.5 billion years.

Decay series, some more elaborate than others, are known for artificial nuclides. Positron emission and $K$ capture are observed in some of these series as well as $\alpha$ and $\beta$ emission. Following are three examples of simple two-step disintegration series of artificial nuclides.

$$^{20}_{8}\text{O} \xrightarrow[14 \text{ sec}]{\beta^-} {}^{20}_{9}\text{F} \xrightarrow[11 \text{ sec}]{\beta^-} {}^{20}_{10}\text{Ne}$$

$$^{30}_{16}\text{S} \xrightarrow[1.4 \text{ sec}]{\beta^+} {}^{30}_{15}\text{P} \xrightarrow[2.6 \text{ min}]{\beta^+} {}^{30}_{14}\text{Si}$$

$$^{76}_{36}\text{Kr} \xrightarrow[10 \text{ hours}]{ec} {}^{76}_{35}\text{Br} \xrightarrow[16.5 \text{ hours}]{\beta^+ \text{ or } ec} {}^{76}_{34}\text{Se}$$

**21.5 NUCLEAR REACTIONS**

In 1915 Ernest Rutherford reported that the following transformation occurs when $\alpha$ particles from $^{214}_{84}\text{Po}$ are passed through nitrogen.

$$^{14}_{7}\text{N} + {}^{4}_{2}\text{He} \rightarrow {}^{17}_{8}\text{O} + {}^{1}_{1}\text{H}$$

This was the first artificial transmutation of one element into another to be reported; in the years following, thousands of such nuclear transformations have been studied. It is assumed that the projectile (in this case, an $\alpha$ particle) forms a compound nucleus with the target ($^{14}_{7}\text{N}$) and that the compound nucleus very rapidly ejects a subsidiary particle ($^{1}_{1}\text{H}$) to form the product nucleus ($^{17}_{8}\text{O}$). Other projectiles, such as neutrons, deuterons ($^{2}_{1}\text{H}$), protons, and ions of low atomic number, are used in addition to $\alpha$ particles.

Particle–particle reactions are usually classified according to the type of projectile employed and the subsidiary particle ejected. Thus, the preceding reaction is called an ($\alpha$, p) reaction, and the complete transformation is indicated by the notation $^{14}_{7}\text{N}$ ($\alpha$, p) $^{17}_{8}\text{O}$. Examples of several of the more common types of nuclear transformations are listed in Table 21.2.

The first artificial, radioactive nuclide produced was made by the ($\alpha$, n) reaction

$$^{27}_{13}\text{Al} + {}^{4}_{2}\text{He} \rightarrow {}^{30}_{15}\text{P} + {}^{1}_{0}\text{n}$$

The product, $^{30}_{15}\text{P}$, decays by positron emission,

$$^{30}_{15}\text{P} \rightarrow {}^{30}_{14}\text{Si} + {}^{0}_{1}e$$

Except for the nature of the product there is no difference between nuclear reactions that produce stable nuclides and those that yield radioactive nuclides.

Table 21.2 Examples of nuclear reactions

| TYPE | REACTION | RADIOACTIVITY OF PRODUCT NUCLIDE |
|---|---|---|
| $(\alpha, n)$ | $^{75}_{33}As + ^{4}_{2}He \rightarrow ^{78}_{35}Br + ^{1}_{0}n$ | $\beta^{+}$ |
| $(\alpha, p)$ | $^{106}_{46}Pd + ^{4}_{2}He \rightarrow ^{109}_{47}Ag + ^{1}_{1}H$ | stable |
| $(p, n)$ | $^{7}_{3}Li + ^{1}_{1}H \rightarrow ^{7}_{4}Be + ^{1}_{0}n$ | ec |
| $(p, \gamma)$ | $^{14}_{7}N + ^{1}_{1}H \rightarrow ^{15}_{8}O + \gamma$ | $\beta^{+}$ |
| $(p, \alpha)$ | $^{9}_{4}Be + ^{1}_{1}H \rightarrow ^{6}_{3}Li + ^{4}_{2}He$ | stable |
| $(d, p)$ | $^{31}_{15}P + ^{2}_{1}H \rightarrow ^{32}_{15}P + ^{1}_{1}H$ | $\beta^{-}$ |
| $(d, n)$ | $^{209}_{83}Bi + ^{2}_{1}H \rightarrow ^{210}_{84}Po + ^{1}_{0}n$ | $\alpha$ |
| $(n, \gamma)$ | $^{59}_{27}Co + ^{1}_{0}n \rightarrow ^{60}_{27}Co + \gamma$ | $\beta^{-}$ |
| $(n, p)$ | $^{45}_{21}Sc + ^{1}_{0}n \rightarrow ^{45}_{20}Ca + ^{1}_{1}H$ | $\beta^{-}$ |
| $(n, \alpha)$ | $^{27}_{13}Al + ^{1}_{0}n \rightarrow ^{24}_{11}Na + ^{4}_{2}He$ | $\beta^{-}$ |

Projectile particles that bear a positive charge are repelled by target nuclei; this is particularly true of the heavier nuclei, which have high charges. Consequently, only a small number of nuclear transformations can be brought about by the positive particles emitted by radioactive sources. Various particle accelerators are used to give protons, deuterons, $\alpha$ particles, and other cationic projectiles sufficiently high kinetic energies to overcome the electrostatic repulsions of the target nuclei. The **cyclotron** (Figure 21.5) is one such instrument.

The ion source is located between two hollow D-shaped plates ($D_1$ and $D_2$), called dees, that are separated by a gap. The dees are enclosed in an evacuated chamber located between the poles of a powerful electromagnet (not shown in the figure). A high-frequency generator keeps the dees oppositely charged. Under the influence of the magnetic and electrical fields the ions move from the source in a circular path. Each time they reach the gap between the dees, the polarity of the dees is reversed. Thus, the positively charged particles are pushed out of a positive dee and attracted into a negative dee. Each time they traverse the gap,

Figure 21.5 Path of a particle in a cyclotron.

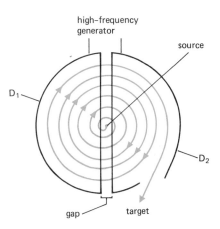

high–frequency generator

source

$D_1$

$D_2$

gap

target

**Figure 21.6** Schematic representation of a linear accelerator.

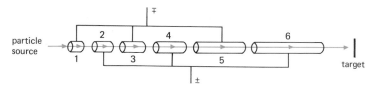

therefore, they are accelerated. Because of this, the particles travel an ever-increasing spiral path; eventually they penetrate a window in the instrument and, moving at extremely high speed, impinge on a target.

The **linear accelerator** (Figure 21.6) operates in much the same way except that no magnetic field is employed. The particles are accelerated through a series of tubes enclosed in an evacuated chamber. A positive ion from the source is attracted into tube 1, which is negatively charged. At this time, the odd-numbered tubes have negative charges and the even-numbered tubes have positive charges. As the particle emerges from tube 1, the charges of the tubes are reversed so that the even-numbered tubes are now negatively charged. The particle is repelled out of tube 1 (now positive) and attracted into tube 2 (now negative); as a result, it is accelerated.

Each time the particle leaves one tube to enter another, the charges of the tubes are reversed. Since the polarity of the tubes is reversed at a constant time interval, and since the speed of the particle increases constantly, each tube must be longer than the preceding one. The accelerated particles leave the last tube at high speed and strike the target.

Neutrons are particularly important projectiles because they bear no charge and therefore are not repelled by the positive charge of the target nuclei. A mixture of beryllium and an $\alpha$ emitter (such as $^{222}_{86}Rn$) is a convenient neutron source.

$$^{9}_{4}Be + ^{4}_{2}He \rightarrow ^{12}_{6}C + ^{1}_{0}n$$

This reaction was used by James Chadwick in his experiments that characterized the neutron (1932). The bombardment of beryllium by accelerated deuterons from a cyclotron is a more intense source of neutrons.

$$^{9}_{4}Be + ^{2}_{1}H \rightarrow ^{10}_{5}B + ^{1}_{0}n$$

A very important source of neutrons is the nuclear reactor (Section 21.6).

Neutrons from nuclear reactions are known as **fast neutrons;** they cause reactions in which a subsidiary particle is ejected [such as (n, $\alpha$) and (n, p) reactions]. **Slow neutrons,** or **thermal neutrons,** are produced when the neutrons derived from a nuclear reaction are passed through a moderator (such as carbon, paraffin, hydrogen, deuterium, or oxygen). Through collisions with the nuclei of the moderator, the kinetic energies of the neutrons are decreased to values approximating those of ordinary gas molecules. Bombardments using slow neutrons bring about (n, γ) reactions, which are also called **neutron-capture reactions** since no subsidiary particle is ejected.

$$^{34}_{16}S + ^{1}_{0}n \rightarrow ^{35}_{16}S + \gamma$$

Isotopes of practically every element have been prepared by this type of reaction.

Nuclear reactions have been used to prepare isotopes belonging to elements that do not exist in nature or that exist in extremely minute concentrations. Thus, isotopes of technetium and astatine have been prepared by the reactions

$$^{96}_{42}Mo + ^{2}_{1}H \rightarrow ^{97}_{43}Tc + ^{1}_{0}n$$

$$^{209}_{83}Bi + ^{4}_{2}He \rightarrow ^{211}_{85}At + 2^{1}_{0}n$$

The elements following uranium in the periodic classification are called the **transuranium elements;** none of these elements are naturally occurring, but many have been made by nuclear reactions. Some of these reactions use targets of artificial nuclides, and in these cases the final products are therefore the result of syntheses consisting of several steps. Examples of these preparations are

$$^{238}_{92}U + ^{1}_{0}n \rightarrow ^{239}_{92}U + \gamma$$

$$^{239}_{92}U \rightarrow ^{239}_{93}Np + ^{0}_{-1}e$$

$$^{239}_{93}Np \rightarrow ^{239}_{94}Pu + ^{0}_{-1}e$$

$$^{239}_{94}Pu + ^{2}_{1}H \rightarrow ^{240}_{95}Am + ^{1}_{0}n$$

$$^{239}_{94}Pu + ^{4}_{2}He \rightarrow ^{242}_{96}Cm + ^{1}_{0}n$$

In addition to the common projectiles, ions of elements of low atomic number are used in some bombardment reactions.

$$^{238}_{92}U + ^{12}_{6}C \rightarrow ^{244}_{98}Cf + 6 ^{1}_{0}n$$

$$^{238}_{92}U + ^{14}_{7}N \rightarrow ^{246}_{99}Es + 6 ^{1}_{0}n$$

$$^{238}_{92}U + ^{16}_{8}O \rightarrow ^{250}_{100}Fm + 4 ^{1}_{0}n$$

$$^{252}_{98}Cf + ^{10}_{5}B \rightarrow ^{257}_{103}Lr + 5 ^{1}_{0}n$$

**21.6 NUCLEAR FISSION AND FUSION**

Nuclear fission reactions are more famous and infamous than particle–particle or particle-capture transformations. In a **fission** process a heavy nucleus is split into nuclei of lighter elements and several neutrons. The fission of a given nuclide results in more than one set of products. Two possible reactions for the slow neutron induced fission of $^{235}_{92}U$ are

$$^{235}_{92}U + ^1_0n \rightarrow ^{93}_{36}Kr + ^{140}_{56}Ba + 3\,^1_0n$$

$$^{235}_{92}U + ^1_0n \rightarrow ^{90}_{38}Sr + ^{144}_{54}Xe + 2\,^1_0n$$

Heavy nuclei have much larger neutron/proton ratios than nuclei of moderate mass (Figure 21.1). The neutrons released in the fission process lower the neutron/proton ratios of the product nuclei; nevertheless, these nuclei are generally radioactive, and further adjustment of the neutron/proton ratios occurs, usually through $\beta$ emission. Certain fissions can be induced by protons, deuterons, or $\alpha$ particles, but the most important are those that are brought about by neutrons.

The **binding energy** of a nucleus may be considered to be the energy *required* to pull the nucleons of the nucleus apart or the energy *released* by the hypothetical formation of the nucleus by the condensation of the individual nucleons. The binding energy of a nucleus may be calculated from the difference between the sum of the masses of the constituent nucleons and the mass of the corresponding nucleus (Section 2.9). For $^{35}_{17}Cl$ this mass difference is 0.320 u, and therefore

$$0.320 \text{ u} \times 931 \text{ MeV/u} = 298 \text{ MeV}$$

The magnitude of the binding energy of a given nucleus indicates the stability of that nucleus toward radioactive decay. For the purpose of comparison, the values are usually given in terms of binding energy per nucleon, and the largest values are characteristic of the most stable nuclei. The binding energy per nucleon in the case of $^{35}_{17}Cl$ is

$$\frac{298 \text{ MeV}}{35 \text{ nucleons}} = 8.5 \text{ MeV/nucleon}$$

In Figure 21.7 mass number is plotted against binding energy per nucleon for the nuclides. Inspection of the curve shows that nuclides of intermediate mass have larger values of binding energy per nucleon than the heavier nuclides. Thus, the fission of $^{235}_{92}U$ produces lighter nuclei with higher binding energy/nucleon values, and energy is liberated. The sum of the masses of the products of a fission reaction is less than the sum of the masses of the reactants. A typical fission of a single $^{235}_{92}U$ nucleus releases approximately 200 MeV.

The fission of each $^{235}_{92}U$ nucleus is induced by a single neutron and produces several neutrons; in the overall process an average of approximately 2.5 neutrons are produced per fission. If each neutron causes the fission of another $^{235}_{92}U$ nucleus, an explosively rapid chain reaction ensues. Using a reproduction factor of 2 for simplicity, the first fission would cause 2 first-generation fissions. Each of these would cause 2 fissions — a total of 4 second-generation fissions; these would be followed by 8, 16, 32, . . . , fissions in succeeding generations. In the $n$th generation, $2^n$ fissions would occur. Since each fission is extremely rapid, an explosion

Figure 21.7 Binding energy per nucleon versus mass number.

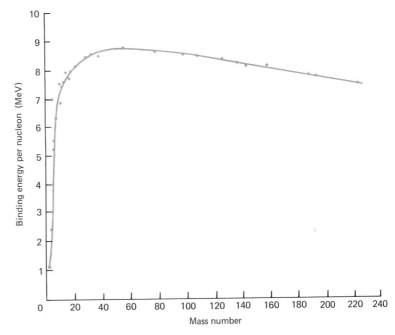

results. If 200 MeV is released by each fission, a tremendous amount of energy is liberated.

In a small amount of $^{235}_{92}U$ undergoing fission, many of the neutrons produced are lost from the surface of the mass before they can bring about nuclear reactions. If the size of the fissionable material exceeds a certain **critical mass,** however, the neutrons are captured before they can leave the mass and an explosive chain reaction results. An atomic bomb is detonated by bringing together two pieces of fissionable material, each of subcritical size, together into one piece of supercritical size. The reaction is started by a stray neutron.

A number of nuclei, such as $^{238}_{92}U$, $^{231}_{91}Pa$, $^{237}_{93}Np$, and $^{232}_{90}Th$, undergo fission with fast, but not with slow, neutrons. Slow neutrons, however, can induce fission in $^{233}_{92}U$, $^{235}_{92}U$ and $^{239}_{94}Pu$. The latter two isotopes are those commonly employed as nuclear fuels. The isotope $^{235}_{92}U$, which comprises only about 0.7% of natural uranium, was employed in the construction of the first atomic bomb. The separation of $^{235}_{92}U$ from the principal uranium isotope, $^{238}_{92}U$, was carried out in many ways, the most successful being the separation of gaseous $^{235}UF_6$ from $^{238}UF_6$ by thermal diffusion through porous barriers (Section 6.12). The isotope $^{239}_{94}Pu$ does not occur in nature but may be prepared from $^{238}_{92}U$ by the series of nuclear reactions given in Section 21.5.

The controlled fission of nuclear fuels in a **nuclear reactor** serves as a source of energy or as a source of neutrons and $\gamma$ radiation for scientific research and the production of artificial isotopes. In a nuclear reactor, cylinders containing at least a critical mass of nuclear fuel ($^{235}_{92}$U, $^{233}_{92}$U, $^{239}_{94}$Pu, natural U, or natural U enriched with $^{235}_{92}$U) are surrounded by a moderator (graphite, water, or heavy water). The moderator serves to slow down the neutrons produced by the fission, thus increasing the possibility of their capture.

The neutron reproduction factor must be maintained at a value close to 1. If it falls below this value, the reaction will eventually stop, and if it increases much above this value, the reaction may become explosively violent. Control rods of cadmium, boron steel, or other materials, which may be inserted into the reactor to any desired depth, serve to capture excess neutrons and control the rate of the reaction.

$$^{113}_{48}\text{Cd} + ^{1}_{0}\text{n} \rightarrow ^{114}_{48}\text{Cd} + \gamma$$

$$^{10}_{5}\text{B} + ^{1}_{0}\text{n} \rightarrow ^{11}_{5}\text{B} + \gamma$$

The $^{114}_{48}$Cd and $^{11}_{5}$B isotopes are not radioactive. Provision is made for the insertion of research samples into the reactor.

Early reactors that used natural uranium as a fuel and graphite blocks as a moderator were large — cubes on the order of 20 to 25 feet on an edge. The use of enriched fuels and moderators other than graphite has reduced the size required considerably. Breeder reactors produce a fissionable isotope (e.g., $^{239}_{94}$Pu from $^{238}_{92}$U) within the reactor and thus provide for the maintenance of a supply of nuclear fuel. Nuclear reactors that are designed as sources of energy employ a circulating coolant to remove the heat from the reactor to the outside where it is used in power production. All nuclear reactors must be heavily shielded.

Nuclear **fusion** is a process in which very light nuclei are fused into a heavier nucleus. The curve of Figure 21.7 shows that such processes should liberate more energy than fission processes. The hydrogen bomb is based on nuclear fusion. Extremely high activation energies are required for fusion reactions, and in actual practice, a fission bomb is used to supply the high temperatures required. The energy of the sun is believed to be derived from the conversion of hydrogen nuclei into helium by nuclear fusion. Reactions such as the following are postulated.

$$^{1}_{1}\text{H} + ^{1}_{1}\text{H} \rightarrow ^{2}_{1}\text{H} + ^{0}_{1}e$$

$$^{2}_{1}\text{H} + ^{1}_{1}\text{H} \rightarrow ^{3}_{2}\text{He} + \gamma$$

$$^{3}_{2}\text{He} + ^{3}_{2}\text{He} \rightarrow ^{4}_{2}\text{He} + 2\,^{1}_{1}\text{H}$$

The liquid drop model of the nucleus interprets fusion and fission in terms of the forces of repulsion and attraction between the nucleons in the nucleus. The size of the nucleus is critical in determining the total effect of these two opposing forces. The nucleons on the surface of the nucleus are subjected to fewer forces of attraction than the nucleons that are surrounded on all sides by other nucleons in the center of the nucleus. Therefore, the total effect of the attractive forces is increased when the surface area is decreased. One large drop has a smaller surface area than several small drops comprising the same amount of nuclear fluid.

On the other hand, the electrostatic forces of repulsion between protons tend to disrupt a large drop into smaller ones. Whether the process of fusion or fission occurs depends upon the balance between these two effects, which do not vary with the size of the drop in a parallel manner. Since the surface area of a sphere is $4\pi r^2$, the increase or decrease in the energy of attraction brought about by a change in surface area is proportional to the square of the radius, $r^2$ (or $A^{2/3}$). The coulombic forces of repulsion, however, are proportional to the square of the nuclear charge divided by the radius, $Z^2/r$ (or $Z^2/A^{1/3}$). For light nuclei the increase in the effect of the attractive forces brought about by the reduction in nuclear surface area more than offsets the increase in the effect of the forces of repulsion, and fusion occurs. For heavy nuclei the reverse applies; fission occurs since it leads to a reduction in the effect of the forces of repulsion that is of more significance than the accompanying reduction in the effect of the forces of attraction.

## 21.7 USES OF ISOTOPES

A large number of uses have been developed for the isotopes that are the products of the processes run in nuclear reactors. Thickness gauges have been developed in which a radioactive source is placed on one side of the material to be tested (cigarettes, metal plates, and so forth) and a counting device on the other. The amount of radiation reaching the counter is a measure of the thickness of the material.

The effectiveness of lubricating oils is measured in an engine constructed from metal into which radioactive isotopes have been incorporated. After the engine has been run for a fixed time period, the oil is withdrawn and tested for the presence of radioactive particles accumulated through engine wear.

When a single pipeline is used to transfer more than one petroleum derivative, a small amount of a radioactive isotope is placed in the last portion of one substance to signal its end and the start of another. The radiation may be used to activate an automatic valve system so that the liquids are diverted into different tanks.

The use of radiocarbon in determining the age of materials of plant origin has been mentioned. Photosynthesis has been studied by tracing the absorption of $CO_2$ containing ${}^{14}_{6}C$ by plants. In this way much has been learned about the conversion of $CO_2$ into sugars, starches, and cellulose.

Radioactive isotopes have found many uses in medical research, therapy, and diagnosis. The radiations from $^{60}_{27}Co$, a $\beta$ and $\gamma$ emitter, are used in cancer therapy. The isotope $^{131}_{53}I$ is used in the diagnosis and treatment of thyroid disorders since this gland concentrates ingested iodine.

Radioactive tracers have found wide use in chemical studies. The following structure of the thiosulfate ion

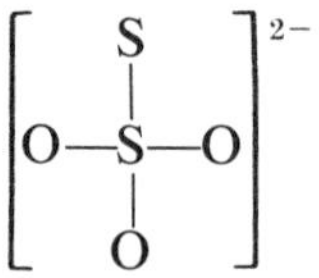

$$\left[\begin{array}{c} S \\ | \\ O{-}S{-}O \\ | \\ O \end{array}\right]^{2-}$$

is indicated by studies using $^{35}_{16}S$. The thiosulfate ion is prepared by heating a sulfite with the radioactive sulfur.

$$^{35}S(s) + SO_3^{2-}(aq) \rightarrow [^{35}SSO_3]^{2-}(aq)$$

Upon the acidification of this thiosulfate solution, the radioactive sulfur is quantitatively precipitated; none is found in the resulting $SO_2$ gas.

$$[^{35}SSO_3]^{2-}(aq) + 2H^+(aq) \rightarrow {}^{35}S(s) + SO_2(g) + H_2O$$

In addition, upon the decomposition of silver thiosulfate derived from this ion, all the radioactive sulfur ends in the silver sulfide.

$$Ag_2[^{35}SSO_3](s) + H_2O \xrightarrow{\text{heat}} Ag_2{}^{35}S(s) + SO_4^{2-}(aq) + 2H^+(aq)$$

These results indicate that the two sulfur atoms of the thiosulfate ion are not equivalent and that the proposed structure is likely.

Radioactive isotopes have been used to study reaction rates and mechanisms as well as the action of catalysts. By using radioactive tracers it becomes possible to follow the progress of tagged molecules and atoms through a chemical reaction.

For example, the mechanism of the reaction between the sulfite ion and the chlorate ion

$$SO_3^{2-} + ClO_3^- \rightarrow SO_4^{2-} + ClO_2^-$$

has been shown to proceed by the exchange of an oxygen atom. The $SO_3^{2-}$ ion used contained ordinary oxygen atoms, but the $ClO_3^-$ ion was prepared with $^{18}_8O$, which is not radioactive but which may be detected by a mass spectrograph. The $SO_4^{2-}$ ion produced by the reaction contained $^{18}_8O$ in an amount that would indicate that one $^{18}_8O$ atom had been added to each $SO_3^{2-}$ ion.

The mechanism proposed, therefore, is a Lewis nucleophilic displacement of $ClO_2^-$ by $SO_3^{2-}$ on one of the oxygen atoms of the $ClO_3^-$ ion. The $^{18}_8O$ atoms are marked with asterisks.

$$\left[\begin{array}{c} O \\ | \\ O-S \\ | \\ O \end{array}\right]^{2-} + \left[\begin{array}{c} O^* \\ | \\ O^*-Cl-O^* \end{array}\right]^{-} \rightarrow \left[\begin{array}{c} O \quad\quad O^* \\ | \quad\quad\quad | \\ O-S \cdots O^* \cdots Cl-O^* \\ | \\ O \end{array}\right]^{3-} \rightarrow$$

$$\left[\begin{array}{c} O \\ | \\ O-S-O^* \\ | \\ O \end{array}\right]^{2-} + \left[\begin{array}{c} O^* \\ | \\ Cl-O^* \end{array}\right]^{-}$$

Activation analysis is the determination of the quantity of an element in a sample by bombarding the sample with suitable nuclear projectiles and measuring the intensity of the radioactivity induced in the element being investigated. The induced activity is not influenced by the chemical bonding of the element. Neutrons are the most frequently employed projectiles. Techniques have been developed for the determination of more than 50 elements. This method is particularly valuable for the determination of elements present in extremely low concentrations.

Determinations of very small vapor pressures and very low solubilities may be made conveniently by using tagged materials. For example, the solubility of water in benzene may be determined by using water that has been enriched with the radioactive $^3_1H$ and measuring the activity of the water-saturated benzene.

## Problems

**21.1** Briefly describe the exchange theory of nuclear bonding.

**21.2** Define the following terms: (a) nuclide, (b) nucleon, (c) critical mass, (d) thermal neutron, (e) fission, (f) fusion, (g) transuranium element.

**21.3** Write equations for the following examples of radioactive decay: (a) alpha emission by $^{218}_{85}At$, (b) beta emission by $^{198}_{79}Au$, (c) positron emission by $^{25}_{13}Al$, (d) electron capture by $^{108}_{47}Ag$.

**21.4** Write equations for the following examples of radioactive decay: (a) alpha emission by $^{181}_{78}Pt$, (b) beta emission by $^{25}_{11}Na$, (c) positron emission by $^{62}_{29}Cu$, (d) electron capture by $^{37}_{18}Ar$.

**21.5** The nuclide $^{192}_{78}Pt$ decays to $^{188}_{76}Os$ by alpha emission. The mass of $^{192}_{78}Pt$ is 191.9614 u, and the mass of $^{188}_{76}Os$ is 187.9560 u. Calculate the energy released in this process.

**21.6** When $^{221}_{87}Fr$ decays, 6.34 MeV and 6.12 MeV $\alpha$ particles are emitted. What is the energy of and the source of the $\gamma$ radiation accompanying this alpha emission?

**21.7** The disintegration energy is

7.03 MeV for the $\beta$ decay of $^{20}_{9}$F to $^{20}_{10}$Ne (mass, 19.99244 u). What is the mass of $^{20}_{9}$F?

**21.8** The nuclide $^{21}_{11}$Na (mass, 20.99883 u) decays to $^{21}_{10}$Ne (mass, 20.99395 u) by positron emission. What is the energy released by this radioactive decay process?

**21.9** The energy released by the decay of $^{31}_{16}$S by positron emission is 5.40 MeV. The daughter nuclide $^{31}_{15}$P has a mass of 30.97376 u. What is the mass of $^{31}_{16}$S?

**21.10** The nuclide $^{37}_{18}$Ar (mass, 36.96678 u) decays to $^{37}_{17}$Cl (mass, 36.96590 u). Is it energetically possible for the process to occur by positron emission, or must it occur by electron capture?

**21.11** The disintegration energy for the decay of $^{51}_{24}$Cr by electron capture is 0.75 MeV, and the mass of the nuclide produced $^{51}_{23}$V is 50.9440 u. What is the mass of $^{51}_{24}$Cr?

**21.12** Why is electron capture accompanied by the production of X rays?

**21.13** The nuclide $^{76}_{35}$Br has a half-life of 16.5 hours. How much of a 0.0100 g sample remains at the end of 1.00 day?

**21.14** The half-life of $^{112}_{47}$Ag is 3.20 hours. How long will it take for 75.0% of a sample to disappear?

**21.15** The rate of decay of $^{18}_{9}$F is such that 10.0% of the original quantity remains after 369 minutes. (a) What is the rate constant for this radioactive disintegration? (b) What is the half-life of $^{18}_{9}$F?

**21.16** The half-life of $^{65}_{30}$Zn is 243 days. What is the activity in curies of a 0.000100 g sample of $^{65}_{30}$Zn?

The mass of $^{65}_{30}$Zn to three significant figures is 64.9 u.

**21.17** A sample of a radioactive material initially gives 2500 counts/minute and 15 minutes later gives 2400 counts/minute. What is the half-life of the nuclide?

**21.18** The half-life of $^{59}_{26}$Fe is 45 days. (a) How many atoms are in a sample that has an activity of 0.75 curie? (b) What is the weight of the sample?

**21.19** The carbon from the heartwood of a giant sequoia tree gives 10.8 $^{14}_{6}$C counts per minute per gram of carbon, whereas the wood from the outer portion of the tree gives 15.3 $^{14}_{6}$C counts per minute per gram of carbon. How old is the tree?

**21.20** Starting with $^{154}_{68}$Er, the successive steps in an artificial decay chain are: $\alpha$, $\beta^{+}$, ec, $\alpha$, $\alpha$. What are the daughter members of the chain?

**21.21** One of the naturally occurring decay series is that of the nuclide $^{232}_{90}$Th. The particles successively emitted in one route are: $\alpha$, $\beta$, $\beta$, $\alpha$, $\alpha$, $\alpha$, $\beta$, $\alpha$, $\beta$, $\alpha$. Determine in order the daughter members of the chain.

**21.22** It is believed that a $^{237}_{93}$Np disintegration series existed in nature at one time. However, the members of the series (with the exception of the stable nuclide that ends the series) have virtually disappeared through radioactive decay in the time since the earth was created. The particles successively emitted in the $^{237}_{93}$Np series are: $\alpha$, $\beta$, $\alpha$, $\alpha$, $\beta$, $\alpha$, $\alpha$, $\alpha$, $\beta$, $\alpha$, $\beta$. Determine in order the daughter members of the chain.

# B

## NOTES ON MATHEMATICAL OPERATIONS

An **exponent** is a superscript added to a **base** to indicate a mathematical operation that is to be performed on the base. In the expression $a^n$ the exponent is $n$ and the base is $a$. The following types of exponents are frequently encountered.

1. *Exponent is a positive integer.* In the expression $a^n$ the exponent $n$ is the number of times that the base $a$ is to be taken as a factor in the expansion; therefore, $(n - 1)$ is the number of times the base is to be multiplied by itself. Hence,

$$a^4 = a \times a \times a \times a$$

2. *Exponent is a negative integer.* The expression $a^{-n}$ is the reciprocal of $a^n$. For example,

$$a^{-4} = \frac{1}{a^4} = \frac{1}{a \times a \times a \times a}$$

3. *Exponent is a fraction of the type* $1/n$. The value of $n$ is the index of a root of the base. Thus,

$$a^{1/2} = \sqrt{a}$$
$$a^{1/3} = \sqrt[3]{a}$$

4. *Exponent is a fraction of the type* $m/n$. This exponent indicates two operations (those of parts 1 and 3). Hence, $a^{m/n}$ is $\sqrt[n]{a^m}$, and

$$a^{3/2} = \sqrt{a^3} = \sqrt{a \times a \times a}$$

5. *Exponent is zero.* Provided the base is not zero, the value of the expression is unity. Thus,

$$a^0 = 1 \qquad (a \neq 0)$$

**21.23** The half-life of $^{237}_{93}\text{Np}$ is $2.1 \times 10^6$ years, and the age of earth is approximately $4.5 \times 10^9$ years. What fraction of the quantity of $^{237}_{93}\text{Np}$ that was formed when the earth was created is now present?

**21.24** Write equations for the following induced nuclear reactions: (a) $^{82}_{35}\text{Br}(n, \gamma)$, (b) $^{10}_{5}\text{B}(n, \alpha)$, (c) $^{35}_{17}\text{Cl}(n, p)$, (d) $^{7}_{3}\text{Li}(p, n)$, (e) $^{130}_{52}\text{Te}(d, 2n)$, (f) $^{43}_{20}\text{Ca}(\alpha, p)$, (g) $^{237}_{93}\text{Np}(\alpha, n)$, (h) $^{238}_{92}\text{U}(^{22}_{10}\text{Ne}, 4n)$.

**21.25** Write equations for the following induced nuclear reactions: (a) $^{10}_{5}\text{B}(\alpha, n)$, (b) $^{6}_{3}\text{Li}(n, \alpha)$, (c) $^{7}_{3}\text{Li}(d, n)$, (d) $^{12}_{6}\text{C}(p, \gamma)$, (e) $^{96}_{42}\text{Mo}(p, n)$, (f) $^{75}_{33}\text{As}(d, p)$, (g) $^{45}_{21}\text{Sc}(\alpha, p)$, (h) $^{51}_{23}\text{V}(d, 2n)$.

**21.26** Following are listed some transuranium nuclides and the types of induced nuclear reactions that have been used to prepare them. In each case what isotope was used as the starting material? (a) $(d, 6n)$ $^{231}_{93}\text{Np}$, (b) $(d, 3n)$ $^{243}_{97}\text{Bk}$, (c) $(\alpha, ^3_1\text{H})$ $^{250}_{99}\text{Es}$, (d) $(^{12}_{6}\text{C}, 4n)$ $^{245}_{99}\text{Es}$, (e) $(^{16}_{8}\text{O}, 5n)$ $^{249}_{100}\text{Fm}$, (f) $(\alpha, 2n)$ $^{255}_{101}\text{Md}$, (g) $(^{22}_{10}\text{Ne}, 4n)$ $^{256}_{102}\text{No}$, (n) $(^{10}_{5}\text{B}, 3n)$ $^{257}_{103}\text{Lr}$

**21.27** An isotope of element 105 was prepared by bombarding a $^{249}_{98}\text{Cf}$ target with $^{15}_{7}\text{N}$ nuclei. Four neutrons were emitted when a $^{249}_{98}\text{Cf}$ nucleus absorbed a $^{15}_{7}\text{N}$ nu-cleus. Write an equation for the transformation.

**21.28** Calculate the binding energy per nucleon of $^{235}_{92}\text{U}$ (mass, 235.0439 u). The masses of the proton, neutron, and electron are 1.007277 u, 1.008665 u, and 0.0005486 u, respectively.

**21.29** (a) Calculate the energy released by the fission:

$$^{235}_{92}\text{U} + {}^{1}_{0}\text{n} \rightarrow {}^{94}_{38}\text{Sr} + {}^{139}_{54}\text{Xe} + 3{}^{1}_{0}\text{n}$$

The atomic masses are: $^{235}_{92}\text{U}$, 235.0439 u; $^{94}_{38}\text{Sr}$, 93.9154 u; $^{139}_{54}\text{Xe}$, 138.9179 u. The mass of the neutron is 1.0087 u. (b) What percent of the total mass of the starting materials is converted into energy?

**21.30** (a) Calculate the energy released by the fusion:

$$^{1}_{1}\text{H} + {}^{2}_{1}\text{H} \rightarrow {}^{3}_{2}\text{He}$$

The atomic masses are: $^{1}_{1}\text{H}$, 1.0078 u; $^{2}_{1}\text{H}$, 2.0141 u; $^{3}_{2}\text{He}$, 3.0160 u. (b) What percent of the total mass of the starting materials is converted into energy?

**21.31** Each of the following nuclides represents one of the products of a neutron-induced fission of $^{235}_{92}\text{U}$. In each case another nuclide and three neutrons are also produced. Identify the other nuclides. (a) $^{148}_{58}\text{Ce}$, (b) $^{95}_{37}\text{Rb}$, (c) $^{134}_{52}\text{Te}$.

## VALUES OF SOME CONSTANTS AND CONVERSION FACTORS

| PHYSICAL CONSTANTS | | |
|---|---|---|
| CONSTANT | SYMBOL | VALUE |
| Avogadro's number | $N$ | $6.0222 \times 10^{23}$/mol |
| Bohr radius | $a_0$ | $5.2917 \times 10^{-9}$ cm |
| electron rest mass | $m$ | $9.1096 \times 10^{-28}$ g |
| | | $5.4859 \times 10^{-4}$ u |
| electronic charge, unit charge | $e$ | $1.6022 \times 10^{-19}$ coulomb |
| | | $4.8033 \times 10^{-10}$ $g^{1/2}cm^{3/2}$/sec (or esu) |
| Faraday | $F$ | $9.6487 \times 10^4$ coulombs/mol |
| | | 23.061 kcal/V mol |
| gas constant | $R$ | $8.2056 \times 10^{-2}$ liter atm/°K mol |
| | | 8.3143 J/°K mol |
| | | 1.9872 cal/°K mol |
| molar volume, ideal gas at STP | — | 22.4136 liters |
| neutron rest mass | — | $1.6749 \times 10^{-24}$ g |
| | | 1.008665 u |
| Planck's constant | $h$ | $6.6262 \times 10^{-27}$ erg sec |
| | | $6.6262 \times 10^{-34}$ J sec |
| proton rest mass | — | $1.6726 \times 10^{-24}$ g |
| | | 1.007277 u |
| speed of light in vacuum | $c$ | $2.9979 \times 10^{10}$ cm/sec |

| CONVERSION FACTO | | |
|---|---|---|
| UNIT | ABBREVIATION | DE |
| ångstrom | Å | 10 |
| atmosphere | atm | 76 |
| | | 1. |
| calorie | cal | 4. |
| curie | Ci | 3. |
| electron volt | eV | 1. |
| | | 2. |
| electrostatic unit | esu | 3 |
| erg | erg | 1 |
| | | 2 |
| joule | J | 1 |
| | | ( |
| Kelvin temperature scale | °K | ° |
| | | |
| torr (or mm of mercury) | torr | |
| unified atomic mass unit | u | |

Some properties of exponents are summarized in the following equations.

1. $a^m a^n = a^{m+n}$  Thus, $a^4 a^2 = a^6$
2. $(a^m)^n = a^{mn}$  Thus, $(a^4)^2 = a^8$
3. $(ab)^n = a^n b^n$  Thus, $(ab)^3 = a^3 b^3$
4. $\dfrac{a^m}{a^n} = a^{m-n}$  Thus, $\dfrac{a^5}{a^2} = a^3;\ \dfrac{a^2}{a^5} = a^{-3} = \dfrac{1}{a^3}$

5. $\dfrac{a^n}{a^n} = 1$  Thus, $\dfrac{a^3}{a^3} = 1$

**Scientific notation** is used to simplify the handling of very large or very small numbers. When scientific notation is employed, a value is expressed in the form

$$a \times 10^n$$

where $a$ is a number with one digit to the left of the decimal point, and the exponent $n$ is a positive or negative integer. A value can be converted into this form by moving the decimal point until there is one digit to the left of it. For each place the decimal point is moved to the *left*, $n$ is *increased* by one; for each place the decimal point is moved to the *right*, $n$ is *decreased* by one. For example

$$1{,}904 = 1.904 \times 10^3$$

$$0.000053 = 5.3 \times 10^{-5}$$

1. When numbers expressed in scientific notation are *multiplied*, the exponents of 10 are added.

$$(3.0 \times 10^5)(4.0 \times 10^{-2}) = (3.0 \times 4.0)(10^5 \times 10^{-2})$$
$$= 12 \times 10^3$$
$$= 1.2 \times 10^4$$

2. In *division*, the powers of 10 are subtracted.

$$\frac{6.89 \times 10^{-7}}{3.36 \times 10^3} = \left(\frac{6.89}{3.36}\right)\left(\frac{10^{-7}}{10^3}\right)$$
$$= 2.05 \times 10^{-10}$$

3. To take a *square root*, we divide the exponent of 10 by 2. If necessary, the number is rewritten so that the exponent of 10 is perfectly divisible by 2.

$$(2.21 \times 10^{-7})^{1/2} = (22.1 \times 10^{-8})^{1/2}$$
$$= 4.70 \times 10^{-4}$$

4. When a *cube root* is taken, the exponent is divided by 3.

$$(1.86 \times 10^8)^{1/3} = (186 \times 10^6)^{1/3}$$
$$= 5.71 \times 10^2$$

5. When numbers expressed in scientific notation are *added* or *subtracted*, the powers of 10 must be the same.

$$(6.25 \times 10^3) + (3.0 \times 10^2) = (6.25 \times 10^3) + (0.30 \times 10^3)$$
$$= 6.55 \times 10^3$$

**B.2 LOGARITHMS**   The **logarithm** of a number is the power to which a base must be raised in order to secure the number. Common logarithms (abbreviated log) employ the base 10. If

$$a = 10^n$$

$$\log a = n$$

and therefore

$$\log 1000 = \log 10^3 = 3$$

$$\log 0.01 = \log 10^{-2} = -2$$

The logarithm of a number that is merely 10 raised to a power is the exponent of 10. The logarithm of a number such as 3.540, however, cannot be determined by inspection. The logarithms of numbers from 1 to 10 can be obtained from the table of logarithms found in Appendix C. Decimal points are omitted from this table. Each of the numbers listed is assumed to have a decimal point following the first digit; each of the logarithms should have a decimal point preceding the value listed. The logarithm of 3.540 is 0.5490.

The logarithm of a number greater than 10 or less than 1 can be obtained in the following way. The number is expressed in scientific notation; for example,

$$3.540 \times 10^{12}$$

Since logarithms are exponents and since $a^m a^n = a^{m+n}$, the logarithm of this value is obtained by adding the logarithm of 3.540 to the logarithm of $10^{12}$. Hence,

$$\log (3.540 \times 10^{12}) = \log 3.540 + \log 10^{12}$$
$$= 0.5490 + 12$$
$$= 12.5490$$

Another example is

$$\log (2.00 \times 10^{-5}) = \log 2.00 + \log 10^{-5}$$
$$= 0.310 + (-5)$$
$$= -4.699$$

Notice that the number of digits that follow the decimal point in the recorded logarithm is equal to the number of significant figures in the original value.

Sometimes it is necessary to find an **antilogarithm,** the number that corresponds to a given logarithm. In such instances the procedure used to find a logarithm is reversed. The given logarithm is written in two parts: a decimal fraction (called a **mantissa**) and a positive or negative whole number (called a **characteristic**). For example,

antilog (3.740) = antilog (0.740 + 3)

The antilogarithm of the mantissa (0.740) is obtained by finding the number corresponding to this logarithm in the table (it is 5.50), and the antilogarithm of the characteristic (3) is merely $10^3$. Therefore,

antilog (3.740) = $5.50 \times 10^3$

or

$3.740 = \log (5.50 \times 10^3)$

All the mantissas recorded in a table of logarithms are *positive.* This fact must be taken into account when an antilogarithm of a negative number is found. For example, to take the antilogarithm of $-3.158$, we must write the number in such a way that the mantissa is positive. Thus,

antilog $(-3.158)$ = antilog $(0.842 - 4)$
$= 6.95 \times 10^{-4}$

or

$-3.158 = \log (6.95 \times 10^{-4})$

Since logarithms are exponents, mathematical operations involving logarithms follow the rules for the use of exponents. When each of the following operations is performed, the logarithms of the values involved are found, the logarithms are treated as indicated, and the antilogarithm of the result is secured as the final answer.

1. *Multiplication.*    $\log (ab) = \log a + \log b$
2. *Division.*    $\log (a/b) = \log a - \log b$
3. *Extraction of a root.*    $\log (a^{1/n}) = \dfrac{1}{n} \log a$
4. *Raising to a power.*    $\log (a^n) = n \log a$

Logarithms that employ the base 10 are called common logarithms. **Natural logarithms** (abbreviated ln) employ the base $e$, where

$e = 2.71828 \ldots$

The relation between natural logarithms and common logarithms is

$\ln a = 2.303 \log a$

Thus, to find the natural logarithm of 6.040, we multiply the common logarithm of 6.040 by 2.303.

$$\ln 6.040 = 2.303 \log 6.040$$
$$= 2.303(0.7810)$$
$$= 1.7986$$

Logarithms can be used to evaluate an expression of the type $e^n$, where $e$ is the base of natural logarithms. Since

$$\ln a = 2.303 \log a$$

$$\log a = \frac{\ln a}{2.303}$$

and since

$$\ln e^n = n$$

$$\log e^n = \frac{n}{2.303}$$

Therefore,

$$e^n = \text{antilog } \frac{n}{2.303}$$

For example, the value of $e^{2.209}$ can be found in the following way.

$$e^{2.209} = \text{antilog } \frac{2.209}{2.303} = \text{antilog } 0.9590$$

$$= 9.100$$

**B.3 QUADRATIC EQUATIONS**

An algebraic equation in the form

$$ax^2 + bx + c = 0$$

is called a **quadratic equation** in one variable. An equation of this type has two solutions given by the **quadratic formula**

$$x = \frac{-b \pm \sqrt{b^2 - 4ac}}{2a}$$

When the quadratic formula is used to find the answer to a chemical problem, two solutions are obtained; one of them, however, must be discarded because it represents a physical impossibility.

Assume that $x$ in the equation

$$x^2 + 0.50\, x - 0.15 = 0$$

is the number of moles of a gas that dissociate under a given set of conditions. The values of the coefficients are: $a = 1$, $b = 0.50$, and $c = -0.15$; and

$$x = \frac{-0.50 \pm \sqrt{(0.50)^2 - 4(1)(-0.15)}}{2(1)}$$

$$x = \frac{-0.50 \pm 0.92}{2}$$

$$x = +0.21, -0.71$$

The value $-0.71$ is discarded because a negative amount of a substance is physically impossible.

# LOGARITHMS

| | 0 | 1 | 2 | 3 | 4 | 5 | 6 | 7 | 8 | 9 |
|---|---|---|---|---|---|---|---|---|---|---|
| 10 | 0000 | 0043 | 0086 | 0128 | 0170 | 0212 | 0253 | 0294 | 0334 | 0374 |
| 11 | 0414 | 0453 | 0492 | 0531 | 0569 | 0607 | 0645 | 0682 | 0719 | 0755 |
| 12 | 0792 | 0828 | 0864 | 0899 | 0934 | 0969 | 1004 | 1038 | 1072 | 1106 |
| 13 | 1139 | 1173 | 1206 | 1239 | 1271 | 1303 | 1335 | 1367 | 1399 | 1430 |
| 14 | 1461 | 1492 | 1523 | 1553 | 1584 | 1614 | 1644 | 1673 | 1703 | 1732 |
| 15 | 1761 | 1790 | 1818 | 1847 | 1875 | 1903 | 1931 | 1959 | 1987 | 2014 |
| 16 | 2041 | 2068 | 2095 | 2122 | 2148 | 2175 | 2201 | 2227 | 2253 | 2279 |
| 17 | 2304 | 2330 | 2355 | 2380 | 2405 | 2430 | 2455 | 2480 | 2504 | 2529 |
| 18 | 2553 | 2577 | 2601 | 2625 | 2648 | 2672 | 2695 | 2718 | 2742 | 2765 |
| 19 | 2788 | 2810 | 2833 | 2856 | 2878 | 2900 | 2923 | 2945 | 2967 | 2989 |
| 20 | 3010 | 3032 | 3054 | 3075 | 3096 | 3118 | 3139 | 3160 | 3181 | 3201 |
| 21 | 3222 | 3243 | 3263 | 3284 | 3304 | 3324 | 3345 | 3365 | 3385 | 3404 |
| 22 | 3424 | 3444 | 3464 | 3483 | 3502 | 3522 | 3541 | 3560 | 3579 | 3598 |
| 23 | 3617 | 3636 | 3655 | 3674 | 3692 | 3711 | 3729 | 3747 | 3766 | 3784 |
| 24 | 3802 | 3820 | 3838 | 3856 | 3874 | 3892 | 3909 | 3927 | 3945 | 3962 |
| 25 | 3979 | 3997 | 4014 | 4031 | 4048 | 4065 | 4082 | 4099 | 4116 | 4133 |
| 26 | 4150 | 4166 | 4183 | 4200 | 4216 | 4232 | 4249 | 4265 | 4281 | 4298 |
| 27 | 4314 | 4330 | 4346 | 4362 | 4378 | 4393 | 4409 | 4425 | 4440 | 4456 |
| 28 | 4472 | 4487 | 4502 | 4518 | 4533 | 4548 | 4564 | 4579 | 4594 | 4609 |
| 29 | 4624 | 4639 | 4654 | 4669 | 4683 | 4698 | 4713 | 4728 | 4742 | 4757 |
| 30 | 4771 | 4786 | 4800 | 4814 | 4829 | 4843 | 4857 | 4871 | 4886 | 4900 |
| 31 | 4914 | 4928 | 4942 | 4955 | 4969 | 4983 | 4997 | 5011 | 5024 | 5038 |
| 32 | 5051 | 5065 | 5079 | 5092 | 5105 | 5119 | 5132 | 5145 | 5159 | 5172 |
| 33 | 5185 | 5198 | 5211 | 5224 | 5237 | 5250 | 5263 | 5276 | 5289 | 5302 |
| 34 | 5315 | 5328 | 5340 | 5353 | 5366 | 5378 | 5391 | 5403 | 5416 | 5428 |
| 35 | 5441 | 5453 | 5465 | 5478 | 5490 | 5502 | 5514 | 5527 | 5539 | 5551 |
| 36 | 5563 | 5575 | 5587 | 5599 | 5611 | 5623 | 5635 | 5647 | 5658 | 5670 |
| 37 | 5682 | 5694 | 5705 | 5717 | 5729 | 5740 | 5752 | 5763 | 5775 | 5786 |
| 38 | 5798 | 5809 | 5821 | 5832 | 5843 | 5855 | 5866 | 5877 | 5888 | 5899 |
| 39 | 5911 | 5922 | 5933 | 5944 | 5955 | 5966 | 5977 | 5988 | 5999 | 6010 |
| 40 | 6021 | 6031 | 6042 | 6053 | 6064 | 6075 | 6085 | 6096 | 6107 | 6117 |
| 41 | 6128 | 6138 | 6149 | 6160 | 6170 | 6180 | 6191 | 6201 | 6212 | 6222 |
| 42 | 6232 | 6243 | 6253 | 6263 | 6274 | 6284 | 6294 | 6304 | 6314 | 6325 |
| 43 | 6335 | 6345 | 6355 | 6365 | 6375 | 6385 | 6395 | 6405 | 6415 | 6425 |
| 44 | 6435 | 6444 | 6454 | 6464 | 6474 | 6484 | 6493 | 6503 | 6513 | 6522 |
| 45 | 6532 | 6542 | 6551 | 6561 | 6571 | 6580 | 6590 | 6599 | 6609 | 6618 |
| 46 | 6628 | 6637 | 6646 | 6656 | 6665 | 6675 | 6684 | 6693 | 6702 | 6712 |
| 47 | 6721 | 6730 | 6739 | 6749 | 6758 | 6767 | 6776 | 6785 | 6794 | 6803 |
| 48 | 6812 | 6821 | 6830 | 6839 | 6848 | 6857 | 6866 | 6875 | 6884 | 6893 |
| 49 | 6902 | 6911 | 6920 | 6928 | 6937 | 6946 | 6955 | 6964 | 6972 | 6981 |
| 50 | 6990 | 6998 | 7007 | 7016 | 7024 | 7033 | 7042 | 7050 | 7059 | 7067 |
| 51 | 7076 | 7084 | 7093 | 7101 | 7110 | 7118 | 7126 | 7135 | 7143 | 7152 |
| 52 | 7160 | 7168 | 7177 | 7185 | 7193 | 7202 | 7210 | 7218 | 7226 | 7235 |
| 53 | 7243 | 7251 | 7259 | 7267 | 7275 | 7284 | 7292 | 7300 | 7308 | 7316 |
| 54 | 7324 | 7332 | 7340 | 7348 | 7356 | 7364 | 7372 | 7380 | 7388 | 7396 |

| | 0 | 1 | 2 | 3 | 4 | 5 | 6 | 7 | 8 | 9 |
|---|---|---|---|---|---|---|---|---|---|---|
| 55 | 7404 | 7412 | 7419 | 7427 | 7435 | 7443 | 7451 | 7459 | 7466 | 7474 |
| 56 | 7482 | 7490 | 7497 | 7505 | 7513 | 7520 | 7528 | 7536 | 7543 | 7551 |
| 57 | 7559 | 7566 | 7574 | 7582 | 7589 | 7597 | 7604 | 7612 | 7619 | 7627 |
| 58 | 7634 | 7642 | 7649 | 7657 | 7664 | 7672 | 7679 | 7686 | 7694 | 7701 |
| 59 | 7709 | 7716 | 7723 | 7731 | 7738 | 7745 | 7752 | 7760 | 7767 | 7774 |
| 60 | 7782 | 7789 | 7796 | 7803 | 7810 | 7818 | 7825 | 7832 | 7839 | 7846 |
| 61 | 7853 | 7860 | 7868 | 7875. | 7882 | 7889 | 7896 | 7903 | 7910 | 7917 |
| 62 | 7924 | 7931 | 7938 | 7945 | 7952 | 7959 | 7966 | 7973 | 7980 | 7987 |
| 63 | 7993 | 8000 | 8007 | 8014 | 8021 | 8028 | 8035 | 8041 | 8048 | 8055 |
| 64 | 8062 | 8069 | 8075 | 8082 | 8089 | 8096 | 8102 | 8109 | 8116 | 8122 |
| 65 | 8129 | 8136 | 8142 | 8149 | 8156 | 8162 | 8169 | 8176 | 8182 | 8189 |
| 66 | 8195 | 8202 | 8209 | 8215 | 8222 | 8228 | 8235 | 8241 | 8248 | 8254 |
| 67 | 8261 | 8267 | 8274 | 8280 | 8287 | 8293 | 8299 | 8306 | 8312 | 8319 |
| 68 | 8325 | 8331 | 8338 | 8344 | 8351 | 8357 | 8363 | 8370 | 8376 | 8382 |
| 69 | 8388 | 8395 | 8401 | 8407 | 8414 | 8420 | 8426 | 8432 | 8439 | 8445 |
| 70 | 8451 | 8457 | 8463 | 8470 | 8476 | 8482 | 8488 | 8494 | 8500 | 8506 |
| 71 | 8513 | 8519 | 8525 | 8531 | 8537 | 8543 | 8549 | 8555 | 8561 | 8567 |
| 72 | 8573 | 8579 | 8585 | 8591 | 8597 | 8603 | 8609 | 8615 | 8621 | 8627 |
| 73 | 8633 | 8639 | 8645 | 8651 | 8657 | 8663 | 8669 | 8675 | 8681 | 8686 |
| 74 | 8692 | 8698 | 8704 | 8710 | 8716 | 8722 | 8727 | 8733 | 8739 | 8745 |
| 75 | 8751 | 8756 | 8762 | 8768 | 8774 | 8779 | 8785 | 8791 | 8797 | 8802 |
| 76 | 8808 | 8814 | 8820 | 8825 | 8831 | 8837 | 8842 | 8848 | 8854 | 8859 |
| 77 | 8865 | 8871 | 8876 | 8882 | 8887 | 8893 | 8899 | 8904 | 8910 | 8915 |
| 78 | 8921 | 8927 | 8932 | 8938 | 8943 | 8949 | 8954 | 8960 | 8965 | 8971 |
| 79 | 8976 | 8982 | 8987 | 8993 | 8998 | 9004 | 9009 | 9015 | 9020 | 9025 |
| 80 | 9031 | 9036 | 9042 | 9047 | 9053 | 9058 | 9063 | 9069 | 9074 | 9079 |
| 81 | 9085 | 9090 | 9096 | 9101 | 9106 | 9112 | 9117 | 9122 | 9128 | 9133 |
| 82 | 9138 | 9143 | 9149 | 9154 | 9159 | 9165 | 9170 | 9175 | 9180 | 9186 |
| 83 | 9191 | 9196 | 9201 | 9206 | 9212 | 9217 | 9222 | 9227 | 9232 | 9238 |
| 84 | 9243 | 9248 | 9253 | 9258 | 9263 | 9269 | 9274 | 9279 | 9284 | 9289 |
| 85 | 9294 | 9299 | 9304 | 9309 | 9315 | 9320 | 9325 | 9330 | 9335 | 9340 |
| 86 | 9345 | 9350 | 9355 | 9360 | 9365 | 9370 | 9375 | 9380 | 9385 | 9390 |
| 87 | 9395 | 9400 | 9405 | 9410 | 9415 | 9420 | 9425 | 9430 | 9435 | 9440 |
| 88 | 9445 | 9450 | 9455 | 9460 | 9465 | 9469 | 9474 | 9479 | 9484 | 9489 |
| 89 | 9494 | 9499 | 9504 | 9509 | 9513 | 9518 | 9523 | 9528 | 9533 | 9538 |
| 90 | 9542 | 9547 | 9552 | 9557 | 9562 | 9566 | 9571 | 9576 | 9581 | 9586 |
| 91 | 9590 | 9595 | 9600 | 9605 | 9609 | 9614 | 9619 | 9624 | 9628 | 9633 |
| 92 | 9638 | 9643 | 9647 | 9652 | 9657 | 9661 | 9666 | 9671 | 9675 | 9680 |
| 93 | 9685 | 9689 | 9694 | 9699 | 9703 | 9708 | 9713 | 9717 | 9722 | 9727 |
| 94 | 9731 | 9736 | 9741 | 9745 | 9750 | 9754 | 9759 | 9763 | 9768 | 9773 |
| 95 | 9777 | 9782 | 9786 | 9791 | 9795 | 9800 | 9805 | 9809 | 9814 | 9818 |
| 96 | 9823 | 9827 | 9832 | 9836 | 9841 | 9845 | 9850 | 9854 | 9859 | 9863 |
| 97 | 9868 | 9872 | 9877 | 9881 | 9886 | 9890 | 9894 | 9899 | 9903 | 9908 |
| 98 | 9912 | 9917 | 9921 | 9926 | 9930 | 9934 | 9939 | 9943 | 9948 | 9952 |
| 99 | 9956 | 9961 | 9965 | 9969 | 9974 | 9978 | 9983 | 9987 | 9991 | 9996 |

# APPENDIX  D

## STANDARD ELECTRODE POTENTIALS AT 25°C[a]

ACID SOLUTION

| Half Reaction | $\mathscr{E}°$ (volts) |
|---|---|
| $Li^+ + e^- \rightleftharpoons Li$ | −3.045 |
| $K^+ + e^- \rightleftharpoons K$ | −2.925 |
| $Rb^+ + e^- \rightleftharpoons Rb$ | −2.925 |
| $Cs^+ + e^- \rightleftharpoons Cs$ | −2.923 |
| $Ra^{2+} + 2e^- \rightleftharpoons Ra$ | −2.916 |
| $Ba^{2+} + 2e^- \rightleftharpoons Ba$ | −2.906 |
| $Sr^{2+} + 2e^- \rightleftharpoons Sr$ | −2.888 |
| $Ca^{2+} + 2e^- \rightleftharpoons Ca$ | −2.866 |
| $Na^+ + e^- \rightleftharpoons Na$ | −2.714 |
| $Ce^{3+} + 3e^- \rightleftharpoons Ce$ | −2.483 |
| $Mg^{2+} + 2e^- \rightleftharpoons Mg$ | −2.363 |
| $Be^{2+} + 2e^- \rightleftharpoons Be$ | −1.847 |
| $Al^{3+} + 3e^- \rightleftharpoons Al$ | −1.662 |
| $Mn^{2+} + 2e^- \rightleftharpoons Mn$ | −1.180 |
| $Zn^{2+} + 2e^- \rightleftharpoons Zn$ | −0.7628 |
| $Cr^{3+} + 3e^- \rightleftharpoons Cr$ | −0.744 |
| $Ga^{3+} + 3e^- \rightleftharpoons Ga$ | −0.529 |
| $Fe^{2+} + 2e^- \rightleftharpoons Fe$ | −0.4402 |
| $Cr^{3+} + e^- \rightleftharpoons Cr^{2+}$ | −0.408 |
| $Cd^{2+} + 2e^- \rightleftharpoons Cd$ | −0.4029 |
| $PbSO_4 + 2e^- \rightleftharpoons Pb + SO_4^{2-}$ | −0.3588 |

[a] Data from A. J. de Bethune and N. A. Swendeman Loud, "Table of Electrode Potentials and Temperature Coefficients," pp. 414–424 in *Encyclopedia of Electrochemistry* (C. A. Hampel, editor), Van Nostrand Reinhold, New York, 1964, and from A. J. de Bethune and N. A. Swendeman Loud, *Standard Aqueous Electrode Potentials and Temperature Coefficients,* 19 pp., C. A. Hampel, publisher, Skokie, Illinois, 1964.

| Half Reaction | $\mathscr{E}°$ (volts) |
|---|---|
| $Tl^+ + e^- \rightleftharpoons Tl$ | $-0.3363$ |
| $Co^{2+} + 2e^- \rightleftharpoons Co$ | $-0.277$ |
| $H_3PO_4 + 2H^+ + 2e^- \rightleftharpoons H_3PO_3 + H_2O$ | $-0.276$ |
| $Ni^{2+} + 2e^- \rightleftharpoons Ni$ | $-0.250$ |
| $Sn^{2+} + 2e^- \rightleftharpoons Sn$ | $-0.136$ |
| $Pb^{2+} + 2e^- \rightleftharpoons Pb$ | $-0.126$ |
| $2H^+ + 2e^- \rightleftharpoons H_2$ | $0.0000$ |
| $S + 2H^+ + 2e^- \rightleftharpoons H_2S$ | $+0.142$ |
| $Sn^{4+} + 2e^- \rightleftharpoons Sn^{2+}$ | $+0.15$ |
| $SO_4^{2-} + 4H^+ + 2e^- \rightleftharpoons H_2SO_3 + H_2O$ | $+0.172$ |
| $AgCl + e^- \rightleftharpoons Ag + Cl^-$ | $+0.2222$ |
| $Cu^{2+} + 2e^- \rightleftharpoons Cu$ | $+0.337$ |
| $H_2SO_3 + 4H^+ + 4e^- \rightleftharpoons S + 3H_2O$ | $+0.450$ |
| $Cu^+ + e^- \rightleftharpoons Cu$ | $+0.521$ |
| $I_2 + 2e^- \rightleftharpoons 2I^-$ | $+0.5355$ |
| $MnO_4^- + e^- \rightleftharpoons MnO_4^{2-}$ | $+0.564$ |
| $O_2 + 2H^+ + 2e^- \rightleftharpoons H_2O_2$ | $+0.6824$ |
| $Fe^{3+} + e^- \rightleftharpoons Fe^{2+}$ | $+0.771$ |
| $Hg_2^{2+} + 2e^- \rightleftharpoons 2Hg$ | $+0.788$ |
| $Ag^+ + e^- \rightleftharpoons Ag$ | $+0.7991$ |
| $2NO_3^- + 4H^+ + 2e^- \rightleftharpoons N_2O_4 + 2H_2O$ | $+0.803$ |
| $Hg^{2+} + 2e^- \rightleftharpoons Hg$ | $+0.854$ |
| $2Hg^{2+} + 2e^- \rightleftharpoons Hg_2^{2+}$ | $+0.920$ |
| $NO_3^- + 4H^+ + 3e^- \rightleftharpoons NO + 2H_2O$ | $+0.96$ |
| $Br_2 + 2e^- \rightleftharpoons 2Br^-$ | $+1.0652$ |
| $O_2 + 4H^+ + 4e^- \rightleftharpoons 2H_2O$ | $+1.229$ |
| $MnO_2 + 4H^+ + 2e^- \rightleftharpoons Mn^{2+} + 2H_2O$ | $+1.23$ |
| $Tl^{3+} + 2e^- \rightleftharpoons Tl^+$ | $+1.25$ |
| $Cr_2O_7^{2-} + 14H^+ + 6e^- \rightleftharpoons 2Cr^{3+} + 7H_2O$ | $+1.33$ |
| $Cl_2 + 2e^- \rightleftharpoons 2Cl^-$ | $+1.3595$ |
| $Au^{3+} + 2e^- \rightleftharpoons Au^+$ | $+1.402$ |
| $PbO_2 + 4H^+ + 2e^- \rightleftharpoons Pb^{2+} + 2H_2O$ | $+1.455$ |
| $Au^{3+} + 3e^- \rightleftharpoons Au$ | $+1.498$ |
| $Mn^{3+} + e^- \rightleftharpoons Mn^{2+}$ | $+1.51$ |
| $MnO_4^- + 8H^+ + 5e^- \rightleftharpoons Mn^{2+} + 4H_2O$ | $+1.51$ |
| $Ce^{4+} + e^- \rightleftharpoons Ce^{3+}$ | $+1.61$ |
| $2HOCl + 2H^+ + 2e^- \rightleftharpoons Cl_2 + 2H_2O$ | $+1.63$ |
| $PbO_2 + SO_4^{2-} + 4H^+ + 2e^- \rightleftharpoons PbSO_4 + 2H_2O$ | $+1.682$ |
| $Au^+ + e^- \rightleftharpoons Au$ | $+1.691$ |
| $MnO_4^- + 4H^+ + 3e^- \rightleftharpoons MnO_2 + 2H_2O$ | $+1.695$ |
| $H_2O_2 + 2H^+ + 2e^- \rightleftharpoons 2H_2O$ | $+1.776$ |
| $Co^{3+} + e^- \rightleftharpoons Co^{2+}$ | $+1.808$ |
| $S_2O_8^{2-} + 2e^- \rightleftharpoons 2SO_4^{2-}$ | $+2.01$ |

| Half Reaction | $\mathscr{E}°$ (volts) |
|---|---|
| $O_3 + 2H^+ + 2e^- \rightleftharpoons O_2 + H_2O$ | +2.07 |
| $F_2 + 2e^- \rightleftharpoons 2F^-$ | +2.87 |

**ALKALINE SOLUTION**

| Half Reaction | $\mathscr{E}°$ (volts) |
|---|---|
| $Al(OH)_4^- + 3e^- \rightleftharpoons Al + 4OH^-$ | −2.33 |
| $Zn(OH)_4^{2-} + 2e^- \rightleftharpoons Zn + 4OH^-$ | −1.215 |
| $Fe(OH)_2 + 2e^- \rightleftharpoons Fe + 2OH^-$ | −0.877 |
| $2H_2O + 2e^- \rightleftharpoons H_2 + 2OH^-$ | −0.82806 |
| $Cd(OH)_2 + 2e^- \rightleftharpoons Cd + 2OH^-$ | −0.809 |
| $S + 2e^- \rightleftharpoons S^{2-}$ | −0.447 |
| $CrO_4^{2-} + 4H_2O + 3e^- \rightleftharpoons Cr(OH)_3 + 5OH^-$ | −0.13 |
| $NO_3^- + H_2O + 2e^- \rightleftharpoons NO_2^- + 2OH^-$ | +0.01 |
| $O_2 + 2H_2O + 4e^- \rightleftharpoons 4OH^-$ | +0.401 |
| $NiO_2 + 2H_2O + 2e^- \rightleftharpoons Ni(OH)_2 + 2OH^-$ | +0.490 |
| $HO_2^- + H_2O + 2e^- \rightleftharpoons 3OH^-$ | +0.878 |

## EQUILIBRIUM CONSTANTS AT 25°C

**I IONIZATION CONSTANTS**

MONOPROTIC ACIDS

| acetic | $HC_2H_3O_2 \rightleftharpoons H^+ + C_2H_3O_2^-$ | $1.8 \times 10^{-5}$ |
|---|---|---|
| benzoic | $HC_7H_5O_2 \rightleftharpoons H^+ + C_7H_5O_2^-$ | $6.0 \times 10^{-5}$ |
| chlorous | $HClO_2 \rightleftharpoons H^+ + ClO_2^-$ | $1.1 \times 10^{-2}$ |
| cyanic | $HOCN \rightleftharpoons H^+ + OCN^-$ | $1.2 \times 10^{-4}$ |
| formic | $HCHO_2 \rightleftharpoons H^+ + CHO_2^-$ | $1.8 \times 10^{-4}$ |
| hydrazoic | $HN_3 \rightleftharpoons H^+ + N_3^-$ | $1.9 \times 10^{-5}$ |
| hydrocyanic | $HCN \rightleftharpoons H^+ + CN^-$ | $4.0 \times 10^{-10}$ |
| hydrofluoric | $HF \rightleftharpoons H^+ + F^-$ | $6.7 \times 10^{-4}$ |
| hypobromous | $HOBr \rightleftharpoons H^+ + OBr^-$ | $2.1 \times 10^{-9}$ |
| hypochlorous | $HOCl \rightleftharpoons H^+ + OCl^-$ | $3.2 \times 10^{-8}$ |
| nitrous | $HNO_2 \rightleftharpoons H^+ + NO_2^-$ | $4.5 \times 10^{-4}$ |

POLYPROTIC ACIDS

| arsenic | $H_3AsO_4 \rightleftharpoons H^+ + H_2AsO_4^-$ | $K_1 = 2.5 \times 10^{-4}$ |
|---|---|---|
| | $H_2AsO_4^- \rightleftharpoons H^+ + HAsO_4^{2-}$ | $K_2 = 5.6 \times 10^{-8}$ |
| | $HAsO_4^{2-} \rightleftharpoons H^+ + AsO_4^{3-}$ | $K_3 = 3 \times 10^{-13}$ |
| carbonic | $CO_2 + H_2O \rightleftharpoons H^+ + HCO_3^-$ | $K_1 = 4.2 \times 10^{-7}$ |
| | $HCO_3^- \rightleftharpoons H^+ + CO_3^{2-}$ | $K_2 = 4.8 \times 10^{-11}$ |
| hydrosulfuric | $H_2S \rightleftharpoons H^+ + HS^-$ | $K_1 = 1.1 \times 10^{-7}$ |
| | $HS^- \rightleftharpoons H^+ + S^{2-}$ | $K_2 = 1.0 \times 10^{-14}$ |
| oxalic | $H_2C_2O_4 \rightleftharpoons H^+ + HC_2O_4^-$ | $K_1 = 5.9 \times 10^{-2}$ |
| | $HC_2O_4^- \rightleftharpoons H^+ + C_2O_4^{2-}$ | $K_2 = 6.4 \times 10^{-5}$ |
| phosphoric | $H_3PO_4 \rightleftharpoons H^+ + H_2PO_4^-$ | $K_1 = 7.5 \times 10^{-3}$ |
| | $H_2PO_4^- \rightleftharpoons H^+ + HPO_4^{2-}$ | $K_2 = 6.2 \times 10^{-8}$ |
| | $HPO_4^{2-} \rightleftharpoons H^+ + PO_4^{3-}$ | $K_3 = 1 \times 10^{-12}$ |

| | | |
|---|---|---|
| phosphorous (diprotic) | $H_3PO_3 \rightleftharpoons H^+ + H_2PO_3^-$ | $K_1 = 1.6 \times 10^{-2}$ |
| | $H_2PO_3^- \rightleftharpoons H^+ + H_2PO_3^{2-}$ | $K_2 = 7 \times 10^{-7}$ |
| sulfuric | $H_2SO_4 \rightarrow H^+ + HSO_4^-$ | strong |
| | $HSO_4^- \rightleftharpoons H^+ + SO_4^{2-}$ | $K_2 = 1.3 \times 10^{-2}$ |
| sulfurous | $SO_2 + H_2O \rightleftharpoons H^+ + HSO_3^-$ | $K_1 = 1.3 \times 10^{-2}$ |
| | $HSO_3^- \rightleftharpoons H^+ + SO_3^{2-}$ | $K_2 = 5.6 \times 10^{-8}$ |

## BASES

| | | |
|---|---|---|
| ammonia | $NH_3 + H_2O \rightleftharpoons NH_4^+ + OH^-$ | $1.8 \times 10^{-5}$ |
| aniline | $C_6H_5NH_2 + H_2O \rightleftharpoons C_6H_5NH_3^+ + OH^-$ | $4.6 \times 10^{-10}$ |
| dimethylamine | $(CH_3)_2NH + H_2O \rightleftharpoons (CH_3)_2NH_2^+ + OH^-$ | $7.4 \times 10^{-4}$ |
| hydrazine | $N_2H_4 + H_2O \rightleftharpoons N_2H_5^+ + OH^-$ | $9.8 \times 10^{-7}$ |
| methylamine | $CH_3NH_2 + H_2O \rightleftharpoons CH_3NH_3^+ + OH^-$ | $5.0 \times 10^{-4}$ |
| pyridine | $C_5H_5N + H_2O \rightleftharpoons C_5H_5NH^+ + OH^-$ | $1.5 \times 10^{-9}$ |
| trimethylamine | $(CH_3)_3N + H_2O \rightleftharpoons (CH_3)_3NH^+ + OH^-$ | $7.4 \times 10^{-5}$ |

## II SOLUBILITY PRODUCTS

### BROMIDES

| | |
|---|---|
| $PbBr_2$ | $4.6 \times 10^{-6}$ |
| $Hg_2Br_2$ | $1.3 \times 10^{-22}$ |
| $AgBr$ | $5.0 \times 10^{-13}$ |

### CARBONATES

| | |
|---|---|
| $BaCO_3$ | $1.6 \times 10^{-9}$ |
| $CdCO_3$ | $5.2 \times 10^{-12}$ |
| $CaCO_3$ | $4.7 \times 10^{-9}$ |
| $CuCO_3$ | $2.5 \times 10^{-10}$ |
| $FeCO_3$ | $2.1 \times 10^{-11}$ |
| $PbCO_3$ | $1.5 \times 10^{-15}$ |
| $MgCO_3$ | $1 \times 10^{-15}$ |
| $MnCO_3$ | $8.8 \times 10^{-11}$ |
| $Hg_2CO_3$ | $9.0 \times 10^{-17}$ |
| $NiCO_3$ | $1.4 \times 10^{-7}$ |
| $Ag_2CO_3$ | $8.2 \times 10^{-12}$ |
| $SrCO_3$ | $7 \times 10^{-10}$ |
| $ZnCO_3$ | $2 \times 10^{-10}$ |

### CHLORIDES

| | |
|---|---|
| $PbCl_2$ | $1.6 \times 10^{-5}$ |
| $Hg_2Cl_2$ | $1.1 \times 10^{-18}$ |
| $AgCl$ | $1.7 \times 10^{-10}$ |

### CHROMATES

| | |
|---|---|
| $BaCrO_4$ | $8.5 \times 10^{-11}$ |
| $PbCrO_4$ | $2 \times 10^{-16}$ |
| $Hg_2CrO_4$ | $2 \times 10^{-9}$ |
| $Ag_2CrO_4$ | $1.9 \times 10^{-12}$ |
| $SrCrO_4$ | $3.6 \times 10^{-5}$ |

### FLUORIDES

| | |
|---|---|
| $BaF_2$ | $2.4 \times 10^{-5}$ |
| $CaF_2$ | $3.9 \times 10^{-11}$ |
| $PbF_2$ | $4 \times 10^{-8}$ |
| $MgF_2$ | $8 \times 10^{-8}$ |
| $SrF_2$ | $7.9 \times 10^{-10}$ |

### HYDROXIDES

| | |
|---|---|
| $Al(OH)_3$ | $5 \times 10^{-33}$ |
| $Ba(OH)_2$ | $5.0 \times 10^{-3}$ |
| $Cd(OH)_2$ | $2.0 \times 10^{-14}$ |
| $Ca(OH)_2$ | $1.3 \times 10^{-6}$ |
| $Cr(OH)_3$ | $6.7 \times 10^{-31}$ |
| $Co(OH)_2$ | $2.5 \times 10^{-16}$ |
| $Co(OH)_3$ | $2.5 \times 10^{-43}$ |
| $Cu(OH)_2$ | $1.6 \times 10^{-19}$ |
| $Fe(OH)_2$ | $1.8 \times 10^{-15}$ |

| | | SULFATES | |
|---|---|---|---|
| $Fe(OH)_3$ | $6 \times 10^{-38}$ | | |
| $Pb(OH)_2$ | $4.2 \times 10^{-15}$ | $BaSO_4$ | $1.5 \times 10^{-9}$ |
| $Mg(OH)_2$ | $8.9 \times 10^{-12}$ | $CaSO_4$ | $2.4 \times 10^{-5}$ |
| $Mn(OH)_2$ | $2 \times 10^{-13}$ | $PbSO_4$ | $1.3 \times 10^{-8}$ |
| $Hg(OH)_2$ (HgO) | $3 \times 10^{-26}$ | $Ag_2SO_4$ | $1.2 \times 10^{-5}$ |
| $Ni(OH)_2$ | $1.6 \times 10^{-16}$ | $SrSO_4$ | $7.6 \times 10^{-7}$ |
| AgOH ($Ag_2O$) | $2.0 \times 10^{-8}$ | | |
| $Sr(OH)_2$ | $3.2 \times 10^{-4}$ | | |
| $Sn(OH)_2$ | $3 \times 10^{-27}$ | | |
| $Zn(OH)_2$ | $4.5 \times 10^{-17}$ | SULFIDES | |

IODIDES

| | | $Bi_2S_3$ | $1.6 \times 10^{-72}$ |
|---|---|---|---|
| | | CdS | $1.0 \times 10^{-28}$ |
| | | CoS | $5 \times 10^{-22}$ |
| $PbI_2$ | $8.3 \times 10^{-9}$ | CuS | $8 \times 10^{-37}$ |
| $Hg_2I_2$ | $4.5 \times 10^{-29}$ | FeS | $4 \times 10^{-19}$ |
| AgI | $8.5 \times 10^{-17}$ | PbS | $7 \times 10^{-29}$ |
| | | MnS | $7 \times 10^{-16}$ |
| | | HgS | $1.6 \times 10^{-54}$ |
| OXALATES | | NiS | $3 \times 10^{-21}$ |
| | | $Ag_2S$ | $5.5 \times 10^{-51}$ |
| $BaC_2O_4$ | $1.5 \times 10^{-8}$ | SnS | $1 \times 10^{-26}$ |
| $CaC_2O_4$ | $1.3 \times 10^{-9}$ | ZnS | $2.5 \times 10^{-22}$ |
| $PbC_2O_4$ | $8.3 \times 10^{-12}$ | | |
| $MgC_2O_4$ | $8.6 \times 10^{-5}$ | | |
| $Ag_2C_2O_4$ | $1.1 \times 10^{-11}$ | | |
| $SrC_2O_4$ | $5.6 \times 10^{-8}$ | | |

| | | MISCELLANEOUS | |
|---|---|---|---|
| PHOSPHATES | | $NaHCO_3$ | $1.2 \times 10^{-3}$ |
| | | $KClO_4$ | $8.9 \times 10^{-3}$ |
| $Ba_3(PO_4)_2$ | $6 \times 10^{-39}$ | $K_2[PtCl_6]$ | $1.4 \times 10^{-6}$ |
| $Ca_3(PO_4)_2$ | $1.3 \times 10^{-32}$ | $AgC_2H_3O_2$ | $2.3 \times 10^{-3}$ |
| $Pb_3(PO_4)_2$ | $1 \times 10^{-54}$ | AgCN | $1.6 \times 10^{-14}$ |
| $Ag_3PO_4$ | $1.8 \times 10^{-18}$ | AgCNS | $1.0 \times 10^{-12}$ |
| $Sr_3(PO_4)_2$ | $1 \times 10^{-31}$ | | |

**III INSTABILITY CONSTANTS**

| | | | |
|---|---|---|---|
| $AlF_6^{3-}$ | $1.4 \times 10^{-20}$ | $Cu(NH_3)_4^{2+}$ | $4.7 \times 10^{-15}$ |
| $Al(OH)_4^-$ | $1.3 \times 10^{-34}$ | $Cu(CN)_2^-$ | $1 \times 10^{-16}$ |
| $Al(OH)^{2+}$ | $7.1 \times 10^{-10}$ | $Cu(OH)^+$ | $1 \times 10^{-8}$ |
| $Cd(NH_3)_4^{2+}$ | $7.5 \times 10^{-8}$ | $Fe(CN)_6^{4-}$ | $1 \times 10^{-35}$ |
| $Cd(CN)_4^{2-}$ | $1.4 \times 10^{-19}$ | $Fe(CN)_6^{3-}$ | $1 \times 10^{-42}$ |
| $Cr(OH)^{2+}$ | $5 \times 10^{-11}$ | $Pb(OH)^+$ | $1.5 \times 10^{-8}$ |
| $Co(NH_3)_6^{2+}$ | $1.3 \times 10^{-5}$ | $HgBr_4^{2-}$ | $2.3 \times 10^{-22}$ |
| $Co(NH_3)_6^{3+}$ | $2.2 \times 10^{-34}$ | $HgCl_4^{2-}$ | $1.1 \times 10^{-16}$ |
| $Cu(NH_3)_2^+$ | $1.4 \times 10^{-11}$ | $Hg(CN)_4^{2-}$ | $4 \times 10^{-42}$ |

| | | | |
|---|---|---|---|
| $HgI_4^{2-}$ | $5.3 \times 10^{-31}$ | $Ag(S_2O_3)_3^{5-}$ | $9.9 \times 10^{-15}$ |
| $Ni(NH_3)_4^{2+}$ | $1 \times 10^{-8}$ | $Zn(NH_3)_4^{2+}$ | $3.4 \times 10^{-10}$ |
| $Ni(NH_3)_6^{2+}$ | $1.8 \times 10^{-9}$ | $Zn(CN)_4^{2-}$ | $1.2 \times 10^{-18}$ |
| $Ag(NH_3)_2^{+}$ | $6.0 \times 10^{-8}$ | $Zn(OH)_4^{2-}$ | $3.6 \times 10^{-16}$ |
| $Ag(CN)_2^{-}$ | $1.8 \times 10^{-19}$ | $Zn(OH)^{+}$ | $4.1 \times 10^{-5}$ |
| $Ag(S_2O_3)_2^{3-}$ | $5 \times 10^{-14}$ | | |

# F

## ANSWERS TO SELECTED NUMERICAL PROBLEMS

CHAPTER 2  **2.3** 72.1% $^{85}Rb$, 27.9% $^{87}Rb$.  **2.5** 24.3 u.  **2.7** $3.51 \times 10^{-12}$ cal [or $1.47 \times 10^{-11}$ J].  **2.9** 28.976 u.  **2.11** $5.45 \times 10^{14}$/sec, $3.61 \times 10^{-12}$ erg; $6.67 \times 10^{14}$/sec, $4.42 \times 10^{-12}$ erg.  **2.14** (a) $4.97 \times 10^{-12}$ erg; (b) $1.31 \times 10^{-12}$ erg;  (c) $5.52 \times 10^{14}$/sec, 5430 Å.  **2.17** $n = 3$ to $n = 2$.  **2.22** $Z = 43$, Tc.  **2.24** $3.8 \times 10^{18}$/sec, 0.79 Å.  **2.28** (a) $6.6 \times 10^{-19}$ cm/sec;  (b) $7.3 \times 10^8$ cm/sec.  **2.34** $n = 4$, $Cr^{2+}$.

CHAPTER 3  **3.21** 43%.  **3.23** 1.27 D.

CHAPTER 5  **5.5** 60.6% W.  **5.8** 32.7 g Be.  **5.9** $C_3O_2$.  **5.10** $C_2H_4O$.  **5.18** (a) 0.180 mol C, 0.360 mol H;  (b) $CH_2$;  (c) 2.52 g.  **5.26** (a) 2.43 g $H_2O$, 0.135 mol $H_2O$;  (b) 0.0450 mol $VF_3$;  (c) 3 mol;  (d) $VF_3 \cdot 3H_2O$.  **5.31** (a) 45.2 g;  (b) 87.4%.  **5.36** 65.7% BaO.  **5.39** $-327$ kcal/mol [or $-1368$ kJ/mol].  **5.45** $-310.61$ kcal/mol [or $-1299.6$ kJ/mol].  **5.48** $-49.0$ kcal [or $-206$ kJ].  **5.54** $-240.2$ kcal/mol [or $-1008.1$ kJ/mol].

CHAPTER 6  **6.1** $5.04 \times 10^{13}$ atoms.  **6.6** (a) 4.10 atm;  (b) 182°K (or $-91$°C);  (c) 364°K (or 91°C).  **6.8** 351 ml.  **6.12** 5.03 liters.  **6.15** 5.60 atm.  **6.17** 120.  **6.22** 15.0 liters $CH_4$, 22.5 liters $O_2$, 15.0 liters $NH_3$, 45.0 liters $H_2O$.  **6.27** 13.3 liters, $p_{N_2O} = 0.50$ atm, $p_{NO_2} = 0.50$ atm.  **6.29** 1.06 liters $H_2$.  **6.34** (a) $CH_2$;  (b) $C_4H_8$.  **6.37** 0.0300 mol $NH_3$, 1.51 g $Mg_3N_2$.  **6.41** 140 ml.  **6.47** 16.0.  **6.49** (a) 22.4 atm, (b) 19.8 atm.

**CHAPTER 7**   **7.3** 8.30 kcal/mol [or 34.7 kJ/mol].   **7.8** 4.3 kcal/mol [or 18 kJ/mol].
**7.22** 50.9.   **7.23** $a = 2.88$ cm.   **7.27** 1.71 Å.   **7.31** 3.85 Å, 3.33 Å.
**7.38 (a)** −217 kcal/mol [or −909 kJ/mol];   **(b)** 10.6% high.

**CHAPTER 8**   **8.26** 45.0.   **8.30 (a)** 17.0;   **(b)** 8.00;   **(c)** 8.50 g $H_2O_2$.

**CHAPTER 9**   **9.11 (a)** $9.63 \times 10^{-4}$ mol;   **(b)** $4.81 \times 10^{-4}$ mol.   **9.15 (a)** $N_2$,
0.800 atm; $O_2$, 0.200 atm;   **(b)** $X_{N_2} = 1.49 \times 10^{-5}$, $X_{O_2} = 0.80 \times 10^{-5}$;
**(c)** −0.0024°C.   **9.17** $mx [1 − (y/100)] = M$.   **9.20** 15.3 g, 8.32 ml.
**9.25** 0.0500$M$.   **9.27** 0.753$N$.   **9.31 (a)** 53.0;   **(b)** 73.9% $Na_2CO_3$.
**9.37 (a)** 0.113$N$;   **(b)** 0.0188$M$.   **9.40 (a)** 0.265 atm;   **(b)** 0.479.
**9.49** 186.   **9.53** 4.02 atm.

**CHAPTER 10**   **10.6** 1.46 g Bi.   **10.12** 922 liters $Cl_2$.   **10.17** 0.0506 amp.   **10.19** 9.96
$\times 10^{-6}$ mol/liter.   **10.23 (a)** +0.740 V;   **(b)** $Cd + Cu^{2+} \rightarrow Cd^{2+} + Cu$;
**(c)** Cu electrode.   **10.36 (a)** Pt | $H_2(g)$ | $OH^-$ ‖ $H^+$ | $H_2(g)$ | Pt;
**(b)** +0.828 V;   **(c)** −19.1 kcal [or −79.9 kJ].   **10.40** 0.253$M$.
**10.44** +1.19 V.

**CHAPTER 12**   **12.4** pure $N_2$, 1.250 g/liter at STP; $N_2$ from air, 1.257 g/liter at STP.

**CHAPTER 13**   **13.7 (a)** Rate $= k$ [A][B];   **(b)** 0.156/$M$ min.   **13.10 (a)** $5.14 \times$
$10^{-3}$/atm sec;   **(b)** $3.38 \times 10^{-1}$ liter/mol sec.   **13.13** $9.3 \times 10^{-3}$
liter/mol sec.   **13.23** 4.09 atm.   **13.26** $[H_2] = [I_2] = 0.319$ mol/liter,
$[HI] = 2.362$ mol/liter.   **13.33** $2.19 \times 10^{-10}$ mol/liter, $6.70 \times 10^{-9}$ atm.
**13.36 (a)** +58.6 kcal [or +245 kJ];   **(b)** $4.85 \times 10^{-13}$ atm.

**CHAPTER 14**   **14.1 (a)** −326.1 kcal/mol [or −1364 kJ/mol];   **(b)** −326.7 kcal [or
−1387 kJ];   **(c)** −66.4 kcal/mol [or −278 kJ/mol].   **14.4 (a)** −17.15
kcal [or −71.53 kJ];   **(b)** −15.97 kcal [or −66.57 kJ].   **14.6** −44.20
kcal/mol [or −184.79 kJ/mol].   **14.10 (a)** −36 kcal [or −150 kJ];
**(b)** −37.9 kcal [or −159 kJ].   **14.14 (a)** $2Cu(s) + Cl_2(g) \rightarrow 2CuCl(s)$;
**(b)** +0.137 V.   **14.18 (a)** +0.2943 V;   **(b)** −13.57 kcal [or −56.79 kJ];
**(c)** −29.11 cal/°K [or −121.7 J/°K];   **(d)** −21.66 kcal [or −90.61 kJ].
**14.21 (a)** +24.6 cal/°K [or +103 J/°K];   **(b)** −140.4 kcal [or −587.6
kJ];   **(c)** −3.044 V.   **14.31** 49.6 cal/°K mol [or 208 J/°K mol].
**14.34** −261.7 kcal [or −1095 kJ].   **14.37** $5.1 \times 10^{-4}$.   **14.45** $5.9 \times 10^9$.

**CHAPTER 16**  **16.1** $3.3 \times 10^{-2}$.  **16.5** $0.082M$, $0.112$ mol.  **16.9** $[H^+] = [NO_2^-] = 9.3 \times 10^{-3}M$; $[HNO_2] = 0.19M$.  **16.12** $0.0025M$.  **16.13** $[H^+] = 1.5 \times 10^{-5}M$.  **16.19** **(a)** 4.14;  **(b)** 1.08;  **(c)** 10.52;  **(d)** 12.62.  **16.23** $0.430$ V.  **16.34** $4.5 \times 10^{-2}M$.  **16.38** $[NH_4^+]/[NH_3] = 0.11$.  **16.40** 4.14.  **16.42** $[H_3AsO_4] = 0.30M$, $[H^+] = [H_2AsO_4^-] = 8.7 \times 10^{-3}M$, $[HAsO_4^{2-}] = 5.6 \times 10^{-8}M$, $[AsO_4^{3-}] = 1.9 \times 10^{-18}M$.

**CHAPTER 17**  **17.1** $2.0 \times 10^{-14}$.  **17.4** $2.1 \times 10^{-4}$ mol/liter.  **17.8** $8.6 \times 10^{-8}$.  **17.12** $3.3 \times 10^{-13}$ mol.  **17.14** $[Na^+] = 0.16M$, $[Cl^-] = 0.30M$, $[C_2O_4^{2-}] = 2.1 \times 10^{-7}M$, $[Ba^{2+}] = 0.07M$.  **17.17** **(a)** $BaSO_4$;  **(b)** $1.2 \times 10^{-5}M$.  **17.26** $0.47M$.  **17.29** $0.086M$.  **17.31** $7.6 \times 10^{-11}M$.  **17.38** 8.18.  **17.40** $4.3 \times 10^{-7}$.  **17.44** 5.82.  **17.49** **(a)** 3.92;  **(b)** 8.46;  **(c)** 12.16.  **17.53** **(a)** $4.2 \times 10^{15}$;  **(b)** $[Cu^{2+}] = 0.050M$, $[Ag^+] = 3.5 \times 10^{-9}M$.

**CHAPTER 18**  **18.26** **(a)** $+0.0592$ V;  **(b)** $+0.770$ V.  **18.27** **(a)** $Fe^{2+}$, $FeO$;  **(b)** $Fe_3O_4$, $\frac{1}{2}$ octahedral, $\frac{1}{8}$ tetrahedral;  **(c)** $Fe_2O_3$, $\frac{2}{9}$.

**CHAPTER 19**  **19.24** **(a)** 5;  **(b)** high-spin.  **19.25** $1.3 \times 10^{-34}$.

**CHAPTER 21**  **21.5** 2.6 MeV.  **21.7** 19.99999 u.  **21.14** 6.38 hours.  **21.19** 2890 years.  **21.28** 7.59 MeV/nucleon.

# THE MODERNIZED

# metric system

## The International System of Units-SI

is a modernized version of the metric system established by international agreement. It provides a logical and interconnected framework for all measurements in science, industry, and commerce. Officially abbreviated SI, the system is built upon a foundation of seven base units, plus two supplementary units, which appear on this chart along with their definitions. All other SI units are derived from these units. Multiples and submultiples are expressed in a decimal system. Use of metric weights and measures was legalized in the United States in 1866, and since 1893 the yard and pound have been defined in terms of the meter and the kilogram. The base units for time, electric current, amount of substance, and luminous intensity are the same in both the customary and metric systems.

Courtesy of the National Bureau of Standards, U.S. Department of Commerce

## SEVEN BASE UNITS

**meter**-m
LENGTH

**kilogram**-kg
MASS

**second**-s
TIME

**ampere**-A
ELECTRIC CURRENT

**kelvin**-k
TEMPERATURE

**mole** -mol
AMOUNT OF SUBSTANCE

**candela** -cd
LUMINOUS INTENSITY

## TWO SUPPLEMENTARY UNITS

**radian** -rad
PLANE ANGLE